Handbook of Natural Computing

Main Editor
Grzegorz Rozenberg

Editors
Thomas Bäck
Joost N. Kok

Handbook of
Natural Computing

Volume 3

With 734 Figures and 75 Tables

 Springer

Editors
Grzegorz Rozenberg
LIACS
Leiden University
Leiden, The Netherlands
and
Computer Science Department
University of Colorado
Boulder, USA

Joost N. Kok
LIACS
Leiden University
Leiden, The Netherlands

Thomas Bäck
LIACS
Leiden University
Leiden, The Netherlands

ISBN 978-3-540-92909-3   ISBN 978-3-540-92910-9 (eBook)
ISBN 978-3-540-92911-6 (print and electronic bundle)
DOI 10.1007/978-3-540-92910-9
Springer Heidelberg Dordrecht London New York

Library of Congress Control Number: 2010933716

Printed on acid-free paper

Springer is part of Springer Science+Business Media (www.springer.com)

Preface

Natural Computing is the field of research that investigates human-designed computing inspired by nature as well as computing taking place in nature, that is, it investigates models and computational techniques inspired by nature, and also it investigates, in terms of information processing, phenomena taking place in nature.

Examples of the first strand of research include neural computation inspired by the functioning of the brain; evolutionary computation inspired by Darwinian evolution of species; cellular automata inspired by intercellular communication; swarm intelligence inspired by the behavior of groups of organisms; artificial immune systems inspired by the natural immune system; artificial life systems inspired by the properties of natural life in general; membrane computing inspired by the compartmentalized ways in which cells process information; and amorphous computing inspired by morphogenesis. Other examples of natural-computing paradigms are quantum computing and molecular computing, where the goal is to replace traditional electronic hardware, by, for example, bioware in molecular computing. In quantum computing, one uses systems small enough to exploit quantum-mechanical phenomena to perform computations and to perform secure communications more efficiently than classical physics, and, hence, traditional hardware allows. In molecular computing, data are encoded as biomolecules and then tools of molecular biology are used to transform the data, thus performing computations.

The second strand of research, computation taking place in nature, is represented by investigations into, among others, the computational nature of self-assembly, which lies at the core of the nanosciences; the computational nature of developmental processes; the computational nature of biochemical reactions; the computational nature of bacterial communication; the computational nature of brain processes; and the systems biology approach to bionetworks where cellular processes are treated in terms of communication and interaction, and, hence, in terms of computation.

Research in natural computing is genuinely interdisciplinary and forms a bridge between the natural sciences and computer science. This bridge connects the two, both at the level of information technology and at the level of fundamental research. Because of its interdisciplinary character, research in natural computing covers a whole spectrum of research methodologies ranging from pure theoretical research, algorithms, and software applications to experimental laboratory research in biology, chemistry, and physics.

Computer Science and Natural Computing

A preponderance of research in natural computing is centered in computer science. The spectacular progress in Information and Communication Technology (ICT) is highly supported by the evolution of computer science, which designs and develops the instruments needed for this progress: computers, computer networks, software methodologies, etc. As ICT has such a tremendous impact on our everyday lives, so does computer science.

However, there is much more to computer science than ICT: it is the science of information processing and, as such, a fundamental science for other disciplines. On one hand, the only common denominator for research done in such diverse areas of computer science is investigating various aspects of information processing. On the other hand, the adoption of Information and Information Processing as central notions and thinking habit has been an important development in many disciplines, biology and physics being prime examples. For these scientific disciplines, computer science provides not only instruments but also a way of thinking.

We are now witnessing exciting interactions between computer science and the natural sciences. While the natural sciences are rapidly absorbing notions, techniques, and methodologies intrinsic to information processing, computer science is adapting and extending its traditional notion of computation, and computational techniques, to account for computation taking place in nature around us. Natural Computing is an important catalyst for this two-way interaction, and this handbook constitutes a significant record of this development.

The Structure of the Handbook

Natural Computing is both a well-established research field with a number of classical areas, and a very dynamic field with many more recent, novel research areas. The field is vast, and so it is quite usual that a researcher in a specific area does not have sufficient insight into other areas of Natural Computing. Also, because of its dynamic development and popularity, the field constantly attracts more and more scientists who either join the active research or actively follow research developments.

Therefore, the goal of this handbook is two-fold:

(i) to provide an authoritative reference for a significant and representative part of the research in Natural Computing, and
(ii) to provide a convenient gateway to Natural Computing for motivated newcomers to this field.

The implementation of this goal was a challenge because this field and its literature are vast — almost all of its research areas have an extensive scientific literature, including specialized journals, book series, and even handbooks. This implies that the coverage of the whole field in reasonable detail and within a reasonable number of pages/volumes is practically impossible.

Thus, we decided to divide the presented material into six areas. These areas are by no means disjoint, but this division is convenient for the purpose of providing a representative picture of the field — representative with respect to the covered research topics and with respect to a good balance between classical and emerging research trends.

Each area consists of individual chapters, each of which covers a specific research theme. They provide necessary technical details of the described research, however they are self-contained and of an expository character, which makes them accessible for a broader audience. They also provide a general perspective, which, together with given references, makes the chapters valuable entries into given research themes.

This handbook is a result of the joint effort of the handbook editors, area editors, chapter authors, and the Advisory Board. The choice of the six areas by the handbook editors in consultation with the Advisory Board, the expertise of the area editors in their respective areas, the choice by the area editors of well-known researchers as chapter writers, and the peer-review for individual chapters were all important factors in producing a representative and reliable picture of the field. Moreover, the facts that the Advisory Board consists of 68 eminent scientists from 20 countries and that there are 105 contributing authors from 21 countries provide genuine assurance for the reader that this handbook is an authoritative and up-to-date reference, with a high level of significance and accuracy.

Handbook Areas

The material presented in the handbook is organized into six areas: Cellular Automata, Neural Computation, Evolutionary Computation, Molecular Computation, Quantum Computation, and Broader Perspective.

Cellular Automata

Cellular automata are among the oldest models of computation, dating back over half a century. The first cellular automata studies by John von Neumann in the late 1940s were biologically motivated, related to self-replication in universal systems. Since then, cellular automata gained popularity in physics as discrete models of physical systems, in computer science as models of massively parallel computation, and in mathematics as discrete-time dynamical systems. Cellular automata are a natural choice to model real-world phenomena since they possess several fundamental properties of the physical world: they are massively parallel, homogeneous, and all interactions are local. Other important physical constraints such as reversibility and conservation laws can be added as needed, by properly choosing the local update rule. Computational universality is common in cellular automata, and even starkly simple automata are capable of performing arbitrary computation tasks. Because cellular automata have the advantage of parallelism while obeying natural constraints such as locality and uniformity, they provide a framework for investigating realistic computation in massively parallel systems. Computational power and the limitations of such systems are most naturally investigated by time- and space-constrained computations in cellular automata. In mathematics — in terms of symbolic dynamics — cellular automata are viewed as endomorphisms of the full shift, that is, transformations that are translation invariant and continuous in the product topology. Interesting questions on chaotic dynamics have been studied in this context.

Neural Computation

Artificial neural networks are computer programs, loosely modeled after the functioning of the human nervous system. There are neural networks that aim to gain understanding of biological neural systems, and those that solve problems in artificial intelligence without necessarily creating a model of a real biological system. The more biologically oriented neural networks model the real nervous system in increasing detail at all relevant levels of information processing: from synapses to neurons to interactions between modules of interconnected neurons. One of the major challenges is to build artificial brains. By reverse-engineering the mammalian brain in silicon, the aim is to better understand the functioning of the (human) brain through detailed simulations. Neural networks that are more application-oriented tend to drift further apart from real biological systems. They come in many different flavors, solving problems in regression analysis and time-series forecasting, classification, and pattern recognition, as well as clustering and compression. Good old multilayered perceptrons and self-organizing maps are still pertinent, but attention in research is shifting toward more recent developments, such as kernel methods (including support vector machines) and Bayesian techniques. Both approaches aim to incorporate domain knowledge in the learning process in order to improve prediction performance, e.g., through the construction of a proper kernel function or distance measure or the choice of an appropriate prior distribution over the parameters of the neural network. Considerable effort is devoted to making neural networks efficient so that large models can be learned from huge databases in a reasonable amount of time. Application areas include, among many others, system dynamics and control, finance, bioinformatics, and image analysis.

Evolutionary Computation

The field of evolutionary computation deals with algorithms gleaned from models of organic evolution. The general aim of evolutionary computation is to use the principles of nature's processes of natural selection and genotypic variation to derive computer algorithms for solving hard search and optimization tasks. A wide variety of instances of evolutionary algorithms have been derived during the past fifty years based on the initial algorithms, and we are now witnessing astounding successes in the application of these algorithms: their fundamental understanding in terms of theoretical results; understanding algorithmic principles of their construction; combination with other techniques; and understanding their working principles in terms of organic evolution. The key algorithmic variations (such as genetic algorithms, evolution strategies, evolutionary programming, and genetic programming) have undergone significant developments over recent decades, and have also resulted in very powerful variations and recombinations of these algorithms. Today, there is a sound understanding of how all of these algorithms are instances of the generic concept of an evolutionary search approach. Hence the generic term "evolutionary algorithm" is nowadays being used to describe the generic algorithm, and the term "evolutionary computation" is used for the field as a whole. Thus, we have observed over the past fifty years how the field has integrated the various independently developed initial algorithms into one common principle. Moreover, modern evolutionary algorithms benefit from their ability to adapt and self-adapt their strategy parameters (such as mutation rates, step sizes, and search distributions) to the needs of the task at hand. In this way, they are robust and flexible metaheuristics for problem-solving even without requiring too much special expertise from their users. The feature of self-adaptation illustrates the ability of evolutionary principles to work on different levels at the same time, and therefore provides a nice demonstration of the universality of the evolutionary principle for search and optimization tasks. The widespread use of evolutionary computation reflects these capabilities.

Molecular Computation

Molecular computing is an emergent interdisciplinary field concerned with programming molecules so that they perform a desired computation, or fabricate a desired object, or control the functioning of a specific molecular system. The central idea behind molecular computing is that data can be encoded as (bio)molecules, e.g., DNA strands, and tools of molecular science can be used to transform these data. In a nutshell, a molecular program is just a collection of molecules which, when placed in a suitable substrate, will perform a specific function (execute the program that this collection represents). The birth of molecular computing is often associated with the 1994 breakthrough experiment by Leonard Adleman, who solved a small instance of a hard computational problem solely by manipulating DNA strands in test tubes. Although initially the main effort of the area was focused on trying to obtain a breakthrough in the complexity of solving hard computational problems, this field has evolved enormously since then. Among the most significant achievements of molecular computing have been contributions to understanding some of the fundamental issues of the nanosciences. One notable example among them is the contribution to the understanding of self-assembly, a central concept of the nanosciences. The techniques of molecular programming were successfully applied in experimentally constructing all kinds of molecular-scale objects or devices with prescribed functionalities. Well-known examples here are self-assembly of Sierpinski triangles, cubes, octahedra, DNA-based logic circuits, DNA "walkers" that move along a track, and autonomous molecular motors. A complementary approach to understanding bioinformation and computation is through studying the information-processing capabilities of cellular organisms. Indeed, cells and nature "compute" by "reading" and "rewriting" DNA through processes that modify DNA (or RNA) sequences. Research into the computational abilities of cellular organisms has the potential to uncover the laws governing biological information, and to enable us to harness the computational power of cells.

Quantum Computation

Quantum computing has been discussed for almost thirty years. The theory of quantum computing and quantum information processing is simply the theory of information processing with a classical notion of information replaced by its quantum counterpart. Research in quantum computing is concerned with understanding the fundamentals of information processing on the level of physical systems that realize/implement the information. In fact, quantum computing can be seen as a quest to understand the fundamental limits of information processing set by nature itself. The mathematical description of quantum information is more complicated than that of classical information — it involves the structure of Hilbert spaces. When describing the structure behind known quantum algorithms, this reduces to linear algebra over complex numbers. The history of quantum algorithms spans the last fifteen years, and some of these algorithms are extremely interesting, and even groundbreaking — the most remarkable are Shor's factorization in polynomial time and Grover's search algorithm. The nature of quantum information has also led to the invention of novel cryptosystems, whose security is not based on the complexity of computing functions, but rather on the physical properties of quantum information. Quantum computing is now a well-established discipline, however implementation of a large-scale quantum computer continues to be extremely challenging, even though quantum information processing primitives, including those allowing secure cryptography, have been demonstrated to be practically realizable.

Broader Perspective

In contrast to the first five areas focusing on more-established themes of natural computing, this area encompasses a perspective that is broader in several ways. First, the reader will find here treatments of certain well-established and specific techniques inspired by nature (e.g., simulated annealing) not covered in the other five areas. Second, the reader will also find application-centered chapters (such as natural computing in finance), each covering, in one chapter, a collection of natural computing methods, thus capturing the impact of natural computing as a whole in various fields of science or industry. Third, some chapters are full treatments of several established research fields (such as artificial life, computational systems biology, evolvable hardware, and artificial immune systems), presenting alternative perspectives and cutting across some of the other areas of the handbook, while introducing much new material. Other elements of this area are fresh, emerging, and novel techniques or perspectives (such as collision-based computing, nonclassical computation), representing the leading edge of theories and technologies that are shaping possible futures for both natural computing and computing in general. The contents of this area naturally cluster into two kinds (sections), determined by the essential nature of the techniques involved. These are "Nature-Inspired Algorithms" and "Alternative Models of Computation". In the first section, "Nature-Inspired Algorithms", the focus is on algorithms inspired by natural processes realized either through software or hardware or both, as additions to the armory of existing tools we have for dealing with well-known practical problems. In this section, we therefore find application-centered chapters, as well as chapters focusing on particular techniques, not otherwise dealt with in other areas of the handbook, which have clear and proven applicability. In the second section, "Alternative Models of Computation", the emphasis changes, moving away from specific applications or application areas, toward more far-reaching ideas. These range from developing computational approaches and "computational thinking" as fundamental tools for the new science of systems biology to ideas that take inspiration from nature as a platform for suggesting entirely novel possibilities of computing.

Handbook Chapters

In the remainder of this preface we will briefly describe the contents of the individual chapters. These chapter descriptions are grouped according to the handbook areas where they belong and given in the order that they appear in the handbook. This section provides the reader with a better insight into the contents, allowing one to design a personal roadmap for using this handbook.

Cellular Automata

This area is covered by nine chapters.

The first chapter, "Basic Concepts of Cellular Automata", by Jarkko J. Kari, reviews some classical results from the theory of cellular automata, relations between various concepts of injectivity and surjectivity, and some basic dynamical system concepts related to chaos in cellular automata. The classical results discussed include the celebrated Garden-of-Eden and Curtis–Hedlund–Lyndon theorems, as well as the balance property of surjective cellular automata. All these theorems date back to the 1960s. The results are provided together with examples that illustrate proof ideas. Different variants of sensitivity to initial conditions and mixing properties are introduced and related to each other. Also undecidability results concerning cellular automata are briefly discussed.

A popular mathematical approach is to view cellular automata as dynamical systems in the context of symbolic dynamics. Several interesting results in this area were reported as early as 1969 in the seminal paper by G.A. Hedlund, and still today this research direction is among the most fruitful sources of theoretical problems and new results. The chapter "Cellular Automata Dynamical Systems", by Alberto Dennunzio, Enrico Formenti, and Petr Kůrka, reviews some recent developments in this field. Recent research directions considered here include subshifts attractors and signal subshifts, particle weight functions, and the slicing construction. The first two concern one-dimensional cellular automata and give precise descriptions of the limit behavior of large classes of automata. The third one allows one to view two-dimensional cellular automata as one-dimensional systems. In this way combinatorial complexity is decreased and new results can be proved.

Programming cellular automata for particular tasks requires special techniques. The chapter "Algorithmic Tools on Cellular Automata", by Marianne Delorme and Jacques Mazoyer, covers classical algorithmic tools based on signals. Linear signals as well as signals of nonlinear slope are discussed, and basic transformations of signals are addressed. The chapter provides results on using signals to construct functions in cellular automata and to implement arithmetic operations on segments. The methods of folding the space–time, freezing, and clipping are also introduced.

The time-complexity advantage gained from parallelism under the locality and uniformity constraints of cellular automata can be precisely analyzed in terms of language recognition. The chapter "Language Recognition by Cellular Automata", by Véronique Terrier, presents results and questions about cellular automata complexity classes and their relationships to other models of computations. Attention is mainly directed to real-time and linear-time complexity classes, because significant benefits over sequential computation may be obtained at these low time complexities. Both parallel and sequential input modes are considered. Separate complexity classes are given also for cellular automata with one-way communications and two-way communications.

The chapter "Computations on Cellular Automata", by Jacques Mazoyer and Jean-Baptiste Yunès, continues with the topic of algorithmic techniques in cellular automata. This chapter uses the basic tools, such as signals and grids, to build natural implementations of common algorithms in cellular automata. Examples of implementations include real-time multiplication of integers and the prime number sieve. Both parallel and sequential input and output modes are discussed, as well as composition of functions and recursion.

The chapter "Universalities in Cellular Automata", by Nicolas Ollinger, is concerned with computational universalities. Concepts of universality include Turing universality (the ability to compute any recursive function) and intrinsic universality (the ability to simulate any other cellular automaton). Simulations of Boolean circuits in the two-dimensional case are explained in detail in order to achieve both kinds of universality. The more difficult one-dimensional case is also discussed, and seminal universal cellular automata and encoding techniques are presented in both dimensions. A detailed chronology of important papers on universalities in cellular automata is also provided.

A cellular automaton is reversible if every configuration has only one previous configuration, and hence its evolution process can be traced backward uniquely. This naturally corresponds to the fundamental time-reversibility of the microscopic laws of physics. The chapter "Reversible Cellular Automata", by Kenichi Morita, discusses how reversible cellular automata are defined, as well as their properties, how they are designed, and their computing abilities. After providing the definitions, the chapter surveys basic properties of injectivity and surjectivity. Three design methods of reversible cellular automata are provided: block rules, partitioned, and second-order cellular automata. Then the computational power of reversible cellular automata is discussed. In particular, simulation methods of irreversible cellular automata, reversible Turing machines, and some other universal systems are given to clarify universality of reversible cellular automata. In spite of the strong constraint of reversibility, it is shown that reversible cellular automata possess rich information processing capabilities, and even very simple ones are computationally universal.

A conservation law in a cellular automaton is a statement of the invariance of a local and additive energy-like quantity. The chapter "Conservation Laws in Cellular Automata", by Siamak Taati, reviews the basic theory of conservation laws. A general mathematical framework for formulating conservation laws in cellular automata is presented and several characterizations are summarized. Computational problems regarding conservation laws (verification and existence problems) are discussed. Microscopic explanations of the dynamics of the conserved quantities in terms of flows and particle flows are explored. The related concept of dissipating energy-like quantities is also discussed.

The chapter "Cellular Automata and Lattice Boltzmann Modeling of Physical Systems", by Bastien Chopard, considers the use of cellular automata and related lattice Boltzmann methods as a natural modeling framework to describe and study many physical systems composed of interacting components. The theoretical basis of the approach is introduced and its potential is illustrated for several applications in physics, biophysics, environmental science, traffic models, and multiscale modeling. The success of the technique can be explained by the close relationship between these methods and a mesoscopic abstraction of many natural phenomena.

Neural Computation

This area is covered by ten chapters.

Spiking neural networks are inspired by recent advances in neuroscience. In contrast to classical neural network models, they take into account not just the neuron's firing rate, but also the time moment of spike firing. The chapter "Computing with Spiking Neuron Networks", by Hélène Paugam-Moisy and Sander Bohte, gives an overview of existing approaches to modeling spiking neural neurons and synaptic plasticity, and discusses their computational power and the challenge of deriving efficient learning procedures.

Image quality assessment aims to provide computational models to predict the perceptual quality of images. The chapter "Image Quality Assessment — A Multiscale Geometric Analysis-Based Framework and Examples", by Xinbo Gao, Wen Lu, Dacheng Tao, and Xuelong Li, introduces the fundamentals and describes the state of the art in image quality assessment. It further proposes a new model, which mimics the human visual system by incorporating concepts such as multiscale analysis, contrast sensitivity, and just-noticeable differences. Empirical results clearly demonstrate that this model resembles subjective perception values and reflects the visual quality of images.

Neurofuzzy networks have the important advantage that they are easy to interpret. When applied to control problems, insight about the process characteristics at different operating regions can be easily obtained. Furthermore, nonlinear model predictive controllers can be developed as a nonlinear combination of several local linear model predictive controllers that have analytical solutions. Through several applications, the chapter "Nonlinear Process Modelling and Control Using Neurofuzzy Networks", by Jie Zhang, demonstrates that neurofuzzy networks are very effective in the modeling and control of nonlinear processes.

Similar to principal component and factor analysis, independent component analysis is a computational method for separating a multivariate signal into additive subcomponents. Independent component analysis is more powerful: the latent variables corresponding to the subcomponents need not be Gaussian and the basis vectors are typically nonorthogonal. The chapter "Independent Component Analysis", by Seungjin Choi, explains the theoretical foundations and describes various algorithms based on those principles.

Neural networks has become an important method for modeling and forecasting time series. The chapter "Neural Networks for Time-Series Forecasting", by G. Peter Zhang, reviews some recent developments (including seasonal time-series modeling, multiperiod forecasting, and ensemble methods), explains when and why they are to be preferred over traditional forecasting models, and also discusses several practical data and modeling issues.

Support vector machines have been extensively studied and applied in many domains within the last decade. Through the so-called kernel trick, support vector machines can efficiently learn nonlinear functions. By maximizing the margin, they implement the principle of structural risk minimization, which typically leads to high generalization performance. The chapter "SVM Tutorial — Classification, Regression and Ranking", by Hwanjo Yu and Sungchul Kim, describes these underlying principles and discusses support vector machines for different learning tasks: classification, regression, and ranking.

It is well known that single-hidden-layer feedforward networks can approximate any continuous target function. This still holds when the hidden nodes are automatically and randomly generated, independent of the training data. This observation opened up many possibilities for easy construction of a broad class of single-hidden-layer neural networks. The chapter "Fast Construction of Single-Hidden-Layer Feedforward Networks", by Kang Li, Guang-Bin Huang, and Shuzhi Sam Ge, discusses new ideas that yield a more compact network architecture and reduce the overall computational complexity.

Many recent experimental studies demonstrate the remarkable efficiency of biological neural systems to encode, process, and learn from information. To better understand the experimentally observed phenomena, theoreticians are developing new mathematical approaches and tools to model biological neural networks. The chapter "Modeling Biological Neural Networks", by Joaquin J. Torres and Pablo Varona, reviews some of the most popular models of neurons and neural networks. These not only help to understand how living systems perform information processing, but may also lead to novel bioinspired paradigms of artificial intelligence and robotics.

The size and complexity of biological data, such as DNA/RNA sequences and protein sequences and structures, makes them suitable for advanced computational tools, such as neural networks. Computational analysis of such databases aims at exposing hidden information that provides insights that help in understanding the underlying biological principles. The chapter "Neural Networks in Bioinformatics", by Ke Chen and Lukasz A. Kurgan, focuses on proteins. In particular it discusses prediction of protein secondary structure, solvent accessibility, and binding residues.

Self-organizing maps is a prime example of an artificial neural network model that both relates to the actual (topological) organization within the mammalian brain and at the same time has many practical applications. Self-organizing maps go back to the seminal work of Teuvo Kohonen. The chapter "Self-organizing Maps", by Marc M. Van Hulle, describes the state of the art with a special emphasis on learning algorithms that aim to optimize a predefined criterion.

Evolutionary Computation

This area is covered by thirteen chapters.

The first chapter, "Generalized Evolutionary Algorithms", by Kenneth De Jong, describes the general concept of evolutionary algorithms. As a generic introduction to the field, this chapter facilitates an understanding of specific instances of evolutionary algorithms as instantiations of a generic evolutionary algorithm. For the instantiations, certain choices need to be made, such as representation, variation operators, and the selection operator, which then yield particular instances of evolutionary algorithms, such as genetic algorithms and evolution strategies, to name just a few.

The chapter "Genetic Algorithms — A Survey of Models and Methods", by Darrell Whitley and Andrew M. Sutton, introduces and discusses (including criticism) the standard genetic algorithm based on the classical binary representation of solution candidates and a theoretical interpretation based on the so-called schema theorem. Variations of genetic algorithms with respect to solution representations, mutation operators, recombination operators, and selection mechanisms are also explained and discussed, as well as theoretical models of genetic algorithms based on infinite and finite population size assumptions and Markov chain theory concepts. The authors also critically investigate genetic algorithms from the perspective of identifying their limitations and the differences between theory and practice when working with genetic algorithms. To illustrate this further, the authors also give a practical example of the application of genetic algorithms to resource scheduling problems.

The chapter "Evolutionary Strategies", by Günter Rudolph, describes a class of evolutionary algorithms which have often been associated with numerical function optimization and continuous variables, but can also be applied to binary and integer domains. Variations of evolutionary strategies, such as the $(\mu+\lambda)$-strategy and the (μ,λ)-strategy, are introduced and discussed within a common algorithmic framework. The fundamental idea of self-adaptation of strategy parameters (variances and covariances of the multivariate normal distribution used for mutation) is introduced and explained in detail, since this is a key differentiating property of evolutionary strategies.

The chapter "Evolutionary Programming", by Gary B. Fogel, discusses a historical branch of evolutionary computation. It gives a historical perspective on evolutionary programming by describing some of the original experiments using evolutionary programming to evolve finite state machines to serve as sequence predictors. Starting from this canonical evolutionary programming approach, the chapter also presents extensions of evolutionary programming into continuous domains, where an attempt towards self-adaptation of mutation step sizes has been introduced which is similar to the one considered in evolutionary strategies. Finally, an overview of some recent applications of evolutionary programming is given.

The chapter "Genetic Programming — Introduction, Applications, Theory and Open Issues", by Leonardo Vanneschi and Riccardo Poli, describes a branch of evolutionary algorithms derived by extending genetic algorithms to allow exploration of the space of computer programs. To make evolutionary search in the domain of computer programs possible, genetic programming is based on LISP S-expression represented by syntax trees, so that genetic programming extends evolutionary algorithms to tree-based representations. The chapter gives an overview of the corresponding representation, search operators, and technical details of genetic programming, as well as existing applications to real-world problems. In addition, it discusses theoretical approaches toward analyzing genetic programming, some of the open issues, as well as research trends in the field.

The subsequent three chapters are related to the theoretical analysis of evolutionary algorithms, giving a broad overview of the state of the art in our theoretical understanding. These chapters demonstrate that there is a sound theoretical understanding of capabilities and limitations of evolutionary algorithms. The approaches can be roughly split into convergence velocity or progress analysis, computational complexity investigations, and global convergence results.

The convergence velocity viewpoint is represented in the chapter "The Dynamical Systems Approach — Progress Measures and Convergence Properties", by Silja Meyer-Nieberg and Hans-Georg Beyer. It demonstrates how the dynamical systems approach can be used to analyze the behavior of evolutionary algorithms quantitatively with respect to their progress rate. It also provides a complete overview of results in the continuous domain, i.e., for all types of evolution strategies on certain objective functions (such as sphere, ridge, etc.). The chapter presents results for undisturbed as well as for noisy variants of these objective functions, and extends the approach to dynamical objective functions where the goal turns into optimum tracking. All results are presented by means of comparative tables, so the reader gets a complete overview of the key findings at a glance.

The chapter "Computational Complexity of Evolutionary Algorithms", by Thomas Jansen, deals with the question of optimization time (i.e., the first point in time during the run of an evolutionary algorithm when the global optimum is sampled) and an investigation of upper bounds, lower bounds, and the average time needed to hit the optimum. This chapter presents specific results for certain classes of objective functions, most of them defined over binary search spaces, as well as fundamental limitations of evolutionary search and related results on the "no free lunch" theorem and black box complexity. The chapter also discusses the corresponding techniques for analyses, such as drift analysis and the expected multiplicative distance decrease.

Concluding the set of theoretical chapters, the chapter "Stochastic Convergence", by Günter Rudolph, addresses theoretical results about the properties of evolutionary algorithms concerned with finding a globally optimal solution in the asymptotic limit. Such results exist for certain variants of evolutionary algorithms and under certain assumptions, and this chapter summarizes the existing results and integrates them into a common framework. This type of analysis is essential in qualifying evolutionary algorithms as global search algorithms and for understanding the algorithmic conditions for global convergence.

The remaining chapters in the area of evolutionary computation report some of the major current trends.

To start with, the chapter "Evolutionary Multiobjective Optimization", by Eckart Zitzler, focuses on the application of evolutionary algorithms to tasks that are characterized by multiple, conflicting objective functions. In this case, decision-making becomes a task of identifying good compromises between the conflicting criteria. This chapter introduces the concept and a variety of state-of-the-art algorithmic concepts to use evolutionary algorithms for approximating the so-called Pareto front of solutions which cannot be improved in one objective without compromising another. This contribution presents all of the required formal concepts, examples, and the algorithmic variations introduced into evolutionary computation to handle such types of problems and to generate good approximations of the Pareto front.

The term "memetic algorithms" is used to characterize hybridizations between evolutionary algorithms and more classical, local search methods (and agent-based systems). This is a general concept of broad scope, and in order to illustrate and characterize all possible instantiations, the chapter "Memetic Algorithms", by Natalio Krasnogor, presents an algorithmic engineering approach which allows one to describe these algorithms as instances of generic patterns. In addition to explaining some of the application areas, the chapter presents some theoretical remarks, various different ways to define memetic algorithms, and also an outlook into the future.

The chapter "Genetics-Based Machine Learning", by Tim Kovacs, extends the idea of evolutionary optimization to algorithmic concepts in machine learning and data mining, involving applications such as learning classifier systems, evolving neural networks, and genetic fuzzy systems, to mention just a few. Here, the application task is typically a data classification, data prediction, or nonlinear regression task — and the quality of solution candidates is evaluated by means of some model quality measure. The chapter covers a wide range of techniques for applying evolutionary computation to machine learning tasks, by interpreting them as optimization problems.

The chapter "Coevolutionary Principles", by Elena Popovici, Anthony Bucci, R. Paul Wiegand, and Edwin D. de Jong, deals with a concept modeled after biological evolution in which an explicit fitness function is not available, but solutions are evaluated by running them against each other. A solution is evaluated in the context of the other solutions, in the actual population or in another. Therefore, these algorithms develop their own dynamics, because the point of comparison is not stable, but coevolving with the actual population. The chapter provides a fundamental understanding of coevolutionary principles and highlights theoretical concepts, algorithms, and applications.

Finally, the chapter "Niching in Evolutionary Algorithms", by Ofer M. Shir, describes the biological principle of niching in nature as a concept for using a single population to find, occupy, and keep multiple local minima in a population. The motivation for this approach is to find alternative solutions within a single population and run of evolutionary algorithms, and this chapter discusses approaches for niching, and the application in the context of genetic algorithms as well as evolutionary strategies.

Molecular Computation

This area is covered by eight chapters.

The chapter "DNA Computing — Foundations and Implications", by Lila Kari, Shinnosuke Seki, and Petr Sosík, has a dual purpose. The first part outlines basic molecular biology notions necessary for understanding DNA computing, recounts the first experimental demonstration of DNA computing by Leonard Adleman in 1994, and recaps the 2001 milestone wet laboratory experiment that solved a 20-variable instance of 3-SAT and thus first demonstrated the potential of DNA computing to outperform the computational ability of an unaided human. The second part describes how the properties of DNA-based information, and in particular the Watson–Crick complementarity of DNA single strands, have influenced areas of theoretical computer science such as formal language theory, coding theory, automata theory, and combinatorics on words. More precisely, it explores several notions and results in formal language theory and coding theory that arose from the problem of the design of optimal encodings for DNA computing experiments (hairpin-free languages, bond-free languages), and more generally from the way information is encoded on DNA strands (sticker systems, Watson–Crick automata). Lastly, it describes the influence that properties of DNA-based information have had on research in combinatorics on words, by presenting several natural generalizations of classical concepts (pseudopalindromes, pseudoperiodicity, Watson–Crick conjugate and commutative words, involutively bordered words, pseudoknot bordered words), and outlining natural extensions in this context of two of the most fundamental results in combinatorics of words, namely the Fine and Wilf theorem and the Lyndon–Schützenberger result.

The chapter "Molecular Computing Machineries — Computing Models and Wet Implementations", by Masami Hagiya, Satoshi Kobayashi, Ken Komiya, Fumiaki Tanaka, and Takashi Yokomori, explores novel computing devices inspired by the biochemical properties of biomolecules. The theoretical results section describes a variety of molecular computing models for finite automata, as well as molecular computing models for Turing machines based on formal grammars, equality sets, Post systems, and logical formulae. It then presents molecular computing models that use structured molecules such as hairpins and tree structures. The section on wet implementations of molecular computing models, related issues, and applications includes: an enzyme-based DNA automaton and its applications to drug delivery, logic gates and circuits using DNAzymes and DNA tiles, reaction graphs for representing various dynamics of DNA assembly pathways, DNA whiplash machines implementing finite automata, and a hairpin-based implementation of a SAT engine for solving the 3-SAT problem.

The chapter "DNA Computing by Splicing and by Insertion–Deletion", by Gheorghe Păun, is devoted to two of the most developed computing models inspired by DNA biochemistry: computing by splicing, and computing by insertion and deletion. DNA computing by splicing was defined by Tom Head already in 1987 and is based on the so-called splicing operation. The splicing operation models the recombination of DNA molecules that results from cutting them with restriction enzymes and then pasting DNA molecules with compatible ends by ligase enzymes. This chapter explores the computational power of the splicing operation showing that, for example, extended splicing systems starting from a finite language and using finitely many splicing rules can generate only the family of regular languages, while extended splicing systems starting from a finite language and using a regular set of rules can generate all recursively enumerable languages. Ways in which to avoid the impractical notion of a regular infinite set of rules, while maintaining the maximum computational power, are presented. They include using multisets and adding restrictions on the use of rules such as permitting contexts, forbidding contexts, programmed splicing systems, target languages, and double splicing. The second model presented, the insertion–deletion system, is based on a finite set of axioms and a finite set of contextual insertion rules and contextual deletion rules. Computational power results described here include the fact that insertion–deletion systems with context-free insertion rules of words of length at most one and context-free deletion rules of words of unbounded length can generate only regular languages. In contrast, for example, the family of insertion–deletion systems where the insertion contexts, deletion contexts, and the words to be inserted/deleted are all of length at most one, equals the family of recursively enumerable languages.

The chapter "Bacterial Computing and Molecular Communication", by Yasubumi Sakakibara and Satoshi Hiyama, investigates attempts to create autonomous cell-based Turing machines, as well as novel communication paradigms that use molecules as communication media. The first part reports experimental research on constructing *in vivo* logic circuits as well as efforts towards building *in vitro* and *in vivo* automata in the framework of DNA computing. Also, a novel framework is presented to develop a programmable and autonomous *in vivo* computer in a bacterium. The first experiment in this direction uses DNA circular strands (plasmids) together with the cell's protein-synthesis mechanism to execute a finite state automaton in *E. coli*. Molecular communication is a new communication paradigm that proposes the use of molecules as the information medium, instead of the traditional electromagnetic waves. Other distinctive features of molecular communication include its stochastic nature, its low energy consumption, the use of an aqueous transmission medium, and its high compatibility with biological systems. A molecular communication system starts with a sender (e.g., a genetically modified or an artificial cell) that generates molecules, encodes information onto the molecules (called information molecules), and emits the information molecules into a propagation environment (e.g., aqueous solution within and between cells). A molecular propagation system (e.g., lipid bilayer vesicles encapsulating the information molecules) actively transports the information molecules to an appropriate receiver. A receiver (e.g., a genetically modified or an artificial cell) selectively receives the transported information molecules, and biochemically reacts to the received information molecules, thus "decoding" the information. The chapter describes detailed examples of molecular communication system designs, experimental results, and research trends.

The chapter "Computational Nature of Gene Assembly in Ciliates", by Robert Brijder, Mark Daley, Tero Harju, Nataša Jonoska, Ion Petre, and Grzegorz Rozenberg, reviews several approaches and results in the computational study of gene assembly in ciliates. Ciliated protozoa contain two functionally different types of nuclei, the macronucleus and the micronucleus. The macronucleus contains the functional genes, while the genes of the micronucleus are not functional due to the presence of many interspersing noncoding DNA segments. In addition, in some ciliates, the coding segments of the genes are present in a permuted order compared to their order in the functional macronuclear genes. During the sexual process of conjugation, when two ciliates exchange genetic micronuclear information and form two new micronuclei, each of the ciliates has to "decrypt" the information contained in its new micronucleus to form its new functional macronucleus. This process is called gene assembly and involves deleting the noncoding DNA segments, as well as rearranging the coding segments in the correct order. The chapter describes two models of gene assembly, the intermolecular model based on the operations of circular insertion and deletion, and the intramolecular model based on the three operations of "loop, direct-repeat excision", "hairpin, inverted-repeat excision", and "double-loop, alternating repeat excision". A discussion follows of the mathematical properties of these models, such as the Turing machine computational power of contextual circular insertions and deletions, and properties of the gene assembly process called invariants, which hold independently of the molecular model and assembling strategy. Finally, the template-based recombination model is described, offering a plausible hypothesis (supported already by some experimental data) about the "bioware" that implements the gene assembly.

The chapter "DNA Memory", by Masanori Arita, Masami Hagiya, Masahiro Takinoue, and Fumiaki Tanaka, summarizes the efforts that have been made towards realizing Eric Baum's dream of building a DNA memory with a storage capacity vastly larger than the brain. The chapter first describes the research into strategies for DNA sequence design, i.e., for finding DNA sequences that satisfy DNA computing constraints such as uniform melting temperature, avoidance of undesirable Watson–Crick bonding between sequences, preventing secondary structures, avoidance of base repeats, and absence of forbidden sequences. Various implementations of memory operations, such as access, read, and write, are described. For example, the "access" to a memory word in Baum's associative memory model, where a memory word consists of a single-stranded portion representing the address and a double-stranded portion representing the data, can be implemented by using the Watson–Crick complement of the address fixed to a solid support. In the Nested Primer Molecular Memory, where the double-stranded data is flanked on both sides by address sequences, the data can be retrieved by Polymerase Chain Reaction (PCR) using the addresses as primer pairs. In the multiple hairpins DNA memory, the address is a catenation of hairpins and the data can be accessed only if the hairpins are opened in the correct order by a process called DNA branch migration. After describing implementations of writable and erasable hairpin memories either in solution or immobilized on surfaces, the topic of *in vivo* DNA memory is explored. As an example, the chapter describes how representing the digit 0 by regular codons, and the digit 1 by wobbled codons, was used to encode a word into an essential gene of *Bacillus subtilis*.

The chapter "Engineering Natural Computation by Autonomous DNA-Based Biomolecular Devices", by John H. Reif and Thomas H. LaBean, overviews DNA-based biomolecular devices that are autonomous (execute steps with no external control after starting) and programmable (the tasks executed can be modified without entirely redesigning the DNA nanostructures). Special attention is given to DNA tiles, roughly square-shaped DNA nanostructures that have four "sticky-ends" (DNA single strands) that can specifically bind them to other tiles via Watson–Crick complementarity, and thus lead to the self-assembly of larger and more complex structures. Such tiles have been used to execute sequential Boolean computation via linear DNA self-assembly or to obtain patterned 2D DNA lattices and Sierpinski triangles. Issues such as error correction and self-repair of DNA tiling are also addressed. Other described methods include the implementation of a DNA-based finite automaton via disassembly of a double-stranded DNA nanostructure effected by an enzyme, and the technique of whiplash PCR. Whiplash PCR is a method that can achieve state transitions by encoding both transitions and the current state of the computation on the same DNA single strand: The free end of the strand (encoding the current state) sticks to the appropriate transition rule on the strand forming a hairpin, is then extended by PCR to a new state, and finally is detached from the strand, this time with the new state encoded at its end. The technique of DNA origami is also described, whereby a scaffold strand (a long single DNA strand, such as from the sequence of a virus) together with many specially designed staple strands (short single DNA strands) self-assemble by folding the scaffold strand — with the aid of the staples — in a raster pattern that can create given arbitrary planar DNA nanostructures. DNA-based molecular machines are then described such as autonomous DNA walkers and programmable DNA nanobots (programmable autonomous DNA walker devices). A restriction-enzyme-based DNA walker consists of a DNA helix with two sticky-ends ("feet") that moves stepwise along a "road" (a DNA nanostructure with protruding "steps", i.e., single DNA strands).

The chapter "Membrane Computing", by Gheorghe Păun, describes theoretical results and applications of membrane computing, a branch of natural computing inspired by the architecture and functioning of living cells, as well as from the organization of cells in tissues, organs, or other higher-order structures. The cell is a hierarchical structure of compartments, defined by membranes, that selectively communicate with each other. The computing model that abstracts this structure is a membrane system (or P system, from the name of its inventor, Gheorghe Păun) whose main components are: the membrane structure, the multisets of objects placed in the compartments enveloped by the membranes, and the rules for processing the objects and the membranes. The rules are used to modify the objects in the compartments, to transport objects from one compartment to another, to dissolve membranes, and to create new membranes. The rules in each region of a P system are used in a maximally parallel manner, nondeterministically choosing the applicable rules and the objects to which they apply. A computation consists in repeatedly applying rules to an initial configuration of the P system, until no rule can be applied anymore, in which case the objects in a priori specified regions are considered the output of the computation. Several variants of P systems are described, including P systems with symport/antiport rules, P systems with active membranes, splicing P systems, P systems with objects on membranes, tissue-like P systems, and spiking neural P systems. Many classes of P systems are able to simulate Turing machines, hence they are computationally universal. For example, P systems with symport/antiport rules using only three objects and three membranes are computationally universal. In addition, several types of P systems have been used to solve NP-complete problems in polynomial time, by a space–time trade-off. Applications of P systems include, among others, modeling in biology, computer graphics, linguistics, economics, and cryptography.

Quantum Computation

This area is covered by six chapters.

The chapter "Mathematics for Quantum Information Processing", by Mika Hirvensalo, contains the standard Hilbert space formulation of finite-level quantum systems. This is the language and notational system allowing us to speak, describe, and make predictions about the objects of quantum physics. The chapter introduces the notion of quantum states as unit-trace, self-adjoint, positive mappings, and the vector state formalism is presented as a special case. The physical observables are introduced as complete collections of mutually orthogonal projections, and then it is discussed how this leads to the traditional representation of observables as self-adjoint mappings. The minimal interpretation, which is the postulate connecting the mathematical objects to the physical world, is presented. The treatment of compound quantum systems is based mostly on operative grounds. To provide enough tools for considering the dynamics needed in quantum computing, the formalism of treating state transformations as completely positive mappings is also presented. The chapter concludes by explaining how quantum versions of finite automata, Turing machines, and Boolean circuits fit into the Hilbert space formalism.

The chapter "Bell's Inequalities — Foundations and Quantum Communication", by Časlav Brukner and Marek Żukowski, is concerned with the nature of quantum mechanics. It presents the evidence that excludes two types of hypothetical deterministic theories: neither a nonlocal nor a noncontextual theory can explain quantum mechanics. This helps to build a true picture of quantum mechanics, and is therefore essential from the philosophical point of view. The Bell inequalities show that nonlocal deterministic theories cannot explain the quantum mechanism, and the Kochen–Specker theorem shows that noncontextual theories are not possible as underlying theories either. The traditional Bell theorem and its variants, GHZ and CHSH among them, are presented, and the Kochen–Specker theorem is discussed. In this chapter, the communication complexity is also treated by showing how the violations of classical locality and noncontextuality can be used as a resource for communication protocols. Stronger-than quantum violations of the CHSH inequality are also discussed. They are interesting, since it has been shown that if the violation of CHSH inequality is strong enough, then the communication complexity collapses into one bit (hence the communication complexity of the true physical world seems to settle somewhere between classical and stronger-than quantum).

The chapter "Algorithms for Quantum Computers", by Jamie Smith and Michele Mosca, introduces the most remarkable known methods that utilize the special features of quantum physics in order to gain advantage over classical computing. The importance of these methods is that they form the core of designing discrete quantum algorithms. The methods presented and discussed here are the quantum Fourier transform, amplitude amplification, and quantum walks. Then, as specific examples, Shor's factoring algorithm (quantum Fourier transform), Grover search (amplitude amplification), and element distinctness algorithms (quantum random walks) are presented. The chapter not only involves traditional methods, but it also contains discussion of continuous-time quantum random walks and, more importantly, an extensive presentation of an important recent development in quantum algorithms, viz., tensor network evaluation algorithms. Then, as an example, the approximate evaluation of Tutte polynomials is presented.

The chapter "Physical Implementation of Large-Scale Quantum Computation", by Kalle-Antti Suominen, discusses the potential ways of physically implementing quantum computers. First, the DiVincenzo criteria (requirements for building a successful quantum computer) are presented, and then quantum error correction is discussed. The history, physical properties, potentials, and obstacles of various possible physical implementations of quantum computers are covered. They involve: cavity QED, trapped ions, neutral atoms and single electrons, liquid-form molecular spin, nuclear and electron spins in silicon, nitrogen vacancies in diamond, solid-state qubits with quantum dots, superconducting charge, flux and phase quantum bits, and optical quantum computing.

The chapter "Quantum Cryptography", by Takeshi Koshiba, is concerned with quantum cryptography, which will most likely play an important role in future when quantum computers make the current public-key cryptosystems unreliable. It gives an overview of classical cryptosystems, discusses classical cryptographic protocols, and then introduces the quantum key distribution protocols BB84, B92, and BBM92. Also protocol OTU00, not known to be vulnerable under Shor's algorithm, is presented. In future, when quantum computers are available, cryptography will most probably be based on quantum protocols. The chapter presents candidates for such quantum protocols: KKNY05 and GC01 (for digital signatures). It concludes with a discussion of quantum commitment, oblivious transfer, and quantum zero-knowledge proofs.

The complexity class BQP is the quantum counterpart of the classical class BPP. Intuitively, BQP can be described as the class of problems solvable in "reasonable" time, and, hence, from the application-oriented point of view, it will likely become the most important complexity class in future, when quantum computers are available. The chapter "BQP-Complete Problems", by Shengyu Zhang, introduces the computational problems that capture the full hardness of BQP. In the very fundamental sense, no BQP-complete problems are known, but the promise problems (the probability distribution of outputs is restricted by promise) bring us as close as possible to the "hardest" problems in BQP, known as BQP-complete promise problems. The chapter discusses known BQP-complete promise problems. In particular, it is shown how to establish the BQP-completeness of the Local Hamiltonian Eigenvalue

Sampling problem and the Local Unitary Phase Sampling problem. The chapter concludes with an extensive study showing that the Jones Polynomial Approximation problem is a BQP-complete promise problem.

Broader Perspective

This area consists of two sections, "Nature-Inspired Algorithms" and "Alternative Models of Computation".

Nature-Inspired Algorithms

This section is covered by six chapters.

The chapter "An Introduction to Artificial Immune Systems", by Mark Read, Paul S. Andrews, and Jon Timmis, provides a general introduction to the field. It discusses the major research issues relating to the field of Artificial Immune Systems (AIS), exploring the underlying immunology that has led to the development of immune-inspired algorithms, and focuses on the four main algorithms that have been developed in recent years: clonal selection, immune network, negative selection, and dendritic cell algorithms; their use in terms of applications is highlighted. The chapter also covers evaluation of current AIS technology, and details some new frameworks and methodologies that are being developed towards more principled AIS research. As a counterpoint to the focus on applications, the chapter also gives a brief outline of how AIS research is being employed to help further the understanding of immunology.

The chapter on "Swarm Intelligence", by David W. Corne, Alan Reynolds, and Eric Bonabeau, attempts to demystify the term Swarm Intelligence (SI), outlining the particular collections of natural phenomena that SI most often refers to and the specific classes of computational algorithms that come under its definition. The early parts of the chapter focus on the natural inspiration side, with discussion of social insects and stigmergy, foraging behavior, and natural flocking behavior. Then the chapter moves on to outline the most successful of the computational algorithms that have emerged from these natural inspirations, namely ant colony optimization methods and particle swarm optimization, with also some discussion of different and emerging such algorithms. The chapter concludes with a brief account of current research trends in the field.

The chapter "Simulated Annealing", by Kathryn A. Dowsland and Jonathan M. Thompson, provides an overview of Simulated Annealing (SA), emphasizing its practical use. The chapter explains its inspiration from the field of statistical thermodynamics, and then overviews the theory, with an emphasis again on those aspects that are important for practical applications. The chapter then covers some of the main ways in which the basic SA algorithm has been modified by various researchers, leading to improved performance for a variety of problems. The chapter briefly surveys application areas, and ends with several useful pointers to associated resources, including freely available code.

The chapter "Evolvable Hardware", by Lukáš Sekanina, surveys this growing field. Starting with a brief overview of the reconfigurable devices used in this field, the elementary principles and open problems are introduced, and then the chapter considers, in turn, three main areas: extrinsic evolution (evolving hardware using simulators), intrinsic evolution (where the evolution is conducted within FPGAs, FPTAs, and so forth), and adaptive hardware (in which real-world adaptive hardware systems are presented). The chapter finishes with an overview of major achievements in the field.

The first of two application-centered chapters, "Natural Computing in Finance — A Review", by Anthony Brabazon, Jing Dang, Ian Dempsey, Michael O'Neill, and David Edelman, provides a rather comprehensive account of natural computing applications in what is, at the time of writing (and undoubtedly beyond), one of the hottest topics of the day. This chapter introduces us to the wide range of different financial problems to which natural computing methods have been applied, including forecasting, trading, arbitrage, portfolio management, asset allocation, credit risk assessment, and more. The natural computing areas that feature in this chapter are largely evolutionary computing, neural computing, and also agent-based modeling, swarm intelligence, and immune-inspired methods. The chapter ends with a discussion of promising future directions.

Finally, the chapter "Selected Aspects of Natural Computing", by David W. Corne, Kalyanmoy Deb, Joshua Knowles, and Xin Yao, provides detailed accounts of a collection of example natural computing applications, each of which is remarkable or particularly interesting in some way. The thrust of this chapter is to provide, via such examples, an idea of both the significant impact that natural computing has already had, as well as its continuing significant promise for future applications in all areas of science and industry. While presenting this eclectic collection of marvels, the chapter also aims at clarity and demystification, providing much detail that helps see how the natural computing methods in question were applied to achieve the stated results. Applications covered include Blondie24 (the evolutionary neural network application that achieves master-level skill at the game of checkers), the design of novel antennas using evolutionary computation in conjunction with developmental computing, and the classic application of learning classifier systems that led to novel fighter-plane maneuvers for the USAF.

Alternative Models of Computation

This section is covered by seven chapters.

The chapter "Artificial Life", by Wolfgang Banzhaf and Barry McMullin, traces the roots, raises key questions, discusses the major methodological tools, and reviews the main applications of this exciting and maturing area of computing. The chapter starts with a historical overview, and presents the fundamental questions and issues that Artificial Life is concerned with. Thus the chapter surveys discussions and viewpoints about the very nature of the differences between living and nonliving systems, and goes on to consider issues such as hierarchical design, self-construction, and self-maintenance, and the emergence of complexity. This part of the chapter ends with a discussion of "Coreworld" experiments, in which a number of systems have been studied that allow spontaneous evolution of computer programs. The chapter moves on to survey the main theory and formalisms used in Artificial Life, including cellular automata and rewriting systems. The chapter concludes with a review and restatement of the main objectives of Artificial Life research, categorizing them respectively into questions about the origin and nature of life, the potential and limitations of living systems, and the relationships between life and intelligence, culture, and other human constructs.

The chapter "Algorithmic Systems Biology — Computer Science Propels Systems Biology", by Corrado Priami, takes the standpoint of computing as providing a philosophical foundation for systems biology, with at least the same importance as mathematics, chemistry, or physics. The chapter highlights the value of algorithmic approaches in modeling, simulation, and analysis of biological systems. It starts with a high-level view of how models and experiments can be tightly integrated within an algorithmic systems biology vision, and then deals in turn with modeling languages, simulations of models, and finally the postprocessing of results from biological models and how these lead to new hypotheses that can then re-enter the modeling/simulation cycle.

The chapter "Process Calculi, Systems Biology and Artificial Chemistry", by Pierpaolo Degano and Andrea Bracciali, concentrates on the use of process calculi and related techniques for systems-level modeling of biological phenomena. This chapter echoes the broad viewpoint of the previous chapter, but its focus takes us towards a much deeper understanding of the potential mappings between formal systems in computer science and systems interpretation of biological processes. It starts by surveying the basics of process calculi, setting out their obvious credentials for modeling concurrent, distributed systems of interacting parts, and mapping these onto a "cells as computers" view. After a process calculi treatment of systems biology, the chapter goes on to examine process calculi as a route towards artificial chemistry. After considering the formal properties of the models discussed, the chapter ends with notes on some case studies showing the value of process calculi in modeling biological phenomena; these include investigating the concept of a "minimal gene set" prokaryote, modeling the nitric oxide-cGMP pathway (central to many signal transduction mechanisms), and modeling the calyx of Held (a large synapse structure in the mammalian auditory central nervous system).

The chapter on "Reaction—Diffusion Computing", by Andrew Adamatzky and Benjamin De Lacy Costello, introduces the reader to the concept of a reaction—diffusion computer. This is a spatially extended chemical system, which processes information via transforming an input profile of ingredients (in terms of different concentrations of constituent ingredients) into an output profile of ingredients. The chapter takes us through the elements of this field via case studies, and it shows how selected tasks in computational geometry, robotics, and logic can be addressed by chemical implementations of reaction—diffusion computers. After introducing the field and providing a treatment of its origins and main achievements, a classical view of reaction—diffusion computers is then described. The chapter moves on to discuss varieties of reaction—diffusion processors and their chemical constituents, covering applications to the aforementioned tasks. The chapter ends with the authors' thoughts on future developments in this field.

The chapter "Rough—Fuzzy Computing", by Andrzej Skowron, shifts our context towards addressing a persistent area of immense difficulty for classical computing, which is the fact that real-world reasoning is usually done in the face of inaccurate, incomplete, and often inconsistent evidence. In essence, concepts in the real world are vague, and computation needs ways to address this. We are hence treated, in this chapter, to an overarching view of rough set theory, fuzzy set theory, their hybridization, and applications. Rough and fuzzy computing are broadly complementary approaches to handling vagueness, focusing respectively on capturing the level of distinction between separate objects and the level of membership of an object in a set. After presenting the basic concepts of rough computing and fuzzy computing in turn, in each case going into some detail on the main theoretical results and practical considerations, the chapter goes on to discuss how they can be, and have been, fruitfully combined. The chapter ends with an overview of the emerging field of "Wisdom Technology" (Wistech) as a paradigm for developing modern intelligent systems.

The chapter "Collision-Based Computing", by Andrew Adamatzky and Jérôme Durand-Lose, presents and discusses the computations performed as a result of spatial localizations in systems that exhibit dynamic spatial patterns over time. For example, a collision may be between two gliders in a cellular automaton, or two separate wave fragments within an excitable chemical system. This chapter introduces us to the basics of collision-based computing and overviews collision-based computing schemes in 1D and 2D cellular automata as well as continuous excitable media. Then, after some theoretical foundations relating to 1D cellular automata, the chapter presents a collision-based implementation for a 1D Turing machine and for cyclic tag systems. The chapter ends with discussion and presentation of "Abstract Geometrical Computation", which can be seen as collision-based computation in a medium that is the continuous counterpart of cellular automata.

The chapter "Nonclassical Computation — A Dynamical Systems Perspective", by Susan Stepney, takes a uniform view of computation, in which inspiration from a dynamical systems perspective provides a convenient way to consider, in one framework, both classical discrete systems and systems performing nonclassical computation. In particular, this viewpoint presents a way towards computational interpretation of physical embodied systems that exploit their natural dynamics. The chapter starts by discussing "closed" dynamical systems, those whose dynamics involve no inputs from an external environment, examining their computational abilities from a dynamical systems perspective. Then it discusses continuous dynamical systems and shows how these too can be interpreted computationally, indicating how material embodiment can give computation "for free", without the need to explicitly implement the dynamics. The outlook then broadens to consider open systems, where the dynamics are affected by external inputs. The chapter ends by looking at constructive, or developmental, dynamical systems, whose state spaces change during computation. These latter discussions approach the arena of biological and other natural systems, casting them as computational, open, developmental, dynamical systems.

Acknowledgements

This handbook resulted from a highly collaborative effort. The handbook and area editors are grateful to the chapter writers for their efforts in writing chapters and delivering them on time, and for their participation in the refereeing process.

We are indebted to the members of the Advisory Board for their valuable advice and fruitful interactions. Additionally, we want to acknowledge David Fogel, Pekka Lahti, Robert LaRue, Jason Lohn, Michael Main, David Prescott, Arto Salomaa, Kai Salomaa, Shinnosuke Seki, and Rob Smith, for their help and advice in various stages of production of this handbook. Last, but not least, we are thankful to Springer, especially to Ronan Nugent, for intense and constructive cooperation in bringing this project from its inception to its successful conclusion.

Leiden; Edinburgh; Nijmegen;   Grzegorz Rozenberg (Main Handbook Editor)
Turku; London, Ontario   Thomas Bäck (Handbook Editor and Area Editor)
October 2010   Joost N. Kok (Handbook Editor and Area Editor)
                                                      David W. Corne (Area Editor)
                                                      Tom Heskes (Area Editor)
                                                      Mika Hirvensalo (Area Editor)
                                                      Jarkko J. Kari (Area Editor)
                                                      Lila Kari (Area Editor)

Editor Biographies

Prof. Dr. Grzegorz Rozenberg

Prof. Rozenberg was awarded his Ph.D. in mathematics from the Polish Academy of Sciences, Warsaw, and he has since held full-time positions at Utrecht University, the State University of New York at Buffalo, and the University of Antwerp. Since 1979 he has been a professor at the Department of Computer Science of Leiden University, and an adjunct professor at the Department of Computer Science of the University of Colorado at Boulder, USA. He is the founding director of the Leiden Center for Natural Computing.

Among key editorial responsibilities over the last 30 years, he is the founding Editor of the book series Texts in Theoretical Computer Science (Springer) and Monographs in Theoretical Computer Science (Springer), founding Editor of the book series Natural Computing (Springer), founding Editor-in-Chief of the journal Natural Computing (Springer), and founding Editor of Part C (Theory of Natural Computing) of the journal Theoretical Computer Science (Elsevier). Altogether he's on the Editorial Board of around 20 scientific journals.

He has authored more than 500 papers and 6 books, and coedited more than 90 books. He coedited the "Handbook of Formal Languages" (Springer), he was Managing Editor of the "Handbook of Graph Grammars and Computing by Graph Transformation" (World Scientific), he coedited "Current Trends in Theoretical Computer Science" (World Scientific), and he coedited "The Oxford Handbook of Membrane Computing" (Oxford University Press).

He is Past President of the European Association for Theoretical Computer Science (EATCS), and he received the Distinguished Achievements Award of the EATCS "in recognition of his outstanding scientific contributions to theoretical computer science". Also he is a cofounder and Past President of the International Society for Nanoscale Science, Computation, and Engineering (ISNSCE).

He has served as a program committee member for most major conferences on theoretical computer science in Europe, and among the events he has founded or helped to establish are the International Conference on Developments in Language Theory (DLT), the International

Conference on Graph Transformation (ICGT), the International Conference on Unconventional Computation (UC), the International Conference on Application and Theory of Petri Nets and Concurrency (ICATPN), and the DNA Computing and Molecular Programming Conference.

In recent years his research has focused on natural computing, including molecular computing, computation in living cells, self-assembly, and the theory of biochemical reactions. His other research areas include the theory of concurrent systems, the theory of graph transformations, formal languages and automata theory, and mathematical structures in computer science.

Prof. Rozenberg is a Foreign Member of the Finnish Academy of Sciences and Letters, a member of the Academia Europaea, and an honorary member of the World Innovation Foundation. He has been awarded honorary doctorates by the University of Turku, the Technical University of Berlin, and the University of Bologna. He is an ISI Highly Cited Researcher.

He is a performing magician, and a devoted student of and expert on the paintings of Hieronymus Bosch.

Prof. Dr. Thomas Bäck

Prof. Bäck was awarded his Ph.D. in Computer Science from Dortmund University in 1994, for which he received the Best Dissertation Award from the Gesellschaft für Informatik (GI). He has been at Leiden University since 1996, where he is currently full Professor for Natural Computing and the head of the Natural Computing Research Group at the Leiden Institute of Advanced Computer Science (LIACS).

He has authored more than 150 publications on natural computing technologies. He wrote a book on evolutionary algorithms, "Evolutionary Algorithms in Theory and Practice" (Oxford University Press), and he coedited the "Handbook of Evolutionary Computation" (IOP/Oxford University Press).

Prof. Bäck is an Editor of the book series Natural Computing (Springer), an Associate Editor of the journal Natural Computing (Springer), an Editor of the journal Theoretical Computer Science (Sect. C, Theory of Natural Computing; Elsevier), and an Advisory Board member of the journal Evolutionary Computation (MIT Press). He has served as program chair for all major conferences in evolutionary computation, and is an Elected Fellow of the International Society for Genetic and Evolutionary Computation for his contributions to the field.

His main research interests are the theory of evolutionary algorithms, cellular automata and data-driven modelling, and applications of these methods in medicinal chemistry, pharmacology and engineering.

Prof. Dr. Joost N. Kok

Prof. Kok was awarded his Ph.D. in Computer Science from the Free University in Amsterdam in 1989, and he has worked at the Centre for Mathematics and Computer Science in Amsterdam, at Utrecht University, and at the Åbo Akademi University in Finland. Since 1995 he has been a professor in computer science, and since 2005 also a professor in medicine at Leiden University. He is the Scientific Director of the Leiden Institute of Advanced Computer Science, and leads the research clusters Algorithms and Foundations of Software Technology.

He serves as a chair, member of the management team, member of the board, or member of the scientific committee of the Faculty of Mathematics and Natural Sciences of Leiden University, the ICT and Education Committee of Leiden University, the Dutch Theoretical Computer Science Association, the Netherlands Bioinformatics Centre, the Centre for Mathematics and Computer Science Amsterdam, the Netherlands Organisation for Scientific Research, the Research Foundation Flanders (Belgium), the European Educational Forum, and the International Federation for Information Processing (IFIP) Technical Committee 12 (Artificial Intelligence).

Prof. Kok is on the steering, scientific or advisory committees of the following events: the Mining and Learning with Graphs Conference, the Intelligent Data Analysis Conference, the Institute for Programming and Algorithms Research School, the Biotechnological Sciences Delft–Leiden Research School, and the European Conference on Machine Learning and Principles and Practice of Knowledge Discovery in Databases. And he has been a program committee member for more than 100 international conferences, workshops or summer schools on data mining, data analysis, and knowledge discovery; neural networks; artificial intelligence; machine learning; computational life science; evolutionary computing, natural computing and genetic algorithms; Web intelligence and intelligent agents; and software engineering.

He is an Editor of the book series Natural Computing (Springer), an Associate Editor of the journal Natural Computing (Springer), an Editor of the journal Theoretical Computer Science (Sect. C, Theory of Natural Computing; Elsevier), an Editor of the Journal of Universal

Computer Science, an Associate Editor of the journal Computational Intelligence (Wiley), and a Series Editor of the book series Frontiers in Artificial Intelligence and Applications (IOS Press).

His academic research is concentrated around the themes of scientific data management, data mining, bioinformatics, and algorithms, and he has collaborated with more than 20 industrial partners.

Advisory Board

Area Editors

Cellular Automata

Jarkko J. Kari
University of Turku
Turku
Finland
jkari@utu.fi

Neural Computation

Tom Heskes
Radboud Universiteit Nijmegen
Nijmegen
The Netherlands
t.heskes@science.ru.nl

Joost N. Kok
LIACS
Leiden University
Leiden
The Netherlands
joost@liacs.nl

Evolutionary Computation

Thomas Bäck
LIACS
Leiden University
Leiden
The Netherlands
baeck@liacs.nl

Molecular Computation

Lila Kari
University of Western Ontario
London
Canada
lila@csd.uwo.ca

Quantum Computation

Mika Hirvensalo
University of Turku
Turku
Finland
mikhirve@utu.fi

Broader Perspective

David W. Corne
Heriot-Watt University
Edinburgh
UK
dwcorne@macs.hw.ac.uk

Table of Contents

Volume 1

Volume 2

Volume 3

Volume 4

List of Contributors

Andrew Adamatzky
Department of Computer Science
University of the West of England
Bristol
UK
andrew.adamatzky@uwe.ac.uk

Paul S. Andrews
Department of Computer Science
University of York
UK
psa@cs.york.ac.uk

Masanori Arita
Department of Computational Biology,
Graduate School of Frontier Sciences
The University of Tokyo
Kashiwa
Japan
arita@k.u-tokyo.ac.jp

Wolfgang Banzhaf
Department of Computer Science
Memorial University of Newfoundland
St. John's, NL
Canada
banzhaf@cs.mun.ca

Hans-Georg Beyer
Department of Computer Science
Fachhochschule Vorarlberg
Dornbirn
Austria
hans-georg.beyer@fhv.at

Sander Bohte
Research Group Life Sciences
CWI
Amsterdam
The Netherlands
s.m.bohte@cwi.nl

Eric Bonabeau
Icosystem Corporation
Cambridge, MA
USA
eric@icosystem.com

Anthony Brabazon
Natural Computing Research and
Applications Group
University College Dublin
Ireland
anthony.brabazon@ucd.ie

Andrea Bracciali
Department of Computing Science and
Mathematics
University of Stirling
UK
braccia@cs.stir.ac.uk

Robert Brijder
Leiden Institute of Advanced Computer
Science
Universiteit Leiden
The Netherlands
robert.brijder@uhasselt.be

Časlav Brukner
Faculty of Physics
University of Vienna
Vienna
Austria
caslav.brukner@univie.ac.at

Anthony Bucci
Icosystem Corporation
Cambridge, MA
USA
anthony@icosystem.com

Ke Chen
Department of Electrical and Computer
Engineering
University of Alberta
Edmonton, AB
Canada
kchen1@ece.ualberta.ca

Seungjin Choi
Pohang University of Science and
Technology
Pohang
South Korea
seungjin@postech.ac.kr

Bastien Chopard
Scientific and Parallel Computing Group
University of Geneva
Switzerland
bastien.chopard@unige.ch

David W. Corne
School of Mathematical and Computer
Sciences
Heriot-Watt University
Edinburgh
UK
dwcorne@macs.hw.ac.uk

Mark Daley
Departments of Computer Science and
Biology
University of Western Ontario
London, Ontario
Canada
daley@csd.uwo.ca

Jing Dang
Natural Computing Research and
Applications Group
University College Dublin
Ireland
jing.dang@ucd.ie

Edwin D. de Jong
Institute of Information and Computing
Sciences
Utrecht University
The Netherlands
dejong@cs.uu.nl

Kenneth De Jong
Department of Computer Science
George Mason University
Fairfax, VA
USA
kdejong@gmu.edu

Benjamin De Lacy Costello
Centre for Research in Analytical, Material
and Sensor Sciences, Faculty of Applied
Sciences
University of the West of England
Bristol
UK
ben.delacycostello@uwe.ac.uk

Kalyanmoy Deb
Department of Mechanical Engineering
Indian Institute of Technology
Kanpur
India
deb@iitk.ac.in

Pierpaolo Degano
Dipartimento di Informatica
Università di Pisa
Italy
degano@di.unipi.it

Marianne Delorme
Laboratoire d'Informatique Fondamentale
de Marseille (LIF)
Aix-Marseille Université and CNRS
Marseille
France
delorme.marianne@orange.fr

Ian Dempsey
Pipeline Financial Group, Inc.
New York, NY
USA
ian.dempsey@pipelinefinancial.com

Alberto Dennunzio
Dipartimento di Informatica
Sistemistica e Comunicazione, Università
degli Studi di Milano-Bicocca
Italy
dennunzio@disco.unimib.it

Kathryn A. Dowsland
Gower Optimal Algorithms, Ltd.
Swansea
UK
k.a.dowsland@btconnect.com

Jérôme Durand-Lose
LIFO
Université d'Orléans
France
jerome.durand-lose@univ-orleans.fr

David Edelman
School of Business
UCD Michael Smurfit Graduate Business
School
Dublin
Ireland
david.edelman@ucd.ie

Gary B. Fogel
Natural Selection, Inc.
San Diego, CA
USA
gfogel@natural-selection.com

Enrico Formenti
Département d'Informatique
Université de Nice-Sophia Antipolis
France
enrico.formenti@unice.fr

Xinbo Gao
Video and Image Processing System Lab,
School of Electronic Engineering
Xidian University
China
xbgao@ieee.org

Shuzhi Sam Ge
Social Robotics Lab
Interactive Digital Media Institute, The
National University of Singapore
Singapore
elegesz@nus.edu.sg

Masami Hagiya
Department of Computer Science
Graduate School of Information Science
and Technology
The University of Tokyo
Tokyo
Japan
hagiya@is.s.u-tokyo.ac.jp

Tero Harju
Department of Mathematics
University of Turku
Finland
harju@utu.fi

Mika Hirvensalo
Department of Mathematics
University of Turku
Finland
mikhirve@utu.fi

Satoshi Hiyama
Research Laboratories
NTT DOCOMO, Inc.
Yokosuka
Japan
hiyama@nttdocomo.co.jp

Guang-Bin Huang
School of Electrical and Electronic
Engineering
Nanyang Technological University
Singapore
egbhuang@ntu.edu.sg

Thomas Jansen
Department of Computer Science
University College Cork
Ireland
t.jansen@cs.ucc.ie

Nataša Jonoska
Department of Mathematics
University of South Florida
Tampa, FL
USA
jonoska@math.usf.edu

Jarkko J. Kari
Department of Mathematics
University of Turku
Turku
Finland
jkari@utu.fi

Lila Kari
Department of Computer Science
University of Western Ontario
London
Canada
lila@csd.uwo.ca

Sungchul Kim
Data Mining Lab, Department of Computer
Science and Engineering
Pohang University of Science and
Technology
Pohang
South Korea
subright@postech.ac.kr

Joshua Knowles
School of Computer Science and
Manchester Interdisciplinary
Biocentre (MIB)
University of Manchester
UK
j.knowles@manchester.ac.uk

Satoshi Kobayashi
Department of Computer Science
University of Electro-Communications
Tokyo
Japan
satoshi@cs.uec.ac.jp

Ken Komiya
Interdisciplinary Graduate School of
Science and Engineering
Tokyo Institute of Technology
Yokohama
Japan
komiya@dis.titech.ac.jp

Takeshi Koshiba
Graduate School of Science and
Engineering
Saitama University
Japan
koshiba@mail.saitama-u.ac.jp

Tim Kovacs
Department of Computer Science
University of Bristol
UK
kovacs@cs.bris.ac.uk

Natalio Krasnogor
Interdisciplinary Optimisation Laboratory,
The Automated Scheduling, Optimisation
and Planning Research Group, School of
Computer Science
University of Nottingham
UK
natalio.krasnogor@nottingham.ac.uk

Lukasz A. Kurgan
Department of Electrical and Computer Engineering
University of Alberta
Edmonton, AB
Canada
lkurgan@ece.ualberta.ca

Petr Kůrka
Center for Theoretical Studies
Academy of Sciences and Charles University in Prague
Czechia
kurka@cts.cuni.cz

Thomas H. LaBean
Department of Computer Science and Department of Chemistry and Department of Biomedical Engineering
Duke University
Durham, NC
USA
thomas.labean@duke.edu

Kang Li
School of Electronics, Electrical Engineering and Computer Science
Queen's University
Belfast
UK
k.li@ee.qub.ac.uk

Xuelong Li
Center for OPTical IMagery Analysis and Learning (OPTIMAL), State Key Laboratory of Transient Optics and Photonics
Xi'an Institute of Optics and Precision Mechanics, Chinese Academy of Sciences
Xi'an, Shaanxi
China
xuelong_li@opt.ac.cn

Wen Lu
Video and Image Processing System Lab,
School of Electronic Engineering
Xidian University
China
luwen@mail.xidian.edu.cn

Jacques Mazoyer
Laboratoire d'Informatique Fondamentale de Marseille (LIF)
Aix-Marseille Université and CNRS
Marseille
France
mazoyerj2@orange.fr

Barry McMullin
Artificial Life Lab, School of Electronic Engineering
Dublin City University
Ireland
barry.mcmullin@dcu.ie

Silja Meyer-Nieberg
Fakultät für Informatik
Universität der Bundeswehr München
Neubiberg
Germany
silja.meyer-nieberg@unibw.de

Kenichi Morita
Department of Information Engineering,
Graduate School of Engineering
Hiroshima University
Japan
morita@iec.hiroshima-u.ac.jp

Michele Mosca
Institute for Quantum Computing and Department of Combinatorics & Optimization
University of Waterloo and St. Jerome's University and Perimeter Institute for Theoretical Physics
Waterloo
Canada
mmosca@iqc.ca

Nicolas Ollinger
Laboratoire d'informatique fondamentale
de Marseille (LIF)
Aix-Marseille Université, CNRS
Marseille
France
nicolas.ollinger@lif.univ-mrs.fr

Michael O'Neill
Natural Computing Research and
Applications Group
University College Dublin
Ireland
m.oneill@ucd.ie

Hélène Paugam-Moisy
Laboratoire LIRIS – CNRS
Université Lumière Lyon 2
Lyon
France
and
INRIA Saclay – Ile-de-France
Université Paris-Sud
Orsay
France
helene.paugam-moisy@univ-lyon2.fr
hpaugam@lri.fr

Gheorghe Păun
Institute of Mathematics of the Romanian
Academy
Bucharest
Romania
and
Department of Computer Science and
Artificial Intelligence
University of Seville
Spain
gpaun@us.es
george.paun@imar.ro

Ion Petre
Department of Information Technologies
Åbo Akademi University
Turku
Finland
ipetre@abo.fi

Riccardo Poli
Department of Computing and Electronic
Systems
University of Essex
Colchester
UK
rpoli@essex.ac.uk

Elena Popovici
Icosystem Corporation
Cambridge, MA
USA
elena@icosystem.com

Corrado Priami
Microsoft Research
University of Trento Centre for
Computational and Systems Biology
(CoSBi)
Trento
Italy
and
DISI
University of Trento
Trento
Italy
priami@cosbi.eu

Mark Read
Department of Computer Science
University of York
UK
markread@cs.york.ac.uk

John H. Reif
Department of Computer Science
Duke University
Durham, NC
USA
reif@cs.duke.edu

Alan Reynolds
School of Mathematical and Computer
Sciences
Heriot-Watt University
Edinburgh
UK
a.reynolds@hw.ac.uk

Grzegorz Rozenberg
Leiden Institute of Advanced Computer
Science
Universiteit Leiden
The Netherlands
and
Department of Computer Science
University of Colorado
Boulder, CO
USA
rozenber@liacs.nl

Günter Rudolph
Department of Computer Science
TU Dortmund
Dortmund
Germany
guenter.rudolph@tu-dortmund.de

Yasubumi Sakakibara
Department of Biosciences and Informatics
Keio University
Yokohama
Japan
yasu@bio.keio.ac.jp

Lukáš Sekanina
Faculty of Information Technology
Brno University of Technology
Brno
Czech Republic
sekanina@fit.vutbr.cz

Shinnosuke Seki
Department of Computer Science
University of Western Ontario
London
Canada
sseki@csd.uwo.ca

Ofer M. Shir
Department of Chemistry
Princeton University
NJ
USA
oshir@princeton.edu

Andrzej Skowron
Institute of Mathematics
Warsaw University
Poland
skowron@mimuw.edu.pl

Jamie Smith
Institute for Quantum Computing and
Department of Combinatorics &
Optimization
University of Waterloo
Canada
ja5smith@iqc.ca

Petr Sosík
Institute of Computer Science
Silesian University in Opava
Czech Republic
and
Departamento de Inteligencia Artificial
Universidad Politécnica de Madrid
Spain
petr.sosik@fpf.slu.cz

Susan Stepney
Department of Computer Science
University of York
UK
susan.stepney@cs.york.ac.uk

Kalle-Antti Suominen
Department of Physics and Astronomy
University of Turku
Finland
kalle-antti.suominen@utu.fi

Andrew M. Sutton
Department of Computer Science
Colorado State University
Fort Collins, CO
USA
sutton@cs.colostate.edu

Siamak Taati
Department of Mathematics
University of Turku
Finland
siamak.taati@gmail.com

Masahiro Takinoue
Department of Physics
Kyoto University
Kyoto
Japan
takinoue@chem.scphys.kyoto-u.ac.jp

Fumiaki Tanaka
Department of Computer Science
Graduate School of Information Science
and Technology
The University of Tokyo
Tokyo
Japan
fumi95@is.s.u-tokyo.ac.jp

Dacheng Tao
School of Computer Engineering
Nanyang Technological University
Singapore
dacheng.tao@gmail.com

Véronique Terrier
GREYC, UMR CNRS 6072
Université de Caen
France
veroniqu@info.unicaen.fr

Jonathan M. Thompson
School of Mathematics
Cardiff University
UK
thompsonjm1@cardiff.ac.uk

Jon Timmis
Department of Computer Science and
Department of Electronics
University of York
UK
jtimmis@cs.york.ac.uk

Joaquin J. Torres
Institute "Carlos I" for Theoretical and
Computational Physics and Department of
Electromagnetism and Matter Physics,
Facultad de Ciencias
Universidad de Granada
Spain
jtorres@ugr.es

Marc M. Van Hulle
Laboratorium voor Neurofysiologie
K.U. Leuven
Leuven
Belgium
marc@neuro.kuleuven.be

Leonardo Vanneschi
Department of Informatics, Systems and
Communication
University of Milano-Bicocca
Italy
vanneschi@disco.unimib.it

Pablo Varona
Departamento de Ingeniería Informática
Universidad Autónoma de Madrid
Spain
pablo.varona@uam.es

Darrell Whitley
Department of Computer Science
Colorado State University
Fort Collins, CO
USA
whitley@cs.colostate.edu

R. Paul Wiegand
Institute for Simulation and Training
University of Central Florida
Orlando, FL
USA
wiegand@ist.ucf.edu

Xin Yao
Natural Computation Group, School of
Computer Science
University of Birmingham
UK
x.yao@cs.bham.ac.uk

Takashi Yokomori
Department of Mathematics, Faculty of
Education and Integrated Arts and
Sciences
Waseda University
Tokyo
Japan
yokomori@waseda.jp

Hwanjo Yu
Data Mining Lab, Department of Computer
Science and Engineering
Pohang University of Science and
Technology
Pohang
South Korea
hwanjoyu@postech.ac.kr

Jean-Baptiste Yunès
Laboratoire LIAFA
Université Paris 7 (Diderot)
France
jean-baptiste.yunes@liafa.jussieu.fr

G. Peter Zhang
Department of Managerial Sciences
Georgia State University
Atlanta, GA
USA
gpzhang@gsu.edu

Jie Zhang
School of Chemical Engineering and
Advanced Materials
Newcastle University
Newcastle upon Tyne
UK
jie.zhang@newcastle.ac.uk

Shengyu Zhang
Department of Computer Science and
Engineering
The Chinese University of Hong Kong
Hong Kong S.A.R.
China
syzhang@cse.cuhk.edu.hk

Eckart Zitzler
PHBern – University of Teacher Education,
Institute for Continuing Professional
Education
Bern
Switzerland
eckart.zitzler@phbern.ch
eckart.zitzler@tik.ee.ethz.ch

Marek Żukowski
Institute of Theoretical Physics and
Astrophysics
University of Gdansk
Poland
marek.zukowski@univie.ac.at
fizmz@univ.gda.pl

Molecular Computation

Lila Kari

33 DNA Computing — Foundations and Implications*

Lila Kari[1] · *Shinnosuke Seki*[2] · *Petr Sosík*[3,4]
[1]Department of Computer Science, University of Western Ontario,
London, Canada
lila@csd.uwo.ca
[2]Department of Computer Science, University of Western Ontario,
London, Canada
sseki@csd.uwo.ca
[3]Institute of Computer Science, Silesian University in Opava,
Czech Republic
petr.sosik@fpf.slu.cz
[4]Departamento de Inteligencia Artificial, Universidad Politécnica de
Madrid, Spain

* This work was supported by The Natural Sciences and Engineering Council of Canada Discovery Grant and Canada
Research Chair Award to L.K.

G. Rozenberg et al. (eds.), *Handbook of Natural Computing*, DOI 10.1007/978-3-540-92910-9_33,
© Springer-Verlag Berlin Heidelberg 2012

Abstract

DNA computing is an area of natural computing based on the idea that molecular biology processes can be used to perform arithmetic and logic operations on information encoded as DNA strands. The first part of this review outlines basic molecular biology notions necessary for understanding DNA computing, recounts the first experimental demonstration of DNA computing (Adleman's 7-vertex Hamiltonian Path Problem), and recounts the milestone wet laboratory experiment that first demonstrated the potential of DNA computing to outperform the computational ability of an unaided human (20 variable instance of 3-SAT).

The second part of the review describes how the properties of DNA-based information, and in particular the Watson–Crick complementarity of DNA single strands, have influenced areas of theoretical computer science such as formal language theory, coding theory, automata theory and combinatorics on words. More precisely, we describe the problem of DNA encodings design, present an analysis of intramolecular bonds, define and characterize languages that avoid certain undesirable intermolecular bonds, and investigate languages whose words avoid even imperfect bindings between their constituent strands. We also present another, vectorial, representation of DNA strands, and two computational models based on this representation: sticker systems and Watson–Crick automata. Lastly, we describe the influence that properties of DNA-based information have had on research in combinatorics on words, by enumerating several natural generalizations of classical concepts of combinatorics of words: pseudopalindromes, pseudoperiodicity, Watson–Crick conjugate and commutative words, involutively bordered words, pseudoknot bordered words. In addition, we outline natural extensions in this context of two of the most fundamental results in combinatorics of words, namely Fine and Wilf's theorem and the Lyndon–Schutzenberger result.

1 Introduction

DNA computing is an area of natural computing based on the idea that molecular biology processes can be used to perform arithmetic and logic operations on information encoded as DNA strands. The aim of this review is twofold. First, the fundamentals of DNA computing, including basics of DNA structure and bio-operations, and two historically important DNA computing experiments, are introduced. Second, some of the ways in which DNA computing research has impacted fields in theoretical computer science are described.

The first part of this chapter outlines basic molecular biology notions necessary for understanding DNA computing (❷ Sect. 2), recounts the first experimental demonstration of DNA computing (❷ Sect. 3), as well as the milestone wet laboratory experiment that first demonstrated the potential of DNA computing to outperform the computational ability of an unaided human (❷ Sect. 4).

The second part of the chapter describes how the properties of DNA-based information, and, in particular, the Watson–Crick (WK) complementarity of DNA single strands have influenced areas of theoretical computer science such as formal language theory, coding theory, automata theory, and combinatorics on words. More precisely, ❷ Sect. 5 summarizes notions and results in formal language theory and coding theory arising from the problem of design of optimal encodings for DNA computing experiments: ❷ Section 5.1 describes the problem of DNA encodings design, ❷ Sect. 5.2 consists of an analysis of intramolecular bonds

(bonds within a given DNA strand), ❯ Sect. 5.3 defines and characterizes languages that avoid certain undesirable intermolecular bonds (bonds between two or more DNA strands), and ❯ Sect. 5.4 investigates languages whose words avoid even imperfect bindings between their constituent strands.

❯ Section 6 contains another representation (vectorial) of DNA strands and two computational models based on this representation: sticker systems and Watson–Crick automata. After a brief description of the representation of DNA partial double strands as two-line vectors, and of the sticking operation that combines them, ❯ Sect. 6.1 describes basic sticker systems, sticker systems with complex structures (❯ Sect. 6.1.1), and observable sticker systems (❯ Sect. 6.1.2). ❯ Section 6.2 investigates the accepting counterpart of the generative sticker systems devices: Watson–Crick automata and their properties.

❯ Section 7 describes the influence that properties of DNA-based information have had on research in combinatorics on words, by enumerating several natural generalizations of classical concepts of combinatorics of words: pseudo-palindromes, pseudo-periodicity, Watson–Crick conjugate and commutative words, involutively bordered words, pseudoknot bordered words. In addition, this section outlines natural extensions in this context of two of the most fundamental results in combinatorics of words, namely Fine and Wilf's theorem and the Lyndon–Schützenberger result.

❯ Section 8 presents general thoughts on DNA-based information, bioinformation, and biocomputation.

2   A Computer Scientist's Guide to DNA

In this section, a brief description of the basic molecular biology notions of DNA structure and DNA-based bio-operations used in DNA computing is given. For further details, the reader is referred to Turner et al. (2000); Drlica (1996); Calladine and Drew (1997); Gonick and Wheelis (1991); and Lewin (2007).

A **DNA (deoxyribonucleic acid)** molecule is a linear polymer. The monomer units of DNA are nucleotides (abbreviated *nt*), and the polymer is known as a "polynucleotide." There are four different kinds of nucleotides found in DNA, each consisting of a nitrogenous *base* (**A**denine, *A*, **C**ytosine, *C*, **G**uanine, *G*, or **T**hymine, *T*), and a sugar-phosphate unit. The abbreviation *N* stands for any nucleotide. The bases are relatively flat molecules and can be divided into purine bases (adenine and guanine) that have two carbon-containing rings in their structure, and smaller pyrimidine bases (cytosine and thymine) that have one carbon-containing ring in their structure.

A sugar-phosphate unit consists of a deoxyribose sugar and one to three phosphate groups. Together, the base and the sugar comprise a *nucleoside*. The sugar-phosphate units are linked together by strong *covalent bonds* to form the backbone of the DNA single strand (ssDNA). A DNA strand consisting of *n* nucleotides is sometimes called an *n*-mer. An *oligonucleotide* is a short DNA single strand, with 20 or fewer nucleotides. Since nucleotides may differ only by their bases, a DNA strand can be viewed simply as a word over the four-letter alphabet $\{A, C, G, T\}$.

A DNA single strand has an orientation, with one end known as the $5'$ end, and the other end known as the $3'$ end, based on their chemical properties. By convention, a word over the DNA alphabet represents the corresponding DNA single strand in the $5'$ to $3'$ orientation, that is, the word ACGTCGACTAC stands for the DNA single strand $5'$-ACGTCGACTAC-$3'$.

◘ Fig. 1
DNA structure: **(a)** DNA's sugar-phosphate backbone, **(b)** DNA bases, **(c)** Watson–Crick (WK) complementarity between bases A and T of two DNA single strands of opposite orientation, **(d)** Watson–Crick complementarity between bases C and G of two DNA single strands of opposite orientation.

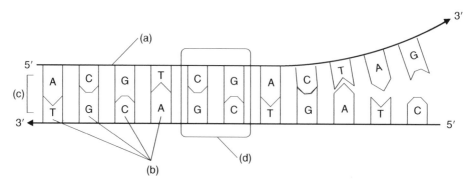

A crucial feature of DNA single strands is their **Watson–Crick (WK) complementarity**: A (a purine) is complementary to T (a pyrimidine), and G (a purine) is complementary to C (a pyrimidine). Two complementary DNA single strands with opposite orientation bind to each other by weak *hydrogen bonds* between their individual bases as follows: A binds with T through two hydrogen bonds, while G binds with C through three hydrogen bonds. Thus, two Watson–Crick complementary single strands form a stable DNA double strand (dsDNA) resembling a helical ladder, with the two backbones at the outside, and the bases paired by hydrogen bonding and stacked on each other, on the inside. For example, the DNA single strand 5′-ACGTCGACTAC- 3′ will bind to the DNA single strand 5′-GTAGTCGACGT-3′ to form the 11 base-pair long (abbreviated as 11bp) double strand

$$5' - ACGTCGACTAC - 3'$$
$$3' - TGCAGCTGATG - 5'.$$

❯ *Figure 1* schematically illustrates this DNA double strand, omitting the double helix structure, for clarity. If v denotes a DNA single strand over the alphabet $\{A, C, G, T\}$, then by $\overrightarrow{v}$ we will denote its Watson–Crick complement.

Another molecule that can be used for computation is *RNA, ribonucleic acid*. RNA is similar to DNA, but differs from it in three main aspects: RNA is usually single-stranded while DNA is usually double-stranded, RNA nucleotides contain the sugar ribose, while DNA nucleotides contain the sugar deoxyribose, and in RNA the base Uracil, U, substitutes for thymine, T, which is present in DNA.

The *genome* of an organism is the totality of its genetic information encoded in DNA. It consists of *chromosomes*, which, in turn, consist of genes. A *gene* is a segment of DNA that holds the information encoding a coherent set of functions necessary to build and maintain cells, and pass genetic traits to offspring. A gene comprises *coding* subsequences (*exons*) that determine what the gene does, and *noncoding* subsequences (*introns*). When a gene is active, the coding and noncoding sequences are copied in a process called *transcription*, producing an RNA copy of the gene's information. This piece of RNA can then direct the *translation* of the catenation of the coding sequences of this gene into *proteins* via the *genetic code*. The genetic code maps each 3-letter RNA segment (called *codon*) into an amino acid. Several designated triplets, the start codon (AUG), and the stop (UAA, UAG, UGA) codons, signal the initiation, respectively, the termination of a translation. There are 20 different standard amino acids. Some of them are encoded by one codon, while others are encoded by several "synonymous" codons. A protein is a sequence over the 20-letter alphabet of amino acids. Proteins are essential parts of organisms and participate in every process within cells having, for example, catalytical, structural, or mechanical functions.

To encode, for example, English text using DNA, one can choose an encoding scheme mapping the Latin alphabet onto strings over $\{A, C, G, T\}$, and proceed to synthesize the obtained information-encoding strings as DNA single strands. In a hypothetical example, one could encode the letters of the English alphabet as $\mathbf{A} \rightarrow ACA$, $\mathbf{B} \rightarrow ACCA$, $\mathbf{C} \rightarrow ACCCA$, $\mathbf{D} \rightarrow AC^4A$, etc., wherein the ith letter of the alphabet is represented by AC^iA, that is, a single strand of DNA consisting of i repetitions of C flanked by one A at the beginning and another A at the end. Under this encoding, the text "To be or not to be" becomes the DNA single strand represented by

$$\overbrace{AC^{20}A}^{T}\,\overbrace{AC^{15}A}^{O}\,\overbrace{AC^2A}^{B}\,\overbrace{AC^5A}^{E}\,\overbrace{AC^{15}A}^{O}\,\overbrace{AC^{18}A}^{R}\,\overbrace{AC^{14}A}^{N}\,\overbrace{AC^{15}A}^{O}\,\overbrace{AC^{20}A}^{T}\,\overbrace{AC^{20}A}^{T}\,\overbrace{AC^{15}A}^{O}\,\overbrace{AC^2A}^{B}\,\overbrace{AC^5A}^{E}$$

that can be readily synthesized.

Indeed, **DNA synthesis** is the most basic **bio-operation** used in DNA computing. DNA solid-state synthesis is based on a method by which the initial nucleotide is bound to a solid support, and successive nucleotides are added step by step, from the $3'$ to the $5'$ direction, in a reactant solution (❷ *Fig. 2*).

While the above encoding example is purely hypothetical, DNA strands of lengths of up to 100 nucleotides can be readily synthesized using fully automated *DNA synthesizers*. The result is a small test tube containing a tiny, dry, white mass of indefinite shape containing a homogeneous population of DNA strands that may contain 10^{18} identical molecules of DNA. In bigger quantities, dry DNA resembles tangled, matted white thread.

Using this or other DNA synthesis methods, one can envisage encoding any kind of information as DNA strands. There are several reasons to consider such a DNA-based memory as an alternative to all the currently available implementations of memories. The first is the extraordinary information-encoding density that can be achieved by using DNA strands. According to Reif et al. (2002), 1 g of DNA, which contains 2.1×10^{21} DNA nucleotides, can store approximately 4.2×10^{21} bits. Thus, DNA has the potential to store data on the order of 10^{10} more compactly than conventional storage technologies. In addition, the robustness of DNA data ensures the maintenance of the archived information over extensive periods of time (Bancroft et al. 2001; Cox 2001; Smith et al. 2003).

For the purposes of DNA computing, after encoding the input data of a problem on DNA strands, DNA bio-operations can be utilized for computations, see Kari (1997), Păun et al. (1998), and Amos (2005). The bio-operations most commonly used to control DNA computations and DNA robotic operations are described below.

DNA single strands with opposite orientation will join together to form a double helix in a process based on the Watson–Crick complementarity and called **base-pairing** (**annealing, hybridization, renaturation**), illustrated in ❷ *Fig. 3*. The reverse process – a double-stranded helix coming apart to yield its two constituent single strands – is called **melting** or **denaturation**. As the name suggests, melting is achieved by raising the temperature, and annealing by lowering it. Each DNA double-strand denatures at a specific temperature, called its *melting*

◘ Fig. 2

DNA solid-state synthesis. The initial nucleotide is bound to a solid support, and successive nucleotides are added step by step, from the 3′ to the 5′ direction, in a reactant solution. (From Kari 1997.)

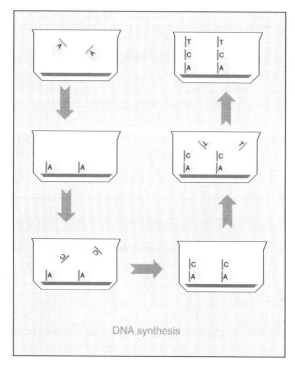

DNA synthesis

temperature. The melting temperature is defined as the temperature at which half of the DNA strands are double-stranded, and half are single-stranded. It depends on both the length of the DNA sequence, and its specific base-composition (SantaLucia 1998).

Cutting DNA double strands at specific sites can be accomplished with the aid of specific enzymes, called *restriction enzymes (restriction endonucleases)*. Each restriction enzyme recognizes a specific short sequence of DNA, known as a *restriction site*. Any double-stranded DNA that contains the restriction site within its sequence is cut by the enzyme at that location, according to a specific pattern. Depending on the enzyme, the cutting operation leaves either two "blunt-ended" DNA double strands or, more often, two DNA strands that are double-stranded but have single-stranded overhangs known as "sticky-ends," ❯ *Fig. 4*. Hundreds of restriction enzymes are now known, and a large number are commercially available. They usually recognize sites ranging in size from 4 to 8 bp, the recognition sites are often pseudo-palindromic (see ❯ Sect. 7 for a definition of pseudopalindrome).

Another enzyme, called *DNA ligase*, can repair breaks in a double-stranded DNA backbone, and can covalently rejoin annealed complementary ends in the reverse of a restriction enzyme reaction, to create new DNA molecules. The process of thus pasting together compatible DNA strands is called **ligation**.

Separation of DNA strands by size is possible by using a technique called *gel electrophoresis*. The DNA molecules, which are negatively charged, are placed in "wells" situated at one

◻ Fig. 3

Melting (separating DNA double strands into their constituent single strands), and annealing (the reformation of double-stranded DNA from thermally denatured DNA). Raising the temperature causes a DNA double strand to "melt" into its constituent Watson–Crick complementary single strands. Decreasing the temperature has the opposite effect: two DNA single strands that are Watson–Crick complementary will bind to each other to form a DNA double strand. (From Kari 1997.)

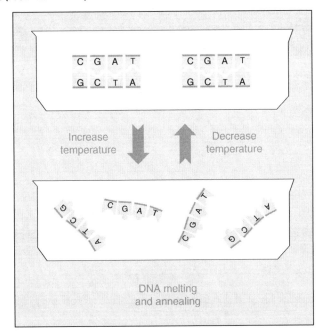

side of an agarose or polyacrylamide gel. Then an electric current is applied to the gel, with the negative pole at the side with the wells, and the positive pole at the opposite side. The DNA molecules will be drawn toward the positive pole, with the larger molecules traveling more slowly through the gel. After a period, the molecules will spread out into distinct bands according to size, ❷ *Fig. 5*. The gel method at constant electric field is capable of separating by size DNA molecules as long as 50,000 base pairs, with a resolution of better than 1% the size of the DNA.

Extraction of DNA single strands that contain a target sequence, *v*, from a heterogeneous solution of DNA single strands, can be accomplished by *affinity purification*, ❷ *Fig. 6*. A DNA probe is a single-stranded DNA molecule used in laboratory experiments to detect the presence of a complementary sequence among a mixture of other singled-stranded DNA molecules.

The extraction process begins by synthesizing probes, that is, strands $\overrightarrow{v}$, Watson–Crick complementary to *v*, and attaching them to a solid support, for example, magnetic beads. Then, the heterogeneous solution of DNA strands is passed over the beads. Those strands containing *v* are "detected" by becoming annealed to $\overrightarrow{v}$, and are thus retained. Strands not containing *v* pass through without being retained. The solid medium, for example, magnetic

■ **Fig. 4**

DNA cutting (digestion) by a restriction enzyme. A hypothetical restriction enzyme (dark gray) recognizes the restriction site *CATC* on a DNA double-strand (light gray), and cuts the two backbones of the DNA strand, as shown. Under most conditions, Watson–Crick complementarity of four base-pairs is not sufficient to keep the strands together, and the DNA molecule separates into two fragments. The result of the digestion is thus a *CATC* − 3′ sticky end overhang, and a 3′ − *GTAG* sticky end overhang. The reverse process of ligation can restore the original strand by bringing together strands with compatible sticky ends by means of Watson–Crick base-pairing, and using the enzyme ligase that repairs the backbone breaks (nicks) that had been introduced by the restriction endonuclease. (From Kari 1997.)

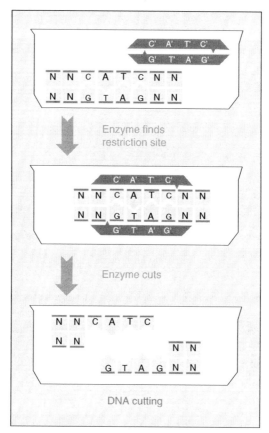

beads, can then be removed from the mixture, washed, and the target DNA molecules can be released from the entrapment.

DNA replication is accomplished by a process called *polymerase chain reaction*, or *PCR*, that involves the *DNA polymerase* enzyme, ❷ *Fig. 7*.

The PCR replication reaction requires a guiding DNA single strand called *template*, and an oligonucleotide called *primer*, that is annealed to the template. The primer is required to initiate the synthesis reaction of the polymerase enzyme. The DNA polymerase enzyme catalyzes DNA synthesis by successively adding nucleotides to one end of the primer.

▣ Fig. 5

Separation of DNA strands by size (length) by using gel electrophoresis. The DNA samples are placed in *wells* (slots) near one end of the gel slab. The power supply is switched on and the DNA, which is highly negatively charged, is allowed to migrate toward the positive electrode (right side of the gel) in separate *lanes* or *tracks*. DNA fragments will move through the gel at a rate which is dependent on their size and shape. After a while, the DNA molecules spread out into distinct bands and the use of a control "ladder" that contains DNA strands of incremental lengths allows the determination of the lengths of the DNA molecules in the samples. (From Kari 1997.)

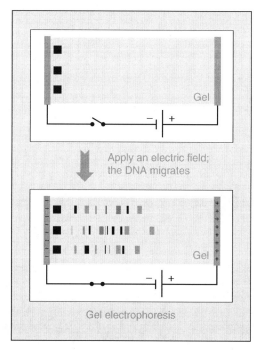

The primer is thus extended at its 3′ end, in the direction 5′ to 3′ only, until the desired strand is obtained that starts with the primer and is complementary to the template. (Note that DNA chemical synthesis and enzymatic DNA replication proceed in opposite directions, namely 3′ − > 5′ and 5′ − > 3′, respectively).

If two primers are used, the result is the exponential multiplication of the subsequence of the template strand that is flanked by the two primers, in a process called *amplification*, schematically explained below. For the purpose of this explanation, if x is a string of letters over the DNA alphabet $\{A, C, G, T\}$, then $\bar{x}$ will denote its simple complement, for example, $\overline{AACCTTGG} = TTGGAACC$.

One can now assume that one desires to amplify the subsequence between x and y from the DNA double strand $\frac{5'-\alpha x\beta y\delta-3'}{3'-\bar{\alpha}\bar{x}\bar{\beta}\bar{y}\bar{\delta}-5'}$, where α, x, β, y, and δ are DNA segments. Then one uses as primers the strand x and the Watson–Crick complement $\overrightarrow{y}$ of y. After heating the solution and thus melting the double-stranded DNA into its two constituent strands, the solution is cooled and the Watson–Crick complement of y anneals to the "top" strand, while x anneals to the "bottom" strand. The polymerase enzyme extends the 3′ ends of both primers into the 5′ to 3′

◨ **Fig. 6**

Extraction of strands that contain a target sequence, $v = 5' - GAT - 3'$, by affinity purification. (The 3nt pattern *GAT* is for illustration purposes only, in practice a longer DNA subsequence would be needed for the process of extraction to work.) The Watson–Crick complement of v, namely $3' - CTA - 5'$, is attached to a solid support, for example, magnetic beads. The heterogeneous solution of DNA strands is poured over. The DNA strands that contain the target sequence v as a subsequence will attach to the complement of the target by virtue of Watson–Crick complementarity. Washing off the solution will result in retaining the beads with the attached DNA strands containing the target sequence. (From Kari 1997.)

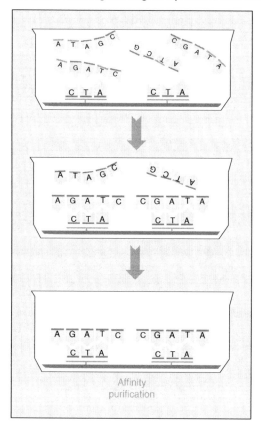

Affinity purification

direction, producing partially double-stranded molecules $\frac{5'-\alpha x\beta y\delta-3'}{3'-\bar{\alpha}\bar{x}\bar{\beta}\bar{y}-5'}$ and $\frac{5'-x\beta y\delta-3'}{3'-\bar{\alpha}\bar{x}\bar{\beta}\bar{y}\delta-5'}$. In a similar fashion, the next heating–cooling cycle will result in the production of the additional strands $5' - x\beta y - 3'$ and $3' - \bar{x}\bar{\beta}\bar{y} - 5'$. These strands are Watson–Crick complementary and will, from now on, be produced in excess of the other strands, since both are replicated during each cycle. At the end, an order of 2^n copies of the desired subsequences flanked by x and y will be present in solution, where n is the number of the heating–cooling cycles, typically 20 to 30.

A biocomputation consists of a succession of *bio-operations* (Daley and Kari 2002), such as the ones described in this section. The DNA strands representing the output of the biocomputation can then be **sequenced** (**read out**) using an automated sequencer. One sequencing method uses special chemically modified nucleotides (dideoxyribonucleoside

◻ **Fig. 7**

DNA strand replication using one primer and the enzyme DNA polymerase. Given a template DNA strand 5′ − *ATAGAGTT* − 3′ to replicate, first a primer 3′ − *TCA* − 5′ (a short DNA sequence, usually 10–15 nt long, that is complementary to a portion of the template) is added to the solution. DNA polymerase (dark gray) extends the primer in the 5′ to 3′ direction, until the template becomes fully double stranded. Repeating the process by heating the solution to denaturate the double-strands and then cooling it to allow annealing of the primer, will thus lead to producing many copies of the portion of the complement of the template strand that starts with the primer. The idea is used in polymerase chain reaction (PCR) that uses two primers, DNA polymerase, dNTPs, and thermal cycling to produce exponentially many copies of the subsequence of a template strand that is flanked by the primers. dNTP, *deoxyribonucleoside triphosphate*, stands for any of the nucleotides dATP, dTTP, dCTP, and dGTP. Each nucleotide consists of a base, plus sugar (which together form a *nucleoside*), plus triphosphate. dNTPs are the building blocks from which the DNA polymerases synthesizes a new DNA strand. (From Kari 1997.)

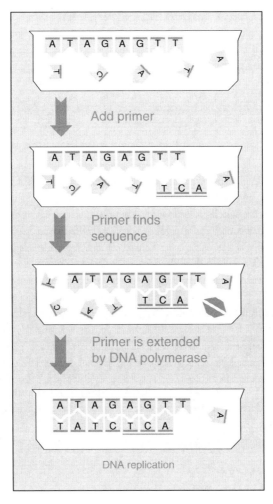

triphosphates – ddNTPs), that act as "chain terminators" during PCR, as follows. A sequencing primer is annealed to the DNA template that one wishes to read. A DNA polymerase then extends the primer. The extension reaction is split into four tubes, each containing a different chain terminator nucleotide, mixed with standard nucleotides. For example, tube C would contain chemically modified C (ddCTP), as well as the standard nucleotides (dATP, dGTP, dCTP, and dTTP). Extension of the primer by the polymerase then produces all prefixes ending in G of the complement of the original strand. A separation of these strands by length using gel electrophoresis allows the determination of the position of all Gs (complements of Cs). Combining the results obtained in this way for all four nucleotides allows the reconstruction of the original sequence.

Some of the novel features of DNA-encoded information and bio-operations have been used for the first time for computational purposes in the breakthrough proof-of-principle experiment in DNA computing reported by Adleman in 1994.

3 The First DNA Computing Experiment

The practical possibilities of encoding information in a DNA sequence and of performing simple bio-operations were used by Leonard Adleman in 1994 (Adleman 1994) to perform the first experimental DNA computation that solved a seven-vertex instance of an NP-complete problem, namely the directed Hamiltonian path problem (HPP).

A directed graph G with designated vertices v_{start} and v_{end} is said to have a Hamiltonian path if and only if there exists a sequence of compatible directed edges $e_1, e_2, \ldots, e_z$ (i.e., a directed path) that begins at v_{start}, ends at v_{end} and enters every other vertex exactly once.

The following (nondeterministic) algorithm solves the problem:

Input. A directed graph G with n vertices and designated vertices v_{start} and v_{end}.
Step 1. Generate random paths through the graph.
Step 2. Keep only those paths that begin with v_{start} and end with v_{end}.
Step 3. Keep only those paths that enter exactly n vertices.
Step 4. Keep only those paths that enter all of the vertices of the graph at least once.
Output. If any paths remain, output "YES"; otherwise output "NO."

Below, we describe Adleman's bio-algorithm which solves the seven-vertex instance of the HPP illustrated in ❷ *Fig. 8*, where $v_{start} = 0$ and $v_{end} = 6$.

To encode the *input* to the problem, that is the vertices and directed edges of the graph, each vertex of the graph was encoded into a carefully chosen 20-mer single strand of DNA. Then, for each oriented edge of the graph, a DNA sequence was designed and synthesized, consisting of the second half of the sequence encoding the source vertex and the first half of the sequence encoding the target vertex. Exceptions were the edges that started with v_{start}, and the ones that ended in v_{end}, for which the DNA encoding consisted of the full sequence of the source vertex followed by the first half of the target vertex, respectively, the second half of the source vertex followed by the full sequence of the target vertex.

For example, the DNA sequences for the vertex 3 and the oriented edges $2 \rightarrow 3$ and $3 \rightarrow 4$ were encoded respectively as

$O_3 = 5' - GCTATTCGAGCTTAAAGCTA - 3'$
$O_{2 \rightarrow 3} = 5' - GTATATCCGAGCTATTCGAG - 3'$
$O_{3 \rightarrow 4} = 5' - CTTAAAGCTAGGCTAGGTAC - 3'.$

◻ Fig. 8

The seven-vertex instance of the Hamiltonian path problem (HPP) solved by Adleman, who used solely molecular biology tools to obtain the answer. This was the first ever experimental evidence that DNA computing is possible. (From Adleman 1994.)

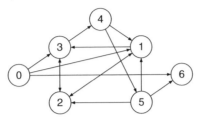

To implement *Step 1*, one mixed together in a test tube multiple copies of each of the encodings of the Watson–Crick complements $\overrightarrow{O_i}$ of all the vertices, together with the encodings for all the directed edges, for a ligation reaction. The complements of the vertices served as splints and brought together sequences associated to compatible edges. For example, the edges $O_{2\rightarrow3}$ and $O_{3\rightarrow4}$ were brought together by the Watson–Crick complement $\overrightarrow{O_3}$ as follows:

$$
\begin{array}{c}
\overbrace{}^{\text{Edge}2\rightarrow3} \qquad\qquad \overbrace{}^{\text{Edge}3\rightarrow4} \\
5' - GTATATCCGAGCTATTCGAG\ CTTAAAGCTAGGCTAGGTAC - 3' \\
3' - \underbrace{CGATAAGCTCGAATTTCGAT}_{\text{Complement of vertex 3}} - 5'
\end{array}
$$

Hence, the Watson–Crick complementarity and the combined ligation reaction resulted in the formation of DNA molecules encoding random paths through the graph. Out of these, the next steps had to find and discard the paths that were not Hamiltonian.

To implement *Step 2* (keep only paths that start with v_{start} and end with v_{end}), the product of *Step 1* was amplified by PCR with primers O_0 and $\overrightarrow{O_6}$. Thus, only those molecules encoding paths that begin with the start vertex 0 and end with the end vertex 6 were amplified.

For implementing *Step 3* (keep only paths of the correct length), gel electrophoresis was used, allowing separation of DNA strands by length. Since any Hamiltonian path, if it exists, has to pass through all the seven vertices of the graph, only DNA double strands of length $7 \times 20 = 140$ were retained.

Step 4 (keep only paths that pass through each vertex at least once) was accomplished by iteratively using affinity purification. After generating single-stranded DNA from the double-stranded product of the preceding step, one attached a sequence $\overrightarrow{O_i}$ to magnetic beads, and the heterogeneous solution of "candidate paths" was passed over the beads. Those strands containing O_i annealed to the complementary sequence and were hence retained. These strands represented paths that pass through the vertex i. This process was repeated successively with $\overrightarrow{O_1}$, $\overrightarrow{O_2}$, $\overrightarrow{O_3}$, $\overrightarrow{O_4}$, and $\overrightarrow{O_5}$.

To obtain the *Output* (are there any paths left?), the presence of a molecule encoding a Hamiltonian path was checked. This was done by amplifying the result of *Step 4* by PCR with primers O_0 and $\overrightarrow{O_6}$, and then reading out the DNA sequence of the amplified molecules. Note that the final PCR was not required for computational purposes: First, it was employed to make sure that, after several filtering steps, the amount of DNA was above the detection threshold, and second to verify the correctness of the answer.

The entire computation required approximately 7 days of wet lab work, and was carried out in approximately 1/50 of a teaspoon of solution (Adleman 1998). It was the first proof-of-concept experiment that DNA computation was possible. From a practical point of view, Adleman's approach had both advantages and disadvantages. On one hand, *Step 1*, which is the cause of the time-complexity exponential blowup in the classical electronic implementation of the algorithm, took only one time-step in Adleman's bio-algorithm. On the other hand, the amount of space needed for the generation of the full solution space grows exponentially relative to the problem size. Indeed, a scaled-up input of size $n = 200$ for the HPP would require, in this brute force approach, an amount of DNA whose weight would be greater than that of the Earth (Hartmanis 1995).

The construction of a DNA pool that contains the full-solution space has been avoided in subsequent bio-algorithms proposed for solving HPP. For example, in Morimoto et al. (1997) a bio-algorithm was proposed that stepwise generated only the possible paths. In this approach, all paths were stepwise extended from v_{start} to v_{end} as follows. After letting many molecules representing v_{start} attach to a surface, $n + 1$ repetitions of the following step found the Hamiltonian path: "Extend each path by one adjacent vertex; Remove paths which contain the same vertex twice." In addition to being space and time efficient (each extension step could be as quick as 30 min), this bio-algorithm avoided the laborious step of separation of strands by length.

4   Beyond Unaided Human Computation

We present another significant milestone in DNA computing research, the first experiment that demonstrated that DNA computing devices can exceed the computational power of an unaided human. Indeed, in 2002, an experiment was reported (Braich et al. 2002), that solved a 20-variable instance of the NP-complete 3-SAT problem, wherein the answer to the problem was found after an exhaustive search of more than 1 million (2^{20}) possible solution candidates.

The input to a 3-SAT problem is a Boolean formula in three-conjunctive-normal-form (3-CNF), that is, a Boolean formula that consists of conjunctions of disjunctive clauses, where each disjunctive clause is the disjunction of at most three literals (a literal is either a Boolean variable or its negation). This formula is called *satisfiable* if there exists a truth value assignment to its variables that satisfies it, that is, that makes the whole formula true. Thus, the output to the 3-SAT problem is "yes" if such a satisfying truth value assignment exists, and "no" otherwise.

The input formula for this experiment was the 20-variable, 24-clause, 3-CNF formula:

$$\Phi = (\overline{x_3} \vee \overline{x_{16}} \vee x_{18}) \wedge (x_5 \vee x_{12} \vee \overline{x_9}) \wedge (\overline{x_{13}} \vee \overline{x_2} \vee x_{20}) \wedge (x_{12} \vee \overline{x_9} \vee \overline{x_5})$$
$$\wedge (x_{19} \vee \overline{x_4} \vee x_6) \wedge (x_9 \vee x_{12} \vee \overline{x_5}) \wedge (\overline{x_1} \vee x_4 \vee \overline{x_{11}}) \wedge (x_{13} \vee \overline{x_2} \vee \overline{x_{19}})$$
$$\wedge (x_5 \vee x_{17} \vee x_9) \wedge (x_{15} \vee x_9 \vee \overline{x_{17}}) \wedge (\overline{x_5} \vee \overline{x_9} \vee \overline{x_{12}}) \wedge (x_6 \vee x_{11} \vee x_4)$$
$$\wedge (\overline{x_{15}} \vee \overline{x_{17}} \vee x_7) \wedge (\overline{x_6} \vee x_{19} \vee x_{13}) \wedge (\overline{x_{12}} \vee \overline{x_9} \vee x_5) \wedge (x_{12} \vee x_1 \vee x_{14})$$
$$\wedge (x_{20} \vee x_3 \vee x_2) \wedge (x_{10} \vee \overline{x_7} \vee \overline{x_8}) \wedge (\overline{x_5} \vee x_9 \vee \overline{x_{12}}) \wedge (x_{18} \vee \overline{x_{20}} \vee x_3)$$
$$\wedge (\overline{x_{10}} \vee \overline{x_{18}} \vee \overline{x_{16}}) \wedge (x_1 \vee \overline{x_{11}} \vee \overline{x_{14}}) \wedge (x_8 \vee \overline{x_7} \vee \overline{x_{15}}) \wedge (\overline{x_8} \vee x_{16} \vee \overline{x_{10}})$$

where, for a Boolean variable x_i, $\overline{x_i}$ denotes the negation of x_i, $1 \le i \le 20$. The formula Φ was designed so as to have a unique satisfying truth assignment, namely $x_1 = F, x_2 = T, x_3 = F, x_4 = F,$

$x_5 = F, x_6 = F, x_7 = T, x_8 = T, x_9 = F, x_{10} = T, x_{11} = T, x_{12} = T, x_{13} = F, x_{14} = F, x_{15} = T, x_{16} = T, x_{17} = T, x_{18} = F, x_{19} = F, x_{20} = F.$

The DNA computing experiment that solved the problem (Braich et al. 2002), was based on the following nondeterministic algorithm.

Input: A Boolean formula Φ in 3-CNF.

Step 1: Generate the set of all possible truth value assignments.

Step 2: Remove the set of truth value assignments that make the first clause false.

Step 3: Repeat Step 2 for all the clauses of the input formula.

Output: The remaining (if any) truth value assignments.

To implement this algorithm, the input data was encoded as follows. Every variable $x_k, k = 1, \ldots, 20$, was associated with two distinct 15-mer DNA single strands. One of them, denoted by X_k^T, represented true (T), while the second, denoted by X_k^F, represented false (F).

Below are some examples of the particular 15-mer sequences – none of which contained the nucleotide G – synthesized and used in the experiment:

$$X_2^T = ATT\ TCC\ AAC\ ATA\ CTC, \quad X_2^F = AAA\ CCT\ AAT\ ACT\ CCT,$$
$$X_3^T = TCA\ TCC\ TCT\ AAC\ ATA, \quad X_3^F = CCC\ TAT\ TAA\ TCA\ ATC.$$

Using these 15-mer encodings for the two truth values of all the 20 variables, the library consisting of all possible 2^{20} truth assignments was assembled using the mix-and-match combinatorial synthesis technique of Faulhammer et al. (2000). In brief, oligonucleotides for X_{20}^T and X_{20}^F were synthesized separately, then mixed together. The mixture was divided in half and the result put in two separate test tubes. The synthesis was restarted separately, with sequences X_{19}^T and X_{19}^F, respectively. In principle, the process can be repeated until the desired library is obtained. In practice, two half-length libraries were created separately, and then linked together using a polymerase-chain extension similar to that in Stemmer et al. (1995). Each *library strand* encoding a truth assignment was thus represented by a 300-mer DNA strand consisting of the ordered catenation of twenty 15-mer value sequence, one for each variable, as follows:

$$X_1 X_2 \ldots X_{20}, \text{ where } \alpha_i \in \{X_i^T, X_i^F\}, 1 \le i \le 20$$

The biocomputation *wet-ware* essentially consisted of a glass module filled with a gel containing the library, as well as 24 glass *clause modules*, one for each of the 24 clauses of the formula. Each clause module was filled with gel containing probes (immobilized DNA single strands) designed to bind only library strands encoding truth assignments satisfying that clause.

The strands were moved between the modules with the aid of gel electrophoresis, that is, by applying an electric current that resulted in the migration of the negatively charged DNA strands through the gel.

The protocol started with the library passing through the first clause module, wherein library strands containing the truth assignments satisfying the first clause (i.e., library strands containing sequences X_3^F, or X_{16}^F, or X_{18}^T) were captured by the immobilized probes, while library strands that did not satisfy the first clause (i.e., library strands containing sequences X_3^T, and X_{16}^T, and X_{18}^F) continued into a buffer reservoir. The captured strands were then released by raising the temperature, and used as input to the second clause module, etc. At the end, only the strand representing the truth assignment that satisfied all 24 clauses remained.

The output strand was PCR amplified with primer pairs corresponding to all four possible true–false combinations of assignments for the first and last variable x_1 and x_{20}. None except the primer pair $(X_1^F, \overrightarrow{X_{20}^F})$ showed any bands, thus indicating two truth values of the satisfying assignment, namely $x_1 = F$ and $x_{20} = F$. The process was repeated for each of the variable pairs (x_1, x_k), $2 \leq k \leq 19$, and, based on the lengths of the bands observed, value assignments were given to the variables. These experimentally derived values corresponded to the unique satisfying assignment for the formula Φ, concluding thus the experiment.

One of the remarkable features of this benchmark DNA computing experiment was that the sole bio-operation that was used (except during input and output) was Watson–Crick complementarity-based annealing and melting.

Generally, it is believed that DNA computers that use a brute-force search algorithm for SAT are limited to 60 to 70 variables (Lipton 1995). Several other algorithms that do not use brute force, such as the breadth-first search algorithm (Yoshida and Suyama 2000), and random walk algorithms (Liu et al. 2005; Diaz et al. 2001) have been proposed. With the breadth-first search algorithm, the capacity of a DNA computer can be theoretically increased to about 120 variables (Yoshida and Suyama 2000). A recent example of this approach that avoids the generation of the full solution space is a solution to the SAT problem using a DNA computing algorithm based on ligase chain reaction (LCR) (Wang et al. 2008). This bio-algorithm can solve an n-variable m-clause SAT problem in m steps, and the computation time required is $O(3m + n)$. Instead of generating the full-solution DNA library, this bio-algorithm starts with an empty test tube and then generates solutions that partially satisfy the SAT formula. These partial solutions are then extended step by step by the ligation of new variables using DNA ligase. Correct strands are amplified and false strands are pruned by a ligase chain reaction (LCR) as soon as they fail to satisfy the conditions.

The two DNA computing experiments described in ❷ Sects. 3 and ❷ 4 are historically significant instances belonging to a vast and impressive array of often astonishing DNA computing experiments, with potential applications to, for example, nanorobotics, nanocomputing, bioengineering, bio-nanotechnology, and micromedicine. Several of these significant experiments are described in the chapters ❷ Molecular Computing Machineries — Computing Models and Wet Implementations, ❷ Bacterial Computing and Molecular Communication, ❷ DNA Memory, and ❷ Engineering Natural Computation by Autonomous DNA-Based Biomolecular Devices of this handbook.

5 DNA Complementarity and Formal Language Theory

The preceding two sections described two milestone DNA computing experiments whose main "computational engine" was the Watson–Crick complementarity between DNA strands. Watson–Crick complementarity can now be studied from a theoretical point of view, and the impact this notion has had on theoretical computer science will be described. This section focuses mainly on the formal language theoretical and coding theoretical approach to DNA-encoded information, and highlights some of the theoretical concepts and results that emerged from these studies.

The idea of formalizing and investigating DNA or RNA molecules and their interactions by using the apparatus of formal language theory is a natural one. Indeed, even though the processes involving DNA molecules are driven by complex biochemical reactions, the primary information is encoded in DNA sequences. Therefore, even without the inclusion of all the thermodynamic parameters that operate during DNA–DNA interactions, formal language theory and coding theory are capable of providing a uniform and powerful framework for an effective study of DNA-encoded information.

To briefly describe the problem of encoding information as DNA strands for DNA computing experiments, one of the main differences between electronic information and DNA-encoded information is that the former has a fixed address and is reusable, while the latter is not. More precisely, DNA strands float freely in solution in a test tube and, if one of the input strands for a bio-operation has become involved in another, unprogrammed, hybridization, that input strand is unavailable for the bio-operation at hand, compromising thus the final result of the experiment. Thus, a considerable effort has been dedicated to finding "optimal" DNA sequences for encoding the information on DNA so as to prevent undesirable interactions and favor only the desirable programmed ones.

Standard biological methods evaluating and predicting hybridization between DNA molecules conditions rely on thermodynamical parameters such as the free energy ΔG (intuitively, the energy needed to melt DNA bonds), and the melting temperature of DNA molecules. For the calculation of these quantities, not only the WK complementarity between single bases is important, but also the WK complementarity (or lack thereof) of their neighboring bases with their counterparts: the *nearest-neighbor model* (SantaLucia 1998) has been frequently used for this purpose. However, in many natural processes, the WK complementarity alone plays a crucial role. Furthermore, together with various similarity metrics, the WK complementarity has been often used to obtain an approximate characterization of DNA hybridization interactions.

This section is devoted to describing methods of discrete mathematics and formal language theory that allow for rapid and mathematically elegant characterization of (partially) complementary DNA molecules, their sets, and set-properties relevant for potential cross-hybridizations. For other approaches to this problem and to the problem of design of molecules for DNA computing, the reader is referred to the chapter ❷ DNA Memory in this book. The section is organized as follows. ❷ Section 5.1 describes the problem of optimal encoding of information for DNA computing experiments that was the initial motivation for this research, and also introduces the basic definitions and notation. ❷ Section 5.2 describes the problem of intramolecular hybridization (hybridization within one molecule, resulting, e.g., in hairpin formation) and results related to hairpin languages and hairpin-free languages. ❷ Section 5.3 describes the problem of intermolecular hybridization (interaction between two or more DNA molecules) and the theoretical concepts and results motivated by this problem. ❷ Section 5.4 investigates properties of languages that guarantee that even undesirable imperfect bonds between DNA strands are avoided.

5.1 DNA Encoding: Problem Setting and Notation

In the process of designing DNA computational experiments, as well as in many general laboratory techniques, special attention is paid to the design of an initial set of "good" DNA strands. Mauri and Ferretti (2004), Sager and Stefanovic (2006), and others distinguish two elementary subproblems of the encoding sequence design:

- *Positive* design problem: Design a set of input DNA molecules such that there is a sequence of reactions that produces the correct result.
- *Negative* design problem: Design a set of input DNA molecules that do not interact in undesirable ways, that is, do not produce incorrect outputs, and/or do not consume molecules necessary for other, programmed, interactions.

❏ **Fig. 9**

Types of undesired (a) intramolecular and (b), (c) intermolecular hybridizations.

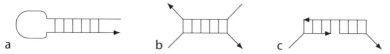

a   b   c

The positive design problem is usually highly related to a specific experiment and it is reported to be hard to find a general framework for its solution. In contrast, the negative design problem can be solved on a general basis by construction of a library of molecules that do not allow for undesired mutual hybridizations. According to Sager and Stefanovic (2006), the following conditions must be guaranteed: (1) no strand forms any undesired secondary structure such as hairpin loops (❷ *Fig. 9a*), (2) no string in the library hybridizes with any string in the library, and (3) no string in the library hybridizes with the complement of any string in the library (❷ *Fig. 9b* or *c*). Many laboratory techniques stress the importance of a unified framework for the negative design. An example is the multiplex PCR in which a set of PCR primers is used simultaneously in a single test tube and mutual bonds between primers must be prevented.

A related issue often studied together with the problem of unwanted hybridization is the uniqueness of the oligonucleotides used in experiments. More precisely, one requires that individual oligonucleotides in a mixture differ substantially from each other such that they (and eventually also longer sequences produced by their catenation) can be easily distinguished. Such a property in the mathematical sense is typical for *codes*, hence many authors adopted this naming convention for sets of oligonucleotides, usually of a fixed length, whose elements are then called *DNA codewords*.

A wide spectrum of methods devoted to the DNA encoding design exists. Thermodynamical methods such as those in Deaton et al. (2003), and Dirks and Pierce (2004) provide the most precise results but are computationally the most expensive. Experimental studies trying to construct DNA codes in vitro with the help of the PCR operation can be found, for example, in Chen et al. (2006), and Deaton et al. (2006). An opposite approach relying solely on the WK complementarity allows for fastest but least-precise methods (see, e.g., Head (2000), Jonoska and Mahalingam (2004), and Kari et al. (2005b)). Approximation methods trying to capture key aspects of the nearest neighbor thermodynamic model represent an intermediate step between these two methodologies. Various discrete metrics based often on Hamming or Levenshtein distance have been studied, for example, in Dyachkov et al. (2006, 2008), and Garzon et al. (1997, 2006). The reader is referred to monographs Amos (2005), Ignatova et al. (2008), and Păun et al. (1998) for an overview.

In the sequel, we focus on the characterization of DNA hybridization and unwanted bonds by means of the concepts of formal language and coding theory. Simple examples of construction of DNA codes are also given.

To describe DNA bonds formally, we represent the single-stranded DNA molecules by strings over the *DNA alphabet* $\Delta = \{A, C, T, G\}$, and their mutual reactions can be reduced to formal manipulation of these strings. Therefore, some formal language prerequisites are necessary. For further details the reader is referred to Hopcroft and Ullman (1979), Choffrut and Karhumäki (1997), and Salomaa (1973).

An *alphabet* is a finite and nonempty set of symbols. In the sequel a fixed non-singleton alphabet Σ shall be used as a generalization of the natural DNA alphabet Δ. The set of all words over Σ is denoted by Σ^*. This set includes the *empty word* λ. The length of a word $w \in \Sigma^*$ is denoted by $|w|$. For an $x \in \Sigma^+$, $|w|_x$ denotes the number of occurrences of x within w. For a nonnegative integer n and a word w, we use w^n to denote the word that consists of n concatenated copies of w. A word v is a *subword* of w if $w = xvy$ for some words x and y. In this case, if $|x| + |y| > 0$ then v is a *proper subword*. By Sub(w) we denote the set of all subwords of w. For a positive integer k, we use Sub$_k(w)$ to denote the set of subwords of length k of w. We say that $u \in \Sigma^*$ is a prefix of a word $v \in \Sigma^*$, and denote it by $u \leq v$, if $v = ut$ for some $t \in \Sigma^*$. Two words u, v are said to be *prefix comparable*, denoted by $u \sim_p v$ if one of them is a prefix of the other. In a similar manner, u is said to be a suffix of v if $v = su$ for some $s \in \Sigma^*$. By Pref(u) (Suff(u)), we denote the sets of all prefixes (respectively suffixes) of u.

The relation of *embedding order* over words is defined as follows: $u \leq_e w$ iff

$$u = u_1 u_2 \cdots u_n, \quad w = v_1 u_1 v_2 u_2 \cdots v_n u_n v_{n+1}$$

for some integer n with $u_i, v_j \in \Sigma^*$.

A language L is a set of words, or equivalently a subset of Σ^*. A language is said to be λ-free if it does not contain the empty word. If n is a nonnegative integer, we write L^n for the language consisting of all words of the form $w_1 \ldots w_n$ such that each w_i is in L, and $L^{\geq n}$ for the language consisting of all catenations of at least n words from L. We also write L^* for the language $L^0 \cup L^1 \cup L^2 \cup \ldots$, and L^+ for the language $L^* \setminus \{\lambda\}$. The set Sub($L$) = $\bigcup_{w \in L}$ Sub(w) we call the set of all subwords of L. The families of regular, linear, context-free, and recursively enumerable languages are denoted by REG, LIN, CF, and RE, respectively.

Many approaches to the construction of DNA encoding are based on the assumption that the set of molecules in a test tube (*tube language*) L is equal to, or a subset of, K^+, where K is a finite language whose elements are called *codewords*. In general, K might contain codewords of different lengths. In many cases, however, the set K consists of words of a certain fixed length l. In this case, we shall refer to K as a *code of length l*.

A mapping $\alpha : \Sigma^* \to \Sigma^*$ is called a *morphism (antimorphism)* of Σ^* if $\alpha(uv) = \alpha(u)\alpha(v)$ (respectively $\alpha(uv) = \alpha(v)\alpha(u)$) for all $u, v \in \Sigma^*$. Note that a morphism or an antimorphism of Σ^* are completely defined if we define their values on the letters of Σ.

If for a morphism α, $\alpha(a) \neq \lambda$ for each $a \in V$, then α is said to be λ-free. A *projection* associated to Σ is a morphism $pr_\Sigma : (V \cup \Sigma)^* \to \Sigma^*$ such that $pr_\Sigma(a) = a$ for all $a \in \Sigma$ and $pr_\Sigma(b) = \lambda$ otherwise. A morphism $h : V^* \to \Sigma^*$ is called a *coding* if $h(a) \in \Sigma$ for all $a \in V$ and *weak coding* if $h(a) \in \Sigma \cup \{\lambda\}$ for all $a \in V$. For a family of languages FL, one can denote by Cod(FL) (respectively wcod(FL)) the family of languages of the form $h(L)$, for $L \in FL$ and h a coding (respectively weak coding).

The *equality set* of two morphisms $h_1, h_2 : V^* \to \Sigma^*$ is defined as

$$\mathrm{EQ}(h_1, h_2) = \{w \in V^* \mid h_1(w) = h_2(w)\}$$

An *involution* $\theta : \Sigma \to \Sigma$ of Σ is a mapping such that θ^2 is equal to the identity mapping, that is, $\theta(\theta(x)) = x$ for all $x \in \Sigma$. It follows then that an involution θ is bijective and $\theta = \theta^{-1}$. The identity mapping is a trivial example of an involution. An involution of Σ can be extended to either a morphism or an antimorphism of Σ^*. For example, if the identity of Σ is extended to a morphism of Σ^*, we obtain the identity involution of Σ^*. However, if the identity of Σ is extended to an antimorphism of Σ^* we obtain instead the mirror-image involution of Σ^* that maps each word u into u^R where

$$u = a_1 a_2 \ldots a_k, \quad u^R = a_k \ldots a_2 a_1, a_i \in \Sigma, 1 \leq i \leq k$$

A word w that is equal to its reverse w^R is called a *palindrome*. If we consider the DNA alphabet Δ, then the mapping $\tau : \Delta \rightarrow \Delta$ defined by $\tau(A) = T, \tau(T) = A, \tau(C) = G, \tau(G) = C$ can be extended in the usual way to an antimorphism of Δ^* that is also an involution of Δ^*. This involution formalizes the notion of Watson–Crick complementarity and will therefore be called the *DNA involution* (Kari et al. 2002).

5.2 Intramolecular Bond (Hairpin) Analysis

In this section, we focus on mathematical properties of DNA hairpins and their importance in DNA encodings. A *DNA hairpin* is a particular type of DNA secondary structure illustrated in ❷ *Fig. 10*.

Hairpin-like secondary structures play an important role in insertion/deletion operations with DNA. Hairpin-freeness is crucial in the design of primers for the PCR reaction. Among numerous applications of hairpins in DNA computing only the Whiplash PCR computing techniques (Rose et al. 2002) and the DNA RAM (see the chapter ❷ DNA Memory in this handbook) are mentioned. The reader is referred, for example, to Kijima and Kobayashi (2006), Kobayashi (2005), and Mauri and Ferretti (2004) for a characterization and design of tube languages with or without hairpins. Coding properties of hairpin-free languages were studied in Jonoska et al. (2002) and Jonoska and Mahalingam (2004). A language-theoretical characterization of hairpins and hairpin languages was given in Păun et al (2001). Hairpins have also been studied in the context of bio-operations occurring in single-celled organisms. For example, the operation of hairpin inversion was defined as one of the three molecular operations that accomplish gene assembly in ciliates (Daley et al. 2003, 2004; Ehrenfeucht et al. 2004). Applications of hairpin structures in biocomputing and bio-nanotechnology are discussed in the chapter ❷ Molecular Computing Machineries — Computing Models and Wet Implementations in this handbook.

The following definition formally specifies hairpin as a structure described in ❷ *Fig. 10*, whose stem consists of at least k base pairs. This condition is motivated by the fact that a hairpin with shorter stem is less stable. An oligonucleotide that does not satisfy this condition is said to be hairpin-free.

Definition 1 (Jonoska et al. 2002; Kari et al. 2006) *Let θ be a (morphic or antimorphic) involution of Σ^* and k be a positive integer. A word $u \in \Sigma^*$ is said to be θ-k-hairpin-free or simply hp(θ,k)-free if $u = xvy\theta(v)z$ for some $x,v,y,z \in \Sigma^*$ implies $|v| < k$.*

We denote by $hpf(\theta,k)$ the set of all hp(θ,k)-free words in Σ^*. The complement of $hpf(\theta,k)$ is the set of all hairpin-forming words over Σ and is denoted by $hp(\theta,k) = \Sigma^* \setminus hpf(\theta,k)$. Observe that $hp(\theta,k+1) \subseteq hp(\theta,k)$ for all $k>0$. A language L is said to be θ-k-hairpin-free or simply hp(θ,k)-free if $L \subseteq hpf(\theta,k)$.

■ **Fig. 10**

A single-stranded DNA molecule forming a hairpin loop.

◻ **Fig. 11**

An NFA accepting the language $hp(\theta,2)$ over the alphabet $\{a,b\}$, where the antimorphism is defined as $\theta(a)=b$ and $\theta(b)=a$.

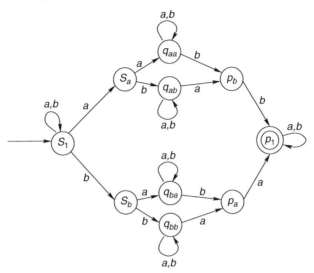

Example 1 Let $\theta = \tau$, the DNA involution over Δ^*. Then

$$hpf(\theta,1) = \{A, C\}^* \cup \{A, G\}^* \cup \{T, C\}^* \cup \{T, G\}^*$$

In the version of Definition 1 given in Jonoska et al. (2002), a θ-k-hairpin-free language was called θ-subword-k-code. The authors focused on their coding properties and relations to other types of codes. A restriction on the length of the loop of a hairpin was also considered: $1 \leq |y| \leq m$ for some $m \geq 1$. Most of the results mentioned in this section remain valid if this additional restriction is considered.

Theorem 1 (Păun et al. 2001) *The languages hp(0,k) and hpf (0,k), k $\geq$ 1, are regular.*

❯ *Figure 11* illustrates a nondeterministic finite automaton (NFA) accepting the language $hp(0,2)$ over the alphabet $\{a,b\}$ where $\theta(a) = b$ and $\theta(b) = a$. Given the above characterization of $hp(0,k)$ and $hpf(0,k)$, the following result is rather immediate.

Theorem 2 (Kari et al. 2006) *The following problem is decidable in linear (or cubic, respectively) time with respect to (w.r.t.) |M| :*

Input: A nondeterministic regular (pushdown, respectively) automaton M
Output: Yes/No depending on whether L(M) is hp(0,k)-free

The *maximality problem* of hairpin-free languages is stated as follows: can a given language $L \subseteq \Sigma^*$, satisfying a certain property (e.g., to be a hairpin-free language), be still extended without loss of this property? Formally, a language $L \subseteq \Sigma^*$ satisfying a property $\mathcal{P}$ is said to be maximal with respect to $\mathcal{P}$ iff $L \cup \{w\}$ does not satisfy $\mathcal{P}$ for any $w \in M \setminus L$, where M is a fixed library of available strands.

Theorem 3 (Kari et al. 2006) *The following problem is decidable in time* $\mathcal{O}(|M_1| \cdot |M_2|)$ *(or* $\mathcal{O}(|M_1| \cdot |M_2|^3)$, *respectively):*

Input: A positive integer k, a deterministic finite (pushdown, respectively) automaton M_1 accepting a hp(0,k)-free language, and an NFA M_2
Output: Yes/No depending on whether there is a word $w \in L(M_2) \setminus L(M_1)$ such that $L(M_1) \cup \{w\}$ is hp(0,k)-free

For hairpin-free languages, it is relatively straightforward to solve the *optimal negative design problem*: to construct a set of hairpin-free DNA words of a given size, where the words can be chosen from a certain library. All the hairpin-free sets are subsets of $hpf\,(0,k)$. For example, if the length of the desired DNA words equals a constant ℓ, the optimal hairpin-free set is simply $hpf\,(0,k) \cap \Sigma^\ell$. Due to Theorem 1, the set $hpf\,(0,k)$ is regular and hence can be accepted by a finite automaton. The size of the automaton, however, grows exponentially with respect to k.

Theorem 4 (Kari et al. 2006) *Consider the DNA alphabet $\Delta = \{A,C,T,G\}$ and the DNA involution τ.*

(i) *The size of a minimal NFA accepting $hp(\tau,k)$ is at most 15×4^k. The number of its states is between 4^k and 3×4^k.*
(ii) *The number of states of either a minimal deterministic finite automaton (DFA) or an NFA accepting $hpf\,(\tau,k)$ is between $2^{2^{k-1}}$ and $2^{3 \times 2^{2k}}$.*

The reader is referred to Kari et al. (2006) for a generalization of the above theorem for the case of an arbitrary alphabet and arbitrary involution. The construction of the automaton is illustrated in ❱ *Fig. 11* for the case of alphabet $\{a,b\}$ and an antimorphism 0, where $0(a) = b$ and $0(b) = a$.

Problems analogous to Theorems 1–4 have been studied also in the case of *scattered hairpins* and *hairpin frames* that represent more complex but rather common types of intramolecular hybridization. The definition of scattered hairpins covers structures like the one described in ❱ *Fig. 12*.

Definition 2 (Kari et al. 2006) *Let 0 be an involution of Σ^* and let k be a positive integer. A word $u = wy$, for $u,w,y \in \Sigma^*$, is 0-k-scattered-hairpin-free or simply shp(0,k)-free if for all $t \in \Sigma^*$, $t \leq_e w$, $0(t) \leq_e y$ implies $|t| < k$.*

◘ **Fig. 12**
An example of a scattered hairpin – a word in $shp(\tau,11)$.

◻ Fig. 13
An example of a hairpin frame.

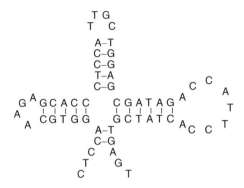

Similarly, the following definition of hairpin frames characterizes secondary structures containing several complementary sequences such as that in ❷ *Fig. 13.*

Definition 3 (Kari et al. 2006) *The pair* $(v,\theta(v))$ *in a word u of the form* $u = xvy\theta(v)z$, *for x,v,y,z* $\in \Sigma^*$, *is called an* hp-pair *of u. The sequence of hp-pairs* $(v_1,\theta(v_1))$, $(v_2,\theta(v_2))$, ..., $(v_j,\theta(v_j))$ *of the word u in the form*:

$$u = x_1 v_1 y_1 \theta(v_1) z_1 x_2 v_2 y_2 \theta(v_2) z_2 \cdots x_j v_j y_j \theta(v_j) z_j$$

is called an hp-frame *of degree j of u or simply an* hp(j)-frame *of u.*

Several other studies have been published, focusing on formal-language aspects of the hairpin formation. A complete characterization of the syntactic monoid of the language consisting of all hairpin-free words over a given alphabet was given in Kari et al. (2007). Manea et al. (2009) described formal language operations of hairpin completion and reduction and studied their closure and other mathematical properties. DNA trajectories – a new formal tool for description of scattered hairpins – were presented in Domaratzki (2007), where also complexity of the set of hairpin-free words described by a set of DNA trajectories and closure properties of hairpin language classes were studied. Hairpin finite automata with the ability to apply the hairpin inversion operation to the remaining part of the input were introduced in Bordihn et al. (2007). The authors studied the power of hairpin-inspired operations and the resulting language classes. Finally, Kari and Seki (2009) focused on a related secondary structure based on intramolecular bonds – *pseudoknots* (see also Dirks and Pierce (2004) for more information on pseudoknots). The authors provided mathematical formalization of pseudoknots and obtained several properties of pseudoknot-bordered and -unbordered words.

5.3 How to Avoid DNA Intermolecular Bonds

In this section, several properties of a tube language $L \subseteq \Sigma^+$, which prohibit various types of undesired hybridizations between two DNA strands, are formally characterized. Many authors assume, for simplicity, that hybridization occurs only between those parts of single-stranded DNA molecules that are perfectly complementary. The following language properties have been considered in Hussini et al. (2003) and Kari et al. (2002, 2003).

(A) 0-**Nonoverlapping**: $L \cap \theta(L) = \emptyset$
(B) 0-**Compliant**: $\forall w \in L, \; x, y \in \Sigma^*, \; w, x\theta(w)y \in L \Rightarrow xy = \lambda$
(C) 0-**p-Compliant**: $\forall w \in L, \; y \in \Sigma^*, \; w, \theta(w)y \in L \Rightarrow y = \lambda$
(D) 0-**s-Compliant**: $\forall w \in L, \; y \in \Sigma^*, \; w, y\theta(w) \in L \Rightarrow y = \lambda$
(E) **Strictly 0-compliant**: both 0-compliant and 0-nonoverlapping
(F) 0-**Free**: $L^2 \cap \Sigma^+ \theta(L) \Sigma^+ = \emptyset$
(G) 0-**Sticky-free**: $\forall w \in \Sigma^+, \; x, y \in \Sigma^*, \; wx, y\theta(w) \in L \Rightarrow xy = \lambda$
(H) 0-**3'-Overhang-free**: $\forall w \in \Sigma^+, \; x, y \in \Sigma^*, \; wx, \theta(w)y \in L \Rightarrow xy = \lambda$
(I) 0-**5'-Overhang-free**: $\forall w \in \Sigma^+, \; x, y \in \Sigma^*, \; xw, y\theta(w) \in L \Rightarrow xy = \lambda$
(J) 0-**Overhang-free**: both 0-3'-overhang-free and 0-5'-overhang-free

For convenience, we agree to say that a language L containing the empty word has one of the above properties if $L \setminus \{\lambda\}$ has that property. Observe that (F) avoids situations like ❯ *Fig. 9c*, while other properties exclude special cases of ❯ *Fig. 9b*.

In Jonoska and Mahalingam (2004), a 0-nonoverlapping language was said to be *strictly 0*. Generally, if any other property holds in conjunction with (A), we add the qualifier *strictly*. This notation has already been used for the property (E). Both *strict* and *non-strict* properties turn out to be useful in certain situations.

For example, a common way to check for the presence of a certain single-stranded molecule w is to add to the solution its complement $\tau(w)$, and use enzymes to destroy any molecules that are not fully double stranded. Simultaneously, we want to prevent all other hybridizations except w and $\tau(w)$. This condition is equivalent to testing whether the whole solution, including w and $\tau(w)$, is non-strictly bond-free (exact matches are allowed).

Further properties have been defined in Jonoska and Mahalingam (2004) for a language L. Observe that the property (K) below avoids bonds like those in ❯ *Fig. 9a*, with a restricted length of the loop part:

(K) $0(k, m_1, m_2)$-**Subword compliant**: $\forall u \in \Sigma^k, \; \Sigma^* u \Sigma^m 0(u) \Sigma^* \cap L = \emptyset$ for $k > 0, m_1 \leq m \leq m_2$
(L) 0-**k-Code**: $\text{Sub}_k(L) \cap \text{Sub}_k(\theta(L)) = \emptyset, k > 0$

The property (L) was also considered implicitly in Baum (1998) and Feldkamp et al. (2002). In particular, Baum (1998) considered tube languages of the form $(sZ)^+$ satisfying (L), where s is a fixed word of length k and Z is a code of length k – the notation sZ represents the set of all words sz such that z is in Z.

The following property was defined for $0 = I$, the identity relation, in Head (2000). A language L is called

(M) **solid** if
    1. $\forall x, y, u \in \Sigma^*, \; u, xuy \in L \Rightarrow xy = \lambda$ and
    2. $\forall x, y \in \Sigma^*, \; u \in \Sigma^+, \; xu, uy \in L \Rightarrow xy = \lambda$

L is *solid relative* to a language $M \subseteq \Sigma^*$ if 1 and 2 above must hold only when there are p, $q \in \Sigma^*$ such that $pxuyq \in M$. L is called *comma-free* if it is solid relative to L^*. Solid languages were also used in Kari et al. (2003) as a tool for constructing error-detecting tube languages that were invariant under bio-operations.

▢ Fig. 14
Classes of tube languages free of certain types of undesired hybridization.

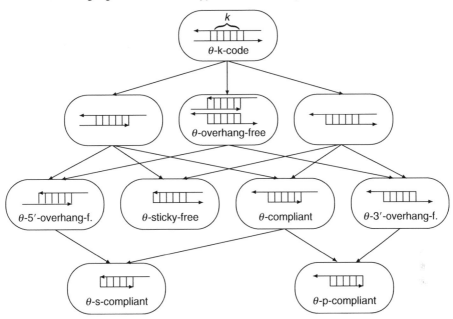

❯ *Figure 14* shows the hierarchy of some of the above language properties. Arrows stand for inclusion relations among language classes satisfying these properties.

Example 2 (Kari et al. 2005b) Consider the language $L = \{A^n T^n \mid n \geq 1\} \subset \Delta^+$, and the DNA involution τ. Observe that $\tau(L) = L$. We can deduce that L is:

- Neither τ-nonoverlapping, nor τ-k-code for any $k \geq 1$
- Not τ-compliant, as for $w = A^n T^n$, $x = A$, $y = T$ we have $w, x\tau(w)y \in L$
- τ-p-Compliant, as $w, \theta(w)y \in L$ implies $w = A^n T^n$, $y = \lambda$; similarly, L is τ-s-compliant
- Not τ-free, as $A^n T^n A^m T^m$, n, $m > 1$, is both in L^2 and in $\Delta^+ L \Delta^+$
- Not τ-sticky-free, as for $w = y = A^n$. $x = T^n$ we have $wx, y\tau(w) \in L$
- τ-$3'$-Overhang-free, as $wx, \tau(w)y \in L$ implies $w = A^n T^m$, $x = T^{n-m}$, $y = T^{m-n}$ and hence $xy = \lambda$; similarly, L is τ-$5'$-overhang-free and hence τ-overhang-free
- Not $\theta(k, m_1, m_2)$-subword compliant for any k, m_1, m_2

For further details and relations between the above listed DNA language properties the reader is referred to Jonoska and Mahalingam (2004) and Kari et al. (2003, 2005b).

To establish a common framework allowing us to handle various types of unwanted hybridization in a uniform way, it is necessary to introduce the generalizing concept of word operations on trajectories. Consider a *trajectory alphabet* $V = \{0, 1\}$ and assume $V \cap \Sigma = \emptyset$. We call any string $t \in V^*$ a *trajectory*. A trajectory is essentially a syntactical condition that specifies how a binary word operation is applied to the letters of its two operands. Let $t \in V^*$ be a trajectory and let α, β be two words over Σ.

Definition 4 (Mateescu et al. 1998) *The shuffle of α with β on a (fixed) trajectory t, denoted by $\alpha \sqcup\!\sqcup_t \beta$, is the following binary word operation:*

$$\alpha \sqcup\!\sqcup_t \beta = \{\alpha_1 \beta_1 \dots \alpha_k \beta_k \mid \alpha = \alpha_1 \dots \alpha_k, \beta = \beta_1 \dots \beta_k, t = 0^{i_1} 1^{j_1} \dots 0^{i_k} 1^{j_k},$$
$$\text{where } |\alpha_m| = i_m \text{ and } |\beta_m| = j_m \text{ for all } m, 1 \le m \le k \}.$$

Example 3 Let $\alpha = a_1 a_2 \dots a_8$, $\beta = b_1 b_2 \dots b_5$ and assume that $t = 0^3 1^2 0^3 10101$. The shuffle of α and β on the trajectory t is:

$$\alpha \sqcup\!\sqcup_t \beta = \{a_1 a_2 a_3 b_1 b_2 a_4 a_5 a_6 b_3 a_7 b_4 a_8 b_5 \}$$

Observe that the result of the operation is generally a set of words, though in the case of shuffle on trajectory it is always a singleton or the empty set. Notice also that $\alpha \sqcup\!\sqcup_t \beta = \emptyset$ if $|\alpha| \ne |t|_0$ or $|\beta| \ne |t|_1$.

A *set of trajectories* is any set $T \subseteq V^*$. The *shuffle of α with β on the set T, denoted by $\alpha \sqcup\!\sqcup_T \beta$, is*

$$\alpha \sqcup\!\sqcup_T \beta = \bigcup_{t \in T} \alpha \sqcup\!\sqcup_t \beta \tag{1}$$

The shuffle on (sets of) trajectories generalizes several traditional word operations. Let, for example, $T = 0^* 1^*$. Then $\sqcup\!\sqcup_T = \cdot$, the operation of catenation.

To characterize the properties of tube languages described above, we define formally a *property* $\mathcal{P}$ as a mapping $\mathcal{P} : 2^{\Sigma^*} \to \{\text{true, false}\}$. We say that a language L has (or satisfies) the property $\mathcal{P}$ if $\mathcal{P}(L) = \text{true}$. The next definition introduces a general concept of *bond-free property* that covers surprisingly many types of undesired bonds studied in the literature.

Definition 5 (Kari et al. 2005b) *Consider a language property $\mathcal{P}$. Let there be binary word operations $\Diamond_{\text{lo}}$, $\Diamond_{\text{up}}$ and an involution θ such that for an arbitrary $L \subseteq \Sigma^*$, $\mathcal{P}(L) = \text{true}$ iff*

(i) $\forall w \in \Sigma^+, x, y \in \Sigma^*$ $((w \Diamond_{\text{lo}} x) \cap L \ne \emptyset, (\theta(w) \Diamond_{\text{up}} y) \cap L \ne \emptyset) \Rightarrow xy = \lambda$, *then $\mathcal{P}$ is called a* bond-free property (of degree 2)

(ii) $\forall w, x, y \in \Sigma^*$ $((w \Diamond_{\text{lo}} x) \cap L \ne \emptyset, (\theta(w) \Diamond_{\text{up}} y) \cap L \ne \emptyset) \Rightarrow w = \lambda$, *then $\mathcal{P}$ is called a* strictly bond-free property (of degree 2)

If not stated otherwise, we assume in the sequel that $\Diamond_{\text{lo}} = \sqcup\!\sqcup_{T\text{lo}}$ and $\Diamond_{\text{up}} = \sqcup\!\sqcup_{T\text{up}}$ for some sets of trajectories $T_{\text{lo}}, T_{\text{up}}$. Intuitively, w and $\theta(w)$ represent two complementary oligonucleotides. Then $w \sqcup\!\sqcup_{T\text{lo}} x$ and $\theta(w) \sqcup\!\sqcup_{T\text{up}} y$ represent two DNA single strands, which could form a double-stranded DNA molecule with blunt ends, depicted in ❯ *Fig. 14*. The bond-free property $\mathcal{P}$ guarantees that $w \sqcup\!\sqcup_{T\text{lo}} x$ and $\theta(w) \sqcup\!\sqcup_{T\text{up}} y$ with nonempty x, y (i), or w (ii), cannot simultaneously exist in L.

Theorem 5 (Kari et al. 2005b)

(i) *The language properties (B), (C), (D), (G), (H), (I), (M.1), (M.2) are bond-free properties.*

(ii) *The language properties (A), strictly (B)–(D), strictly (G)–(I), (L), strictly (L) are strictly bond-free properties.*

Moreover, in both cases, the associated sets of trajectories $T_{\text{lo}}, T_{\text{up}}$ are regular.

Proof Let θ be an antimorphism and let the sets of trajectories T_{lo}, T_{up} corresponding to the listed bond-free properties be

(A) $T_{lo} = T_{up} = 0^+$
(B) $T_{lo} = 0^+$, $T_{up} = 1^*0^+1^*$
(C) $T_{lo} = 0^+$, $T_{up} = 0^+1^*$
(D) $T_{lo} = 0^+$, $T_{up} = 1^*0^+$
(G) $T_{lo} = 0^+1^*$, $T_{up} = 1^*0^+$
(H) $T_{lo} = 0^+1^*$, $T_{up} = 0^+1^*$
(I) $T_{lo} = 1^*0^+$, $T_{up} = 1^*0^+$
(L) $T_{lo} = T_{up} = 1^*0^k1^*$
(L) strictly: $T_{lo} = T_{up} = 1^*0^k1^* \cup 0^+$

If θ is a morphism, one can similarly define

(M.1) $T_{lo} = 0^*$, $T_{up} = 1^*0^*1^*$
(M.2) $T_{lo} = 1^*0^+$, $T_{up} = 0^+1^*$

Consider, for example, the property (H), θ-3'-overhang-freedom. Then $w \sqcup_{T_{lo}} x = \{wx\}$ and $\theta(w) \sqcup_{T_{up}} y = \{\theta(w)y\}$. The relations in ❷ Definition 5 (i) take the form $wx \in L, \theta(w)y \in L$ that corresponds to the definition of (H) above. The proofs of the other mentioned properties are analogous.

Observe that T_{lo}, T_{up} for a certain property correspond to the "shape" of the bonds prohibited in languages satisfying the property. The above theorem allows for a general characterization of bond-free properties via a solution of an *unique* language inequation in Kari et al. (2005b). As a consequence, the following result can be obtained.

Theorem 6 (Kari et al. 2005b) *Let $\mathcal{P}$ be a (strictly) bond-free property associated with regular sets of trajectories T_{lo}, T_{up}. Then the following problem is decidable in quadratic time with respect to $|A|$:*

Input: an NFA A
Output: Yes/No depending on whether L(A) satisfies $\mathcal{P}$

By ❷ Theorem 5, the above result applies to the properties (A) – (D), (G) – (J), (M), strictly (B) – strictly (D), strictly (G) – strictly (J), (L), strictly (L), in the case of regular languages, on one hand. On the other hand, for a given context-free language L it is undecidable whether it satisfies certain bond-free properties, for example, (B) and (F).

Theorem 7 (Hussini et al. 2003) *The following problem is undecidable.*

Input: A bond-free property $\mathcal{P}$ associated with regular sets of trajectories T_{lo}, T_{up}, and a context-free language L
Output: Yes/No depending on whether $\mathcal{P}(L) = \text{true}$

An important problem studied in the literature is the *optimal negative design problem*: How to construct a non-crosshybridizing set (i.e., a set of single-stranded molecules which do not mutually hybridize) of a certain required size, given a fixed library of available molecules. In general, even to decide whether such a *finite* set exists is an NP-complete problem and its equivalence with the maximal independent set problem can be easily shown

(Deaton et al. 2003). Various heuristics were used to find a nearly optimal solution (Phan and Garzon 2008). Here we focus on a similar but in some cases easier *maximality problem* defined formally in ❯ Sect. 5.2.

Theorem 8 (Kari et al. 2005b) *Let $M \subseteq \Sigma^+$ be a regular set of words, and $L \subseteq M$ a regular language satisfying a bond-free property $\mathcal{P}$.*

(a) *Let θ be an antimorphism and let $\mathcal{P}$ be one of the properties (B), (C), (D), (G), strictly (B) – strictly (D), strictly (G), (L), strictly (L), or*
(b) *Let θ be a morphism and let $\mathcal{P}$ be one of the properties (B), (C), (D), (H), (I), strictly (B) – strictly (D), strictly (H), strictly (I), (L), strictly (L).*

Then there is an algorithm deciding whether there is a $w \in M \setminus L$ such that $L \cup \{w\}$ satisfies $\mathcal{P}$.

Algorithms deciding the maximality of these properties can require an exponential time with respect to the size of an NFA accepting L. The same holds when we want to construct a maximal regular set of DNA strands satisfying these bond-free properties. In two important cases, however, a polynomial time can be achieved (Kari et al. 2005b): (1) for maximality of *regular* nonoverlapping sets satisfying the property (A), and (2) for maximality of *finite* θ-compliant sets satisfying the property (B).

5.4 Preventing Imperfect DNA–DNA Bonds

In ❯ Sects. 5.2 and ❯ 5.3, properties of DNA codes are presented based on the assumption that hybridization binds only two perfectly WK complementary single-stranded DNA molecules. In reality, however, thermodynamical laws allow for hybridization also in cases of some "roughly" complementary molecules with certain irregularities in the WK complementarity sense. This section describes properties of languages that ensure that even such imperfect bindings can be described and eventually prevented.

As already mentioned, the negative design problem is rather computationally expensive when using thermodynamical methods. Various approximative methods using discrete similarity metrics have been therefore studied. In Marathe et al. (2001) and Tulpan et al. (2003), the authors considered codes K of length k satisfying the following property ($H(u,v)$ is the Hamming distance, i.e., the number of mismatches at corresponding positions, between words u and v of the same length).

$\mathbf{X}[d,k]$: If u and v are any codewords in K then $H(u, \tau(v)) > d$.

In fact the above property is studied in conjunction with the uniqueness property $H(K) > d$ – here $H(K)$ is the smallest Hamming distance between any two different words in K.

Garzon et al. (1997) introduced the H-measure based on Hamming distance for two words x and y of length k and explained how this measure can be used to encode instances of the HPP. This measure was also used in Garzon et al. (2006) to search optimal codes for DNA computing using the shuffle operation on DNA strands. A similar measure was defined in Arita (2004) and Arita and Kobayashi (2002) extending the work of Frutos et al. (1997) and was applied to codes of length k whose words can be concatenated in arbitrary ways. Thus, the tube language here was $L = K^+$. The code K satisfied certain uniqueness conditions. In particular, the tube language $L = K^+$. satisfied the following property.

$\mathbf{Y}[d,k]$: If x is a subword of L of length k and v is a codeword in K then $H(x, \tau(v)) > d$.

This property was considered also in Reif et al. (2002) for tube languages of the form $K_1 K_2 \ldots K_m$, where each K_i is a certain code of length k.

Finally, Kari et al. (2005a) introduced the following property of a tube language L, motivated by the fact that the above defined properties **X**, **Y** still allow for certain types of undesired bonds between DNA codewords.

Z[d,k]: If x and y are any subwords of L of length k then $H(x,\tau(y)) > d$.

The reader can observe that the property **X** is a generalization of (A) from the previous section. Similarly, **Y** is a generalization of (B) and **Z** generalizes (L). Note that any set L satisfying property **Z**[d,k] also satisfies **Y**[d,k]. Further relations among different bond-free properties using similarity measures were studied in Kari et al. (2005a).

The choice of the Hamming distance in the condition $H(x,\tau(y)) \le d$ for *similarity* between words is a very natural one and has attracted a lot of interest in the literature. One might argue, however, that parts of two DNA molecules could form a stable bond even if they have different lengths. In **❷** *Fig. 15*, for example, the bound parts of the two molecules have lengths 10 and 9. Such hybridizations (and even more complex ones) were addressed in Andronescu et al. (2003). Based on this observation, the condition for two subwords x and y to bind together should be

$$|x|, |y| \ge k \quad \text{and} \quad Lev(x, \tau(y)) \le d$$

The symbol $Lev(u, v)$ denotes the *Levenshtein distance* between the words u and v – this is the smallest number of substitutions, insertions, and deletions of symbols required to transform u into v.

To establish a general framework that would cover both the similarity measure H, Lev and possibly also others, Kari et al. (2005a) considered a general binary relation γ on words over Σ, that is, a subset of $\Sigma^* \times \Sigma^*$. The expression "(u,v) is in γ" can be rephrased as "$\gamma(u,v)$ is true" when γ is viewed as a logic predicate. A binary relation is called *rational* if it can be realized by a finite transducer.

Definition 6 (Kari et al. 2005a) *A binary relation* sim *is called a* similarity *relation with parameters* (t, l), *where t and l are nonnegative integers, if the following conditions are satisfied.*

(i) *If* sim (u,v) *is true then* abs$(|u|-|v|) \le t$, *where abs is the absolute value function.*
(ii) *If* sim (u,v) *is true and* $|u|,|v| > l$ *then there are proper subwords x and y of u and v, respectively, such that* sim (x,y) *is true.*

We can interpret the above conditions as follows: (1) the lengths of two similar words cannot be too different and (2) if two words are similar and long enough, then they contain two similar proper subwords.

The notation $H_{d,k}$ shall be used for the relation "$|u|,|v| \ge k$ and $H(u,v) \le d$", and $Lev_{d,k}$ for "$|u|,|v| \ge k$ and $Lev(u,v) \le d$". It is evident that the $H_{d,k}$ is an example of a rational similarity

❏ Fig. 15

Two DNA molecules in which the parts 5′ – *AAGCGTTCGA* – 3′ and 5′ – *TCGGACGTT* – 3′ bind together although these parts have different lengths.

$$5'-\text{v A A G C G T T C G A w } -3'$$
$$\backslash \; \backslash \; | \; | \; | \quad | \; | \; |$$
$$3'-\text{z T T G C A G G C T y } -5'$$

relation with parameters $(0, k)$. It is also easy to show that $Lev_{d,k}$ is a rational similarity relation as well, with parameters $(d, d + k)$.

Based on the above definition, for any similarity relation $sim(\cdot, \cdot)$ between words and for every involution θ, we define the following property of a language L with strong mathematical properties.

P$[\theta, sim]$: If x and y are any nonempty subwords of L then $sim(x, \theta(y))$ is false.

Any language satisfying **P**$[\theta, sim]$ is called a (θ, sim)-*bond-free language*. Although this property seems to be quite general and covering many possible situations, it can be shown that it is only a special case of the strict bond-freeness defined in ❷ Sect. 5.3.

Theorem 9   (Kari et al. 2005a) $P[\theta, sim]$ *is a strictly bond-free property.*

Proof   The mappings sim_L and sim_R are defined as follows:

$$sim_L(y) = \{x \in \Sigma^* \mid sim(x, y)\}$$
$$sim_R(x) = \{y \in \Sigma^* \mid sim(x, y)\}$$

Recall that a language L is (θ, sim)-bond-free iff

$$\forall x_1, y_1, x_2, y_2 \in \Sigma^*, \ w_1, w_2 \in \Sigma^+$$
$$(x_1 w_1 y_1, \ x_2 w_2 y_2 \in L) \Rightarrow \text{ not } sim(w_1, \theta(w_2)) \qquad \text{iff}$$
$$\forall x_1, y_1, x_2, y_2 \in \Sigma^*, \ w_1, w_2 \in \Sigma^+$$
$$(x_1 w_1 y_1, \ x_2 \theta(w_2) y_2 \in L) \Rightarrow \text{ not } sim(w_1, w_2) \qquad \text{iff}$$
$$\forall x_1, w_1, y_1, x_2, w_2, y_2 \in \Sigma^*$$
$$(x_1 w_1 y_1, \ x_2 \theta(w_2) y_2 \in L, w_2 \in sim_R(w_1)) \Rightarrow (w_1 = \lambda \text{ or } w_2 = \lambda) \qquad \text{iff}$$
$$\forall x_1, y_1, x_2, y_2, w \in \Sigma^*$$
$$(\{x_1 w y_1\} \cap L \neq \emptyset, \ \{x_2\} \cdot \theta(sim_R(w) \cap \Sigma^+) \cdot \{y_2\} \cap L \neq \emptyset) \Rightarrow w = \lambda \quad \text{iff}$$
$$\forall x, y, w \in \Sigma^*$$
$$(w \sqcup_T x \cap L \neq \emptyset, \ \theta(sim_R(w)) \sqcup_T y \cap L \neq \emptyset) \Rightarrow w = \lambda$$

where $T = 1^* 0^+ 1^*$.

Therefore, an expression corresponding to the definition of strictly bond-free property has been obtained. Notice that results analogous to ❷ Theorem 9 could be proved also for the properties **X**$[d,k]$, **Y**$[d,k]$, and **Z**$[d,k]$. Observe, furthermore, that the operation on words w and y defined as $\theta(sim_R(w)) \sqcup_T y$ is "almost" $\sqcup_T$, and hence some results from ❷ Sect. 5.3 are applicable in the case of (θ, sim)-bond-free languages, provided that the relation sim is "reasonable."

Theorem 10   (Kari et al. 2005a) *Let* sim *be a rational relation. The following problem is decidable in quadratic time with respect to* $|A|$.

Input: An NFA A
Output: YES/NO, depending on whether $L(A)$ *is a* (θ, sim)-*bond-free language*

For the case where sim is one of the similarity relations $H_{d,k}$ or $Lev_{d,k}$, the algorithm runs at time $\mathcal{O}(dk|A|^2)$ (or $\mathcal{O}(dk^2|A|^2)$, respectively). The (θ, sim)-bond-freeness remains decidable even in the case of context-free tube languages, although the existence of a polynomial-time algorithm cannot be guaranteed.

Two problems related to the design of large sets of bond-free molecules, the *optimal negative design problem* and the *maximality problem* have been addressed, too. Both were formally specified in previous sections. The optimal negative design problem remains NP complete even for *finite* tube languages in the case of various similarity metrics. The problem of maximality of *regular* $(0, \text{sim})$-bond-free languages has been shown decidable in Kari et al. (2005a), although the existence of a polynomial-time algorithm cannot be generally guaranteed. However, rather surprisingly, in the important Hamming case such an algorithm exists. One can consider languages that are subsets of $(\Sigma^k)^+$, for some positive integer k. We call such languages *k-block languages*. Naturally, any regular k-block language can be represented by a special type of lazy DFA (Wood 1987), which we call a *k-block DFA*.

Theorem 11 (Kari et al. 2005a) *Let d be fixed to be either 0 or 1. The following problem is computable in a polynomial time.*

Input: k-block DFA A such that $L(A)$ is a $(0, H_{d,k})$-bond-free subset of $(\Sigma^k)^+$

Output: YES/NO, depending on whether the language $L(A)$ is maximal with that property. Moreover, if $L(A)$ is not maximal, output a minimal-length word $w \in (\Sigma^k)^+ \setminus L(A)$ such that $L(A) \cup \{w\}$ is a $(0, H_{d,k})$-bond-free subset of $(\Sigma^k)^+$.

In particular, the time complexity $t(|A|)$ is bounded as follows:

$$t(|A|) = \begin{cases} O(k|A|^3), & \text{if } k \text{ is odd and } d = 0 \\ O(|A|^6), & \text{if } k \text{ is even and } d = 0 \\ O(k^3|A|^6), & \text{if } d = 1 \end{cases}$$

The concept of $(0, \text{sim})$-bond-free languages is quite general and could cover also subtler similarity measures than the Hamming or Levenshtein distance. Consider the original nearest-neighbor thermodynamical approach to the hybridization problem (SantaLucia 1998). The calculation of $\Delta G_{\min}$, the minimum free energy among the free energies of all possible secondary substructures that may be formed by the examined DNA sequences, is frequently used to determine the most likely secondary structure that will actually form. Assume secondary substructures of a size limited from above (say, of at most 25 bp), which is reasonable from the practical point of view. One can consider two DNA sequences *similar* if and only if they contain subsequences satisfying the condition $\Delta G_{\min} \geq B$, where B is a threshold value for hybridization energy (the "all or nothing" hybridization model). Such a similarity relation obviously fulfills the conditions of ❷ Definition 6 and, furthermore, it is rational (even finite). Therefore, for a fixed value of B, the hybridization analysis can possibly benefit from the above mentioned results and rapid algorithms.

We conclude with two examples of a construction of DNA codes in the Hamming case. The first example is based on the method of *k-templates* proposed originally in Arita and Kobayashi (2002). This method allows us to produce codes that are subsets of $(\Sigma^k)^+$.

Theorem 12 (Kari et al. 2005a) *Let I be a nonempty subset of $\{1, \ldots, k\}$ of cardinality $m = \lfloor k/2 \rfloor + 1 + \lfloor (d + (k \text{ rem } 2))/2 \rfloor$. Then the language K^+ is $(\tau, H_{d,k})$-bond-free, where*

$$K = \{v \in \Sigma^k \mid \text{if } i \in I \text{ then } v[i] \in \{A, C\}\}$$

Observe that the size of the code K is $2^m 4^{k-m}$. An advantage of the method of $\bullet$ Theorem 12 is that we can construct $(\tau, H_{d,k})$-bond-free languages with a large ratio d/k.

Another method is based on the operation of *subword closure* $K^\otimes$ of a set $K \subseteq \Sigma^k$

$$K^\otimes = \{w \in \Sigma^* \mid |w| \geq k, \mathrm{Sub}_k(w) \subseteq K\}$$

Denote further $K^\oplus \stackrel{\text{def}}{=} K^\otimes \cap (\Sigma^k)^+$. The following theorem characterizes all maximal $(\tau, H, _{d,k})$-bond-free subsets of $(\Sigma^k)^+$ and $\Sigma^k \Sigma^*$.

Theorem 13   (Kari et al. 2005a) *The class of all maximal $(\tau, H_{d,k})$-bond-free subsets of $(\Sigma^k)^+$ is finite and equal to*

$$\{K^\oplus \mid K \text{ is a maximal } (\tau, H_{d,k})\text{-bond-free subset of } \Sigma^k\}.$$

The above theorem holds also for subsets of $\Sigma^k \Sigma^*$ if one replaces $K^\oplus$ with $K^\otimes$. As a consequence, if one constructs a maximal finite subset K of Σ^k satisfying $\tau(K) \cap H_d(K) = \emptyset$, then the language $K^\oplus$ is a maximal $(\tau, H_{d,k})$-bond-free subset of $(\Sigma^k)^+$. Another implication of $\bullet$ Theorem 13 is that all maximal $(\tau, H_{d,k})$-bond-free subsets of $(\Sigma^k)^+$ or $\Sigma^k \Sigma^*$ are regular.

To conclude, in $\bullet$ Sect. 5 several basic types of DNA interaction based on Watson–Crick complementarity are characterized, using the apparatus of formal language and automata theory. Besides its value as a contribution to theoretical computer science, the main application of this research is the description and construction of sets of DNA molecules (called also DNA codes) that are free of certain types of unwanted binding interactions. These codes are especially useful in DNA computing and many other laboratory techniques, which assume that an undesirable hybridization between DNA molecules does not occur. It has been shown that a uniform mathematical characterization exists for many types of DNA bonds, both perfect and imperfect with respect to the WK complementarity principle. This characterization, in turn, implies the existence of effective algorithms for manipulation and construction of these DNA codes.

6   Vectorial Models of DNA-Based Information

In $\bullet$ Sect. 5, we saw that formal language theory is a natural tool for modeling, analyzing, and designing "good" DNA codewords that control DNA strand inter- and intramolecular interactions based on Watson–Crick complementarity. This was achieved by formalizing a DNA single-strand in the $5'$-$3'$ orientation as a linear word over the DNA alphabet $\{A, C, G, T\}$. A limitation of this representation of DNA single strand as words is that it does not model double-stranded DNA molecules, partially double-stranded DNA molecules (such as DNA strands with sticky ends), or interactions between DNA single strands that may lead to double strands.

This section offers an alternative natural way of representing DNA strands, namely as vectors. A (single, partially double-stranded, or fully double-stranded) DNA molecule is modeled namely by a vector whose first component is a word over the DNA alphabet representing the "top" strand, and the second component is a word over the DNA alphabet representing the "bottom" strand. Using this representation, one can naturally model the operations of annealing and ligation.

We now introduce two computational models that use this representation of DNA strands: a language generating device called *sticker system* ($\bullet$ Sect. 6.1) (Freund et al. 1999; Kari et al.

1998; Păun and Rozenberg 1998), and its automata counterpart, *Watson–Crick automata* (❯ Sect. 6.2) (Freund et al. 1999). A summary of essential results on these topics can be found in Păun et al. (1998).

We start by introducing a vectorial formalism of the notions of DNA single strand, double strand, and partial double strand, as well of the bio-operations of annealing (hybridization) and ligation. Other notions and notations that are used here were defined in ❯ Sect. 5.1.

The notion of Watson–Crick complementarity is formalized by a symmetric relation. A relation $\rho \subseteq \Sigma \times \Sigma$ is said to be *symmetric* if for any $a, b \in \Sigma$, $(a, b) \in \rho$ implies $(b, a) \in \rho$. In order to define a symmetric relation ρ, it suffices to specify one of (a, b) and (b, a) as long as we explicitly note that ρ is symmetric.

DNA strands are modeled by 2×1 vectors, wherein the first row corresponds to the "top" DNA strand and the second row corresponds to the "bottom" DNA strand. In this formalism, DNA double strands are modeled as 2×1 vectors in square brackets where the top word is in relation ρ with the bottom word, while DNA single strands are modeled as 2×1 vectors in round brackets where one of the rows is the empty word. More concretely, we write $\begin{bmatrix} a \\ b \end{bmatrix}_\rho$ if two letters a, b are in the relation ρ. Whenever ρ is clear from context, the subscript ρ is omitted. We now define an alphabet of double-stranded columns

$$\Sigma_d = \begin{bmatrix} \Sigma \\ \Sigma \end{bmatrix}_\rho \cup \begin{pmatrix} \Sigma \\ \lambda \end{pmatrix} \cup \begin{pmatrix} \lambda \\ \Sigma \end{pmatrix}$$

where $\begin{bmatrix} \Sigma \\ \Sigma \end{bmatrix}_\rho = \left\{ \begin{bmatrix} a \\ b \end{bmatrix} \middle| a, b \in \Sigma, (a, b) \in \rho \right\}$, $\begin{pmatrix} \Sigma \\ \lambda \end{pmatrix} = \left\{ \begin{pmatrix} a \\ \lambda \end{pmatrix} \middle| a \in \Sigma \right\}$, and $\begin{pmatrix} \lambda \\ \Sigma \end{pmatrix} = \left\{ \begin{pmatrix} \lambda \\ b \end{pmatrix} \middle| b \in \Sigma \right\}$.

The *Watson–Crick domain* associated to Σ and ρ is the set $\mathrm{WK}_\rho(\Sigma)$ defined as $\mathrm{WK}_\rho(\Sigma) = \begin{bmatrix} \Sigma \\ \Sigma \end{bmatrix}_\rho^*$. An element $\begin{bmatrix} a_1 \\ b_1 \end{bmatrix} \begin{bmatrix} a_2 \\ b_2 \end{bmatrix} \cdots \begin{bmatrix} a_n \\ b_n \end{bmatrix} \in \mathrm{WK}_\rho(\Sigma)$ can be written as $\begin{bmatrix} a_1 a_2 \cdots a_n \\ b_1 b_2 \cdots b_n \end{bmatrix}$ succinctly. Note that $\begin{pmatrix} w_1 \\ w_2 \end{pmatrix}$ means no more than a pair of words w_1, w_2, whereas $\begin{bmatrix} w_1 \\ w_2 \end{bmatrix}$ imposes that $|w_1| = |w_2|$ and their corresponding letters are complementary in the sense of the relation ρ. Elements of $\mathrm{WK}_\rho(\Sigma)$ are called *complete double-stranded sequences* or *molecules*. Moreover, we denote $\mathrm{WK}_\rho^+(\Sigma) = \mathrm{WK}_\rho(\Sigma) \setminus \{(\lambda, \lambda)\}$.

Note that elements of $\mathrm{WK}_\rho(\Sigma)$ are fully double-stranded. In most DNA computing experiments, for example, Adleman's first experiment, partially double-stranded DNA, that is, DNA strands with sticky ends, are essential. To introduce sticky ends in the model, let $S(\Sigma) = \begin{pmatrix} \lambda \\ \Sigma^* \end{pmatrix} \cup \begin{pmatrix} \Sigma^* \\ \lambda \end{pmatrix}$ be the set of sticky ends. Then we define a set $W_\rho(\Sigma)$ whose elements are molecules with sticky ends at both sides as $W_\rho(\Sigma) = L_\rho(\Sigma) \cup R_\rho(\Sigma) \cup LR_\rho(\Sigma)$, where

$$L_\rho(\Sigma) = S(\Sigma) \mathrm{WK}_\rho(\Sigma)$$
$$R_\rho(\Sigma) = \mathrm{WK}_\rho(\Sigma) S(\Sigma)$$
$$LR_\rho(\Sigma) = S(\Sigma) \mathrm{WK}_\rho^+(\Sigma) S(\Sigma)$$

Note that unlike an element in $L_\rho(\Sigma)$ or $R_\rho(\Sigma)$, elements of $LR_\rho(\Sigma)$ must have at least one "column" $\begin{bmatrix} a \\ b \end{bmatrix}$. Any element of $W_\rho(\Sigma)$ with at least a position $\begin{bmatrix} a \\ b \end{bmatrix}$, $a \neq \lambda$, $b \neq \lambda$, is called a *well-started (double stranded) sequence.* Thus, $LR_\rho(\Sigma)$ is equivalent to the set of all well-started sequences.

Annealing and ligation of DNA molecules can be modeled as a partial operation among elements of $W_\rho(\Sigma)$. A well-started molecule can be prolonged to the right or to the left with another molecule, provided that their sticky ends match. We define "*sticking y to the right of x*" operation, denoted by $\mu_r(x, y)$. In a symmetric way, $\mu_\ell(y, x)$ (sticking y to the left of x) is defined. Let $x \in LR_\rho(\Sigma)$ and $y \in W_\rho(\Sigma)$. Being well-started, $x = x_1 x_2 x_3$ for some $x_1, x_3 \in S(\Sigma)$, $x_2 \in WK_\rho^+(\Sigma)$. Then $\mu_r(x, y)$ is defined as follows (also see ❷ *Fig. 16*):

Case A If y is single-stranded, that is, $y \in S(\Sigma)$, we have the following cases: for $r, p \geq 0$,

1. If $\quad x_3 = \begin{pmatrix} a_1 \cdots a_r \\ \lambda \end{pmatrix} \quad$ and $\quad y = \begin{pmatrix} a_{r+1} \cdots a_{r+p} \\ \lambda \end{pmatrix}, \quad$ then $\quad \mu_r(x, y) = x_1 x_2$
$\begin{pmatrix} a_1 \cdots a_r a_{r+1} \cdots a_{r+p} \\ \lambda \end{pmatrix}.$

2. If $x_3 = \begin{pmatrix} a_1 \cdots a_r \\ \lambda \end{pmatrix}\begin{pmatrix} a_{r+1} \cdots a_{r+p} \\ \lambda \end{pmatrix}$, $y = \begin{pmatrix} \lambda \\ b_1 \cdots b_r \end{pmatrix}$, and $(a_i, b_i) \in \rho$ for $1 \leq i \leq r$,

then $\mu_r(x, y) = x_1 x_2 \begin{bmatrix} a_1 \cdots a_r \\ b_1 \cdots b_r \end{bmatrix}\begin{pmatrix} a_{r+1} \cdots a_{r+p} \\ \lambda \end{pmatrix}.$

3. If $x_3 = \begin{pmatrix} a_1 \cdots a_r \\ \lambda \end{pmatrix}$, $y = \begin{pmatrix} \lambda \\ b_1 \cdots b_r \end{pmatrix}\begin{pmatrix} \lambda \\ b_{r+1} \cdots b_{r+p} \end{pmatrix}$, and $(a_i, b_i) \in \rho$ for $1 \leq i \leq r$,

then $\mu_r(x, y) = x_1 x_2 \begin{bmatrix} a_1 \cdots a_r \\ b_1 \cdots b_r \end{bmatrix}\begin{pmatrix} \lambda \\ b_{r+1} \cdots b_{r+p} \end{pmatrix}.$

4. The counterparts of cases A1 – A3 where the roles of upper and lower strands are reversed.

Case B If y is well-started (partially double-stranded), that is, $y = y_1 y_2 y_3$ for some $y_1, y_3 \in S(\Sigma)$ and $y_2 \in WK_\rho^+(\Sigma)$: for $r \geq 0$,

1. If $x_3 = \begin{pmatrix} a_1 \cdots a_r \\ \lambda \end{pmatrix}$, $y_1 = \begin{pmatrix} \lambda \\ b_1 \cdots b_r \end{pmatrix}$, and $(a_i, b_i) \in \rho$ for $1 \leq i \leq r$, then $\mu_r(x, y) = x_1 x_2 \begin{bmatrix} a_1 \cdots a_r \\ b_1 \cdots b_r \end{bmatrix} y_2 y_3.$

2. The counterpart of case B1 when the roles of upper and lower strands are reversed.

◘ **Fig. 16**

Sticking operations: Prolongation of a well-started molecule to the right.

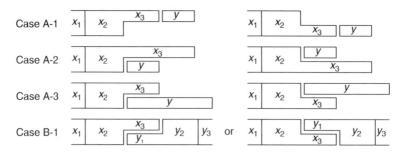

If none of these cases applies, $\mu_r(x,y)$ is undefined. Note that in all cases, r can be 0, that is, the system is allowed to prolong the blunt ends of molecules. Moreover, we have

$$\mu_r\left(x,\begin{pmatrix}\lambda\\\lambda\end{pmatrix}\right) = \mu_\ell\left(\begin{pmatrix}\lambda\\\lambda\end{pmatrix},x\right) = x \text{ for any } x \in \mathrm{LR}_\rho(\Sigma).$$

Note also that this vectorial model of DNA molecules allows – unlike its linear counterpart presented in the previous section – the differentiation between single-stranded DNA molecules, double-stranded DNA molecules, and DNA molecules with sticky ends, as well as for modeling of DNA–DNA interactions such as annealing and ligation.

6.1 Sticker Systems

Sticker systems are formal models of molecular interactions occurring in DNA computing, based on the sticking operation. Several variants of sticker systems have been defined in the literature. In this section, the simple regular sticker system, which is the most realistic variant but which is weak in terms of generating capacity, is described. We also introduce two practical ways to strengthen this variant: by using some complex DNA structures (❯ Sect. 6.1.1), and by observing the sticker system externally (❯ Sect. 6.1.2). Both enhance the computational power of the sticker system to Turing universality.

A sticker system prolongs given well-started DNA molecules, both to the left and to the right, by using the sticking operation, so as to turn them into complete double-stranded molecules. This system is an appropriate model of molecular interaction occurring, for example, in Adleman's 1994 experiment, and was proposed under the name *bidirectional sticker system* (Freund et al. 1998).

A (bidirectional) *sticker system over a relation* ρ is a 4-tuple $\gamma = (\Sigma, \rho, A, P)$, where Σ is an alphabet endowed with the symmetric relation $\rho \subseteq \Sigma \times \Sigma$, $A \subseteq \mathrm{LR}_\rho(\Sigma)$ is a finite subset of well-started sequences (axioms), and P is a finite subset of $W_\rho(\Sigma) \times W_\rho(\Sigma)$. Starting from an axiom in A, the system prolongs it with using a pair in P as follows:

$$x \Rightarrow_\gamma w \text{ iff } w = \mu_r(\mu_\ell(y,x),z) \text{ for some } (y,z) \in P$$

In other words, $x \Rightarrow_\gamma w$ iff sticking y to the left and z to the right of x results in w. The reflexive and transitive closure of $\Rightarrow_\gamma$ is denoted by $\Rightarrow_\gamma^*$. A sequence $x_1 \Rightarrow_\gamma x_2 \Rightarrow_\gamma \ldots \Rightarrow_\gamma x_k$, $x_1 \in A$, is called a *computation* in γ (of length $k-1$). A computation as above is *complete* if $x_k \in \mathrm{WK}_\rho^+(\Sigma)$.

The set of all molecules over Σ generated by complete computations in γ is defined as $LM(\gamma) = \{w \in \mathrm{WK}_\rho^+(\Sigma) \mid x \Rightarrow_\gamma^* w, x \in A\}$. We can also consider the sticker systems as a generator of languages of strings rather than double-stranded molecules. To this aim, the following language is associated with $LM(\gamma)$

$$L(\gamma) = \left\{ w \in \Sigma^* \;\middle|\; \begin{bmatrix} w \\ w' \end{bmatrix}_\rho \in LM(\gamma) \text{ for some } w' \in \Sigma^* \right\}$$

A language L is called a *sticker language* if there exists a sticker system γ such that $L(\gamma) = L$.

A sticker system $\gamma = (\Sigma, \rho, A, P)$ is said to be *simple* (respectively *regular*) if for each pair $(y,z) \in P$, $y,z \in S(\Sigma)$ (respectively $y = \lambda$). A simple regular system extends either the upper or lower strand, one at a time (hence the attribute "simple"), and only to the right (hence the attribute "regular"). Thus, a simple regular sticker system can be rewritten as 5-tuple $(\Sigma, \rho, A,$

■ **Fig. 17**

The idea of the proof of ❯ Theorem 14. For any sticker system γ over a relation ρ, one can construct a sticker system γ′ over the identity relation, such that $L(\gamma)=L(\gamma')$. The newly constructed sticker system γ′, based on *id*, simulates the process of γ to generate $(x,y)\in WK_\rho^+(\Sigma)$ where $x=x_0x_1\ldots x_n$ and $y=y_0y_1\ldots y_m$, by generating $(x,z)\in WK_{id}^+(\Sigma)$, where $z=z_0z_1\ldots z_m$.

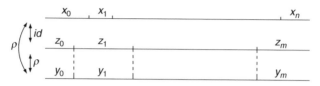

D_u, and D_ℓ), where D_u and D_ℓ are finite subsets of $\left(\dfrac{\Sigma^*}{\lambda}\right)$ and $\left(\dfrac{\lambda}{\Sigma^*}\right)$, respectively. The family of languages generated by simple regular sticker systems is denoted by SRSL(n), where n means "no-restriction." This variant is the most precise and realistic model of the annealing/ligation-based hybridization occurring in Adleman's experiment. In fact, since being proposed in Kari et al. (1998), this type of sticker system has been intensively investigated.

A normal form for sticker systems with respect to the relation ρ was introduced in Hoogeboom and van Vugt (2000) for simple regular variants, and in Kuske and Weigel (2004) for general sticker systems. It says that the identity relation *id* suffices to generate any sticker language.

Theorem 14 (Hoogeboom and van Vugt 2000; Kuske and Weigel 2004) *For a sticker system γ over a relation ρ, one can construct a sticker system γ′ over the identity relation id such that* $L(\gamma) = L(\gamma')$.

Proof The ideas proposed in Hoogeboom and van Vugt (2000) and Kuske and Weigel (2004) are essentially the same; they work for arbitrary bidirectional sticker systems. Here we present their proofs applied to *regular* (unidirectional) sticker systems to suggest the fact that the identity relation suffices also for Watson–Crick automata introduced later.

Let $\gamma = (\Sigma, \rho, A, P_r)$ be a regular sticker system. We construct a regular sticker system $\gamma'=(\Sigma, id, A', P_r')$, ❯ *Fig. 17*, where

$$A' = \left\{\left(\begin{array}{c}x_0\\z_0\end{array}\right) \;\middle|\; \left(\begin{array}{c}x_0\\y_0\end{array}\right) \in A \text{ for some } y_0 \text{ such that } \left[\begin{array}{c}z_0\\y_0\end{array}\right]_\rho\right\}$$

$$P_r' = \left\{\left(\begin{array}{c}x_i\\z_i\end{array}\right) \;\middle|\; \left(\begin{array}{c}x_i\\y_i\end{array}\right) \in P_r \text{ for some } y_i \text{ such that } \left[\begin{array}{c}z_i\\y_i\end{array}\right]_\rho\right\}$$

Assume that $x \in L(\gamma)$, that is, there is a word y such that $(x,y) \in WK_\rho^+(\Sigma)$, $x = x_0x_1\ldots x_n$ and $y=y_0y_1\ldots y_m$, where $(x_0,y_0) \in A$, and $(x_j,y_j)\in P_r'$ $(1 \leq j \leq m)$. Due to the symmetric property of ρ, x can also be written as the catenation of words z_0 and $z_1,\ldots,z_m \in \Sigma^*$ such that $(z_k,y_k) \in WK_\rho(\Sigma)$ $(0 \leq k \leq m)$. According to the definition of A', $(x_0,z_0) \in A'$, and $(x_j,z_j) \in P_r'$ $(1 \leq j \leq m)$. As a result, $\left[\begin{array}{c}x_0x_1\cdots x_n\\z_0z_1\cdots z_m\end{array}\right]_{id} = \left[\begin{array}{c}x\\x\end{array}\right]_{id} \in LM(\gamma')$, and hence $x \in L(\gamma')$. The proof that $L(\gamma')\subseteq L(\gamma)$ is similar.

As mentioned in the previous proof, this theorem proved to be valid for general sticker systems in Kuske and Weigel (2004). Moreover, in the paper, the authors show that an analogous result holds even for Watson–Crick automata, that is, the identity relation suffices for WK-automata. From a historical viewpoint of the theory of computation, the normal forms for grammars and acceptors have proved to be useful tools. Analogously, ❯ Theorem 14 will be useful for several proofs in the rest of this section.

The simple regular sticker system is one of the most "natural" computational models for annealing/ligation-based hybridization. Kari et al. (1998) initiated an investigation into the generative capacity of general sticker systems, including the simple regular variant, and the investigation continued in Freund et al. (1998) and Păun and Rozenberg (1998). The conclusion was that some classes of sticker systems can even characterize the recursively enumerable languages. On the contrary, the simple regular variant turned out to be quite weak.

Theorem 15   (Kari et al. 1998; Păun et al. 1998) SRSL $(n) \subsetneq$ REG = Cod (SRSL (n)).

Thus, this "natural variant" of sticker systems has no more generative power than finite automata, even with the aid of encoding.

In Kari et al. (1998), the notion of *fair computation* was proposed. Let γ be a simple regular sticker system. A complete computation in γ is said to be *fair* if through the computation, the number of extensions occurring on the upper strand is equal to the number of extensions occurring on the lower strand. A language L is called a *fair sticker language* if there exists a simple regular sticker system γ such that L is the set of all words that are generated by fair computations in γ. The family of fair sticker languages is denoted by SRSL(f).

It is known that REG $\subsetneq$ Cod(SRSL(f)) and also that we cannot obtain characterizations of RE starting from languages in SRSL(f) and using an arbitrary generalized sequential machine (*gsm*) mapping, including a coding (see Păun et al. (1998)). Hence the question of whether SRSL(f) is included in CF (or even in LIN) arose. The answer to this question was obtained in Hoogeboom and van Vugt (2000). First, their example can be introduced to show that SRSL $(f) \not\subseteq$ LIN.

Example 4   Let $\gamma = (\{a,b,c,d\}, \rho, A, D_u, D_\ell)$ be a simple regular sticker system, where $\rho = \{(a,a),(b,b),(b,c),(d,d)\}$, $A = \left\{ \begin{bmatrix} d \\ d \end{bmatrix} \right\}$, $D_u = \left\{ \begin{pmatrix} aa \\ \lambda \end{pmatrix}, \begin{pmatrix} b \\ \lambda \end{pmatrix} \right\}$, and $D_\ell = \left\{ \begin{pmatrix} \lambda \\ a \end{pmatrix}, \begin{pmatrix} \lambda \\ bc \end{pmatrix} \right\}$. This is a technical modification of Example 2 from Hoogeboom and van Vugt (2000) in order to make an axiom well started. Then $LM_n(\gamma) = \begin{bmatrix} d \\ d \end{bmatrix} \left\{ \begin{bmatrix} aa \\ aa \end{bmatrix}, \begin{bmatrix} bb \\ bc \end{bmatrix} \right\}^*$, and hence $L_n(\gamma) = d\{aa, bb\}^*$, $L_f(\gamma) = \{x \in L_n(\gamma) \mid \#_a(x) = \#_b(x)\}$. The pumping lemma for linear languages can be used to prove that $L_f(\gamma)$ is not linear.

Theorem 16   (Hoogeboom and van Vugt 2000) SRSL(f) $\subsetneq$ Cod (SRSL(f)) $\subsetneq$ CF.

Kari et al. (1998) imposed an additional constraint on fair computations called *coherence*, the use of which leads to a representation of RE. In the rest of this section, two new approaches are introduced to the problem of how to obtain characterizations of RE by using sticker systems augmented with more practical assumptions. Regarding the generative capacity and further topics about sticker systems, the reader is referred to the thorough summary in chapter 4 of Păun et al. (1998).

6.1.1 Sticker Systems with Complex Structures

Sakakibara and Kobayashi (2001) proposed a novel use of stickers, that involved the formation of *DNA hairpins*. In ❯ Sect. 5, a hairpin was modeled as a linear word $w_1 a_1 \ldots a_n w_2 b_n \ldots b_1 w_3$, where $(a_i, b_i) \in \rho$, $1 \leq i \leq n$. Here we represent the same hairpin vectorially as $\begin{pmatrix} \langle w_2 \rangle \\ w_1 \mid w_3 \end{pmatrix} \in \begin{pmatrix} \Sigma^* \\ \Sigma^* \end{pmatrix}$ as shown in ❯ *Fig. 18*. This hairpin-shaped molecule may stick to other molecules by its two sticky ends w_1 and w_3 or by its loop part w_2. The sets of this type and inverted type of hairpins are denoted by $T_u(\Sigma)$ and $T_\ell(\Sigma)$, respectively, that is,

$$T_u(\Sigma) = \left\{ \begin{pmatrix} \langle w_2 \rangle \\ w_1 \mid w_3 \end{pmatrix} \;\middle|\; w_1, w_2, w_3 \in \Sigma^* \right\}, T_\ell(\Sigma) = \left\{ \begin{pmatrix} w_1 \mid w_3 \\ \langle w_2 \rangle \end{pmatrix} \;\middle|\; w_1, w_2, w_3 \in \Sigma^* \right\},$$

The operation of "sticking a hairpin $x = \begin{pmatrix} \langle w_2 \rangle \\ w_1 \mid w_3 \end{pmatrix}$ onto a single-stranded molecule y" is defined whenever $y = y_1 y_2 y_3$, and y_2 is complementary to $w_1 w_3$ as follows:

$$\mu(y, x) = \left(\begin{pmatrix} \lambda \\ y_1 \end{pmatrix} \begin{bmatrix} \langle w_2 \rangle \\ w_1 \mid w_3 \\ y_2 \end{bmatrix} \begin{pmatrix} \lambda \\ y_3 \end{pmatrix} \right).$$

Moreover, this operation $\mu(y, x)$ is extended to the general case; for a molecule,

$$z = \left(\begin{pmatrix} \lambda \\ z_0 \end{pmatrix} \begin{bmatrix} \langle u_1 \rangle \\ t_1 \mid v_1 \\ z_1 \end{bmatrix} \begin{bmatrix} \langle u_2 \rangle \\ t_2 \mid v_2 \\ z_2 \end{bmatrix} \cdots \begin{bmatrix} \langle u_n \rangle \\ t_n \mid v_n \\ z_n \end{bmatrix} \begin{pmatrix} \lambda \\ z_{n+1} z_{n+2} \end{pmatrix} \right)$$

such that $w_1 w_3$ is complementary to z_{n+1}, $\mu(z, x)$ is defined as

$$\mu(z, x) = \left(\begin{pmatrix} \lambda \\ z_0 \end{pmatrix} \begin{bmatrix} \langle u_1 \rangle \\ t_1 \mid v_1 \\ z_1 \end{bmatrix} \begin{bmatrix} \langle u_2 \rangle \\ t_2 \mid v_2 \\ z_2 \end{bmatrix} \cdots \begin{bmatrix} \langle u_n \rangle \\ t_n \mid v_n \\ z_n \end{bmatrix} \begin{bmatrix} \langle w_2 \rangle \\ w_1 \mid w_3 \\ z_{n+1} \end{bmatrix} \begin{pmatrix} \lambda \\ z_{n+2} \end{pmatrix} \right).$$

Thus, this sticking operation forms "multiple hairpins" structures as illustrated in ❯ *Fig. 18* (Middle). The set of molecules with this type of structures and the set of molecules with inverted structures are denoted by $\mathrm{TW}_u(\Sigma)$ and $\mathrm{TW}_\ell(\Sigma)$, respectively, that is,

◘ **Fig. 18**
(Left) A hairpin that the word $w_1 a_1 \ldots a_n w_2 b_n \ldots b_1 w_3$ may form if $(a_i, b_i) \in \rho$, $1 \leq i \leq n$; the sticky ends w_1, w_3 or the hairpin loop w_2 can bind to other molecules; (Middle) two hairpins stick to the lower strand via their sticky ends, leaving sticky ends at both ends of the lower strand; (Right) two "complete" molecules can bind together via their hairpin loops, if they are matched as shown.

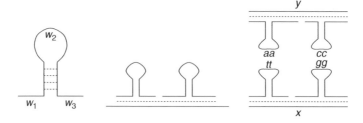

$$\mathrm{TW}_u(\Sigma) = \begin{pmatrix} \lambda \\ \lambda \\ \Sigma^* \end{pmatrix} \left(\begin{bmatrix} \langle \Sigma^* \rangle \\ \Sigma^* \\ \Sigma^* \end{bmatrix} \right)^* \begin{pmatrix} \lambda \\ \lambda \\ \Sigma^* \end{pmatrix}, \quad \mathrm{TW}_\ell(\Sigma) = \begin{pmatrix} \Sigma^* \\ \lambda \\ \lambda \end{pmatrix} \left(\begin{bmatrix} \Sigma^* \\ \Sigma^* \\ \langle \Sigma^* \rangle \end{bmatrix} \right)^* \begin{pmatrix} \Sigma^* \\ \lambda \\ \lambda \end{pmatrix}.$$

Similarly, the hybridization operation $\mu(z,x)$ is defined for $z \in \mathrm{TW}_\ell(\Sigma)$ and $x \in T_\ell(\Sigma)$. Note that $\begin{pmatrix} \lambda \\ \lambda \\ x \end{pmatrix} \in \mathrm{TW}_\ell(\Sigma)$ or $\begin{pmatrix} x \\ \lambda \\ \lambda \end{pmatrix} \in \mathrm{TW}_u(\Sigma)$ can be regarded as a word $x \in \Sigma^*$. Hence we will give also a word as a first argument of μ in the following. Now we can define another type of "complete" molecules, namely elements in the set $\mathrm{TWK}(\Sigma) = \mathrm{TWK}_u(\Sigma) \cup \mathrm{TWK}_\ell(\Sigma)$, where

$$\mathrm{TWK}_u(\Sigma) = \left(\begin{bmatrix} \langle \Sigma^* \rangle \\ \Sigma^* \\ \Sigma^* \end{bmatrix} \right)^*, \quad \mathrm{TWK}_\ell(\Sigma) = \left(\begin{bmatrix} \Sigma^* \\ \Sigma^* \\ \langle \Sigma^* \rangle \end{bmatrix} \right)^*.$$

Second, we consider another sticking operation for the loop (w_2 of x) of a hairpin, which is illustrated in ❷ *Fig. 18* (Right). Consider two complete molecules $x \in \mathrm{TWK}_u(\Sigma)$ and $y \in \mathrm{TWK}_\ell(\Sigma)$ defined as

$$x = \left(\begin{bmatrix} \langle s_1 \rangle \\ r_1|t_1 \\ x_1 \end{bmatrix} \begin{bmatrix} \langle s_2 \rangle \\ r_2|t_2 \\ x_2 \end{bmatrix} \cdots \begin{bmatrix} \langle s_n \rangle \\ r_n|t_n \\ x_n \end{bmatrix} \right), \quad y = \left(\begin{bmatrix} y_1 \\ u_1|w_1 \\ \langle v_1 \rangle \end{bmatrix} \begin{bmatrix} y_2 \\ u_2|w_2 \\ \langle v_2 \rangle \end{bmatrix} \cdots \begin{bmatrix} y_n \\ u_n|w_n \\ \langle v_n \rangle \end{bmatrix} \right).$$

When s_i is complementary to v_i $(1 \leq i \leq n)$, $\phi(x,y)$ is defined as

$$\phi(x,y) = \left(\begin{bmatrix} y_1 \\ u_1|w_1 \\ \langle v_1 \rangle \\ \langle s_1 \rangle \\ r_1|t_1 \\ x_1 \end{bmatrix} \begin{bmatrix} y_2 \\ u_2|w_2 \\ \langle v_2 \rangle \\ \langle s_2 \rangle \\ r_2|t_2 \\ x_2 \end{bmatrix} \cdots \begin{bmatrix} y_n \\ u_n|w_n \\ \langle v_n \rangle \\ \langle s_n \rangle \\ r_n|t_n \\ x_n \end{bmatrix} \right).$$

Thus, for two complete molecules $x \in \mathrm{TWK}_u(\Sigma)$ and $y \in \mathrm{TWK}_\ell(\Sigma)$, $\phi(x,y)$ is well defined to be an element of the set

$$\mathrm{DTWK}(\Sigma) = \left(\begin{bmatrix} \Sigma^* \\ \Sigma^* \\ \langle \Sigma^* \rangle \\ \langle \Sigma^* \rangle \\ \Sigma^* \\ \Sigma^* \end{bmatrix} \right)^*.$$

A *sticker system with complex structures* is a 4-tuple $\gamma = (\Sigma, \rho, D_u, D_\ell)$, where D_u and D_ℓ are finite subsets of $T_u(\Sigma)$ and $T_\ell(\Sigma)$, respectively. For molecules $x, y \in \mathrm{TW}_u(\Sigma)$, we write $x \Rightarrow_u y$ if $y = \mu(x, v)$ for some $v \in D_u$. Analogously, for $x', y' \in \mathrm{TW}_\ell(\Sigma)$, we write $x' \Rightarrow_\ell y'$ if $y' = \mu(x', v')$ for some $v' \in D_\ell$. The reflexive and transitive closures of these operations are denoted by $\Rightarrow_u^*$ and $\Rightarrow_\ell^*$.

A sequence $x_1 \Rightarrow_\alpha x_2 \Rightarrow_\alpha \cdots \Rightarrow_\alpha x_k$, with $x_1 \in \Sigma^*$ and $\alpha \in \{u, \ell\}$, is called a computation in γ. (In this context, x_1, the start of the computation, can be regarded as either a word in Σ^* or an element of $\mathrm{TW}_\alpha(\Sigma)$.) A computation $x_1 \Rightarrow_\alpha^* x_k$ is said to be *complete* when $x_k \in \mathrm{TWK}_\alpha(\Sigma)$.

Suppose that one has two complete computations $x \Rightarrow_u^* y$ and $x \Rightarrow_\ell^* z$ for a word $x \in \Sigma^*$. When $\phi(y, z)$ becomes a complete matching, that is, $\phi(y, z) \in \mathrm{DTWK}(\Sigma)$, this computation process is said to be *successful*. The following language is associated with γ:

$$L(\gamma) = \{x \in (\Sigma \setminus \{\#\})^* \mid x\# \Rightarrow_u^* y, y \in \mathrm{TWK}_u(\Sigma)$$
$$x\# \Rightarrow_\ell^* z, z \in \mathrm{TWK}_\ell(\Sigma), \text{ and } \phi(y, z) \in \mathrm{DTWK}(\Sigma)\}.$$

Thus, one can consider this system as a language-accepting device. The family of languages accepted by sticker systems with complex structures is denoted by SLDT.

Now it can be shown that the use of hairpins enables one to characterize RE based on the following lemma.

Lemma 1 (Kari et al. 1998) *For each recursively enumerable language $L \subseteq \Sigma^*$, there exist two λ-free morphisms $h_1, h_2 : \Sigma_2^* \to \Sigma_1^*$, a regular language $R \subseteq \Sigma_1^*$, and a projection $pr_\Sigma : \Sigma_1^* \to \Sigma^*$ such that $L = pr_\Sigma(h_1(\mathrm{EQ}(h_1, h_2)) \cap R)$.*

Theorem 17 (Sakakibara and Kobayashi 2001) *Every recursively enumerable language is the weak coding of a language in the family SLDT.*

Proof Let $L \in \mathrm{RE}$. Due to ❷ Lemma 1, L can be obtained from $h_1(\mathrm{EQ}(h_1, h_2)) \cap R$ by a projection, where $h_1, h_2 : \Sigma_2^* \to \Sigma_1^*$ are λ-free morphisms, and $R \in \mathrm{REG}$. Hence it suffices to construct a sticker system with complex structures γ that accepts an *encoding* of $h_1(\mathrm{EQ}(h_1, h_2)) \cap R$. Consider a complete deterministic finite automaton $M = (Q, \Sigma_1, \delta, q_0, F)$ for R, where $Q = \{q_0, q_1, \ldots, q_m\}$. Any word $w = b_1 b_2 \ldots b_k$ ($b_i \in \Sigma_1$) is encoded uniquely as $q_{l_0} b_1 q_{l_1} q_{l_1} b_2 q_{l_2} \cdots q_{l_{k-1}} b_k q_{l_k} q_{l_k}$, where $q_{l_0} = q_0, q_{l_1}, \ldots, q_{l_k} \in Q$ such that $\delta(q_{l_{j-1}}, b_j) = q_{l_j}$ for $1 \le j \le k$. So $w \in R$ iff $q_{l_k} \in F$.

Let $\Sigma_2 = \{a_1, \ldots, a_n\}$, and for each a_i, let $h_1(a_i) = c_1 c_2 \cdots c_{k_i}$ ($c_j \in \Sigma_1$). Note that for an arbitrary state in Q, there is a unique transition on M by $h_1(a_1)$ because M is a complete deterministic automaton. Thus, a set of encodings of all such transitions is defined for each $a_i \in \Sigma_2$ as follows:

$$T_1(h_1(a_i)) = \bigcup_{q_{l_0} \in Q} \{q_{l_0} b_1 q_{l_1} \cdots q_{l_{k_i-1}} b_{k_i} q_{l_{k_i}} \mid \delta(q_{l_{j-1}}, b_j) = q_{l_j}, 1 \le j \le k_i\}$$

Following the same idea, $T_2(h_2(a_i))$ is defined for each $a_i \in \Sigma_2$. Now γ is constructed as $(\Sigma_1 \cup \Sigma_2 \cup Q \cup \{\#\}, id, D_u \cup D_\ell)$, where

$$D_\ell = \left\{ \binom{t_2}{\langle a_i \rangle} \mid t_2 \in T_2(h_2(a_i)) \right\} \cup \left\{ \binom{q_f \#}{\langle \# \rangle} \mid q_f \in F \right\}$$

$$D_u = \left\{ \binom{\langle a_i \rangle}{t_1} \mid t_1 \in T_1(h_1(a_i)) \right\} \cup \left\{ \binom{\langle \# \rangle}{q_f \#} \mid q_f \in F \right\}$$

❷ *Figure 19* illustrates the idea of how γ recognizes the language $h_1(\mathrm{EQ}(h_1, h_2))$ (the encoding mentioned above for R is omitted for clarity).

By this construction, $L(\gamma)$ is the set of encodings of words $u \in R$ for which there exists a word $w \in \Sigma_2^*$ with $h_1(w) = h_2(w) = u$. Projection being a weak coding, there exists a weak coding h such that $h(L(\gamma)) = L$.

❏ **Fig. 19**
A brief sketch of the proof for ❷ Theorem 17. Two hairpins, one with sticky end $h_1(a_i)$ and the other with sticky end $h_2(a_i)$, binding together via their matching loops, can be abstracted as a finite control (square) labeled by a_i, with two heads (bifurcated arrows). The control checks for each i, $1 \leq i \leq n$, if the single-stranded part of the upper strand begins with $h_2(a_i)$, and the lower strand begins with $h_1(a_i)$; if so, then the corresponding hairpins are stuck. This process is repeated until either this system cannot proceed in this way anymore or it generates a complete matching of two complete molecules. In fact this system is essentially equivalent to Watson–Crick automata, which will be introduced in ❷ Sect. 6.2.

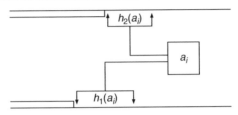

6.1.2 Observable Sticker Systems

Another idea to strengthen a computational system is to let someone observe and report how the system works step by step. This composite system, inspired by the common practice of observing the progress of a biology or chemistry experiment, has been introduced in Cavaliere and Leupold (2004) to "observe" membrane systems. There, a finite automaton observed the change of configurations of a "computationally weak" membrane system (with context-free power). Surprisingly, this composite system proved to be universal. Following this idea, many computations were "observed": splicing systems, derivations of grammars and string-rewriting systems, and also sticker systems. For the details of these observations as well as the formal definition of computation by observation in general, the readers are referred to Cavaliere and Leupold (2004), Cavaliere (2008), and references thereof.

The idea of observing sticker systems was introduced in Alhazov and Cavaliere (2005). Informally, an *observable sticker system* is composed of a "computationally weak" sticker system and an external observer. Observing the computation of a sticker system, starting from an axiom, the observer notes – at each computational step – the current configuration, and processes it according to its own rules, producing an output. The catenation of all the outputs thus produced by the observer during a complete computation, constitutes a word in the language of the observable sticker system. The collection of all words thus obtained is the language generated by this observable sticker system.

Formally speaking, configurations of a sticker system, which are elements of $\Sigma_d{}^*$, are observed so that an observer is implemented as a finite automaton that works on elements of $\Sigma_d{}^*$. This automaton is defined as a 6-tuple $O = (Q, \Sigma_d, \Delta, q_0, \delta, \sigma)$, where a finite set of states Q, an input alphabet Σ_d, the initial state $q_0 \in Q$, and a complete transition function $\delta : \Sigma_d \times Q \rightarrow Q$ are defined as usual for conventional finite automata: whereas Δ is an output alphabet and σ is a labeling function $Q \rightarrow \Delta \cup \{\bot, \lambda\}$, $\bot$ being a special symbol.

An *observable (simple regular) sticker system* is a pair $\phi = (\gamma, O)$ for a simple regular sticker system γ and an observer O. For a computation $c : x_0 \Rightarrow_\gamma x_1 \Rightarrow_\gamma \ldots \Rightarrow_\gamma x_k$ ($x_i \in R_\rho(\Sigma)$), $O(c)$ is defined as $O(x_0)O(x_1)\ldots O(x_k)$. The language $L(\phi)$ generated by ϕ is defined as $L(\phi) = \{O(c) \mid c$ is a complete computation by $\gamma\}$.

Theorem 18 (Alhazov and Cavaliere 2005) *There exists an observable simple regular sticker system which generates a non-context-free language.*

This theorem shows how stronger very restricted sticker systems can get with the aid of the observer (cf. ❷ Theorem 15). A natural question that follows is of how to get universality within the framework of observable sticker systems. The next theorem proves that observers with the capability to discard any "bad" evolution endow simple regular sticker systems with universality.

The symbol $\perp \notin \Delta$ makes it possible for an observer O to distinguish bad evolutions by a sticker system γ from good ones in a way that the observation of bad evolutions leads to the output of $\perp$. For the observable sticker system $\phi = (\gamma, O)$, one can weed out any word that contains $\perp$ from $L(\phi)$ by taking $\widehat{L}(\phi) = L(\phi) \cap \Delta^*$.

Theorem 19 (Alhazov and Cavaliere 2005) *For each $L \in$ RE, there exists an observable simple regular sticker system $\phi = (\gamma, O)$ such that $\widehat{L}(\phi) = L$.*

Due to the fact that recursive languages are closed under intersection with regular languages, ❷ Theorem 19 has the following result as its corollary.

Corollary 1 (Alhazov and Cavaliere 2005) *There exists an observable simple regular sticker system $\phi = (\gamma, O)$ such that $L(\phi)$ is a non-recursive language.*

6.2 Watson–Crick Automata

While sticker systems *generate* complete double-stranded molecules by using the sticking operation, their *accepting* counterparts, the Watson–Crick automata, parse a given complete double-stranded molecule and determine whether the input is accepted or not. A Watson–Crick automaton is equipped with a finite state machine with two heads. This machine has its heads read the respective upper and lower strands of a given complete double-stranded molecule simultaneously, and changes its state accordingly. The basic idea of how Watson–Crick automata work was described in ❷ *Fig. 19.*

In parallel to the research on sticker systems, these biologically inspired automata have been intensively investigated within the last decade. Early studies including, for example, Freund et al. (1999) and Martin-Víde et al. (1998) investigate variants of WK-automata, relationships among them with respect to generative capacity, and universal Watson–Crick automata, topics summarized in chapter 5 of Păun et al. (1998). A more recent survey Czeizler and Czeizler (2006c) includes results on complexity measures (Păun and Păun 1999) and Watson–Crick automata systems (Czeizler and Czeizler 2006a,b). Other studies on WK-automata comprise Watson–Crick ω-automata (Petre 2003), the role of the complementarity relation (Kuske and Weigel 2004) (see ❷ Theorem 1), local testability and regular reversibility (Sempere 2007, 2008), 5'→3' sensing Watson–Crick finite automata (Nagy 2008), and deterministic Watson–Crick automata (Czeizler et al. 2008a).

In this section, we present recent results in this field, such as studies of deterministic WK-automata, and the role of the complementarity relation. In particular, deterministic WK-automata are essential for the design of efficient molecular parsers.

A (*nondeterministic*) *Watson–Crick (finite) automaton over a symmetric relation* $\rho \subseteq \Sigma \times \Sigma$ is a 6-tuple $M = (\Sigma, \rho, Q, q_0, F, \delta)$, where Σ, Q, q_0, and F are defined in the same manner as for finite automata. The *transition function* δ is a mapping $\delta : Q \times \Sigma^* \times \Sigma^* \to 2^Q$ such that $\delta\left(q, \begin{pmatrix} w_1 \\ w_2 \end{pmatrix}\right) \neq \emptyset$ only for *finitely many* pairs $(q, w_1, w_2) \in Q \times \Sigma^* \times \Sigma^*$. We can replace the transition function by rewriting rules, by denoting $q\begin{pmatrix} w_1 \\ w_2 \end{pmatrix} \to q'$ instead of $q' \in \delta\left(q, \begin{pmatrix} w_1 \\ w_2 \end{pmatrix}\right)$. Transitions in M are defined as follows. For $q, q' \in Q$ and $\begin{pmatrix} x_1 \\ x_2 \end{pmatrix}, \begin{pmatrix} w_1 \\ w_2 \end{pmatrix}, \begin{pmatrix} y_1 \\ y_2 \end{pmatrix} \in \begin{pmatrix} \Sigma^* \\ \Sigma^* \end{pmatrix}$ such that $\begin{bmatrix} x_1 w_1 y_1 \\ x_2 w_2 y_2 \end{bmatrix} \in \mathrm{WK}_\rho(\Sigma)$, we write $\begin{pmatrix} x_1 \\ x_2 \end{pmatrix} q \begin{pmatrix} w_1 \\ w_2 \end{pmatrix} \begin{pmatrix} y_1 \\ y_2 \end{pmatrix} \Rightarrow \begin{pmatrix} x_1 \\ x_2 \end{pmatrix} \begin{pmatrix} w_1 \\ w_2 \end{pmatrix} q' \begin{pmatrix} y_1 \\ y_2 \end{pmatrix}$ iff $q' \in \delta\left(q, \begin{pmatrix} w_1 \\ w_2 \end{pmatrix}\right)$. If $\Rightarrow^*$ is the reflexive and transitive closure of $\Rightarrow$, then the language accepted by M is

$$L(M) = \left\{ w_1 \in \Sigma^* \,\middle|\, q_0 \begin{bmatrix} w_1 \\ w_2 \end{bmatrix} \Rightarrow^* \begin{bmatrix} w_1 \\ w_2 \end{bmatrix} q_f \text{ for some} \right.$$

$$\left. q_f \in F \text{ and } w_2 \in \Sigma^* \text{ such that } \begin{bmatrix} w_1 \\ w_2 \end{bmatrix} \in \mathrm{WK}_\rho(\Sigma) \right\}.$$

Other languages are also considered in Freund et al. (1999) and Păun et al. (1998) such as control words associated to computations — but they are not introduced here. By convention, as suggested also in Păun and Păun (1999), in this section we consider two languages differing only by the empty word λ as identical.

A WK-automaton $M = (\Sigma, \rho, Q, q_0, F, \delta)$ is said to be *stateless* if $Q = F = \{s_0\}$; *all-final* if $Q = F$; *simple* if for any rewriting rule $q\begin{pmatrix} w_1 \\ w_2 \end{pmatrix} \to q'$, either w_1 or w_2 is λ; *1-limited* if for any transition $q\begin{pmatrix} w_1 \\ w_2 \end{pmatrix} \to q'$, we have $|w_1 w_2| = 1$. By AWK, NWK, FWK, SWK, and 1WK, we denote the families of languages accepted by WK-automata that are arbitrary (A), stateless (N, no-state), all-final (F), simple (S), and 1-limited (1). When two restrictions are imposed at the same time, both of the corresponding symbols are used to identify the family.

As customary in automata theory, normal forms for WK-automata are available. For example, we can convert any WK-automaton into a 1-limited one without changing the language accepted, or the symmetric relation over which the original WK-automaton is defined (Martin-Víde et al. 1998). Another normal form, standardizing the symmetric relation, is as follows:

Theorem 20 (Kuske and Weigel 2004) *For any WK-automaton M, we can construct a WK-automaton M_{id} over the identity relation id with $L(M) = L(M_{id})$.*

A 1-limited WK-automaton over the identity relation is equivalent to a *two-head finite automaton*. Therefore the next theorem follows, where TH denotes the family of languages accepted by two-head finite automata.

Theorem 21 (Păun et al. 1998) 1WK = SWK = 1SWK = AWK = TH.

Czeizler et al. (2008a) proposed three criteria of determinism. A WK-automaton M is said to be *weakly deterministic* if at any point of computation by M, there is at most one possibility to continue the computation; and *deterministic* if for any pair of transition rules $q\binom{u}{v} \rightarrow q'$ and $q\binom{u'}{v'} \rightarrow q''$, we have $u \not\sim_p u'$ or $v \not\sim_p v'$. Clearly, a deterministic WK-automaton is weakly deterministic. Moreover, a deterministic WK-automaton over a symmetric relation ρ is said to be *strongly deterministic* if ρ is the identity. The families of languages accepted by *weakly deterministic*, *deterministic*, and *strongly deterministic* WK-automata are denoted by wdAWK, dAWK, and sdAWK, respectively. The symbol "A" can be replaced with N, F, S, 1, or their combination as in the nondeterministic case.

Proposition 1 (Czeizler et al. 2008a) sdAWK $\subseteq$ dAWK $\subseteq$ wdAWK $\subseteq$ AWK.

The generative power of finite automata or Turing machines remains unchanged by bringing in determinism, while the determinism strictly weakens pushdown automata (Hopcroft and Ullman 1979). Thus, a natural question is whether the relation in ❶ Proposition 1 includes some strict inclusion or all of the families are the same.

Czeizler et al. (2008a) proved that dAWK = d1WK using a similar proof technique for AWK = 1WK. As mentioned above, the technique keeps the symmetric relation unchanged. As such, sdAWK = sd1WK follows immediately. Following the same reasoning to prove AWK = TH (❶ Theorem 21), we can see that sdAWK is equivalent to the family dTH of languages accepted by *deterministic two-head finite automata*. It is known that there exists a language in TH\dTH, for example, $L' = \{w \in \Sigma^* \mid w \neq w^R\}$. This means that nondeterministic WK-automata are strictly more powerful than strongly deterministic ones. In fact, the following stronger result holds.

Theorem 22 (Czeizler et al. 2008a) sdAWK $\subsetneqq$ dAWK.

Proof It suffices to prove that $L' \in$ dAWK. We prove the statement only for the case when Σ is binary, but the statement holds for arbitrary finite alphabets.

Let $M = (\Sigma \cup \{c, d_a, d_b\}, \rho, Q, q_0, F, \delta)$ be a WK-automaton, where $\rho = \{(a, a), (a, d_a), (b, b), (b, d_b), (a, c), (b, c)\}$, $Q = \{q_0, q_f, q_a, q_b\}$, $F = \{q_f\}$, and δ consists of the following rules:

$$q_0\binom{\lambda}{x} \rightarrow q_0, \quad q_0\binom{\lambda}{d_x} \rightarrow q_x, \text{with } x \in \{a, b\}$$

$$q_x\binom{y}{z} \rightarrow q_x, \text{ with } x, y, z \in \{a, b\}$$

$$q_x\binom{zy}{c} \rightarrow q_f, \text{ with } x, y, z \in \{a, b\}, \ x \neq y$$

$$q_f\binom{x}{\lambda} \rightarrow q_f, \text{ with } x \in \{a, b\}$$

It is clear that M is deterministic. Let $w = w_1 w_2 \ldots w_n$ with $w_i \in \Sigma$. If $w \neq w^R$, then there exists a position k in the first half of w such that $w_k \neq w_{n-k+1}$. The characters d_a, d_b are used as a marker of this position. When M runs on the input $\begin{pmatrix} w_1 \cdots w_{k-1} w_k w_{k+1} \cdots w_{n-1} w_n \\ w_1 \cdots w_{k-1} d_{w_k} w_{k+1} \cdots w_{n-1} c \end{pmatrix}$, it accepts w. On the other hand, M does not accept any palindrome regardless of what complement one chooses; thus $L(M) = \{w \in \{a, b\}^+ \mid w \neq w^R\}$.

This contrasts with the nondeterministic case where the complementarity relation does not play any active role (❷ Theorem 20). Though it is natural now to ask if dAWK $\subseteq$ wdAWK is strict or not, this question remains open. Note that there exists a weakly deterministic WK-automaton that is not deterministic.

❷ Proposition 1 and ❷ Theorem 22 conclude that strong determinism strictly weakens WK-automata. However, deterministic WK-automata are still more powerful than finite automata. Since the construction of a strongly deterministic WK-automaton that simulates a given deterministic finite automaton is straightforward, in the following a stronger result is included.

Theorem 23   REG $\subseteq$ sdF1WK.

Proof   In order to simulate a deterministic finite automaton $A = (Q, \Sigma, q_0, F, \delta)$ with $Q = \{q_0, q_1, \ldots, q_n\}$, we construct a strongly deterministic all-final 1-limited WK-automaton $M = (\Sigma, id, Q', q_{0,0}, Q', \delta')$, where $Q' = \{q_{i,j}, \overline{q_{i,j}} \mid 0 \leq i, j \leq n\}$, and for $a \in \Sigma$, $0 \leq i, j \leq n$,

$$\delta'\left(q_{i,j}, \begin{pmatrix} a \\ \lambda \end{pmatrix}\right) = \begin{cases} q_{k,j} & \text{if } \delta(q_i, a) = q_k \text{ and } q_k \notin F \\ \overline{q_{k,j}} & \text{if } \delta(q_i, a) = q_k \text{ and } q_k \in F \end{cases}$$

$$\delta'\left(\overline{q_{i,j}}, \begin{pmatrix} \lambda \\ a \end{pmatrix}\right) = \begin{cases} \overline{q_{i,k}} & \text{if } \delta(q_j, a) = q_k \text{ and } q_k \notin F \\ q_{i,k} & \text{if } \delta(q_j, a) = q_k \text{ and } q_k \in F \end{cases}$$

The recognition of a sequence $\begin{bmatrix} w \\ w \end{bmatrix}$ consists of two identical simulations of recognition process of w on A over the upper and lower strands. Current states of the automaton A working on upper and lower strands are recorded as first and second subscripts of states in Q'. The overline of states indicates that M is now in the simulation over the lower strand. One switches these two phases every time the simulated automaton reaches some final state of A. We can see easily that $L(M) = L(A) \cup \{\lambda\}$.

From ❷ Theorem 23, it is clear that WK-automata are more powerful than finite automata. Additional results on the generative capacity of WK-automata can be found in, for example, Păun et al. (1998). Moreover, it was shown in Czeizler et al. (2008a) and Păun and Păun (1999) that WK-automata recognize some regular languages in a less space-consuming manner. Usage of space is measured by the *state complexity* (for more details, see Păun and Păun (1999) and Yu (2002)). It is well-known that the state complexity of some families of finite languages is unbounded when one considers the finite automata recognizing them. In other words, for any $k \geq 1$, there is a finite language that cannot be recognized by any finite automaton with at most $k-1$ states. On the contrary, any finite language can be recognized by a WK-automaton with two states (Czeizler et al. 2008a).

Determinism is one of the most essential properties for the design of an efficient parser. Due to the time-space trade-off, the stronger the determinism is, the more space-consuming WK-automata get. For example, it was shown in Czeizler et al. (2008a) that in order to recognize a finite language $L_k = \{a, aa, \ldots, a^{k-1}\}$, the strongly deterministic WK-automaton needs at least k states, while for any k, two states are enough once the strong determinism is changed to determinism. Little is known about the state complexity with respect to the (strongly, weakly) deterministic WK-automata.

As a final result on this topic, the following undecidability property about determinism of WK-automata is mentioned.

Theorem 24 (Czeizler et al. 2008a) *It is undecidable whether a given WK-automaton is weakly deterministic.*

7 DNA Complementarity and Combinatorics on Words

❯ Sections 5 and ❯ 6 described new concepts and results in formal language theory and automata theory that attest to the influence that the notion of DNA-based information and its main aspect, the Watson–Crick complementarity, has had on formal language theory, coding theory, and automata theory. This section is devoted to describing some of the ideas, notions, and results that DNA-based information has brought to another area of theoretical computer science, namely combinatorics on words. Indeed, the "equivalence" from the informational point of view between a DNA single strand and its Watson–Crick complement led to several interesting generalizations of fundamental notions such as bordered word, palindrome, periodicity, conjugacy, and commutativity. Moreover, these generalizations led to natural extensions of two of the most fundamental results in combinatorics on words: the Fine and Wilf theorem and the Lyndon–Schützenberger equation. This section presents some of these concepts and results in combinatorics of words motivated by the Watson–Crick complementarity property of DNA-based information.

In the following, θ will denote an antimorphic involution. For further details on combinatorics on words, the reader is referred to Choffrut and Karhumäki (1997).

As mentioned in ❯ Sect. 2, recognition sites of enzymes are often palindromic in a biological sense. Conventionally speaking, a palindrome is a word that reads the same way from the left and from the right, such as "racecar" in English. In contrast, in molecular biology terms, a palindromic DNA sequence is one that is equal to its WK complementary sequence. For example, *TGGATCCA* is palindromic in this sense. The biological palindromic motif is herein modeled as *θ-palindrome* (or, more generally, *pseudopalindrome*) defined as follows: A word w is a θ-palindrome if $w = \theta(w)$. θ-palindromes were investigated intensively from a theoretical computer science perspective, see for example, in de Luca and Luca (2006) and Kari and Mahalingam (2010).

The properties of θ-sticky-freeness and θ-overhang-freeness motivate the generalization of the notions of *border, commutativity,* and *conjugacy* of words into θ-*border*, θ-*commutativity*, and θ-*conjugacy* of words. A word $w \in \Sigma^+$ is *bordered* if there exists a word $v \in \Sigma^+$ satisfying $w = vx = yv$ for some $x, y \in \Sigma^+$. Kari and Mahalingam (2007) proposed an extended notion called the θ-borderedness of words as w is θ-*bordered* if $w = vx = y\theta(v)$ for some $v, x, y \in \Sigma^+$. Note that θ-sticky-free languages do not contain any θ-bordered word. The *pseudoknot-bordered-word* proposed in Kari and Seki (2009) is a further extension of the notion of θ-bordered word. A word $w \in \Sigma^+$ is *pseudoknot-bordered* if $w = uvx = y\theta(u)\theta(v)$. A pseudoknot-bordered word models the crossing dependency occurring in DNA and RNA pseudoknots. A word $v \in \Sigma^+$ is said to be a *conjugate* of another word $u \in \Sigma^+$ if $v = yx$ and $u = xy$ for some $x, y \in \Sigma^*$, that is, $ux = xv$ holds. This notion was extended in Kari and Mahalingam (2008) as follows: a word $u \in \Sigma^+$ is a *θ-conjugate* of a word $v \in \Sigma^+$ if $ux = \theta(x)v$ for some $x \in \Sigma^+$. In this case, either $v = \theta(u)$ or $u = \theta(x)z$, $v = zx$ for some $z \in \Sigma^+$; hence a language that contains both u and v cannot be strictly θ-sticky-free.

In contrast with the above two notions, the *θ-commutativity of words* introduced in Kari and Mahalingam (2008) has a purely theoretical significance, being a mathematical tool for obtaining results involving WK complementarity. Two words $u, v \in \Sigma^+$ are said to *commute* if $uv = vu$ holds. For two words u, v, we say that u θ-commutes with v if $uv = \theta(v)u$. This equation is a special case of conjugacy equations. Thus, by applying a known result in combinatorics on words, one can deduce the following results.

Theorem 25 (Kari and Mahalingam 2008) *For an antimorphic involution θ, and two words $u, v \in \Sigma^+$, if $uv = \theta(v)u$ holds, then $u = r(tr)^i$, $v = (tr)^j$ for some $i \geq 0$, $j \geq 0$, and θ-palindromes $r, t \in \Sigma^*$ such that rt is primitive.*

This theorem relates the three important notions: θ-commutativity of words, θ-palindrome, and primitivity.

Another interesting perspective is the informational equivalence between the two strands of a DNA double helix. Indeed, the two constituent single strands of a DNA double strand are "equivalent" with respect to the information they encode, and thus we can say that w and $\theta(w)$ are equivalent in this sense. Practical applications of this idea include an extended Lempel–Ziv algorithm for DNA molecules proposed in Grumbach and Tahi (1993).

A word $w \in \Sigma^+$ is called *primitive* if it cannot be written as a power of another word; that is, $w = u^n$ implies $n = 1$ and $w = u$. For a word $w \in \Sigma^+$, the shortest $u \in \Sigma^+$ such that $w = u^n$ for some $n \geq 1$ is called the *primitive root* of the word w and is denoted by $\rho(w)$. It is well known that for each word $w \in \Sigma^*$, there exists a unique primitive word $t \in \Sigma^+$ such that $\rho(w) = t$, that is, $w = t^n$ for some $n \geq 1$, see for example, Choffrut and Karhumäki (1997). In Czeizler et al. (2008b), the primitivity was extended to θ-primitivity for an antimorphic involution θ. A word $u \in \Sigma^+$ is said to be *θ-primitive* if there does not exist any word $t \in \Sigma^+$ such that $u \in \{t, \theta(t)\}^{\geq 2}$. For a word $w \in \Sigma^+$, we define the *θ-primitive root* of w, denoted by $\rho_\theta(w)$, as the shortest word $t \in \Sigma^+$ such that $w \in t\{t, \theta(t)\}^*$. Note that if w is θ-primitive, then $\rho_\theta(w) = w$. Note also that θ-primitivity can be defined also for θ being a morphic involution. However, as it is meant as a model of WK complementarity, we will continue to assume that θ is antimorphic. For counterparts of the following results when θ is a morphic involution see Czeizler et al. (2008b).

We start by looking at some basic properties of θ-primitive words.

Proposition 2 (Czeizler et al. 2008b) *If a word $w \in \Sigma^+$ is θ-primitive, then it is also primitive. Moreover, the converse is not always true.*

Proof By definition, it is clear that θ-primitive words are primitive. Remark that if there exists $a \in \Sigma$ satisfying $a \neq \theta(a)$, then the word $a\theta(a)$ is not θ-primitive, but is primitive.

If a word $w \in \Sigma^+$ is in $\{t, \theta(t)\}^+$ for some word $t \in \Sigma^+$ and t in turn is in $\{s, \theta(s)\}^+$ for some word $s \in \Sigma^+$, then we have that $w \in \{s, \theta(s)\}^+$. Thus, we have the following result.

Proposition 3   (Czeizler et al. 2008b) *The 0-primitive root of a word is 0-primitive, and hence primitive.*

It is a well-known fact that any conjugate of a primitive word is primitive. This fact is heavily employed in obtaining fundamental results including the Fine and Wilf theorem and solutions to the Lyndon–Schützenberger equation. In contrast, a conjugate of a θ-primitive word need not be θ-primitive. For instance, for the DNA involution τ defined in ❷ Sect. 5, the word $w = GCTA$ is τ-primitive, while its conjugate $w' = AGCT = AG\tau(AG)$ is not.

Proposition 4   (Czeizler et al. 2008b) *The class of θ-primitive words is not closed under circular permutations.*

An essential property of primitive words is that a primitive word cannot be equal to its conjugate. In other words, for a primitive word u, the equation $uu = xuy$ implies that either x or y is empty. Thus, u^i and u^j, with $i,j \geq 1$, cannot overlap non-trivially on a sequence longer than $|u|$. This is not the case when considering overlap between $\alpha(v,\theta(v)), \beta(v,\theta(v)) \in \{v,\theta(v)\}^+$ for some θ-primitive word $v \in \Sigma^+$. For instance, for the DNA involution τ and a τ-primitive word $v = CCGGAT$, $v^2 = CCGG \cdot \tau(v) \cdot AT$ holds. Nevertheless, an analogous result for θ-primitive words was obtained.

Theorem 26   (Kari et al. 2010) *Let $v \in \Sigma^+$ be a θ-primitive word. Neither $v\theta(v)$ nor $\theta(v)v$ can be a proper infix of a word in $\{v,\theta(v)\}^3$.*

Furthermore, Czeizler et al. completely characterized all such nontrivial overlaps with the set of all solutions of the corresponding equation (Czeizler et al. 2009).

Theorem 27   (Czeizler et al. 2009) *Let $v \in \Sigma^+$ be a θ-primitive word. The only possible proper overlaps of the form $\alpha(v,\theta(v)) \cdot x = y \cdot \beta(v,\theta(v))$ with $\alpha(v,\theta(v)),\beta(v,\theta(v)) \in \{v,\theta(v)\}^+$, $x,y \in \Sigma^+$ and $|x|,|y| < |v|$ are given in ❷ Table 1 (modulo a substitution of v by $\theta(v)$) together with the characterization of their sets of solutions.*

❑ Table 1

Characterization of possible proper overlaps of the form $\alpha(v,\theta(v)) \cdot x = y \cdot \beta(v,\theta(v))$. For the last three equations, $n \geq 0$, $m \geq 1$, $r,t \in \Sigma^+$ such that $r = \theta(r)$, $t = \theta(t)$, and rt is primitive. Note that the fourth and fifth equations are the same up to the antimorphic involution θ

Equation	Solution
$v^i x = y\theta(v)^i, i \geq 1$	$v = yp, x = \theta(y), p = \theta(p)$, and whenever $i \geq 2$, $y = \theta(y)$
$vx = yv$	$v = (pq)^{j+1}p, x = qp, y = pq$ for some $p,q \in \Sigma^+, j \geq 0$
$v\theta(v)x = yv\theta(v),$	$v = (pq)^{j+1}p, x = \theta(pq), y = pq$, with $j \geq 0$, $qp = \theta(qp)$
$v^{i+1}x = y\theta(v)^i v, i \geq 1$	$v = r(tr)^{n+m}r(tr)^n, x = (tr)^m r(tr)^n, y = r(tr)^{n+m}$
$v\theta(v)^i x = yv^{i+1}, i \geq 1$	$v = (rt)^n r(rt)^{m+n}r, y = (rt)^n r(rt)^m, x = (rt)^{m+n}r$
$v\theta(v)^i x = yv\theta(v), i \geq 2$	$v = (rt)^n r(rt)^{m+n}r, y = (rt)^n r(rt)^m, x = (tr)^m r(tr)^n$

We now shift one's attention to extensions of two essential results in combinatorics on words. The following theorem is known as the *Fine and Wilf theorem* (Fine and Wilf 1965), in its form for words (Choffrut and Karhumäki 1997; Lothaire 1983). It illustrates a fundamental periodicity property of words. Its concise proof is available, in, for example, Choffrut and Karhumäki (1997). As usual, $\gcd(n, m)$ denotes the *greatest common divisor of n and m*.

Theorem 28 *Let $u, v \in \Sigma^*$, $n = |u|$, and $m = |v|$. If a power of u and a power of v have a common prefix of length at least $n + m - \gcd(n, m)$, then $\rho(u) = \rho(v)$. Moreover, the bound $n + m - \gcd(n, m)$ is optimal.*

A natural question is whether one can obtain an extension of this result when, instead of taking powers of two words u and v, one looks at a word in $u\{u, \theta(u)\}^*$ and a word in $v\{v, \theta(v)\}^*$. The answer is yes. Note that without loss of generality, one can suppose that the two words start with u and v because θ is an involution. Czeizler, Kari, and Seki provided extensions of ❷ Theorem 28 in two forms (❷ Theorems 29 and ❷ 30) (Czeizler et al. 2008b). As illustrated in the following example, the bound given by ❷ Theorem 28 is not sufficient anymore.

Example 5 (Czeizler et al. 2008b) Let $\theta : \{a, b\}^* \rightarrow \{a, b\}^*$ be the mirror image mapping defined as follows: $\theta(a) = a$, $\theta(b) = b$, and $\theta(w_1 \ldots w_n) = w_n \ldots w_1$, where $w_i \in \{a, b\}$ for all $1 \leq i \leq n$. Obviously, θ is an antimorphic involution on $\{a, b\}^*$. Let $u = (ab)^k b$ and $v = ab$. Then, u^2 and $v^k \theta(v)^{k+1}$ have a common prefix of length $2|u| - 1 > |u| + |v| - \gcd(|u|, |v|)$. Nevertheless, u and v do not have the same θ-primitive root, that is $\rho_\theta(u) \neq \rho_\theta(v)$.

In the following, $\mathrm{lcm}(n, m)$ denotes the *least common multiple of n and m*.

Theorem 29 (Czeizler et al. 2008b) *Let $u, v \in \Sigma^+$, and $\alpha(u, \theta(u)) \in u\{u, \theta(u)\}^*$, $\beta(v, \theta(v)) \in v\{v, \theta(v)\}^*$ be two words sharing a common prefix of length at least $\mathrm{lcm}(|u|, |v|)$. Then, there exists a word $t \in \Sigma^+$ such that $u, v \in t\{t, \theta(t)\}^*$, i.e., $\rho_\theta(u) = \rho_\theta(v)$. In particular, if $\alpha(u, \theta(u)) = \beta(v, \theta(v))$, then $\rho_\theta(u) = \rho_\theta(v)$.*

This theorem provides us with an alternative definition of the θ-primitive root of a word.

Corollary 2 (Czeizler et al. 2008b) *For any word $w \in \Sigma^+$ there exists a unique θ-primitive word $t \in \Sigma^+$ such that $w \in t\{t, \theta(t)\}^*$, i.e., $\rho_\theta(w) = t$.*

Corollary 3 (Czeizler et al. 2008b) *Let $u, v \in \Sigma^+$ be two words such that $\rho(u) = \rho(v) = t$. Then, $\rho_\theta(u) = \rho_\theta(v) = \rho_\theta(t)$.*

Next, another bound is provided for this extended Fine and Wilf theorem, which is in many cases much shorter than the bound given in ❷ Theorem 29. As noted before, due to ❷ Proposition 4, it is intuitive that one cannot use the concise proof technique based on the fact that a conjugate of a primitive word is primitive. The proof given in Czeizler et al. (2008b) involves rather technical case analyses.

Theorem 30 (Czeizler et al. 2008b) *Given two words $u,v \in \Sigma^+$ with $|u| > |v|$, if there exist two words $\alpha(u,\theta(u)) \in u\{u,\theta(u)\}^*$ and $\beta(v,\theta(v)) \in v\{v,\theta(v)\}^*$ having a common prefix of length at least $2|u| + |v| - gcd(|u|,|v|)$, then $\rho_\theta(u) = \rho_\theta(v)$.*

This section is concluded with an extension of another fundamental result related to the periodicity on words, namely the solution to the *Lyndon–Schützenberger equation*. The equation is of the form $u^\ell = v^n w^m$ for some $\ell, n, m \geq 2$ and $u,v,w \in \Sigma^+$. Lyndon and Schützenberger proved that this equation implies $\rho(u) = \rho(v) = \rho(w)$ (Lyndon and Schützenberger 1962). A concise proof when u,v,w belong to a free semigroup can be found in, for example, Harju and Nowotka (2004).

Incorporating the idea of θ-periodicity, the Lyndon–Schützenberger equation has been extended in Czeizler et al. (2009) in the following manner. Let $u,v,w \in \Sigma^+$, θ be an antimorphic involution over Σ, and ℓ, n, m be integers ≥ 2. Let $\alpha(u,\theta(u)) \in \{u,\theta(u)\}^\ell$, $\beta(v,\theta(v)) \in \{v,\theta(v)\}^n$, and $\gamma(w,\theta(w)) \in \{w,\theta(w)\}^m$. The *extended Lyndon–Schützenberger equation* is

$$\alpha(u, \theta(u)) = \beta(v, \theta(v))\gamma(w, \theta(w))$$

In Czeizler et al. (2009), the authors investigated the problem of finding conditions on ℓ, n, m such that if the equation holds, then $u,v,w \in \{t,\theta(t)\}^+$ for some $t \in \Sigma^+$. If such t exists, we say that (ℓ, n, m) *imposes θ-periodicity on u,v,w*. The original condition $\ell, n, m \geq 2$ is not enough for this extension as shown below.

Example 6 (Czeizler et al. 2009) Let $\Sigma = \{a,b\}$ and θ be the mirror image. Take now $u = a^k b^2 a^{2k}$, $v = \theta(u)^l a^{2k} b^2 = (a^{2k} b^2 a^k)^l a^{2k} b^2$, and $w = a^2$, for some $k, l \geq 1$. Then, although $\theta(u)^{l+1} u^{l+1} = v^2 w^k$, there is no word $t \in \Sigma^+$ with $u,v,w \in \{t,\theta(t)\}^+$.

Example 7 (Czeizler et al. 2009) Consider again $\Sigma = \{a,b\}$ and the mirror image θ, and take $u = b^2(aba)^k$, $v = u^l b = (b^2(aba)^k)^l b$, and $w = aba$ for some $k, l \geq 1$. Then, although $u^{2l+1} = v\theta(v)w^k$, there is no word $t \in \Sigma^+$ with $u,v,w \in \{t,\theta(t)\}^+$.

Example 8 Let $\Sigma = \{a,b\}$ and θ be the mirror image. Let $v = a^{2m} b^{2j}$ and $w = aa$ for some $m \geq 1$ and $j \geq 1$. Then $v^n w^m = (a^{2m} b^{2j})^n a^{2m}$. This is θ-palindrome of even length and hence it can be written as $u\theta(u)$ for some $u \in \Sigma^+$. Clearly $\rho_\theta(w) = a$ and $\rho_\theta(v) \neq a$ because v contains b. Hence $(2, n, m)$ is not enough to impose the θ-periodicity.

Thus, once either n or m is 2, it is not always the case that there exists a word $t \in \Sigma^+$ such that $u,v,w \in \{t,\theta(t)\}^+$. On the other hand, it was proved that (ℓ, n, m) imposes θ-periodicity on u,v,w if $\ell \geq 5$, $n,m \geq 3$.

Theorem 31 (Czeizler et al. 2009) *For words $u,v,w \in \Sigma^+$ and $\ell, n, m \geq 2$, let $\alpha(u,\theta(u)) \in \{u,\theta(u)\}^\ell$, $\beta(v,\theta(v)) \in \{v,\theta(v)\}^n$, and $\gamma(w,\theta(w)) \in \{w,\theta(w)\}^m$. If $\alpha(u,\theta(u)) = \beta(v,\theta(v))\gamma(w,\theta(w))$ holds and $\ell \geq 5$, $n,m \geq 3$, then $u,v,w \in \{t,\theta(t)\}^+$ for some $t \in \Sigma^+$.*

This theorem requires rather complex case analyses, too, and hence the proof is omitted here. However, if ℓ, n, m are "big" enough, one can easily see that there is much room to employ the extended Fine and Wilf theorem.

◻ Table 2

Summary of the extended Lyndon–Schützenberger equation results

l	*n*	*m*	*θ*-Periodicity	Proved by
≥ 5	≥ 3	≥ 3	Yes	❷ Theorem 31
4	≥ 3	≥ 3	?	
3	≥ 3	≥ 3	?	
≥ 3	2	≥ 2	No	❷ Examples 6 and ❷ 7
≥ 3	≥ 2	2	No	
2	≥ 2	≥ 2	No	❷ Example 8

This section is concluded with ❷ *Table 2*, which summarizes the results obtained so far on the extended Lyndon–Schützenberger equation.

8 Conclusions

In this chapter we described how information can be encoded on DNA strands and how bio-operations can be used to perform computational tasks. In addition, we presented examples of the influence that fundamental properties of DNA-encoded information, especially the Watson–Crick complementarity, have had on various areas of theoretical computer science such as formal language theory, coding theory, automata theory, and combinatorics on words.

A few final remarks are in order regarding DNA-encoded data and the bio-operations that are used to act on it. The descriptions of DNA structure and DNA bio-operations point to the fact that DNA-encoded information is very different from electronically encoded information, and bio-operations also differ from electronic computer operations. A fundamental difference arises, for example, from the fact that in electronic computing data interaction is fully controlled, while in a test-tube DNA-computer, free-floating data-encoding DNA single strands can interact because of Watson–Crick complementarity. Another difference is that, in DNA computing, a bio-operation usually consumes both operands. This implies that, if one of the operands is either involved in an illegal binding or has been consumed by a previous bio-operation, it is unavailable for the desired computation. Yet another difference is that, while in electronic computing a bit is a single individual element, in DNA experiments, each submicroscopic DNA molecule is usually present in millions of identical copies. The bio-operations operate in a massively parallel fashion on all identical strands and this process is governed by the laws of chemistry and thermodynamics, with the output obeying statistical laws.

Differences like the ones mentioned above point to the fact that a fresh approach is needed when employing as well as when theoretically investigating bioinformation and biocomputation, and they offer a wealth of problems to explore at this rich intersection between molecular biology and computer science.

Examples of such fascinating research areas and topics include models and wet implementations of molecular computing machineries, membrane computing, DNA computing by splicing and insertion–deletion operations, bacterial computing and communication, DNA memory, DNA computing by self-assembly, and computational aspects of gene assembly in ciliates, as described in, for example, the chapter ❷ Computational Nature of Gene Assembly in Ciliates of this handbook.

References

Adleman L (1994) Molecular computation of solutions to combinatorial problems. Science 266(5187): 1021–1024

Adleman L (1998) Computing with DNA. Sci Am 279:54–61

Alhazov A, Cavaliere M (2005) Computing by observing bio-systems: the case of sticker systems. In: Feretti C, Mauri G, Zandron C (eds) Proceedings of DNA computing 10, Milan, Italy, June 2004. Lecture notes in computer science, vol 3384. Springer-Verlag, Berlin, pp 1–13

Amos M (2005) Theoretical and experimental DNA computation. Springer-Verlag, Berlin

Andronescu M, Dees D, Slaybaugh L, Zhao Y, Condon A, Cohen B, Skiena S (2003) Algorithms for testing that sets of DNA words concatenate without secondary structure. In: Hagiya M, Ohuchi A (eds) Proceedings of DNA computing 8, Sapporo, Japan, June 2002. Lecture notes in computer science, vol 2568. Springer-Verlag, Berlin, pp 182–195

Arita M (2004) Writing information into DNA. In: Jonoska N, Păun G, Rozenberg G (eds) Aspects of molecular computing. Lecture notes in computer science, vol 2950. Springer-Verlag, Berlin, pp 23–35

Arita M, Kobayashi S (2002) DNA sequence design using templates. New Generation Comput 20:263–277

Bancroft C, Bowler T, Bloom B, Clelland C (2001) Long-term storage of information in DNA. Science 293: 1763–1765

Baum E (1998) DNA sequences useful for computation. In: Landweber L, Baum E (eds) DNA based computers II. DIMACS series in discrete mathematics and theoretical computer science, vol 44. American Mathematical Society, Providence, RI, pp 235–246

Bordihn H, Holzer M, Kutrib M (2007) Hairpin finite automata. In: Harju T, Karhumäki J, Lepistö A (eds) Developments in language theory, Turku, Finland, July 2007. Lecture notes in computer science, vol 4588. Springer-Verlag, Berlin, pp 108–119

Braich R, Chelyapov N, Johnson C, Rothemund P, Adleman L (2002) Solution of a 20-variable 3-SAT problem on a DNA computer. Science 296:499–502

Calladine C, Drew H (1997) Understanding DNA: the molecule and how it works, 2nd edn. Academic Press, London

Cavaliere M (2008) Computing by observing: a brief survey. In: Logic and theory of algorithms. Lecture notes in computer science, vol 5028. Springer, Berlin, pp 110–119

Cavaliere M, Leupold P (2004) Evolution and observation – a new way to look at membrane systems. In: Martin-Vide C, Mauri G, Păun G,

Rozenberg G, Salomaa A (eds) Membrane computing. Lecture notes in computer science, vol 2933. Springer, Berlin, pp 70–87

Chen J, Deaton R, Garzon M, Kim J, Wood D, Bi H, Carpenter D, Wang YZ (2006) Characterization of non-crosshybridizing DNA oligonucleotides manufactured in vitro. Nat Comput 5(2):165–181

Choffrut C, Karhumäki J (1997) Combinatorics of words. In: Rozenberg G, Salomaa A (eds) Handbook of formal languages, vol 1. Springer-Verlag, Berlin-Heidelberg-New York, pp 329–438

Cox J (2001) Long-term data storage in DNA. Trends Biotechnol 19:247–250

Czeizler E, Czeizler E (2006a) On the power of parallel communicating Watson-Crick automata systems. Theor Comput Sci 358:142–147

Czeizler E, Czeizler E (2006b) Parallel communicating Watson-Crick automata systems. Acta Cybern 17:685–700

Czeizler E, Czeizler E (2006c) A short survey on Watson-Crick automata. Bull EATCS 89:104–119

Czeizler E, Czeizler E, Kari L, Salomaa K (2008a) Watson-Crick automata: determinism and state complexity. In: DCFS'08: Proceedings of descriptional complexity of formal systems, University of Prince Edward Island, Charlottetown, PE, Canada, July 2008, pp 121–133

Czeizler E, Kari L, Seki S (2008b) On a special class of primitive words. In: MFCS 2008: Proceedings of mathematical foundations of theoretical computer science, Torun, Poland, August 2008. Lecture notes in computer science, vol 5162. Springer, Berlin-Heidelberg, pp 265–277

Czeizler E, Czeizler E, Kari L, Seki S (2009) An extension of the Lyndon Schützenberger result to pseudoperiodic words. In: Diekert V, Nowotka D (eds) DLT'09: Proceedings of developments in language theory, Stuttgart, Germany, June 2009. Lecture notes in computer science, vol 5583. Springer-Verlag, Berlin

Daley M, Kari L (2002) DNA computing: Models and implementations. Comments Theor Biol 7:177–198

Daley M, Ibarra O, Kari L (2003) Closure and decidability properties of some language classes with respect to ciliate bio-operations. Theor Comput Sci 306 (1):19–38

Daley M, Kari L, McQuillan I (2004) Families of languages defined by ciliate bio-operations. Theor Comput Sci 320:51–69

de Luca A, Luca AD (2006) Pseudopalindrome closure operators in free monoids. Theor Comput Sci 362:282–300

Deaton R, Chen J, Bi H, Rose J (2003) A software tool for generating non-crosshybridizing libraries of DNA oligonucleotides. In: Hagiya M, Ohuchi A (eds) Proceedings of DNA computing 8, Sapporo, Japan, June 2002. Lecture notes in computer science, vol 2568. Springer-Verlag, Berlin, pp 252–261

Deaton R, Chen J, Kim J, Garzon M, Wood D (2006) Test tube selection of large independent sets of DNA oligonucleotides. In: Chen J, Jonoska N, Rozenberg G (eds) Nanotechnology: science and computation, Springer-Verlag, Berlin, pp 147–161

Diaz S, Esteban J, Ogihara M (2001) A DNA-based random walk method for solving k-SAT. In: Condon A, Rozenberg G (eds) Proceedings of DNA computing 6, Leiden, the Netherlands, June 2000. Lecture notes in computer science, vol 2054. Springer-Verlag, Berlin, pp 209–219

Dirks R, Pierce N (2004) An algorithm for computing nucleic acid base-pairing probabilities including pseudoknots. J Comput Chem 25:1295–1304

Domaratzki M (online 2007) Hairpin structures defined by DNA trajectories. Theory Comput Syst DOI 10.1007/s00224-007-9086-6

Drlica K (1996) Understanding DNA and gene cloning: a guide for the curious. Wiley, New York

Dyachkov A, Macula A, Pogozelski W, Renz T, Rykov V, Torney D (2006) New t-gap insertion-deletion-like metrics for DNA hybridization thermodynamic modeling. J Comput Biol 13(4):866–881

Dyachkov A, Macula A, Rykov V, Ufimtsev V (2008) DNA codes based on stem similarities between DNA sequences. In: Garzon M, Yan H (eds) Proceedings of DNA computing 13, Memphis, TN, June 2007. Lecture notes in computer science, vol 4848. Springer-Verlag, Berlin, pp 146–151

Ehrenfeucht A, Harju T, Petre I, Prescott D, Rozenberg G (2004) Computation in living cells: gene assembly in ciliates. Natural Computing Series. Springer-Verlag, Berlin

Faulhammer D, Cukras A, Lipton R, Landweber L (2000) Molecular computation: RNA solutions to chess problems. Proc Nat Acad Sci U S A 97:1385–1389

Feldkamp U, Saghafi S, Banzhaf W, Rauhe H (2002) DNASequenceGenerator – a program for the construction of DNA sequences. In: Jonoska N, Seeman N (eds) Proceedings of DNA computing 7, Tampa, FL, June 2001. Lecture notes in computer science, vol 2340. Springer-Verlag, Berlin, pp 23–32

Fine N, Wilf H (1965) Uniqueness theorem for periodic functions. Proc Am Math Soc 16(1):109–114

Freund R, Păun G, Rozenberg G, Salomaa A (1998) Bidirectional sticker systems. In: Altman R, Dunker A, Hunter L, Klein T (eds) Pacific symposium on biocomputing, vol 98. World Scientific, Singapore, pp 535–546

Freund R, Păun G, Rozenberg G, Salomaa A (1999) Watson-Crick finite automata. DIMACS Ser Discrete Math Theor Comput Sci 48:297–327

Frutos A, Liu Q, Thiel A, Sanner A, Condon A, Smith L, Corn R (1997) Demonstration of a word design strategy for DNA computing on surfaces. Nucleic Acids Res 25(23):4748–4757

Garzon M, Neathery P, Deaton R, Murphy R, Franceschetti D, Stevens Jr S (1997) A new metric for DNA computing. In: Koza J, Deb K, Dorigo M, Vogel D, Garzon M, Iba H, Riolo R (eds) Proceedings of genetic programming 1997, Stanford University, July 1997. Morgan Kaufmann, San Francisco, CA, pp 479–490

Garzon M, Phan V, Roy S, Neel A (2006) In search of optimal codes for DNA computing. In: Mao C, Yokomori T (eds) Proceedings of DNA computing 12, Seoul, Korea, June 2006. Lecture notes in computer science, vol 4287. Springer-Verlag, Berlin, pp 143–156

Gonick L, Wheelis M (1991) The cartoon guide to genetics, updated edn. Collins, New York

Grumbach S, Tahi F (1993) Compression of DNA sequences. In: Proceedings of IEEE symposium on data compression. IEEE Computer Society Press, San Francisco, CA, June 1993. pp 340–350

Harju T, Nowotka D (2004) The equation $x^i = y^j z^k$ in a free semigroup. Semigroup Forum 68:488–490

Hartmanis J (1995) On the weight of computations. Bull EATCS 55:136–138

Head T (2000) Relativised code concepts and multi-tube DNA dictionaries. In: Calude C, Păun G (eds) Finite versus infinite: contributions to an eternal dilemma. Springer-Verlag, London, pp 175–186

Hoogeboom H, van Vugt N (2000) Fair sticker languages. Acta Inform 37:213–225

Hopcroft J, Ullman J (1979) Introduction to automata theory, languages, and computation. Addison-Wesley, Reading, MA

Hussini S, Kari L, Konstantinidis S (2003) Coding properties of DNA languages. Theor Comput Sci 290 (3):1557–1579

Ignatova Z, Martínez-Pérez I, Zimmermann KH (2008) DNA computing models. Springer-Verlag, Berlin

Jonoska N, Mahalingam K (2004) Languages of DNA based code words. In: Chen J, Reif J (eds) Proceedings of DNA computing 9, Madison, WI, June 2003. Lecture notes in computer science, vol 2943. Springer, Berlin, pp 61–73

Jonoska N, Kephart D, Mahalingam K (2002) Generating DNA code words. Congressus Numerantium 156:99–110

Kari L (1997) DNA computing: the arrival of biological mathematics. Math Intell 19(2):9–22

Kari L, Kitto R, Thierrin G (2002) Codes, involutions and DNA encoding. In: Brauer W, Ehrig H, Karhumäki J, Salomaa A (eds) Formal and natural computing. Lecture notes in computer science, vol 2300. Springer-Verlag, Berlin, pp 376–393

Kari L, Konstantinidis S, Losseva E, Wozniak G (2003) Sticky-free and overhang-free DNA languages. Acta Inform 40:119–157

Kari L, Konstantinidis S, Sosík P (2005a) Bond-free languages: formalizations, maximality and construction methods. Int J Foundations Comput Sci 16 (5):1039–1070

Kari L, Konstantinidis S, Sosík P (2005b) On properties of bond-free DNA languages. Theor Comput Sci 334 (1–3):131–159

Kari L, Konstantinidis S, Losseva E, Sosík P, Thierrin G (2006) A formal language analysis of DNA hairpin structures. Fundam Inform 71(4):453–475

Kari L, Mahalingam K (2007) Involutively bordered words. Int J Foundations Comput Sci 18(5): 1089–1106

Kari L, Mahalingam K (2008) Watson-Crick conjugate and commutative words. In: Garzon M, Yan H (eds) Proceedings of DNA computing 13, Memphis, TN, June 2007. Lecture notes in computer science, vol 4848. Springer-Verlag, Berlin, pp 273–283

Kari L, Mahalingam K (2010) Watson-Crick palindromes in DNA computing, Natural Comput 9(2):297–316

Kari L, Mahalingam K, Thierrin G (2007) The syntactic monoid of hairpin-free languages. Acta Inform 44 (3–4):153–166

Kari L, Masson B, Seki S (2010) Properties of pseudo-primitive words and their applications. CoRR abs/ 1002.4084

Kari L, Păun G, Rozenberg G, Salomaa A, Yu S (1998) DNA computing, sticker systems, and universality. Acta Inform 35:401–420

Kari L, Seki S (2009) On pseudoknot-bordered words and their properties. J Comput Syst Sci 75(2): 113–121

Kijima A, Kobayashi S (2006) Efficient algorithm for testing structure freeness of finite set of biomolecular sequences. In: Carbone A, Pierce N (eds) Proceedings of DNA 11, London, ON, Canada, June 2005. Lecture notes in computer science, vol 3892. Springer-Verlag, Berlin, pp 171–180

Kobayashi S (2005) Testing structure-freeness of regular sets of biomolecular sequences. In: Feretti C, Mauri G, Zandron C (eds) Proceedings of DNA computing 10, Milan, Italy, June 2004. Lecture notes in computer science, vol 3384. Springer-Verlag, Berlin, pp 192–201

Kuske D, Weigel P (2004) The role of the complementarity relation in Watson-Crick automata and sticker systems. In: Calude C, Claude E, Dinneen M (eds) DLT 2004: Proceedings of development in language theory, Auckland, New Zealand, December 2004. Lecture notes in computer science, vol 3340. Springer-Verlag, Berlin Heidelberg, pp 272–283

Lewin B (2007) Genes IX. Johns and Bartlett Publishers, Sudbury, MA

Lipton R (1995) Using DNA to solve NP-complete problems. Science 268:542–545

Liu W, Gao L, Zhang Q, Xu G, Zhu X, Liu X, Xu J (2005) A random walk DNA algorithm for the 3-SAT problem. Curr Nanosci 1:85–90

Lothaire M (1983) Combinatorics on words. Encyclopedia of mathematics and its applications, vol 17. Addison-Wesley, Reading, MA

Lyndon R, Schützenberger M (1962) The equation $a^m = b^n c^p$ in a free group. Mich Math J 9:289–298

Manea F, Mitrana V, Yokomori T (2009) Two complementary operations inspired by the DNA hairpin formation: completion and reduction. Theor Comput Sci 410(4–5):417–425

Marathe A, Condon A, Corn R (2001) On combinatorial DNA word design. J Comput Biol 8(3):201–220

Martin-Víde C, Păun G, Rozenberg G, Salomaa A (1998) Universality results for finite H systems and for Watson-Crick automata. In: Păun G (ed) Computing with bio-molecules. Theory and experiments. Springer, Berlin, pp 200–220

Mateescu A, Rozenberg G, Salomaa A (1998) Shuffle on trajectories: syntactic constraints. Theor Comput Sci 197:1–56

Mauri G, Ferretti C (2004) Word design for molecular computing: a survey. In: Chen J, Reif J (eds) Proceedings of DNA computing 9, Madison, WI, June 2003. Lecture notes in computer science, vol 2943. Springer, Berlin, pp 37–46

Morimoto N, Arita M, Suyama A (1997) Stepwise generation of Hamiltonian path with molecules. In: Lundh D, Olsson B, Narayanan A (eds) Proceedings of bio-computing and emergent computation, Skovde, Sweden, September 1997. World Scientific, Singapore. pp 184–192

Nagy B (2008) On 5'→3' sensing Watson-Crick finite automata. In: Garzon M, Yan H (eds) Proceedings of DNA computing 13, Memphis, TN, June 2007. Lecture notes in computer science, vol 4848. Springer-Verlag, Berlin, pp 256–262

Petre E (2003) Watson-Crick ω-automata. J Automata, Lang Combinatorics 8:59–70

Phan V, Garzon M (online 2008) On codeword design in metric DNA spaces. Nat Comput DOI 10.1007/ s11047-008-9088-6

Păun A, Păun M (1999) State and transition complexity of Watson-Crick finite automata. In: Ciobanu G, Păun G (eds) FCT'99, Iasi, Romania, August–September 1999. Lecture notes in computer science, vol 1684. Springer-Verlag, Berlin Heidelberg, pp 409–420

Păun G, Rozenberg G (1998) Sticker systems. Theor Comput Sci 204:183–203

Păun G, Rozenberg G, Salomaa A (1998) DNA computing: new computing paradigms. Springer, Berlin

Păun G, Rozenberg G, Yokomori T (2001) Hairpin languages. Int J Foundations Comput Sci 12 (6):837–847

Reif J, LaBean T, Pirrung M, Rana V, Guo B, Kingsford C, Wickham G (2002) Experimental construction of very large scale DNA databases with associative search capability. In: Jonoska N, Seeman N (eds) Proceedings of DNA computing 7, Tampa, FL, June 2001. Lecture notes in computer science, vol 2340. Springer-Verlag, Berlin, pp 231–247

Rose J, Deaton R, Hagiya M, Suyama A (2002) PNA-mediated whiplash PCR. In: Jonoska N, Seeman N (eds) Proceedings of DNA computing 7, Tampa, FL, June 2001. Lecture notes in computer science, vol 2340. Springer-Verlag, Berlin, pp 104–116

Sager J, Stefanovic D (2006) Designing nucleotide sequences for computation: a survey of constraints. In: Carbone A, Pierce N (eds) Proceedings of DNA computing 11, London, ON, Canada. Lecture notes in computer science, vol 3892. Springer-Verlag, Berlin, pp 275–289

Sakakibara Y, Kobayashi S (2001) Sticker systems with complex structures. Soft Comput 5:114–120

Salomaa A (1973) Formal languages. Academic, New York

SantaLucia J (1998) A unified view of polymer, dumbbell and oligonucleotide DNA nearest-neighbor thermodynamics. Proc Nat Acad Sci U S A 95(4):1460–1465

Sempere J (2007) On local testability in Watson-Crick finite automata. In: Proceedings of the international workshop on automata for cellular and molecular computing, tech report of MTA SZTAKI, Budapest, Hungary, August 2007. pp 120–128

Sempere J (2008) Exploring regular reversibility in Watson-Crick finite automata. In: AROB 2008: Proceedings of 13th international symposium on artificial life and robotics, Beppu, Japan, January–February 2008. pp 505–509

Smith G, Fiddes C, Hawkins J, Cox J (2003) Some possible codes for encrypting data in DNA. Biotechnol Lett 25:1125–1130

Stemmer W, Crameri A, Ha K, Brennan T, Heyneker H (1995) Single-step assembly of a gene and entire plasmid from large numbers of oligodeoxyribonucleotides. GENE 164:49–53

Tulpan D, Hoos H, Condon A (2003) Stochastic local search algorithms for DNA word design. In: Hagiya M, Ohuchi A (eds) Proceedings of DNA computing 8, Sapporo, Japan, June 2002. Lecture notes in computer science, vol 2568. Springer-Verlag, Berlin, pp 229–241

Turner P, McLennan A, Bates A, White M (2000) Instant notes in molecular biology, 2nd edn. Garland, New York

Wang X, Bao Z, Hu J, Wang S, Zhan A (2008) Solving the SAT problem using a DNA computing algorithm based on ligase chain reaction. Biosystems 91: 117–125

Wood D (1987) Theory of computation. Harper & Row, New York

Yoshida H, Suyama A (2000) Solution to 3-SAT by breadth-first search. In: Winfree E, Gifford D (eds) DNA based computers V. DIMACS series in discrete mathematics and theoretical computer science, vol 54. American Mathematical Society, Providence, RI, pp 9–22

Yu S (2002) State complexity of finite and infinite regular languages. Bull EATCS 76:142–152

34 Molecular Computing Machineries — Computing Models and Wet Implementations

Masami Hagiya[1] · *Satoshi Kobayashi*[2] · *Ken Komiya*[3] · *Fumiaki Tanaka*[4] ·
Takashi Yokomori[5]
[1]Department of Computer Science, Graduate School of Information
Science and Technology, The University of Tokyo, Tokyo, Japan
hagiya@is.s.u-tokyo.ac.jp
[2]Department of Computer Science, University of
Electro-Communications, Tokyo, Japan
satoshi@cs.uec.ac.jp
[3]Interdisciplinary Graduate School of Science and Engineering, Tokyo
Institute of Technology, Yokohama, Japan
komiya@dis.titech.ac.jp
[4]Department of Computer Science, Graduate School of Information
Science and Technology, The University of Tokyo, Tokyo, Japan
fumi95@is.s.u-tokyo.ac.jp
[5]Department of Mathematics, Faculty of Education and Integrated Arts
and Sciences, Waseda University, Tokyo, Japan
yokomori@waseda.jp

G. Rozenberg et al. (eds.), *Handbook of Natural Computing*, DOI 10.1007/978-3-540-92910-9_34,
© Springer-Verlag Berlin Heidelberg 2012

Abstract

This chapter presents both the theoretical results of molecular computing models mainly from the viewpoint of computing theory and the biochemical implementations of those models in wet lab experiments. Selected topics include a variety of molecular computing models with computabilities ranging from finite automata to Turing machines, and the associated issues of molecular implementation, as well as some applications to logical controls for circuits and medicines.

1   Introduction

Molecular computing machineries (molecular machineries, in short), the primary theme of this chapter, are novel computing devices inspired by the biochemical nature of biomolecules. In the early days it was understood that molecular machineries would replace silicon-based hardware with biological hardware through the development of molecular-based computers. However, it is now recognized that molecular machineries should not substitute for existing computers but complement them.

One may be surprised at how early the term "molecular computers" appeared in the literature. In fact, one can go back to an article of *Biofizika* in 1973 where an idea of cell molecular computers is discussed (Vaintsvaig and Liberman 1973). Since 1974, M. Conrad, one of the pioneering researchers in this area, has been conducting consistent research on information-processing capabilities using macromolecules (such as proteins) (Conrad 1985), and he edited a special issue entitled *Molecular Computing Paradigms* in Conrad (1992). These initial efforts in the short history of molecular computing have been succeeded by Head's pioneering work on splicing systems and languages in Head (1987) whose theoretical analysis on splicing phenomena is highly appreciated, in that it was the first achievement of mathematical analysis using biochemical operations of DNA recombination. It is also worth mentioning that, prior to Head's work, one can find some amount of study dealing with DNA/RNA molecules in the framework of formal language theory, such as Eberling and Jimenez-Montano (1980), Jimenez-Montano (1984), and Brendel and Busse (1984). Then, all of these initial works were eventually evoked by Adleman's groundbreaking attempt in 1994 (Adleman 1994), which decisively directed this new research area of *molecular computing*.

This chapter presents both the theoretical results of molecular computing models mainly from the viewpoint of computing theory and the biochemical implementations of those models in wet lab experiments. Selection of topics included here might reflect the authors' taste and it is intended to give concise and suggestive expositions for each chosen topic.

The chapter is organized as follows. This section ends by providing the rudiments of the theory of formal languages and computation. ❷ Section 2 presents a variety of molecular computing models that are classified into two categories: the first is for molecular computing models with *Turing computability* and the second is for the molecular models of *finite automata*. Among the various ideas proposed for realizing Turing computability within the framework of molecular computing theory, four typical approaches to achieving universal computability were selected. These are based on formal grammars, equality sets, Post systems, and logical formulae. It is hoped that succinct expositions of those models may help the reader to gain some new insight into the theory of molecular computing models.

Most of the molecular computing models developed so far involve both the use of structural molecules with various complexity and sophisticated control of them.

❯ Section 3 is devoted to those *computing models with structured molecules* in which hairpin structures, tree structures, and more involved structures are focused on, among others.

In ❯ Sect. 4 attention is turned to the wet implementation issues of the molecular computing models and their applications as well. The lineup of the chosen topics is as follows:

- Enzyme-based implementation of *DNA finite automata* and its application to drug delivery
- Molecular design and implementation of *logic gates and circuits* using DNAzymes and *DNA tiles*
- Abstract models called *reaction graphs* for representing various reaction dynamics of DNA assembly pathways
- DNA implementation of finite automata with *whiplash machines*
- Hairpin-based implementation of *SAT engines* for solving the 3-SAT problem

The results mentioned above are just some successful examples in the area of molecular computing machineries that have been realized through devoted efforts and collaboration between biochemists and computer scientists. It is hoped that they will be able to provide a good springboard for advanced beginners in this area and for senior researchers in other areas.

1.1 Basic Notions and Notations

The basic notions and notations required to understand this chapter are summarized below. Then standard notions and notations are assumed from the theory of formal languages (see, e.g., Hopcroft et al. 2001, Rozenberg and Salomaa 1997, and Salomaa 1985).

For a set X, $|X|$ denotes the cardinality of X. The power set of X (i.e., the set of all subsets of X) is denoted by $\mathscr{P}(X)$.

An *alphabet* is a nonempty finite set of symbols. For an alphabet V, V^* denotes the set of all strings (of finite length) over V, where the empty string is denoted by ε (or λ). By V^+ one denotes the set $V^* - \{\varepsilon\}$. A *language* over V is a subset of V^*. For a string $x = a_1 a_2 \ldots a_n$ over V, the *mirror image* of $x \in V^*$, denoted by $mi(x)$ (or by x^R), is defined by $a_n \ldots a_2 a_1$.

A collection of languages over an alphabet V is called a *family of languages* (*language family*) or a *class of languages* (*language class*) over V.

A *phrase-structure grammar* (or *Chomsky grammar*) is a quadruple $G = (N, T, S, P)$, where N and T are disjoint alphabets of nonterminals and terminals, respectively, $S(\in N)$ is the initial symbol, and P is a finite subset of $(N \cup T)^* N (N \cup T)^* \times (N \cup T)^*$. An element (u, v) of P, called a rule, is denoted by $u \rightarrow v$.

For strings $x, y \in (N \cup T)^*$, let

$$x \underset{G}{\Longrightarrow} y \overset{\mathrm{def}}{\Longleftrightarrow} \exists x_1, x_2 \in (N \cup T)^* \text{ and } u \rightarrow v \in P \text{ such that } x = x_1 u x_2, \ y = x_1 v x_2.$$

A language generated by G is defined as $L(G) = \{x \in T^* | S \underset{G}{\overset{*}{\Longrightarrow}} x\}$

A phrase-structure grammar $G = (N, T, S, P)$ is

1. *Context-sensitive* if for $u \rightarrow v \in P$ there exist $u_1, u_2 \in (N \cup T)^*$, $A \in N$, $x \in (N \cup T)^+$ such that $u = u_1 A u_2$, $v = u_1 x u_2$
2. *Context-free* if for $u \rightarrow v \in P$, $u \in N$
3. *(Context-free) linear* if for $u \rightarrow v \in P$, $u \in N$, and $v \in T^* N T^* \cup T^*$
4. *Regular* if for $u \rightarrow v \in P$, $u \in N$, and $v \in T \cup T N \cup \{\varepsilon\}$

RE, CS, CF, LIN, and *REG* denote the families of languages generated by phrase-structure grammars, context-sensitive grammars, context-free grammars, (context-free) linear grammars, and regular grammars, respectively. *RE* comes from the alternative name of *recursively enumerable* language. The following inclusions (called the *Chomsky hierarchy*) are proved.

$$REG \subset LIN \subset CF \subset CS \subset RE$$

A (nondeterministic) *Turing machine* (TM) is a construct $M = (Q, V, T, p_0, F, \delta)$, where Q, V are finite sets of states and a tape alphabet, respectively. V contains a special symbol B (a blank symbol), $T (\subseteq V - \{B\})$ is an input alphabet, $p_0 (\in Q)$ is the initial state, $F (\subseteq Q)$ is a set of final states, and δ is a function from $Q \times V$ to $\mathscr{P}(Q \times V \times \{L, R\})$ acting as follows: for $p, p' \in Q, a, b \in V, d \in \{L, R\}$, if $(p', b, d) \in \delta(p, a)$, then M changes its state from p to p', rewrites a as b, and moves the head to the left (if $d = L$) or the right (if $d = R$). M is called *deterministic* (denoted by DTM) if for each $(p, a) \in Q \times V, |\delta(p, a)| \le 1$.

A configuration in M is a string of the form: xpy, where $x \in V^*, y \in V^* (V - \{B\}) \cup \{\varepsilon\}$, $p \in Q$, and the head of M takes its position at the leftmost symbol of y. A binary relation $\vdash$ on the set of configurations in M is defined by

$$\begin{cases} xpay \vdash xbqy \iff (q, b, R) \in \delta(p, a) \\ xp \vdash xbq \iff (q, b, R) \in \delta(p, B) \\ xpay \vdash xqby \iff (q, b, L) \in \delta(p, a) \\ xcp \vdash xqcb \iff (q, b, L) \in \delta(p, B) \end{cases}$$

where $a, b, c \in V, x, y \in V^*, p, q \in Q$. Let $\vdash^*$ be the reflexive, transitive closure of $\vdash$. A language $L(M)$ recognized by M is defined as

$$L(M) = \{w \in T^* \mid p_0 w \vdash^* xpy, \text{ for some } x, y \in V^*, p \in F\}$$

that is, the set of input strings such that, starting with the initial state, they drive their configurations in M to a final state. It is known that the family of languages recognized by Turing machines is equal to the family *RE*.

As subfamilies of *RE*, there are many to be noted. Among others, two families are of importance: **NP** and **P** are the families of languages recognized in polynomial-time, that is, in $f(n)$ steps by TMs and DTMs, respectively, where f is a polynomial function defined over the set of nonnegative integers and n is the size of the input string. It remains open whether **P** = **NP** or not, while the inclusion **P** $\subseteq$ **NP** is clear from the definition. In particular, a subfamily of **NP** called *NP-complete* is of special interest, because NP-complete languages are representative of the hardness in computational complexity of **NP** in the sense that any language in **NP** is no harder than the NP-complete languages. Further, by **NSPACE**$(f(n))$ and **DSPACE**$(f(n))$, the families of languages recognized in $f(n)$ space (memory) are denoted by TMs and DTMs, respectively.

Due to Watson–Crick complementarity, not only single-stranded molecules but also completely hybridized double-stranded molecules can be regarded as strings over an alphabet {A,C,G,T}. Therefore, it is natural to employ the formal framework of the classical formal language theory, in order to study and formulate notions in molecular computing.

Given an alphabet Σ, a *morphism* h over Σ is a mapping defined by $h(a) \in \Sigma^*$ for each $a \in \Sigma$. Any h is extended to $h: \Sigma^* \to \Sigma^*$ by $h(\varepsilon) = \varepsilon$ and $h(xa) = h(x)h(a)$ for each $x \in \Sigma^*, a \in \Sigma$. A *weak coding* over Σ is a morphism h such that $h(a)$ is in $\Sigma \cup \{\varepsilon\}$ for each $a \in \Sigma$.

An *involution* over Σ is a morphism h such that $h(h(a)) = a$ for each $a \in \Sigma$. Any involution h over Σ such that $h(a) \neq a$ for all $a \in \Sigma$ is called a *Watson–Crick morphism* over Σ.

In analogy to DNA molecules (where A and T (C and G) are complementary pairs), for $x, y \in \Sigma^*$, if $h(x) = y$ (and therefore, $h(y) = x$), then it is said that x and y are *complementary* to each other, and denoted by $\overline{x} = y$ ($\overline{y} = x$). An alphabet $\overline{\Sigma} = \{\overline{a} \mid a \in \Sigma\}$ is called the *complementary alphabet* of Σ.

2 Molecular Machine Models

Since Adleman's groundbreaking work on the DNA implementation of computing a small instance of the directed Hamiltonian path problem (Adleman 1994), numerous research papers in this new computation paradigm have been published. In fact, Adleman's model has been extensively studied by many researchers seeking to generalize his technique to solve larger classes of problems (Lipton 1995; Lipton and Baum 1996), provide abstract DNA computer models with Turing computability (Adleman 1996; Beaver 1995; Rothemund 1995; Winfree et al. 1996), and so forth.

In this section, the focus is on computing models of molecular machines with Turing computability and others, restricted primarily to finite automata.

2.1 Molecular Models with Turing Computability

2.1.1 Grammar Approach

The Turing machine or its grammatical equivalent (like phrase-structure grammar) is one of the most common computing models when one wants to prove the universal computability of a new computing device. Among others, most interesting is the normal form theorems proved by Geffert (1991), which tell us that each recursively enumerable language can be generated by a phrase-structure grammar with only five nonterminals $\{S, A, B, C, D\}$ (S is the initial nonterminal) and with all of the context-free rules being of the form $S \to v$ and two extra (non context-free) rules $AB \to \varepsilon$ and $CD \to \varepsilon$. Thus, the feature of this normal form is of great importance in that *most* of the computation processes can be carried out without checking any context-sensitivity requirement.

Geffert derived an interesting normal form theorem for phrase-structure grammars which can be conveniently rephrased here with some modifications.

Proposition 1 (Geffert 1991) *Each recursively enumerable language L over Σ can be generated by a phrase-structure grammar with $G = (\{S, A, C, G, T\}, \Sigma, R \cup \{TA \to \varepsilon, CG \to \varepsilon\}, S)$, where $R = R_1 \cup R_2 \cup R_3$ and*

$$R_1 = \{S \to z_a Sa \mid a \in \Sigma \text{ and } z_a \text{ is in } \{T, C\}^*\} \text{ (called Type-(i) rules)}$$
$$R_2 = \{S \to uSv \mid u \text{ is in } \{T, C\}^* \text{ and } v \text{ is in } \{A, G\}^*\} \text{ (called Type-(ii) rules)}$$
$$R_3 = \{S \to u\} \text{ (called Type-(iii) rule), where } u \text{ is in } \{T, C\}^*$$

The following properties of the phrase-structure grammar G in normal form in Proposition 1 are quite useful for deriving the computing model discussed later.

1. It is not until a Type-(*iii*) rule has been applied that two extra cancellation rules (TA $\to \varepsilon$ and CG $\to \varepsilon$) are available.

2. Mixed use of Type-(i) and Type-(ii) rules leads to sentential forms for which no chance to get terminal strings remains.

3. Any occurrence of $x (\in \Sigma)$ in any substring (except for a suffix) of a sentential form blocks a terminal string.

4. A successful derivation process can be divided into three phases: At the first phase (*Generation*), Type-(i) rules are used. The second phase (*Extension*) consists of using Type-(ii) rules and ends up with one application of the Type-(iii) rule. Finally, the third phase (*Test*) erases well-formed pairings (TA or CG) using two extra rules only.

5. There is only one occurrence of either TA or CG in any sentential form to which cancellation rules can be applied.

6. The *Test* phase can only be started up with the unique position where either TA exclusively or CG appears immediately after one application of the Type-(iii) rule.

Grammar-Based Computation of Self-Assembly: (GCS)

We can now present a way of DNA computation, GCS, which is based on the features of Geffert normal form grammars in ❷ Proposition 1. GCS (this is referred to as YAC in the original paper (Yokomori 2000)), grammar-based computation of self-assembly, has the following set of basic components assembled:

1. For each Type-(ii) rule: $S \rightarrow uSv$, construct a component with the shape illustrated in ❷ *Fig. 1b*, and let it be called a *Type-(ii)* component.

2. For the unique Type-(iii) rule: $S \rightarrow u$, construct a component with the shape illustrated in ❷ *Fig. 1c*, where a connector is attached as an endmarker. This is called a *Type-(iii)* component.

3. Finally, for a given input string $w = b_1 \ldots b_n (\in \Sigma^*)$ to be recognized, a component in the shape shown in ❷ *Fig. 1a* is prepared, where Z_w of the component in (a) is the binary coding of the input w and $\overline{Z_w}$ denotes the complementarity sequence of Z_w.

Under the presence of *ligase*, all these components with sufficiently high concentration in a single pot of a test tube are provided. (See ❷ *Fig. 1*.)

Finally, the following operation is used:

- d-Detect: Given a test tube that may contain double-stranded strings, return "*yes*" if it contains (at least) one *completely hybridized* double-stranded string, and "*no*" otherwise.

❷ *Figure 1* illustrates a computation schema for GCS, where a random pool is a multi-set of basic components with sufficiently high concentration that can produce all possible assembly necessary for recognizing an input string. Besides the operation d-Detect, GCS only requires two basic operations to execute its computation process: *annealing* and *melting*.

In summary, one has the following:

Theorem 1 (Yokomori 2000) *Any recursively enumerable language can be recognized by the GCS computation schema.*

(Note that, in contrast, GCS is based on grammatical formulation, its functional behavior is close to an acceptor rather than a generative grammar.)

The structural simplicity of computation assembly units in GCS is expected to facilitate its implementation at least in its design schema. In particular, the two groups of four DNAs (A,G:

◘ Fig. 1

Basic components and a successful computation using the grammar-based computation process (GCS): A large quantity of molecules consisting of all basic components are put into a single test tube. Then, hybridization of those molecules produces candidate complex molecules for successful computations. By first melting the complexes and then annealing them, a desired double-stranded molecule is detected using the w-probe attached with magnetic beads when the input w is in L.

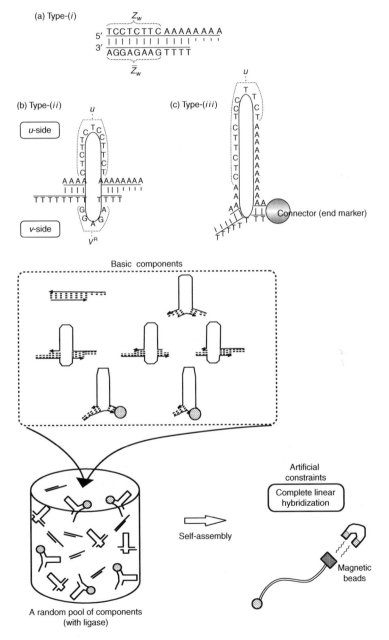

purine and C,T: pyrimidine) can be readily applied both to encoding the computational information of a language into its GCS assembly units and to checking the completeness of the computation results in the final process with its complementary property. On the other hand, as for the biomolecular feasibility of the GCS, it still remains at the conceptual level, and more molecular biological investigation is needed before the model becomes truly feasible.

2.1.2 Equality Checking Approach

In the theory of formal languages and computation, a simple computational principle called *equality checking* can often play an important role. The equality checking principle (EC principle) is simple enough for humans to understand and for machines of any kind to implement. In particular, it seems very suitable for biomolecular machines, because this EC principle only requires the task of equality checking of two memories, represented as strings of symbols, just like a completely hybridized *double-stranded* DNA *sequence.*

Based on the EM approach, one can present an abstract model of DNA computing, DNA-EC, which has the universal computability of Turing machines.

Equality Machines

An *equality machine* (Engelfriet and Rozenberg 1980) is an abstract acceptor shown in ❯ *Fig. 2.* Formally, an equality machine (EM) M is a structure $(Q, \Sigma, \Delta, \delta, q_0, F)$, where Q is the finite set of states, $q_0 \in Q$ is the initial state, $F \subseteq Q$ is the set of final states, Σ is the input alphabet, Δ is the output alphabet, and δ is a finite set of transition rules of the form: $(p, a) \rightarrow (q, u, i)$, where $p, q \in Q$, $a \in \Sigma \cup \{\varepsilon\}$, $u \in \Delta^*$, and $i \in \{1, 2\}$.

A configuration of M is of the form $(p, w, (v_1, v_2))$, where $p \in Q$, $w \in \Sigma^*$, and $v_1, v_2 \in \Delta^*$. If $(p, a) \rightarrow (q, u, i) \in \delta$, then one writes $(p, aw, (v_1, v_2)) \vdash (q, w, (v_1', v_2'))$, where $v_1' = v_1 u$ and $v_2' = v_2$

☐ Fig. 2

Equality machine EM: the machine consists of a one-way *input* tape, a *finite control*, and two *write-only* tapes for its memory. With the initial empty memory tapes, the machine starts to read the input from the left and changes its state, then extends one of the two memory tapes by concatenating an output string, according to the transition rules specified. At the end of computation, where the input string has been fully read, the machine accepts it only if the contents of the two memory tapes are identical.

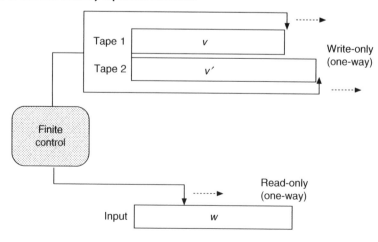

(if $i = 1$), and $v_1' = v_1$ and $v_2' = v_2 u$ (if $i = 2$). The reflexive and transitive closure of $\vdash$ is denoted by $\vdash^*$. The *language accepted* by M is $L(M) = \{w \in \Sigma^* \,|\, (q_0, w, (\varepsilon, \varepsilon)) \vdash^* (q, \varepsilon, (v, v))$ for some $q \in F$ and $v \in \Delta^*\}$. An EM M is *deterministic* if δ is a mapping from $Q \times \Sigma$ into $Q \times (\Delta \times \{1,2\})^*$.

The classes of languages accepted by nondeterministic and deterministic EMs are denoted by $\mathscr{L}_N(EM)$ and $\mathscr{L}_D(EM)$, respectively.

Very interesting results from Engelfriet and Rozenberg (1980) are summed up here which motivated the results in this section.

Proposition 2 (Engelfriet and Rozenberg 1980) *The following relations hold true:*

1. $\mathscr{L}_N(EM) = RE$
2. $\mathscr{L}_D(EM) \subseteq \mathbf{DSPACE}(\log n)$ *and* $REG \subset \mathscr{L}_D(EM) \subset \mathbf{NSPACE}(n) = CS$
3. $\mathscr{L}_D(EM)$ *is incomparable to CF but not disjoint.*

DNA-EC: A Model of DNA Computing

Following the notations and definitions in Rooß and Wagner (1996), the DNA computation model called DNA-EC is described, which is based on the characterization results given in the previous section.

In the *test tube computation* discussed below, T denotes a test tube containing a set of strings over some fixed alphabet Σ, and its contents are denoted by $I(T)$. Further, it is assumed that $I(T) \subseteq \Sigma^*$ unless stated otherwise.

The DNA-EC requires the following two set *test* operations:

- EM (*Emptiness Test*): Given a test tube T, return "*yes*" if $I(T)$ contains a string (i.e., $I(T) \neq \emptyset$) and "*no*" if it contains none.
- EQ (*Equivalence Test*): Given a test tube T (where $I(T) \subseteq D^*$) that may contain double-stranded strings, return "*yes*" if $I(T)$ contains (at least) one complete double-stranded string, and "*no*" otherwise. (Note that D is the alphabet for representing double strands.)

Let T, T_1, T_2 be set variables for test tubes used in DNA-EC. ❷ *Table 1* shows a collection of *set* operations used in DNA-EC, where $a, b, c, d \in \Sigma$, $u, v, w \in \Sigma^*$.

Let $\mathcal{O}$ and $\mathcal{T}$ be collections of set operations and set test operations, respectively, introduced above. Then, DNA$(\mathcal{O}, \mathcal{T})$-EC is the class of languages recognized by DNA-EC, which allows the use of set operations from $\mathcal{O}$ and of set test operations from $\mathcal{T}$.

❑ **Table 1**

Set operations (from Rooß and Wagner 1996)

Name	Syntax	$I(T)$ after the Operation
UN(union)	$T := T_1 \cup T_2$	$I(T_1) \cup I(T_2)$
LC(left cut)	$T := a \backslash T_1$	$\{u \,|\, au \in I(T_1)\}$
LA(left adding)	$T := a \cdot T_1$	$\{a\} \cdot I(T_1)$
EX(extraction)	$T := \mathrm{Ex}(T_1, w)$	$I(T_1) \cap \Sigma^* \{w\} \Sigma^*$
RE(replacement)	$T := \mathrm{Re}(T_1, u, v)$	$\{xvy \,|\, xuy \in I(T_1)\}$
RV(reversal)	$T := T_1^R$	$\{u^R \,|\, u \in I(T)\}$ (u^R is the reversal of u)

EX is essentially the same as the one often referred to as *Separation* in the literature

As for the computational power of DNA-EC, it is proved that given an EM $M = (Q, \Sigma, \Delta, \delta, q_0, F)$ and $w \in \Sigma^*$, the computation process of M with w is simulated within the DNA-EC model in a variety of ways. The idea for the simulation is simply to keep only the difference between the contents of Tapes 1 and 2 in the configuration of w in the computing schema dealing with a sequence of DNAs as a single string. The hybridization property of DNA complementarity can also be readily used for checking two memories in the case of manipulating *double-stranded strings* within the schema. The obtained results include the following:

Theorem 2 (Yokomori and Kobayashi 1999) *The following equalities hold:*

1. DNA({UN,LA,RV,RE}, {EM})-EC = DNA({UN,RE}, {EM})-EC = RE
2. DNA({UN,LA,LC,EX}, {EQ})-EC

2.1.3 Post System Approach

Among a variety of computing models based on string rewriting, a classical Post system will be introduced below that has a unique position, somewhere between formal grammars and logical systems. Because of the simplicity of its components and functional behavior, a Post system can sometimes provide a clear picture of the essence of the computation.

A Lesson From Post Systems

A rewriting system called the *general regular Post system* (GRP system) is a quadruple $G = (V, \Sigma, P, A)$, where V and $\Sigma(\subseteq V)$ are alphabets, P is a finite set of rules of the form of either $uX \to wX$ or $aX \to Xb$ ($X \notin V$: variable, $u, w \in V^*, a, b \in V$), $A(\subset V^+)$ is a finite set of axioms.

Given strings $\alpha, \beta \in V^*$, define a binary relation $\Longrightarrow$ as follows:

$$\alpha \Longrightarrow \beta \overset{\text{def}}{\Longleftrightarrow} \begin{cases} \exists uX \to wX \in P, \delta \in V^* \ [\alpha = u\delta, \beta = w\delta] \quad \text{or} \\ \exists aX \to Xb \in P, \delta \in V^* \ [\alpha = a\delta, \beta = \delta b] \end{cases}$$

Let $\Longrightarrow^*$ be the reflexive and transitive closure of $\Longrightarrow$. Then, a language generated by G is defined as

$$L(G) = \{w \in \Sigma^* | \ \exists u \in A \text{ such that } u \Longrightarrow^* w\}$$

It is known that the family of languages generated by GRP systems coincides with the family of recursively enumerable languages (Salomaa 1985). The source of the universal computing capability of GRP systems comes from the use of two types of rules $\alpha X \to \beta X$ (say, type 1 rule) and $aX \to Xb$ (type 2 rule). The generative capability between these two types of rule is interesting.

Theorem 3 (Post 1943) *GRP systems with type 1 rules alone can generate only regular languages, while type 1 rules together with type 2 rules enable GRP systems to generate any recursively enumerable language.* (This fact gives the base of a technique called the "rotate-simulate method" used to prove that a computing model has universal computability.)

This result seems to be useful in analyzing the computing power of rewriting systems. As seen in the subsequent discussion, this observation leads to new types of computing models with universal computability based on splicing operations.

Multi-test Tube Models

We note that Type 1 rules of GRP systems can be generally simulated by some devices of existing rewriting models of computation. Therefore, in order to achieve Turing computability, one has to answer the question: "Is there any way to carry out a rotation rule $aX \rightarrow Xb$ within the computing system?" One possible way is to consider an extended model of test tube systems studied in Csuhaj-Varju et al. (1996) and Kobayashi and Sakakibara (1998).

An *elementary formal system* (EFS) is a triple $E = (D, \Sigma, M)$, where D is a finite set of predicate symbols, Σ is a finite alphabet, and M is a set of logic formulas. Then, F in M is an *H-form* iff it is of the form:

$$P(x_1 u_1 v_2 y_2) \leftarrow Q(x_1 u_1 u_2 x_2) \ R(y_1 v_1 v_2 y_2) \text{ or } P(w) \leftarrow$$

where x_1, x_2, y_1, y_2 are all distinct variables, u_1, u_2, v_1, v_2 and w are in Σ^*.

An EFS $E = (D, \Sigma, M)$ is called *H-form EFS* iff every F in M is an H-form. Further, an H-form EFS with a single predicate P, that is, an H-form EFS $E = (\{P\}, \Sigma, M)$ is called a *simple H-form EFS*.

Given a predicate P and EFS $E = (\{P\}, \Sigma, M)$, define

$$L(E, P) = \{w \in \Sigma^* | P(w) \text{ is provable from } E\}$$

It is known that a simple H-form EFS is equivalent to an H-system in its computing capability.

Theorem 4 *For any H-system $S = (\Sigma, R, A)$, there effectively exists a simple H-form EFS $E = (\{P\}, \Sigma, M)$ such that $L(E, P) = L(S)$. Conversely, for any simple H-form EFS $E = (\{P\}, \Sigma, M)$, one can construct an H-system $S = (\Sigma, R, A)$ such that $L(S) = L(E, P)$.*

Therefore, in order to get the computing power beyond H-systems, it is necessary to consider an H-form EFS with more than one predicate symbol, which leads to the idea of "multi-splicing models" where more than one test tube is available for splicing operations.

Given an $n > 0$, by HE_n the family of languages defined by H-form EFSs is denoted with n predicate symbols. Further, let $HE_* = \cup_{i \geq 1} HE_i$.

Then, the following has been proved.

Theorem 5
1. $H(FIN, FIN) = HE_1 \subset HE_2 \subseteq HE_3 \subseteq \ldots \subseteq HE_{13} = \ldots = HE_* = RE$.
2. HE_2 *contains non-regular languages.*
3. HE_3 *contains non-context-free languages.*

Thus, 13 test tubes are sufficient to generate any recursively enumerable language in multi-splicing models.

The way to simulate an H-form EFS $E = (D, R, A)$ by a multi-splicing model is outlined below, where $|D| = 2$, that is, two test tubes T_1 and T_2 are available (Kobayashi and Sakakibara 1998). It suffices to show how to simulate $aX \rightarrow Xb$ using splicing operations in two test tubes. For a given string $w = ax$, make $w' = BaxE$, where B and E are new symbols of marker. Then, delete a prefix Ba from w' in T_1 and transfer xE to T_2 that contains b. Then, replace E in xE with b, resulting in xb.

It remains open to solve how many test tubes is minimally necessary to generate any recursively enumerable language. See Csuhaj-Varju et al. (1996) for related discussion.

2.1.4 Logic Approach

Horn clause logic is one of the most important logical systems in mathematical logic. Its significance was first pointed out by the logician Alfred Horn (Horn 1951). Horn clause logic is a mathematical basis of a descriptive programming language, PROLOG, and is known to have Turing computability (Tarnlund 1977). Therefore, it is natural to construct a molecular computing device based on Horn clause computation (Kobayashi et al. 1997). Kobayashi proposed to use Horn clause logic as an underlying computational framework of a molecular computing machine (Kobayashi 1999). His idea is to implement a reduction process of a subclass of a Horn program, called a *simple* Horn program, with DNA molecules.

For a predicate symbol A and a sequence $X_1, \ldots, X_k$ of constants or variables, $A(X_1, \ldots, X_k)$ is called a *simple atom*. If every X_k is a constant, $A(X_1, \ldots, X_k)$ is called a *simple ground atom*. A *simple Horn clause* is a formula of the form $\forall X_1, \ldots, X_m (F \leftarrow F_1 \wedge \cdots \wedge F_n)$, where F and F_i $(1 \leq i \leq n)$ are simple atoms. For a simple Horn clause $r : F \leftarrow F_1, \ldots, F_n$, F and F_i $(i = 1, \ldots, n)$ are called a *head* and a *body* of r, respectively. A *simple Horn program* is a finite set of simple Horn clauses.

A substitution θ is a mapping from V to $V \cup C$. An *expression* means a constant, a variable, a sequence of atoms, or a clause. For an expression E and a substitution θ, $E\theta$ stands for the result of applying θ to E, which is obtained by simultaneously replacing each occurrence of a variable X in E by the corresponding variable or constant $X\theta$.

Let H be a simple Horn program and I be a set of simple ground atoms. Then, an *immediate consequence operator*, $T_H(I)$ is defined as follows:

$$T_H(I) = \{F \mid F_1, \ldots, F_k \in I \wedge A \leftarrow B_1, \ldots, B_k \in H \wedge \exists \theta \ s.t. \ (B_i\theta = F_i \wedge A\theta = F)\}$$

Let Σ be a finite alphabet. A *decision problem* is a function from Σ^* to $\{0, 1\}$. Let Γ be an alphabet (possibly infinite) for representing simple Horn programs and Γ^* be the set of all strings of *finite* length over Γ. A function π from Σ^* to Γ^* is said to be an *encoding function for problem instances* if for every $w \in \Sigma^*$, $\pi(w)$ is a finite set of simple Horn clauses and π can be computed by a deterministic Turing machine in polynomial time. Let sz be a function from Σ^* to the set of all positive integers such that for every $w \in \Sigma^*$, $sz(w) \leq |w|$ and $sz(w)$ can be computed in polynomial time with respect to $|w|$ by a deterministic Turing machine. A family $\{H_i\}_{i \geq 1}$ of simple Horn programs *computes a decision problem*, f, *with encoding function*, π, *and size function*, sz, if there exists a predicate symbol A such that for every $w \in \Sigma^*$ with $sz(w) = n$, $f(w) = 1$ holds iff $A \in T_{H_{sz(w)} \cup \pi(w)}(\emptyset)$.

It is said that a function ϕ from $\{1\}^+$ to Γ^* *generates* a family $\{H_i\}_{i \geq 1}$ of simple Horn programs if for every positive integer i, $\phi(1^i) = H_i$ holds. A family $\{H_i\}_{i \geq 1}$ of simple Horn programs is said to be *uniform* if it is generated by some function, which can be computed by a deterministic Turing machine in polynomial time.

Then, one can show the following theorem:

Theorem 6 *For any decision problem P in **NP**, there exists a uniform family of simple Horn programs which computes P in polynomial steps.*

By the above theorem, efficient DNA implementation of the operation T_H for a simple Horn program enables us to construct a DNA computer that solves problems in **NP** in a polynomial number of biolab steps. One idea to implement T_H is given by Kobayashi, in which the whiplash polymerase chain reaction (PCR) technique is used to (1) check the equality of arguments in a body of a clause and (2) copy an argument from a body to a head. (For details, refer to ❯ Sect. 4.4 in this chapter.)

For representing ground atoms of a given simple Horn program, it is necessary to design sequences of the following oligos:

1. Oligos for representing constant symbols
2. For each predicate symbol A with k-arguments and for each i ($1 \leq i \leq k$), two different oligos A_i and A_i', where A_i and A_i' are called a *pre-address* and a *post-address* of A at the ith position

Each ground atom $A(c_1, \ldots, c_k)$ is represented as a single-stranded DNA consisting of k segments, each of which contains data c_i of each argument of A. Each segment is composed of three regions, *pre-address*, *data*, and *post-address*. The data region is an oligo representing a constant. Further, there exist stopper sequences in whiplash PCR at each of the segment ends, and at each end of the pre-address, data, and post-address regions. (See ❯ Fig. 3a.)

The procedure for computing $T_H(I)$ is as follows. First, construct a set of ground instances of all possible bodies of rules in H. This is accomplished by concatenating ground atoms in DNA representation of I by using appropriate complementary sequences (see ❯ Fig. 3b). Then, check the equality constraint among arguments specified by the rule, which is implemented by using whiplash PCR with two different stopper sequences in a sophisticated manner. In this way, one can extract correct ground instances of all possible bodies of rules. Then, copy arguments from the body to the head by using whiplash PCR once again. For instance, in ❯ Fig. 3b, the arguments c_1 and c_4 are copied to the head, which results in the DNA

◻ **Fig. 3**

Representation of atoms and their ligation: (a) DNA representation of a ground atom $A(c_1,\ldots,c_k)$. (b) Concatenation of atoms $B(c_1,c_2)$, $C(c_2,c_3)$, $D(c_3,c_4)$ for applying a clause $A(X,W) \leftarrow B(X,Y) \wedge C(Y,Z) \wedge D(Z,W)$. After checking the equivalence of corresponding arguments (the second of B and the first of C, the second of C and the first of D), one makes a new DNA sequence representing $A(c_1,c_4)$ by using the whiplash PCR technique.

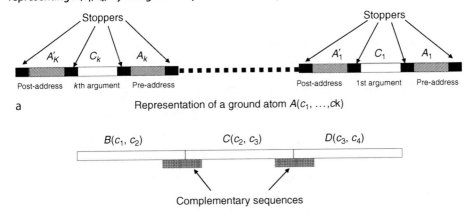

representation of a ground atom $A(c_1, c_4)$. Refer to Kobayashi (1999) for details. In conclusion, from ❷ Theorem 6 and the above discussion, one has:

Claim 1 (1) *There effectively exists a bio-procedure based on the simple Horn clause computation model with Turing computability.*
(2) *For any decision problem P in* **NP**, *there effectively exists a bio-procedure of a polynomial number of steps for computing P based on the simple Horn clause computation model.*

Other approaches to the DNA implementation of Horn clause computation were also proposed by Uejima et al. (2001).

2.2 Molecular Finite State Machines

A *finite automaton* (FA) is one of the most popular concepts in theoretical computer science, and, hence, it is of great use to establish a simple implementation method for realizing an FA in the formulation of molecular computing. In particular, it is strongly encouraged to consider the problem of how simply one can construct a molecular FA using DNAs in the framework of *self-assembly computation*.

Formally, a *nondeterministic finite automaton* (NFA) is a construct $M = (Q, \Sigma, \delta, p_0, F)$, where Q is a set of states, Σ is the input alphabet, $p_0 (\in Q)$ is the initial state, $F (\subseteq Q)$ is a set of final states, and δ is defined as a function from $Q \times \Sigma$ to $\mathcal{P}(Q)$ (the power set of Q). (If δ is restricted to being a function from $Q \times \Sigma$ to Q, then M is said to be *deterministic* and is denoted by DFA.)

2.2.1 Bulge Loop Approach

The first method introduced here, proposed by Gao et al. (1999), consists of three basic phases: (1) *encoding* the transition rules of a given NFA with an input string onto DNA molecules, (2) *hybridization* and *ligation* which carry out the simulation of the state transition, and (3) *extracting* the results to decide the acceptance or rejection of the input string.

As an example, consider an NFA M with the initial state 0 and two final states (0:2) and (0:3) given as a transition graph in ❷ *Fig. 4a*. For the phase (1), the set of six states is encoded by DNA sequences of length 4 as follows:

State 0	atat	State (0:3)	ttat
State (1:3)	gctg	State (2:3)	ctca
State (0:2)	tatt	State (1:2)	gtcg

The start molecule is constructed as a double-stranded molecule with a sticky end shown in ❷ *Fig. 4b*. Each transition rule $p \rightarrow^a q$ (where p, q are states, and a is in $\{0, 1\}$) is encoded as a molecule with a so-called "bulge loop" structure. For example, two transition rules: $0 \rightarrow^0 (0:3)$ and $0 \rightarrow^0 (0:2)$ are encoded as in ❷ *Fig. 4c*, while two transition rules: $0 \rightarrow^1 (1:3)$ and $0 \rightarrow^1 (1:2)$ are encoded as in ❷ *Fig. 4d*. All of the other transition rules have to be encoded in the same principle on structured molecules with bulge loops. (The idea behind the design principle of molecular encoding will be explained below.)

◻ Fig. 4
NFA for divisibility by 2 or 3: Coding transition rules.

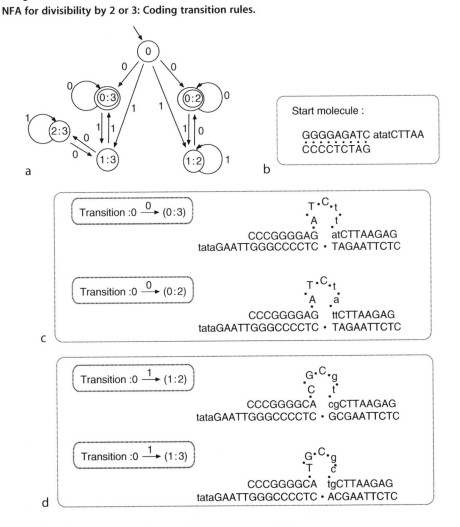

A computation in M is simulated in three steps as follows:

(*Step 1*): At the beginning of the reaction, one has the start molecule having the atatCTTAA overhang. Suppose that input 0 will be processed. Then, there are two possibilities (two molecules) that have a tata hangover, which matches atat in the hangover of the current state (of molecule). After hybridization and ligation, the remaining unreacted molecules are all washed away and collected. At this moment, the new possible states (0:3) and (0:2) from 0 with input 0 are covered by the loop part and do not show up yet. (The molecule potentially representing the state (0:3) is illustrated in ❷ *Fig. 5a.*)

(*Step 2*): Add a restriction enzyme *Sma*I which recognizes $\frac{CCCGGG}{GGGCCC}$ and cut at the middle between C and G, providing two parts, one of which contains the bulge loop. (See ❷ *Fig. 5b.*)

(*Step 3*): Add a restriction enzyme *Eco*RI which recognizes $\frac{CTTAAG}{GAATTC}$ on the portion with the bulge loop and cut it as shown in ❷ *Fig. 5c*, providing a molecule representing a new state (0:3), ready for the next new transition.

◨ **Fig. 5**

Simulating transition with input 0 from the state 0.

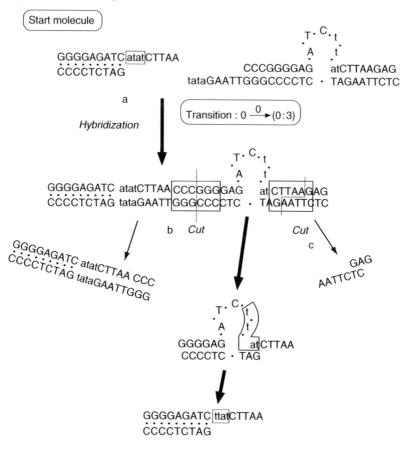

The "bulge loop" structure functions to prevent the new state sequence in it from reaching the wrong blunt end reaction during ligation. The *EcoRI* recognition site is used for breaking the bulge loop to expose the next state (i.e., (0:3) or (0:2) in this example) for the subsequent reaction. That is, Steps (1)–(3) are repeated for as many input symbols as necessary. In this manner, it is seen that the nondeterministic transitions in *M* can be simulated by a sequence of chemical reactions with a certain chemical control.

This implementation method has several advantages: (1) The size of molecules representing "state" remains unchanged, no matter how long the computation process. (2) The reactions proceed in a cyclic way that facilitates automated implementation. (3) The manner of designing molecules representing states and transitions used in the section can be standardized in the sense that once an input alphabet and the number of states are given, a library of molecules that represent all possible transitions with inputs could be created, independent of machine specificity.

Finally, it would be interesting and instructive to compare this work with an implementation method proposed in Benenson et al. (2001) where a different type of restriction enzyme and more sophisticated encoding techniques are used; this will be described later.

2.2.2 Length-Based Approach

A new molecular implementation method based on *length-only encoding* is proposed, which leads to a very simple-molecular implementation technique to solve graph problems. Here, for example, an effective molecular implementation method for a nondeterministic finite automaton based on *one pot self-assembly* computing is demonstrated (Yokomori et al. 2002).

In the following, the focus is on ε-free regular languages, that is, regular languages not containing the empty string ε. Therefore, in any NFA M considered here, the initial state p_0 is not a final state in F of M. The following result is easily shown, but is of crucial importance to attain our objective. Given any DFA, there effectively exists an equivalent NFA M such that (1) M has the unique final state q_f, and (2) there is neither a transition into the initial state q_0 nor a transition from the final state q_f.

Thus, in what follows, *one may assume that a finite automaton M is an NFA satisfying the properties* (1) *and* (2) *above.*

Now, a simple molecular implementation for NFA is presented, which one may call *NFA Pot* and its computing schema is illustrated in ❷ *Fig. 6.* NFA Pot has the following advantages:

1. In order to encode with DNAs each state transition of M, no essential design technique of encoding is required. Only the *length* of each DNA sequence involved is important.
2. Self-assembly due to hybridization of complementary DNAs is the only mechanism for carrying out the computation process.
3. An input w is accepted by M iff a completely hybridized double strand $[c(w)/\overline{c(w)}]$ is detected from the pot, where $c(w)$ is a DNA sequence representing w and the "overline" version denotes its complementary sequence.

Without loss of generality, one may assume that the state set Q of M is represented as $\{0, 1, 2, \ldots, m\}$ (for some $m \geq 1$) and in particular, "0" and "m" denote the initial and the final states, respectively.

Using an example, it is shown, how one can implement an NFA Pot for M given an NFA M.

Let one consider an NFA M (in fact, M is a DFA) in ❷ *Fig. 7,* where an input string $w = abba$ is a string to be accepted by $M = (\{0,1,2,3\}, \{a,b\}, \delta, 0, \{3\})$, where $\delta(0, a) = 1$,

❑ Fig. 6

Finite automaton pot.

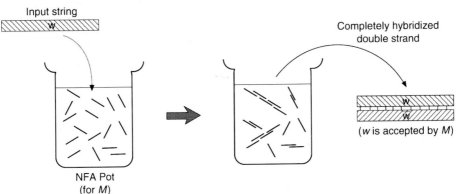

(*w* is accepted by *M*)

◻ Fig. 7

FA and its DNA implementation.

Input string (to be accepted) w = a b b a

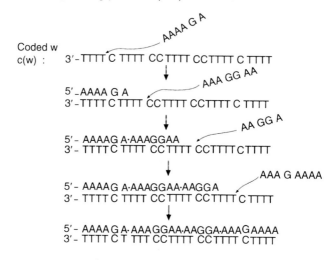

$\delta(1, b) = 2$, $\delta(2, b) = 1$, $\delta(1, a) = 3$. (Note that M satisfies properties (1) and (2) in the assumption previously mentioned and that the state 3 is the unique final state.)

In order to encode each symbol a_k from $\Sigma = \{a_1, \ldots, a_n\}$, each is associated with an oligo GGG...G of length k. Furthermore, each transition $\delta(i, a_k) = j$ is encoded as follows:

$$5'- \overbrace{\text{AAA} \cdots \text{A}}^{m-i} \overbrace{\text{GGG} \cdots \text{G}}^{k} \overbrace{\text{AAA} \cdots \text{A}}^{j}\text{-}3'$$

The idea is the following. Two consecutive valid transitions $\delta(i, a_k) = j$ and $\delta(j, a_{k'}) = j'$ are implemented by concatenating two corresponding encoded molecules, that is,

$$5'- \overbrace{\text{AAA} \cdots \text{A}}^{m-i} \overbrace{\text{GGG} \cdots \text{G}}^{k} \overbrace{\text{AAA} \cdots \text{A}}^{j}\text{-}3' \text{ and } 5'- \overbrace{\text{AAA} \cdots \text{A}}^{m-j} \overbrace{\text{GGG} \cdots \text{G}}^{k'} \overbrace{\text{AAA} \cdots \text{A}}^{j'}\text{-}3'$$

together make

$$5'- \overbrace{\text{AAA} \cdots \text{A}}^{m-i} \overbrace{\text{GGG} \cdots \text{G}}^{k} \overbrace{\text{AAA} \cdots \text{A}}^{m} \overbrace{\text{GGG} \cdots \text{G}}^{k'} \overbrace{\text{AAA} \cdots \text{A}}^{j'}\text{-}3'$$

Thus, an oligo $\overbrace{\text{AAA} \cdots \text{A}}^{m}$ plays the role of a "joint" between two transitions and it guarantees that the two are valid in M.

In order to get the result of the recognition task of an input w, one application of the "**Detect**" operation is applied to the pot, that is, the input w is accepted by M iff the operation *detects* a completely hybridized double strand $[c(w)/\overline{c(w)}]$, where $c(w)$ is a DNA sequence encoding w and the "overline" version denotes its complementary sequence. Specifically, for an input string $w = a_{i_1} \cdots a_{i_n}$, $c(w)$ is encoded as:

$$3'\text{-}\overbrace{\text{TTT}\cdots\text{T}}^{m}\overbrace{\text{CCC}\cdots\text{C}}^{i_1}\overbrace{\text{TTT}\cdots\text{T}}^{m}\overbrace{\text{CCC}\cdots\text{C}}^{i_2} \cdots \overbrace{\text{TTT}\cdots\text{T}}^{m}\overbrace{\text{CCC}\cdots\text{C}}^{i_n}\overbrace{\text{TTT}\cdots\text{T}}^{m}\text{-}5'$$

❯ *Figure 7* also illustrates the self-assembly process of computing $w = abba$ in the NFA Pot for M (although M is actually a DFA).

A method for molecular implementation is outlined to solve the directed Hamiltonian path problem (DHPP) which seems simpler than any other ever proposed. Since this implementation is also based on the "self-assembly in a one-pot computation" paradigm, one may call it *DHPP Pot* .

The DHPP Pot introduced has several advantages that are common with and similar to the implementation of NFA Pot. That is,

1. In order to encode each directed edge of a problem instance of DHPP, no essential design technique of encoding is required. Only the *length* of each DNA sequence involved is important.
2. Self-assembly due to hybridization of complementary DNAs is the only mechanism of carrying out the computation process.
3. An instance of DHPP has a solution path, p, iff a completely hybridized double strand $[c(p)/\overline{c(p)}]$ of a certain fixed length is detected from the pot, where $c(p)$ is a DNA sequence representing p and the "overline" version denotes its complementary sequence.

As seen above, the implementation methods discussed have a common distinguished feature in molecular encoding, that is, *no essential design technique is necessary* in that only the length of the DNA sequences involved in the encoding is critical. It should be remarked that this aspect of this noncoding implementation could apply to a large class of problems that are formulated in terms of graph structures. (It would be easy to list a number of problems associated with graphs.)

One can also propose a general schema for solving graph problems in the framework of the self-assembly molecular computing paradigm where the *length-only-encoding* technique may be readily applied. In fact, some of the graph-related optimization problems such as the traveling salesman problems can be handled in the schema mentioned above.

In the framework of self-assembly molecular computing, it is of great use to explore the computing capability of the self-assembly schema with "double occurrence checking (doc)" as a screening mechanism, because many of the important NP-complete problems can be formulated into this computing schema. It is shown by constructing DHPP Pot that at the sacrifice of the *nonlinear* (*exponential*) length increase of strands, the "doc" function can be replaced with the length-only encoding technique. It is, however, strongly encouraged to develop more practical methods for dealing with the doc function.

Finally, the biological implementations based on the method of length-only encoding (discussed here) can be found in Kuramochi and Sakakibara (2005) and in the corresponding chapter ❯ Bacterial Computing and Molecular Communication.

2.3 Bibliographic Notes

Various types of abstract models for DNA computing with Turing computability have been presented. These models are classified into various principles of computation, such as a specific grammar called "Geffert normal form" (Geffert 1991), the equality machine (or equality checking) in Engelfriet and Rozenberg (1980), Post systems (Post 1943), and logical formulations (Tarnlund 1977). In relation to the approach by equality checking discussed in ❷ Sect. 2.1.2, one can find a series of intensive works on Watson–Crick automata (Freund et al. 1999b) that share the computing principle with equality machines.

Many attempts to discover grammatical DNA computation models with Turing computability have already been made (Freund et al. 1999a; Păun 1996a, b; etc.), based on the *splicing* operation (originally proposed by Head (1987) or on H-systems (refer to the chapter ❷ DNA Computing by Splicing and by Insertion–Deletion for details). In nature, there exist DNA molecules, such as bacterial DNA and plasmids, that form circular structures. This inspires a new type of splicing models of computing where "circular splicing" produces a mixture of linear and circular molecules. One can find several works on such an extended H-system, called circular splicing system, in Siromoney et al. (1992), Pixton (1995), and Yokomori et al. (1997), in which the type 2 rule of the GPS system can be simulated by using circular molecules within the framework of splicing computing.

A further extension of splicing systems is studied in which tree structural molecules are considered to be spliced (Sakakibara and Ferretti 1999). The language family generated by tree splicing systems is shown to be the family of context-free languages (refer to ❷ Sect. 3.2). Recently, Csuhaj-Varju and Verlan proposed a new model of test tube systems with Turing computability, based on splicing, where the length-only separation of strings functions filtering in communication (Csuhaj-Varju and Verlan 2008). More models of extended splicing systems can be found in Păun (1998), while lab experimental work on splicing systems is reported in Laun and Reddy (1997).

Kari et al. (2005) propose and investigate another type of grammatical model for DNA computation based on insertion/deletion schemes that are inspired by pioneering works on contextual grammars due to S. Marcus (1969). The reference book by Păun et al. (1998) is one of the best sources for details of many topics mentioned above (including splicing systems, Watson–Crick automata, insertion/deletion systems, and others).

Reif (1995) proposes a parallel molecular computation model called parallel associative memory (PAM), which can *theoretically* and efficiently simulate a nondeterministic Turing machine and a parallel random access machine (PRAM), where a PA-Match operation, similar to the joint operation in relational database, is employed in an effective manner. Due to the complexity of the molecular biological transformation procedure used to realize a PAM program, it seems more study is needed to fill in the technological gaps between the two computation procedures at different levels.

Rooß and Wagner (1996) propose a formal framework called "DNA-Pascal" to study the computational powers of various abstract models for DNA computing and derive several interesting relationships between proposed models and known computational complexity classes within a polynomial time DNA computation. A recently published book (Ignatove et al. 2008) on DNA computing models contains broader topics in the area and partly overlaps with this chapter.

3 Computing Models with Structured Molecules

The molecules used in the first experiment in DNA computing reported in Adleman (1994) were linear non-branching structures of DNA. One of the important developments in DNA computing is based on the use of more involved molecules, which perform specific computations essentially through the process of self-assembly.

This section, introduces some of the computing models based on such structural DNA molecules.

3.1 Computing by Hairpin Structures

Molecules with *hairpin* structure(s) form a natural extension of linear non-branched molecules, because a hairpin results from a linear molecule that folds onto itself. This structural feature of molecules is recognized as a convenient and useful tool for many applications in DNA computing based on self-assembly.

This naturally brings about the following question: what sort of problems can be computed by a schema using hairpin structures? Investigating this question leads one to consider sets of string molecules which contain complementary sequences of potential hairpin formations, that is, the set of strings containing a pair of complementary subsequences.

In Păun et al. (2001) several types of such "hairpin languages" are considered and their language theoretic complexity (using the Chomsky hierarchy) is investigated.

When forming a hairpin molecule, it is necessary that the annealed sequence is "long enough" in order to ensure the stability of the construct, so it is always imposed that this sequence is, at least, of length, k, for a given k.

Let, then, k be a positive integer. The *unrestricted hairpin language* (of degree k) is defined by

$$uH_k = \{zvwxy \mid z, v, w, x, y \in V^*, x = mi(h(v)), \text{ and } |x| \geq k\}$$

By imposing restrictions on the location where the annealing may occur, one gets various sublanguages of uH_k (❷ *Fig. 8* illustrates the hairpin constructions corresponding to these languages):

◘ **Fig. 8**
(Left) Hairpin constructions corresponding to eight languages ; (right) a hairpin formation to be removed in solving the directed Hamiltonian path problem (DHPP).

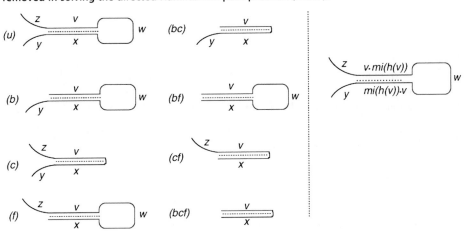

$bH_k = \{vwxy \mid v, w, x, y \in V^*, \text{cond}(x, v; k)\}$, $bfH_k = \{vwx \mid v, w, x \in V^*, \text{cond}(x, v; k)\}$,

$cH_k = \{zvxy \mid z, v, x, y \in V^*, \text{cond}(x, v; k)\}$, $cfH_k = \{zvx \mid z, v, x \in V^*, \text{cond}(x, v; k)\}$,

$fH_k = \{zvwx \mid z, v, w, x \in V^*, \text{cond}(x, v; k)\}$, $bcfH_k = \{vx \mid v, x \in V^*, \text{cond}(x, v; k)\}$,

$bcH_k = \{vxy \mid v, x, y \in V^*, \text{cond}(x, v; k)\}$, where $\text{cond}(x, v; k)$ means "$x = mi(h(v)) \wedge |x| \geq k$."

The computation schemata mentioned above is in fact embodied in the literature, for example, one subtracts a language uH_k, for some k, from a (regular) language in ❯ Sect. 4.5, while one intersects a given (linear) language with bcH_1 in GCS in ❯ Sect. 2.2.2. Therefore, let one consider languages of the form αH_k and $V^* - \alpha H_k$, for $\alpha \in \{u, b, c, f, bc, bf, cf, bcf\}$. As for the languages of the form αH_k, the following are obtained.

Theorem 7

(1) All languages αH_k, for $\alpha \in \{u, b, f, c, bf\}$ and $k \geq 1$, are regular.

(2) The languages $bcH_k, cfH_k, bcfH_k$, $k \geq 1$, are linear, but not regular

In order to see that those in (2) are not regular, one has only to consider the alphabet $V = \{a, b\}$, with a, b complementary to each other, and to intersect these languages with the regular language $a^+ b^+$, which leads to obtaining the following three languages: $\{a^n b^m \mid m \geq n \geq k\}$, $\{a^n b^m \mid n \geq m \geq k\}$, and $\{a^n b^n \mid n \geq k\}$ (in this order), and none of these languages is regular.

The next consequence is of interest from the viewpoint of DNA computing by hairpin removal.

Theorem 8
Let $\mathcal{F}$ be a family of languages which is closed under intersection with regular languages. If $L \in \mathcal{F}$, then $L - \alpha H_k \in \mathcal{F}$, for all $\alpha \in \{u, b, c, f, bf\}$ and all $k \geq 1$.

3.1.1 Complements of Hairpin Languages

In turn, one can consider languages bcH_k, cfH_k and $bcfH_k$. By (2) of ❯ Theorem 7, one knows already that the complements of bcH_k, cfH_k, and $bcfH_k$ are not necessarily regular. However, the languages $bcfH_k, k \geq 1$, are not "very non-regular" in the sense that for each $k \geq 1$, one has $V^* - bcfH_k$ as a linear language.

For more details, the next results are obtained:

Theorem 9

(1) If L is a regular language, then $L - bcfH_k$, $k \geq 1$, is a linear language which does not have to be regular.

(2) If L is a context-free language, then $L - bcfH_k$, $k \geq 1$, is a context-sensitive language which does not have to be context-free.

On the other hand, for languages bcH_k and cfH_k, the situation is more involved.

Theorem 10

(1) The languages $V^* - bcH_k$, $V^* - cfH_k$, and $k \geq 1$ are not context-free.

(2) If L is a regular language, then the languages $L - bcH_k$ and $L - cfH_k$, $k \geq 1$ are context-sensitive, but not necessarily context-free.

It would be interesting to investigate the place of the languages $L - bcH_k$ and $L - cfH_k$ with respect to families of languages intermediate between the context-free and the context-sensitive families (such as matrix or other regulated rewriting families (Dassow and Păun 1989)).

As an example of the hairpin computation schema discussed here, one can find a solution for the directed Hamiltonian path problem as well.

3.1.2 A New Hairpin-Based Operation: Hairpin Completion

Another interesting aspect of the computability involved in hairpin structures is brought about when a biological phenomenon known as *lengthening DNA by polymerases* is introduced into the DNA computation process. More specifically, consider the following hypothetical biological situation: one is given one single-stranded DNA molecule, z, such that either a prefix or a suffix of z is Watson–Crick complementary to a subword of z. Then the prefix or suffix of z and the corresponding subword of z get annealed by complementary base pairing and then z is lengthened by DNA polymerases up to a complete hairpin structure. The mathematical expression of this hypothetical situation leads to a new formal operation in DNA computing theory; starting from a single word, one can generate a set of words by this formal operation called *hairpin completion*. One can also consider the inverse operation of hairpin completion, namely hairpin reduction. That is, the *hairpin reduction* of a word x consists of all words y such that x can be obtained from y by hairpin completion. These operations are schematically illustrated in ❷ *Fig. 9*.

However, attention is focussed *only* on hairpin completion here.

Recall that for any $\alpha \in V^*$, $\overline{\alpha}^R$ denotes the complement of the mirror image of α. For any $w \in V^+$ and $k \geq 1$, the *k-hairpin completion* of w, denoted by $(w \rightarrow_k)$, is defined as follows.

$$(w \rightarrow_k) = \{\overline{\gamma}^R w \mid w = \alpha \beta \overline{\alpha}^R \gamma, |\alpha| = k, \alpha, \beta, \gamma \in V^+\}$$
$$(w \rightarrow_k) = \{w \overline{\gamma}^R \mid w = \gamma \alpha \beta \overline{\alpha}^R, |\alpha| = k, \alpha, \beta, \gamma \in V^+\}$$
$$(w \rightarrow_k) = (w \rightarrow_k) \cup (w \rightarrow_k)$$

Then, the hairpin completion is naturally extended to languages by

$$(L \rightarrow_k) = \bigcup_{w \in L} (w \rightarrow_k)$$

◻ Fig. 9
Hairpin completion and hairpin reduction: For each $k \geq 1$, hairpin completion (hairpin reduction) requires the condition of $|\alpha| = k$.

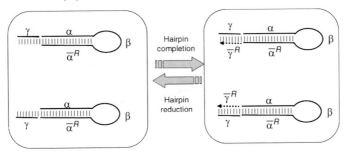

Note that all these phenomena here are considered in an idealized way. For instance, polymerase is allowed to extend in either end (3' or 5') despite that, due to the greater stability of 3' when attaching new nucleotides, DNA polymerase can act continuously only in the $5' \rightarrow 3'$ direction. However, polymerase can also act in the opposite direction, but in short "spurts" (Okazaki fragments). Moreover, in order to have a "stable" hairpin structure, the subword α should be sufficiently long.

For a class of languages $\mathscr{F}$ and an integer $k \geq 1$ one denotes by

$$WCOD(\mathscr{F} \rightarrow_k) = \{h(L \rightarrow_k) \mid L \in \mathscr{F} \text{ and } h \text{ is a weak coding}\}$$

the weak-coding image of the class of the hairpin completion of languages in $\mathscr{F}$.

The following characterizations concerning the class of linear context-free languages are interesting:

Theorem 11 (Cheptea et al. 2006) (i) *For any integer $k \geq 1$, $LIN = WCOD(REG \rightarrow_k)$ holds.* (Manea et al. 2009a) (ii) *For any integer $k \geq 1$, $WCOD(LIN \rightarrow_k)$ is a family of mildly context-sensitive languages.*

It should be remarked that this is the characterization result of a class of mildly context-sensitive languages (Joshi and Schabes 1997) that is based on a bio-inspired operation (hairpin completion) and weak codings starting from a simple class of linear languages in the Chomsky hierarchy. It remains open whether a similar result holds for the class of languages by hairpin reduction.

In contrast to its simplicity, hairpin computation as the underlying schema for DNA computing seems to be quite promising.

3.2 Computing by Tree Splicing

In their study on splicing systems, Sakakibara and Ferretti extended the notion of splicing on linear strings, proposed by Tom Head, to splicing on tree-like structures which resemble some natural molecular structures such as RNA structures at the secondary structure level (Sakakibara and Ferretti 1999).

Consider a finite alphabet V and trees over V. By V^T one denotes the set of all trees over V. A *splicing rule on tree structures over V* is given by a pair of tree substructures as in ❷ *Fig. 10*, where u_i's ($i = 1,...,4$) are subtrees over V, λs are null trees, and # is a special symbol not

❑ Fig. 10

Splicing rule on tree structures.

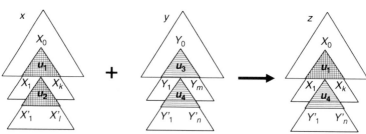

□ Fig. 11

Splicing on tree structures: Let x, y be trees which contain $u_1(\lambda,\ldots,u_2(\lambda,\ldots,\lambda),\ldots,\lambda)$ and $u_3(\lambda,\ldots,u_4(\lambda,\ldots,\lambda),\ldots,\lambda)$ as a part of the tree, respectively. Then, one can apply the rule in ❷ *Fig. 10* to x and y to generate a new tree z, which is obtained by replacing the subtree of x under u_2 with that of y under u_4.

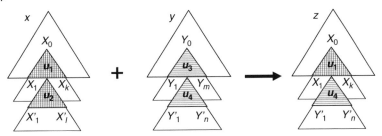

contained in the alphabet V indicating a single node in a tree. The symbol \$ is a separator located between the two tree substructures.

A splicing rule r takes two trees $x, y \in V^T$ as inputs and generates a tree $z \in V^T$, written $(x, y) \vdash_r z$. ❷ *Figure 11* illustrates an application of a splicing rule of ❷ *Fig. 10*.

A *splicing scheme on tree structures* is a pair $\gamma = (V, R)$, where V is an alphabet and R is a set of splicing rules on tree structures over V. A pair $\mathscr{S} = (\gamma, L)$ for a splicing scheme γ and a tree language $L \subseteq V^T$ is called a *splicing scheme on tree structures* (or HT system). For an HT system $\mathscr{S} = (\gamma, L)$, the tree language generated by $\mathscr{S}$ is defined as follows:

$$\gamma(L) = \{z \in V^T \mid (x, y)\vdash_r z \text{ for some } x, y \in L, r \in R\}$$

Iterated splicing is also defined as follows:

$$\begin{cases} \gamma^0(L) = L, \\ \gamma^{i+1}(L) = \gamma^i(L) \cup \gamma(\gamma^i(L)), \text{ and} \\ \gamma^*(L) = \bigcup_{i \geq 0} \gamma^i(L). \end{cases}$$

For a class of tree languages F_1 and a class of sets of splicing rules F_2, the set of all tree languages $\gamma^*(L)$ is denoted by $HT(F_1\, F_2)$, such that $\gamma = (V, R)$, $L \in F_1$, and $R \in F_2$.

Let FIN_T be the class of finite tree languages, and FIN_R be the class of finite sets of splicing rules. Then, $HT(FIN_T, FIN_R)$ has a close relation to the class of context-free languages in the following sense.

Theorem 12 *For any context-free grammar G, the set of derivation trees of G is in the class HT (FIN_T, FIN_R).*

3.3 Computing by PA Matching

Reif (1999) introduced an operation, called *parallel associative matching* (*PA Matching*), in order to provide a theoretical basis for parallel molecular computing. The operation PA Matching, denoted by ⋈, is defined as follows.

Let Σ be a DNA alphabet $\Sigma = \{\sigma_0, \sigma_1, ..., \sigma_{n-1}, \bar{\sigma}_0, \bar{\sigma}_1, ..., \bar{\sigma}_{n-1}\}$ of $2n$ distinct symbols where for each i, the symbols $\sigma_i, \bar{\sigma}_i$ are called (Watson–Crick) *complements*. Note that the DNA alphabet is extended theoretically to have symbols more than four. Consider another finite alphabet A and an encoding function E from A^+ to Σ^+. Thus, for a string $\alpha \in A^s$ of length s for some positive integer s, $E(\alpha)$ is an encoded DNA sequence of α. For a pair $\alpha, \beta \in A^s$, let $E(\alpha, \beta)$ denote an encoding of the ordered paired strings (α, β). (For details of how to encode a pair of strings, refer to Reif (1999).)

For each $\alpha, \beta, \beta', \gamma \in A^s$, the *PA-Match* operation $\bowtie$ is defined to be $E(\alpha, \beta) \bowtie E(\beta', \gamma) = E(\alpha, \gamma)$ when $\beta = \beta'$, and the operation $\bowtie$ yields no value if $\beta \neq \beta'$.

Reif introduced a formal parallel computation model, called the PAM model, in which the following five operations are allowed to be executed in one step:

1. Merge: Given test tubes T_1, T_2, produce the union $T_1 \cup T_2$.
2. Copy: Given a test tube T_1, produce a test tube T_2 containing the same contents of T_1.
3. Detect: Given a test tube T, say "yes" if T is not empty and say "no" otherwise.
4. Separation: Given a test tube T and a string $\alpha \in \Sigma^+$ produce a test tube T_1 consisting of all strings of T, which contain α as a substring, and also a test tube T_2 consisting of all strings of T, which do not contain α.
5. PA-Match: Given test tubes T_1, T_2, apply the operation $\bowtie$ to yield the test tube $T_1 \bowtie T_2$.

Reif showed a surprising simulation result of a nondeterministic Turing machine based on PAM model.

Theorem 13 *Nondeterministic Turing machine computation with space bound s and time bound $2^{O(s)}$ can be simulated by the PAM model using $O(s\log s)$ PAM operations that are not PA-Match and $O(s)$ PA-Match operations, where strings of length $O(s)$ over the alphabet size $O(s)$ (corresponding to DNA of length $O(s\log s)$ base-pairs) are used in the simulation.*

Motivated by Reif's work, Kobayashi et al. studied the formal properties of the PA-Match operation, where the operation $\bowtie$ was extended in order to deal with string pairs of different length (Kobayashi et al. 2001).

Informally speaking, the operation consists of cutting two strings in two segments such that the prefix of one of them matches the suffix of another, removing these two matching pieces, and pasting the remaining parts. Formally, given an alphabet V, a subset X of V^+, and two strings $u, v \in V^+$, one defines

$$PAm_X(u, v) = \{wz \mid u = wx, v = xz, \text{ for } x \in X, \text{ and } w, z \in V^*\}$$

The operation is naturally extended to languages over V by

$$PAm_X(L_1, L_2) = \bigcup_{u \in L_1, v \in L_2} PAm_X(u, v)$$

When $L_1 = L_2 = L$, $PAm_X(L)$ is written instead of $PAm_X(L_1, L_2)$. Since one shall only deal either with finite sets X or with $X = V^+$, one uses the notation $fPAm$ for finite PA-matching and the notation PAm for arbitrary PA-matching PAm_{V^+}.

Theorem 14
(1) *The families REG and RE are closed under PAm.*
(2) *The families LIN, CF, and CS are not closed under PAm.*

They further investigate the iterated version of the PA-match operation which is defined as follows. For a language $L \subseteq V^*$ and a finite set $X \subseteq V^+$, one defines:

$$\begin{cases} PAm_X^0(L) = L, \\ PAm_X^{k+1}(L) = PAm_X^k(L) \cup PAm_X(PAm_X^k(L)), \; k \geq 0, \\ PAm_X^*(L) = \bigcup_{k \geq 0} PAm_X^k(L). \end{cases}$$

When X is finite, the iterated PA-matching operation is denoted by $fPAm^*$; in the case $X = V^*$, the corresponding operation is denoted by PAm^*. Then, one has:

Theorem 15
(1) *The families REG and RE are closed under both $fPAm^*$ and PAm^*.*
(2) *The family LIN is not closed under $fPAm^*$ and PAm^*.*
(3) *The family CF is closed under $fPAm^*$ but it is not closed under PAm^*.*
(4) *The family CS is closed neither under $fPAm^*$ nor under PAm^*.*

As shown above, if one extends the operation $\bowtie$ to deal with string pairs of different length, then the computational power of the PA-Matching operation is rather weak, because one cannot escape from regularity when one starts from regular languages. In this sense, the length condition of $\bowtie$ is important in Reif's results.

3.4 Bibliographic Notes

Needless to say, it is of great importance in molecular computing theory to investigate the computational properties of structural molecules. In fact, there is a variety of structural complexity of computing molecules which range from 1-D linear to 3-D structures such as a duplex with hairpin molecules (Eng 1997), molecular graphs (Jonoska et al. 1998), quite sophisticated structures like double crossover molecules (Winfree et al. 1996) and DNA tiles (LaBean et al. 2000; Lagoudakis and LaBean 2000). Among others, Jonoska et al. used the 3-D graph structure of molecules to solve two NP-complete problems (SAT and 3-vertex-colorability problems). Further, Winfree et al. investigated the computational capability of various types of linear tile assembly models (Winfree et al. 2000).

Kari et al. studied hairpin sets and characterized the complexity and decidability properties of those sets (Kari et al. 2005b). They also extended their study to *scattered* hairpin structures in which the stem allows any number of unbounded regions (Kari et al. 2005a). Furthermore, the notion of DNA trajectory is proposed to describe the bonding between two separate DNA strands, and is used to study *bond-free* properties of DNA code design (Kari et al. 2005c), while the notion is also adopted to model the scattered hairpin structures (Doramatzki 2006).

On the other hand, hairpin molecules find many applications in DNA computing. For example, the use of hairpin formation of single-stranded DNA molecules in solving 3-SAT problems is demonstrated in Sakamoto et al. (2000) (for details, refer to ❷ Sect. 4.5). Also, as previously seen in ❷ Sect. 2.2.2, a molecular computing schema GCS uses the power of hairpin formation in finalizing its computation process (Yokomori 2000), and so forth. These two computation models share a common feature, which follows a general computational schema called "generation and test" strategy and is described as follows: (i) first, prepare the initial random pool of all possible structured molecules, (ii) then extract only target molecules with (or without) hairpin formation from the pool. In fact, this paradigm is formalized and intensively studied under the formal language theoretic terms of "computing by carving" in Păun (1999).

4   Molecular Implementations and Applications

In this section, several wet lab experimental works of molecular computing machineries based on structured DNA molecules are introduced. Those molecular devices involve a variety of ideas in controlling molecular behaviors, such as utilizing specific restriction enzymes, deoxyribozymes, sophisticatedly designed DNA tiles, and so forth.

4.1   DNA Finite Automata

4.1.1   Enzyme-Based DNA Automata

An interesting wet lab experimental work has been reported from Israel's group led by Shapiro, where an autonomous computing machine that comprises DNAs with DNA manipulating enzymes and behaves as a finite automaton was designed and implemented (Benenson et al. 2001). The hardware of the automaton consists of a restriction nuclease and ligase, while the software and input are encoded by double-stranded DNAs.

Encoding DFA onto DNA

Using an example DFA M given in ❷ *Fig. 12a*, it will be demonstrated how to encode state transition rules of M onto DNA molecules.

Before going into the details, however, it is important to observe how a restriction enzyme *Fok*I works on double-stranded molecules to recognize the site and to cut them into two portions, which is illustrated in ❷ *Fig. 12b*. A restriction enzyme *Fok*I has its recognition site $\frac{GGATG}{CCTAC}$ and cuts out a double-stranded DNA molecule as shown in ❷ *Fig. 12b*. That is, splicing by *Fok*I takes place nine DNAs to the right of the recognition site on the upper strand and 13 DNAs to the right of the recognition site on the lower strand.

A transition molecule detects the current state and symbol and determines the next state. It consists of (1) a pair (state, symbol) detector, (2) a *Fok*I recognition site, and (3) a spacer; all these three together determine the location of the *Fok*I cleavage site inside the next symbol encoding, which in turn defines the next state: For example, 3-base-pair spacers maintain the current state, and 5-base-pair spacers transfer s_0 to s_1, etc. Also, a special molecule for detecting the "acceptance" of an input is prepared as shown in ❷ *Fig. 13c*. (Note that *the details of the actual design of molecules for implementing the automaton M is not given here*; rather only the principle of encoding the molecules and computation will be outlined.)

Taking this fact into account, encoding *transition molecules* for the transition rules of DFA is designed as follows. First, there are two points to be mentioned in designing transition molecules: One is that neither a state nor symbol alone is encoded but a pair (state, symbol) is taken into consideration to be encoded, and the other is that a sticky end of 6-base length encoding (state, symbol) is sophisticatedly designed such that the length 4 prefix of the sticky end represents (s_1, x), while the length 4 suffix of the same sticky end represents (s_0, x), where x

☐ **Fig. 12**

(a) DFA *M* with states s_0, s_1 and input symbols *a*, *b*. The initial state is s_0, which is the unique final state as well. *M* accepts the set of strings containing an even number of *b*'s. (b) How a restriction enzyme *Fok*I works on molecules.

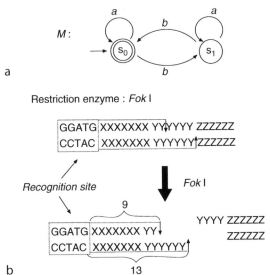

is in $\{a, b, t\}$, and *t* is the special symbol, not in the input alphabet, for the endmarker (see ❯ *Fig. 13a*). Reflecting this design philosophy, the structures of the transition molecules are created as shown in ❯ *Fig. 13b*, where molecules for two transition rules: $s_0 \rightarrow^a s_0$ and $s_0 \rightarrow^b s_1$ are selectively described. The transition molecule $m_{s_0,a}$ for $s_0 \rightarrow^a s_0$ consists of the recognition site of *Fok*I, the center base-pairs of length 3 and the sticky end CCGA, because its complementary sequence GGCT is designed to represent the pair (s_0, a) as seen from the table. A transition $s_0 \rightarrow^b s_1$ is encoded by a similar principle onto the transition molecule $m_{s_0,b}$ shown in ❯ *Fig. 13b*, where a noteworthy difference from $m_{s_0,a}$ is that it has the center base-pairs of length 5. In this principle of designing transition molecules, a transition molecule has the center base-pairs of length 5 iff the transition changes its state (from s_0 to s_1 or vice versa). Thus the distinct length of the center base-pairs plays a crucial role.

Computing Process

The molecular automaton processes the input as shown in ❯ *Fig. 14*, where input *w* is assumed to be of the form $ab \ldots (t)$, where *t* is the endmarker. First, the input *w* is encoded into the form of the molecule shown in ❯ *Fig. 14a*, where seven *X* base-pairs represent a "spacer" consisting of arbitrary bases.

The input molecule is cleaved by *Fok*I, exposing a 4-base sticky end that encodes the pair of the initial state and the first input symbol, that is, (s_0, a) in the running example. The computation process is performed in a cyclic manner: In each cycle, the sticky end of an applicable transition molecule ligates to the sticky end of the (remaining) input molecule (obtained immediately after cleaved by the last transition), detecting the current state and the input symbol (❯ *Fig. 14b*). (Note that it is assumed at the beginning of computation that all

◘ **Fig. 13**

States and symbols encoding.

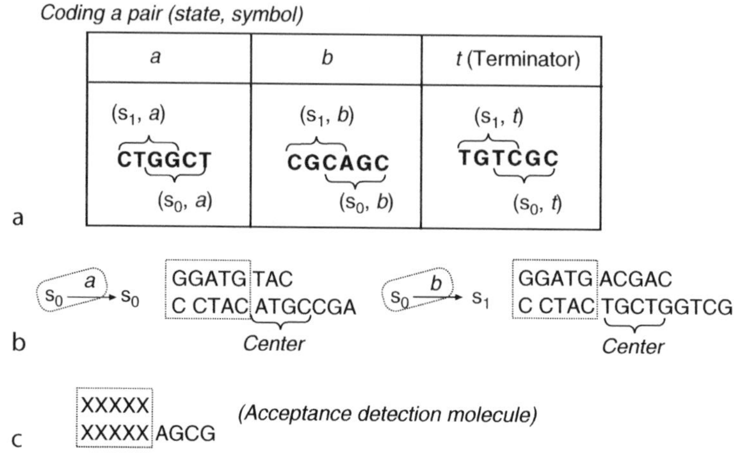

Coding a pair (state, symbol)

	a	b	t (Terminator)
	(s_1, a)	(s_1, b)	(s_1, t)
	C T G G C T	C G C A G C	T G T C G C
	(s_0, a)	(s_0, b)	(s_0, t)

a

b

$s_0 \xrightarrow{a} s_0$

GGATG	TAC
C CTAC	ATGCCGA

Center

$s_0 \xrightarrow{b} s_1$

GGATG	ACGAC
C CTAC	TGCTGGTCG

Center

c

XXXXX	
XXXXX	AGCG

(Acceptance detection molecule)

of the possible transition molecules are contained in a solution pot together with the initial molecule and the acceptance detect molecules.)

At this moment, since the transition molecule for (s_0, a), shown in ❷ *Fig. 13b*, can get access to the remaining input molecule on the right-hand side in ❷ *Fig. 14b*, it produces the hybridized molecule shown in ❷ *Fig. 14c*. Therefore, the resulting molecule is again cleaved by *Fok*I, exposing a new 4-base sticky end (❷ *Fig. 14d*). The design of the transition molecule ensures that the 6-base-pair long encodings of the input symbols a and b are cleaved by *Fok*I at only two different "frames" as seen above, the prefix frame of 4-base-pair length for s_1, and the suffix frame of 4-base-pair length for s_0. Thus, the next restriction site and the next state are exactly determined by the current state and the length of the center base-pairs in an applicable transition molecule. In the example, since the sticky end of the remaining input molecule can match that of the transition molecule $m_{s_0,b}$ for (s_0, b) (❷ *Fig. 14e*) and the molecule $m_{s_0,b}$ exists in the solution pot, so that these two can hybridize to form the double-stranded molecule shown in ❷ *Fig. 14f*.

The computation process proceeds until no transition molecule matches the exposed sticky end of the rest of the input molecule, or until the special terminator symbol is cleaved, forming an output molecule having the sticky end encoding the final state (see ❷ *Fig. 14g*). The latter means that the input string w is accepted by the automaton M.

In summary, the implemented automaton consists of a restriction nuclease and ligase for its hardware, and the software and input are encoded by double-stranded DNA, and programming amounts to choosing the appropriate software molecules. Upon mixing solutions containing these components, the automaton processes the input molecule by a cascade cycle of restriction, hybridization, and ligation, producing a detectable output molecule that encodes the final state of the automaton.

In the implementation, a total amount of 10^{12} automata sharing the same software ("transition molecules") run independently and parallel on inputs in 120 μℓ solution at room temperature at a combined rate of 10^9 transitions per second with a transition fidelity greater than 99.8%, consuming less than 10^{10} W. (For detailed reports of the experimental results, refer to Benenson et al. (2001).)

◘ **Fig. 14**
Computing process of input *w=ab(t)*.

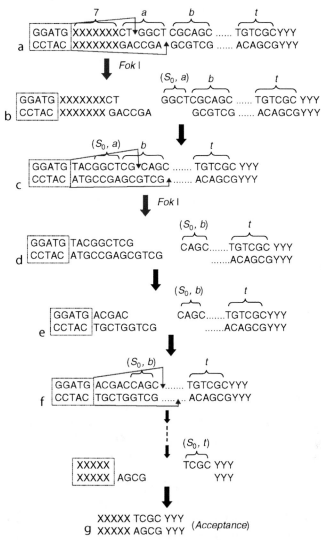

4.1.2 Application to Drug Delivery

As an application of the molecular finite automata, Benenson et al. (2004) proposed an idea to use their molecular automata for carrying out a diagnosis in vivo, and demonstrated an in vitro experiment, which shows the feasibility of logical control of gene expression for drug delivery. An excellent review article on the work by Benenson et al. is given by Condon (2004).

In order to carry out a medical diagnosis, it is necessary to achieve in vivo checking of a medical statement such as "if a certain series of diagnostic conditions are true, then the drug is administered." Such an *if–then* mechanism is critically essential in computing models and in diagnostic process as well. It is assumed that each condition is described in the form: "a certain type of gene expression is regulated by either high or low level of concentrations of particular mRNA (called *indicator molecule*) for the gene."

❯ *Figure 15* outlines the idea used in the diagnostic molecular automaton (*DMA*) experiments by Benenson et al. Suppose that a certain disease involves four conditions of gene expression to be checked before administering a drug for the disease, and each gene-i (for $i = 1, \ldots, 4$) as its label is encoded into a structured molecule (called the *computation molecule*) with a hairpin in which the drug is enclosed (❯ *Fig. 15a*). Note that a diagnostic rule is of the form: if the conditions for gene-1 through gene-4 are all true, then the drug is released. The *DMA* (based on the automaton described in the previous section) can carry out the checking of each diagnostic condition in one transition step, and is constructed in such a way that it has two states "Yes" and "No" and three transitions: Yes → Yes if indicator molecules are present, Yes → No if indicator molecules are absent, and No → No in any other situation. Note that once the transition Yes → No occurs, the state of *DMA* remains unchanged, that is, it keeps staying in No no matter what the situation of the indicator molecule is. Therefore, starting with the initial state Yes, the *DMA* behaves as follows: after each transition, the state is either Yes indicating that the corresponding conditions tested *so far* are all true, or No otherwise (i.e., at least one of the conditions tested is false). Thus, if the *DMA* can successfully make four transitions, then the drug is released.

In ❯ *Fig. 15*, at the initial state (State 0), a hairpin structural molecule (computation molecule) is designed so that the diagnostic condition is the conjunction of four logical statements on gene expression, each of which is true for gene-i if the level of a particular indicator molecule for gene-i is either high or low (depending upon the requirement on each gene expression), for each $i = 1, \ldots, 4$. In order to make a transition from State $(i-1)$ to State i, the *DMA* requires a specific molecule, say a *bridge molecule*, which is supposed to be created from a team of molecules designed for each gene-i ($i = 1, \ldots, 4$) under a certain condition.

Suppose, for example, the *DMA* is in State 3 and is going to effect a transition to State 4, which must diagnose a high level of indicator molecule for gene-4. The team of molecules to do the task comprises *transition molecules* 4A, 4B, and 4A′, where 4B and 4A′ initially form a binding structure, 4A and 4A′ contain a sequence of oligonucleotides and its complementary one, respectively, and annealing the two can provide a bridge molecule containing the recognition site for the restriction enzyme *Fok*I (❯ *Fig. 15b*). In addition to the team molecules, when an indicator molecule for gene-4 is present in the solution, the scenario for the *DMA* to complete a successful transition to State 4 goes as follows. Transition molecule 4B is designed so that it prefers to pair with the indicator molecule for gene-4 rather than transition molecule 4A′ (current partner), because 4B shares a longer complementary region with indicator molecule 4 than 4A′. Therefore, after leaving transition molecule 4A′, transition molecule 4B settles in to pair with the indicator molecule for gene-4. On the other hand, transition molecules 4A and 4A′ get together forming a structure due to their mutual affinity. As a result, this structured molecule can participate as a bridge molecule in the successful transition to gene-4 which is carried out by *Fok*I.

It should be noted that the *DMA* is stochastic in overall behavior, because the transition is strictly dependent upon the level of concentration of indicator molecules in each diagnostic condition, and there remain some problems to be solved before the demonstrated mechanism will work in vivo as well. However, without question, this is the first achievement to prove the great potential of developing medical diagnostic molecular devices for antisense therapies.

◨ **Fig. 15**
DNA automaton for logical control of drug delivery: **(a)** At the initial state (State 0), a hairpin molecule (computation molecule) is designed so that a hairpin may maintain the drug molecule that is guarded by four specific sequences gene-*i* (*i* = 1, . . . , 4) each of which corresponds to one of four conjunctive logical statements in a diagnostic rule for the disease. Each State (*i* − 1) is associated with a team of molecules (given in **b** of this figure for *i* = 4), and in the presence of the indicator molecule for gene-*i*, the team can produce a specific molecule (bridge molecule) which is necessary for cleaving the guarded gene-*i*, resulting in a successful transition to State *i*.
(b) A team of molecules needed for a transition from State 3 to State 4 comprises transition molecules 4A, 4B, and 4A′. In the presence of the indicator molecule for gene-4, members in the team can be relocated into the new formation in which a bridge molecule (involving the recognition site of *Fok*I) is created to contribute to a transition to State 4.

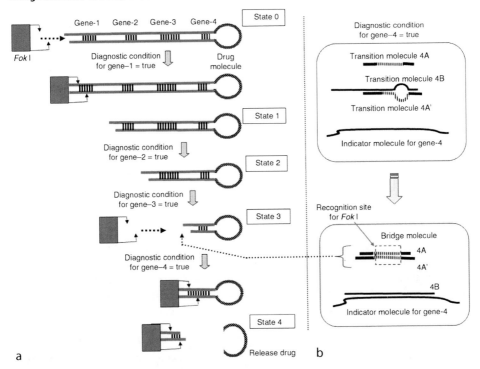

a   b

4.2   Molecular Gates and Circuits

4.2.1   DNAzymes

Stojanovic et al. established the use of DNAzymes to implement logic gates by DNA (Stojanovic et al. 2002). Just as a ribozyme refers to an enzyme made of RNA, a DNAzyme (deoxyribozyme) is a DNA molecule that has some catalytic function. ❷ *Figure 16* shows the DNAzyme named E6. The strand forming a loop structure is the DNAzyme that partially hybridizes with the target strand and cleaves it at the site shown by the arrow. The base denoted as rA is a single ribonucleotide (RNA nucleotide) inserted in the target strand. The enzyme cuts the phosphodiester bond at this site.

◻ **Fig. 16**

DNAzyme named E6. The strand containing a ribonucleotide denoted as rA is cleaved at the site shown by the arrow.

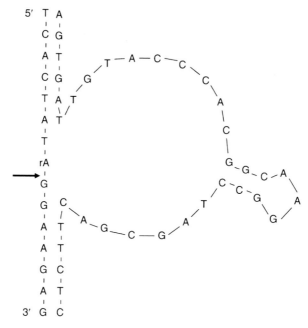

After the target strand is cleaved, the partially double-stranded structure becomes unstable and the two pieces that arise from the target strand come apart. They can be regarded as outputs of the logic gate. If a fluorescent group and a quencher group are attached at the terminals of the target strand, an increase of fluorescence is observed when the strand is cleaved.

They introduced various modifications into the DNAzyme in order to control its catalytic function. Existence or non-existence of some strands affects the structure of such a modification and turns the catalytic function on or off. They can be regarded as inputs to the logic gate implemented by the modified DNAzyme.

❯ *Figure 17a* shows a schematic figure of the logic gate $A \wedge \neg B$. The modified enzyme implementing this gate has two more loops in addition to the loop of the original DNAzyme. It has the catalytic function if the loop denoted by $\overline{I_A}$ is opened while the loop denoted by $\overline{I_B}$ is intact. If the strand I_A which is complementary to the loop $\overline{I_A}$ is present, the loop is opened and the DNAzyme gains its catalytic function (❯ *Figure 17b*). On the other hand, if the strand I_B is present, the loop $\overline{I_B}$ is opened and the catalytic function is lost because the structure of the original loop is changed (❯ *Figure 17c*). Simply mixing this gate and the similar gate $B \wedge \neg A$ together results in the disjunction $(A \wedge \neg B) \vee (B \wedge \neg A)$, which is the XOR function of A and B.

Using logic gates implemented DNAzymes, Stojanovic and Stefanovic programmed a strategy to play the game of tic-tac-toe for demonstrating the expressiveness of DNAzyme-based logic circuits (Stojanovic and Stefanovic 2003).

A move of a human player is represented by one of nine single strands of DNA, each corresponding to one of the nine positions on the board of tic-tac-toe. A logic circuit made of

◻ **Fig. 17**
The schematic figure of the gate $A \wedge \neg B$. The DNAzyme is functional only if A is present and B is not.

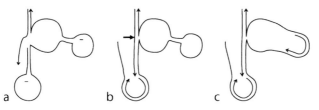

a   b   c

◻ **Fig. 18**
A move in tic-tac-toe. The next move of the DNA side is at the upper right position.

DNAzymes is placed at each position. For each move of the human player, the corresponding single strand is given to all the circuits. The programmed strategy shows the next move by making the circuit at the position of the move return output 1. More concretely, fluorescence emitted by the cleaved strand is increased at the position of the next move.

In ❷ *Fig. 18*, two strands corresponding to the two moves of the human player (denoted by O) are put to all the nine positions. Then the circuit at the upper right position is triggered and gives output 1, or fluorescence at the position is increased.

In this work by Stojanovic and Stefanovic, all logic circuits at the nine positions are disjunctive normal forms and implemented by simply mixing the gates of all disjuncts. Each disjunct is a conjunction consisting of at most three literals. Each conjunction with three literals is of the form $A \wedge B \wedge \neg C$. This kind of conjunction can be implemented by a modified DNAzyme shown in ❷ *Fig. 19*, which has three additional loops.

DNAzymes or ribozymes are used to implement logic circuits in many applications other than that of Stojanovic and Stefanovic as mentioned above. For example, Win and Smolke use a hammerhead ribozyme to control translation from mRNA to protein (Win and Smolke 2008).

They insert a hammerhead ribozyme with some additional structures into an untranslated region of a gene. As in the upper panel of ❷ *Fig. 20*, the hammerhead ribozyme cleaves itself at the site shown by the arrow. The lower panel of the same figure shows additional structures attached at the loops of the ribozyme. These structures contain sensors and logic gates. Sensors are implemented by so-called aptamers, which are RNA molecules that change their structure if they bind to a certain small molecule. Logic gates are implemented by loop structures similar to those in Stojanovic's logic gates.

After the gene is transcribed, the hammerhead ribozyme in the mRNA transcript becomes active under some specific conditions depending on the additional structures, and cleaves the transcript itself. As a result, the rate of translation from the transcript is decreased. In this way,

◻ Fig. 19
A schematic figure of a three-literal conjunction. A conjunction of the form $A \wedge B \wedge \neg C$ can be implemented in this way.

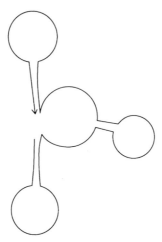

◻ Fig. 20
A hammerhead ribozyme. The upper panel shows the original structure of the ribozyme, which cleaves itself at the site shown by the arrow. The lower panel shows the additional structures containing sensors and logic gates.

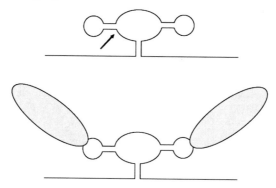

it is possible to control the expression of the gene by a combination of conditions that can be detected by the sensors.

4.2.2 Strand Displacements

As shown in the previous section, many bio-molecular machineries and logic gates have been proposed. Such molecular systems sophisticatedly utilized various restriction enzymes or DNAzymes (RNAzymes) to control their state transitions. However, enzyme-based logic gates have a disadvantage in that it is difficult to cascade from one gate to another one because of design restrictions.

In this section, logic gates without enzymes (or enzyme-free logic gates) are introduced, which were developed by Seelig et al. (2006b). The driving force of Seelig's logic gates is a branch migration rather than enzymatic reactions. A branch migration, which is also called "strand displacement" or "oligonucleotide replacement," is the process in which a single strand hybridizes to a dangling end of duplex and then removes the resident strand from the duplex, resulting in the formation of a new duplex (see ❷ *Fig. 21*). The dangling end used to promote a branch migration is called a toehold, the length of which dominantly affects the kinetics of the reaction (Yurke and Mills 2003). The process of branch migration is regarded as a random walk (❷ *Fig. 21a–c*); the branching point moves left or right when two strands competitively hybridize to their complementary strand. If the resident strand wins the competition, the state comes back to ❷ *Fig. 21a*. In contrast, once the resident strand is removed from the duplex, the reaction stops because the new duplex does not have a toehold region (❷ *Fig. 21d*).

By using this branch migration reaction, various logic gates were successfully constructed (❷ *Figs. 22* and ❷ *23*). The reaction pathways of these gates are as follows:

- **AND gate** – First, by the branch migration, the input1 hybridizes to the toehold of the output complex (❷ *Fig. 22b*) and removes one single strand from the complex, resulting in

■ **Fig. 21**
Branch migration.

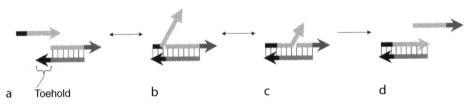

■ **Fig. 22**
AND gate.

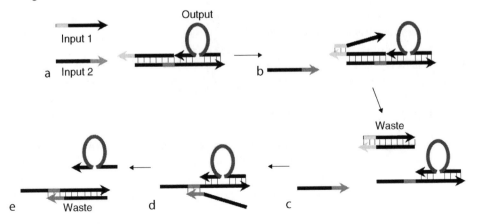

□ Fig. 23
Various gates.

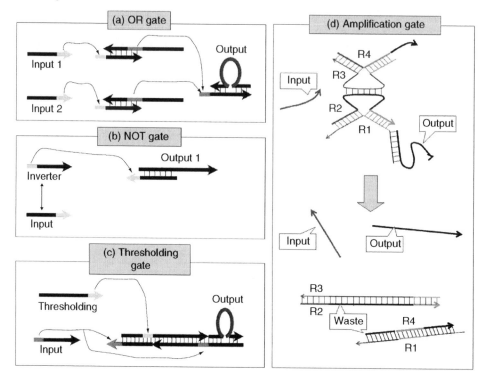

the exposed toehold region for input2 (❷ *Fig. 22c*). Then, input2 removes the second single strand from the complex (❷ *Fig. 22d, e*), producing a single output strand. Thus, only if both input strands exist, is an output released as a single strand.

- **OR gate** – After either input1 or input2 hybridizes to the corresponding intermediate gate, either of their output strands produces the final output strand (❷ *Fig. 23a*). Therefore, the output strand is produced if either input1 or input2 exists.

- **NOT gate** – This gate consists of the output complex and the inverter strand that is complementary to the input strand (❷ *Fig. 23b*). The inverter strand hybridizes to the toehold of the complex and produces the single output strand. In contrast, if the input exists, it hybridizes to the inverter and prevents the branch migration between the inverter and output complex. Thus, this works as a NOT gate.

- **Thresholding gate** – This gate, which consists of a "thresholding" strand and output complex, can be used to remove the extra single strand that was produced unexpectedly (❷ *Fig. 23c*). The output complex has three toeholds. The first and third toeholds are complementary to the subsequence of input, while the second toehold corresponds to the thresholding strand. Thus, the output strand is released only if two inputs and one thresholding strand complete their branch migrations. Due to this stoichiometry, if input concentrations are twice as high as the thresholding strand, most gates will produce the output (see the supporting online material of Seelig et al. (2006b) for the theoretical consideration). Therefore, the threshold value can be controlled by the concentration of the thresholding strand.

- **Amplification gate** – Although the previous thresholding gate can remove the extra single strand, the concentration of the output does not exceed half of the input concentration. To restore the signal, this amplification gate is used (❷ *Fig. 23d*). This gate produces input and output as a single strand after the multiple-stage branch migrations. Note that the input strand can react again with the other amplification gate. Thus, as long as amplification gates remain, input strands produce the output. Therefore, by combining the thresholding and amplification gates, one can exclude incorrect signals without deleting correct ones wrongly. For the detailed reaction pathway, please refer to Seelig et al. (2006a).

These enzyme-free logic gates have a great advantage in the cascading of gates. Because the output of each gate is a single strand, it can be easily used as the input of the other gates.

Here, readers may wonder if the above logic gates must be redesigned and reconstructed when one selects the other sequences as inputs. This problem can be solved with the following converter, which converts an input strand to another output strand (❷ *Fig. 24*). By two-step branch migration, the input strand is converted into the output strand, whose sequence is not relevant to the input sequence. Thus, logic gates shown above can accept any input sequences by combining appropriate converters unless the sequences unfortunately cross-hybridize with each other.

With these logic gates shown above, one can construct the more complex logic circuits. In fact, Seelig et al. achieved the logic circuit including 11 gates and six inputs (Seelig et al. 2006b).

Based on Seelig's logic gates, Tanaka et al. developed a DNA comparator, which is a DNA machine for comparing DNA concentrations (Tanaka et al. 2009). The purpose of the DNA comparator is to determine whether the concentration of the target strand is higher than that of the reference strand. If the concentration of the target is higher, the DNA comparator outputs a single-strand, which may be an input strand to another molecular logic gate.

A DNA comparator consists of three double strands and two inputs (see ❷ *Fig. 25*). One of the two inputs is a target strand whose concentration one wants to compare, while the other is a reference strand whose concentration is the standard. Two of three double strands have bulge loops in the middle of the sequences and toehold structures, whose two inputs can hybridize to, at the 5′-end of the opposite sequences. The DNA comparator was designed to work as follows: First, two inputs *reference* and *target* hybridize to the toeholds of *cp_B1* and *cp_B2*, respectively. Then, *reference* and *cp_B1* form a double strand, while *target* and *cp_B2* also form a double strand by the branch migration reaction. After that, *B1* and *B2* are released as single strands and immediately form a double strand because the loop domains of *B1* and *B2* are complementary to each other. Thus, only the extra amount of *B2* can hybridize to the toehold of *cp_out*, resulting in branch migration. As a result, *output* is released as a single

❏ **Fig. 24**
Converter based on branch migration.

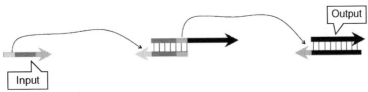

◫ **Fig. 25**
DNA comparator.

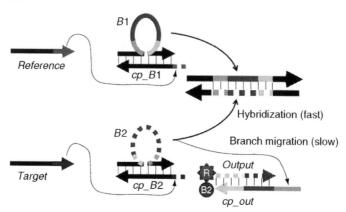

strand, which can be monitored by the fluorescent intensity of a fluophor. The key point is that the hybridization between *B1* and *B2* is much faster than the branch migration between *B2* and *output&cp_out*. Thus, if the concentration of *B2* is lower than that of *B1*, almost all *B2* will probably hybridize to *B1* rather than the toehold of *cp_out*. Therefore, only if the concentration of *target* is higher than that of *reference* can *B2* hybridize to the toehold of *cp_out* and remove *output* from *cp_out* by branch migration.

4.2.3 DNA Tiles

Self-assembly of tiles has a powerful computational ability as shown by Wang (1961, 1962), and it even has universal computability, because it can simulate cellular automata and therefore Turing machines. More precisely, Wang showed that it is undecidable whether a given set of tiles can completely fill out the entire plane. Winfree reformulated this result in the context of DNA tiles (Winfree 1998). In this section, the focus is on the use of DNA tiles for implementing logic gates.

❷ *Figure 26* shows tiles for computing XOR and constructing Sierpinski's triangle. The four tiles in the top row implement the XOR function. The two edges facing down accept two inputs to XOR and the other two edges facing up give the (same) output of XOR. The three tiles in the bottom row form the boundary of Sierpinski's triangle.

In the mathematical model of tile assembly known as aTAM (abstract tile assembly model), proposed by Winfree, tiles self-assemble under the condition that only edges with the same label can bind together. Moreover, each label has its own binding strength. A tile can bind to a tile complex (aggregate) only if the sum of the strengths of the binding edges is greater than or equal to the predefined threshold called the temperature.

In this example, labels 0 and 1 have binding strength 1, while label 2 has binding strength 2. The temperature is set to 2. Therefore, as in ❷ *Fig. 27*, beginning with the corner tile (the middle tile in the bottom row of ❷ *Fig. 26*), the boundary of Sierpinski's triangle is formed. The tiles for XOR then bind to the boundary and construct Sierpinski's triangle as in ❷ *Fig. 28*.

◘ **Fig. 26**

Tiles for Sierpinski's triangle. The upper tiles compute the XOR function. The lower tiles form the boundary of Sierpinski's triangle.

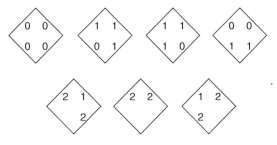

◘ **Fig. 27**

The boundary of Sierpinski's triangle.

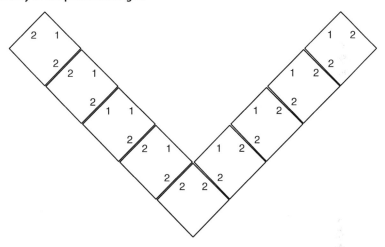

❷ *Figure 29* shows an actual implementation of tiles by DNA (Rothemund et al. 2004). This structure, called a double crossover molecule, consists of four single strands of DNA. In other words, it consists of two double strands that exchange their single strands at two positions. This rectangular structure has four sticky ends, through which one instance of the structure can hybridize with four other instances.

Consequently, it works as a tile with four edges as mentioned above, and as in ❷ *Fig. 30*, the self-assembly of DNA tiles forms a planar structure. Sticky ends of a DNA tile correspond to edges of an abstract tile, and different sequences of sticky ends correspond to different edge labels. Their length roughly corresponds to their binding strength.

More complex computations can be performed by tiling (Mao 2000). For example, ❷ *Fig. 31* shows the addition of two binary numbers. The full-adder tile in the upper panel receives the corresponding bits of two numbers and the carry bit through its upper, lower, and right edges. It propagates its carry bit through its left edge. Addition of two three-bit numbers is performed as in the lower panel of the figure. The full-adder tile can be implemented by a triple crossover molecule, which has six sticky ends.

❑ Fig. 28
Sierpinski's triangle.

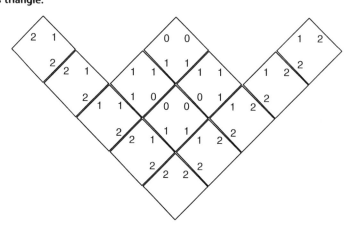

❑ Fig. 29
A double crossover molecule, also called a DNA tile. It consists of four single strands of DNA and forms a rectangular planar structure.

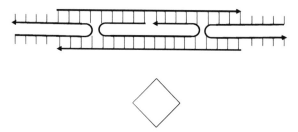

4.3 Reaction Graphs

Many bio-molecular machineries and logic gates have been constructed successfully (Seelig et al. 2006a, b). As mentioned in ❷ Sect. 4.2.2, DNA logic circuits without enzymes have attracted attention recently (Stojanovic et al. 2002; Stojanovic and Stefanovic 2003). However, such logic circuits are sometimes complicated because multiple sequences react with each other in various ways such as assembly, disassembly, and branch migration. Furthermore, sequences often consist of many domains, where some domains may be toeholds, and others may be loops of hairpin structures. This complexity leads to difficulties in understanding or designing the reaction dynamics of logic circuits.

To simply describe the self-assembly pathways, Peng Yin and his colleagues proposed the abstract model named "reaction graph," which mainly focuses on circuits consisting of versatile DNA hairpin motifs (Yin et al. 2008a). With the reaction graph, the reaction dynamics can be simply described by nodes including input/output ports and arrows between nodes.

A simple example is shown in ❷ *Fig. 32*. ❷ *Figure 32a* shows the reaction dynamics. First, the sequence *I* hybridizes to the toehold region of hairpin *A* and then opens the hairpin by the branch-migration reaction (❷ *Fig. 32a* (1)). Second, the complex structure *A ·I* hybridizes to

◘ **Fig. 30**

Self-assembly of DNA tiles. Through sticky ends, a DNA tile can hybridize with four other DNA tiles.

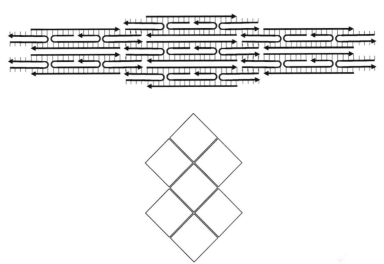

◘ **Fig. 31**

Addition by tiles. The upper panel shows a full-adder tile, and the lower panel shows addition of two three-bit numbers.

	x	
c-out	z	c-in
	y	

	x2		x1		x0	
	x2		x1		x0	
c-out	z2		z1		z0	0
	y2		y1		y0	
	y2		y1		y0	

the toehold region of B and then opens the hairpin (❷ *Fig. 32a* (2)). Third, the sequence B removes the sequence I from A by branch migration (❷ *Fig. 32a* (3)). Because the removed I is a single strand, it can hybridize to the toehold of hairpin A again. Thus, the sequence I works as a catalyst to promote the hybridization between sequences A and B. (Note that without I, sequences A and B cannot hybridize to each other, because the complementary regions are blocked by their hairpin structures.)

This reaction explained above can be represented by the reaction graph shown in ❷ *Fig. 32c*. Each sequence is represented by a node (see ❷ *Fig. 32b*). Each node has some ports (input/output ports), which correspond to domains in each sequence. Input ports correspond to the domains which can receive the upstream outputs and are represented by triangles, while output ports are the domains which can be inputs for downstream hairpins and are

■ **Fig. 32**

Simple example from Yin et al. (2008a) (slightly changed).

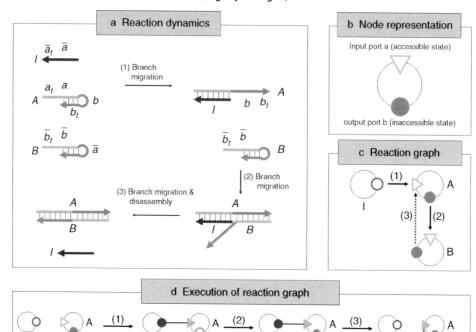

represented by circles. Input and output ports have two states, accessible or inaccessible. If a domain is a single-stranded region, the corresponding port is accessible (represented by open triangles/circles) because the single-stranded region can hybridize with other sequences. In contrast, if a domain is a double-stranded region, the corresponding port is inaccessible (represented by filled triangles/circles). A solid arrow from an output port to an input port represents that the output port can hybridize to the input port if both ports are accessible. Furthermore, a dashed arrow from an output port to an input port represents that the output port can hybridize to the input port and remove the hybridized sequence from the input port. For example, the solid arrow denoted by (1) in ❷ *Fig. 32c* represents that the sequence *I* can hybridize to the toehold domain of *A*, while the dashed arrow denoted by (3) represents that *B* can hybridize to *A* and remove *I* from *A*.

   ❷ *Figure 32d* is the transition of the reaction graph shown in ❷ *Fig. 32c*. The thin lines represent the hydrogen bonds between input and output ports. As readers can check easily, the reaction graph correctly represents the reaction dynamics: When *I* hybridizes to *A*, the output port of *I* and the input port of *A* are flipped to inaccessible states and connected by a thin line (❷ *Fig. 32d* (1)). Then, the output port of *A* is flipped to accessible state and hybridizes to the input port of *B*. Both ports are flipped to inaccessible states and connected by a thin line, while the output port of *B* is flipped to accessible (❷ *Fig. 32d* (2)). The output port of *B* removes *I* from *A* and hybridizes to the input port of *A*, when both ports are flipped to inaccessible and connected by a thin line. Finally, the system outputs *I* with the accessible output port and the *A* ·*B* complex with inaccessible input/output ports (❷ *Fig. 32d* (3)).

❷ *Figure 33* shows a more complicated example from Yin et al. (2008a). This example may be somewhat confusing. Complex *A* ·*B* is produced by the hybridization between *A* and *B* promoted by *I*. Furthermore, *A* ·*B* promotes the hybridization between *C* and *D*, while *C* ·*D* does the hybridization between *A* and *B* simultaneously. Thus, the reaction dynamics are complicated and cannot be easily understood. With the reaction graph, however, the system can be simplified as shown in ❷ *Fig. 33b*:

- Sequence *I* hybridizes to the input port of *A* and opens the hairpin, resulting in the accessible output port of *A* (1).
- Then, the output port of *A* hybridizes to the input port of *B* (2a), when one output port of *B*, denoted by a black circle, is flipped to accessible, removes *I* from *A*, and hybridizes to the input port of *A* (2b).
- At the same time, the other output port of *B*, denoted by a gray circle, is flipped to accessible. This accessible output port can hybridize to the input port of *C*, resulting in the accessible output port of *C* (3).

◻ **Fig. 33**
More complicated example from Yin et al. (2008a) (slightly changed).

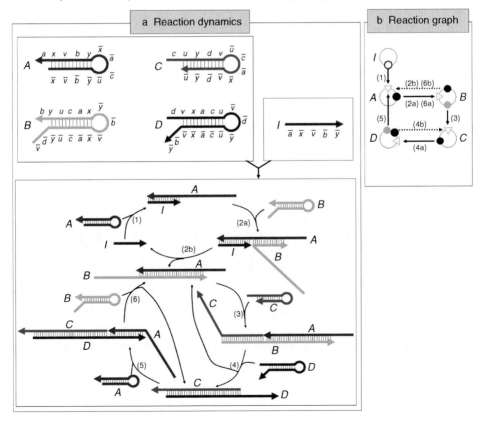

- The accessible output port of C hybridizes to the input port of D, when two output ports of D are flipped to accessible (4a).
- The accessible output port of D, denoted by a black circle, removes the complex $A \cdot B$ from C and hybridizes to the input port of C (4b), while the gray output port of D hybridizes to the input port of A (5).
- Then, the accessible output port of A hybridizes to the input port of B (6a), when the accessible black output port of B removes the complex $C \cdot D$ from A (6b).
- As a result, the system promotes the formation of complexes $A \cdot B$ and $C \cdot D$.

Readers can find some other examples of reaction graphs in Yin et al. (2008a), including three- or four-arm junctions, dendritic growth structures, and DNA walkers with two legs.

4.4 Whiplash Machines

4.4.1 Single-Molecule DNA State Machine

A distinctive DNA-based computing machine, called the *Whiplash machine*, has been continuously studied by a Japanese research group; it is a finite state machine implemented by encoding with a single DNA molecule. The parallel execution of multiple programs with different inputs in one reaction pot is realized by *Whiplash PCR* (WPCR), which is the recursive, self-directed polymerase extension of single-stranded DNA (ssDNA), proposed by Hagiya et al. (2000) and Sakamoto et al. (1999).

DNA Implementation of a Finite State Machine

The Whiplash machine utilizes the facts: a single-stranded DNA can form a *hairpin structure* by autonomously hybridizing the complementary sequences, and polymerase extension can generate a new DNA sequence as a reading out operation. Note that the rate of the computing process of the Whiplash machine is independent of the DNA concentration since hairpin formation is the intramolecular reaction.

Suppose that a finite state machine M having transition rules for the three-state, two-step transitions, $a \rightarrow b \rightarrow c$. Then, the DNA implementation of M is carried out as follows.

(1) *Designing transition rules*: WPCR involves a sophisticated coding design of a ssDNA sequence (❯ *Fig. 34a*). The set of transition rules are encoded in the form of catenated *rule blocks*, each encoding a state transition from state x to state y by an adjacent pair of sequences, $5'\text{-}\bar{y}\,\bar{x}\text{-}3'$. Here, $\bar{x}$ denotes the complementary sequence of x. A rule block also contains at its $5'$ side the *stopper*, which is commonly implemented by a short DNA sequence, AAA. Specifically, states a, b, c are encoded by sequences consisting of C, G, and T. The *transition rule region* contains rule blocks, $a \rightarrow b$ and $b \rightarrow c$. The $3'$ most sequence forms the *head*, which represents the Whiplash machine's current state.

(2) *Carrying out state transition*: By annealing, the complementary sequences a and $\bar{a}$ come to hybridize, then a polymerase extension of the head takes place and the extension will be terminated automatically when the polymerase encounters the stopper, where the polymerization buffer is assumed to miss T (deoxythymidine triphosphate). The first transition, $a \rightarrow b$ is completed (❯ *Fig. 34b* (the upper panel)). Then, by denaturation, the DNA again becomes the single-stranded form (❯ *Fig. 34b* (the lower panel)). The $3'$ end of the ssDNA retains the current state b, renewed by reading out the complementary sequence $\bar{b}$ in the rule block $a \rightarrow b$.

■ **Fig. 34**

Computing process of finite state machine by Whiplash PCR.

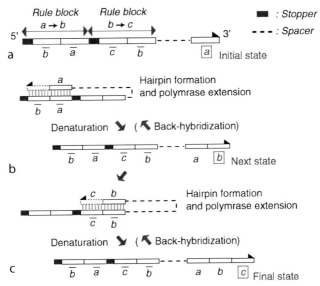

(3) *Successive transitions*: The process mentioned above can be repeated to perform further simulation of the state transition of *M*. After denaturation, the head representing the current state *b* is now ready for performing another simulation of the next transition, $b \to c$ (❷ *Fig. 34c*).

Eight successive state transitions were so far achieved in the wet experiment (Komiya et al. 2006). As each DNA may be encoded with a distinct set of transition rules and a distinct initial state, parallel computation is achievable by the production of a combinatorial mixture of ssDNA. What is called the "multiple-instruction, multiple-data (MIMD)" computing paradigm is implemented by biomolecules. A solution to a 6-city instance of the directed Hamiltonian path problem using the Whiplash machine was demonstrated by Komiya et al. (2006). In addition, the WPCR model equipped with input/output interface was proposed in Komiya et al. (2001). Note that, in WPCR, a record of computation is encoded at the ssDNA's 3′ end as a string of catenated state sequences. In ❷ *Fig. 34c*, this corresponds to the string, 5′-*abc*-3′. Implementation of evolutionary computation and its application to a novel protein evolution technique, taking advantage of the *computation record region*, were proposed (Rose et al. 2002b, 2003; Wood et al. 2002).

Back-Hybridization

Although the Whiplash machine is highly versatile in principle, it is known to suffer from an efficiency problem due to its systematic tendency towards self-inhibition, a problem called *back-hybridization*. Consider again the two-step transitions, shown in ❷ *Fig. 34*. The second polymerase extension requires the hybridization of the head sequence *b* with the sequence $\bar{b}$ of rule block $b \to c$ (❷ *Fig. 34c*). The efficiency of this process, however, is compromised by the ability of the ssDNA to form the alternative hairpin shown in ❷ *Fig. 34b*, which is no longer extendable and energetically more favorable than the planned, extendable hairpin. For multiple rounds of WPCR implementation, the number of distinct back-hybridized configurations increases as the computation continues. Back-hybridization is a serious obstacle to practical application of the Whiplash machine (Rose et al. 2002a). In the next section, an optimally reengineered architecture of the Whiplash machine to overcome this efficiency problem is described.

4.4.2 Efficient and Controllable Whiplash Machine

An extended architecture of the WPCR, *displacement whiplash PCR* (DWPCR), was proposed by Rose et al. (2006). In DWPCR, the new *rule protect* operation, which is the primer-directed conversion of each implemented rule-block to double-stranded DNA, accompanied by opening the back-hybridized hairpins by strand displacement, is introduced. DWPCR implements the same series of operations applied in standard WPCR, adding only a protection step, where primer annealing, extension, and strand displacement occur at the targeted rule block, following each round of transition. A rule block contains an additional sequence at its 3′ side for primer binding, represented by a black rectangle in ❷ *Fig. 35*.

DWPCR supports isothermal operation at physiological temperatures and is expected to attain a near-ideal efficiency by abolishing back-hybridized hairpins, and thus open the door for potential biological applications. Note that the use of rule-protect for controlling computation is being investigated (Komiya et al. 2009). Operation timing can be controlled

❑ **Fig. 35**

Computing process of displacement Whiplash PCR.

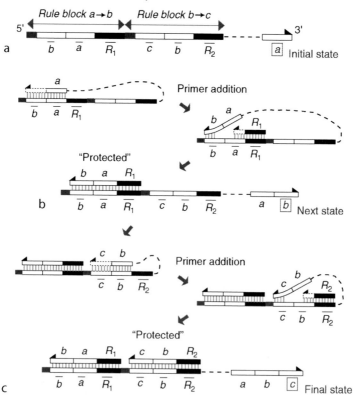

by primer addition as an external signal for regulation. Furthermore, as transition rules may also be deactivated via rule-protect prior to their implementation, the transition rules to be implemented can be switched dependent on primer addition. Consequently, the extended Whiplash machine performs successive state transitions, following both a computational program encoded with an ssDNA molecule and a regulatory program implemented by a series of additions of primers, as a signal-dependent self-directed operation.

4.5 SAT Engines

The satisfiability problem (SAT, in short) is one of the well-known NP-complete problems where, given a Boolean formula, one has to decide whether or not the formula is satisfiable (i. e., there exists a truth value assignment under which the truth value of the formula is "true").

As its name suggests, the *SAT Engine* (Sakamoto et al. 2000) has been developed for solving the SAT problem, in which the mechanism of hairpin formation is successfully employed to detect and remove many "inconsistent assignments" to a given Boolean formula.

Using an example, the idea of the SAT algorithm via DNA molecules is outlined. Consider a Boolean formula F with three variables, consisting of five clauses:

$$F = C_1 \cdot C_2 \cdots C_5$$
$$= (a \lor b \lor c) \land (a \lor \neg b \lor c) \land (\neg a \lor b \lor c) \land (\neg a \lor b \lor \neg c) \land (\neg a \lor \neg b \lor \neg c)$$

where a variable x and its negation $\neg x$ are called *literal*. A truth value assignment is a mapping, which for each variable assigns 1 or 0 (true or false). A formula F is *satisfiable* iff there exists a truth value assignment for which the value of F is true. In the running example, since there is a truth value assignment $(a, b, c) = (1, 1, 0)$ which leads to $F = 1$, F is satisfiable. If each clause C_i in F contains at most k literals, then F is called an instance of the k-SAT problem.

4.5.1 DNA SAT Algorithm

First, let one consider a string $u_F = t_1 t_2 \ldots t_5$ over the alphabet $\Gamma = \{a, \neg a, b, \neg b, c, \neg c\}$, where each t_i is an arbitrary literal chosen from C_i, for each $i = 1, 2, \ldots, 5$. A string u_F is called a *literal string* of F. For example, $aab \neg c \neg c$ and $acbb \neg c$ are literal strings of F. Each literal t_i chosen from C_i can be regarded as the value for which $C_i = 1$. (In the example, a literal string $aabb \neg c$ is interpreted as $(a, b, c) = (1, 1, 0)$ and it makes F consistently true, while a literal string $acbb \neg c$ is inconsistent with a variable c, that is, it contains both c and $\neg c$, failing to make F true.) A literal string of F is satisfiable if it can make F true, and it is unsatisfiable otherwise.

Now, the SAT algorithm comprises the following three steps;

(Step 1): For a formula F with n clauses, make the set $LS(F)$ of all literal strings of F whose length are all n. (In case of the 3-SAT problem, $LS(F)$ consists of at most 3^n literal strings.)

(Step 2): Remove from $LS(F)$ any literal string that is unsatisfiable.

(Step 3): After removing in Step 2, if there is any literal string remaining, then that is a satisfiable one, so that the formula F turns out to be satisfiable. Otherwise, F is not.

These can be implemented using biomolecular experimental techniques in the following manner:

(Step 1): By ligating each DNA molecule representing each literal, make all elements of the set $LS(F)$ in parallel.

◼ **Fig. 36**

Encoding a formula onto ssDNA and hairpin formation.

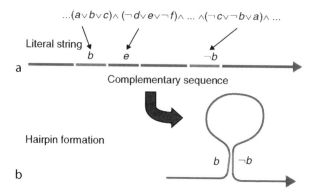

◼ **Fig. 37**

SAT algorithmic schema.

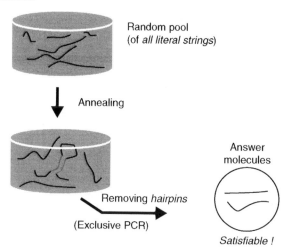

(Step 2): A variable x and its negation $\neg x$ are encoded so that these two are just complementary. Therefore, a literal string containing x and $\neg x$ is encoded into a molecule containing two subsequences which are complementary. (See ❯ *Fig. 36a*.) Thus, an encoded molecule for an unsatisfiable literal string can form a *hairpin structure*. (See ❯ *Fig. 36b*.)

(Step 3): Remove all DNA molecules that form hairpin structures. If there remains any molecule, then detect it and read the sequence for decoding the variable assignment. (See ❯ *Fig. 37*)

4.5.2 Implementation and Experiments

A wet lab experiment was carried out by Sakamoto et al. (2000) where the following Boolean formula was considered

$$F = (a \vee b \vee \neg c) \wedge (a \vee c \vee d) \wedge (a \vee \neg c \vee \neg d)$$
$$\wedge (\neg a \vee \neg c \vee d) \wedge (a \vee \neg c \vee e) \wedge (a \vee d \vee \neg f)$$
$$\wedge (\neg a \vee c \vee d) \wedge (a \vee c \vee \neg d) \wedge (\neg a \vee \neg c \vee \neg d) \wedge (\neg a \vee c \vee \neg d)$$

The total number of literal strings of F is $3^{10}(= 59,049)$ among which only 24 literal strings are satisfiable. (In fact, these 24 literal strings are for the identical truth value assignment.)

In their experiment, each literal was conceptually encoded onto a single-stranded DNA molecule (called *literal DNA*) of length 30 as follows:

a: TTGGTTGATAGACCCAGGATCGAGTGCCAT

b: TTGGCATAAGTTGCCAGGCAGGAACTCCAT

c: TTGGAACGTAGTACCAGGAGTCTCCTCCAT

d: TTGGTGATACGGACCAGGCCTTCTTACCAT

e: TTGGTTGCCCTCTCCAGGTGACTAATCCAT

f: TTGGACTGCTCATCCAGGACTGAAGACCAT,

while the negative counterpart $\neg x$ of x (x in $\Gamma = \{ a, b, c, d, e, f \}$) was encoded as the complementary sequence of x.

(Step 1) was actually carried out by concatenating with ligase all literal DNAs for x and $\neg x$ ($x \in \Gamma$), where in an actual implementation, each literal DNA was designed as a dsDNA with sticky ends at either side, playing the role of a linker for concatenation. (This is because ligase used for concatenation does not act on ssDNAs but on dsDNAs.) Another purpose of a linker is to ensure that the resultant of concatenating literal DNAs contains exactly one literal from each clause. A final pot of a random pool consisting of all molecules for $LS(F)$ was created autonomously. The experimental result of (Step 1) is reported to have been successful.

In (Step 2), each dsDNA in the pot was first heated to separate into two ssDNAs, one of which was washed out. Then, the remaining ssDNAs are now ready to form *hairpin structures* if they are DNAs encoding unsatisfiable assignments. In other words, if there exist ssDNAs that do not form hairpin structures, then it means that there exists truth value assignments for a given F (i.e., the formula F is satisfiable). Note that each literal DNA contains a recognition site of a restriction enzyme *Bst*NI, and this hidden sequence will play an important role, as seen below.

In (Step 3), by using a restriction enzyme *Bst*NI, the hairpin structure formed in (Step 2) is cut out, so that all DNAs containing hairpin structures lose a chance to survive until the next procedure in which the application of PCR will multiply the remaining DNAs for sequencing them.

In practice, 6 literal strings were finally obtained which include

$$bc\neg d\neg ae\neg f\neg ac\neg a\neg a \text{ and } bc\neg d\neg ae\neg f\neg a\neg d\neg a\neg a$$

It is reported that all those literal strings were for the unique assignment:

$$(a, b, c, d, e, f) = (0, 1, 1, 0, 1, 0)$$

which provides, in fact, a correct solution to the example formula.

4.6 Bibliographic Notes

Research on computing Boolean circuits using DNA is well established and dates back to Ogihara and Ray (1999) and Amos et al. (1998). Their methods, however, require that for each logic gate an external experimental operation is applied to a test tube. On the other hand, this section focused on molecular computing machineries that can compute a series of logic gates in a self-controlled way without human intervention in the sense of autonomous molecular computers advocated by Hagiya (1999).

As DNA automata are implemented by enzymes, logic gates can also be implemented by enzymes. In particular, an autonomous molecular computer Reverse-transcription-and-TRanscription-based Autonomous Computing System (RTRACS) proposed and implemented in Nitta and Suyama (2004) and Takinoue et al. (2008) has some unique features. Modeled after the retroviral replication mechanism, RTRACS utilizes not only DNA/RNA molecules but also reverse transcriptase and other enzymatic substrates, and the computation is autonomously carried out under an isothermal environment. An in vitro experiment of simulating an AND gate in Takinoue et al. (2008) demonstrates a potentially high computational capability of the integrated circuits of the modularized basic units of RTRACS.

More and more efforts are being made to implement in vivo logic circuits as proposed by Benenson et al. in ❯ Sect. 4.1.2. Smolke's RNA computer mentioned in ❯ Sect. 4.2.1 is one such effort. Benenson also proposed to use small interfering RNAs (siRNAs) for implementing logic circuits (Rinaudo et al. 2007). They both use RNA to regulate gene expression. Implementing logic gates is also a main research topic in the field of synthetic biology. Logic circuits are implemented by modifying genetic circuits made of genes and regulatory proteins that regulate gene expression (Endy 2005). Artificial genetic circuits are not covered in this section, as synthetic biology is considered out of the scope of this chapter.

As is also mentioned in the bibliographic notes of ❯ Sect. 3, self-assembly of DNA tiles has been one of the main topics in DNA computing research (Jonoska and McColm 2009; Winfree 1998; Winfree et al. 1996, 2000), though logic gates implemented by DNA tiling is briefly touched on in ❯ Sect. 4.2.3.

The machineries based on DNAzymes and those based on strand displacements, which have been extended to reaction graphs in ❯ Sect. 4.3, are being applied to implement more typical molecular machines, including walkers and transporters (Bath and Turberfield 2007).

References

Adleman L (1994) Molecular computation of solutions to combinatorial problems. Science 266:1021–1024

Adleman L (1996) On constructing a molecular computer. DNA based computers. Series in mathematics and theoretical computer science, vol 27. American Mathematical Society, Providence, RI, pp 1–22

Amos M, Dunne PE, Gibbons A (1998) DNA simulation of Boolean circuits. In: Koza et al. (eds) Proceedings of the third annual conference on genetic programming, University of Wisconsin, Madison, WI, July 1998. Morgan Kaufmann, San Francisco, CA, pp 679–683

Arita M, Hagiya M, Suyama A (1997) Joining and rotating data with molecules. In: IEEE international

conference on evolutionary computation, Indianapolis, IN, June 1996, IEEE Service Center, pp 243–248

Bath J, Turberfield JA (2007) DNA nanomachines. Nat Nanotechnol 2:275–284

Beaver D (1995) A universal molecular computer. In: Lipton J, Baum B (eds) DNA based computers. DIMACS series in discrete mathematics and theoretical computer science, vol 27. American Mathematics Society, Providence, RI, pp 29–36

Benenson Y, Adar R, Paz-Elizur T, Livneh Z, Shapiro E (2003) DNA molecule provides a computing machine with both data and fuel. Proc Natl Acad Sci USA 100(5):2191–2196

Benenson Y, Gil B, Ben-Dor U, Adar R, Shapiro E (2004) An autonomous molecular computer for logical control of gene expression. Nature 429: 423–429

Benenson Y, Paz-Elizur T, Adar R, Keinan E, Livneh Z, Shapiro E (2001) Programmable and autonomous computing machine made of biomolecules. Nature 414:430–434

Brauer W, Ehrig H, Karhumäki J, Salomaa A (eds) (2002) Formal and natural computing. Lecture notes in computer science, vol 2300. Springer, Berlin

Brendel V, Busse HG (1984) Genome structure described by formal languages. Nucl Acids Res 12:2561–2568

Cheptea D, Martin-Vide C, Mitrana V (2006) A new operation on words suggested by DNA biochemistry: hairpin completion. In: Proceedings of transgressive computing, Universidad de Granada, Spain, April 2006, 216–228

Condon A (2004) Automata make antisense. Nature 429:351–352

Conrad M (1985) On design principles for a molecular computer. Comm ACM 28(5):464–480

Conrad M (1992) Molecular computing paradigms. IEEE Comput 25(11):6–9

Csuhaj-Varju E, Kari L, Păun Gh (1996) Test tube distributed systems based on splicing. Comput AI 15:211–232

Csuhaj-Varju E, Verlan S (2008) On length-separating test tube systems. Nat Comput 7:167–181

Dassow J, Păun Gh (1989) Regulated rewriting in formal language theory. Springer, Berlin

Doramatzki M (2006) Hairpin structures defined by DNA trajectories. In: DNA 12: Proceedings of 12th international meeting on DNA computing, Seoul, Korea, June 2006. Lecture notes in computer science, vol 4287. Springer, Berlin, pp 182–194

Eberling W, Jimenez-Montano MA (1980) On grammars, complexity and information measures of biological macromolecules. Math Biosci 52:53–72

Endy D (2005) Foundations for engineering biology. Nature 438:449–453: (vol 438 – 24 November 2005 – doi:10.1038/nature04342.)

Eng T (1997) Linear DNA self-assembly with hairpins generates the equivalent of linear context-free grammars. DNA based computers III. DIMACS series in discrete mathematics and theoretical computer science, vol 48. American Mathematical Society, Providence, RI, pp 289–296

Engelfriet J, Rozenberg G (1980) Fixed point languages, equality languages, and representation of recursively enumerable languages. J ACM, 27(3):499–518

Freund R, Kari L, Păun Gh (1999a) DNA computing based on splicing: the existence of universal computers. Technical report, Fachgruppe Informatik, Tech. Univ. Wien, 1995, and Theory Comput Syst 32:69–112

Freund R, Păun Gh, Rozenberg G, Salomaa A (1999b) Watson-Crick finite automata. DNA based computers III. DIMACS series in discrete mathematics and theoretical computer science, vol 48. American Mathematical Society, Providence, RI, pp 297–327

Gao Y, Garzon M, Murphy RC, Rose JA, Deaton R, Franceschetti DR, Stevens SE Jr (1999) DNA implementation of nondeterminism. DNA based computers III. DIMACS series in discrete mathematics and theoretical computer science, vol 48. American Mathematical Society, Providence, RI, pp 137–148

Geffert V (1991) Normal forms for phrase-structure grammars. RAIRO Theor Inform Appl 25:473–496

Hagiya M (1999) Perspectives on molecular computing. New Generation Comput 17:131–151

Hagiya M (2001) From molecular computing to molecular programming. In: DNA6: Proceedings of sixth international meeting on DNA based computers, Leiden, the Netherlands, June 2000. Lecture notes in computer science, vol 2054. Springer, Berlin, pp 89–102

Hagiya M, Arita M, Kiga D, Sakamoto K, Yokoyama S (2000) Towards parallel evaluation and learning of Boolean μ-formulas with molecules. In: Rubin H, Wood D (eds). DNA based computers III, DIMACS series in discrete mathematics, vol 48. American Mathematical Society, Providence, RI, pp 57–72

Hagiya M, Ohuchi A (2002) Preliminary proceedings of the eighth international meeting on DNA based computers, Hokkaido University, Hokkaido, Japan, June 10–13, 2002

Head T (1987) Formal language theory and DNA: an analysis of the generative capacity of specific recombinant behaviors. Bull Math Biol 49:737–759

Head T, Păun Gh, Pixton D (1997) Language theory and molecular genetics. Generative mechanisms suggested by DNA recombination In: Rozenberg G, Salomaa A (eds) Handbook of formal languages, vol 2. Springer, Berlin, pp 295–360

Hopcroft JE, Motwani R, Ullman JD (2001) Introduction to automata theory, languages, and computation, 2nd edn. Addison-Wesley, Reading, MA

Horn A (1951) On sentences which are true of direct unions of algebras. J Symbolic Logic 16:14–21

Ignatove Z, Martinez-Perez I, Zimmermann K-H (2008) DNA Computing Models. Springer, New York

Jimenez-Montano MA (1984) On the syntactic structure of protein sequences and the concept of grammar complexity, Bull Math Biol 46:641–659

Jonoska N, Karl SA, Saito M (1998) Three dimensional DNA structures in computing. In: Kari L (ed) Proceedings of 4th DIMACS meeting on DNA based computers, University of Pennsylvania, Philadelphia, PA, June 16–19, 1998, American Mathematics Society, pp 189–200

Jonoska N, McColm GL (2009) Complexity classes for self-assembling flexible tiles. Theor Comput Sci 410:332–346

Joshi AK, Schabes Y (1997) Tree-adjoining grammars. In Rozenberg G, Salomaa A (eds) Handbook of formal languages, vol. 3. Springer, Berlin, pp 69–123

Kari L (1996) DNA computers: tomorrow's reality. Tutorial Bull EATCS 59:256–266

Kari L, Konstantinidis S, Losseva E, Sosik P, Thierrin G (2005a) Hairpin structures in DNA words. In: DNA11: Proceedings of the 11th international meeting on DNA computing, London, ON, Canada, June 2005, Lecture notes in computer science, vol 3892. Springer, Berlin, pp 267–277

Kari L, Konstantinidis S, Sosik P, Thierrin G (2005b) On hairpin-free words and languages. In: deFelice C, Restivo A (eds) Proceedings of the 9th international conference on developments in language theory, Palermo, Italy, July 2005. Lecture notes in computer science, vol 3572. Springer, Berlin, pp 296–307

Kari L, Rozenberg G (2008) The many facets of natural computing, C. ACM: 51(10):72–83

Kari L, Konstantinidis S, Sosik P (2005c) On properties of bond-free DNA languages, Theor Comput Sci. 334:131–159

Kobayashi S (1999) Horn clause computation with DNA molecules. J Combinatorial Optimization 3:277–299

Kobayashi S, Mitrana V, Păun G, Rozenberg G (2001) Formal properties of PA-matching. Theor Comput Sci 262:117–131

Kobayashi S, Sakakibara Y (1998) Multiple splicing systems and the universal computability. Theor Comput Sci 264:3–23

Kobayashi S, Yokomori T, Sanpei G, Mizobuchi K (1997) DNA implementation of simple Horn clause computation. In: IEEE international conference on evolutionary computation, Indianapolis, IN, April 1997, IEEE Service Center, pp 213–217

Komiya K, Rose JA (2009) Experimental validation of signal dependent operation in Whiplash PCR. In: Goel A, Simmel FC (eds) DNA computing. 14th international workshop on DNA-based computers, Prague, Czech Republic, June 2008. Lecture notes in computer science, vol 5347, pp 1–10

Komiya K, Sakamoto K, Gouzu H, Yokohama S, Arita M, Nishikawa A, Hagiya M (2001) Successive state transitions with I/O interface by molecules. In: Condon A, Rozenberg G (eds) DNA computing. 6th international workshop on DNA-based computers, Leiden, the Netherlands, June 2000. Lecture notes in computer science, vol 2054, pp 17–26

Komiya K, Sakamoto K, Kameda A, Yamamoto M, Ohuchi A, Kiga D, Yokoyama S, Hagiya M (2006) DNA polymerase programmed with a hairpin DNA incorporates a multiple-instruction architecture into molecular computing. Biosystems 83:18–25

Kuramochi J, Sakakibara Y (2005) Intensive *in vitro* experiments of implementing and executing finite automata in test tube. In: DNA11: Proceedings of

11th international workshop on DNA computers, London, Canada, June 2005. Lecture notes in computer science, vol 3892, pp 193–202

LaBean TH, Winfree E, Reif JH (2000) Experimental progress in computation by self-assembly of DNA tilings. In: Winfree E, Gifford DK (eds) DNA based computers V. DIMACS series in discrete mathematics and theoretical computer science, vol 54. American Mathematical Society, Providence, RI, pp 123–140

Lagoudakis MG, LaBean TH (2000) 2D DNA self-assembly for satisfiability. In: Winfree E, Gifford DK (eds) DNA based computers V. DIMACS series in discrete mathematics and theoretical computer science, vol 54. American Mathematical Society, Providence, RI, pp 141–154

Laun E, Reddy K (1997) Wet splicing systems. In: Proceedings of 3rd DIMACS meeting on DNA based computers, University of Pennsylvania, Philadelphia, PA, June 23–25, 1997, pp 115–126

Lipton RJ (1995) DNA solution of hard computational problems. Science 268:542–545

Lipton RJ, Baum EB (eds) (1996) DNA based computers. Series in mathematics and theoretical computer science, vol 27. American Mathematical Society, Providence, RI

Manea F, Mitrana V (2007) Hairpin completion versus hairpin reduction. In: Computation in Europe, CiE 2007, Siena, Italy, June 2007. Lecture notes in computer science, vol 4497. Springer, Berlin, pp 532–541

Manea F, Mitrana V, Yokomori T (2008) Some remarks on the hairpin completion. In: Proceedings of 12th international conference on AFL, Hungary, May 2008. (to appear in Int J Found Comput Sci)

Manea F, Mitrana V, Yokomori T (2009a) Two complementary operations inspired by the DNA hairpin formation: completion and reduction. Theor Comput Sci 410:417–425

Manea F, Martín-Vide C, Mitrana V (2009b) On some algorithmic problems regarding the hairpin completion. Discrete Appl Math 157:2143–2152

Mao C, LaBean TH, Relf JH, Seeman NC (Sep 2000) Logical computation using algorithmic self-assembly of DNA triple-crossover molecules. Nature 407 (6803):493–496, Sep 2000

Marcus S (1969) Contextual grammars. Revue Roum Math Pures Appl 14(10):473–1482

Nitta N, Suyama A (2004) Autonomous biomolecular computer modeled after retroviral replication. In: Chen J, Reif J (eds) DNA9: Proceedings of 9th international meeting on DNA-based computers, Madison, WI, June 2003. Lecture notes in computer science, vol 2943, pp 203–212

Ogihara M, Ray A (1999) Simulating Boolean circuits on a DNA computer. Algorithmica 25:239–250

Păun Gh (1996a) Five (plus two) universal DNA computing models based on the splicing operation.

In: Proceedings of 2nd DIMACS workshop on DNA based computers, Princeton, NJ, June 1996, pp 67–86

Păun Gh (1996b) Regular extended H systems are computationally universal. J Automata, Lang Combinatorics 1(1):27–36

Păun Gh (1999) (DNA) Computing by carving. Soft Comput 3(1):30–36

Păun Gh, Rozenberg G, Salomaa A (1998) DNA computing: new computing paradigms. Springer, Berlin

Păun Gh, Rozenberg G, Yokomori T (2001) Hairpin languages. Int J Found Comput Sci 12(6):837–847

Pixton D (1995) Linear and circular splicing systems. In: Proceedings of 1st international symposium on intelligence in neural and biological systems, Herndon, VA, May 1995. IEEE, Washington, DC, pp 38–45

Post E (1943) Formal reductions of the general combinatorial decision problem. Am J Math 65:197–215

Reif J (1995) Parallel molecular computation. In: SPAA'95: 7th annual ACM symposium on parallel algorithms and architectures, Santa Barbara, CA, July 1995, pp 213–223

Reif J (1999) Parallel biomolecular computation: models and simulations. Algorithmica 25:142–176

Rinaudo K, Bleris L, Maddamsetti R, Subramanian S, Weiss R, Benenson Y (2007) A universal RNAi-based logic evaluator that operates in mammalian cells. Nat Biotechnol 25:795–801. (Published online: May 21, 2007 – doi:10.1038/nbt1307.)

Rooß D, Wagner K (1996) On the power of DNA-computing. Infor Comput 131:95–109

Rose JA, Deaton RJ, Hagiya M, Suyama A (2002a) Equilibrium analysis of the efficiency of an autonomous molecular computer. Phys Rev E 65:021910

Rose JA, Hagiya M, Deaton RJ, Suyama A (2002b) A DNA-based in vitro genetic program. J Biol Phys 28:493–498

Rose JA, Komiya K, Yaegashi S, Hagiya, M (2006) Displacement Whiplash PCR: optimized architecture and experimental validation. In: Mao C, Yokomori T (eds) DNA computing. 12th international workshop on DNA-based computers, Seoul, Korea, June 2006. Lecture notes in computer science, vol 4287, pp 393–403

Rose JA, Takano M, Hagiya M, and Suyama A (2003) A DNA computing-based genetic program for in vitro protein evolution via constrained pseudomodule shuffling. J Genet Programming Evolvable Mach 4:139–152

Rothemund PWK (1995) A DNA and restriction enzyme implementation of Turing machines. In: Lipton J, Baum B (eds) DNA based computers. DIMACS series in discrete mathematics and theoretical computer science, vol 27. American Mathematical Society, Providence, RI, pp 75–119

Rothemund PWK, Papadakis N, Winfree E (Dec 2004) Algorithmic self-assembly of DNA Sierpinski triangles. PLoS Biol 2(12):e424

Rozenberg G, Salomaa A (eds) (1997) Handbook of formal languages, 3 volumes. Springer, Berlin

Salomaa A (1985) Computation and automata. Cambridge University Press, Cambridge

Sakakibara Y, Ferretti C (1997) Splicing on tree-like structures. In: Proceedings of 3rd DIMACS meeting on DNA based computers, University of Pennsylvania, Philadelphia, PA, June 23–25, 1997, pp 348–358. Also, in Theor Comput Sci 185:15–45, 1999

Sakakibara Y, Kobayashi S (2001) Sticker systems with complex structures. Soft Comput 5:114–120

Sakakibara Y, Suyama A (2000) Intelligent DNA chips: logical operation of gene expression profiles on DNA computers. In: Genome Informatics 2000: Proceedings of 11th workshop on genome informatics, Tokyo, Japan, December 2000. Universal Academy Press, Tokyo, Japan, pp 33–42

Sakamoto K, Gouzu H, Komiya K, Kiga D, Yokoyama S, Yokomori T, Hagiya M (2000) Molecular computation by DNA hairpin formation. Science 288:1223–1226

Sakamoto K, Kiga D, Komiya K, Gouzu H, Yokoyama S, Ikeda S, Sugiyama H, Hagiya M (1999) State transitions by molecules. BioSystems 52(1–3):81–91

Seelig G, Yurke B, Winfree E (2006a) Catalyzed relaxation of a metastable DNA fuel. J Am Chem Soc 128 (37):12211–12220

Seelig G, Soloveichik D, Zhang DY, Winfree E (Dec 2006b) Enzyme-free nucleic acid logic circuits. Science 314(5805):1585–1588

Shih W (Feb 2008) Biomolecular assembly: dynamic DNA. Nat Mater 7(2):98–100

Siromoney R, Subramanian KB, Rajkumar Dare V (1992) Circular DNA and splicing systems. In: Proceedings of Parallel Image Analysis, Ube, Japan, December 1992. Lecture notes in computer science, vol 654. Springer, Berlin, pp 260–273

Smith WD (1995) DNA computers in vitro and vivo. In: Lipton J, Baum B (eds) DNA based computers. DIMACS series in discrete mathematics and theoretical computer science, vol 27. American Mathematics Society, Providence, RI, pp 121–185

Stojanovic MN, Mitchell TE, Stefanovic D (Apr 2002) Deoxyribozyme-based logic gates. J Am Chem Soc 124(14):3555–3561

Stojanovic MN, Semova S, Kolpashchikov D, Macdonald J, Morgan C, Stefanovic D (May 2005) Deoxyribozyme-based ligase logic gates and their initial circuits. J Am Chem Soc 127(19):6914–6915

Stojanovic MN, Stefanovic D (Sep 2003) A deoxyribozyme-based molecular automaton. Nat Biotechnol 21(9):1069–1074

Takinoue M, Kiga D, Shohda K, Suyama A (2008) Experiments and simulation models of a basic computation element of an autonomous molecular computing system. Phys Rev E 78:041921

Tanaka F, Tsuda T, Hagiya M (2009) Towards DNA comparator: the machine that compares DNA concentrations. In: Goel A, Simmel FC (eds) DNA computing, 14th international workshop on DNA-based computers, Prague, Czech Republic, June 2008. Lecture notes in computer science, vol 5347, pp 11–20

Tarnlund S (1977) Horn clause computability. BIT 17:215–226

Uejima H, Hagiya M, Kobayashi S (2001) Horn clause computation by self-assembly of DNA molecules. In: Proceedings of 7th international workshop on DNA-based computers, Tampa, FL, June 2001. Lecture notes in computer science, vol 2340, pp 308–320

Vaintsvaig MN, Liberman EA (1973) Formal description of cell molecular computer. Biofizika 18:939–942

Venkataraman S, Dirks RM, Rothemund PWK, Winfree E, Pierce NA (2007) An autonomous polymerization motor powered by DNA hybridization. Nat Nanotechnol 2:490–494

Wang H (1961) Proving theorems by pattern recognition-II. Bell Syst Tech J 40:1–41

Wang H (1962) Dominoes and the AEA case of the decision problem. In: Proceedings of the symposium on mathematical theory of automata. New York, April 1962. Polytechnic Institute of Brooklyn, Brooklyn, New York, pp 23–55

Win MN, Smolke CD (Oct 2008) Higher-order cellular information processing with synthetic RNA devices. Science 322(5900):456–460

Winfree E (1998) Algorithmic self-assembly of DNA. Ph.D. thesis, California Institute of Technology

Winfree E, Eng T, Rozenberg G (2000) String tile models for DNA computing by self-assembly. In: Proceedings of the 6th international meeting on DNA based computers. Leiden University, Leiden, the Netherlands, June 13–17, 2000, pp 65–84

Winfree E, Yang X, Seeman NC (1996) Universal computation via self-assembly of DNA: some theory and experiments DNA based computers II. DIMACS series in discrete mathematics and theoretical computer science, vol 44. American Mathematical Society, Providence, RI, pp 191–213

Wood D, Bi H, Kimbrough S, Wu D-J, Chen J (2002) DNA starts to learn poker. In: Jonoska N, Seeman N (eds) DNA computing, 7th international workshop on DNA-based computers Tampa, FL, June 2001. Lecture notes in computer science, vol 2340, pp 22–32

Yin P, Choi HMT, Calvert CR, Pierce NA (Jan 2008a) Programming biomolecular self-assembly pathways. Nature 451(7176):318–322

Yin P, Hariadi RF, Sahu S, Choi HMT, Park SH, LaBean TH, Reif JH (2008b) Programming DNA tube circumferences. Science 321:824–826

Yokomori T (1999) Computation = self-assembly+ conformational change: toward new computing paradigms. In: DLT'99: Proceedings of 4th international conference on developments in language theory, Aachen, Germany, July 1999, pp 21–30

Yokomori T (2000) YAC: yet another computation model of self-assembly. In: Winfree E, Gifford DK (eds) DNA based computers V. DIMACS series in discrete mathematics and theoretical computer science, vol 54. American Mathematical Society, Providence, RI, pp 155–169

Yokomori T, Kobayashi S (1999) DNA-EC: a model of DNA computing based on equality checking. DNA based computers III. DIMACS series in discrete mathematics and theoretical computer science, vol 48. American Mathematical Society, Providence, RI, pp 347–360

Yokomori T, Kobayashi S, Ferretti C (1997) On the power of circular splicing systems and DNA computability. In: Proceedings of IEEE international conference on evolutionary computation, Indianapolis, IN, April 1997, IEEE Service Center, pp 219–224

Yokomori T, Sakakibara Y, Kobayashi S (2002) A magic pot: self-assembly computation revisited. In: Brauer W, Ehrig H, Karhumaki J, Salomaa A (eds) Formal and natural computing. Lecture notes in computer science, vol 2300. Springer, Berlin, pp 418–429

Yurke B, Mills AP Jr (2003) Using DNA to power nanostructures. Genet Programming Evolvable Mach 4:111–122

Zhang DY, Turberfield AJ, Yurke B. Winfree E (Nov 2007) Engineering entropy-driven reactions and networks catalyzed by DNA. Science 318(5853):1121–1125

35 DNA Computing by Splicing and by Insertion–Deletion

Gheorghe Păun
Institute of Mathematics of the Romanian Academy, Bucharest, Romania
Department of Computer Science and Artificial Intelligence,
University of Seville, Spain
gpaun@us.es
george.paun@imar.ro

G. Rozenberg et al. (eds.), *Handbook of Natural Computing*, DOI 10.1007/978-3-540-92910-9_35,
© Springer-Verlag Berlin Heidelberg 2012

Abstract

This chapter is devoted to two of the most developed theoretical computing models inspired by DNA biochemistry, computing by splicing (a formal operation with strings that models the recombination of DNA molecules under the influence of restriction enzymes and ligase) and by insertion–deletion. Only basic ideas and results are presented, as well as a comprehensive – although not complete – list of titles where further information can be found.

1   Introduction

DNA computing was mainly developed after the 1994 Adleman experiment of solving a small instance of the Hamiltonian path problem in a test tube, making use of standard lab operations, but from a theoretical point of view, one may say that everything started much earlier. For instance, as early as in 1987, Head introduced the so-called splicing operation (Head 1987), an operation with strings that models the recombination of DNA molecules cut by restriction enzymes and pasted together by means of ligase. This initiated a powerful research direction in computability (mainly formal language and automata theory, see Păun et al. (1998), but recently also in terms of complexity, see, e.g., Loos et al. (2008) and Loos and Ogihara (2007)), *DNA computing by splicing*. Actually, the "prehistory" of DNA computing can be pushed further, for instance, considering the operations of inserting and of deleting strings, considered in linguistics and formal languages well before – only Galiukschov (1981) and Marcus (1969) are mentioned – and whose study was continued in the new framework of handling DNA molecules.

In what follows, both of these directions of research in theoretical DNA computing, computing by splicing and computing by insertion–deletion, will be briefly overviewed. Only basic notions are given; a few typical proofs (mainly proof ideas) and a series of results will be recalled – without attempting to be complete from this last point of view, because, on the one hand, there exists a large number of results and, on the other hand, the research in these areas is still active, so that the existing open problems are under continuous examination and there is continuous progress. In particular, only a few basic references are provided and the reader can refer to Head et al. (1997), Jonoska et al. (2004), and Păun et al. (1998, 2009) for further details and references; many PhD theses devoted to (theoretical) DNA computing can also be useful in this respect – some of them are mentioned in the bibliography of this chapter.

2   Some Prerequisites

It is assumed that the reader is familiar with elementary facts from language and automata theory, so that only a few notions and notations are recalled here. If necessary, any of the many monographs in this area may be consulted. Only Harrison (1978) and Salomaa (1973) and the handbook by Rozenberg and Salomaa (1997) are mentioned as sources of comprehensive information.

For an alphabet V, V^* is the set of all strings over V, the empty string, denoted by λ, included. The length of $x \in V^*$ is denoted by $|x|$. A mapping $h : V \to U^*$ extended to $h : V^* \to U^*$ by $h(\lambda) = \lambda$ and $h(uv) = h(u)h(v)$ for all $u, v \in V^*$ is called a *morphism*. A morphism $h : V^* \to U^*$ is called a *coding* if $h(a) \in U$ for all $a \in V$ and a *weak coding* if

$h(a) \in U \cup \{\lambda\}$. A weak coding $h: V \to U$ with $U \subseteq V$ such that $h(a) = a$ for $a \in U$, and $h(a) = \lambda$ for $a \in V - U$ is called a *projection* (on U).

A Chomsky grammar (also called type-0 grammar) is a quadruple $G = (N, T, S, P)$, where N is the nonterminal alphabet, T is the terminal alphabet, S is the axiom, and R is the set of rewriting rules. If these rules are of the following forms,

- $AB \to CD$, where $A, B, C, D \in N$ (type 1: context-sensitive rules)
- $A \to BC$, where $A, B, C \in N$ (type 2: context-free rules)
- $A \to a$, where $A \in N$ and $a \in T \cup \{\lambda\}$ (type 3: terminal rules; they may be erasing rules)

then the grammar is in the *Kuroda normal form*.

A type-0 grammar $G = (\{S, A, B, C, D\}, T, S, P)$ with P containing rules of the forms $S \to uSv$, $S \to x$, with $u, v, x \in (T \cup \{A, B, C, D\})^*$, as well as the non-context-free rules $AB \to \lambda$, $CD \to \lambda$, is said to be in the *Geffert normal form*.

The families of finite, regular, linear, context-free, context-sensitive, and recursively enumerable languages are denoted by *FIN, REG, LIN, CF, CS*, and *RE*, respectively.

A family of languages closed under union, concatenation, λ-free morphisms, inverse morphisms, intersection with regular languages, and Kleene closure is called an AFL (abstract family of languages); an AFL closed under arbitrary morphisms is called a full-AFL.

The Dyck language over an alphabet $V_n = \{a_i, b_i | 1 \leq i \leq n\}$ is the language generated by the context-free grammar $G_n = (\{S\}, V_n, S, P_n)$ with the following rules

$$P_n = \{S \to SS, \ S \to \lambda\} \cup \{S \to a_iSb_i \mid 1 \leq i \leq n\}$$

Every context-free language L can be written in the form $L = h(D_n \cap R)$, where h is a projection, D_n, $n \geq 1$, is a Dyck language, and R is a regular language. This result is known as the *Chomsky–Schützenberger characterization of context-free languages*.

A deterministic finite automaton is a construct $M = (K, V, q_0, F, \delta)$, where K is the finite set of states, V is the alphabet of the automaton, q_0 is the initial state, $F \subseteq K$ is the set of terminal states, and $\delta: K \times V \to K$ is the next-state mapping. Finite automata characterize the family, *REG*, of languages generated by Chomsky regular grammars.

3 Computing by Splicing

As mentioned in ❷ Sect. 1, the splicing operation was introduced in Head (1987), as an abstraction of a biochemical operation (actually, a sequence of cut- and paste-operations) with DNA molecules. Here this abstraction process is not recalled, but we refer to the seminal paper and to Păun et al. (1998) for details. The operation proposed in Head (1987) was further generalized in Păun (1996a), formulated as a "pure" operation with strings and formal languages, and it was then mainly considered in this form in the subsequent mathematical developments. In particular, it was this version of the splicing operation that was used for defining a computing model, called an *H system* in Păun et al. (1996), with H coming from the name of the originator of this approach, Tom Head. Later, further variations were introduced, for example, in Pixton (2000) and Loos (2006). In what follows, the DNA computing is presented by splicing based on the operation as defined in Păun (1996), hence in terms of strings and languages, abstracting away such important features of the DNA molecules, always present in the background of the theoretical developments, such as the use of only four letters (representing nucleotides), the double stranded structure, the Watson–Crick complementarity, the bidirectionality, the palindromicity of the restriction sites of many enzymes, and so on. The use of the other variants of splicing, in particular, comparisons between them, can be found in a series of papers; some of them are also included in the bibliography of this chapter.

3.1 The Splicing Operation

One can consider an arbitrary finite alphabet V. A splicing rule over V is a string $u_1\#u_2\$u_3\#u_4$, where u_1, u_2, u_3, u_4 are strings over V and $\#$, $\$$ are special symbols not in V. (The idea is that (u_1, u_2), (u_3, u_4) represent the restriction sites of two enzymes, which cut DNA molecules in such a way that they produce sticky ends which match, hence the fragments produced by these enzymes can be recombined; the first enzyme cuts molecules – strings here – in between u_1 and u_2, and the second enzyme cuts in between u_3 and u_4.)

For a splicing rule $r = u_1\#u_2\$u_3\#u_4$ and four strings x, y, z, w over V, we write

$$(x, y) \models_r (z, w) \text{ iff } x = x_1 u_1 u_2 x_2, \ y = y_1 u_3 u_4 y_2,$$
$$z = x_1 u_1 u_4 y_2, \ w = y_1 u_3 u_2 x_2$$
$$\text{for some } x_1, x_2, y_1, y_2 \in V^*$$

We say that we *splice* x, y at the *sites* $u_1 u_2$, $u_3 u_4$, respectively, and the result consists of the strings z, w. This is called the 2-splicing, because both strings obtained by recombination are taken as the result of the operation. From a mathematical point of view, in many contexts it is sufficient/more elegant to consider only the first string, z, as the result of the splicing – and this is called the 1-splicing. The idea is that in most cases, we work with a set of splicing rules and then, together with the rule $u_1\#u_2\$u_3\#u_4$ we can also consider the symmetric rule, $u_3\#u_4\$u_1\#u_2$, and, by splicing the same strings x, y, the first rule produces the first result, z, and the second one produces the second result, w. In what follows it will be specified which of the two operations is used. Most of the results (for instance, the regularity and the universality results) are similar for the two versions of the splicing, but there are cases where a difference can be found – see, for example, Verlan and Zizza (2003).

In particular, we use $\vdash$ to denote the 1-splicing and $\models$ for the 2-splicing operation. When only "splicing" is referred to, it will be clear from the context which type of splicing is meant.

These operations can be extended to languages in a natural way. One can first introduce the notion of an H scheme, as a pair $\sigma = (V, R)$, where V is an alphabet and $R \subseteq V^*\#V^* \$V^*\#V^*$ is a set of splicing rules. Note that in this formulation R is a language, hence we can consider its complexity with respect to various language hierarchies. For instance, if $R \in FL$, for a given family of languages, FL, then we say that the H scheme σ is *of FL type*.

Now, for a given H scheme $\sigma = (V, R)$ and a language $L \subseteq V^*$, we define

$$\sigma_1(L) = \{z \in V^* \mid (x, y) \vdash_r z, \text{ for some } x, y \in L, r \in R\}$$

and then, for two families FL_1, FL_2 of languages, we denote

$$S_1(FL_1, FL_2) = \{\sigma_1(L) \mid L \in FL_1 \text{ and } \sigma = (V, R) \text{ with } R \in FL_2\}$$

This is the one-step splicing operation, extended to languages and performed with respect to sets of splicing rules. The operation can be iterated, in the natural way: for an H scheme $\sigma = (V, R)$ and a language $L \subseteq V^*$ we define

$$\sigma_1^0(L) = L$$
$$\sigma_1^{i+1}(L) = \sigma_1^i(L) \cup \sigma_1(\sigma_1^i(L)), \ i \geq 0$$
$$\sigma_1^*(L) = \bigcup_{i \geq 0} \sigma_1^i(L)$$

Consequently, $\sigma_1^*(L)$ is the closure of L under the splicing with respect to σ, that is, the smallest language L' that contains L, and is closed under the splicing with respect to σ, that is to say, $\sigma_1(L') \subseteq L'$.

For two families of languages, FL_1, FL_2, we define

$$H_1(FL_1, FL_2) = \{\sigma_1^*(L) \mid L \in FL_1 \text{ and } \sigma = (V, R) \text{ with } R \in FL_2\}$$

Of course, the same definitions can be considered for the 2-splicing and then we obtain the families $S_2(FL_1, FL_2)$, $H_2(FL_1, FL_2)$, respectively.

3.2 The Power of the Splicing Operation

We examine now the size of families $S_1(FL_1, FL_2)$, $H_1(FL_1, FL_2)$, for FL_1, FL_2 one of the following families: *FIN, REG, LIN, CF, CS, RE*.

For the non-iterated splicing, the following theorem synthesizes the best results known in the framework specified above; proofs and references can be found in Păun et al. (1998) and Head et al. (1997).

Theorem 1 *The relations in* ❯ *Table 1 hold, where at the intersection of the row marked with* FL_1 *with the column marked with* FL_2 *there appear either the family* $S_1(FL_1, FL_2)$, *or two families* FL_3, FL_4 *such that* $FL_3 \subset S_1(FL_1, FL_2) \subset FL_4$. *These families* FL_3, FL_4 *are the best possible estimations among the six families considered here.*

In the case of iterated splicing, things are more difficult, especially concerning the characterization of the families $H_1(FL, FIN)$, $FL \in \{FIN, REG\}$. Several proofs of the regularity of the languages in these families were given, some of them rather complex and some of them extended to abstract families of languages.

❏ **Table 1**
The size of families $S_1(FL_1, FL_2)$

	FIN	REG	LIN	CF	CS	RE
FIN	FIN	FIN	FIN	FIN	FIN	FIN
REG	REG	REG	REG, LIN	REG, CF	REG, RE	REG, RE
LIN	LIN, CF	LIN, CF	RE	RE	RE	RE
CF	CF	CF	RE	RE	RE	RE
CS	RE	RE	RE	RE	RE	RE
RE	RE	RE	RE	RE	RE	RE

Proofs of the following result, called in Păun et al. (1998) *the Regularity Preserving Lemma*, can be found in Culik II and Harju (1991), Pixton (1996) (recalled in Păun et al. (1998)), and Manca (2000):

Lemma 1 $H_1(REG, FIN) \subseteq REG$.

A stronger result has been presented in Pixton (2000); the proof is recalled in Head et al. (1997).

Lemma 2 *If FL is a full AFL, then* $H_1(FL, FIN) \subseteq FL$.

Thus, when using a finite set of rules (remember: this corresponds to a finite set of restriction enzymes), we cannot compute too much, we remain inside the computing competence of finite automata. Much more can be computed if we use a set of rules of the next level in the hierarchy of language families considered above, that is, a regular set. This result is presented, called in Păun et al. (1998) *the Basic Universality Lemma*, with a proof, mainly because the proof idea, the so-called rotate-and-simulate technique, is useful in many related contexts.

Lemma 3 *Every language* $L \in RE$, $L \subseteq T^*$, *can be written in the form* $L = L' \cap T^*$ *for some* $L' \in H_1(FIN, REG)$.

Proof Consider a Chomsky type-0 grammar $G = (N, T, S, P)$, denote $U = N \cup T \cup \{B\}$, where B is a new symbol, and construct the H scheme $\sigma = (V, R)$, where

$$V = N \cup T \cup \{X, X', B, Y, Z\} \cup \{Y_\alpha \mid \alpha \in U\}$$

and R contains the following groups of rules:

$$
\begin{aligned}
\textit{Simulate} : 1.\ & Xw\#uY\$Z\#vY, & \text{for } u \to v \in P, w \in U^* \\
\textit{Rotate} : 2.\ & Xw\#\alpha Y\$Z\#Y_\alpha, & \text{for } \alpha \in U, w \in U^* \\
3.\ & X'\alpha\#Z\$X\#wY_\alpha, & \text{for } \alpha \in U, w \in U^* \\
4.\ & X'w\#Y_\alpha\$Z\#Y, & \text{for } \alpha \in U, w \in U^* \\
5.\ & X\#Z\$X'\#wY, & \text{for } w \in U^* \\
\textit{Terminate} : 6.\ & \#ZY\$XB\#wY, & \text{for } w \in T^* \\
7.\ & \#Y\$XZ\#
\end{aligned}
$$

Consider also the language

$$L_0 = \{XBSY,\ ZY,\ XZ\} \cup \{ZvY \mid u \to v \in P\}$$
$$\cup \{ZY_\alpha,\ X'\alpha Z \mid \alpha \in U\}$$

We obtain $L = \sigma_1^*(L_0) \cap T^*$.

Indeed, one can examine the work of σ, namely, the possibilities to obtain a string in T^*.

No string in L_0 is in T^*. All rules in R involve a string containing the symbol Z, but this symbol will not appear in the string produced by splicing. Therefore, at each step, we have to use a string in L_0 and, excepting the case of using the string $XBSY$ in L_0, a string produced at a previous step.

The symbol B is a marker for the beginning of the sentential forms of G simulated by σ.

By rules in group 1 we can simulate the rules in P. Rules in groups 2–5 move symbols from the right-hand end of the current string to the left-hand end, thus making possible the simulation of rules in P at the right-hand end of the string produced by σ. However, because B is always present and marks the place where the string of G begins, it is known in each moment which is that string. Namely, if the current string in σ is of the form $\beta_1 w_1 B w_2 \beta_2$, for some β_1, β_2 markers of types X, X', Y, Y_α with $\alpha \in U$, and $w_1, w_2 \in (N \cup T)^*$, then $w_2 w_1$ is a sentential form of G.

We start from $XBSY$, hence from the axiom of G, marked to the left hand with B and bracketed by X, Y.

Let us see how the rules 2–5 work. Take a string $Xw\alpha Y$, for some $\alpha \in U$, $w \in U^*$. By a rule of type 2 we get

$$(Xw|\alpha Y, Z|Y_\alpha) \vdash XwY_\alpha$$

The symbol Y_α memorizes the fact that α has been erased from the right-hand end of $w\alpha$. No rule in R can be applied to XwY_α, excepting the rules of type 3:

$$(X'\alpha|Z, X|wY_\alpha) \vdash X'\alpha wY_\alpha$$

Note that the same symbol α removed at the previous step is now added in the front of w. Again, we have only one way to continue, namely, by using a rule of type 4. We get

$$(X'\alpha w|Y_\alpha, Z|Y) \vdash X'\alpha wY$$

If we now use a rule of type 7, removing Y, then X' (and B) can never be removed, and hence the string cannot be turned to a terminal one. A rule of type 5 must be used:

$$(X|Z, X'|\alpha wY) \vdash X\alpha wY$$

We have started from $Xw\alpha Y$ and obtained $X\alpha wY$, a string with the same end markers. These steps can be iterated as long as we want, so any circular permutation of the string between X and Y can be produced. Moreover, what we obtain are exactly the circular permutations and nothing more (for instance, at every step, there will still be one and only one occurrence of B).

To every string XwY one can also apply a rule of type 1, providing w ends with the left-hand member of a rule in P. Any rule of P can be simulated in this way, at any place we want in the corresponding sentential form of G, by preparing the string as above, using rules in groups 2–5.

Consequently, for every sentential form w of G there is a string $XBwY$, produced by σ, and, conversely, if $Xw_1 Bw_2 Y$ is produced by σ, then $w_2 w_1$ is a sentential form of G.

The only way to remove the symbols not in T from the strings produced by σ is by using rules in groups 6, 7. More precisely, the symbols XB can only be removed in the following conditions: (1) Y is present (hence the work is blocked if one uses first rule 7, removing Y: the string cannot participate in any further splicing, and it is not terminal), (2) besides one occurrence of B, the current string bracketed by X, Y consists of terminal symbols only, and (3) the symbol B is adjacent to X, in the right hand of it. After removing X and B, one can remove Y, too, and what one obtains is a string in T^*. From the previous discussion, it is clear that such a string is in $L(G)$, hence $\sigma_1^*(L_0) \cap T^* \subseteq L(G)$. Conversely, each string in $L(G)$ can be produced in this way, hence $L(G) \subseteq \sigma_1^*(L_0) \cap T^*$. We have the equality $L(G) = \sigma_1^*(L_0) \cap T^*$, which completes the proof.

◻ **Table 2**

The size of families $H_1(FL_1, FL_2)$

	FIN	REG	LIN	CF	CS	RE
FIN	FIN, REG	FIN, RE	FIN, RE	FIN, RE	FIN, RE	FIN, RE
REG	REG	REG, RE	REG, RE	REG, RE	REG, RE	REG, RE
LIN	LIN, CF	LIN, RE	LIN, RE	LIN, RE	LIN, RE	LIN, RE
CF	CF	CF, RE	CF, RE	CF, RE	CF, RE	CF, RE
CS	CS, RE	CS, RE	CS, RE	CS, RE	CS, RE	CS, RE
RE	RE	RE	RE	RE	RE	RE

We also recall, without a proof, a result concerning the limitations of families $H_1(FL_1, FL_2)$.

Lemma 4 *Let FL be a family of languages closed under intersection with regular languages and restricted morphisms. For every $L \subseteq V^*$, $L \notin FL$, and $c, d \notin V$, the language $L' = (dc)^* L(dc)^* \cup c(dc)^* L(dc)^* d$ does not belong to the family $H_1(FL, RE)$.*

Based on the previous results, we get the following synthesis theorem.

Theorem 2 *The relations in ❯ Table 2 hold, where at the intersection of the row marked with FL_1 with the column marked with FL_2 there appear either the family $H_1(FL_1, FL_2)$, or two families FL_3, FL_4 such that $FL_3 \subset H_1(FL_1, FL_2) \subset FL_4$. These families FL_3, FL_4 are the best possible estimations among the six families considered here.*

3.3 Extended Splicing Systems

In the sections above the power of splicing operations is considered without explicitly introducing a computing model. This can be done now, by defining the main device investigated in DNA computing by splicing, the H systems; these devices are presented directly in the extended version, with a terminal alphabet used for selecting the result of computations out of all results of splicing operations.

An *extended H system* is a quadruple $\gamma = (V, T, A, R)$, where V is an alphabet, $T \subseteq V$, $A \subseteq V^*$, and $R \subseteq V^* \# V^* \$ V^* \# V^*$, where #, $ are special symbols not in V.

We call V the alphabet of γ, T is the *terminal* alphabet, A is the set of *axioms*, and R is the set of splicing rules. Therefore, we have an *underlying H scheme*, $\sigma = (V, R)$, augmented with a given subset of V and a set of axioms. When $T = V$ we say that γ is a *non-extended* H system.

The *language generated* by γ is defined by $L(\gamma) = \sigma_1^*(A) \cap T^*$, where σ is the underlying H scheme of γ.

For two families of languages, FL_1, FL_2, we denote by $EH_1(FL_1, FL_2)$ the family of languages $L(\gamma)$ generated by extended H systems $\gamma = (V, T, A, R)$, with $A \in FL_1$, $R \in FL_2$.

A number of the results from the previous section can be reformulated in terms of extended H systems. Moreover, we have the following inclusion (we recall its simple proof from Păun et al. (1998) in order to have an example of an extended H system).

Lemma 5 $REG \subseteq EH_1(FIN, FIN)$.

Proof Consider a regular grammar $G = (N, T, S, P)$ (hence with rules in P of the forms $A \to aB$, $A \to a$, for $A, B \in N$ and $a \in T$). We construct the H system

$$\gamma = (N \cup T \cup \{Z\}, T, A_1 \cup A_2 \cup A_3, R_1 \cup R_2)$$
$$A_1 = \{S\}$$
$$A_2 = \{ZaY \mid X \to aY \in P, X, Y \in N, a \in T\}$$
$$A_3 = \{ZZa \mid X \to a \in P, X \in N, a \in T\}$$
$$R_1 = \{\#X\$Z\#aY \mid X \to aY \in P, X, Y \in N, a \in T\}$$
$$R_2 = \{\#X\$ZZ\#a \mid X \to a \in P, X \in N, a \in T\}$$

The work of γ can be now examined. If a string ZxX is spliced, possibly one from A_2 (this is the case if $x = c \in T$ and $U \to cX \in P$) using a rule in R_1, then we get a string of the form $ZxaY$. The symbol Z cannot be eliminated, hence no terminal string can be obtained if one continues to use the resulting string as the first term of a splicing. On the other hand, a string ZxX with $|x| \geq 2$ cannot be used as the second term of a splicing. Thus, the only way to obtain a terminal string is to start from S, to use splicings with respect to rules in R_1 an arbitrary number of times, and to end with a rule in R_2. Always the first term of a splicing is that obtained by a previous splicing and the second one is from A_2 or from A_3 (at the last step). This corresponds to a derivation in G, hence we have $L(\gamma) = L(G)$.

Combining now the previous results, we get the following theorem.

Theorem 3 *The relations in* ❯ *Table 3 hold, where at the intersection of the row marked with FL_1 with the column marked with FL_2 there appear either the family $EH_1(FL_1, FL_2)$, or two families FL_3, FL_4 such that $FL_3 \subset EH_1(FL_1, FL_2) \subseteq FL_4$. These families FL_3, FL_4 are the best possible estimations among the six families considered here.*

It is worth noting that the only family that is not equal to a family in the Chomsky hierarchy is $EH_1(LIN, FIN)$.

Two of the relations summarized in ❯ *Table 3* are central for the DNA computing based on splicing, $EH_1(FIN, FIN) = REG$ and $EH_1(FIN, REG) = RE$.

Therefore, using a finite extended H system, that is, a system with a finite set of axioms and a finite set of splicing rules, we only obtain the computing power of finite automata.

◻ **Table 3**
The generative power of extended H systems

	FIN	REG	LIN	CF	CS	RE
FIN	REG	RE	RE	RE	RE	RE
REG	REG	RE	RE	RE	RE	RE
LIN	LIN, CF	RE	RE	RE	RE	RE
CF	CF	RE	RE	RE	RE	RE
CS	RE	RE	RE	RE	RE	RE
RE	RE	RE	RE	RE	RE	RE

However, working with a regular set of rules that is not finite is not of much interest from a practical point of view since we cannot physically realize an infinite "computer" (in this case, an infinite number of restriction enzymes).

3.4 Universal H Systems

The results in ❷ *Table 3* hold also for 2-splicing – which is important from a "practical" point of view, as in a test tube both results of a splicing operation are equally possible. From a "practical" point of view, the difficulty suggested by the equalities $EH_2(FIN, FIN) = REG$, $EH_2(FIN, REG) = RE$ can also be avoided, and the idea is to use a finite set of rules, but not to apply them freely, as in a usual extended H system, but impose some control on the strings that can be handled by a specific rule. In this way, characterizations of the computing power of Turing machines are obtained by means of finite H systems. The proofs are constructive, in most cases based on the rotate-and-simulate technique from the proof of Basic Universality Lemma, hence, starting from a universal type-0 grammar, we get a universal H system (a system that simulates any particular H system after introducing a coding of this particular system in the axioms of the universal one). That is why *universality results* are spoken about even when explicitly dealing with results which, mathematically speaking, provide characterizations of Turing computability (of RE languages).

There are many controls that can be imposed to a finite extended H system in order to increase its computing power from the power of finite automata to the power of Turing machines. (This shows that the splicing operation needs very little additional power – more technically: context-sensitivity – in order to reach the power of Turing machines.)

Several of these controls are mentioned here without entering into technical details: (1) permitting contexts (each rule has associated to it some symbols and the rule can be applied to splice two strings only if these strings contain the respective symbols), (2) forbidding contexts (as above, but the associated symbols should not be present in the spliced strings), (3) target languages, with two possibilities, local targets and a global target (in the first case, a language is associated with each rule and the results of a splicing operation must be a string in the language associated with the used rule; in the global case, the target languages associated with different rules are identical), (4) programmed H systems (a next rule mapping is considered, which controls the sequence of rules used), (5) double splicing (similar to the previous restriction, but only couples of consecutive rules are specified; the strings resulting after applying the first rule in a couple are spliced by the second rule), (6) using multisets (counting the copies of strings present in any step of the computation). Also considered were other computing devices based on splicing, using suggestions from the regulated rewriting area (Dassow and Păun 1989), from grammar systems area (Csuhaj-Varju et al. 1994), or from membrane computing (Păun et al. 2009) (the handbook (Păun et al. 2009) contains a comprehensive chapter devoted to membrane systems using string objects processed by splicing). Details and source references for all these ways to get universality by means of controlled finite H systems can be found in Păun et al. (1998).

Therefore, universal computing devices based on the splicing operation can be constructed – at the theoretical level. Implementing such "computers" is, of course, another story. From the computing theory point of view, important insights have been obtained; the universality itself, in any of the cases mentioned above, shows that the computability can be "reconstructed" on the basis of an operation, splicing, which is much different from the usual operation of rewriting (replacing a short substring of a longer string) that is used in most, if not all, classic computing models: Turing machines, Chomsky grammars, Thue and Post systems, and Markov algorithms. This is significant, for instance, in view of the question "what does it mean to compute in a natural way?": at the level of DNA, nature mainly "computes" by means of splicing. (In the next section, it will also be seen that the insertion–deletion operation can lead to universality, but such operations are not so far from rewriting.)

3.5 Further Developments

There are many other directions of research and many results related to the splicing operation. One idea is to have variations of the basic operations, starting with those considered in Head (1987) and Pixton (2000) and continuing with many others, either closer to the biological reality or motivated mathematically. A well-investigated class of H systems is that of simple H systems, whose rules are of the form $r_a = a\#\$a\#$ or $r'_a = \#a\$\#a$, where a is a symbol. Then, we can consider multiple splicing (crossingover) taking place in each step, or we can pass to circular strings; this latter case is both well motivated and interesting and still raises a series of problems that were not solved. Indeed, as we mentioned above, there are open problems also related to the relationship between regular languages and languages in $H_1(FIN, FIN)$ (characterization, decidability, etc.). The splicing was also considered for other data structures than strings, such as trees, arbitrary graphs, arrays of various forms, and even for multisets of symbols.

Furthermore, the architecture of the computing device based on splicing can be of different forms, not only H systems, which directly correspond to formal grammars. Grammar systems and membrane systems (P systems) based on splicing are mentioned above; for instance, test-tube systems, time-varying H systems, two-level systems, and more recently, networks of language processors using splicing were also considered.

On the other hand, many other problems than the computing power were investigated. Descriptional complexity was considered from the very beginning: number of axioms and of rules, length of axioms and size of rules (the maximal length of the strings u_1, u_2, u_3, u_4 in splicing rules, or the vector of lengths of the four strings). Recently, also complexity classes based on H systems were introduced. Universality was proved since the very beginning, but also concrete universal H systems were produced. Splicing was related to Schützenberger recombination of strings by means of *constants* and to other combinatorics on word notions.

Of course, splicing was tested also in laboratory, but this is beyond the scope of this chapter.

For all these topics and for further ones the reader can find information in various titles listed in the bibliography.

4 Computing by Insertion–Deletion

As in the case of splicing, details are not recalled about the way insertion and deletion operations are realized in terms of DNA biochemistry, but a formal language presentation is followed, directly introducing the basic notions related to this approach and then recalling some of the results known in this area. As above, only a sample proof will be sketched.

4.1   Insertion and Deletion Operations

An *ins-del system* is a construct $\gamma = (V, T, A, I, D)$, where V is an alphabet, $T \subseteq V$ (terminal symbols), A is a finite language over V (axioms), and I, D are finite sets of triples of the form (u, α, v), where u, v, α are strings over V; the triples in I are called insertion rules, those in D are called deletion rules, and they are used as follows.

For $x, y \in V^*$ we define

$$x \Longrightarrow_{ins} y \text{ iff } x = x_1 u v x_2, y = x_1 u w v x_2 \text{ for some } (u, w, v) \in I, x_1, x_2 \in V^*$$
$$x \Longrightarrow_{del} y \text{ iff } x = x_1 u w v x_2, y = x_1 u v x_2 \text{ for some } (u, w, v) \in D, x_1, x_2 \in V^*$$

The reflexive and transitive closure of $\Longrightarrow_\alpha$, $\alpha \in \{ins, del\}$ is denoted by $\Longrightarrow_\alpha^*$. When this is clear from the context, the subscript *ins* or *del* is omitted.

For a system γ as above we define the language

$$L(\gamma) = \{w \in T^* \mid z \Longrightarrow^* w, \quad z \in A\}$$

Thus, we start from an axiom of γ, proceed through a finite number of insertion and deletion operations, and we select the strings consisting of terminal symbols – very similar to the way the language generated by a grammar is defined. Actually, there are papers where an ins-del system has the rules presented in a rewriting style: $(u, \lambda/\alpha, v)$ or even $uv \rightarrow u\alpha v$ for insertion rules and $(u, \alpha/\lambda, v)$ or even $u\alpha v \rightarrow uv$ for deletion rules.

The complexity of an ins-del system is evaluated in terms of the complexity of its rules, that is, the length of strings used as contexts and the length of inserted or deleted strings. More precisely, an ins-del system $\gamma = (V, T, A, I, D)$ is *of weight* $(n, m; p, q)$ if

$$n = max\{|\alpha| \,|\, (u, \alpha, v) \in I\}$$
$$m = max\{|u| \,|\, (u, \alpha, v) \in I \text{ or } (v, \alpha, u) \in I\}$$
$$p = max\{|\alpha| \,|\, (u, \alpha, v) \in D\}$$
$$q = max\{|u| \,|\, (u, \alpha, v) \in D \text{ or } (v, \alpha, u) \in D\}$$

We denote by $INS_n^m \, DEL_p^q$, for $n, m, p, q \geq 0$, the family of languages $L(\gamma)$ generated by ins-del systems of weight $(n', m'; p', q')$ such that $n' \leq n, m' \leq m, p' \leq p, q' \leq q$. When one of the parameters n, m, p, q is not bounded, we replace it by $*$. Thus, the family of all ins-del languages is $INS_*^* \, DEL_*^*$. Because the insertion-deletion of the empty string changes nothing, when $n = 0$ we also suppose that $m = 0$, and when $p = 0$ we also suppose that $q = 0$. The meaning of INS_0^0 is that no insertion rule is used, and the meaning of DEL_0^0 is that no deletion rule is used. When $m = 0$ or $q = 0$, we have a context-free insertion or deletion, respectively; if both these parameters are zero, then we get a context-free ins-del system.

This section will be concluded with a few simple examples.

The ins-del system

$$\gamma = (\{a, b\}, \{a, b\}, \{ab\}, \{(a, ab, b)\}, \emptyset)$$

clearly generates the (linear nonregular) language $L(\gamma) = \{a^n b^n \mid n \geq 1\}$. If the set of insertion rules is replaced with $I = \{(\lambda, ab, \lambda)\}$, then the generated language is the Dyck language over the alphabet $\{a, b\}$, excepting the empty string.

As a more complex ins-del system, the following one from Margenstern et al. (2005) can be recalled:

$$\gamma = (\{S, S', a, b\}, \{a, b\}, \{S\}, \{(\lambda, S'aSb, \lambda), \ (\lambda, S'ab, \lambda)\}, \{(\lambda, SS', \lambda)\})$$

Because of the selection of terminal strings, symbols S, S' should not be present in the strings that form the language $L(\gamma)$; such symbols can be removed only by the deletion rule (λ, SS', λ), which means that always S' is introduced in the right hand of a symbol S. Initially, S is alone; assume that we have a string $w_1 S w_2$ (initially, $w_1 = w_2 = \lambda$). We have to use the insertion rule $(\lambda, S'aSb, \lambda)$ such that we get $w_1 SS'aSbw_2$, otherwise SS' cannot be removed. In this way, we pass to the string $w_1 aSbw_2$. This process can be repeated. When using the insertion rule $(\lambda, S'ab, \lambda)$, the obtained string, $w_1 SS'abw_2$, can be turned to a terminal one. Consequently, we get $L(\gamma) = \{a^n b^n \mid n \geq 1\}$.

4.2 The Power of Ins-Del Systems

Continuing the previous examples, we recall now a counterpart of ❯ Lemma 5 for ins-del systems.

Lemma 6 $REG \subset INS_*^* DEL_0^0$.

Proof Let L be a regular language and let $M = (K, V, q_0, F, \delta)$ be the minimal deterministic finite automaton recognizing L.

For each $w \in V^*$, we define the mapping $\rho_w : K \to K$ by

$$\rho_w(q) = q' \text{ iff } (q, w) \vdash^* (q', \lambda), \ q, q' \in K$$

Obviously, if $x_1, x_2 \in V^*$ are such that $\rho_{x_1} = \rho_{x_2}$, then for every $u, v \in V^*$, ux_1v is in L if and only if ux_2v is in L.

The set of mappings from K to K is finite. Hence the set of mappings ρ_w as above is finite. Let n_0 be their number. We construct the ins-del system $\gamma = (V, V, A, I, \emptyset)$ with

$$A = \{w \in L \mid |w| \leq n_0 - 1\}$$
$$I = \{(w, v, \lambda) \mid |w| \leq n_0 - 1, 1 \leq |v| \leq n_0, |wv| \leq n_0, \text{ and } \rho_w = \rho_{wv}\}$$

From the definition of mappings ρ_w and the definitions of A, I, it follows immediately that $L(\gamma) \subseteq L$.

Assume that the converse inclusion is not true and let $x \in L - L(\gamma)$ be a string of minimal length with this property. Thus $x \notin A$. Hence $|x| \geq n_0$. Let $x = zz'$ with $|z| = n_0$ and $z' \in V^*$. If $z = a_1 a_2 \ldots a_{n_0}$, then it has $n_0 + 1$ prefixes, namely, $\lambda, a_1, a_1 a_2, \ldots, a_1 \ldots a_{n_0}$. There are only n_0 different mappings ρ_w. Therefore there are two prefixes u_1, u_2 of z such that $u_1 \neq u_2$ and $\rho_{u_1} = \rho_{u_2}$. With no loss in generality it may be assumed that $|u_1| < |u_2|$. By substituting u_2 by u_1 we obtain a string x' that is also in L. As $|x'| < |x|$ and x was of minimal length in $L - L(\gamma)$, we obtain $x' \in L(\gamma)$. However, $|u_2| - |u_1| \leq |u_2| \leq n_0$, so if $u_2 = u_1 u_3$, then (u_1, u_3, λ) is an insertion rule in I. This implies that $x' \Longrightarrow_{ins} x$, that is, $x \in L(\gamma)$, a contradiction. In conclusion, $L \subseteq L(\gamma)$.

The strictness of the inclusion is obvious (see, for instance, the first example considered in the end of the previous section).

We recall now a few non-universality results, that is, pointing out classes of ins-del systems, which generate families of languages strictly included in RE (references are omitted for those results that also appear in Păun et al. (1998); the proofs are in general based on simulating type-0 grammars in Kuroda normal form or in Geffert normal form).

Theorem 4 *The following relations hold:*

1. $INS_*^* DEL_0^0 \subset CS$ and $LIN - INS_*^* DEL_0^0 \neq \emptyset$, but $INS_2^2 DEL_0^0$ contains non-semilinear languages.
2. $INS_*^1 DEL_0^0 \subset CF$.
3. $INS_*^0 DEL_1^0 \subset CF$, $INS_2^0 DEL_2^0 \subset CF$, $INS_1^0 DEL_*^0 \subset REG$ (Verlan 2007).
4. $REG \subset INS_*^* DEL_0^0$ and each regular language is the coding of a language in the family $INS_*^1 DEL_0^0$.

There are however a lot of universality results in this area (characterizations of RE languages). Some of them are recalled here.

Theorem 5 *We have $INS_n^m DEL_p^q = RE$ for all the following quadruples $(n,m;p,q)$ (for each of these quadruples we also mention the place where a proof can be found, omitting again the cases whose proof can be found in* Păun et al. (1998)): *(i) (1,2;1,1), (ii) (1,2;2,0), (iii) (2,1;2,0), (iv) (1,1;1,2) – Tanaka and Yokomori (2003), (v) (2,1;1,1) – Tanaka and Yokomori (2003), (vi) (2,0;3,0) – Margenstern et al. (2005), (vii) (3,0;2,0) – Margenstern et al. (2005), (viii) (1,1;2,0), (ix) (1,1;1,1) – Tanaka and Yokomori (2003).*

There are also a series of characterizations of RE languages by means of operations with languages starting from languages generated by ins-del systems. Only one of them is recalled here, together with some details of its proof, namely, the central result from Păun et al. (2008), which gives a Chomsky–Schützenberger-like characterization for RE languages (note that only insertion rules are used).

Theorem 6 *Each language $L \in RE$ can be represented in the form $L = h(L' \cap D)$, where $L' \in INS_3^0 DEL_0^0$, h is a projection, and D is a Dyck language.*

Proof (Sketch) Consider a language $L \subseteq T^*$, generated by a type-0 grammar $G = (N, T, S, P)$ in Kuroda normal form. Assume that the rules of P are labeled in a one-to-one manner with elements of a set $Lab(P)$.

We construct an ins-del system $\gamma = (V \cup \overline{V}, V \cup \overline{V}, \{S\}, I, \emptyset)$, of weight $(3, 0; 0, 0)$, with

$$V = N \cup T \cup Lab(P)$$

and with I containing the following insertion rules.

- Group 1: For each rule $r : AB \to CD$ of type 1 in P we construct the following two insertion rules: (λ, CDr, λ) and $(\lambda, \overline{BA\overline{r}}, \lambda)$
- Group 2: For each rule $r : A \to BC$ of type 2 in P we construct the following two insertion rules: (λ, BCr, λ) and $(\lambda, \overline{A\overline{r}}, \lambda)$
- Group 3: For each rule $r : A \to a$ of type 3 in P we construct the following two insertion rules: $(\lambda, a\overline{a}r, \lambda)$ and $(\lambda, \overline{A\overline{r}}, \lambda)$, where $\overline{\lambda} = \lambda$

For a rule $r : u \to v$ in P we say that two rules (λ, vr, λ) and $(\lambda, \overline{u^R\overline{r}}, \lambda)$ in P' are *r-complementary*, and denote their labels by r_+ and r_-, respectively, (x^R is the reversal of the string x).

We define the projection $h : (V \cup \overline{V})^* \to T^*$ by $h(a) = a$ for all $a \in T$, and $h(a) = \lambda$ otherwise. Let D be the Dyck language over V, taking as pairs $(a, \overline{a})$ for $a \in V$.

We have $L(G) = h(L(\gamma) \cap D)$. To begin with some useful notions are introduced.

For any rule $r : u \to v \in P$, let $U_r(u) = ru\overline{u}^R\overline{r}$; we call this an r-block. Then, this notion is extended to define U-structures as follows:

1. An r-block $U_r(u)$ is a U-structure.
2. If U_1 and U_2 are U-structures, then $U_1 U_2$ is a U-structure.
3. Let α_i, $i = 1, 2, 3$, be U-structures or empty, with at least one α_i being non-empty; consider a string of the form $r\alpha_1 u_1 \alpha_2 u_2 \alpha_3 \overline{u}^R\overline{r}$, where $u = u_1 u_2$ is such that $r : u \to v \in P$. Then, this string, denoted by $U_r(\alpha_1 u_1 \alpha_2 u_2 \alpha_3)$, is a U-structure.
4. Nothing else is a U-structure.

Let us now define a mapping ϕ over $(V \cup \overline{V})$ as follows: For any $a \in V - T$, let $a\overline{a} \sim \lambda$, and for any $a \in T$, let $a\overline{a} \sim a$. Then, one can consider a reduction operation over $(V \cup \overline{V})^*$ by iteratively using the binary relation $\sim$. We define $\phi(w)$ as the string finally obtained as the *irreducible* string in terms of this reduction operation. (Because the symbols from T and from $V - T$ are subject of different "reduction rules," the irreducible string reached when starting from a given string is unique, hence the mapping ϕ is correctly defined.)

The following lemma holds:

Lemma 7 *Let $S \Longrightarrow^{n-1} z_{n-1} (= \alpha u \beta) \Longrightarrow z_n (= \alpha v \beta)$ in G, where $r : u \to v$ is used in the last step. Then, there exists a derivation of γ such that $S \Longrightarrow^{2n} \tilde{z}_n$ and $\phi(\tilde{z}_n) = z_n$.*

The following observations are useful for the proof of the inclusion $h(L(\gamma) \cap D) \subseteq L(G)$.

Observation 1 For a rule $r : u \to v$ in P, let $r_+ : (\lambda, vr, \lambda)$ and $r_- : (\lambda, \overline{u}^R\overline{r}, \lambda)$ be the two r-complementary rules.

1. Any successful derivation of γ requires the use of both r-complementary rules r_+ and r_-.
2. Let $\tilde{z}$ be any sentential form in a derivation of γ that eventually leads to a string in D (we say that such a derivation is *successful*). Then, it must hold that for any prefix α of $\tilde{z}$, $\#_{vr}(\alpha) \geq \#_{\overline{u}^R\overline{r}}(\alpha)$, where $\#_x(\alpha)$ denotes the number of occurrences of a string x in α.
3. Applying insertion rules within a U-structure leads to only invalid strings. Indeed, suppose that r_+ and r_- are applied on some occurrence of u appearing in a U-structure $U_{r'}(\delta_1 u\delta_2) = r'\delta_1 u\delta_2 \overline{u}^R\overline{r'}$, where $r' : u \to v'$. This derives a string $r'\delta_1 vU_r(u)\delta_2 \overline{u}^R\overline{r'}$ that leads to an invalid string (i.e., *not* in D) unless $u = v$. This also occurs in the case where u appears in separate locations in the U-structure.
4. A location in $\tilde{z}$ is called *valid* for two r-complementary rules if it is either immediately before u_1 for r_+ or immediately after u_2 for r_-, by ignoring U-structures in $\tilde{z}$, where $u = u_1 u_2$. Then, applying insertion rules at valid locations only leads to valid strings. This is seen as follows: From 1, 2, 3 above, the locations for r_+ and r_- to be used are restricted to somewhere in the left and right, respectively, of u. In order to derive a valid string from $\tilde{z}$, it is necessary to apply r_+ and r_- to u so that these two rules together with u may eventually lead to forming a U-structure.

Now, the following result holds:

Lemma 8 *Let $S \Longrightarrow^{2n} \tilde{z}$ in γ and $\phi(\tilde{z}) = z(\in (N \cup T)^*)$. Then, one has $S \Longrightarrow^n z$ in G.*

The proof of the theorem can be completed now.

For any $w \in L(G)$, consider a derivation $S \Longrightarrow^* w$. Then, by ❷ Lemma 7 there exists a derivation $S \Longrightarrow^* \tilde{w}$ in γ such that $\phi(\tilde{w}) = w$. Since ϕ deletes only U-structures and elements of $\overline{T}$, this implies that $\tilde{w} \in D$ and $h(\tilde{w}) = w \in T^*$. Thus, $w \in h(L(\gamma) \cap D)$. Hence, we have $L(G) \subseteq h(L(\gamma) \cap D)$.

Conversely, suppose that $S \Longrightarrow^*$ in γ and $\phi(\tilde{w}) = w(\in T^*)$. Then, by ❷ Lemma 8 we have $S \Longrightarrow^* w$ in G. Again, $\phi(\tilde{w}) \in T^*$ implies that $\tilde{w} \in D$ and $h(\tilde{w}) = w$. Thus, we have $h(L(\gamma) \cap D) \subseteq L(G)$.

4.3 Further Remarks

As in the case of computing by splicing, there also are several other developments – and a series of open problems – in the case of computing by insertion–deletion, starting with the linguistically motivated models and results (for instance, related to Marcus contextual grammars), and ending with insertion–deletion operations for data structures other than strings. The bibliography below provides several titles that might be useful to the interested reader.

5 Concluding Remarks

Computing by splicing and by insertion–deletion provides convincing evidence that computer science – more specifically computability theory – has much to learn from the biochemistry of DNA. Whether or not this will prove to be useful to practical computer science is still a matter for further research, but from a theoretical point of view, we stand to gain much insight into computability from reasoning in this framework; in particular, it enables everyone to ask the question: "what does it mean to compute in a *natural way*?"

References

Bonizzoni P, Mauri G (2005) Regular splicing languages and subclasses. Theor Comput Sci 340:349–363

Bonizzoni P, Mauri G (2006) A decision procedure for reflexive regular splicing languages. In: Proceedings of the developments in language theory '06, Santa Barbara, Lecture notes in computer science, vol. 4036. Springer, Berlin, pp 315–326

Bonizzoni P, De Felice C, Mauri G, Zizza R (2003) Regular languages generated by reflexive finite splicing systems. In: Proceedings of the developments in language theory '03, Szeged, Hungary, Lecture notes in computer science, vol. 2710. Springer, Berlin, pp 134–145

Bonizzoni P, De Felice C, Mauri G (2005a) Recombinant DNA, gene splicing as generative devices of formal languages. In: Proceedings of the Computability in Europe '05, Amsterdam, The Netherlands, Lecture notes in computer science, vol. 3536. Springer, Berlin, pp 65–67

Bonizzoni P, De Felice C, Zizza R (2005b) The structure of reflexive regular splicing languages via Schützenberger constants. Theor Comput Sci 334:71–98

Cavaliere M, Jonoska N, Leupold P (2006) Computing by observing DNA splicing. Technical Report 11/2006, Microsoft Center for Computational Biology, Trento

Ceterchi R, Subramanian KG (2003) Simple circular splicing systems. Romanian J Inform Sci Technol 6:121–134

Csuhaj-Varju E, Dassow J, Kelemen J, Păun Gh (1994) Grammar systems. A grammatical approach to distribution and cooperation. Gordon & Breach, London

Culik II K, Harju T (1991) Splicing semigroups of dominoes and DNA. Discrete Appl Math 31:261–277

Dassen R, Hoogebooom HJ, van Vugt N (2001) A characterization of non-iterated splicing with regular rules. In: Martin-Vide C, Mitrana V (eds) Where mathematics, computer science, linguistics and biology meet. Kluwer, Dordrecht, pp 319–327

Dassow J, Păun Gh (1989) Regulated rewriting in formal language theory. Springer, Berlin

Dassow J, Vaszil G (2004) Multiset splicing systems. BioSystems 74:1–7

De Felice C, Fici G, Zizza R (2007) Marked systems and circular splicing. In: Proceedings of the Fundamentals of Computation theory, Budapest, Hungary, Lecture notes in computer science, vol. 4639. Springer, Berlin, pp 238–249

Frisco P (2004) Theory of molecular computing. Splicing and membrane computing. Ph.D. thesis, Leiden University, The Netherlands

Galiukschov BS (1981) Semicontextual grammars (in Russian). Mat logica i mat ling, Tallinn Univ 38–50

Goode E, Pixton D (2001) Semi-simple splicing systems. In: Martin-Vide C, Mitrana V (eds) Where mathematics, computer science, linguistics and biology meet. Kluwer, Dordrecht, pp 343–352

Goode E, Pixton D (2007) Recognizing splicing languages: syntactic monoids and simultaneous pumping. Discrete Appl Math 155:989–1006

Harju T, Margenstern M (2005) Splicing systems for universal Turing machines. In: Proceedings of the DNA Computing '04, Milano, Italy, Lecture notes in computer science, vol. 3384. Springer, Berlin, pp 149–158

Harrison M (1978) Introduction to formal language theory. Addison-Wesley, Reading, MA

Head T (1987) Formal language theory and DNA: an analysis of the generative capacity of specific recombinant behaviors. Bull Math Biol 49:737–759

Head T, Păun Gh, Pixton D (1997) Language theory and molecular genetics. Generative mechanisms suggested by DNA recombination. In: Rozenberg G, Salomaa A (eds) Handbook of formal languages, vol. 2. Springer, Berlin, pp 295–360

Hemalatha S (2007) A study on rewriting P systems, splicing grammar systems and picture array languages. Ph.D. thesis, Anna University, Chennai, India

Jonoska N, Păun Gh, Rozenberg G (eds) (2004) Aspects of molecular computing. Essays dedicated to Tom Head on the occasion of his 70th birthday, Lecture notes in computer science, vol. 2950. Springer, Berlin

Kari L (1991) On insertion and deletion in formal languages. Ph.D. thesis, University of Turku

Kari L, Sosik P (2008) On the weight of universal insertion grammars. Theor Comput Sci 396:264–270

Krassovitskiy A, Rogozhin Y, Verlan S (2007) Further results on insertion-deletion systems with one-sided contexts. Proceedings of the LATA 2007, Tarragona, Spain, Technical Rep. RGML, 36/2008, pp 347–358

Krassovitskiy A, Rogozhin Y, Verlan S (2008) One-sided insertion and deletion: traditional and P systems case. In: Csuhaj-Varju et al. (eds) International workshop on computing with biomolecules. Vienna, Austria, pp 51–63

Loos R (2006) An alternative definition of splicing. Theor Comput Sci 358:75–87

Loos R, Ogihara M (2007) Complexity theory of splicing systems. Theor Comput Sci 386:132–150

Loos R, Malcher A, Wotschke D (2008) Descriptional complexity of splicing systems. Int J Found Comput Sci 19:813–826

Manca V (2000) Splicing normalization and regularity. In: Calude CS, Păun Gh (eds) Finite versus infinite. Contributions to an eternal dilemma, Springer, Berlin, pp 199–215

Marcus S (1969) Contextual grammars. Rev Roum Math Pures Appl 14:1525–1534

Margenstern M, Rogozhin Y, Verlan S (2002) Time-varying distributed H systems of degree 2 can carry out parallel computations. In: Proceedings of the DNA computing '02, Sapporo, Japan, Lecture notes in computer science, vol. 2568. Springer, Berlin, pp 326–336

Margenstern M, Rogozhin Y, Verlan S (2004) Time-varying distributed H systems with parallel computations: the problem is solved. In: Proceedings of the DNA Computing '04, Madison, Wisconsin, Lecture notes in computer science, vol. 2943. Springer, Berlin, pp 48–53

Margenstern M, Păun Gh, Rogozhin Y, Verlan S (2005) Context-free insertion-deletion systems. Theor Comput Sci 330:339–348

Mateescu A, Păun Gh, Rozenberg G, Salomaa A (1998) Simple splicing systems. Discrete Appl Math 84:145–163

Matveevici A, Rogozhin Y, Verlan S (2007) Insertion-deletion systems with one-sided contexts. In: Proceedings of the machines, computations, and universality '07, Orleans, France, LNCS 4664, Springer, 2007, 205–217

Păun Gh (1996a) On the splicing operation. Discrete Appl Math 70:57–79

Păun Gh (1996b) Regular extended H systems are computationally universal. J Auto Lang Comb 1:27–36

Păun Gh (1997) Marcus contextual grammars. Kluwer, Boston, MA

Păun A (2003) Unconventional models of computation: DNA and membrane computing. Ph.D. thesis, University of Western Ontario, Canada

Păun Gh, Rozenberg G, Salomaa A (1996) Computing by splicing. Theor Comput Sci 168:321–336

Păun Gh, Rozenberg G, Salomaa A (1998) DNA computing. New computing paradigms. Springer, Berlin

Păun Gh, Pérez-Jiménez MJ, Yokomori T (2008) Representations and characterizations of languages in Chomsky hierarchy by means of insertion-deletion systems. Int J Found Comput Sci 19:859–871

Păun Gh, Rozenberg G, Salomaa A (eds) (2009) Handbook of membrane computing. Oxford University Press, Oxford, UK

Penttonen M (1974) One-sided and two-sided contexts in phrase structure grammars. Inform Control 25:371–392

Pixton D (1996) Regularity of splicing languages. Discrete Appl Math 69:101–124

Pixton D (2000) Splicing in abstract families of languages. Theor Comput Sci 234:135–166

Rozenberg G, Salomaa A (eds) (1997) Handbook of formal languages, 3 vol. Springer, Berlin

Salomaa A (1973) Formal languages. Academic, New York

Tanaka A, Yokomori T (2003) On the computational power of insertion-deletion systems. In: Proceedings of the DNA Computing '02, Sapporo, Japan, Lecture notes in computer science, vol. 2568. Springer, Berlin, pp 269–280

Thomas DG, Begam MH, David NG (2007) Hexagonal array splicing systems. Ramanujan Math Soc Lect Notes Ser 3:197–207

Verlan S (2007) On minimal context-free insertion-deletion systems. J Auto Lang Comb 12:317–328

Verlan S, Zizza R (2003) 1-splicing vs. 2-splicing: separating results. In: Harju T, Karhumaki J (eds) Proceedings of the WORDS'03, TUCS General Publisher, 27, pp 320–331

Zizza R (2002) On the power of classes of splicing systems. Ph.D. thesis, University of Milano-Bicocca

36 Bacterial Computing and Molecular Communication

Yasubumi Sakakibara[1] · *Satoshi Hiyama*[2]
[1]Department of Biosciences and Informatics, Keio University, Yokohama, Japan
yasu@bio.keio.ac.jp
[2]Research Laboratories, NTT DOCOMO, Inc., Yokosuka, Japan
hiyama@nttdocomo.co.jp

G. Rozenberg et al. (eds.), *Handbook of Natural Computing*, DOI 10.1007/978-3-540-92910-9_36,
© Springer-Verlag Berlin Heidelberg 2012

Abstract

Emerging technologies that enable the engineering of nano- or cell-scale systems using biological and/or artificially synthesized molecules as computing and communication devices have been receiving increased attention. This chapter focuses on "bacterial computing," which attempts to create an autonomous cell-based Turing machine, and "molecular communication," which attempts to create non-electromagnetic-wave-based communication paradigms by using molecules as an information medium. ❯ Section 2 introduces seminal works for constructing in vivo logic circuits, and focuses on research into implementing in vitro and in vivo finite automata in the framework of DNA-based computing. Furthermore, the first experimental development of a programmable in vivo computer that executes a finite-state automaton in bacteria is highlighted. ❯ Section 3 reports on the system design, experimental results, and research trends of molecular communication components (senders, molecular communication interfaces, molecular propagation systems, and receivers) that use bacteria, lipids, proteins, and DNA as communication devices.

1   Introduction

1.1   Brief Introduction to Bacterial Computing

Biological molecules such as DNA, RNA, and proteins are natural devices that store information, activate (chemical) functions, and communicate between systems (such as cells). Research in DNA computing aims to utilize these biological devices to make a computer. One of the ultimate goals is to make an autonomous cell-based Turing machine for genetic and life engineering.

Finite-state automata (machines) is the most basic computational model in the Chomsky hierarchy and is the starting point for building universal DNA computers. Many works have proposed theoretical models of DNA-based computer to develop finite-state automata (Păun et al. 1998), and several experimental research works have attempted to implement finite automata in vitro and in vivo. For example, Benenson et al. (2001, 2003) successfully implemented the two-state finite automata in vitro by sophisticated use of the restriction enzyme (actually, *Fok*I), which is cut outside of its recognition site in a double-stranded DNA. However, their method has some limitations for extending to more than two states. Yokomori et al. (2002) proposed a theoretical framework using a length-encoding technique to implement finite automata on DNA molecules. Theoretically, the length-encoding technique has no limitations to implement finite automata of any larger states. Based on the length-encoding technique, Sakakibara and his group (Kuramochi and Sakakibara 2005) attempted to implement and execute finite automata of a larger number of states in vitro, and carried out intensive laboratory experiments on various finite automata from two states to six states for several input strings.

On the other hand, Sakakibara and his group (Nakagawa et al. 2005) proposed a method using a protein-synthesis mechanism combined with four-base codon techniques to simulate a computation (accepting) process of finite automata in vivo (a codon is normally a triplet of bases, and different base triplets encode different amino acids in protein). The proposed method is quite promising, has several advanced features such as the protein-synthesis process, is very accurate, overcomes the mis-hybridization problem in the self-assembly computation, and further offers an autonomous computation.

Attempts to implement a finite number of states in vivo have also been actively studied in the area of synthetic biology. One of the most significant goals of synthetic biology is the rewiring of genetic regulatory networks to realize controllable systems in a cell. Recently, this technology has been applied to the development of biomolecular computations in vivo. Weiss et al. (2003, 2004) explored in vivo technology to develop genetic circuits and perform logical computations in cells. They developed cellular units that implement the logical functions NOT, IMPLIES, and AND. These units are combined into genetic circuits for cellular logical computation.

The first half of this chapter overviews seminal works of Weiss et al. (2003, 2004) on the construction of in vivo logic circuits, and research by Sakakibara and his group (Kuramochi and Sakakibara 2005; Nakagawa et al. 2005) on implementing finite automata in vitro and in vivo in the framework of DNA-based computing.

First, the length-encoding technique proposed in Yokomori et al. (2002) and Sakakibara and Hohsaka (2003) is introduced to implement finite automata in test tubes. In the length-encoding method, the states and state transition functions of a target finite automaton are effectively encoded into DNA sequences, a computation (accepting) process of finite automata is accomplished by self-assembly of encoded complementary DNA strands, and the acceptance of an input string is determined by the detection of a completely hybridized double-strand DNA. Second, intensive in vitro experiments are shown in which several finite-state automata have been implemented and executed in test tubes. Practical laboratory protocols that combine several in vitro operations such as annealing, ligation, PCR, and streptavidin–biotin bonding are designed and developed to execute in vitro finite automata based on the length-encoding technique. Laboratory experiments are carried out on various finite automata from two states to six states for several input strings. Third, a novel framework to develop a programmable and autonomous in vivo computer using *Escherichia coli (E. coli)* is presented, and in vivo finite-state automata are implemented based on the framework by employing the protein-synthesis mechanism of *E. coli*. The fundamental idea to develop a programmable and autonomous finite-state automata in *E. coli* is that first an input string is encoded into one plasmid, state-transition functions are encoded into the other plasmid, and those two plasmids are introduced into an *E. coli* cell by electroporation. Fourth, a protein-synthesis process in *E. coli* combined with four-base codon techniques is executed to simulate a computation (accepting) process of finite automata, as proposed for in vitro translation-based computations in Sakakibara and Hohsaka (2003). This approach enables one to develop a programmable in vivo computer by simply replacing a plasmid encoding a state-transition function with others. Furthermore, the in vivo finite automata are autonomous because the protein-synthesis process is autonomously executed in the living *E. coli* cell.

1.2 Brief Introduction to Molecular Communication

The second half of this chapter overviews molecular communication that uses molecules (chemical substances such as proteins and deoxyribonucleic acid (DNA)) as information media and offers a new communication paradigm based on biochemical reactions caused by received molecules (Hiyama et al. 2005a, 2008b; Moritani et al. 2007a; Suda et al. 2005).

Molecular communication has received increasing attention as an interdisciplinary research area that spans the fields of nanotechnology, biotechnology, and communication engineering.

Molecular communication is inspired by the observation of specific features of biological molecular systems that living organisms have acquired through the evolutionary process over billions of years. Communication in biological systems is typically done through molecules (i.e., chemical signals). For instance, multicellular organisms including human beings perform maintenance of homeostasis, growth regulation, kinematic control, memory, and learning through intercellular and intracellular communication using signal-transducing molecules such as hormones (Alberts et al. 1997; Pollard and Earnshaw 2004). Molecular communication aims to develop artificially designed and controllable systems that could transmit biochemical information such as phenomena and the status of living organisms that cannot be feasibly carried with traditional communication using electromagnetic waves (i.e., electronic and optical signals).

In molecular communication, a sender generates molecules, encodes information into the molecules (called information molecules), and emits the information molecules to the propagation environment. A propagation system transports the emitted information molecules to a receiver. The receiver, upon receiving the transported information molecules, reacts biochemically according to the information molecules (this biochemical reaction represents decoding of the information).

Molecular communication is a new communication paradigm and is different from the traditional communication paradigm (❷ *Table 1*). Unlike traditional communication which utilizes electromagnetic waves as an information medium, molecular communication utilizes molecules as an information medium. In addition, unlike traditional communication in which encoded information such as voice, text, and video is decoded and regenerated at a receiver, in molecular communication information molecules activate some biochemical reactions at a receiver and recreate phenomena and/or chemical status that a sender transmits. Other features of molecular communication include aqueous environmental communication, the stochastic nature of communication, low energy-consumption communication, and high compatibility with biological systems.

❑ Table 1

Comparisons of key features between traditional communication and molecular communication

Key features	Traditional communication	Molecular communication
Information medium	Electromagnetic waves	Molecules
Signal type	Electronic and optical signals	Chemical signals
Propagation speed	Light speed (3×10^5 km/s)	Slow speed (a few $\mu m/s$)
Propagation distance	Long (ranging from m to km)	Short (ranging from nm to m)
Propagation environment	Airborne and cable medium	Aqueous medium
Encoded information	Voice, text, and video	Phenomena and chemical status
Behavior of receivers	Decoding of digital info	Biochemical reaction
Communication model	Deterministic communication	Stochastic communication
Energy consumption	High	Extremely low

Although the communication speed/distance of molecular communication is slower/shorter than that of traditional communication, molecular communication may carry information that is not feasible to carry with traditional communication (such as the biochemical status of a living organism) between entities to which traditional communication does not apply (such as biological entities). Molecular communication has unique features that are not seen in traditional communication and is not competitive but complementary to traditional communication.

Molecular communication is an emerging research area and its feasibility is being verified through biochemical experiments. Detailed system design, experimental results, and research trends in molecular communication are described in ❥ Sect. 3.

2 Bacterial Computing

2.1 In Vivo Genetic Circuit

Seminal work to develop genetic circuits and perform logical computations in cells was done by Weiss et al. (2003, 2004). They developed cellular units that implement the logical functions NOT, IMPLIES, and AND. These units were combined into genetic circuits for cellular logical computation. These genetic circuits employed messenger RNA (mRNA), DNA-binding proteins, small molecules that interacted with the related proteins, and DNA motifs that regulated the transcription and expression of the proteins. They further attempted to construct a circuit component library so that genetic circuit design is the process of assembling library components into a logical circuit.

2.1.1 AND Gate

Given a library of circuit components, various genetic circuits are designed by combining and assembling a set of library components. For example, the AND logical function was implemented in cells by using RNA polymerase, activator proteins, inducer molecules, and DNA that encodes the RNA polymerase binding domain, the activator binding domain, and an output gene (Weiss et al. 2003). In this genetic circuit, the input and output signals are represented by the concentrations of some mRNAs or small molecules.

There can be several designs to implement the same logical function AND. Sakakibara et al. (2006) developed another circuit design for implementing the AND function and performed an in vivo experiment to run the AND circuit. As basic components, they used RNA polymerase, activator protein LuxR, repressor protein LacI, two inducer molecules, GFP protein, and DNA that encodes the RNA polymerase binding domain, the activator binding domain, the repressor binding domain, and the GFP gene. In this circuit design, the input signals are two inducers AHL and IPTG, and the output signal is GFP protein. This genetic circuit executes the AND logical function as follows. If only the IPTG inducer is present, the activator LuxR is active with IPTG, binds to the DNA promoter and activates RNA polymerase transcription, but the repressor LacI is still active, it keeps binding to the DNA promoter and hence prevents transcription. If only the AHL inducer is present, the inducer binds to the repressor LacI and changes the conformation of the repressor. The conformation change prevents the repressor from binding to the DNA promoter. However, the activator LuxR is inactive without IPTG and hence does not activate the RNA polymerase transcription. Finally, if both inducers AHL and IPTG are present, the inactive repressor LacI does not bind to the

☐ **Fig. 1**

(*Left*) A genetic circuit design for AND function and (*right*) its experimental results.

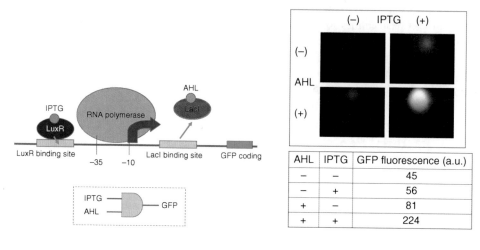

☐ **Fig. 2**

(*Left*) Toggle switch circuit and (*right*) two states using gene expression levels.

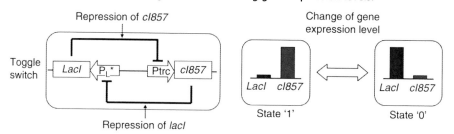

DNA promoter and the activator LuxR activates the RNA polymerase transcription, and therefore RNA polymerase transcribes the GFP gene, yielding the GFP output. The genetic circuit design is illustrated in ❯ *Fig. 1* (left) and the experimental results for executing the AND logical function are shown in ❯ *Fig. 1* (right).

2.1.2 Toggle Switch

In order to realize a finite number of states in vivo, Gardner et al. (2000) proposed the toggle switch mechanism. The toggle switch is composed of two repressors and two constitutive promoters (see ❯ *Fig. 2* (left)). Each promoter is inhibited by the repressor that is transcribed by the opposing promoter. The behavior of the toggle switch and the conditions for bistability to represent two states can be formalized using the following equations:

$$\frac{dV_{cI857}}{dt} = \frac{\alpha_1}{1 + V_{lacI}^{\beta}} - V_{cI857}$$

$$\frac{dV_{lacI}}{dt} = \frac{\alpha_2}{1 + V_{cI857}^{\gamma}} - V_{lacI}$$

where V_{cI857} is the concentration of repressor gene $cI857$, V_{lacI} is the concentration of repressor gene $lacI$, α_1 is the effective rate of synthesis of repressor $cI857$, α_2 is the effective rate of synthesis of repressor $lacI$, β is the cooperativity of repression of promoter Ptrc, and γ is the cooperativity of repression of promoter $P_{L^\cdot}$. The bistability of the toggle switch depends on these parameters. The state where $cI857$ is expressed and $lacI$ is repressed is termed state "1" and the state where $cI857$ is repressed and $lacI$ is expressed is termed state "0," as shown in ❯ *Fig. 2* (right). The state 1 is switched to state 0 by a thermal pulse of $42\,^\circ$C induction and the state 0 is switched to state 1 by a pulse of IPTG induction, which is a molecular compound that triggers transcription of the lac operon.

2.2 Bacterial Finite-State Machine

2.2.1 Preliminaries

The aim of bacteria-based computing is to develop a computing machine in a living system. Sakakibara and his group (Nakagawa et al. 2005) have attempted to simulate a computation process of finite automata in *Escherichia coli (E. coli)* by employing the in vivo protein-synthesis mechanism of *E. coli*. (*Escherichia coli* is a typical bacterium living inside our body, in the large intestine.) This in vivo computation possesses the following two novel features, not found in any previous biomolecular computer. First, an in vivo finite automaton is implemented in a living *E. coli* cell. Second, this automaton increases in number very rapidly according to bacterial growth; one bacterial cell can multiply to over a million cells overnight. This chapter explores the feasibility of in vivo computation.

The main feature of the in vivo computer based on *E. coli* is that first an input string is encoded into one plasmid, state-transition functions are encoded into the other plasmid, and *E. coli* cells are transformed with these two plasmids by electroporation. Second, a protein-synthesis process in *E. coli* combined with four-base codon techniques is executed to simulate a computation (accepting) process of finite automata, which has been proposed for in vitro translation-based computations in Sakakibara and Hohsaka (2003). The successful computations are detected by observing the expressions of a reporter gene linked to mRNA encoding input data. Therefore, when an encoded finite automaton accepts an encoded input string, the reporter gene, *lacZ*, is expressed and hence a blue color is observed. When the automaton rejects the input string, the reporter gene is not expressed and hence no blue color is observed. The in vivo computer system based on *E. coli* is illustrated in ❯ *Fig. 3*.

Thus, this *E. coli*-based computer enables one to develop a programmable and autonomous computer. As known to the authors, this is the first experimental development of an in vivo computer and it has succeeded in executing a finite-state automaton using *E. coli*.

2.2.2 A Framework of a Programmable and Autonomous In Vivo Computer using *E. coli*

Length-Encoding Method to Implement Finite-State Automata

Let $M = (Q, \Sigma, \delta, q_0, F)$ be a (deterministic) finite automaton, where Q is a finite set of states numbered from 0 to k, Σ is an alphabet of input symbols, δ is a state-transition function such that $\delta : Q \times \Sigma \rightarrow Q$, q_0 is the initial state, and F is a set of final states. The length-encoding

◼ **Fig. 3**

The framework of an in vivo computer system based on *E. coli.*

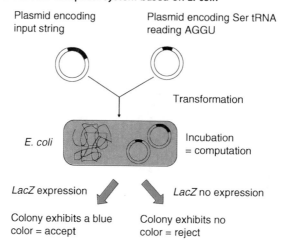

technique (Yokomori et al. 2002) is adopted to encode each state in Q by the length of DNA subsequences.

For the alphabet Σ, each symbol a in Σ is encoded into a single-strand DNA subsequence, denoted $e(a)$, of fixed length. For an input string w on Σ, $w = x_1 x_2 \cdots x_m$ is encoded into the following single-strand DNA subsequence, denoted $e(w)$:

$$5'\text{-}e(x_1) \underbrace{X_1 X_2 \cdots X_k}_{k \text{ times}} e(x_2) \underbrace{X_1 X_2 \cdots X_k}_{k \text{ times}} \cdots e(x_m) \underbrace{X_1 X_2 \cdots X_k}_{k \text{ times}} \text{-}3'$$

where X_i is one of four nucleotides A,C,G,T, and the subsequences $X_1 X_2 \cdots X_k$ are used to encode $k+1$ states of the finite automaton M. For example, when a symbol "1" is encoded into a ssDNA subsequence GCGC and a symbol "0" into GGCC, and we encode three states into TT, a string "1101" is encoded into the following ssDNA sequence:

$$5'\text{-} \overbrace{\text{GCGC}}^{1} \text{TT} \overbrace{\text{GCGC}}^{1} \text{TT} \overbrace{\text{GGCC}}^{0} \text{TT} \overbrace{\text{GCGC}}^{1} \text{TT-}3'$$

In addition, two supplementary subsequences are appended at both ends for PCR primers and probes for affinity purifications with magnetic beads that will be used in the laboratory protocol:

$$5'\text{-} \underbrace{S_1 S_2 \cdots S_s}_{\text{PCR primer}} e(x_1) X_1 X_2 \cdots X_k \cdots e(x_m) X_1 X_2 \cdots X_k \underbrace{Y_1 Y_2 \cdots Y_t}_{\text{probe}} \underbrace{R_1 R_2 \cdots R_u}_{\text{PCR primer}} \text{-}3'$$

For a state-transition function from state q_i to state q_j with input symbol $a \in \Sigma$, the state-transition function $\delta(q_i, a) = q_j$ is encoded into the following complementary single-strand DNA subsequence:

$$3'\text{-} \underbrace{\overline{X}_{i+1} \overline{X}_{i+2} \cdots \overline{X}_k}_{k-i \text{ times}} \overline{e(a)} \underbrace{\overline{X}_1 \overline{X}_2 \cdots \overline{X}_j}_{j \text{ times}} \text{-}5'$$

where $\overline{X}_i$ denotes the complementary nucleotide of X_i, and $\overline{y}$ denotes the complementary sequence of y. Further, two more complementary ssDNA sequences are added for the supplementary subsequences at both ends:

$$3'\text{-}\overline{S}_1\overline{S}_2\cdots\overline{S}_s\text{-}5', \quad 3'\text{-}\underbrace{\overline{Y}_1\overline{Y}_2\cdots\overline{Y}_t\overline{R}_1\overline{R}_2\cdots\overline{R}_u}_{\text{biotinylated}}\text{-}5'$$

where the second ssDNA is biotinylated for streptavidin–biotin bonding.

Now, all those ssDNAs encoding an input string w and encoding state-transition functions and the supplementary subsequences of probes and PCR primers are put together into a test tube. Then, a computation (accepting) process of the finite automata M is accomplished by self-assembly among those complementary ssDNAs, and the acceptance of an input string w is determined by the detection of a completely hybridized double-strand DNA.

The main idea of the length-encoding technique is explained as follows. Two consecutive valid transitions $\delta(h, a_n) = i$ and $\delta(i, a_{n+1}) = j$ are implemented by concatenating two corresponding encoded ssDNAs, that is,

$$3'\text{-}\underbrace{AAA\cdots A}_{k-h}\,\overline{e(a_n)}\,\underbrace{AAA\cdots A}_{i}\text{-}5'$$

and

$$3'\text{-}\underbrace{AAA\cdots A}_{k-i}\,\overline{e(a_{n+1})}\,\underbrace{AAA\cdots A}_{j}\text{-}5'$$

together make

$$3'\text{-}\underbrace{AAA\cdots A}_{k-h}\,\overline{e(a_n)}\,\underbrace{AAA\cdots A}_{k}\,\overline{e(a_{n+1})}\,\underbrace{AAA\cdots A}_{j}\text{-}5'$$

Thus, the subsequence $\underbrace{AAA\cdots A}_{k}$ plays the role of "joint" between two consecutive state-transitions and it guarantees that the two transitions are valid in M.

Designing Laboratory Protocols to Execute Finite Automata in Test Tubes

First, the length-encoding method was applied to in vitro experiments to see the effectiveness of the length-encoding techniques for implementing finite-state automata in vitro and in vivo. In order to practically execute the laboratory in vitro experiments for the length-encoding method, the following experimental laboratory protocol was designed, which is also illustrated in ❯ *Fig. 4*:

0. Encoding: Encode an input string into a long ssDNA, and state-transition functions and supplementary sequences into short pieces of complementary ssDNAs.
1. Hybridization: Put all those encoded ssDNAs together into one test tube, and anneal those complementary ssDNAs to be hybridized.
2. Ligation: Put DNA "ligase" into the test tube and invoke ligations at a temperature of $37\,°C$. When two ssDNAs encoding two consecutive valid state-transitions $\delta(h, a_n) = i$ and $\delta(i, a_{n+1}) = j$ are hybridized at adjacent positions on the ssDNA of the input string, these two ssDNAs are ligated and concatenated.

⬛ **Fig. 4**

The flowchart of the laboratory protocol used to execute in vitro finite automata consists of five steps: hybridization, ligation, denature and extraction by affinity purification, amplification by PCR, and detection by gel-electrophoresis. The acceptance of the input string by the automata is the left case, and the rejection is the right case.

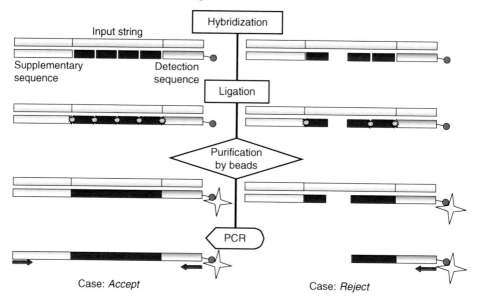

3. Denature and extraction by affinity purification: Denature double-stranded DNAs into ssDNAs and extract concatenated ssDNAs containing the biotinylated probe subsequence by streptavidin–biotin bonding with magnetic beads.
4. Amplification by PCR: Amplify the extracted ssDNAs with PCR primers.
5. Detection by gel-electrophoresis: Separate the PCR products by length using gel-electrophoresis and detect a particular band of the full-length. If the full-length band is detected, that means a completely hybridized double-strand DNA has been formed, and hence the finite automaton "accepts" the input string. Otherwise, it "rejects" the input string. In the laboratory experiments, a "capillary" electrophoresis microchip-based system, called Bioanalyser 2100 (Agilent Technologies), was used in place of conventional gel-electrophoresis. Capillary electrophoresis is of higher resolution and is more accurate than gel electrophoresis using agarose gel.

Simulating Computation Process of Finite Automata Using Four-Base Codons and the Protein-Synthesis Mechanism

Sakakibara and Hohsaka (2003) have proposed a method using the protein-synthesis mechanism combined with four-base codon techniques to simulate a computation (accepting) process of finite automata. An approach to make an in vivo computer is to execute the proposed method in E. coli in order to improve the efficiency of the method and further develop a programmable in vivo computer. The proposed method is described using the example of the simple finite automaton, illustrated in ❷ Fig. 5, which is of two states $\{s_0, s_1\}$,

◻ **Fig. 5**

A simple finite automaton of two states $\{s_0, s_1\}$, defined on one symbol "1", and accepting input strings with even numbers of symbol 1 and rejecting input strings with odd numbers of 1s.

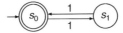

defined as one symbol "1", and accepts input strings with even numbers of symbol 1 and rejects input strings with odd numbers of 1s.

The input symbol "1" is encoded into the four-base subsequence AGGU and an input string is encoded into an mRNA by concatenating AGGU and A alternately, and adding a special sequence AAUUAAC that contains a stop codon at the $3'$-end. This one nucleotide A in between AGGU is used to encode two states $\{s_0, s_1\}$, which is the same technique presented in Yokomori et al. (2002). The stop codon UAA is used for the decision of acceptance or rejection together with a downstream reporter gene. For example, a string "111" is encoded into an mRNA:

The four-base anticodon $(3')\text{UCCA}(5')$ of tRNA encodes the transition rule $s_0 \xrightarrow{1} s_1$, that is a transition from state s_0 to state s_1 with input symbol 1, and the combination of two three-base anticodons $(3')\text{UUC}(5')$ and $(3')\text{CAU}(5')$ encodes the rule $s_1 \xrightarrow{1} s_0$. Furthermore, the encoding mRNA is linked to the *lacZ*-coding RNA subsequence as a reporter gene for the detection of successful computations. Together with these encodings and the tRNAs containing a four-base anticodon $(3')\text{UCCA}(5')$, if a given mRNA encodes an input string with odd numbers of symbol 1, an execution of the in vivo protein-synthesis system stops at the stop codon, which implies that the finite automaton does not accept the input string, and if a given mRNA encodes even numbers of 1s, the translation goes through the entire mRNA and the detection of acceptance is found by the *blue* signal of *lacZ*. Examples of accepting processes are shown in ❷ *Fig. 6* (upper): For an mRNA encoding a string "1111," the translation successfully goes through the entire mRNA and translates the reporter gene of *lacZ* that emits the blue signal (lower). For an mRNA encoding a string "111," the translation stops at the stop codon UAA, does not reach the *lacZ* region, and produces no blue signal.

If the competitive three-base anticodon $(3')\text{UCC}(5')$ comes faster than the four-base anticodon $(3')\text{UCCA}(5')$, the incorrect translation (computation) immediately stops at the following stop codon UAA.

In order to practically execute the laboratory experiments for the in vivo finite automata described in this section, the details of the laboratory protocol are described in Nakagawa et al. (2005).

2.2.3 Programmability and Autonomy

Two important issues to consider when developing DNA-based computers are whether they are *programmable* and *autonomous*.

Programmability means that a program is stored as data (i.e., stored program computer) and any computation can be accomplished by just choosing a stored program. In DNA-based computers, it is required that a program is encoded into a molecule different from the main and

■ **Fig. 6**

Examples of accepting processes: (*Upper*) For an mRNA encoding a string "1111," the translation successfully goes through the mRNA and translates the reporter gene of *lacZ* emitting the blue signal. (*Lower*) For an mRNA encoding a string "111," the translation stops at the stop codon UAG, does not reach the *lacZ* region and produces no blue signal.

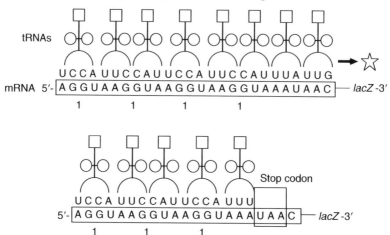

fixed units of the DNA computer, that molecule encoding program can be stored and changed, and that a change of the molecule encoding program accomplishes any computations.

The main features of the in vivo computer enable one to develop a programmable in vivo computer. A plasmid encoding a state-transition function can be simply replaced with another plasmid encoding a different state-transition function, and the *E. coli* cell transformed a new plasmid computes a different finite automaton.

With autonomous DNA computers, once one sets a program and input data and starts a computation, the entire computational process is carried out without any operations from the outside. The in vivo finite automata are autonomous in the sense that the protein-synthesis process that corresponds to a computation (accepting) process of an encoded finite automata is autonomously executed in a living *E. coli* cell and requires no laboratory operations from the outside (❷ *Fig. 7*).

2.3 In Vitro and In Vivo Experiments

2.3.1 In Vitro Experiments

Laboratory in vitro experiments have been carried out on various finite automata for several input strings.

Two-States Automaton with Two Input Strings

The experiments begin with a simple two-states automaton shown in ❷ *Fig. 8* (left) with two input strings, (a) "1101" and (b) "1010." The language accepted by this automaton is $(10)^+$, and hence the automaton accepts the string 1010 and rejects the other string 1101.

◼ **Fig. 7**
A programmable and autonomous in vivo computer system based on _E. coli_.

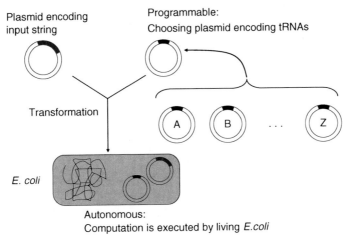

◼ **Fig. 8**
(*Left*) The simple two-states automaton used for this experiment. (*Right*) The results of electrophoresis are displayed in the gel-like image. Lane (**a**) is for the input string 1101, and lane (**b**) for 1010. Since the full-length band (190 mer) is detected only in lane (**b**), the automaton accepts only the input string (**b**) 1010.

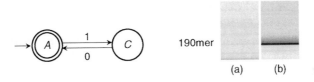

The results of electrophoresis using the Bioanalyser are displayed in the form of the gel-like image (as shown in ❷ *Fig. 8* (right)). For these two input strings, the full-length DNA is of 190 bps (mer). Hence, if a band at a position of 190 mer is detected in the results of the electrophoresis, that means a completely hybridized double-strand DNA is formed, and hence the finite automaton "accepts" the input string.

❷ *Figure 8* (right) clearly shows that the in vitro experiments have successfully identified the correct acceptance of this automaton for two input strings, and hence correctly executed the computation process of the automaton in vitro.

Four-States Automaton with Three Input Strings
The second experiment attempts a four-states automaton, shown in ❷ *Fig. 9* (left) for the three input strings (a) 1101, (b) 1110, and (c) 1010. This four-states automaton accepts the language $(1(0 \cup 1)1)^* \cup (1(0 \cup 1)1)^*0$, and hence it accepts 1110 and 1010 and rejects 1101.

The results are shown in ❷ *Fig. 9* (*right*) in the gel-like image. As in the first experiment, the full-length DNA is of 190 bps (mer). Bands at the position of 190 mer are detected in lane

□ **Fig. 9**

(*Left*) The four-states automaton used for this experiment. (*Right*) The results of electrophoresis are displayed in the gel-like image. Lane (a) is for the input string 1101, lane (b) for 1110, and lane (c) for 1010. Since the full-length band (190 mer) is detected in lanes (b) and (c), the automaton accepts two input strings (b) 1110 and (c) 1010.

(b) and lane (c). Hence, the in vitro experiments have successfully detected that the automaton accepts the two input strings (b) 1110 and (c) 1010.

From Two-States to Six-States Automata with One Input String "111111" of Length 6

The second experiments are five different automata from two states to six states shown in ❷ *Fig. 10* (upper) for one input string "111111" of length 6. The automaton (2) accepts the language $(11)^*$, that is, strings with even numbers of symbol "1", (3) accepts the language $(111)^*$, strings repeating three times 1s, (4) accepts the language $(1111)^*$, strings repeating four times 1s, (5) accepts the language $(11111)^*$, strings repeating five times 1s, (6) accepts the language $(111111)^*$, strings repeating six times 1s. Since 6 is a multiple of 2, 3 and 6, the automata (2), (3) and (6) accept the input string 111111 of length 6.

The results are shown in ❷ *Fig. 10* (lower) in the gel-like image. For the input string 111111, the full-length DNA is of 240 bps (mer). Bands at position 240 mer are detected in lanes (2), (3), and (6) in ❷ *Fig. 10*. Hence, in these in vitro experiments, the automaton (2), (3), and (6) have correctly accepted the input string 111111 and the automaton (4) and (5) have correctly rejected 111111.

2.3.2 In Vivo Experiments

In vivo experiments that followed the laboratory protocols presented in Nakagawa et al. (2005) were done to execute the finite automaton shown in ❷ *Fig. 5*, which is of two states $\{s_0, s_1\}$, defined on one symbol "1," and accepts input strings with even numbers of symbol 1 and rejects input strings with odd numbers of 1s.

Six input strings, "1," "11," "111," "1111," "11111," and "111111," were tested to see whether the method correctly accepts the input strings "11," "1111," "111111," and rejects the strings "1," "111," "11111."

The results are shown in ❷ *Fig. 11*. Blue-colored colonies that indicate the expression of *lacZ* reporter gene were observed only in the plates for the input strings 11, 1111, and 111111. Therefore, the in vivo finite automaton succeeded in correctly computing the six input strings, that is, it correctly accepted the input strings 11, 1111, 111111 of even numbers of symbol "1" and correctly rejected 1, 111, 11111 of odd number of 1s. To the best of our knowledge, this is the first experimental implementation and in vivo computer that succeeded in executing an finite-state automaton in *E. coli*.

☐ **Fig. 10**
(Upper) **Five different automata from two states to six states used for this experiment.**
(Lower) **The results of the electrophoresis are displayed in the gel-like image. Lane (2) is for the automaton (2), (3) for (3), (4) for (4), (5) for (5), and (6) for (6). Since the full-length bands (240 mer) are detected in lanes (2), (3), and (6), the automata (2), (3), and (6) accept the input string 111111.**

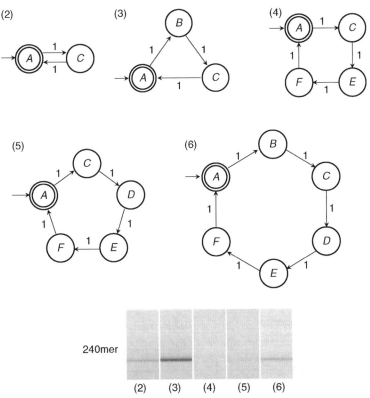

2.4   Further Topics

While a single *E. coli* bacteria cell contains approximately 10^{10} order of active molecules (Weiss et al. 2004), the previous works only utilized a small set of molecules to implement genetic circuits and finite-state automata. The main difficulty is to control such a huge number of molecules and complex wired networks for our purposes to implement bacterial computers. The idea of a "genomic computer" relating to developmental biology and evolution proposed by Istrail et al. (2007) presents some interesting direction to handle hundreds of thousands of regulatory modules that genomes contain for information processing. An animal genome contains many thousands of *cis*-regulatory control modules (called CRMs) that are wired in large networks and constitute genomic computers of information processing capability. Each CRM receives multiple inputs in the form of transcription factors that bind to CRM and execute information processing, such as AND, OR and NOT functions, for their multiple inputs. The idea is very similar to genetic circuits. The interesting features of genomic

◘ **Fig. 11**

Computation by *E. coli* cells with plasmids of the input strings 1, 11, 111, 1111, 11111, 111111. In each panel, the upper plate (part of a LB plate) shows the result in the presence of the suppressor tRNA with UCCU anticodon in the cell, while the lower plate shows the result of the control experiment with no suppressor tRNA expressed. The signs (+) and (−) indicate the theoretical values about the expressions of the *lacZ* reporter gene: (+) means that the cultured *E. coli* cells must express *lacZ* theoretically, and (−) means it must not express. Circles indicate the blue-colored colonies expressing *lacZ*. Therefore, the in vivo finite automaton has correctly computed the six input strings, that is, it correctly accepts the input strings 11, 1111, 111111 of even numbers of symbol "1" and correctly rejects 1, 111, 11111 of odd number of 1s.

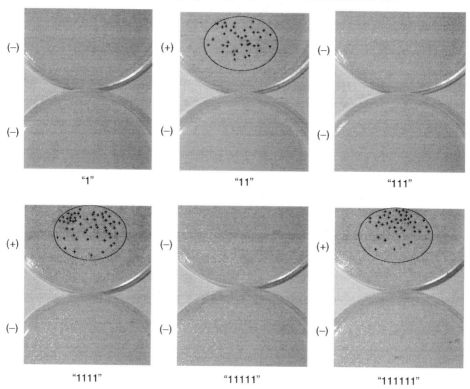

"1"   "11"   "111"

"1111"   "11111"   "111111"

computers are: (1) multiplicity of processors and synchronicity, (2) robustness against system failure, and (3) evolvability especially relating to developmental biology.

Another recent interesting approach to utilizing genomes is the genetic memory proposed by Ham et al. (2008) that employs genome rearrangement operations to develop biological memory. They designed and synthesized an inversion switch using the DNA inversion recombination systems to create a heritable sequential memory switch.

One of the important directions of in vivo computation is multicellular bacterial computation. One necessary and fundamental mechanism for implementing multicellular computation is the communication between cells for sending and receiving the signals. Such intercellular communication was realized by means of the quorum-sensing mechanism, which in nature allows the living bacteria to detect cell density. By using such a quorum sensing mechanism, Sakakibara and his group (Sakakibara et al. 2007) constructed a genetic circuit in *E.coli* that provides a one-dimensional cellular automaton. The main purpose was to extend previous works of implementing finite-state automata and develop in vivo computational systems higher than finite-state machine in terms of the Chomsky hierarchy. The computational capacity of the cellular automaton is equivalent to the universal Turing Machine. CA are multicellular systems in which uniform cells are allocated on a lattice grid. Each cell has a finite number of states. The state transition is determined by the current state and the states of the neighbor cells. Three mechanisms are required to implement CA in bacteria: (1) sending and receiving signals between cells, (2) sustaining the state, and (3) sensing input signals and changing the state following the state transition rules. To encode signals for mechanism (1), small molecules inducing transcription were used. For mechanism (2), the toggle switch circuit was employed to represent a finite number of states using gene expression. If one gene is expressed and the other is not, the state is 1, and, vice versa, 0. For mechanism (3), the state transition functions were implemented as logic gates using transcriptional regulatory proteins that bind to specific signal molecules (chemicals).

3 Molecular Communication

Molecular communication is a new and interdisciplinary research area; as such, it requires research into a number of key areas. Key research challenges in molecular communication include (1) design of a sender that generates molecules, encodes information onto the molecules (called information molecules), and emits the information molecules, (2) design of a molecular propagation system that transports the emitted information molecules from a sender to a receiver, (3) design of a receiver that receives the transported information molecules and biochemically reacts to the received information molecules resulting in decoding of the information, (4) design of a molecular communication interface between a sender and a molecular propagation system and also between the propagation system and a receiver to allow a generic transport of information molecules independent of their biochemical and physical characteristics, (5) mathematical modeling of molecular communication components and systems.

This section describes system design in detail, experimental results, and research trends in molecular communication.

3.1 Overview of the Molecular Communication System

❷ *Figure 12* depicts an overview of a molecular communication system that includes senders, molecular communication interfaces, molecular propagation systems, and receivers.

A sender generates molecules, encodes information into the molecules (called information molecules), and emits the information molecules into a propagation environment. The sender may encode information into the type of the molecules used or the concentration of the molecules used. Possible approaches to creating a sender include genetically modifying eukaryotic cells and artificially constructing biological devices to control the encoding.

◘ Fig. 12

An overview of a molecular communication system that includes senders, molecular communication interfaces, molecular propagation systems, and receivers.

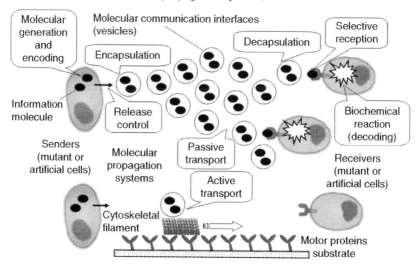

A molecular communication interface acts as a molecular container that encapsulates information molecules to hide the characteristics of the information molecules during the propagation from the sender to a receiver to allow a generic transport of information molecules independent of their biochemical and physical characteristics. Using a lipid bilayer vesicle (Luisi and Walde 2000) is a promising approach to encapsulate the information molecules. Encapsulated information molecules are decapsulated at a receiver.

A molecular propagation system passively or actively transports information molecules (or vesicles that encapsulate information molecules) from a sender to an appropriate receiver through the propagation environment. The propagation environment is an aqueous solution that is typically found within and between cells. Using biological motor systems (motor proteins and cytoskeletal filaments) (Vale 2003) is a promising approach to actively and directionally transport information molecules.

A receiver selectively receives transported and decapsulated information molecules, and biochemically reacts with the received information molecules (this biochemical reaction represents decoding of the information). Possible approaches to creating a receiver are to genetically modify eukaryotic cells and to artificially construct biological devices to control the biochemical reaction.

3.2 Molecular Communication Interface

A vesicle-based communication interface provides a mechanism for transporting different types of information molecules in diverse propagation environments (Moritani et al. 2006a). The vesicle structure (i.e., a lipid bilayer membrane) compartmentalizes information molecules from the propagation environments and provides a generic architecture for transporting diverse types of information molecules independent of their biochemical and physical characteristics. The vesicle structure also protects information molecules from denaturation (e.g., molecular deformation caused by enzymatic attacks or changes in pH) in the propagation environment. Key research issues in implementing the vesicle-based communication interface include how vesicles encapsulate information molecules at a sender and how vesicles decapsulate the information molecules at a receiver.

On these grounds, a molecular communication interface that uses a vesicle embedded with gap junction proteins has been proposed (Moritani et al. 2006b, 2007b). A gap junction is an intercellular communication channel formed between neighboring two cells, and it consists of two docked hemichannels (connexons) constructed from self-assembled six gap junction proteins (connexins) (Kumar and Gilula 1996). When a gap junction is open, molecules whose molecular masses are less than 1.5 kDa can directly propagate through the gap junction channel connecting two cells according to the molecular concentration gradient. A gap junction hemichannel is closed unless two hemichannels are docked.

In the molecular communication interface shown in ❷ *Fig. 13* a sender stores information molecules inside itself and has gap junction hemichannels. When a vesicle with gap junction hemichannels physically contacts the sender, gap junction channels are formed between the sender and the vesicle, and the information molecules are transferred from the sender to the vesicle according to the molecular concentration gradient. When the vesicle detaches from the sender spontaneously, the gap junction hemichannels at the sender and at the vesicle close, and the information molecules transferred from the sender to the vesicle are encapsulated in the vesicle. Encapsulation of information molecules in a vesicle allows a molecular propagation system to transport the information molecules from the sender to a receiver independent of their biochemical and physical characteristics. The receiver also has gap junction hemichannels, and when the transported vesicle physically contacts the receiver, a gap junction channel is formed between the vesicle and the receiver, and the information molecules in the vesicle are transferred into the receiver according to the molecular concentration gradient.

The feasibility of the designed communication interface has been investigated by creating connexin-43 (one of the gap junction proteins) embedded vesicles (Moritani et al. 2006b). Microscopic observations confirmed that calceins (hydrophilic dyes used as model

❏ **Fig. 13**
A schematic diagram of a molecular communication interface using a vesicle embedded with gap junction proteins.

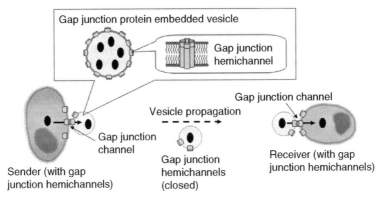

information molecules) were transferred between connexin-43 embedded vesicles and the transferred calceins were encapsulated into the vesicles (Moritani et al. 2007b). This result indicates that the created connexin-43 embedded vesicle (a molecular communication interface) may encapsulate information molecules and receive/transfer information molecules from/into a sender/receiver through gap junctions.

3.3 Molecular Propagation System

The simplest and easiest approach to transport information molecules (or vesicles that encapsulate information molecules) from a sender to a receiver is to use free diffusion. In biological systems, for instance, neurotransmitters such as acetylcholines diffuse at a synapse (around 100 nm gap between a nerve cell and a target cell), cellular slime molds such as amebas exhibit chemotaxis by detecting the molecular concentration gradient of cyclic adenosine monophosphates (cAMPs) diffused from hundreds of micrometers away, and pheromones secreted from insects can affect receiver insects even if they diffuse over a distance of several meters. Generally, the achievable rates of transmitted information molecules at a receiver are low due to Brownian motion and the dilution effect in the propagation environment; however, these observed facts indicate that diffusion-based passive molecular communication is feasible as long as the receiver sensitivity is high.

On the other hand, biological systems have a directional molecular propagation mechanism as well as nondirectional free diffusion. Motor proteins such as kinesins directionally transport cargo such as subcellular organelles and vesicles almost exactly to their destinations in spite of frequent collisions with disturbance molecules (e.g., water molecules, inorganic ions, organic small molecules, and organic macromolecules) within a biological cell. This mechanism is known as active transport and is realized by motor proteins' enzymatic actions that convert chemical energy called adenosine triphosphate (ATP) into mechanical work during their directional walk along filamentous proteins such as microtubules (MTs) (Alberts et al. 1997; Pollard and Earnshaw 2004). Motor proteins are nanometer-scaled actuators and have received increasing attention as engineering materials because they also work in an artificial environment outside of biological cells where aqueous conditions such as temperature and pH are reasonable (Goel and Vogel 2008; Van den Heuvel and Dekker 2007). For instance, as shown in ❷ *Fig. 14*, directional gliding of MTs was demonstrated by immobilizing kinesins onto a glass substrate etched by lithography (Hiratsuka et al. 2001). This mechanism may be applied to a molecular propagation system that uses gliding MTs as cargo transporters of information molecules. The idea is to load information molecules onto gliding MTs at a sender, to transport (glide) the information molecule-loaded MTs from a sender to a receiver, and to unload the transported information molecules from the gliding MTs at a receiver. Major issues in demonstrating this idea are how to load specified information molecules onto gliding MTs at a sender and how to unload the information molecules from the gliding MTs at a designated receiver in a reversible manner. Note that cargo transporters that utilize avidin–biotin (Bachand et al. 2004; Hess et al. 2001) or antigen–antibody (Bachand et al. 2006; Ramachandran et al. 2006) bindings for cargo loading may not be suitable for cargo unloading because of their tight bindings.

On these grounds, a molecular propagation system that uses the reverse geometry of MT motility on kinesins and also uses DNA hybridization and strand exchange has been proposed (Hiyama et al. 2005b, 2008a, c). The propagation system uses DNA hybridization and strand

◘ Fig. 14

Unidirectional movement of MTs along circular tracks. (a) Transmission electron microscopy (TEM) image of the micro-patterned track equipped with arrow-shaped rectifiers. (b–d) Snapshots of the gliding MTs labeled with fluorescent dyes. MTs in the outer and inner circle are gliding clockwise and counterclockwise, respectively. MTs entering the arrowheads from the wrong direction often make 180° turns and move out in the correct direction. Each panel is reproduced with permission from Hiratsuka et al. (2001); copyright 2001, Biophysical Society.

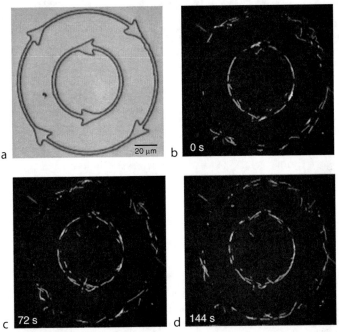

exchange to achieve autonomous and reversible loading/unloading of specified cargos (e.g., vesicles encapsulating information molecules) at a sender/receiver and MT motility to transport the loaded cargo from a sender to a receiver. In order to use DNA hybridization and strand exchange, each gliding MT, cargo, and unloading site (receiver) is labeled with different single-stranded DNAs (ssDNAs). Note that the length of an ssDNA attached to an MT is designed to be shorter than that of the cargo, and the length of an ssDNA attached to a cargo is designed to be as long as that of the unloading site. Cargo are pooled at a given loading site (a given sender) and the ssDNA for the cargo is designed to be either complementary or noncomplementary to that of the MT. When an MT labeled with an ssDNA passes through a given loading site (❷ Fig. 15a–b), a cargo labeled with an ssDNA complementary to that of the MT is selectively loaded onto the gliding MT through DNA hybridization (❷ Fig. 15c), while cargo labeled with a noncomplementary ssDNA remain at the loading site. The cargo loaded onto the MT (i.e., an MT-cargo complex) is transported by MT motility on kinesins toward a given unloading site (❷ Fig. 15d–e). To achieve autonomous unloading at a given unloading site, the ssDNA attached to each unloading site is designed to be either complementary or noncomplementary to that attached to the cargo. When the MT-cargo complex passes

⬛ **Fig. 15**

Microscopic images of cargo loading and transport. Black and white arrows indicate an MT and a cargo, respectively. Each panel is reproduced with permission from Hiyama et al. (2008a); copyright 2008, Wiley-VCH Verlag GmbH & Co. KGaA.

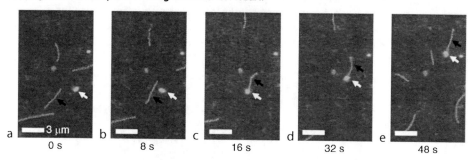

a ▬▬ 3 μm   b ▬▬   c ▬▬   d ▬▬   e ▬▬

   0 s   8 s   16 s   32 s   48 s

through an unloading site, the cargo labeled with an ssDNA complementary to the one attached to the unloading site is selectively unloaded from the gliding MT through DNA strand exchange (Hiyama et al. 2008a, c). These results indicate that gliding MTs may load/unload cargo-vesicles at a sender/receiver through DNA hybridization and strand exchange.

3.4 Senders and Receivers

The most straightforward approach to create senders and receivers in molecular communication is to use living biological cells. In eukaryotic cells, organelles called ribosomes synthesize some proteins that are transported to cell membranes through the vesicle transport between the endoplasmic reticulum (ER) and the Golgi apparatus. Then the transported molecules are secreted into the outside of the cell by exocytosis and diffused away (Alberts et al. 1997; Pollard and Earnshaw 2004). The diffused molecules are selectively captured by cell surface receptors. The received molecules are transduced into intracellular signals that induce cell behavior or are taken inside the cell by endocytosis, resulting in biochemical reactions such as activation of enzymes and expression of genes (Alberts et al. 1997; Pollard and Earnshaw 2004). These facts indicate that it is the fastest way to use living biological cells as senders/receivers in molecular communication, rather than creating senders/receivers artificially, because biological cells inherently have most of the required functionalities as senders/receivers in molecular communication. In that case, the issues to be solved are how to control these inherent functionalities and how to encode information onto molecules.

For instance, *Escherichia coli (E. coli)* bacteria produce intercellular messengers such as acyl-homoserine lactones (AHLs) and respond according to the sensed population of surrounding bacteria. It was demonstrated that genetic manipulations for these mechanisms can enable one to artificially control the fluctuation of bacterial population and bioluminescence (Basu et al. 2005; You et al. 2004). This uses a synthetic multicellular system in which genetically engineered "receiver"cells that produce fluorescent proteins were programmed to form ring-like patterns of differentiation based on chemical gradients of AHL signals that were synthesized by "sender"cells. Other patterns, such as ellipses and clovers, were achieved by

◻ **Fig. 16**
Bacterial pattern formation and molecular communication. Red and green colored portions in each panel represent colonies of "sender"cells and "receiver"cells that received AHL signals sent by sender cells, respectively. Genetically engineered receiver cells formed ring-like patterns such as ellipses (**a**), hearts (**b**), and clovers (**c**) by placing sender cells in different configurations. Each panel is reproduced with permission from Basu et al. (2005); copyright 2005, Macmillan Publishers Ltd.

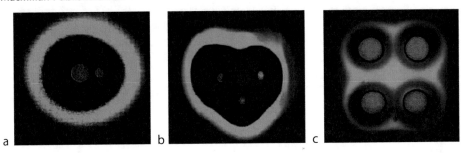

a   b   c

placing senders in different configurations (❷ *Fig. 16*). Such a synthetic multicellular system is an example of a molecular communication system that uses bacteria as molecular senders/receivers.

Genetically altered mutant eukaryotic cells can be also used as molecular senders/receivers. In the process of vesicle transport within eukaryotic cells such as yeasts, export molecules are selected and sorted strictly at the ER and recycled between the ER and the Golgi apparatus (Sato and Nakano 2003). If information is encoded onto the type of molecules or the concentration of molecules, the genetically altered cells, which change the sorting and secreting molecules in response to external stimuli such as temperature and light, may act as a sender in molecular communication.

In contrast, another possible approach to creating senders and receivers in molecular communication is to assemble functional molecules that are designed and synthesized artificially. For instance, a synthetic sender/receiver that uses a giant liposome embedded with gemini-peptide lipids has been proposed (Mukai et al. 2009; Sasaki et al. 2006a). A liposome is an artificially created vesicle that has a lipid bilayer membrane structure similar to cell membranes. The gemini-peptide lipids are composed of two amino acid residues, each having a hydrophobic double-tail and a functional spacer unit connecting to the polar heads of the lipid (❷ *Fig. 17a*). The liposomes embedded with the same type of gemini-peptide lipids in their lipid bilayer membranes assemble in response to specific triggers such as ions and light (❷ *Fig. 17b*) (Iwamoto et al. 2004; Sasaki et al. 2006b). Note that assembled liposomes with gemini-peptide lipids can be dissociated reversibly by applying a complementary trigger (e.g., applying UV-light for liposomal assembly and applying visible light for liposomal dissociation). The reversible liposomal dissociation and assembly mechanisms may be applied to the selective transmission and reception mechanisms of information molecules (or small liposomes that encapsulate information molecules) at a sender and a receiver (Mukai et al. 2009; Sasaki et al. 2006a), respectively (❷ *Fig. 18*). The gemini-peptide lipids are used as a molecular tag. A small liposome embedded with molecular tags acts as a container of information molecules (a molecular container) and a giant liposome embedded with a

◻ **Fig. 17**

Example structures and features of gemini-peptide lipids. (a) Chemical structures of ion-responsive gemini-peptide lipids (GPL type 1 and GPL type 2) and a photo-responsive gemini-peptide lipid (GPL type 3). (b) A schematic diagram of reversible liposomal assembly and dissociation. For instance, liposomes embedded with GPL type 3 assemble/dissociate by exposure to UV-light/visible light in the presence of divalent metal ions due to the photoisomerization of azobenzene-spacer units in the GPL type 3. This figure is reproduced with permission from Iwamoto et al. (2004); copyright 2004, Elsevier Ltd.

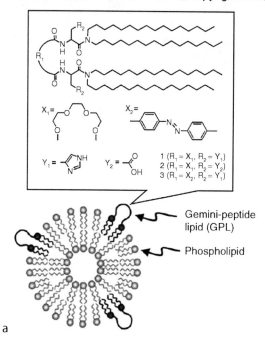

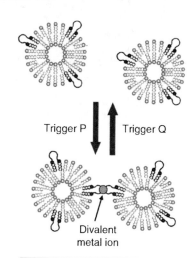

a

b

GPL	Trigger P	Trigger Q
1	Zn^{2+}	K^+
2	Ca^{2+}	K^+
3	Zn^{2+} & UV	Zn^{2+} & VIS

molecular tag acts as a sender/receiver. A sender/receiver is embedded with a specific molecular tag, and a molecular container whose destination is the receiver is also embedded with the same type of molecular tag. When trigger A/B is applied to the sender/receiver and the molecular container, a sender/receiver embedded with a molecular tag that is responsive to the applied trigger transmits/receives the molecular container embedded with the same type of molecular tag (❯ *Fig. 18*). This selective transmission/reception mechanism controlled by specific triggers may lead to the creation of not only unicast-type but also multicast- and broadcast-type molecular communication.

As for the biochemical reaction at a receiver, artificial signal transduction systems have been fabricated on a liposomal membrane (Fukuda et al. 2001; Mukai et al. 2009). The artificial signal transduction system is inspired by biological signal transduction that involves ligand–receptor interaction and G-protein-linked pathways. The system is composed of three molecular components: an artificial receptor, an enzyme, and bilayer-forming lipids (❯ *Fig. 19*) (Fukuda et al. 2001). In this figure, an azobenzene-containing artificial receptor is embedded in the bilayer-forming lipids (dimyristoylphosphatidylcholine: DMPC) and an enzyme (lactate dehydrogenase: LDH) is immobilized on the bilayer-forming lipids. In the

◘ Fig. 18

A schematic diagram of a sender/receiver using a giant liposome embedded with gemini-peptide lipids and molecular capsules using small liposomes embedded with gemini-peptide lipids. By applying a complementary trigger (e.g., visible light/UV-light exposure) to the system, selective transmission/reception of molecular capsules can be performed at a sender/receiver.

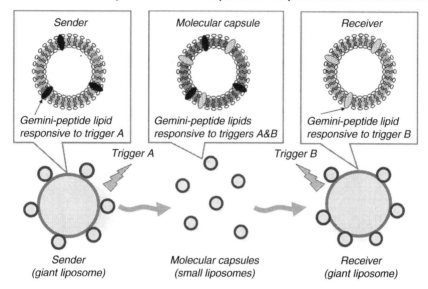

absence of an external signal (1-hydroxy-2-naphthoaldehyde: HNA), Cu^{2+} ions bind to LDH and inhibit LDH activity (corresponding to the off-state). In contrast, in the presence of HNA, Cu^{2+} ions bind to the HNA-receptor complex and activate LDH activity (corresponding to the on-state), resulting in the switching of enzymatic activity through the molecular recognition of the receptor toward the signal. Similar systems can be constructed by using gemini-peptide lipids as photo-responsive molecular tags for liposomal dissociation and assembly. In this case, the photo-responsive molecular tag embedded in the receiver can also act as an artificial receptor in an artificial signal transduction system fabricated on a liposomal membrane (Mukai et al. 2009). The photo-responsive molecular tag drastically changes Cu^{2+} ion-binding affinity through the photoisomerization of azobenzene-spacer units in the GPL type 3. Thus input of a photonic signal (UV-light/visible light exposure) to the molecular tag would be converted to amplified chemical signal output at the enzyme through the translocation of Cu^{2+} ions as a mediator between them. These results indicate that a receiver may biochemically react to the received information molecules by applying a specific trigger.

On the other hand, in the long run it may be possible to create a "white box"-type artificial cell that is only composed of known biological materials and functions. In the field of synthetic biology, some innovative challenges are going on: creating artificial organisms that do not exist or could exist in nature, and creating standardized biological "parts" to turn artisanal genetic engineering into "real" engineering (Bio Fab Group 2006). By combining these technologies, it may be possible to create an artificial cell acting as a sender/receiver in molecular communication from scratch.

■ **Fig. 19**

A schematic diagram of an artificial signal transduction system fabricated on a liposomal membrane. The system is composed of an azobenzene-containing artificial receptor, an enzyme (LDH), and bilayer-forming lipids (DMPC), achieving the switching of enzymatic activity through the molecular recognition of the receptor toward an external signal (HNA). This figure is reproduced with permission from Fukuda et al. (2001); copyright 2001, Elsevier Ltd.

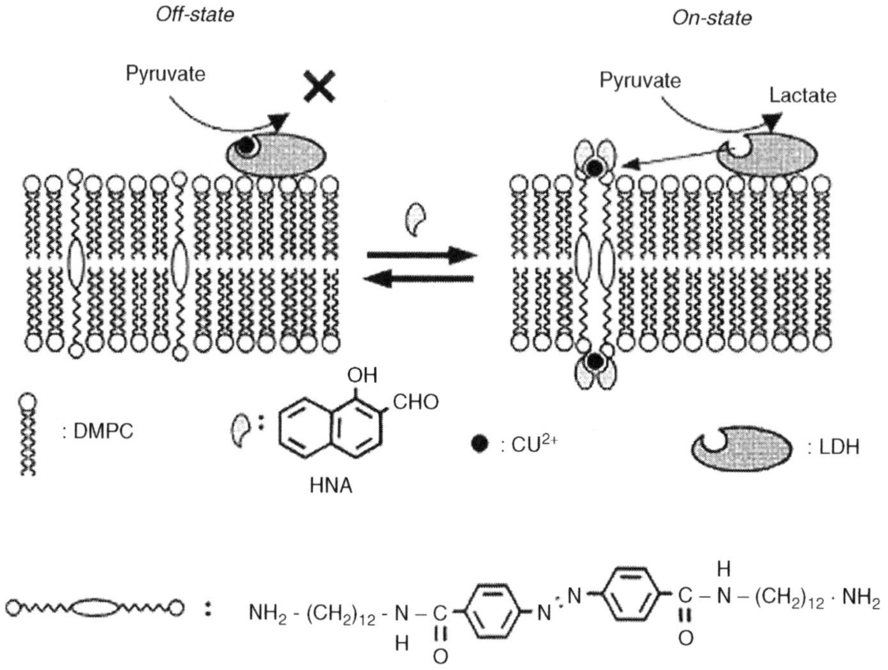

3.5 Research Trends and Future Prospects

Although molecular communication is a new research area, it has received increasing attention in the areas of biophysics, biochemistry, information science, and communication engineering. As summarized in ❷ *Table 2*, various workshops and symposia have been organized worldwide in recent years. The National Science Foundation (NSF) has recognized the importance and impact of molecular communication research, and has already started investigation into funding (NSF 2008). In the middle of 2008, the Institute of Electrical and Electronics Engineers (IEEE) Emerging Technologies Committee launched a new subcommittee to advance the concept of nanoscale networking and molecular communication (IEEE 2008).

Molecular communication research started with experimental approaches to first verify the feasibility, and now it has been extended to theoretical approaches based on information theory. For instance, achievable information rates have been calculated assuming that distinct information molecules are freely diffused (Eckford 2007). Other examples include an information-theoretic model of molecular communication based on cellular signaling (Liu and Nakano 2007), calculating the maximum capacity for the molecular communication channel between a sender and a receiver (Atakan and Akan 2007), and

◘ Table 2

Selected major activities in molecular communication

Date	Venue	Event
Mar. 2005	Miami (FL, USA)	Panel at IEEE INFOCOM (2005)
Oct. 2005	Huntington Beach (CA, USA)	Technical session at IEEE CCW'05
Jan. 2006	Tokyo (Japan)	International symposium
Nov. 2006	Okinawa (Japan)	Symposium at EABS & BSJ'06
Dec. 2006	Cavalese (Italy)	Panel at BIONETICS (2006)
Sep. 2007	Okazaki (Japan)	Domestic workshop
Dec. 2007	Budapest (Hungary)	Workshop at BIONETICS (2007)
Dec. 2007	Yokohama (Japan)	Symposium at biophysics conference
Feb. 2008	Arlington (VA, USA)	NSF workshop (2008)
Sep. 2008	Kawasaki (Japan)	Tutorial at IEICE society conference
Nov. 2008	Hyogo (Japan)	Workshop at BIONETICS (2008)
Aug. 2009	San Francisco (CA, USA)	Workshop at IEEE ICCCN (2009)

mathematical modeling of molecular communication among floating nanomachines (acting as senders/receivers) in an aqueous environment based on probabilistic timed finite-state automata (Wiedermann and Petrů 2008). These mathematical models are also now in the initial phase of research under many constraints and partly nonrealistic assumptions, but they will become sophisticated soon.

In conclusion, molecular communication is an emerging interdisciplinary research area, and, as such, it requires a lot of research effort and collaboration among biophysicists, biochemists, and information scientists. The authors of this chapter hope that many researchers will participate in and contribute to the development of molecular communication.

Acknowledgments

The first author of this book chapter would like to thank many people who executed and supported the research shown in this chapter, especially Junna Kuramochi, Hirotaka Nakagawa, Kensaku Sakamoto, Takahiro Hohsaka, Takashi Yokomori, and Satoshi Kobayashi. This work was supported in part by special coordination funds for Promoting Science and Technology from the Ministry of Education, Culture, Sports, Science and Technology of the Japanese Government, and through a grant from the Bioinformatics Research and Development of the Japanese Science and Technology Agency.

The second author of this chapter would like to thank Prof. Kazuo Sutoh and Associate Prof. Shoji Takeuchi (The University of Tokyo), Prof. Jun-ichi Kikuchi (Nara Institute of Science and Technology), Prof. Kazunari Akiyoshi and Associate Prof. Yoshihiro Sasaki (Tokyo Medical and Dental University), Prof. Tatsuya Suda (University of California, Irvine), and Mr. Yuki Moritani (NTT DOCOMO, Inc.) for their considerable help with the promotion of molecular communication research.

References

Alberts B, Johnson A, Raff M, Walter P, Bray D, Roberts K (1997) Essential cell biology – an introduction to the molecular biology of the cell. Garland, New York

Atakan B, Akan ÖB (2007) An information theoretical approach for molecular communication. In: Proceedings of the bio inspired models of network, information and computing systems, December 2007. Budapest, Hungary

Bachand GD, Rivera SB, Boal AK, Gaudioso J, Liu J, Bunker BC (2004) Assembly and transport of nanocrystal CdSe quantum dot nanocomposites using microtubules and kinesin motor proteins. Nano Lett 4:817–821

Bachand GD, Rivera SB, Carroll-Portillo A, Hess H, Bachand M (2006) Active capture and transport of virus particles using a biomolecular motor-driven, nanoscale antibody sandwich assay. Small 2:381–385

Basu S, Gerchman Y, Collins CH, Arnold FH, Weiss R (2005) A synthetic multicellular system for programmed pattern formation. Nature 434:1130–1134

Benenson Y, Paz-Ellzur T, Adar R, Keinan E, Livneh Z, Shapiro E (2001) Programmable and autonomous computing machine made of biomolecules. Nature 414:430–434

Benenson Y, Adar R, Paz-Ellzur T, Livneh Z, Shapiro E (2003) DNA molecule provides a computing machine with both data and fuel. Proc Natl Acad Sci 100:2191–2196

Bio Fab Group (2006) Engineering life: building a FAB for biology. Sci Am June:44–51

Eckford A (2007) Achievable information rates for molecular communication with distinct molecules. In: Proceedings of the workshop on computing and communications from biological systems: theory and applications, December 2007. Budapest, Hungary

Fukuda K, Sasaki Y, Ariga K, Kikuchi J (2001) Dynamic behavior of a transmembrane molecular switch as an artificial cell-surface receptor. J Mol Catal B Enzym 11:971–976

Gardner TS, Cantor CR, Collins JJ (2000) Construction of a genetic toggle switch in *Escherichia coli*. Nature 403:339–342

Goel A, Vogel V (2008) Harnessing biological motors to engineer systems for nanoscale transport and assembly. Nat Nanotechnol 3:465–475

Ham TS, Lee SK, Keasling JD, Arkin AP (2008) Design and construction of a double inversion recombination switch for heritable sequential genetic memory. PLoS ONE 3(7):e2815

Hess H, Clemmens J, Qin D, Howard J, Vogel V (2001) Light-controlled molecular shuttles made from

motor proteins carrying cargo on engineered surfaces. Nano Lett 1:235–239

van den Heuvel MGL, Dekker C (2007) Motor proteins at work for nanotechnology. Science 317:333–336

Hiratsuka Y, Tada T, Oiwa K, Kanayama T, Uyeda TQP (2001) Controlling the direction of kinesin-driven microtubule movements along microlithographic tracks. Biophys J 81(3):1555–1561

Hiyama S, Moritani Y, Suda T, Egashira R, Enomoto A, Moore M, Nakano T (2005a) Molecular communication. In: Proceedings of the NSTI nanotechnology conference and trade show, vol 3. May 2005. Anaheim, CA, pp 391–394

Hiyama S, Isogawa Y, Suda T, Moritani Y, Sutoh K (2005b) A design of an autonomous molecule loading/transporting/unloading system using DNA hybridization and biomolecular linear motors. In: Proceedings of the European nano systems workshop, December 2005. Paris, France, pp 75–80

Hiyama S, Inoue T, Shima T, Moritani Y, Suda T, Sutoh K (2008a) Autonomous loading, transport, and unloading of specified cargoes by using DNA hybridization and biological motor-based motility. Small 4:410–415

Hiyama S, Moritani Y, Suda T (2008b) A biochemically-engineered molecular communication system (invited paper). In: Proceedings of the international conference on nano-networks, September 2008. Boston, MA

Hiyama S, Takeuchi S, Gojo R, Shima T, Sutoh K (2008c) Biomolecular motor-based cargo transporters with loading/unloading mechanisms on a micro-patterned DNA array. In: Proceedings of the IEEE international conference on micro electro mechanical systems, January 2008, Tucson, AZ, pp 144–147

Hohsaka T, Ashizuka Y, Murakami H, Sisido M (2001a) Five-base codons for incorporation of nonnatural amino acids into proteins. Nucleic Acids Res 29:3646–3651

Hohsaka T, Ashizuka Y, Taira H, Murakami H, Sisido M (2001b) Incorporation of nonnatural amino acids into proteins by using various four-base codons in an *Escherichia coli* in vitro translation system. Biochemistry 40:11060–11064

IEEE Emerging Technical Subcommittee on Nano-Scale, Molecular, and Quantum Networking. http://www.comsoc.org/socsTR/org/operation/comm/subemerging.html. Accessed Sep 2008

Istrail S, De-Leon S, Davidson E (2007) The regulatory genome and the computer. Dev Biol 310:187–195

Iwamoto S, Otsuki M, Sasaki Y, Ikeda A, Kikuchi J (2004) Gemini peptide lipids with ditopic ion-recognition site. Preparation and functions as an inducer for assembling of liposomal membranes. Tetrahedron 60:9841–9847

Kumar NM, Gilula NB (1996) The gap junction communication channel. Cell 84:381–388

Kuramochi J, Sakakibara Y (2005) Intensive in vitro experiments of implementing and executing finite automata in test tube. In: Proceedings of 11th international meeting on DNA based computers. London, Canada, pp 59–67

Liu J-Q, Nakano T (2007) An information theoretic model of molecular communication based on cellular signaling. In: Proceedings of the workshop on computing and communications from biological systems: theory and applications, December 2007, Budapest, Hungary

Luisi PL, Walde P (2000) Giant vesicles. Wiley, New York

Moritani Y, Hiyama S, Suda T (2006a) Molecular communication among nanomachines using vesicles. In: Proceedings of the NSTI nanotechnology conference and trade show, vol 2, May 2006, Boston, MA, pp 705–708

Moritani Y, Nomura S-M, Hiyama S, Akiyoshi K, Suda T (2006b) A molecular communication interface using liposomes with gap junction proteins. In: Proceedings of the bio inspired models of network, information and computing systems, December 2006, Cavalese, Italy

Moritani Y, Hiyama S, Suda T (2007a) Molecular communication – a biochemically-engineered communication system. In: Proceedings of the frontiers in the convergence of bioscience and information technologies, October 2007, Jeju Island, Korea, pp 839–844

Moritani Y, Nomura S-M, Hiyama S, Suda T, Akiyoshi K (2007b) A communication interface using vesicles embedded with channel forming proteins in molecular communication. In: Proceedings of the bio inspired models of network, information and computing systems, December 2007, Budapest, Hungary

Mukai M, Maruo K, Kikuchi J, Sasaki Y, Hiyama S, Moritani Y, Suda T (2009) Propagation and amplification of molecular information using a photo-responsive molecular switch. Supramolecular Chem 21:284–291

Nakagawa H, Sakamoto K, Sakakibara Y (2005) Development of an in vivo computer based on Escherichia coli. In: Proceedings of 11th international meeting on DNA based computers. London, Ontario, pp 68–77

NSF Workshop on Molecular Communication (2008) Biological communications technology. http://netresearch.ics.uci.edu/mc/nsfws08/index.html. Accessed Feb 2008

Panel on Nanoscale Communications. (2005) IEEE INFOCOM'05. http://www.ieee-infocom.org/2005/panels.htm. Accessed Mar 2005

Panel on Nano Scale Communications and Computing. (2006) IEEE/ACM BIONETICS'06. http://www.bionetics.org/2006/. Accessed Dec 2006

Păun Gh, Rozenberg G, Salomaa A (1998) DNA computing. Springer, Heidelberg

Pollard TD, Earnshaw WC (2004) Cell biology (Updated Edition). Saunders, Philadelphia, PA

Ramachandran S, Ernst K-H, Bachand GD, Vogel V, Hess H (2006) Selective loading of kinesin-powered molecular shuttles with protein cargo and its application to biosensing. Small 2:330–334

Sakakibara Y, Hohsaka T (2003) In vitro translation-based computations. In: Proceedings of 9th international meeting on DNA based computers. Madison, WI, pp 175–179

Sakakibara Y, Yugi K, Takagi H (2006) Implementation of AND genetic circuit in Escherichia coli. unpublished manuscript

Sakakibara Y, Nakagawa H, Nakashima Y, Yugi K (2007) Implementing in vivo cellular automata using toggle switch and inter-bacteria communication mechanism. In: Proceedings of workshop on computing and communications from biological systems (in conjunction with IEEE/ACM BIONETICS'07). Budapest, Hungary. http://www.bionetics.org/2007/ccbs.shtml

Sasaki Y, Hashizume M, Maruo K, Yamasaki N, Kikuchi J, Moritani Y, Hiyama S, Suda T (2006a) Controlled propagation in molecular communication using tagged liposome containers. In: Proceedings of the bio inspired models of network, information and computing systems, December 2006, Cavalese, Italy

Sasaki Y, Iwamoto S, Mukai M, Kikuchi J (2006b) Photo- and thermo-responsive assembly of liposomal membranes triggered by a gemini peptide lipid as a molecular switch. J Photoch Photobiol A 183:309–314

Sato K, Nakano A (2003) Oligomerization of a cargo receptor directs protein sorting into COPII-coated transport vesicles. Mol Biol Cell 14:3055–3063

Suda T, Moore M, Nakano T, Egashira R, Enomoto A (2005) Exploratory research on molecular communication between nanomachines. In: Proceedings of the genetic and evolutionary computation conference, June 2005, Washington, DC

Symposium on Molecular Computing and Molecular Communication (2007) New computing and communication systems using biological functions. In: 45th annual meeting of the Biophysical Society of Japan. http://www.tuat.ac.jp/~biophy07/symposium_e.html#sinpo5. Accessed Dec 2007

Vale RD (2003) The molecular motor toolbox for intracellular transport. Cell 112:467–480

Weiss R, Basu S, Hooshangi S, Kalmbach A, Karig D, Mehreja R, Netravali I (2003) Genetic circuit building blocks for cellular computation, communications, and signal processing. Nat Comput 2:47–84

Weiss R, Knight T, Sussman G (2004) Genetic process engineering. In: Amos M (ed) Cellular computing. Oxford University Press, Oxford, UK, pp 42–72

Wiedermann J, Petrů L (2008) Communicating mobile nano-machines and their computational power. In: Proceedings of the international conference on nano-networks, September 2008. Boston, MA

Workshop on Computing and Communications from Biological Systems (2007) Theory and applications (in conjunction with IEEE/ACM BIONETICS'07). http://www.bionetics.org/2007/ccbs.shtml. Accessed Dec 2007

Workshop on Computing and Communications from Biological Systems (2008) Theory and applications (in conjunction with IEEE/ACM BIONETICS'08). http://www.bionetics.org/ccbs.html. Accessed Nov 2008

Workshop on Nano, Molecular, and Quantum Communications (in conjunction with IEEE ICCCN'09) (2009) http://cms.comsoc.org/eprise/main/SiteGen/Nano/Content/Home/NanoCom_369.html. Accessed Aug 2009

Yokomori T, Sakakibara Y, Kobayashi S (2002) A magic pot: self-assembly computation revisited. In: Formal and natural computing, Lecture notes in computer science, vol 2300. Springer, Berlin, pp 418–429

You L, Cox RS III, Weiss R, Arnold FH (2004) Programmed population control by cell-cell communication and regulated killing. Nature 428:868–871

37 Computational Nature of Gene Assembly in Ciliates

Robert Brijder[1] · *Mark Daley*[2] · *Tero Harju*[3] · *Nataša Jonoska*[4] · *Ion Petre*[5] ·
Grzegorz Rozenberg[1,6]
[1]Leiden Institute of Advanced Computer Science, Universiteit Leiden,
The Netherlands
robert.brijder@uhasselt.be
[2]Departments of Computer Science and Biology, University of Western
Ontario, London, Ontario, Canada
daley@csd.uwo.ca
[3]Department of Mathematics, University of Turku, Finland
harju@utu.fi
[4]Department of Mathematics, University of South Florida, Tampa, FL,
USA
jonoska@math.usf.edu
[5]Department of Information Technologies, Åbo Akademi University,
Turku, Finland
ipetre@abo.fi
[6]Department of Computer Science, University of Colorado, Boulder, CO,
USA
rozenber@liacs.nl

G. Rozenberg et al. (eds.), *Handbook of Natural Computing*, DOI 10.1007/978-3-540-92910-9_37,
© Springer-Verlag Berlin Heidelberg 2012

Abstract

Ciliates are a very diverse and ancient group of unicellular eukaryotic organisms. A feature that is essentially unique to ciliates is the nuclear dualism, meaning that they have two functionally different types of nuclei, the macronucleus and the micronucleus. During sexual reproduction a micronucleus is transformed into a macronucleus – this process is called *gene assembly*, and it is the most involved naturally occurring example of DNA processing that is known to us. Gene assembly is a fascinating research topic from both the biological and the computational points of view.

In this chapter, several approaches to the computational study of gene assembly are considered. This chapter is self-contained in the sense that the basic biology of gene assembly as well as mathematical preliminaries are introduced. Two of the most studied molecular models for gene assembly, *intermolecular* and *intramolecular*, are presented and the main mathematical approaches used in studying these models are discussed. The topics discussed in more detail include the string and graph rewriting models, invariant properties, template-based DNA recombination, and topology-based models. This chapter concludes with a brief discussion of a number of research topics that, because of the space restrictions, could not be covered in this chapter.

1 Introduction

The mathematical study of gene assembly in ciliates was initiated in Landweber and Kari (1999, 2002), where it was noted that the DNA rearrangements performed by ciliates have a strong computational appeal. The first results dealt with the computational capabilities of suitably defined models for gene assembly, using classical approaches from theoretical computer science, especially based on formal languages and computability theory. Shortly afterward, a parallel line of research was initiated in Ehrenfeucht et al. (2001b) and in Prescott et al. (2001a), where the focus was to study various properties of the gene assembly process itself, understood as an information processing process that transforms one genetic structure into another.

This research area has witnessed an explosive development, with a large number of results and approaches currently available. Some of them belong to computer science: models based on rewriting systems, permutations, strings, graphs, and formal languages, invariants results, computability results, etc.; see, for example, Ehrenfeucht et al. (2003a). While others, such as template-based DNA recombination, belong to theoretical and experimental biology; see, for example, Ehrenfeucht et al. (2007) and Angeleska et al. (2007).

In this chapter, several approaches and results in the computational study of gene assembly are reviewed. In ❯ Sect. 2, the basic biological details of the gene assembly process as currently understood and experimentally observed are introduced. After mathematical preliminaries in ❯ Sect. 3, two of the most studied molecular models for gene assembly (intermolecular and intramolecular) are introduced in ❯ Sects. 4 and ❯ 5. Several mathematical approaches used in studying these models are also discussed. In ❯ Sect. 6, some properties of the gene assembly process, called invariants, that hold independently of molecular model and assembly strategy, are discussed. In ❯ Sect. 7, models for template-based DNA recombination are presented as a possible molecular implementation of the gene assembly process. The chapter is concluded with a brief discussion in ❯ Sect. 8.

2   The Basic Biology of Gene Arrangement in Ciliated Protozoa

All living cells can be classified as belonging to one of two high-level groups: the *prokaryotes* or the *eukaryotes*. The prokaryotes are defined chiefly by their simple cellular organization: within a prokaryotic cell, there are no compartments or subdivisions, all of the intracellular materials (e.g., enzymes, DNA, food, waste) are contained within a single cellular membrane and are free to intermix. By contrast, eukaryotic cells contain nested membranes with many functionally and morphologically distinct organelles, the most well known being the nucleus that contains the DNA, and DNA processing machinery. All bacteria and archaea are prokaryotes while all higher multicellular organisms are eukaryotes.

The *protozoa* are single-celled eukaryotes of striking complexity; each cell functions as a complete, individual organism capable of advanced behaviors typically associated with multicellular organisms. Protozoa are equipped with extensive sensory capabilities and various species can sense temperature, light, and motion, as well as chemical and magnetic gradients. In addition to locomotion via swimming, some types of ciliated protozoa are able to coordinate legs made from fused cilia to walk along substrates in search of food. Many protozoan species are active hunters and possess sensory apparatus enabling them to identify, hunt, and consume prey.

Protozoa are found in nearly every habitat on Earth, and form a critical portion of the microbial food web. Beyond such ecological significance, the study of protozoa is fundamental to evolutionary inquiry as the protozoa represent a significant portion of eukaryotic evolutionary diversity on Earth – through their evolution they have produced unique features, one of which we study further in this chapter.

The *ciliated protozoa* (phylum Ciliophora) are a particularly interesting group due to their unique, and complex, nuclear morphology and genetics. Where most eukaryotic cells contain only one type of nucleus (sometimes in many copies), ciliate cells contain two functionally different types of nucleus: *Macronucleus* (abbreviated as *MAC*) and *Micronucleus* (*MIC*), each of which may be present in various multiplicities. For details on the many other aspects of ciliate biology which are not touched upon here, the reader can refer to Hausmann and Bradbury (1997) and Prescott (2000).

The diversity within the phylum Ciliophora is staggering and even relatively closely related species have extreme evolutionary distances, for example, the ciliates *Tetrahymena thermophila* and *Paramecium caudatum* have an evolutionary distance roughly equivalent to that between rat and corn (Prescott 1994). In this chapter, the discussion is restricted to ciliates of the subclass Stichotrichia for which the unique features that are discussed in this chapter are especially pronounced.

In Stichotrichia there is a profound difference in genome organization between the MIC and MAC on both a global level (where one considers the organization of chromosomes) and a local level (considering the organization of DNA sequence within individual genes).

On the global level, the MIC is diploid (there are exactly two copies of each chromosome) and composed of long chromosomes containing millions of base pairs of DNA. Each of these chromosomes contains a large number of genes, although these genes account for only a very small portion of the total sequence of the chromosomes (approximately 2–5% in stichotrichs). This long, but genetically sparse, chromosome structure is similar to that found in the nuclei of most other eukaryotes.

The global organization of the MAC provides a stark contrast to that of the MIC: most obviously, the number of copies of chromosomes in MACs is very large. For example, typically more than 1,000 copies in stichotrichs, more than 10,000 in *Stylonichia*, and at the extreme end up to millions of copies of the rDNA-containing chromosome in the stichotrich *Oxytricha trifallax* (also called *Sterkiella histriomuscorum*). Along with much higher number of copies, MAC chromosomes are much shorter than those of the MIC and they contain only 1–3 genes each. Although they are small, these chromosomes are genetically dense – indeed, approximately 85% of a typical chromosome sequence contain genes. Hence, although we have a huge multiplicity of these chromosomes, the total DNA content is still much lower than in the MIC. Thus, the MIC chromosomes are long and "sparse" while the MAC chromosomes are short and "dense."

The difference in global genetic organization between the MAC and the MIC is impressive, but turns out to be much less startling than the difference in the local organization of the genes. On the local level, the MAC is a functional nucleus and, like most eukaryotic nuclei, it carries out the day-to-day genetic "housekeeping" tasks of the cell including the production of proteins from the functional MAC genes. In contrast, the genes of the MIC are not functional (not expressible as proteins) due to the presence of many noncoding sequences that break up the genes. Indeed, the macronuclear forms of ciliate genes are heavily modified from the original micronuclear configurations. MIC genes can be divided into two interleaving types of regions: *macronuclear destined segments* or *MDSs* and *internal eliminated sequences* or *IESs*. The MDSs are the regions of the MIC gene that end up in the functional MAC version of the gene, while the IESs are interspersed nongenetic regions of sequence that do not appear in the MAC version of the gene, see ❷ *Fig. 1*. The MDSs are assembled, via overlapping regions called *pointers* to form the macronuclear genes. Each MDS, with the exception of the first and last, is flanked on either side by one of these pointer regions. The outgoing pointer region on one side, which we depict at the right side of MDS n has identical DNA sequence to the incoming pointer region on the left side of MDS $n + 1$. An illustration of this can be seen in ❷ *Fig. 2*; note that the outgoing pointer of MDS n is identical to the incoming pointer of MDS $(n + 1)$.

In some ciliates, in addition to the appearance of IESs, the MIC genes have the further complication that the MDSs do not appear in the same order as they do in the functional MAC gene. That is, the order of MDSs in the MIC gene is *scrambled* relative to their "orthodox" order in the MAC gene and may even contain segments that are inverted (i.e., rotated 180°). A schematic representation of the gene actin I from *O. trifallax* is shown in ❷ *Fig. 3* (see Prescott and Greslin (1992)). Note that the numbers enumerating MDSs refer to the orthodox order of the MDSs in the macronuclear version of the gene, and a bar is used to denote MDSs that are inverted.

Ciliates reproduce asexually but, during times of environmental stress (e.g., starving due to lack of nutrients) can also undergo the (nonreproductive) sexual activity of *conjugation*. Rather than producing offspring, the purpose of conjugation in ciliates is to give each cell a "genetic facelift" by incorporating new DNA from the conjugating partner cell. The specifics of conjugation are highly species dependent, but almost all ciliate species follow the same basic outline, the generic form of which is illustrated in ❷ *Fig. 4*. Two cells form a cytoplasmic bridge while meiotically dividing their diploid MICs into four haploid MICs. Each cell sends one haploid MIC across the bridge to the conjugating partner. The cell then combines one of its own haploid MICs with the newly arrived haploid MIC from the partner to form a new fused diploid MIC. Also, each cell destroys its old MAC and remaining haploid MICs, the fused MIC divides into two parts, one of which will be the new MIC and the other one will develop into (will be transformed into) the new MAC.

◻ Fig. 1
A diagram of the arrangement of six IESs and seven MDSs in the micronuclear (top) and the macronuclear (bottom) gene encoding βTP protein. During macronuclear development the IESs are excised and the MDSs are ligated (by overlapping of the ends) to yield a macronuclear gene. MDSs are *rectangles*. IESs are *line segments* between *rectangles* (from Prescott and DuBois (1996)).

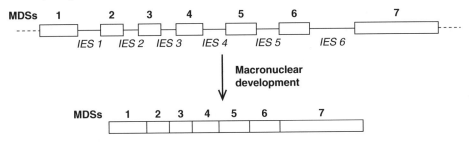

◻ Fig. 2
Schematic representation of MIC gene (*top*) and associated MAC gene (*bottom*) with pointers indicated on the MDSs.

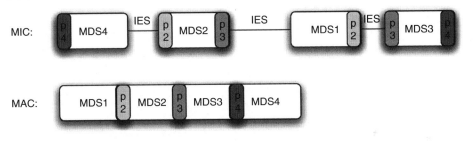

◻ Fig. 3
Schematic representation of the structure of the micronuclear gene-encoding actin protein in the stichotrich *Sterkiella nova*. The nine MDSs are in a scrambled disorder (from Prescott and Greslin (1992)).

The topic of interest in this chapter is this transformation of a new MIC into a new MAC. We focus in particular on the transformations of the individual genes – a process called *gene assembly*. This process is fascinating both from the biological and information processing points of view. Indeed, gene assembly is the most complex example of DNA processing known in nature, underscored by the dramatic difference in the structure, and composition, of the MIC and MAC genomes explained above. To form a functional macronuclear gene, all IESs must be excised, all MDSs must be put in their original (orthodox) order with inverted segments switched to their proper orientation.

⬛ Fig. 4

Conjugation in stichotrichs: **(a)** A stichotrich with two macronuclei and two micronuclei. **(b)** Two stichotrichs have joined and formed a cytoplasmic channel. The two diploid micronuclei have each formed four haploid micronuclei. **(c)** The two cells exchanged haploid micronuclei, and the two organisms have separated. The exchanged haploid micronucleus has fused with a resident haploid micronucleus forming a new diploid micronucleus (half white and half black). **(d)** The new diploid micronucleus has divided by mitosis. The unused haploid micronuclei (six) and the two macronuclei are degenerating. **(e)** One of the new daughter micronuclei has developed into a new macronucleus. The old macronucleus and the unused haploid micronuclei have disappeared. **(f)** Conjugation has been completed. The micronucleus and macronucleus have divided, yielding the appropriate nuclear numbers (in this case, two MICs and two MACs) (from Ehrenfeucht et al. (2003a)).

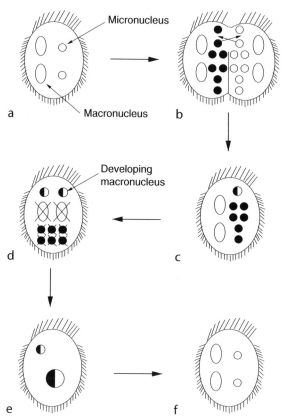

Understanding and investigating the computational nature of the gene assembly process, and its biological implications, becomes the central focus of this chapter. For further details on the biology of gene assembly we refer to Jahn and Klobutcher (2000), Prescott (1999, 2000), Möllenbeck et al. (2008), and Ehrenfeucht et al. (2007) and, in particular to Ehrenfeucht et al. (2003a), which contains chapters explaining basic biology, basic cell biology, and the basic biology of ciliates written specifically for motivated computer scientists.

3 Mathematical Preliminaries

In this section we fix basic mathematical notions and terminology used in this chapter.

3.1 Strings

For an alphabet Σ, let Σ^* denote the set of all strings over Σ. Let Λ denote the *empty string*. For a string $u \in \Sigma^*$, we denote by $|u|$ its length, that is, the number of letters u consists of.

A string u is a *substring* of a string v, if $v = xuy$, for some $x, y \in \Sigma^*$. In this case, we denote $u \leq v$. We say that u is a *conjugate* of v if $v = w_1 w_2$ and $u = w_2 w_1$, for some $w_1, w_2 \in \Sigma^*$.

For an alphabet Σ, let $\overline{\Sigma} = \{\overline{a} \mid a \in \Sigma\}$ be a signed disjoint copy of Σ. The set of all strings over $\Sigma \cup \overline{\Sigma}$ is denoted by $\Sigma^{\circledast} = (\Sigma \cup \overline{\Sigma})^*$. A string $v \in \Sigma^{\circledast}$ is called a *signed string over* Σ. We adopt the convention that $\overline{\overline{a}} = a$ for each letter $a \in \Sigma$.

Let $v \in \Sigma^{\circledast}$ be a signed string over Σ. We say that a letter $a \in \Sigma \cup \overline{\Sigma}$ *occurs* in v, if a or $\overline{a}$ is a substring of v. Let $\mathsf{dom}(v) \subseteq \Sigma$, called the *domain* of v, be the set of the (unsigned) letters that occur in v.

Example 1 For the alphabet $\Sigma = \{2, 3\}$ of pointers, $\overline{\Sigma} = \{\overline{2}, \overline{3}\}$. Here $u = 2\,3\,3 \in \Sigma^* \subseteq \Sigma^{\circledast}$, while $v = 2\,\overline{3}\,\overline{2} \in \Sigma^{\circledast}$ is a signed string over Σ for which $\mathsf{dom}(v) = \{2, 3\}$ although 3 is not a substring of v.

The signing $a \mapsto \overline{a}$ of letters extends to strings in a natural way: for a signed string $u = a_1 a_2 \ldots a_n \in \Sigma^{\circledast}$, with $a_i \in \Sigma \cup \overline{\Sigma}$ for each i, let the *inversion* of u be

$$\overline{u} = \overline{a}_n \overline{a}_{n-1} \ldots \overline{a}_1 \in \Sigma^{\circledast}$$

For any set of strings $S \subseteq \Sigma^{\circledast}$, we denote $\overline{S} = \{\overline{u} \mid u \in S\}$. For two strings $u, v \in \Sigma^{\circledast}$, we say that u and v are *equivalent*, denoted $u \approx v$, if u is a conjugate of either v or $\overline{v}$.

For an alphabet Σ, let $\|.\|$ be the substitution that unsigns the letters: $\|a\| = a = \|\overline{a}\|$. Mapping $\|.\|$ extends to $\Sigma^{\circledast}$ in the natural way. A signed string v over Σ is a *signing* of a string $u \in \Sigma^*$, if $\|v\| = u$. A signed string u, where each letter from Σ occurs exactly once in u, is called a *signed permutation*.

For two alphabets Σ and Δ, a mapping $f : \Sigma^{\circledast} \to \Delta^{\circledast}$ is called a *morphism* if $f(uv) = f(u)f(v)$ and $f(\overline{u}) = \overline{f(u)}$. If $\Delta \subseteq \Sigma$, a morphism $f : \Sigma^* \to \Delta^*$ is called a *projection* if $f(x) = x$ for $x \in \Delta$ and $f(x) = \lambda$, for $x \in \Sigma \setminus \Delta$.

Example 2 For the alphabet $\Sigma = \{2, 3, 4, 5\}$, there are $2^4 \cdot 4! = 384$ signed permutations. The signed strings $2\,3\,4\,5$ and $\overline{4}\,\overline{2}\,5\,3$ are among them.

Throughout this chapter we $\{2, 3, \ldots\}$ the set of *pointers*.

3.2 Graphs

For a finite set V, let $E(V) = \{\{x, y\} \mid x, y \in V, x \neq y\}$ be the set of all unordered pairs of different elements of V. A (*simple*) *graph* is an ordered pair $G = (V, E)$, where V and E are finite sets of *vertices*, and *edges*, respectively. If $e = \{x, y\} \in E$ is an edge, then the vertices x and y are the *ends* of e. In this case, x and y are *adjacent* in G. If $V = \emptyset$, then we denote $G = \emptyset$ and call it the *empty graph*.

For a vertex x in G, let $N_G(x) = \{y \mid \{x, y\} \in E\}$ be the *neighborhood* of x in G. A vertex x is *isolated* in G, if $N_G(x) = \emptyset$.

A *signed graph* $G = (V, E, \sigma)$ consists of a simple graph (V, E) together with a labeling $\sigma : V \rightarrow \{-, +\}$ of the vertices. A vertex x is said to be *positive*, if $\sigma(x) = +$; otherwise x is *negative*. We write $x^{\sigma(x)}$ to indicate that the vertex x has sign $\sigma(x)$.

Let $G = (V, E, \sigma)$ be a signed graph. For a subset $A \subseteq V$, its *induced subgraph* is the signed graph $(A, E \cap E(A), \sigma)$, where σ is restricted to A. The *complement* of G is the signed graph $G^c = (V, E(V) \setminus E, \sigma^c)$, where $\sigma^c(x) = +$ if and only if $\sigma(x) = -$. Also, let $\mathrm{loc}_x(G)$ be the signed graph obtained from G by replacing the subgraph induced by $N_G(x)$ by its complement. For a subset $A \subseteq V$, we denote by $G - A$ the subgraph induced by $V \setminus A$.

A *multigraph* is a (undirected) graph $G = (V, E, \varepsilon)$, where parallel edges are possible. Therefore, E is a finite set of edges and $\varepsilon : E \rightarrow \{\{x, y\} \mid x, y \in V\}$ is the *endpoint mapping*. We allow $x = y$, and therefore edges can be of the form $\{x, x\} = \{x\}$ – an edge of this form should be seen as a "loop" for x.

4 The Intermolecular Model for Gene Assembly

The very first formal model of the ciliate gene assembly process was the *intermolecular model* proposed in Landweber and Kari (1999, 2002). In this section we introduce a string-based formalization of the model and we refer to Landweber and Kari (1999) for the biological details of the model. Given a string *uxvxw*, where *u*, *x*, *v*, *w* are nonempty substrings, the authors defined the following intramolecular operation: *uxvxw* → {*uxw*, ·*vx*} where · denotes that the string *vx* is circular. Intuitively, this models the action of a strand of DNA, *uxvxw*, looping over onto itself and aligning the regions containing the subsequence *x*. With the *x*'s aligned, the DNA strand can undergo recombination, yielding the two products *uxw* and ·*vx*. Likewise, the inverse, intermolecular, operation was also defined: {*uxw*, ·*vx*} → *uxvxw*. These operations are illustrated in ❷ *Fig. 5*

❑ **Fig. 5**

The (reversible) recombination of *uxvxw* ↔ { *uxw*, ·*vx*} in the intermolecular model for gene assembly of Landweber and Kari (1999, 2002).

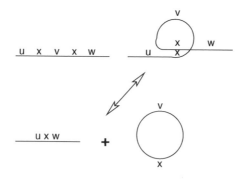

Example 3 Consider a hypothetical micronuclear gene with 4 MDSs where the MDSs come in the order $M_1 M_2 M_4 M_3$. If we denote each MDS by its pair of incoming/outgoing pointers and/or markers, we can then denote the whole micronuclear gene as $\delta = (b, p_2)(p_2, p_3)(p_4, e)(p_3, p_4)$. (For more details on formal representation of ciliate genes, see ❷ Sect. 5.) An assembly strategy for this gene in the intermolecular model is the following (we indicate for each operation the pointer on which the molecule is aligned):

$$\delta \xrightarrow{p_3} (b, p_2)(p_2, p_3, p_4) + \cdot(p_4, e)p_3 \xrightarrow{p_4} (b, p_2)(p_2, p_3, p_4, e)p_3 p_4$$

$$\xrightarrow{p_2} (b, p_2, p_3, p_4, e)p_3, p_4 + \cdot p_2$$

As indicated by the formal notation above, the gene gets assembled with copies of pointers p_3 and p_4 following the assembled gene and a copy of pointer p_2 placed on a separate circular molecule. Note that the notation above ignores all IESs of the micronuclear and of the assembled gene.

The obvious difficulty with the intermolecular model is that it cannot deal with DNA molecules in which a pointer is inverted – this is the case, for example, for the actin I gene in *S. nova*. Nevertheless, we can show that inverted pointers can be handled in this model, provided the input molecule (or its MDS–IES descriptor) is available in two copies. Moreover, *we consider all linear descriptors modulo inversion*. The first assumption is essentially used in research on the intermolecular model, see Kari et al. (1999), Kari and Landweber (1999), and Landweber and Kari (2002). The second assumption is quite natural whenever double-stranded DNA molecules are modeled.

Example 4 Consider the micronuclear actin gene in *Sterkiella nova*, see ❷ *Fig. 3*. Denoting each of its MDSs by the pair of incoming/outgoing pointers and/or markers similarly to our previous example, the gene may be written as

$$\delta = (p_3, p_4)(p_4, p_5)(p_6, p_7)(p_5, p_6)(p_7, p_8)(p_9, e)(\overline{p_3}, \overline{p_2})(b, p_2)(p_8, p_9)$$

Then δ can be assembled in the intermolecular model as follows:

$$\delta \xrightarrow{p_5} (p_3, p_4)(p_4, p_5, p_6)(p_7, p_8)(p_9, e)(\overline{p_3}, \overline{p_2})(b, p_2)(p_8, p_9) + \cdot p_5(p_6, p_7)$$

$$\xrightarrow{p_8} (p_3, p_4)(p_4, p_5, p_6)(p_7, p_8, p_9) + \cdot p_8(p_9, e)(\overline{p_3}, \overline{p_2})(b, p_2) + \cdot p_5(p_6, p_7)$$

$$\xrightarrow{p_4} (p_3, p_4, p_5, p_6)(p_7, p_8, p_9) + \cdot p_4 + \cdot p_8(p_9, e)(\overline{p_3}, \overline{p_2})(b, p_2) + \cdot p_5(p_6, p_7)$$

$$\xrightarrow{p_7} (p_3, p_4, p_5, p_6)p_7 p_5(p_6, p_7, p_8, p_9) + \cdot p_4 + \cdot p_8(p_9, e)(\overline{p_3}, \overline{p_2})(b, p_2)$$

$$\xrightarrow{p_6} (p_3, p_4, p_5, p_6, p_7, p_8, p_9) + \cdot p_6 p_7 p_5 + \cdot p_4 + \cdot p_8(p_9, e)(\overline{p_3}, \overline{p_2})(b, p_2)$$

$$\xrightarrow{p_9} (p_3, p_4, p_5, p_6, p_7, p_8, p_9, e)(\overline{p_3}, \overline{p_2})(b, p_2)p_8 p_9 + \cdot p_6 p_7 p_5 + \cdot p_4$$

Assuming that $\overline{\delta}$ is also available, the assembly continues as follows. Here, for a (circular) string τ, we use 2τ to denote $\tau + \tau$:

$$\delta + \bar{\delta} \longrightarrow \cdots \longrightarrow (p_3, p_4, p_5, p_6, p_7, p_8, p_9, e)(\overline{p_3}, \overline{p_2})(b, p_2)p_8 p_9$$
$$+ \, \overline{p_9}\,\overline{p_8}(\overline{p_2}, \bar{b})(p_2, p_3)(\bar{e}, \overline{p_9}, \overline{p_8}, \overline{p_7}, \overline{p_6}, \overline{p_5}, \overline{p_4}, \overline{p_3})$$
$$+ \, 2 \cdot p_6 p_7 p_5 + 2 \cdot p_4$$
$$\xrightarrow{\;p_2\;} (p_3, p_4, p_5, p_6, p_7, p_8, p_9, e)(\overline{p_3}, \overline{p_2})(b, p_2, p_3)(\bar{e}, \overline{p_9}, \overline{p_8}, \overline{p_7}, \overline{p_6}, \overline{p_5}, \overline{p_4}, \overline{p_3})$$
$$+ \, \overline{p_9}\,\overline{p_8}(\overline{p_2}, \bar{b})p_2 p_8 p_9 + 2 \cdot p_6 p_7 p_5 + 2 \cdot p_4$$
$$\xrightarrow{\;\overline{p_2}\;} (p_3, p_4, p_5, p_6, p_7, p_8, p_9, e)(\overline{p_3}, \overline{p_2}, \bar{b})p_2 p_8 p_9$$
$$+ \, \overline{p_9}\,\overline{p_8}\overline{p_2}(b, p_2, p_3)(\bar{e}, \overline{p_9}, \overline{p_8}, \overline{p_7}, \overline{p_6}, \overline{p_5}, \overline{p_4}, \overline{p_3})$$
$$+ \, 2 \cdot p_6 p_7 p_5 + 2 \cdot p_4$$
$$\xrightarrow{\;\overline{p_3}\;} (p_3, p_4, p_5, p_6, p_7, p_8, p_9, e)\overline{p_3}$$
$$+ \, \overline{p_9}\,\overline{p_8}\overline{p_2}(b, p_2, p_3)(\bar{e}, \overline{p_9}, \overline{p_8}, \overline{p_7}, \overline{p_6}, \overline{p_5}, \overline{p_4}, \overline{p_3}, \overline{p_2}, \bar{b})p_2 p_8 p_9$$
$$+ \, 2 \cdot p_6 p_7 p_5 + 2 \cdot p_4$$
$$\xrightarrow{\;p_3\;} p_3(\bar{e}, \overline{p_9}, \overline{p_8}, \overline{p_7}, \overline{p_6}, \overline{p_5}, \overline{p_4}, \overline{p_3}, \overline{p_2}, \bar{b})p_2 p_8 p_9$$
$$+ \, \overline{p_9}\,\overline{p_8}\overline{p_2}(b, p_2, p_3, p_4, p_5, p_6, p_7, p_8, p_9, e)\overline{p_3} + 2 \cdot p_6 p_7 p_5 + 2 \cdot p_4$$
$$= 2\overline{p_9}\,\overline{p_8}\overline{p_2}(b, p_2, p_3, p_4, p_5, p_6, p_7, p_8, p_9, e)\overline{p_3} + 2 \cdot p_6 p_7 p_5 + 2 \cdot p_4$$

This calculation shows that the gene is eventually assembled, with p_3 preceding the assembled gene and $p_2 p_8 p_9$ following it, and that two circular molecules are excised during the assembly.

We refer to ❯ Sect. 5 for intramolecular assembly strategies of the actin I gene in *S. nova* and to ❯ Sect. 6 for a set of properties that are independent of particular assembly strategies, called invariants. The DNA sequences of all molecules produced by gene assembly are particular invariants of the process. We refer to Harju et al. (2004a) for a detailed discussion on intermolecular assembly strategies and a comparison to intramolecular assemblies.

Consider now contextual versions of the operations. We define rules of the form (p, x, q) (p', x, q') with the intended semantics that p (p', resp.) and q (q', resp.) represent contexts flanking x (x', resp.). These contexts are not directly involved in the recombination process, but instead regulate it: recombination may only take place if x (x', resp.) is flanked by p (p', resp.) and q (q', resp.). Our intramolecular operation now becomes $uxvxw \to \{uxw, \cdot vx\}$, where $u = u'p, v = qv' = v''p', w = q'w'$, with similar constraints for the intermolecular operation.

It was shown in Landweber and Kari (2002) that iterated nondeterministic application of these contextual rules to an initial axiom string yields a computing system with the generative power of a Turing machine.

5 The Intramolecular Model for Gene Assembly

5.1 Molecular Model

The *intramolecular model* was introduced in Ehrenfeucht et al. (2001b) and Prescott et al. (2001a). It consists of a set of three irreversible molecular operations explaining the excision of IESs and the unscrambling and ligation of MDSs during the MIC–MAC development. All three operations postulate the folding of a DNA molecule into a specific pattern that allows the alignment of some pointers. Subsequent DNA recombination on those pointers leads to the MDSs (and IESs) being rearranged. Each of the three operations is described in the following.

5.1.1 Loop, Direct-Repeat Excision (in Short, ld)

The ld *operation* is applicable to a DNA molecule having two occurrences of a pointer, say p, on the same strand. We say in this case that pointer p has a *direct repeat* along the molecule. The molecule is then folded into a *loop* (❷ *Fig. 6a*) so that the two occurrences of p are aligned. Recombination on p is thus facilitated (❷ *Fig. 6b*) and as a result, a linear molecule and a circular molecule are obtained, (❷ *Fig. 6c*). Each of the two molecules has an occurrence of p.

5.1.2 Hairpin, Inverted-Repeat Excision/Reinsertion (in Short, hi)

The hi *operation* is applicable to a DNA molecule having two occurrences of a pointer, say p, on different strands of the molecule. We say in this case that pointer p has an *inverted repeat* along the molecule. Then the molecule is folded into a *hairpin* (❷ *Fig. 7a*) so that the two occurrences of p have a direct alignment. Recombination on p is thus facilitated (❷ *Fig. 7b*), and as a result, a new linear molecule is obtained, where one block of nucleotides has been inverted (❷ *Fig. 7c*).

5.1.3 Double Loop, Alternating Direct-Repeat Excision (in Short, dlad)

The dlad *operation* is applicable to a DNA molecule having two pointers, say p and q, each with two occurrences. All four pointer occurrences should be on the same strand. Moreover, pointer p has one occurrence in-between the two occurrences of q and one occurrence outside

◘ **Fig. 6**
Illustration of the ld-rule. **(a)** The molecule is first folded into a loop, aligning the two direct repeats of a pointer. **(b)** Recombination is facilitated on the two occurrences of the pointer. **(c)** One linear and one circular molecule are produced as a result of the operation (from Harju et al. (2004b)).

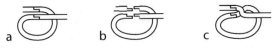

◘ **Fig. 7**
Illustration of the hi-rule. **(a)** The molecule is first folded into a hairpin, aligning the two inverted repeats of a pointer. **(b)** Recombination is facilitated on the two occurrences of the pointer. **(c)** A new linear molecule is produced as a result of the operation (from Harju et al. (2004b)).

◘ **Fig. 8**

Illustration of the dlad-rule. (a) The molecule is first folded into a double loop, simultaneously aligning two pairs of pointer occurrences. (b) Recombination is facilitated on both pointer alignments. (c) A new linear molecule is produced as a result of the operation (from Harju et al. (2004b)).

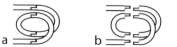

them. (By consequence, the same holds true for pointer q with respect to the two occurrences of pointer p.) The molecule is then folded into a double loop (❷ *Fig. 8a*), so that the two occurrences of p and the two occurrences of q are simultaneously aligned. Recombination events on p and on q are facilitated (❷ *Fig. 8b*), and as a result, a new linear molecule is obtained, see ❷ *Fig. 8c*.

5.1.4 Discussion

The {ld, hi, dlad} model is often referred to as the *intramolecular model*, to stress that in this model, the input to which operations are applied is always a single molecule. (In contrast, the model presented in ❷ Sect. 4 is referred to as the *intermolecular model*.) It is important to note however that ld yields as an output two molecules. As such, for the intramolecular assembly to succeed, that is, to assemble all MDSs, it is essential that all coding blocks remain within one of the molecules produced by ld. This gives two restrictions for applying ld on a pointer p in a successful gene assembly:

(i) No MDSs exist in-between the two occurrences of p. (Equivalently, only one, possibly composite, IES exists in-between the two occurrences of p.) We say in this case that we have a *simple* application of ld. As a result, a (possibly composite) IES is excised as a circular molecule and all MDSs remain on the resulting linear molecule.

(ii) All MDSs are placed in-between the two occurrences of p. We say in this case that we have a *boundary* application of ld. As a result, all MDSs are placed on the resulting circular molecule. The final assembled gene will be a circular molecule. It has been shown in Ehrenfeucht et al. (2001a, 2003a) that using a boundary application of ld can be postponed to the last step of any successful assembly. In this way, in-between the two occurrences of p there is only one (possibly composite) IES, similarly as in the case of simple ld.

The mechanistic details of the alignment and recombination events postulated by the three molecular operations ld, hi, and dlad are not indicated in the original proposal for the intramolecular model. Two mechanisms were later proposed in Prescott et al. (2001b) and in Angeleska et al. (2007). The details of both are discussed in ❷ Sect. 7.

5.2 String and Graph Representations for Ciliate Genes

We present three different formalizations for the gene assembly process: signed permutations, legal strings, and overlap graphs. The first two of these are linear in the sense that the order of the MDSs can be readily seen from the presentation. On the other hand, from the overlap graphs the order of the components is more difficult to capture. The gene assembly process will, however, be different in nature for signed permutations as for legal strings as well as for overlap graphs. Permutations need to be sorted while strings and graphs need to be reduced to empty string and graph, respectively. Another formalization in terms of *descriptors* is given in ❯ Sect. 6.

5.2.1 Signed Permutations and Legal Strings

Each micronuclear arrangement of MDSs and IESs can be represented as a signed permutation over the alphabet $\{M_1, M_2, \ldots, M_\kappa\}$, for an integer $\kappa \geq 1$ corresponding to the number of macronuclear destined sequences that assemble to a functional gene. We identify such a string with a signed permutation over the index set $\{1, 2, \ldots, \kappa\}$.

Example 5 Consider the micronuclear arrangement of the actin I gene of *Sterkiella nova*: $M_3 M_4 M_6 M_5 M_7 M_9 \overline{M_2} M_1 M_8$. This is represented as the signed permutation $\alpha = 3\,4\,6\,5\,7\,9\,\overline{2}\,1\,8$.

The gene assembly process is equivalent to sorting a signed permutation in proper order, that is, in the order $p(p+1) \ldots \kappa 1 \ldots (p-1)$ or its inverse for suitable p (and κ, the number of MDSs in the micronuclear gene). If $p = 1$ here, then the MDSs are linearly ordered; otherwise they are cyclically ordered.

Representation by strings will preserve the order of the MDSs that are coded as "pairs of pointers."

A string $v \in \Sigma^*$ over an alphabet Σ is said to be a *double occurrence string*, if every letter $a \in \mathrm{dom}(v)$ occurs exactly twice in v. A signing of a nonempty double occurrence string is a *legal string*. A letter $a \in \Sigma \cup \overline{\Sigma}$ is *positive* in a legal string $v \in \Sigma^\circledast$, if v contains both a and $\overline{a}$; otherwise, a is *negative* in v.

Example 6 Consider the legal string $u = 2\,4\,3\,\overline{2}\,\overline{5}\,3\,4\,5$ of pointers. Pointers 2 and 5 are positive in u, while 3 and 4 are negative in u. On the other hand, the string $w = 2\,4\,3\,\overline{2}\,5\,3\,5$ is not legal, since 4 has only one occurrence in w.

Let $u = a_1 a_2 \ldots a_n \in \Sigma^\circledast$ be a legal string over Σ with $a_i \in \Sigma \cup \overline{\Sigma}$ for each i. For $a \in \mathrm{dom}(u)$, let $1 \leq i < j \leq n$ be such that $\|a_i\| = a = \|a_j\|$. Then the substring

$$u_{(a)} = a_i a_{i+1} \ldots a_j$$

is called the *a-interval* of u. Two different letters $a, b \in \Sigma$ are said to *overlap* in u, if the a-interval and the b-interval of u overlap: if $u_{(a)} = a_{i_1} \ldots a_{j_1}$ and $u_{(b)} = a_{i_2} \ldots a_{j_2}$, then either $i_1 < i_2 < j_1 < j_2$ or $i_2 < i_1 < j_2 < j_1$.

Example 7 Let $u = 2\,4\,3\,5\,3\,\overline{2}\,6\,\overline{5}\,4\,6$ be a signed string of pointers. The 2-interval of u is the substring $u_{(2)} = 2\,4\,3\,5\,3\,\overline{2}$, and hence pointer 2 overlaps with 4 and 5 but not with 3 or 6. Similarly, for example, $u_{(4)} = 4\,3\,5\,3\,\overline{2}\,6\,\overline{5}\,4$ and hence 4 overlaps with 2 and 6.

A signed permutation will be represented by a legal string using the following substitution $\varrho_\kappa : \{1, 2, \ldots, \kappa\}^\circledast \to \{2, 3, \ldots, \kappa\}^\circledast$

$$\varrho_\kappa(1) = 2, \quad \varrho_\kappa(\kappa) = \kappa, \quad \varrho_\kappa(p) = p\,(p+1) \text{ for } 2 \le p < \kappa$$

and $\varrho_\kappa(\overline{p}) = \overline{\varrho_\kappa(p)}$ for each p with $1 \le p \le \kappa$.

Example 8 Consider the signed permutation α from Example 5. We have

$$\varrho_9(\alpha) = 34\ 45\ 67\ 56\ 78\ 9\ \overline{3}\ \overline{2}\ 2\ 89.$$

5.2.2 Overlap Graphs

The focus now shifts to representations by graphs. We use signed graphs to represent the structure of overlaps of letters in legal strings as follows: let $v \in \Sigma^{\circledast}$ be a legal string. The *overlap graph* of v is the signed graph $G_v = (\text{dom}(v), E, \sigma)$ such that

$$\sigma(x) = \begin{cases} +, & \text{if } x \text{ is positive in } v \\ -, & \text{if } x \text{ is negative in } v \end{cases}$$

and

$$\{x, y\} \in E \iff x \text{ and } y \text{ overlap in } v$$

Example 9 Consider the legal string $v = 3\,4\,\overline{5}\,2\,3\,\overline{2}\,4\,5$ of pointers. Then its overlap graph G_v is given in ❯ *Fig. 9*.

Overlap graphs of double occurrence strings are also known as *circle graphs*.

Example 10 Notice that the mapping $w \mapsto G_w$ of legal strings to overlap graphs is not injective. The following eight legal strings of pointers have the same overlap graph (of one edge): $2\overline{3}23$, $\overline{3}232$, $23\overline{2}3$, $32\overline{3}2$, $\overline{2}3\overline{2}\overline{3}$, $\overline{3}\overline{2}3\overline{2}$, $\overline{2}3\overline{2}3$, $3\overline{2}3\overline{2}$. For a more complicated example, we mention that the strings $v_1 = 23342554$ and $v_2 = 35242453$ define the same overlap graph.

5.2.3 Reduction Graph

Recall that legal strings represent the MIC form of genes. We now introduce a graph, called the *reduction graph*, that represents the MAC form of a gene and the other molecules obtained as results of the assembly, given a legal string (the MIC form of that gene). In this way, the

◼ **Fig. 9**
The overlap graph of the signed string $v = 3\,4\,\overline{5}\,2\,3\,\overline{2}\,4\,5$.

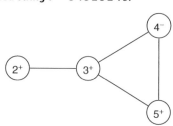

reduction graph represents the end result after recombination on all pointers. First, we define a
2-edge colored graph as a tuple (V, E_1, E_2, f, s, t), where V are the vertices, $s, t \in V$ are called
source and *target*, and $f: V \setminus \{s, t\} \to \Gamma$ is a vertex labeling function with Γ a finite set of vertex
labels. There are two (not necessarily disjoint) sets of undirected edges E_1 and E_2. We let $\mathsf{dom}(G)$
be the range of f, and say that 2-edge colored graphs G and G' are *isomorphic*, denoted $G \approx G'$,
when they are equal up to a renaming of the vertices. However, we require that the labels of the
identified vertices be equal, and that the sources and targets of G and G' be identified.

A reduction graph (Brijder et al. 2006) is a 2-edge colored graph where the two types of
edges E_1 and E_2 are called *reality edges* and *desire edges* respectively. Moreover, each vertex,
except s and t, is labeled by an element of $\Delta_\kappa = \{2, 3, \ldots, \kappa\}$. As an example, consider the
representation of legal string $u = 2\bar{7}473\bar{5}3\bar{4}2656$ over $\Pi_\kappa = \Delta_\kappa \cup \bar{\Delta}_\kappa$ given in ❷ *Fig. 10*. This
legal string will be used as our running example.

A reduction graph $\mathcal{R}_u$ for a legal string u is defined in such a way that (1) each occurrence
of a pointer of u appears twice (in unbarred form) as a label of a vertex in the graph to
represent both sides of the pointer in the representation of u, (2) the reality edges (depicted as
"double edges" to distinguish them from the desire edges) represent the segments between the
pointers, (3) the desire edges represent which segments should be glued to each other when
recombination operations are applied on the corresponding pointers. To enforce this last
requirement, positive pointers are connected by crossing desire edges (cf. pointers 4 and 7 in
❷ *Fig. 11*), while negative pointers are connected by parallel desire edges. The vertices s and
t represent the left end and the right end, respectively. Note that, since the reduction graph is
fixed for a given u, the end product after recombination is fixed as well. The notion of
reduction graph is similar to the breakpoint graph (or reality-and-desire diagram) known
from the theory of sorting by reversal, see, for example, Setubal and Meidanis (1997) and
Pevzner (2000). A formal definition of reduction graph is found, for example, in Brijder et al.
(2006). The reduction graph for string $u = 2\bar{7}473\bar{5}3\bar{4}2656$ is in ❷ *Fig. 11*.

In depictions of reduction graphs, the vertices (except for s and t) will be represented by
their labels, because the exact identities of the vertices are not essential here – we consider
reduction graphs up to isomorphism. Note that the reduction graph is defined for the general
concept of legal strings. Therefore, the reduction graph represents the end product after
recombination of arbitrary sequences of pointers (which by definition come in pairs) – not
only those that correspond to sequences of MDSs.

❏ **Fig. 10**

The representation of $u = 2\bar{7}473\bar{5}3\bar{4}2656$.

| | 2 | | 7 | | 4 | | 7 | | 3 | | 5 | | 3 | | 4 | | 2 | | 6 | | 5 | | 6 | |

❏ **Fig. 11**

The reduction graph for $u = 2\bar{7}473\bar{5}3\bar{4}2656$.

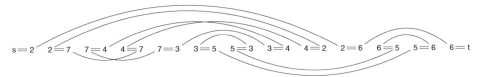

□ Fig. 12
The reduction graph $\mathcal{R}_u$ of u from the running example.

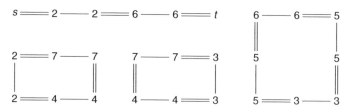

In the running example, the reduction graph $\mathcal{R}_u$ of u is again depicted in ❷ *Fig. 12* – we have only rearranged the vertices.

Each reduction graph has a connected component, called the *linear component*, containing both vertices s and t. The other connected components are called *cyclic components*.

5.3 Mathematical Formalizations of the Intramolecular Model

In this section we formalize the gene assembly process involving the three molecular operations ld, hi, and dlad in the framework of legal strings and overlap graphs.

5.3.1 Assembly Operations on Strings

Recall that each micronuclear MDS structure can be faithfully represented as a signed permutation α, which in turn has a presentation as a legal string $v = \varrho_\kappa(\alpha)$, where κ is the number of the MDSs in the micronuclear gene, and as an overlap graph G_v with κ vertices.

The assembly operations for strings are described first. The rules are (snr) the *string negative rule*, (spr) the *string positive rule*, and (sdr) the *string double rule*. For simplicity we consider only strings of pointers. Recall that we denote $\Delta_\kappa = \{2, 3, \ldots, \kappa\}$ and $\Pi_\kappa = \Delta_\kappa \cup \overline{\Delta}_\kappa$. Below, we assume that $p, q \in \Pi_\kappa$.

- snr_p applies to a legal string of the form $u = u_1 p p u_2$ resulting in

$$snr_p(u_1 p p u_2) = u_1 u_2 \qquad (1)$$

- spr_p applies to a legal string of the form $u = u_1 p u_2 \overline{p} u_3$ resulting in

$$spr_p(u_1 p u_2 \overline{p} u_3) = u_1 \overline{u}_2 u_3 \qquad (2)$$

- $sdr_{p,q}$ applies to a legal string of the form $u = u_1 p u_2 q u_3 p u_4 q u_5$ resulting in

$$sdr_{p,q}(u_1 p u_2 q u_3 p u_4 q u_5) = u_1 u_4 u_3 u_2 u_5 \qquad (3)$$

We define $\text{dom}(\rho)$ for a string reduction rule ρ by $\text{dom}(snr_p) = \text{dom}(spr_p) = \{\|p\|\}$ and $\text{dom}(sdr_{p,q}) = \{\|p\|, \|q\|\}$ for $p, q \in \Pi_\kappa$.

We adopt the following graphical notations for the applications of these operations:

$$u \xrightarrow{\mathsf{snr}_p} \mathsf{snr}_p(u), \quad u \xrightarrow{\mathsf{spr}_p} \mathsf{spr}_p(u), \quad u \xrightarrow{\mathsf{sdr}_{p,q}} \mathsf{sdr}_{p,q}(u)$$

A composition $\varphi = \varphi_n \ldots \varphi_1$ of the above operations φ_i is a *string reduction* of u, if φ is applicable to u. Also, φ is *successful* for u, if $\varphi(u) = \Lambda$, the empty string. Moreover, we define $\mathsf{dom}(\varphi) = \mathsf{dom}(\varphi_1) \cup \mathsf{dom}(\varphi_2) \cup \ldots \cup \mathsf{dom}(\varphi_n)$.

Example 11 The rule snr_2 is applicable to the legal string $u = 5\,2\,2\,3\,\overline{5}\,\overline{4}\,3\,4$: $\mathsf{snr}_2(u) = 5\,3\,\overline{5}\,\overline{4}\,3\,4$. Moreover, we have

$$5\,2\,2\,3\,\overline{5}\,\overline{4}\,3\,4 \xrightarrow{\mathsf{snr}_2} 5\,3\,\overline{5}\,\overline{4}\,3\,4 \xrightarrow{\mathsf{spr}_{\overline{4}}} 5\,3\,\overline{5}\,\overline{3} \xrightarrow{\mathsf{spr}_3} 5\,5 \xrightarrow{\mathsf{snr}_5} \Lambda,$$

and hence φ is successful for u.

The following is the basic universality result for legal strings.

Theorem 1 (Ehrenfeucht et al. 2000; Brijder et al. 2006) *Each legal string has a successful string reduction.*

5.3.2 Assembly operations on graphs

The assembly operations for graphs will now be described. The rules are (gnr) the *graph negative rule*, (gpr) the *graph positive rule*, (gdr) the *graph double rule*.

Let x and y be vertices of a signed graph G.

- gnr_x is applicable to G, if x is isolated and negative. The result is $\mathsf{gnr}_x(G) = G - x$.
- gpr_x is applicable to G, if x is positive. The result is $\mathsf{gpr}_x(G) = \mathsf{loc}_x(G) - x$.
- $\mathsf{gdr}_{x,y}$ is applicable to G, if x and y are adjacent and negative. The result is $\mathsf{gdr}_{x,y}(G) = \mathsf{loc}_x\mathsf{loc}_y\mathsf{loc}_x(G) - \{x,y\}$ obtained by complementing the edges between the sets $N_G(x) \setminus N_G(y)$, $N_G(y) \setminus N_G(x)$, and $N_G(x) \cap N_G(y)$.

Example 12 Consider the overlap graph $G = G_w$ for $w = 3\,\overline{5}\,2\,6\,5\,4\,7\,3\,6\,7\,2\,\overline{4}$; see ❷ *Fig. 13a*. The graph $\mathsf{gpr}_4(G)$ is given in ❷ *Fig. 13b*, and the graph $\mathsf{gdr}_{2,3}(G)$ is given in ❷ *Fig. 13c*.

❑ **Fig. 13**
The graphs (a) $G = G_w$, (b) $\mathsf{gpr}_4(G)$, (c) $\mathsf{gdr}_{2,3}(G)$, where $w = 3\,\overline{5}\,2\,6\,5\,4\,7\,3\,6\,7\,2\,\overline{4}$.

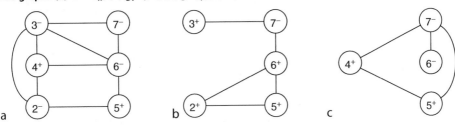

A composition $\varphi = \varphi_n \ldots \varphi_1$ of the above graph operations is called a *graph reduction* for a signed graph G, if φ is applicable to G. Also, φ is *successful*, if $\varphi(G)$ is the empty graph.

Example 13 The overlap graph $G = G_w$ given in ❷ *Fig. 13a* is reduced to the empty graph by the composition $\text{gpr}_5\text{gpr}_6\text{gpr}_7\text{gpr}_4\text{gdr}_{2,3}$.

The above operations are universal for signed graphs.

Theorem 2 (Harju et al. 2004b) *Each signed graph G has a successful graph reduction.*

5.3.3 Equivalence of the Systems

The relation between the systems for strings and graphs is now studied, and we show that there is a correspondence between these operations.

Theorem 3 (Ehrenfeucht et al. 2003b) *Let w be a legal string. Each string reduction $\varphi = \varphi_n \ldots \varphi_1$ for w translates into a graph reduction $\varphi' = \varphi'_n \ldots \varphi'_1$ for the overlap graph G_w by the translation:*

$$\text{snr}_p \mapsto \text{gnr}_p, \quad \text{spr}_p \mapsto \text{gpr}_p, \quad \text{sdr}_{p,q} \mapsto \text{gdr}_{p,q}.$$

Consequently, if φ is successful for w, then φ' is successful for G_w.

The reverse implication, from graphs to strings, is not as straightforward. Recall first that the mapping from the legal strings to the overlap graphs is not injective.

Denote by $p(w)$ the *first occurrence* of p or $\bar{p}$ in w.

Theorem 4 (Ehrenfeucht et al. 2003b) *Let w be a legal string, and let φ be a successful reduction for G_w. Then there exists a permutation $\varphi'' = \varphi_n \ldots \varphi_1$ of φ, which is successful for G_w, and which can be translated to a successful string reduction $\varphi' = \varphi'_n \ldots \varphi'_1$ for w by the following translations:*

$$\text{gnr}_p \mapsto \text{snr}_{p(w)}, \quad \text{gpr}_p \mapsto \text{spr}_{p(w)}, \quad \text{gdr}_{p,q} \mapsto \text{sdr}_{p(w),q(w)}.$$

5.4 Properties of Intramolecular Assemblies

Some properties of intramolecular gene assemblies are discussed in this section. Many of these properties hold in all the mathematical models described in ❷ Sect. 5.3. In each case however, we choose to describe the results only on a level that allows for the simplest or the most elegant formulation. For more results we refer to Ehrenfeucht (2003a) and Brijder (2008).

5.4.1 Nondeterminism and Confluence

It is easy to show on all model levels, from the molecular level to that of graphs, that for a given gene (permutation, string, graph, resp.), there can be more than one strategy to assemble it.

Different assembly reduction strategies have recently also been confirmed experimentally (Möllenbeck et al. 2008). We say that the intramolecular model is *nondeterministic*. Consider, for example, the signed string associated to the actin I gene in *S. nova*: $u = 3\,4\,4\,5\,6\,7\,5\,6\,7\,8\,9\,\overline{3}\,\overline{2}\,2\,8\,9$. There are 3060 different sequential strategies to reduce this string (assemble the gene), see Ehrenfeucht et al. (2003a), of which only two are shown in ❯ *Table 1*. Note in particular that these two strategies differ in the number and type of operations used. On the other hand, both strategies are successful, reducing the input string to the empty string. As shown in Theorems 1 and 2, this is true in general: although several operations may be applicable to a given input, successful strategies for that input exist starting with any of those operations. The biological interpretation is that all (potentially many) assembly strategies of a given micronuclear gene, have the same result: the assembled corresponding macronuclear gene. We call such a model *confluent*. Consequently, ciliates "need not remember" a particular sequential strategy which in turn contributes to the robustness of the gene assembly process.

In ❯ Sect. 6 it will be proved that the assembly strategies of a given gene share a number of other properties beyond yielding the same assembled gene: the number of molecules (linear and circular) produced throughout the assembly, their nucleotide sequence, whether the assembled gene is linear or circular.

5.4.2 The Structure of MAC Genes with By-products

Since legal strings represent the initial configuration (gene in MIC form) and the corresponding reduction graph the end result (the same gene in MAC form and its excised products), it is natural to study the possible forms of reduction graphs. Formally, we now characterize the graphs that are (isomorphic to) reduction graphs.

A graph G isomorphic to a reduction graph must be a 2-edge colored graph (V, E_1, E_2, f, s, t) such that for each p in the range of f, $p \in \Delta$ and there must be exactly four vertices labeled by p. Each vertex must be connected to exactly one (reality) edge from E_1, and each vertex, except s and t, must be connected to exactly one (desire) edge from E_2. Finally, edges from E_2 must connect vertices with a common label. One can call these graphs *abstract reduction graphs*, and let the set of abstract reduction graphs be ARG. It turns out that there are graphs in ARG that are not (isomorphic to) reduction graphs.

To obtain a characterization, one more property of reduction graphs is needed: the ability to linearly order the vertices to resemble its (in general not unique) underlying legal string, as done in ❯ *Fig. 11*. To make this linear order of vertices explicit, we introduce a third set of edges, called *merge edges*, to the reduction graph as done in ❯ *Fig. 14*.

❑ **Fig. 14**
Merge edges are added to the reduction graph of ❯ *Fig. 11*.

Now, when is a set of edges M for $G \in$ ARG a set of merge edges? Like desire edges, they have the properties that (1) the edges connect vertices with a common label and (2) each vertex except s and t is connected to exactly one merge edge. Moreover, M and the set E_2 are disjoint – no desire edge is parallel to a merge edge. Finally, the reality edges and merge edges must allow for a path from s to t passing each vertex once. This last requirement is equivalent to the fact that the reality and merge edges induce a connected graph.

If it is possible to add a set of merge edges to the graph, then it is not difficult to see that the graph is isomorphic to a reduction graph $\mathcal{R}_u$. Indeed, we can identify such a u for this reduction graph by simply considering the alternating path from s to t over the reality and merge edges. The orientation (positiveness or negativeness) of each pointer is determined by the crossing or noncrossing of the desire edges (exactly as we defined the notion of reduction graph).

To characterize reduction graphs we need the notion of a pointer-component graph. Given an abstract reduction graph, a pointer-component graph describes how the labels of that abstract reduction graph are distributed among its connected components.

Definition 1 *Let $G \in$ ARG. The* pointer-component graph *of G, denoted by $\mathcal{PC}_G$, is a multigraph (ζ, E, ε), where ζ is the set of connected components of G, $E = \mathrm{dom}(G)$ and ε is, for $e \in E$, defined by $\varepsilon(e) = \{C \in \zeta \mid C \text{ contains vertices labeled by } e\}$.*

The pointer-component graph of $G = \mathcal{R}_u$ of ❯ *Fig. 12* is given in ❯ *Fig. 15*. We have $\zeta = \{C_1, C_2, C_3, R\}$ where R is the linear component and the other elements are cyclic components of $\mathcal{R}_u$.

It is shown in Brijder and Hoogeboom (2008a) that, surprisingly, $G \in$ ARG has a set of merge edges precisely when the pointer-component graph $\mathcal{PC}_G$ is a connected graph. Therefore, we have the following result.

Theorem 5 (Brijder and Hoogeboom 2008a) *An abstract reduction graph G is isomorphic to a reduction graph iff $\mathcal{PC}_G$ is a connected graph.*

Consider the legal string $u = 2\bar{7}47353\bar{4}2656$ as before and $v = 2\bar{7}42653\bar{4}7356$. It turns out that they have the same reduction graph (up to isomorphism): $\mathcal{R}_u \approx \mathcal{R}_v$ (see ❯ *Fig. 11*). The reason for this is that a reduction graph may have more than one set of merge edges – each one corresponding to a different legal string. Thus, there can be many legal strings giving the same reduction graph. In Brijder and Hoogeboom (2008a) it is shown how for a given legal string u we can obtain precisely the set of all legal strings having the same reduction graph

❑ **Fig. 15**
The graph $\mathcal{PC}_{\mathcal{R}_u}$ of the graph $\mathcal{R}_u$ in ❯ Fig. 12.

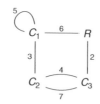

(up to isomorphism). In fact, it turns out that this set is exactly the set of all legal strings obtained by applying compositions of the following string rewriting rules.

For all $p, q \in \Pi_\kappa$ with $\|p\| \neq \|q\|$ we define

- The *dual string positive rule* for p is defined by $\mathrm{dspr}_p(u_1 p u_2 p u_3) = u_1 \bar{p} \bar{u}_2 p u_3$
- The *dual string double rule* for p,q is defined by $\mathrm{dsdr}_{p,q}(u_1 p u_2 q u_3 \bar{p} u_4 \bar{q} u_5) = u_1 p u_4 q u_3 \bar{p} u_2 \bar{q} u_5$

where $u_1, u_2, \ldots, u_5$ are arbitrary (possibly empty) strings over Π_κ. Notice the strong similarities of these rules with the string positive rule and the string double rule. As an example, if we take $u = 2\bar{7}47353\bar{4}2656$ and $v = 2\bar{7}426\bar{5}3\bar{4}7356$ given earlier, then $\mathrm{dsdr}_{4,5}\,\mathrm{dspr}_3(u) = v$ and hence both legal strings indeed have a common reduction graph.

5.4.3 Intermediate Legal Strings

We now show that we can generalize the notion of reduction graph to allow for representations of any intermediate product during the reduction process. In such an intermediate product, some pointers, represented as a subset D of $\mathrm{dom}(u)$, where u is a legal string, have not yet been used in recombination operations, while the other pointers, in $\mathrm{dom}(u) \setminus D$, have already been used in recombination operations. A reduction graph of u with respect to this set D, denoted by $\mathcal{R}_{u,D}$, represents such intermediate product. As before, we simply ignore the pointers in D – they are put as strings on the reality edges that are now directed edges.
❯ *Figure 16* gives an example of $\mathcal{R}_{u,D}$ with $u = 2\bar{7}47353\bar{4}2656$ and $D = \{2, 4\}$ (recall that Λ represents the empty string).

We denote the legal string obtained from a legal string u by removing the pointers from $D \subseteq \mathrm{dom}(u)$ and its barred variants by $\mathrm{rem}_D(u)$. In our example, $\mathrm{rem}_D(u) = \bar{7}7353656$. We define $\mathrm{red}(u, D)$ as the label of the alternating path from s to t. Thus, in our example $\mathrm{red}(u, D) = 2\bar{4}\bar{4}2$. Assuming that gene assembly is intramolecular (all recombination takes place on a single DNA molecule), then the cyclic connected components must have only empty strings as edge labels. In our example, it is easy to obtain an "invalid" intermediate product: take, for example, $D = \{3, 4, 5, 6, 7\}$. Hence, it is not possible to first recombine pointer 2, followed by recombination of the remaining pointers.

Theorem 6 (Brijder et al. 2006) *Let u be a legal string, let φ be a composition of reduction rules with $\mathrm{dom}(\varphi) \subseteq \mathrm{dom}(u)$, and let $D = \mathrm{dom}(u) \setminus \mathrm{dom}(\varphi)$. Then φ is applicable to u iff φ is applicable to $\mathrm{rem}_D(u)$ and $\mathrm{red}(u, D)$ is a legal string with domain D. Moreover, if this is the case, then $\varphi(u) = \mathrm{red}(u, D)$.*

As a consequence of Theorem 6, reductions φ_1 and φ_2 with the same domain have the same effect: $\varphi_1(u) = \varphi_2(u)$ for all legal strings u. Note that in general there are $D \subseteq \mathrm{dom}(u)$ for

⬛ **Fig. 16**
Graph $\mathcal{R}_{u,D}$ with $u = 2\bar{7}47353\bar{4}2656$ and $D = \{2, 4\}$.

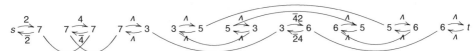

which there is no reduction φ of u with $D = \mathsf{dom}(\varphi(u))$. In our example, $\mathsf{red}(u, D)$ is a legal string with domain D and we have, for example, $(\mathsf{snr}_6\ \mathsf{sdr}_{3,5}\ \mathsf{spr}_{\bar{7}})(u) = 2\bar{4}\bar{4}2 = \mathsf{red}(u, D)$.

5.4.4 Cyclic Components

Since the reduction graph is a representation of the end result after recombination, the cyclic components of a reduction graph represent circular molecules. If we now consider again the intramolecular model of gene assembly, we notice that each such molecule is obtained by loop recombination. Hence, although there can be many different sequences of operations that obtain the fixed end product, the *number* of loop recombination operations (string negative rules in the model) in each such sequence is the same.

Theorem 7 (Brijder et al. 2006) *Let N be the number of cyclic components in the reduction graph of legal string u. Then every successful reduction of u has exactly N string negative rules.*

Example 14 Since $\mathcal{R}_u$ in ❷ *Fig. 12* has three cyclic components, by Theorem 7, every successful reduction φ of u has exactly three string negative rules. For example, $\varphi = \mathsf{snr}_2\ \mathsf{snr}_{\bar{4}}\ \mathsf{spr}_{\bar{7}}\ \mathsf{snr}_6\ \mathsf{sdr}_{3,5}$ is a successful reduction of u. Indeed, φ has exactly three string negative rules. Alternatively, $\mathsf{snr}_6\ \mathsf{snr}_3\ \mathsf{snr}_7\ \mathsf{spr}_2\ \mathsf{spr}_{\bar{5}}\ \mathsf{spr}_4$ is also a successful reduction of u, with a different number of (spr and sdr) operations.

It turns out that the reduction graph also allows for determining *on which pointers* the string negative rules can be applied using the pointer-component graph (Brijder et al. 2008). For convenience, one can denote $\mathcal{PC}_{\mathcal{R}_u}$ by $\mathcal{PC}_u$. Also, one can denote $\mathcal{PC}_u|_D$ as the graph obtained from $\mathcal{PC}_u$ by removing the edges outside D. Finally, for a reduction φ, let $\mathsf{snrdom}(\varphi) \subseteq \mathsf{dom}(\varphi)$ be the (unbarred) pointers used in snr rules in φ.

Theorem 8 (Brijder et al. 2008) *Let u be a legal string, and let $D \subseteq \mathsf{dom}(u)$. There is a successful reduction φ of u with $\mathsf{snrdom}(\varphi) = D$ iff $\mathcal{PC}_u|_D$ is a tree.*

In our running example, we see that $D = \{2, 3, 6\}$ induces a (spanning) tree of $\mathcal{PC}_u$. Therefore, there is a successful reduction φ of u with $\mathsf{snrdom}(\varphi) = D$. Indeed, we have $\mathsf{sdr}_{\bar{4},5}\ \mathsf{spr}_{\bar{7}}(u) = 226336$. It is clear that we can extend $\mathsf{sdr}_{\bar{4},5}\ \mathsf{spr}_{\bar{7}}$ to a successful reduction that applies string negative rules on 2, 3, and 6. Notice that here snr_3 must be applied *before* snr_6. In fact in Brijder et al. (2009) it is shown that the possible orders in which the string negative rules can be applied is also deducible from $\mathcal{PC}_u$ by considering rooted trees.

The results above can be carried over to intermediate products; for example, the number of string negative rules from u to $\varphi(u)$ is fixed and is equal to the number of cyclic components of $\mathcal{R}_{u,D}$.

5.5 Simple and Parallel Gene Assemblies

The general formulation of the intramolecular operations allows for the aligned pointers to be arbitrarily far from each other. We discuss in this section a *simple* variant of the model, where all alignments and folds involved in the operations are *local*. In the simple versions of ld, hi, and dlad, the pointers involved in the recombination are at a minimal distance from each other. It turns out that the simple model is able to explain the successful assemblies of all currently known micronuclear gene sequences, see Cavalcanti et al. (2005), Prescott et al. (2001a), and Langille et al. (2010). In this section we discuss the molecular and the mathematical formulation of the simple model and indicate some interesting properties of the model.

In the second part of this section, a notion of *parallel gene assembly* is discussed. In each (parallel) step of the assembly, we apply a number of well-selected operations simultaneously in such a way that the total number of steps is minimal. In each step, the operations are selected in such a way that their application is independent of the others applied in the same step: All sequential compositions of those operations are applicable to the current graph. Several difficult computational problems arise in this context, including deciding whether a given graph has a parallel assembly of a given length, or deciding whether there are graphs (or even trees) of arbitrarily high parallel complexity.

5.5.1 Simple Gene Assembly

The three intramolecular operations allow in their general formulation that the MDSs participating in an operation may be located anywhere along the molecule. Arguing on the principle of parsimony, a simplified model was discussed already in Prescott et al. (2001a) and then formalized in Harju et al. (2006c), asking that all operations are applied "locally". In the simple model, the restriction is that there is at most one coding block involved in each of the three operations. This idea was then further developed into two separate models. In one of them, which is referred to as the *simple model* (Langille and Petre 2006), both micronuclear, as well as composite MDSs (obtained by splicing of several micronuclear MDSs) may be manipulated in each of the three molecular operations. In the other, called the *elementary model* and introduced in Harju et al. (2006b, 2008c), the model was further restricted so that only *micronuclear*, but not *composite*, MDSs could be manipulated by the molecular operations. Consequently, once two or more micronuclear MDSs are combined into a larger composite MDS, they can no longer be moved along the sequence. In this section, only the simple model is discussed and for details of the elementary model one can refer to Harju et al. (2006b, 2008c), Langille et al. (2010), and Petre and Rogojin (2008).

We already discussed in ❯ Sect. 5.1 that ld must always be *simple* in a successful assembly. As such, the effect of ld is that it will combine two consecutive MDSs into a bigger composite MDS. For example, consider that $M_3 M_4$ is a part of the molecule, that is, MDS M_4 succeeds M_3 being separated by one IES I. Thus, pointer 4 has two occurrences that flank I: one in the end of MDS M_3 and the other one in the beginning of MDS M_4. Then ld makes a fold as in ❯ *Fig. 6a* aligned by pointer 4, IES I is excised as a circular molecule and M_3 and M_4 are combined into a longer coding block as shown in ❯ *Fig. 6c*.

In the case of hi and dlad, the pointers involved can be separated by arbitrarily large sequences; for example, in the actin I gene in *S. nova*, pointer 3 has two occurrences: one in the beginning of M_3 and one, inverted, in the end of M_2. Thus, hi is applicable to this sequence with the hairpin aligned on pointer 3, even though five MDSs separate the two occurrences of pointer 3. Similarly, dlad is applicable to the MDS sequence $M_2 M_8 M_6 M_5 M_1 M_7 M_3 M_{10} M_9 M_4$, with the double loops aligned on pointers 3 and 5. Here the first two occurrences of pointers 3 and 5 are separated by two MDSs (M_8 and M_6) and their second occurrences are separated by four MDSs (M_3, M_{10}, M_9, M_4).

An application of the hi operation on pointer p is *simple* if the part of the molecule that separates the two copies of p in an inverted repeat contains only one MDS and one IES. We have here two cases, depending on whether the first occurrence of p is incoming or outgoing, see ❷ *Fig. 17a*.

An application of dlad on pointers p, q is *simple* if the sequence between the first occurrences of p and q, as well as the sequence between the second occurrences of p and q consist of either one MDS or one IES. We have again two cases, depending on whether the first occurrence of p is incoming or outgoing, see ❷ *Fig. 17b*.

The simple operations can be formalized as operations on signed permutations, signed strings, and signed graphs. We only give here the definitions for the string-based operations, where the mathematical formulation is more concise. For the other formulations, including the relationships among these models, one can refer to Harju et al. (2006c), Langille et al. (2010), and Brijder and Hoogeboom (2008b).

The *simple* hi *operation* for pointer p, denoted sspr_p, is applicable to strings of the form $u = u_1 p u_2 \overline{p} u_3$, where $|u_2| \le 1$, resulting in $\mathsf{sspr}_p(u_1 p u_2 \overline{p} u_3) = u_1 \overline{u_2} u_3$.

The *simple* dlad *operation* for pointers p, q, denoted $\mathsf{ssdr}_{p,q}$ is applicable to strings of the form $u = u_1 p q u_2 p q u_3$, resulting in $\mathsf{ssdr}_{p,q}(u_1 p q u_2 p q u_3) = u_1 u_2 u_3$.

Let ϕ be a composition of snr, sspr, and ssdr operations such that ϕ is applicable to string u. We say that ϕ is a *simple reduction* for u if either $\phi(u) = \Lambda$ (in which case we say that φ is *successful*), or $\phi(u) \ne \Lambda$ and no simple operation is applicable to $\phi(u)$ (in which case we say that ϕ is *unsuccessful*). For example, the string reduction in ❷ *Table 1b* is simple, unlike the one in ❷ *Table 1a*.

The simple model has a number of properties that do not hold for the general model. One of them concerns the length of reduction strategies for a given string. While in the general model a string may have reduction strategies of different lengths, see the example in ❷ *Table 1*, the same is not true in the simple model, see the next result of Langille and Petre (2007). Moreover, if we consider *parallel applications of simple operations* (a notion of Langille and Petre (2007) that is not defined in this chapter), we get a new twist: for any given n there exists a string having maximal parallel reductions of any length between n and $2n$.

◘ **Fig. 17**
The MDS/IES structures where **(a)** *simple hi*-rules and **(b)** *simple dlad*-rules are applicable. The MDSs are indicated by rectangles and their flanking pointers are shown. Between the two MDSs there is only one IES represented by a straight line (From Harju et al. (2006c) and Prescott et al. (2001a)).

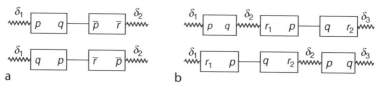

◻ Table 1

Two reduction strategies for the signed string corresponding to the actin I micronuclear gene in S.nova

(a)	(b)
$u_1 = \mathsf{spr}_3(u) = \overline{9}8765\overline{44}\overline{2}289$	$u'_1 = \mathsf{snr}_4(u) = 3567567893\overline{3}289$
$u_2 = \mathsf{snr}_4(u_1) = \overline{9}8765\overline{2}289$	$u'_2 = \mathsf{sdr}_{5,6}(u'_2) = 377893\overline{3}289$
$u_3 = \mathsf{spr}_8(u_2) = \overline{9}225675679$	$u'_3 = \mathsf{snr}_7(u'_2) = 3893\overline{3}289$
$u_4 = \mathsf{spr}_2(u_3) = \overline{9}5675679$	$u'_4 = \mathsf{sdr}_{8,9}(u'_3) = 3\overline{3}\overline{2}2$
$u_5 = \mathsf{sdr}_{5,7}(u_4) = \overline{9}669$	$u'_5 = \mathsf{spr}_2(u'_4) = 3\overline{3}$
$u_6 = \mathsf{snr}_6(u_5) = \overline{9}9$	$u'_6 = \mathsf{spr}_3(u'_5) = \Lambda$
$u_7 = \mathsf{spr}_9(u_6) = \Lambda$	

Theorem 9 (Langille and Petre 2007) *Let u be a signed double occurrence string and ϕ, ψ two reduction strategies for u. Then ϕ and ψ have the same number of operations.*

Regarding the outcome of reduction strategies, the simple model is different from the general model in several respects; for example, there are strings that cannot be reduced in the simple model, unlike in the general model where all strings have reduction strategies. Indeed, no simple operation is applicable to the string $\overline{2}43423$. The following is a result of Langille and Petre (2006).

Theorem 10 (Langille and Petre 2006) *No signed string has both successful and unsuccessful reductions in the simple model.*

On the other hand, the outcome of various strategies for a given string can differ; for example, for $u = 234678567823 45$, $u_1 = \mathsf{ssdr}_{2,3} \circ \mathsf{ssdr}_{7,8}(u) = 465645$, whereas $u_2 = \mathsf{ssdr}_{3,4} \circ \mathsf{ssdr}_{6,7}(u) = 285825$. The strings u_1 and u_2 are, however, identical modulo a relabeling of their letters. This observation can be extended to define a notion of *structure* that can be used to prove that the results of various reductions, although different, always have the same structure. For details, one can refer to Langille and Petre (2006), where the discussion is in terms of signed permutations, rather than signed strings.

5.5.2 Parallel Gene Assembly

The notion of parallelism is usually defined in concurrency theory for processes whose application is independent of each other. In other words, a number of processes can be applied in parallel to a signed graph if they can be (sequentially) applied in any order. Adopting this approach, the following gives the definition of parallel application of the three molecular operations on a signed graph. For a similar discussion, albeit technically more tedious, on the level of signed strings, one can refer to Harju et al. (2006a).

Definition 2 (Harju et al. 2006a) *Let S be a set of k* gnr, gpr, *and* gdr *operations and let G be a signed graph. We say that the rules in S are applicable in parallel to G if for any ordering $\varphi_1, \varphi_2, \ldots, \varphi_k$ of S, the composition $\varphi_k \circ \cdots \circ \varphi_1$ is applicable to G.*

Based on the definition of parallelism, which presumes that the rules are applicable in any possible order, the following theorem shows that the result is always the same regardless of the order in which they are applied.

Theorem 11 (Harju et al. 2006a) *Let G be a signed graph and let S be a set of operations applicable in parallel to G. Then for any two compositions φ and ψ of the operations of S, $\varphi(G) = \psi(G)$.*

Based on Theorem 11, we can write $S(G) = \varphi(G)$ for any set S of operations applicable in parallel to G and any composition φ of these operations. The notion of parallel complexity is defined as follows.

Definition 3 (Harju et al. 2007) *Let G be a signed graph, and let $S_1, \ldots, S_k$ be sets of* gnr, gpr, gdr *operations. If $(S_k \circ \ldots \circ S_1)(G) = \emptyset$, then we say that $S = S_k \circ \ldots \circ S_1$ is a* parallel reduction *for G. In this case the* parallel complexity *of S is $\mathcal{C}(S) = k$. The* parallel complexity *of the signed graph G is:*

$$\mathcal{C}(G) = \min\{\mathcal{C}(S) \mid S \text{ is a parallel reduction strategy for } G\}.$$

Deciding whether a given set of graph operations is applicable in parallel to a given graph turns out to be a difficult problem if gdr operations are involved. When at most two gdr operations are involved, then simple characterizations were given in Harju et al. (2008a). The computational complexity of the general problem was upper bounded in the co-NP class, see Alhazov et al. (2009).

It can be easily verified that the parallel complexity of the graphs corresponding to the currently known micronuclear gene sequences is at most two, see Harju et al. (2008b). However, examples of graphs of higher complexity can be given. For example, the graph with the highest known complexity has 24 vertices and can be reduced in six parallel steps. The tree with the highest known parallel complexity has 12 vertices and can be reduced in five parallel steps. One can refer to Harju et al. (2008b) for more examples. Although the parallel complexity of certain types of graphs (e.g., for uniformly signed trees) is known to be finitely bounded, see Harju et al. (2008a), the general problem is currently open and seems to be very difficult. In particular, it seems to require a characterization for the parallel applicability of arbitrary sets of operations, another open problem. The problem is open even in seemingly simpler cases: for trees, or for negative graphs. The computational complexity of deciding whether the parallel complexity of a given graph is upper bounded by a given contact was placed in the NP^{NP} class in Alhazov et al. (2009). Two algorithms for computing the parallel complexity of signed graphs were given in Alhazov et al. (2009, 2010), both with exponential computational complexity. A visual graph editor including support for computing the parallel complexity of signed graphs can be found in Petre and Skogman (2006).

5.6 Gene Assembly by Folding and Unfolding

In this section, gene assembly is considered from a somewhat more general viewpoint. The section is based on Ehrenfeucht et al. (2002). The molecular operations provided by the gene assembly process each involve one molecule. This observation underlies the model of gene assembly as fold-and-recombine computing paradigm. For convenience, we consider circular graphs to be representations of DNA molecules. Our initial situation is a set of circular DNA molecules represented by bicolored and labeled circular graphs. The fold-and-recombine process is reflected by a two-stage processing of the graphs: (1) fold on vertices representing pointers; (2) unfold using a pairing function. In this setup, gene assembly becomes a dynamic process for recombination graphs.

Graphs with multiple edges and loops are allowed. The vertices denote all pointers of the gene. We consider bicolored graphs, where color 1 is used to indicate an IES and color 2 is used to indicate an MDS. To represent the sequence of nucleotides comprising various (IES or MDS) segments we use a labeling of the edges.

Each edge will be oriented in both directions: $a = (x, y)$ and $\bar{a} = (y, x)$ are *reverse pairs*. Let V be a set of vertices, and consider E as a set of edge symbols such that $E = \{e_1, \ldots, e_n, \bar{e}_1, \ldots, \bar{e}_n\}$ with $\bar{\bar{e}}_i = e_i$. A *(general) bicolored graph* G consists of an *end point map* $\varepsilon_G = \varepsilon: E \rightarrow V \times V$ such that $\varepsilon(\bar{e}) = \overline{\varepsilon(e)}$ for all $e \in E$; a *labeling* $f_G = f: E \rightarrow \Sigma^*$ for an alphabet Σ, with $f(\bar{e}) = \overline{f(e)}$ for all $e \in E$; a *coloring* $h_G = h: E \rightarrow \{1, 2\}$ such that $h(e) = h(\bar{e})$ for all $e \in E$.

For simplicity, we write $e = (x, y)$ for $\varepsilon(e) = (x, y)$. An edge $e = (x, y) \in E$ is a *loop*, if $x = y$. The *valency* $\mathrm{val}_G(x)$ of a vertex x is the number of edges leaving x. A bicolored graph is *even* if its valencies are all even. A bicolored graph G is a *recombination graph*, if $\mathrm{val}_G(x) \in \{2, 4\}$ for all x, and every vertex of valency 4 is balanced: two incident edges have color 1 and the other two have color 2; see ❯ *Fig. 18a*.

For each vertex x in an even bicolored graph G let ψ_x be a bijection that maps incoming edges to outgoing edges respecting inversions, such that $\psi_x(\overline{\psi_x(e)}) = \bar{e}$ and $\psi_x(e) = \bar{e}$ if and only if e is a loop. Then the map $\psi: x \mapsto \psi_x$ is a *pairing*. Each recombination graph G has the *natural pairing* ψ where e and $\psi_x(e)$ have the same color whenever $\mathrm{val}_G(x) = 4$.

5.6.1 Folding and Unfolding

A pair $p = \{x, y\} \in E(V)$ will be called a *pointer* with *ends* x and y. A set P of mutually disjoint pointers is a *pointer set*. The *p-folded graph* $G * p$ is obtained by identifying the ends of p; see ❯ *Fig. 18a, b*. For a pointer set $\{p, q\}$, $G * p * q = G * q * p$. This allows one to define, for a pointer set $P = \{p_1, \ldots, p_m\}$, the *P-folded graph* $G * P$ as $G * p_1 * \ldots * p_m$.

⬛ **Fig. 18**

(a) A recombination graph, where color 1 is represented by thick edge. (b) The *p*-folded graph $G *$ *p* for p = {1, 7}. (c) The ψ-unfolded graph $G \diamond_\psi 1$ with the natural pairing (from Ehrenfeucht et al. (2002)).

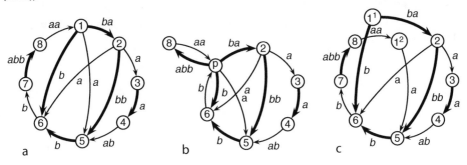

Let G be an even bicolored graph with a pairing ψ. For a vertex x, let $e_{11}, e_{12}, \ldots, e_{m1}, e_{m2}$ be the incoming edges with $\psi_x(e_{i1}) = \bar{e}_{i2}$. In the ψ-*unfolded graph*, the vertex x is replaced by the new vertices $x^1, \ldots, x^m$ and the edges are redirected according to the pairing ψ_x; see ❷ *Fig. 18c*, where the redirection is determined by the colors.

Notice that if G is a recombination graph, so is $G \diamond_\psi x$. Also, if $x \neq y$, then $G \diamond_\psi x \diamond_\psi y = G \diamond_\psi y \diamond_\psi x$. Therefore, we can write $G \diamond_\psi A = G \diamond_\psi x_1 \diamond_\psi \ldots \diamond_\psi x_m$ for a subset $A = \{x_1, \ldots, x_m\}$.

For an even bicolored graph G with a pairing ψ, let $F(G) = \{x \in V_G \mid \mathrm{val}_G(x) \geq 4\}$. Then the graph $G \diamond_\psi F(G)$ is called the ψ-*unfolded graph* of G.

Lemma 1   (Ehrenfeucht et al. 2002) *If G is an even bicolored graph with a pairing ψ, then its ψ-unfolded graph is a disjoint union of cycles.*

Let G be a bicolored graph with a pointer set P, and let ψ be a pairing of the P-folded graph $G * P$. We denote $G \circledast_\psi P = (G * P) \diamond_\psi P$. We shall write $G \circledast P$ for $G \circledast_\psi P$, if $G * P$ is a recombination graph and ψ is its natural pairing.

Lemma 2   (Ehrenfeucht et al. 2002) *Let G be a disjoint union of bicolored cyclic graphs. Let P be a pointer set of G, and let ψ be a pairing of $G * P$. Then $G \circledast_\psi P$ is a disjoint union of bicolored cyclic graphs.*

5.6.2   Assembled Graphs of Genomes

Let G be a bicolored cyclic graph with $V_G = \{x_1, \ldots, x_n\}$ and the edge set $E_G = \{e_1, \ldots, e_n, \bar{e}_1, \ldots, \bar{e}_n\}$, where $e_i = (x_i, x_{i+1})$ and $x_{n+1} = x_1$. A vertex x_i is a *boundary vertex* of G, if $h_G(e_{i-1}) \neq h_G(e_i)$, where $i - 1$ is modulo n. A path π is a *segment*, if its edges have color 1 and the ends of π are boundary vertices. For a disjoint union $G = \sum_{i=1}^m G_i$ of bicolored cyclic graphs G_i, we let its boundary vertex set be the union of corresponding sets of the components.

A pair $\mathcal{G} = (G, P)$ is called a *genome*, if G is a disjoint union of bicolored cyclic graphs and P is a pointer set of G containing boundary vertices only; see ❷ *Fig. 19*.

The *assembled genome* of a genome $\mathcal{G} = (G, P)$ is $A(\mathcal{G}) = (G \circledast P, \emptyset)$, and it is a genome. Each segment of the unfolded graph $\gamma \circledast P$ is a *gene*.

◘ **Fig. 19**

The genome (G, P) with P = { p, q} for p = {2, 9} and q = {5, 8}. The labels correspond to the MDSs M_1, M_2, N_1 and P_1, P_2. The labels corresponding to IESs are omitted (from Ehrenfeucht et al. (2002)).

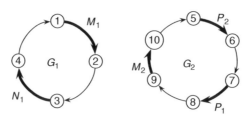

◼ **Fig. 20**
Let $\mathcal{G}$ be the genome of ❯ *Fig. 19*. Then $G * P$ is given in (a). Unfolding gives $G \circledast P$ in (b).
The genes of $\mathcal{G}$ are $g_1 : 1 \to p^1 \to 10$ (with the value M_1M_2), $g_2 : 7 \to q^1 \to 6$ (with the value P_1P_2),
and $g_3 : 3 \to 4$ (with the value N_1) (from Ehrenfeucht et al. (2002)).

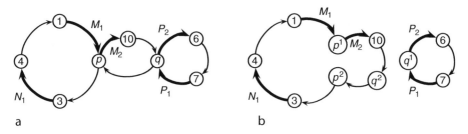

a   b

For $\mathcal{G} = (G, P)$, a sequence $\mathcal{S} = (P_1, P_2, \ldots, P_m)$ of subsets of P is an *assembly strategy* of $\mathcal{G}$, if $\{P_1, \ldots, P_m\}$ is a partition of P. One can show that in a genome $\mathcal{G} = (G, P)$, if $P_1, P_2 \subseteq P$ are disjoint, then $(G \circledast P_1) \circledast P_2 = G \circledast (P_1 \cup P_2) = (G \circledast P_2) \circledast P_1$. This gives the following general invariance property.

Theorem 12 (Ehrenfeucht et al. 2002) *Every assembly strategy* $\mathcal{S} = (P_1, P_2, \ldots, P_m)$ *of a genome produces the same assembled genome* $\mathcal{G} = (G, P)$: $G \circledast P = G \circledast P_1 \circledast P_2 \circledast \ldots \circledast P_m$.

A pointer set $R \subseteq P$ of $\mathcal{G} = (G, P)$ is *intracyclic*, if any two parts g' and g'' of each gene g that lie in the same connected component of G, lie in the same connected component of $G \circledast R$. The unfolding of a genome is illustrated in ❯ *Fig. 20*.

Theorem 13 (Ehrenfeucht et al. 2002) *For each genome* $\mathcal{G}$, *there exists an intracyclic assembly strategy* $\mathcal{S} = (P_1, P_2, \ldots, P_m)$ *such that* $1 \leq |P_i| \leq 2$ *for all* i.

We can also show that the following stronger result holds.

Theorem 14 (Ehrenfeucht et al. 2002) *Let* $\mathcal{G} = (G, P)$ *be a genome for a connected* $G \circledast P$. *Then there is a genome* $\mathcal{G}' = (G', P')$, *where* G' *is connected, such that* $A(\mathcal{G}) = A(\mathcal{G}')$, *and* $\mathcal{G}'$ *has an assembly strategy* $\mathcal{S} = (P_1, P_2, \ldots, P_m)$ *of* $\mathcal{G}'$ *for which* $1 \leq |P_i| \leq 2$ *for all* i, *and each* $G \circledast \cup_{i=1}^{j} P_i$ *is a cyclic graph for each* j.

6 Invariant Properties of Gene Assembly

As discussed already in this chapter, both an intermolecular model and an intramolecular model exist for gene assembly. Moreover, both models are nondeterministic: For a given gene, there may be several assembly strategies and also, a gene may be assembled either on a linear, or on a circular molecule. As such, a natural question is that of *invariants*: What properties of the assembled gene and of the assembly process hold for all assembly strategies of both models? For example, for a given gene, is the set of molecules excised during the assembly an invariant of the process? The same question for whether or not the assembled gene is linear or cyclic, and for whether or not the obtained structure of the IESs is fixed for given gene is also open.

An affirmative answer to these questions was given already in Ehrenfeucht et al. (2001a) for the intramolecular model, showing that these properties are invariants of the intramolecular model. We follow here a presentation of Petre (2006), where the result is given in a stronger form, showing that the properties above are invariants of *any model based on the paradigm of pointer-directed assembly*. This result may also be deduced based on the graph-theoretical framework of Ehrenfeucht et al. (2002). The presentation given in this section is in terms of strings and permutations. One can refer to Petre (2006) for more details, examples, and full proofs.

6.1 Gene Structure

In this section, we introduce a novel formal representation for the gene structures of ciliates, able to track the transformations witnessed by a gene from its micronuclear form to its assembled form. A gene is first represented as a signed permutation over the alphabet of MDSs by denoting their sequence and orientation. Our notation is then extended to denote also the IESs and all the pointers.

It can be recalled that $u, v \in \Sigma^*$ are called *equivalent*, denoted $u \approx v$, if u is a conjugate of either v or $\bar{v}$. Two finite sets $X_1, X_2 \subseteq \Sigma^*$ are called *equivalent*, denoted $X_1 \approx X_2$, if they have the same number of elements and for any $x_i \in X_i$ there is $x_j \in X_j$ such that $x_i \approx x_j$, with $i, j = 1, 2$, $i \neq j$. Intuitively, if u denotes a circular DNA molecule and $u \approx v$, then v denotes the same molecule, potentially starting from a different nucleotide and/or in the reverse direction. Similarly, if X_1 denotes a set of circular molecules and $X_1 \approx X_2$, then X_2 denotes the same set of molecules.

We denote the MDSs of the given gene by letters from the alphabet $\mathbf{M}_n = \{M_1, M_2, \ldots, M_n\}$ in the order they occur in the macronuclear gene, where $n \geq 1$. Thus, the sequence of MDSs in the macronuclear gene is $M_1 M_2 \ldots M_n$. On the other hand, the sequence of MDSs in the micronuclear gene is in general a signed permutation over $\mathbf{M}$.

Example 15 The MDSs of the micronuclear gene actin I in *S. nova may be represented as the signed permutation* $M_3 M_4 M_6 M_5 \, M_7 M_9 \overline{M_2} M_1 M_8$, *see* Prescott and DuBois (1996). *In this gene, MDS M_2 is inverted.*

We call $\mathbf{M}_n$-*descriptor* any signed permutation over $\mathbf{M}_n$. We say that μ is an *assembled* $\mathbf{M}_n$-*descriptor* if μ or $\bar{\mu}$ is of the form $M_i \ldots M_n M_1 \ldots M_{i-1}$, for some $1 \leq i \leq n$.

Consider now the alphabet of IESs $\mathbf{J}_n = \{I_0, I_1, \ldots, I_n\}$. For any $J \subseteq \mathbf{M}_n \cup \mathbf{J}_n$, a signed permutation over J will be called an $\mathbf{M}\mathbf{J}_n$-*descriptor*.

Let π_n be the projection $\pi_n : (\mathbf{M}_n \cup \mathbf{J}_n)^* \to \mathbf{M}_n^*$. We say that $\delta \in (\mathbf{M}_n \cup \mathbf{J}_n)^*$ is an *assembled* $\mathbf{M}\mathbf{J}_n$-descriptor if $\pi_n(\delta)$ is an assembled $\mathbf{M}_n$-descriptor.

We can associate an $\mathbf{M}\mathbf{J}_n$-descriptor to any $\mathbf{M}_n$-descriptor as follows. Let $\mu = \widetilde{M}_{i_1} \widetilde{M}_{i_2} \ldots \widetilde{M}_{i_n}$ be an $\mathbf{M}_n$-descriptor, where $\widetilde{M}_{i_k} \in \{M_{i_k}, \overline{M}_{i_k}\}$ and $i_1, i_2, \ldots, i_n$ is a permutation over $\{1, 2, \ldots, n\}$. Then the $\mathbf{M}\mathbf{J}_n$-*descriptor associated to* μ is $\tau_\mu = I_0 \widetilde{M}_{i_1} I_1 \widetilde{M}_{i_2} I_2 \ldots \widetilde{M}_{i_n} I_n$ – We denote by $I_0, I_1, \ldots, I_n$ the noncoding blocks separating the MDSs. We say in this case that τ_μ is a *micronuclear* $\mathbf{M}\mathbf{J}_n$-descriptor.

Example 16 (i) The $\mathbf{M}\mathbf{J}_9$-descriptor associated to the actin I gene in S.nova, see Example 15, is $I_0 M_3 I_1 M_4 I_2 M_6 I_3 \, M_5 I_4 M_7 I_5 M_9 I_6 \overline{M_2} I_7 M_1 I_8 M_8 I_9$.
(ii) $\delta = I_0 \overline{I_3} \overline{M_3} \overline{M_2} \overline{M_1} I_2 I_1$ is an assembled $\mathbf{M}_3$-descriptor.

The $\mathcal{MI}_n$-descriptors can now be extended to include also the information about the position of pointers in the gene. Consider then the alphabet $\mathcal{P}_n = \{2, 3, \ldots, n\}$ and denote the markers by b and e. Denote $\Sigma_n = \mathcal{M}_n \cup \mathcal{I}_n \cup \mathcal{P}_n \cup \{b, e\}$ and let π_n be the projection $\pi_n : \Sigma_n^{\circledast} \to (\mathcal{M}_n \cup \mathcal{I}_n)^{\circledast}$. We say that $\sigma \in \Sigma_n^{\circledast}$ is a Σ_n-*descriptor* if it has one of the following forms:

(i) $\sigma = \alpha_0 p_1 p_1 \alpha_1 p_2 p_2 \ldots p_k p_k \alpha_k, \; \alpha_0 \alpha_k \neq \Lambda$ or

(ii) $\sigma = p_1 \alpha_1 p_2 p_2 \cdots p_{k-1} p_{k-1} \alpha_k p_1$

where $k \geq 0$, $p_i \in \mathcal{P}_n \cup \overline{\mathcal{P}}_n$, $\alpha_i \in (\Sigma_n \setminus \mathcal{P}_n)^{\circledast}$, for all $0 \leq i \leq k$ and moreover, $\pi_n(\sigma)$ is a $\mathcal{MI}_n$-descriptor. In case (i), we call σ *linear*, and in case (ii) we call it *circular*. We say that σ is *assembled* if $\pi_n(\sigma)$ is an assembled $\mathcal{MI}_n$-descriptor.

Example 17 (i) $\sigma_1 = 22M_2 33M_3 \; eI_2 \overline{I}_3 b M_1 22$ is an assembled circular Σ_3-descriptor.

(ii) $\sigma_2 = I_0 22 M_2 33 I_1 \overline{2}\,\overline{2}\,\overline{M}_1 \; \overline{b} I_2 33 \; M_3 e I_3$ is a linear micronuclear Σ_3-descriptor.

(iii) $\sigma_3 = 22M_2 3 M_3 e I_2 \overline{I}_3 b M_1 22$ is not a Σ_3-descriptor.

Every MDS of micronuclear ciliate genes is flanked at its both ends by a pointer or a marker. We denote the pointers flanking MDS M_i by writing $iiM_i(i+1)(i+1)$. For M_1 and M_n we write $bM_1 22$ and $nnM_n e$, respectively. We use a double letter notation for pointers in order to deal with splicing in a simple way in ❷ Sect. 6.2. Formally, to associate a Σ_n-descriptor to an $\mathcal{MI}_n$-descriptor, consider the morphism $\phi_n : (\mathcal{M}_n \cup \mathcal{I}_n)^{\circledast} \to \Sigma_n^{\circledast}$ defined as follows: $\phi_n(I) = I$, for all $I \in \mathcal{I}_n$; $\phi_n(M_i) = iiM_i(i+1)(i+1)$, for all $2 \leq i \leq n-1$; $\phi_n(M_1) = bM_1 22$ and $f(M_n) = nnM_n e$. For any $\mathcal{MI}_n$-descriptor δ, we say that $\phi_n(\delta)$ is the Σ_n-descriptor *associated* to δ. It can be said that $\phi_n(\delta)$ is a *micronuclear* Σ_n-descriptor if δ is a micronuclear $\mathcal{MI}_n$-descriptor. Note that all micronuclear Σ_n-descriptors are linear.

Example 18 The micronuclear Σ_9-descriptor of the actin I gene in S.nova, see Example 16, is $I_0 33M_3 44 I_1 44 M_4 55 I_2 66 M_6 77 I_3 55 M_5 66 I_4 77 M_7 88 I_5 99 M_9 e I_6 \overline{33M}_2 \; \overline{2}\,\overline{2} I_7 b M_1 22 I_8 88 M_8 99 I_9$.

For any micronuclear Σ_n-descriptor σ and any $p \in \mathcal{P}_n$, σ contains two occurrences from the set $\{pp, \overline{p}\,\overline{p}\}$: pp represents the pointer in the beginning of MDS M_p and at the end of M_{p-1}, while $\overline{p}\,\overline{p}$ is its inversion.

6.2 Invariants

In this section, we give a number of invariants of the gene assembly process: the circularity of the assembled gene (whether or not the gene is assembled on a circular molecule), with the IES-context of the gene (the sequence of IESs preceding and succeeding the assembled gene), but also with the set of molecules excised during assembly. It is worth emphasizing that we establish all these properties based solely on the generic paradigm of pointer-directed assembly, independently of the specificities of either the intra-, or the inter-molecular model.

During the pointer-directed assembly, ciliates allegedly align their DNA molecules along their pointers, and through recombination they sort the MDSs in the orthodox order. It is essential to observe that in this process, the two strands of any pointer p will be separated: One strand will remain with the block preceding the pointer, while the other strand will remain with the block succeeding the pointer. The single strands will then recombine with the complementary strands obtained by separating in a similar way the second occurrence of p in the gene. This splicing on pointers can be formalized by a word-cutting operation defined in the following.

Let σ be a Σ_n-descriptor. If σ is linear, $\sigma = \alpha_0 p_1 \alpha_1 p_2 p_2 \alpha_2 \ldots \alpha_{k-1} p_k p_k \alpha_k$, with $p_i \in \mathcal{P}_n \cup \overline{\mathcal{P}}_n$, $\alpha_i \in (\Sigma_n \setminus \mathcal{P}_n)^*$, then $W_\sigma = \{\alpha_0 p_1, p_k \alpha_k, p_i \alpha_i p_{i+1} \mid 1 \leq i < k\}$. If σ is circular, $\sigma = p_1 \alpha_1 p_2 p_2 \alpha_2 \ldots \alpha_{k-1} p_k p_k \alpha_k p_1$, then $W_\sigma = \{p_i \alpha_i p_{i+1}, p_k \alpha_k p_1 \mid 1 \leq i < k\}$. For any set $S \subseteq \Sigma_n^{*\circledast}$, we denote $W_S = \cup_{\sigma \in S} W_\sigma$. Note that the set W_σ is equivalent to the set of edges of genome graphs (❷ Sect. 5.6) and to the reality edges of reduction graphs (❷ Sect. 5.2).

It is important to note that we do not conjecture that ciliates split their genes by cutting *simultaneously* on each pointer, to yield on the scale of 10^5 MDSs and IESs, followed then by a precise reassembly of all these blocks. Indeed, it is difficult to imagine that such a mechanism would lead to the precise effective assembly that we see in ciliates. Here we merely represent those pointer-delimited coding and noncoding blocks that will be eventually reshuffled to assemble the gene. Our main result states that, given the fixed order in which MDSs must be assembled, the pointer-directed assembly of all the other blocks (IESs) is uniquely determined by the micronuclear structure of the gene.

Example 19   For the Σ_9-descriptor σ in Example 18, we have

$$W_\sigma = \{I_0 3, 3M_3 4, 4I_1 4, 4M_4 5, 5I_2 6, 6M_6 7, 7I_3 5, 5M_5 6, 6I_4 7, 7M_7 8, 8I_5 9,$$
$$9M_9 eI_6 \overline{3}, \overline{3}\,\overline{M_2}\overline{2}, \overline{2}I_7 bM_1 2, 2I_8 8, 8M_8 9, 9I_9\}.$$

Our invariant theorem may be stated now as follows.

Theorem 15   (Petre 2006) *Let σ be a micronuclear Σ_n-descriptor.*

(i) *There exists a set $\mathcal{A}_\sigma$ of Σ_n-descriptors such that*
   (a) $W_\sigma \cup \overline{W_\sigma} = W_{\mathcal{A}_\sigma} \cup \overline{W_{\mathcal{A}_\sigma}}$
   (b) *there exists an assembled Σ_n-descriptor in $\mathcal{A}_\sigma$*
(ii) *For any other set S of Σ_n-descriptors, if S satisfies conditions (a)-(b) above, then $S \approx \mathcal{A}_\sigma$*

Moreover, $\mathcal{A}_\sigma$ consists of exactly one linear Σ_n-descriptor and possibly several circular ones.

Theorem 15 may be stated informally by saying that the final result of gene assembly, including the molecule where the assembled gene is placed, as well as all the other noncoding molecules excised in the process, is unique.

Example 20   Consider the Σ_9-descriptor σ associated to gene actin I in S.nova in Example 18, with W_σ given in Example 19. It follows from Theorem 15 that the results of assembling the gene are

$$\{I_0 33\overline{I}_6 \overline{e}\overline{M_9} \overline{99M_8} \overline{88M_7} \overline{77M_6} \overline{66M_5} \overline{55M_4} \overline{44M_3} \overline{33M_2} \overline{22M}_1 \overline{b}\overline{I}_7 22I_8 88I_5 99I_9, 4I_1 4\}.$$

Thus, the noncoding block $4I_1 4$ is excised as a circular molecule and the gene is assembled linearly in the inverse order from $\overline{M_9}$ to $\overline{M_1}$ with the noncoding block $I_0 33\overline{I}_6$ preceding it and $\overline{I}_7 22I_8 88I_5 99I_9$ succeeding it.

Note that Theorem 15 holds both for the intra-molecular and the inter-molecular models for gene assembly. Consequently, the set of molecules generated by the assembly cannot be used to distinguish between different assembly strategies, either intramolecular, or intermolecular. Instead, to (in)validate either model, one could experimentally identify the sets of molecules generated at various stages of the assembly and verify it against the predictions made by the two models.

Our results hold also in a more general way. We have proved that the final result $\mathcal{A}_\sigma$ of assembling a micronuclear Σ_n-descriptor σ is unique modulo conjugation and inversion. As a matter of fact, our proofs apply unchanged also to the following variant proved in Brijder et al. (2006) for the intramolecular model. Let $P \subseteq \mathcal{P}_n$. There exists a unique (modulo conjugation and inversion) set $\mathcal{A}_{P,\sigma}$ of Σ_n-descriptors with the following property: $M_{p-1}ppM_p \leq \alpha$ for some $\alpha \in \mathcal{A}_{P,\sigma} \cup \overline{\mathcal{A}_{P,\sigma}}$ if and only if $p \in P$. In other words, if the assembly is to be done only on a given set P (that may be different from the total set $\mathcal{P}_n$), then the result is unique. To prove the result, it is enough to replace the morphism ϕ_n in ❷ Sect. 6.1 with a morphism $\phi_{P,n}$ that only inserts pp in case $p \in P$. This extension of Theorem 15 does not contradict the non-determinism of gene assembly: The ciliate may choose to reduce the pointers in any order. The result above only says that after assembling on a *fixed* set of pointers, the result, including the excised molecules, is unique.

7 Template-Guided Recombination

In the previous sections, we have considered models that take an abstract view of the gene assembly process in ciliates. A lower, implementation-oriented, level of abstraction is considered next in which we attempt to begin addressing the question of how the assembly process takes place in vivo. Our quest for the discovery of the "biological hardware" responsible for implementing assembly begins with a simple examination of how MDSs might overlap to fit together, in the correct order, while also removing IESs. As has been noted above, each MDS is flanked by pointer sequences – that is, looking at the level of DNA sequence, there will be a proper suffix of MDS n that is equal to a proper prefix of MDS $n + 1$. Considering this type of structure for an entire gene of several MDSs, a computer scientist will immediately recognize the *linked list data structure*: the core of the MDS is the data while the pointer sequence indicates the "address" of the next data item (MDS). The initial state of MDSs, distributed throughout the MIC, is reminiscent of the nonlinear distribution of linked list data in heap memory. It is worth noting that the ciliate data structure is, in fact, significantly more sophisticated than a classic linked list; whereas each element of a linked list contains two separate components, an area for a data payload and an area for a pointer to the next element, the ciliate version of the linked list actually combines these two elements. Since the pointers always lie within the MDSs, the pointer to the next MDS is also part of the "data" contained in the MDS and is fully integrated into the assembled gene.

Completing the process of gene assembly, starting with the MIC and ending with the MAC, can be seen as implementing a linked-list specification. However, it does not appear that the pointers alone are guiding this implementation process as some of them may be too short to serve as unique pointers to the following MDSs. Yet the pointer sequences are still always present. A natural question to be answered by any suggested implementation of gene assembly is thus: "what role do the pointers play and why are they present?"

A realistic, biologically implementable, model must incorporate a number of principal features. It must be *irreversible*, it must be *self-propagating* or *reusable*, it cannot be sequence specific, that is, it cannot rely on the presence of certain fixed sequences, as a huge variety of pointer sequences are known (instead it must be *configuration specific*), and it must have some mechanism for *identifying the MDS/IES boundary and hence the pointers*. The basic DNA template model of *template-guided recombination* was introduced in Prescott et al. (2001b) to address exactly these requirements that will be clarified in more detail after describing the model.

7.1 DNA Template-Guided Recombination

We consider here a schematic view of DNA as a picket fence with the sugar-phosphate backbones running horizontally along the top and bottom of the strand and the hydrogen bonds running vertically between the backbones. Suppose now that one wishes to assemble two strands of DNA, X and Y, having the sequences $X_1\alpha\beta\delta X_2$ and $Y_1\epsilon\beta\gamma Y_2$, respectively, where $\beta = \beta_1\beta_2$. In order to guide this assembly, we assume the existence of a DNA template T of the form $T_1\alpha\beta\gamma T_2$, which is placed in-between the two target strands X and Y, as seen in ❯ *Fig. 21*. Note that we use the notation $\bar{\alpha}$ to denote the Watson–Crick complement of the DNA sequence α and that, in ❯ *Fig. 21*, the $\alpha\beta$ region of X is now aligned with $\overline{\alpha\beta}$ on the template; similarly, $\beta\gamma$ of Y is aligned with its complement on the template. At the same time, the sequence of δ, beginning from the first nucleotide, must not be complementary to γ and, in a similar way, ϵ must not be complementary to α. It is important to note that X and Y need not be physically disconnected independent strands but may instead be different regions of a single, connected, strand of DNA.

The template-guided recombination now takes place in these principal steps:

- The hydrogen bonds in the $\alpha\beta_1$ region of X and the $\overline{\alpha\beta_1}$ in the template are broken and switch from binding the backbones within X and Y vertically to binding complementary sequence horizontally between X and Y; likewise for the second half of the template. In our fence analogy, the vertical pickets within DNA double-strands are replaced by floors and roofs *across* DNA strands, as pictured in ❯ *Fig. 22*.
- Cuts are now made in the backbones of the roof/floor assembly (at the ends of the roof/ floor structures) to yield the free-standing structure of ❯ *Fig. 23*.
- The same cuts also yield a new copy of the original template strand, shown in ❯ *Fig. 24*, and the strands $Y_1\epsilon\beta_1$ and $\beta_2\delta X_2$ (not pictured) that are left free to float away.

◘ **Fig. 21**
Sequences $X_1\alpha\beta\delta X_2$ and $Y_1\epsilon\beta\gamma Y_2$ stacked with DNA template $T_1\alpha\beta\gamma T_2$ in-between (from Prescott et al. (2001b)).

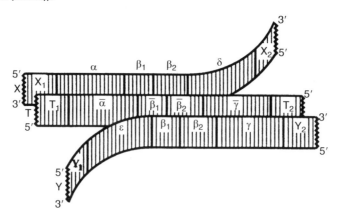

- The roof and floor structures of ❯ *Fig. 23* rotate to align the cut backbones that are then healed (via ligation) yielding the complete double-stranded DNA $X_1\alpha\beta\gamma Y_2$, which is the recombination of the prefix $X_1\alpha\beta_1$ of X and the suffix $\beta_2\gamma Y_2$ of Y – thus X and Y have been recombined.
- Likewise, the roof and floor structures of ❯ *Fig. 24* rotate, align, and ligate to yield $T_1\bar{\alpha}\bar{\beta}\bar{\gamma} T_2$ – thus the template is reconstituted.

◻ **Fig. 22**
Hydrogen bonds switch from being vertical pickets holding individual strands together to forming floors and roofs across strands (from Prescott et al. (2001b)).

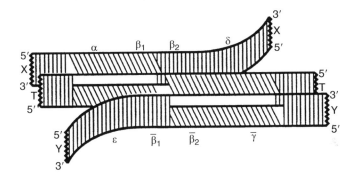

◻ **Fig. 23**
One of the products resulting from cuts made in the backbones of the configuration of ❯ *Fig. 22* (from Prescott et al. (2001b)).

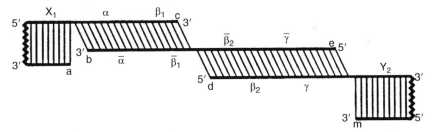

◻ **Fig. 24**
The reconstituted template, resulting from cuts made in the backbones of the configuration of ❯ *Fig. 22* (from Prescott et al. (2001b)).

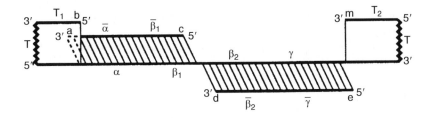

It can be seen clearly that the model, when considered in symbolic terms (as often done in computer science), meets our requirements for a biologically implementable system. The process of template-guided recombination may be described by the following implication:

$$\alpha\beta\delta + \alpha\beta\gamma + \varepsilon\beta\gamma \Rightarrow \alpha\beta\gamma + \alpha\beta\gamma + \varepsilon\beta_1 + \beta_2\delta$$

where $\beta_1\beta_2 = \beta$. On the left side, we have the first term consisting of the required subsequences for X, the second term is the template, and the third term is the required subsequence of Y. On the right side, the first term expresses the form of X recombined with Y, the form of the reconstituted template, and the last two terms are the forms of the "loose ends." Note that we must have the components on the left side present in order for the reaction to take place. If they are present, this equation describes *what* happens, though not how it happens, and what we obtain is the four products on the right side. Note carefully that from this moment on, no three components of the right-hand side have the form required of the three components of the left-hand side – thus the process is irreversible (one-way).

Examining the right side, we see $\alpha\beta\gamma$ twice, demonstrating that the template in this model is self-propagating: we begin with one component $\alpha\beta\gamma$ on the left-hand side and after a single iteration we get two such components. Hence as the process progresses iteratively, we have an explanation of the growth of the number of available templates. Thus even if we begin with only a single copy of the template, we very quickly end up with an "abundance" of templates. This is necessary as we recall from ❷ Sect. 2 on ciliate biology that many copies of the MIC chromosomes are present during the polytene chromosome stage so that multiple copies of each template are required to successfully assemble a full MAC genome.

Further, it is clear that the whole three-step process does not depend on *specific* sequences α, β, γ; indeed, all that matters is the relationship between the sequences that causes the formation of a particular configuration.

Considering this process carefully, the true nature of pointers becomes apparent: Pointers are sequence segments within which the transfer of roofs, and dually the transfer of floors, takes place. Consequently, the most essential backbone cuts of the recombination process will take place in the pointer region. Pointers are records of transfers. Hence, in our scheme, pointers are the regions denoted $\beta = \beta_1\beta_2$ while $\alpha\beta$ and $\beta\gamma$ correspond to MDSs M_i and M_{i+1}.

7.2 RNA Template-Guided Recombination

A variant of DNA template-guided recombination has been proposed in Angeleska et al. (2007), which considers *RNA templates*. Following through the steps of the DNA template model, one can see that a portion of the template strand ends up being incorporated into the final assembled product; this presents no problem for DNA templates but is infeasible with RNA templates since DNA and RNA backbones are incompatible. Rather than proposing a template which sits "in-between" the strands to be assembled, the RNA template model suggests a template that "hangs above" the strands to be recombined, guiding the recombination but never directly participating in it.

Consider DNA strands $\alpha\beta\delta$ and $\varepsilon\beta\gamma$ again containing the MDSs $\alpha\beta$ and $\beta\gamma$ and a double-stranded RNA template molecule $\alpha\beta\gamma$. We stack the two substrate strands in tandem, as in the DNA model, but place the template horizontally above, rather than between, the substrates as in ❷ *Fig. 25*. The $\bar{\beta}$ sequence of the RNA template begins to form hydrogen bonds with the sequence β in one of the substrates; note that while the backbones are incompatible,

◻ **Fig. 25**

Reading *left* to *right*, *top* to *bottom*: The RNA template forms hydrogen bonds with β and $\bar{\beta}$ in the substrate strands, causing the formation of a floor between strands. When the RNA template is removed, a roof structure is now formed. Four cuts are made, followed by backbones swinging back into position and being healed to form the assembled strand (from Angeleska et al. (2007)).

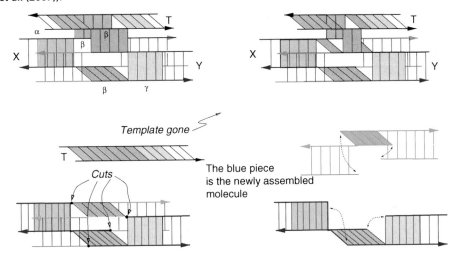

it is certainly possible for RNA and DNA to share hydrogen bonds across two strands. The complementary strand of the template RNA similarly binds to the other substrate strand. With the internal "picket fence" hydrogen bonds of the substrate strands broken, the complementary portions at the bottom of each strand now form a floor of hydrogen bonds. If the RNA template is now removed, the complementary strands at the top of the substrates will form a roof of hydrogen bonds. Cutting the DNA backbones in the four places indicated in ❯ *Fig. 25* and rotating, followed by healing (ligation), the broken backbones similarly to the DNA model yields a correctly assembled DNA strand.

Note carefully that the RNA template is not integrated into the resulting structure; rather, the RNA template served only to break up the hydrogen bonds in the β region of the substrate strands, inducing the formation of a floor of hydrogen bonds between the two substrates, which then induced the formation of a complementary roof structure following the removal of the RNA template.

Yet another similar approach requires only a single-stranded RNA template, which is placed "diagonally" between the two DNA picket fence strands (details are given in Angeleska et al. (2007)).

The development of theoretical models of template-guided recombination has helped to direct biological inquiry into core questions surrounding gene assembly. It is natural, from a biological perspective, to ask where templates might originate. Both the RNA and DNA template-guided models suggest that templates are composed of sequence that is highly similar to that in the old MAC. Recent experimental results in the ciliate *Oxytricha trifallax* (*Sterkiella histriomuscorum*), guided by the insight provided in the theoretical models, support the hypothesis that short MAC-specific RNA templates are involved in gene assembly (Nowacki et al. 2008). We also point out that the appendix of Angeleska et al. (2007) contains a straightforward RNA-template-combination explanation for the molecular operations ld, hi, and dlad introduced in ❷ Sect. 5, which are the basis of the intramolecular models presented in this chapter.

7.3 Template-Guided Recombination on Words and Languages

We move now to consider theoretical research on the template model. Together with combinatorial models discussed in preceding sections, the theoretical work concerning the template model provided both some new insights into the nature of gene assembly and a whole spectrum of novel and interesting notions, models, and results for theoretical computer science.

A formal language theoretic version of the basic *DNA template-guided recombination* (abbreviated TGR) operation of Prescott et al. (2001b) was first studied in Daley and McQuillan (2005b). For words $x, y, z, t \in \Sigma^*$ and natural numbers $n_1, n_2 \geq 1$, we denote by z, the product of the recombination of x and y, guided by template t, by $(x, y) \vdash_{t, n_1, n_2} z$. More specifically, if $x = u_1 \alpha \beta v_1, y = v_2 \beta \gamma u_2, t = \alpha \beta \gamma$ with $\alpha, \beta, \gamma, u_1, u_2, v_1, v_2 \in \Sigma^*$, $|\alpha|, |\gamma| \geq n_1$ and $|\beta| = n_2$, then we may write the TGR product $z = u_1 \alpha \beta \gamma u_2$. If $T, L \subseteq \Sigma^*$ are languages, then $⋔_{T, n_1, n_2}(L)$ is defined by

$$⋔_{T, n_1, n_2}(L) = \{z \ : \ \exists x, y \in L, t \in T \text{ such that } (x, y) \vdash_{t, n_1, n_2} z\}$$

The shorthand notation $⋔_T(L)$ is used whenever n_1, n_2 are understood. We note that the restriction on pointer length, $|\beta| = n_2$, is not as strict as it appears since it has been proven equivalent to the restriction $|\beta| \geq n_2$ in Daley and McQuillan (2005b).

Given the nature of biochemical reactions, it is natural to consider an iterated version of TGR as well: let $⋔_{T, n_1, n_2}^0(L) = L$ and for all $i \geq 1$, let

$$⋔_{T, n_1, n_2}^i(L) = ⋔_{T, n_1, n_2}^{i-1}(L) \cup ⋔_{T, n_1, n_2}(⋔_{T, n_1, n_2}^{i-1}(L))$$

Then we also define $⋔_{T, n_1, n_2}^*(L)$ as

$$⋔_{T, n_1, n_2}^*(L) = \bigcup_{i \geq 0} ⋔_{T, n_1, n_2}^i(L)$$

Finally, let $\mathcal{L}, \mathcal{T}$ be classes of languages and $n_1, n_2 \geq 1$. We define the following closure classes:

$$⋔_{\mathcal{T}, n_1, n_2}(\mathcal{L}) = \{⋔_{T, n_1, n_2}(L) \ : \ T \in \mathcal{T}, L \in \mathcal{L}\}$$
$$⋔_{\mathcal{T}, n_1, n_2}^*(\mathcal{L}) = \{⋔_{T, n_1, n_2}^*(L) \ : \ T \in \mathcal{T}, L \in \mathcal{L}\}$$

The families of finite languages are denoted by FIN and regular languages by REG, and we recall that a family of languages is said to be a full AFL if it is closed under homomorphism, inverse homomorphism, intersection with regular languages, union, concatenation, and Kleene plus.

On initial inspection of TGR, there appears to be a similarity with the well-known model of splicing systems, but this relationship has been shown to be superficial in Daley and McQuillan (2005b). Before the relevant formal result is stated, we recall the basic operational scheme of splicing systems.

A *splicing* scheme or *H scheme* is a pair $\sigma = (\Sigma, R)$ where Σ is an alphabet and $R \subseteq \Sigma^* \# \Sigma^* \$ \Sigma^* \# \Sigma^*$ is a set of splicing rules where $\$, \#$ are not elements of Σ. For a rule $r \in R$, we define the relation $(x,y) \models_r z$ if $r = u_1 \# u_2 \$ u_3 \# u_4$, $x = x_1 u_1 u_2 x_2$, $y = y_1 u_3 u_4 y_2$, $z = x_1 u_1 u_4 y_2$, for some $u_1, u_2, u_3, u_4, y_1, y_2, x_1, x_2 \in \Sigma^*$.

For a language $L \subseteq \Sigma^*$ and an H scheme $\sigma = (\Sigma, R)$, we define $\sigma(L) = \{z \in \Sigma^* : \exists x, y \in L, r \in R$ such that $(x,y) \models_r z\}$ and extend to iterated splicing as follows: let $\sigma^0(L) = L$ and $\sigma^i(L)$ be defined by $\sigma^i(L) = \sigma^{i-1}(L) \cup \sigma(\sigma^{i-1}(L))$ for all $i \geq 1$. Finally, as expected,

$$\sigma^*(L) = \bigcup_{i \geq 0} \sigma^i(L)$$

For classes of languages $\mathcal{L}, \mathcal{R}$, let $H(\mathcal{L}, \mathcal{R}) = \{\sigma^*(L) : L \in \mathcal{L}, \sigma = (\Sigma, R), R \in \mathcal{R}\}$.

Lemma 3 (Daley and McQuillan 2006) *For all $n_1, n_2 \geq 1$, for all full AFLs $\mathcal{L}$:*

$$\mathcal{L} = \pitchfork^*_{\mathrm{FIN}, n_1, n_2}(\mathcal{L}) = H(\mathcal{L}, \mathrm{FIN}),$$

while for all finite languages and templates:

$$\mathrm{FIN} \subset \pitchfork^*_{\mathrm{FIN}, n_1, n_2}(\mathrm{FIN}) \subset H(\mathrm{FIN}, \mathrm{FIN}) \subset \mathrm{REG}$$

Despite this result, it has been shown in Daley and McQuillan (2005b) that every regular language is the coding of a language in $\pitchfork^*_{\mathrm{FIN}}(\mathrm{FIN})$ demonstrating a relatively modest computational power for single-application TGR.

To investigate the computational power of the iterated case, it is necessary to define the notion of a "useful" template; we say that a template $t \in T$ is *useful* on L, n_1, n_2 if there exists $u_1 \alpha \beta v_1, v_2 \beta \gamma u_2 \in \pitchfork^*_{T, n_1, n_2}(L)$ with $|\alpha|, |\gamma| \geq n_1, |\beta| = n_2, u_1, u_2, v_1, v_2 \in \Sigma^*$ and $t = \alpha \beta \gamma$. If every template $t \in T$ is useful on L, n_1, n_2, then we say that T is useful on L, n_1, n_2. The following results, demonstrating the surprisingly limited power of iterated TGR, were shown in Daley and McQuillan (2006):

Theorem 16 (Daley and McQuillan 2006) *Let $n_1, n_2 \geq 1$, $\mathcal{L}$ be a full AFL and $L, T \in \mathcal{L}$. If T is useful on L, n_1, n_2, then $\pitchfork^*_{T, n_1, n_2}(L) \in \mathcal{L}$.*

Corollary 1 (Daley and McQuillan 2006) *Let $n_1, n_2 \geq 1$. For all full AFLs $\mathcal{L}$, $\pitchfork^*_{\mathrm{REG}, n_1, n_2}(\mathcal{L}) = \mathcal{L}$.*

The problem of template equivalence, viz. "Given sets of templates T_1, T_2, are $\pitchfork_{T_1}$ and $\pitchfork_{T_2}$ identical operations?" has been considered in Domaratzki (2007), which gives a characterization of when two sets of templates define the same TGR operation in formal language theoretic terms, leading to the following decidability result.

Theorem 17 (Domaratzki 2007) *Let $n_1, n_2 \geq 1$ and $T_1, T_2 \subseteq \Sigma^*$ ($|\Sigma| \geq 3$) be regular sets of templates. Then it is decidable whether or not $\pitchfork_{T_1, n_1, n_2}(L) = \pitchfork_{T_2, n_1, n_2}(L)$ for all $L \subseteq \Sigma^*$.*

Several variants of TGR have been studied as well, including a computationally universal version with added deletion contexts (Daley and McQuillan 2005a) and a purely intramolecular version that resembles a templated version of the ld operation (Daley et al. 2007) discussed above.

In addition to the very literal formalization of TGR considered in this section, the underlying theoretical model has also inspired work at a more abstract level.

7.4 Covers from Templates

The process of gene descrambling may be abstractly formulated in simple terms as a procedure that takes MDSs from the MIC and connects them, via overlap, to form the MAC. The template-guided recombination model discussed above provides a concrete suggestion of how this process might be implemented; one can view a template, T, as a sort of magnet, which glues together regions of MIC chromosomes containing MDSs to form orthodox MAC genes. Returning to a more abstract level, we now consider the set of all MDSs as our primary object of study. In this view, a template is now a request: "with this set of segments(MDSs), please cover me": and therefore our core question is now "given a set of segments, how can a particular word be covered with these segments?" This leads naturally to the study of various properties of coverings, and the notion of uniqueness of coverings formalized as *scaffolds* presented in Ehrenfeucht and Rozenberg (2006).

An *interval* is a set of integers of the form $\{n, n+1, \ldots, n+m\}$, where $n \in \mathbb{Z}$, and $m \in \mathbb{N}$. For a given alphabet Σ, we define a *segment* as a function $f : A \rightarrow \Sigma$, where A is an interval, and denote the set of all segments over Σ by $\mathbf{S}_\Sigma$. Since f is a function, we alternatively view segments as sets of ordered pairs of the form $(n, f(n))$, called *elements of* f, with n called the *location* of $(n, f(n))$; thus, elements of f are ordered through their locations. This point of view is very convenient as it provides a set-theoretical calculus of segments: we can consider inclusions, union, intersections, differences,... of segments. Also, in this way, a set of segments is a family of sets.

For a set $C \subseteq \mathbf{S}_\Sigma$ and a segment $f \in \mathbf{S}_\Sigma$, we say that C *covers* f if $f = \bigcup C$. Intuitively, f is covered by C if each element of f is present in at least one segment of C, and all elements of all segments of C are present in f. Note that in general an element of f may be present in several segments of C, that is, the segments of C may overlap. One may also have redundant segments in C, that is, segments which cover only elements of f that are already covered by other segments of C. We thus say that a cover C of f is *tight* if for every $z \in C$, $C - \{z\}$ is not a cover of z.

Often covers are chosen from a subset of $\mathbf{S}_\Sigma$. Given such an $F \subseteq \mathbf{S}_\Sigma$, and a cover C of f with $C \subseteq F$, we say that C is a *small cover* of f (with respect to F) if $|C| \leq |Z|$ for all $Z \subseteq F$ covering f. Note that the property of being a small cover is a global property: C is a small cover of f with respect to F if any other cover Z of f, $Z \subseteq F$, has at least as many segments as C. The set of all small covers of f with respect to F is denoted $SC_F(f)$, and the *small index* of f (with respect to F) is the cardinality of the small covers of f (with respect to F). For any small cover C of f with respect to F, we get a natural order $C(1), \ldots, C(m)$ of C, where m is the small index of f, and the order is determined by increasing locations of first elements of the segments of C.

Example 21 Let $\Sigma = \{a, b, c\}$ and $f \in \mathbf{S}_\Sigma$ be defined by

$$f = \{(3, a), (4, b), (5, b), (6, a), (7, c)\}$$

which may be abbreviated as $f = (3, abbac)$ since f begins at location 3.

Consider the sets $F = \{(3, ab), (4, bb), (4, bba), (5, bac), (6, ac)\}$ and $C = \{(3, ab), (4, bb), (6, ac)\}$. It is clear that C covers f and is tight, since no element of C can be removed while still covering f; however, with respect to F, C is not small since $|C| = 3$ and the set $\{(3, ab), (5, bac)\} \subseteq F$ also covers f and has cardinality 2.

From the point of view of the original biological motivation, the segment f represents a descrambled MAC gene while the set F is the collection of MIC gene fragments available for assembly. We now proceed to investigate the structure of f by considering the family of all small covers of f with respect to some fixed F.

Let $f \in S_\Sigma$, $F \subseteq S_\Sigma$ be such that it contains a cover of f, and let m be the small index of f with respect to F. Let $1 \leq i \leq m$. The ith kernel of f with respect to F (denoted $ker_{i,F}(f)$) is defined by

$$
ker_{i,F}(f) = \left(\bigcap_{C \in SC_F(f)} C(i) \right) - \bigcup_{\substack{C \in SC_F(f) \\ j \neq i}} C(j)
$$

We now define the *scaffold* of f (with respect to F) as the set $\{ker_{1,F}(f), \ldots, ker_{m, F}(f)\}$ of all kernels of f.

Theorem 18 (Ehrenfeucht and Rozenberg 2006) *For each $1 \leq i \leq m$, $ker_{i,F}(f)$ is a nonempty segment.*

For any segment f, the choice of F determines a certain natural class of "maximal" subsegments of f; namely, those subsegments that are not strict subsegments of other subsegments. Let $P_F(f)$ be the set of all subsegments of f that belong to F. We say that a segment $g \in P_F(f)$ is *long* (with respect to F) if it is not properly included in any other segment in $P_F(f)$. The set of all long segments of f (with respect to F) is denoted by $LP_F(f)$. Additionally, we call a cover long if it consists solely of long segments.

Attention can now be turned to the study of the canonical class of covers that have both the "long" and "small" property. For each $1 \leq i \leq m$, let $LP_F(f,i) = \{y \in LP_F(f): ker_{i,F} \subseteq y\}$. That is, we categorize the long segments of f with respect to F according to containment of kernels. Since $LP_F(f,i)$ is an ordered set, we let $rt_F(f,i)$ (resp., $lt_F(f,i)$) be the maximal, or rightmost (resp., minimal, or leftmost) element of $LP_F(f,i)$. The following result provides a method for constructing "canonical" small covers of f.

Theorem 19 (Ehrenfeucht and Rozenberg 2006) *The sets $\{rt_F(f,1), \ldots, rt_F(f,m)\}$ and $\{lt_F(f,1), \ldots, lt_F(f,m)\}$ are long small covers of f with respect to F.*

More detailed analysis of the structure of scaffolds is given in Ehrenfeucht and Rozenberg (2006).

7.5 Topology-Based Models

An important question that arises from considering template-guided recombination concerns the three-dimensional structure of DNA undergoing multiple recombination events. Recently, two new approaches to this question have been undertaken. The physical structure of the DNA strand undergoing recombination is directly considered in Angeleska et al. (2007) through the use of virtual knot diagrams. Micronuclear genes are represented in a schematic form that explicitly denotes only the relative locations of the pointer sequences. Consider, for example, the Uroleptus gene *USGI* (Chang et al. 2006) that has the following MDS descriptor: $M_1M_3M_4M_5M_7M_{10}M_{11}M_6M_8M_2M_9$. The corresponding legal string is the following: 2 3 4 4 5 5 6 7 8 10 11 11 6 7 8 9 2 3 9 10. It is now possible to interpret this sequence as a Gauss code. Each symbol in a Gauss code must occur exactly twice and to each code we associate a virtual knot diagram in a similar manner to the construction of the recombination graphs exposed above:

- For each symbol occurring in the code, we place a (disconnected) crossing in the plane and label the crossing with the corresponding symbol. A crossing may be thought of as similar to a vertex with predetermined order of the incident edges of degree 4 in a graph.
- We choose an arbitrary point in the plane to denote as the base point.
- Following a chosen direction, we connect crossings according to the order in the Gauss code. In our example, we would draw arcs from the base point to crossing 2, crossing 2 to crossing 3, crossing 3 to crossing 4, crossing 4 to itself, crossing 4 to crossing 5, and so forth. Each crossing corresponds to a "roof-floor" structure depicted in ❷ *Fig. 26* left.
- We connect the remaining arc leaving the final crossing back to the base point.

Note that during the construction of the virtual knot diagram, it might happen that we have to cross an already sketched arc. This crossing is not labeled and does not correspond to a required pointer-guided homologous recombination, therefore it is called a "virtual crossing." The virtual crossings correspond to a cross-over embedding of the DNA in space when one helix crosses over another.

The virtual knot diagram is now relabeled with each crossing receiving the label of its associated pointer and indicating inverted pointers with a bar. The process of assembly is now reduced to one of *smoothing* the crossings of the virtual knot diagram (see ❷ *Fig. 26* right). The smoothing of a crossing consists of eliminating the crossing by splicing together the arcs of the crossing according to the pointers: if the pointers are not inverted, we splice the arcs together

◻ Fig. 26
Roof-floor structure of a crossing in a virtual knot diagram (from Angeleska et al. (2007)).

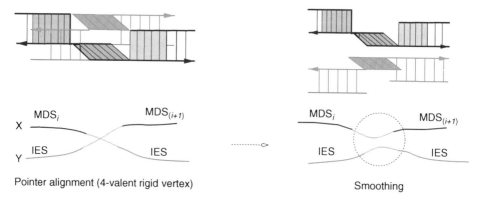

<table>
<tr><td>Pointer alignment (4-valent rigid vertex)</td><td>Smoothing</td></tr>
</table>

following the orientation; if the pointers are inverted, we splice the arcs together opposite to the orientation. The result of a simultaneous smoothing of all crossings is a virtual link diagram with no real crossings remaining; note that the link may be composed of multiple components. If the arcs of the virtual knot diagram are labeled with the respective MDS and IES names (see Angeleska et al. (2007)), then we have the following theorem.

Theorem 20   (Angeleska et al. 2007) *For every labeled virtual knot diagram derived from a representation of a scrambled gene G, there exists a simultaneous smoothing yielding a link with a component C containing a subarc labeled with all MDSs in orthodox macronuclear order.*

Note that the basic mechanism of gene assembly through the smoothing of virtual knot diagrams was already introduced in Ehrenfeucht et al. (2002) in terms of a graph-based model, which was also briefly discussed in ❷ Sect. 5.6. This model is based on the notion of recombination graph representing pointers as vertices, and MDSs and IESs as edges between the vertices standing for their flanking pointers. The recombination graph is first subject to a process of *graph folding* yielding a structure similar to the virtual knot diagram. A subsequent step of *graph unfolding* is similar to that of *smoothing* above and yields a representation of the assembled gene and of all the molecules excised during gene assembly. In this way, the approach of virtual knot diagrams from Angeleska et al. (2007) is a translation of the graph-theoretic approach from Ehrenfeucht et al. (2002) into a topological framework. This allows one to apply a rich set of techniques from both graph theory and knot theory to the investigation of gene assembly.

One significant issue that has not yet been addressed concerns the thermodynamics of template-guided recombination. For template-guided recombination to be implemented as suggested in the theoretical models requires some intricate positioning and biochemical operations; specifically, it requires the juxtaposition of three nucleic acid strands in space followed by a strand branch migration process as is observed in vivo (see, e.g., Dennis et al. (2004) and Zerbib et al. (1998)) and in vitro (see, e.g., Yan et al. (2002) and Yurke et al. (2000)). In order for such a branch migration process to start, it is possibly necessary that some of the cuts in the molecule (Fig. 25 bottom left) appear early in the process. On the other side, it is possible to suppose that the juxtaposition of the molecules is artificially created by as-yet undiscovered enzymatic mechanics. Yet another possibility would be a simple argument indicating the thermodynamic favorability of the process. In the absence of such complex additional enzymatic machinery, template-guided recombination must rely upon diffusion processes within the cell to juxtapose the template and substrate strands. It is a well-known theorem in mathematics that random walks in three-space need never pass through the same point twice, so the likelihood of juxtaposing three molecules, in three-space, seems low. An alternative knot theory–based model presented in Daley et al. (2010) attempts to reduce the complexity of this problem by considering the effects of individual recombinations on the structure of the substrate DNA.

When circular DNA undergoes recombination, it is supercoiled due to twist applied by the recombination machinery (Krasnow et al. 1983; White 1992). Where linear DNA has unbound ends that are free to rotate and relax, thus removing the induced twist, circular DNA cannot relax in this fashion after recombination. Instead, the relaxation of the DNA strand induces supercoiling of the DNA molecule. The connection between linking, twist, and writhe of the DNA has been observed decades ago (White 1992), therefore, multiple recombinations on a closed circle of DNA can lead to a knotted, supercoiled strand.

It is suggested in Daley et al. (2010) that DNA is assembled in a closed circular topology where the twist induced by a template-guided recombination (or multiple recombinations) causes a new, potentially knotted and supercoiled, topology to form. This new topology could facilitate the juxtaposition of the "next" regions of DNA to be assembled.

8   Discussion

In this chapter, we have discussed a number of topics related to research on the computational nature of gene assembly in ciliates, including the two main models for gene assembly. Among others, we have discussed their mathematical formalizations, invariant properties, template-based DNA recombination, and topology-based models for gene assembly. Due to space restrictions, all lines of research could not be discussed; for the sake of completeness, some of the topics that were not covered are now mentioned briefly.

The organization of the micronuclear genes into broken and shuffled MDSs, separated by IESs, is one of the characteristic features of the ciliates. To explain the evolutionary origin of this organization, a somewhat geometrical hypothesis based on a novel proposal for DNA repair is suggested in Ehrenfencht et al. (2007).

Approaches based on formal languages have been introduced for both the intermolecular model and the intramolecular model, leading to very diverse research topics: computability, language equations, closure properties, hierarchies of classes of languages, etc. A typical approach is to consider contextually based applications of string rewriting rules; see, for example, Landweber and Kari (1999) for the intermolecular model. A derivation relation and an axiom are introduced, thus obtaining an acceptance mechanism: Start with a multiset of strings, for example, consisting of several copies of the input string and eventually derive a multiset containing the axiom. It can be proved that such a mechanism is a universal computing device. A similar result can be obtained also based on the intramolecular model, see Ishdorj et al. (2007), where the multisets are replaced with single strings. The idea is to concatenate several copies of the input string rather than having them in a multiset. A generating, rather than accepting, computing device inspired by gene assembly was also considered in Dassow and Vaszil (2006). Defining non-contextual string-based rewriting rules inspired by the molecular operations in either model leads to language operations and to questions related to closure properties, or solutions of language equations, see Daley et al. (2003a), Dassow and Holzer (2005), and Freund et al. (2002), and also Dassow et al. (2002) for a more general framework. Language operations inspired by the template-based DNA recombination were considered in Daley and McQuillan (2005b) and in Domaratzki (2007).

The original, non-contextual, intramolecular and intermolecular operations were generalized to the synchronized insertion and deletion operations on linear strings in Daley et al. (2003a). Let α, β be two nonempty words in an alphabet Σ^*. The synchronized insertion of β into α is defined as $\alpha \oplus \beta = \{ uxvxw \mid \alpha = uxw, \beta = vx, x \in \Sigma^+, u, v, w \in \Sigma^* \}$. While the synchronized deletion of β from α is defined as $\alpha \ominus \beta = \{ uxw \mid \alpha = uxvxw, \beta = vx, x \in \Sigma^+, u, v, w \in \Sigma^* \}$. All language families in the Chomsky hierarchy were shown to be closed under synchronized insertion while only the families of regular and recursively enumerable languages were closed under synchronized deletion. The existence of a solution was shown to be decidable for language equations of the form $L \odot Y = R$ and $X \odot L = R$ where $\odot$ is one of the synchronized insertion or synchronized deletion operations and L, R are regular languages. The same problems are undecidable in the case that L is a context-free language.

More general results considering families of languages defined by reversal-bounded counter machines are also given in Daley et al. (2003a) along with results for a similarly generalized version of the hi operation. Generalized versions of the ld and dlad operations are considered in Daley et al. (2003b) while families of languages defined by closure under these generalized operations are examined in Daley et al. (2004).

Two novel classes of codes based on synchronized insertion were defined and studied in Daley and Domaratzki (2007): the synchronized outfix codes ($\oplus$-codes) defined by the equation $(L \oplus \Sigma^+) \cap L = \emptyset$ and the synchronized hypercodes ($\otimes$-codes) defined by $(L \otimes \Sigma^+) \cap L = \emptyset$ (where $\otimes$ is the transitive closure of $\oplus$, namely synchronized scattered insertion). The $\oplus$-codes and $\otimes$-codes are shown to be completely disjoint from the regularly studied classes of $*$-codes and it is demonstrated that it is decidable if a regular language is an $\oplus$-code while the same property is undecidable for linear context-free languages. It is, however, decidable if an arbitrary context-free language is an $\otimes$-code while this same property is, unsurprisingly, undecidable for context-sensitive languages.

A different computability approach was developed in Alhazov et al. (2008), where the process of gene assembly is used to solve an NP-complete problem. Conceptually, this is important since gene assembly is confluent, see ❯ Sect. 5, that is, it yields a computing device that answers "yes" to all legal inputs. A way around the problem is to make the device highly nondeterministic by extending its set of legal inputs. For example, strings with more than two occurrences of each letter may be allowed. With this modification, a suitably defined model can be introduced to solve the Hamiltonian path problem (HPP) by mimicking gene assembly on an encoding of the input to HPP, see Alhazov et al. (2008). The intermolecular model also leads to solutions to computational problems, see Ishdorj et al. (2008) for a solution to the satisfiability problem. Yet another approach based on Boolean circuits was investigated in Ishdorj and Petre (2008). Algorithmic questions related to finding a gene assembly strategy were considered in Ilie and Solis-Oba (2006).

The topological model of gene assembly with virtual knot diagrams raises new mathematical questions. Considering that a virtual knot diagram could represent the physical structure of the micronuclear DNA at the time of recombination, in Angeleska et al. (2009) it was shown that for every such virtual knot diagram there is an embedding of the molecule in space, and there is smoothing of the vertices (recombination along the pointers) such that the resulting molecule is always unlinked. Further, it was shown that the smoothing guided by the pointers differs from the existent smoothing notions defined earlier for virtual knot diagrams (Kauman 1999). This opens completely new problems on virtual knot diagrams that have not been studied before.

Research on the computational nature of gene assembly is an example of genuinely interdisciplinary research that contributed to both computer science, through a whole range of novel and challenging models of computation, and to biology, by increasing our understanding of the biological nature of gene assembly – it has even led to formulating biological models of this process based on the notion of template-guided recombination.

Acknowledgments

MD and GR acknowledge support by NSF, grant 0622112. IP acknowledges support by the Academy of Finland, grants 108421 and 203667. NJ has been supported in part by the NSF grants CCF 0523928 and CCF 0726396.

References

Alhazov A, Petre I, Rogojin V (2008) Solutions to computational problems through gene assembly. J Nat Comput 7(3):385–401

Alhazov A, Petre I, Rogojin V (2009) The parallel complexity of signed graphs: Decidability results and an improved algorithm. Theor Comput Sci 410 (24–25):2308–2315

Alhazov A, Li C, Petre I (2010) Computing the graph-based parallel complexity of gene assembly. Theor Comput Sci 411(25):2359–2367

Angeleska A, Jonoska N, Saito M, Landweber LF (2007) RNA-template guided DNA assembly. J Theor Biol 248:706–720

Angeleska A, Jonoska N, Saito M (2009) DNA recombination through assembly graphs. Discrete Appl Math 157(14):3020–3037

Brijder R (2008) Gene assembly and membrane systems. PhD thesis, University of Leiden

Brijder R, Hoogeboom H (2008a) The fibers and range of reduction graphs in ciliates. Acta Inform 45:383–402

Brijder R, Hoogeboom HJ (2008b) Extending the overlap graph for gene assembly in ciliates. In: Martín-Vide C, Otto F, Fernau H (eds) LATA 2008: 2nd international conference on language and automata theory and applications, Tarragona, Spain, March 2008. Lecture notes in computer science, vol 5196. Springer, Berlin Heidelberg, pp 137–148

Brijder R, Hoogeboom H, Rozenberg G (2006) Reducibility of gene patterns in ciliates using the breakpoint graph. Theor Comput Sci 356:26–45

Brijder R, Hoogeboom H, Muskulus M (2008) Strategies of loop recombination in ciliates. Discr Appl Math 156:1736–1753

Cavalcanti A, Clarke TH, Landweber L (2005) MDS_IES_DB: a database of macronuclear and micronuclear genes in spirotrichous ciliates. Nucl Acids Res 33:396–398

Chang WJ, Kuo S, Landweber L (2006) A new scrambled gene in the ciliate *Uroleptus*. Gene 368:72–77

Daley M, Domaratzki M (2007) On codes defined by bio-operations. Theor Comput Sci 378(1):3–16

Daley M, McQuillan I (2005a) On computational properties of template-guided DNA recombination. In: Carbone A, Pierce N (eds) DNA 11: Proceedings of 11th international meeting on DNA-based computers, London, Ontario, June 2005. Lecture notes in computer science, vol 3892. Springer, Berlin, Heidelberg, pp 27–37

Daley M, McQuillan I (2005b) Template-guided DNA recombination, Theor Comput Sci 330 (2):237–250

Daley M, McQuillan I (2006) Useful templates and iterated template-guided DNA recombination in ciliates. Theory Comput Syst 39(5):619–633

Daley M, Ibarra OH, Kari L (2003a) Closure properties and decision questions of some language classes under ciliate bio-operations. Theor Comput Sci 306 (1–3): 19–38

Daley M, Ibarra OH, Kari L, McQuillan I, Nakano K (2003b) The ld and dlad bio-operations on formal languages. J Autom Lang Comb 8(3):477–498

Daley M, Kari L, McQuillan I (2004) Families of languages defined by ciliate bio-operations. Theor Comput Sci 320(1):51–69

Daley M, Domaratzki M, Morris A (2007) Intramolecular template-guided recombination. Int J Found Comput Sci 18(6):1177–1186

Daley M, McQuillan I, Stover N, Landweber LF (2010) A simple topological mechanism for gene descrambling in Stichotrichous ciliates. (under review)

Dassow J, Holzer M (2005) Language families defined by a ciliate bio-operation: hierarchies and decidability problems. Int J Found Comput Sci 16(4):645–662

Dassow J, Vaszil G (2006) Ciliate bio-operations on finite string multisets. In: Ibarra OH, Dang Z (eds) 10th international conference on developments in language theory, Santa Barbara, CA, June 2006. Lecture notes in computer science, vol 4036. Springer, Berlin Heidelberg, pp 168–179

Dassow J, Mitrana V, Salomaa A (2002) Operations and languages generating devices suggested by the genome evolution. Theor Comput Sci 270(1–2):701–738

Dennis C, Fedorov A, Käs E, Salomé L, Grigoriev M (2004) RuvAB-directed branch migration of individual Holliday junctions is impeded by sequence heterology. EMBO J 23:2413–2422

Domaratzki M (2007) Equivalence in template-guided recombination, J Nat Comput 7(3):439–449

Ehrenfeucht A, Rozenberg G (2006) Covers from templates. Int J Found Comput Sci 17(2):475–488

Ehrenfeucht A, Petre I, Prescott DM, Rozenberg G (2000) Universal and simple operations for gene assembly in ciliates. In: Mitrana V, Martin-Vide C (eds) Where mathematics, computer science, linguistics and biology meet. Kluwer, Dordrecht, pp 329–342

Ehrenfeucht A, Petre I, Prescott DM, Rozenberg G (2001a) Circularity and other invariants of gene assembly in ciliates. In: Ito M, Păun G and Yu S (eds) Words, semigroups, and transductions. World Scientific, Singapore, pp 81–97

Ehrenfeucht A, Prescott DM, Rozenberg G (2001b) Computational aspects of gene (un)scrambling in ciliates. In: Landweber LF, Winfree E (eds)

Evolution as computation. Springer, Berlin, Heidelberg, New York, pp 216–256

Ehrenfeucht A, Harju T, Rozenberg G (2002) Gene assembly in ciliates through circular graph decomposition. Theor Comput Sci 281:325–349

Ehrenfeucht A, Harju T, Petre I, Prescott DM, Rozenberg G (2003a) Computation in living cells: gene assembly in ciliates. Springer, Berlin, Heidelberg, New York

Ehrenfeucht A, Harju T, Petre I, Prescott DM, Rozenberg G (2003b) Formal systems for gene assembly in ciliates. Theor Comput Sci 292:199–219

Ehrenfeucht A, Prescott DM, Rozenberg G (2007) A model for the origin of internal eliminated segments (IESs) and gene rearrangement in stichotrichous ciliates. J Theor Biol 244(1):108–114

Freund R, Martin-Vide C, Mitrana V (2002) On some operations on strings suggested by gene assembly in ciliates. New Generation Comput 20(3):279–293

Harju T, Petre I, Rozenberg G (2004a) Two models for gene assembly. In: Karhumäki J, Păun G, Rozenberg G (eds) Theory is forever. Springer, Berlin, pp 89–101

Harju T, Petre I, Rozenberg G (2004b) Gene assembly in ciliates: formal frameworks. In: Paun G, Rozenberg G, Salomaa A (eds) Current trends in theoretical computer science (the challenge of the new century). World Scientific, Hackensack, NJ, pp 543–558

Harju T, Li C, Petre I, Rozenberg G (2006a) Parallelism in gene assembly. Nat Comp 5(2):203–223

Harju T, Petre I, Rogojin V, Rozenberg G (2006b) Simple operations for gene assembly. In: Kari L (ed) Proceedings of 11th international meeting on DNA-based computers, London, Ontario, 2005. Lecture notes in computer science, vol 3892. Springer, Berlin, pp 96–111

Harju T, Petre I, Rozenberg G (2006c) Modelling simple operations for gene assembly. In: Chen J, Jonoska N, Rozenberg G (eds) Nanotechnology: science and computation. Springer, Berlin, pp 361–376

Harju T, Li C, Petre I, Rozenberg G (2007) Complexity measures for gene assembly. In: Tuyls K, Westra R, Saeys Y, Now'e A (eds) Proceedings of the knowledge discovery and emergent complexity in bioinformatics workshop, Ghent, Belgium, May 2006. Lecture notes in bioinformatics, vol 4366. Springer, Berlin, Heidelberg, pp 42–60

Harju T, Li C, Petre I (2008a) Graph theoretic approach to parallel gene assembly. Discr Appl Math 156(18): 3416–3429

Harju T, Li C, Petre I (2008b) Parallel complexity of signed graphs for gene assembly in ciliates. Soft Comput Fusion Found Methodol Appl 12(8):731–737

Harju T, Petre I, Rogojin V, Rozenberg G (2008c) Patterns of simple gene assembly in ciliates, Discr Appl Math 156(14):2581–2597

Hausmann K, Bradbury PC (eds) (1997) Ciliates: cells as organisms. Vch Pub, Deerfield Beach, FL

Ilie L, Solis-Oba R (2006) Strategies for DNA self-assembly in ciliates. In: Mao C, Yokomori T (eds) DNA'06: Proceedings of the 12th international meeting on DNA computing, Seoul, Korea, June 2006. Lecture notes in computer science, vol 4287. Springer, Berlin, pp 71–82

Ishdorj T-O, Petre I (2008) Gene assembly models and boolean circuits. Int J Found Comput Sci 19(5):1133–1145

Ishdorj T-O, Petre I, Rogojin V (2007) Computational power of intramolecular gene assembly. Int J Found Comp Sci 18(5):1123–1136

Ishdorj T-O, Loos, R, Petre I (2008) Computational efficiency of intermolecular gene assembly. Fund Informaticae 84(3-4):363–373

Jahn CL, Klobutcher LA (2000) Genome remodeling in ciliated protozoa. Ann Rev Microbiol 56:489–520

Kari L, Landweber LF (1999) Computational power of gene rearrangement. In: Winfree E, Gifford DK (eds) Proceedings of DNA based computers, V. Massachusetts Institute of Technology, 1999 American Mathematical Society, Providence, RI, pp 207–216

Kari L, Kari J, Landweber LF (1999) Reversible molecular computation in ciliates. In: Karhumäki J, Maurer H, Păun G, Rozenberg G (eds), Jewels are forever. Springer, Berlin, Heidelberg, New York, pp 353–363

Kauman LH (1999) Virtual knot theory. Eur J Combinatorics 20:663–690

Krasnow MA, Stasiak A, Spengler SJ, Dean F, Koller T, Cozzarelli NR (1983) Determination of the absolute handedness of knots and catenanes of DNA. Nature 304(5926):559–560

Landweber LF, Kari L (1999) The evolution of cellular computing: Nature's solution to a computational problem. In: Kari L, Rubin H, Wood D (eds) Special issue of Biosystems: proceedings of DNA based computers IV, vol 52(1–3). Elsevier, Amsterdam, pp 3–13

Landweber LF, Kari L (2002) Universal molecular computation in ciliates. In: Landweber LF, Winfree E (eds) Evolution as computation. Springer, Berlin, Heidelberg, New York, pp 257–274

Langille M, Petre I (2006) Simple gene assembly is deterministic. Fund Inf 72:1–12

Langille M, Petre I (2007) Sequential vs. parallel complexity in simple gene assembly. Theor Comput Sci 395(1):24–30

Langille M, Petre I, Rogojin V (2010) Three models for gene assembly in ciliates: a comparison. Comput Sci J Moldova 18(1):1–26

Möllenbeck M, Zhou Y, Cavalcanti ARO, Jönsson F, Higgins BP, Chang W-J, Juranek S, Doak TG, Rozenberg G, Lipps HJ, Landweber LF (2008) The pathway to detangle a scrambled gene. PLoS ONE 3(6):2330

Nowacki M, Vijayan V, Zhou Y, Schotanus K, Doak TG, Landweber LF (2008) RNA-mediated epigenetic programming of a genome-rearrangement pathway. Nature 451:153–158

Petre I (2006) Invariants of gene assembly in stichotrichous ciliates. IT, Oldenbourg Wissenschaftsverlag 48(3):161–167

Petre I, Rogojin V (2008) Decision problems for shuffled genes. Info Comput 206(11):1346–1352

Petre I, Skogman S (2006) Gene assembly simulator. http://combio.abo.fi/simulator/simulator.php

Pevzner P (2000) Computational molecular biology: an algorithmic approach. MIT Press, Cambridge, MA

Prescott DM (1994) The DNA of ciliated protozoa. Microbiol Rev 58(2):233–267

Prescott DM (1999) The evolutionary scrambling and developmental unscrambling of germline genes in hypotrichous ciliates. Nucl Acids Res 27(5):1243–1250

Prescott DM (2000) Genome gymnastics: unique modes of DNA evolution and processing in ciliates. Nat Rev Genet 3:191–198

Prescott DM, DuBois M (1996) Internal eliminated segments (IESs) of oxytrichidae. J Eukariot Microbiol 43:432–441

Prescott DM, Greslin AF (1992) Scrambled actin I gene in the micronucleus of Oxytricha nova. Dev Gen 13(1):66–74

Prescott DM, Ehrenfeucht A, Rozenberg G (2001a) Molecular operations for DNA processing in hypotrichous ciliates. Eur J Protistol 37:241–260

Prescott DM, Ehrenfeucht A, Rozenberg G (2001b) Template-guided recombination for IES elimination and unscrambling of genes in stichotrichous ciliates. J Theor Biol 222:323–330

Setubal J, Meidanis J (1997) Introduction to computational molecular biology. PWS Publishing Company, Boston, MA

White JH (1992) Geometry and topology of DNA and DNA-protein interactions. In: Sumners et al. (eds) New scientific applications of geometry and topology, American Mathematical Society, Providence, RI, pp 17–38

Yan H, Zhang X, Shen Z, Seeman NC (2002) A robust DNA mechanical device controlled by hybridization topology. Nature 415:62–65

Yurke B, Turberfield AJ, Mills AP, Simmel FC Jr (2000) A DNA fueled molecular machine made of DNA. Nature 406:605–608

Zerbib D, Mezard C, George H, West SC (1998) Coordinated actions of RuvABC in Holliday junction processing. J Mol Biol 281(4):621–630

38 DNA Memory

Masanori Arita[1] · *Masami Hagiya*[2] · *Masahiro Takinoue*[3] · *Fumiaki Tanaka*[4]

[1]Department of Computational Biology, Graduate School of Frontier Sciences, The University of Tokyo, Kashiwa, Japan
arita@k.u-tokyo.ac.jp

[2]Department of Computer Science, Graduate School of Information Science and Technology, The University of Tokyo, Tokyo, Japan
hagiya@is.s.u-tokyo.ac.jp

[3]Department of Physics, Kyoto University, Kyoto, Japan
takinoue@chem.scphys.kyoto-u.ac.jp

[4]Department of Computer Science, Graduate School of Information Science and Technology, The University of Tokyo, Tokyo, Japan
fumi95@is.s.u-tokyo.ac.jp

G. Rozenberg et al. (eds.), *Handbook of Natural Computing*, DOI 10.1007/978-3-540-92910-9_38,
© Springer-Verlag Berlin Heidelberg 2012

Abstract

This chapter summarizes the efforts that have been made so far to build a huge memory using DNA molecules. These efforts are targeted at increasing the size of the address space of a molecular memory and making operations on a specified word in the address space more efficient and reliable. The former issue should be solved by careful design of the base sequences of the address portions. The latter issue depends on the architecture of a molecular memory and the available memory operations. Concrete examples of molecular memories described in this chapter are classified into in vitro DNA memory, DNA memory on surfaces, and in vivo DNA memory. This chapter also describes the technology for designing base sequences of DNA molecules.

1 Introduction

Soon after Adleman published his seminal work on DNA computing (Adleman 1994) and Lipton (1995) gave a more general framework for solving nondeterministic polynomial time (NP)-complete problems using DNA, Baum wrote a technical comment in Science on building a huge memory using DNA molecules (Baum 1995). He claimed that it was possible to build an associative memory vastly larger than the brain. More concretely, he wrote that it was not completely implausible to imagine vessels storing 10^{20} words, which is comparable to standard estimates of brain capacity as 10^{14} synapses each storing a few bits. Among various proposals for constructing a molecular memory, he considered a scheme that stores DNA molecules consisting of an address portion and a data portion. In order for the scheme to work as an ordinary random access memory, each address portion should be accessible only by the sequence that is complementary to the sequence of the address portion. He even mentioned the use of an error correcting code to avoid accidental bonding due to approximate match.

This chapter summarizes the efforts that have been made so far toward achieving Baum's dream. Those efforts are targeted at increasing the size of the address space of a molecular memory and making operations on a specified word in the address space more efficient and reliable. As Baum pointed out, the former issue should be solved by careful design of base sequences of address portions. The latter issue depends on the architecture of a molecular memory and available memory operations. Sequence design is also affected by the required memory operations.

The next section describes the technology for designing base sequences of DNA molecules. It is one of the most fruitful technological achievements in the research field of DNA computing, and can be used in various application fields including biotechnology and nanotechnology. The following section then summarizes a series of memory operations that have been devised for constructing various kinds of molecular memories. Concrete examples of molecular memories are then described in the sections that follow, classified into in vitro DNA memory, DNA memory on surfaces, and in vivo DNA memory.

2 Sequence Design

This section describes sequence design in DNA computing. Sequence design is the search for sequences satisfying given constraints or having a high value for a given evaluation function.

First, this section describes the thermodynamics of DNA molecules and the constraints to be satisfied in DNA computing. Then, various strategies for sequence design are described.

2.1 Thermodynamics of DNA Molecules

Deoxyribonucleic acid (DNA) is a polymer consisting of three units: bases, deoxyriboses, and phosphoric acids. In DNA, the bases comprising the nucleotides are adenine (A), thymine (T), cytosine (C), and guanine (G); in RNA, thymine is replaced by uracil (U). The pairs between A and T and between C and G are called Watson–Crick base pairs (Watson and Crick 1953). The former has two hydrogen bonds and the latter has three, making it more stable. Other pairs (called "mismatches") such as G·T are generally less stable than the Watson–Crick base pairs. On the basis of these selective hydrogen bonds, a DNA sequence and its complementary strand form a double-stranded structure in an antiparallel direction at low temperature (Watson and Crick 1953), which is called hybridization. For example, sequence 5′-AGGCTACC-3′ hybridizes with its complement 3′-TCCGATGG-5′ and forms a double-stranded structure. A double strand melts into two single strands at high temperature, which is called denaturation.

A secondary structure of DNA is defined as a set of base pairs. DNA sequences can form various secondary structures consisting of stacking pairs, various loops, and dangling ends (see ❯ *Fig. 1*).

◻ **Fig. 1**
Example of DNA secondary structure. The dots represent the bases, while the thin and dotted lines represent the backbones and hydrogen bonds, respectively.

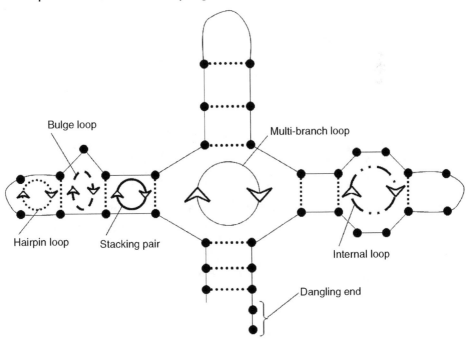

Bulge loop

Multi-branch loop

Hairpin loop   Stacking pair

Internal loop

Dangling end

The stability of double-stranded DNA depends on the temperature, the DNA concentration, and the buffer solution. The free energy change (ΔG) is the commonly used indicator of stability for DNA complexes. ΔG is the free energy change when the sequence folds into the structure from a random coil. In particular, ΔG_T denotes the free energy change at temperature T. The lower the free energy, the more stable the structure, so more sequences form into a structure in solution. The free energy of an entire structure is the sum of the free energies of all the local substructures (Zuker and Stiegler 1981). The minimum free energy (ΔG_{min}) is the minimum value of free energy for any structure. Thus, the structure with the minimum free energy must be the most stable in solution. Stability can be well approximated using the minimum free energy calculation. In fact, the minimum free energy is often used for secondary structure prediction of single-stranded RNA (Zuker and Stiegler 1981). The minimum free energy can be calculated using dynamic programming.

Another indicator for the stability of a double-stranded DNA is the melting temperature, T_m, which is defined as the temperature at which single and double strands are in equilibrium with a 1:1 ratio. A pair of sequences with a high melting temperature is stable, while one with a low melting temperature is unstable. The melting temperature is calculated using (Borer et al. 1974)

$$T_m = \frac{\Delta H}{R \ln(C_t/\alpha) + \Delta S}, \tag{1}$$

where R is the gas constant, C_t is the total oligo concentration. ΔH and ΔS are enthalpy and entropy change from a random coil, respectively. Parameter α is set to one for self-complementary strands and to four for non-self-complementary strands.

Parameters ΔH, ΔS and ΔG are called "thermodynamic parameters" and are calculated on the basis of the nearest-neighbor model (SantaLucia 1998). For example, the ΔG of double-stranded DNA 5'-AGTG-3'/3'-TCAC-5' is calculated using

ΔG (5'-AGTG-3'/3'-TCAC-5') = *initiation* + *symmetry* + ΔG (5'-AG-3'/3'-TC-5') + ΔG (5'-GT-3'/3'-CA-5') + ΔG (5'-TG-3'/3'-AC-5')

where "*initiation*" is a contribution by terminal base pairs and the value differs depending on whether the terminal base pair is an A-T pair or G-C pair. The term "*symmetry*" describes whether the sequence is a palindrome or not. The thermodynamic parameters of a substructure, such as 5'-GT-3'/3'-CA-5', are called "nearest-neighbor parameters" or "nearest-neighbor thermodynamics." Nearest-neighbor parameters are determined by the sequence forming the substructure and must be estimated in advance through thermodynamic analysis. ❷ *Table 1* shows an example of reported nearest-neighbor parameters for Watson–Crick base pairs (Allawi and SantaLucia 1997; Tanaka et al. 2004). If the SantaLucia parameters (in the rightmost column of the table) are used, the ΔG of 5'-AGTG-3'/3'-TCAC-5' is calculated as $(0.98 + 1.03) + 0 + (-1.28) + (-1.44) + (-1.45) = -2.16$ kcal/mol.

All possible nearest-neighbor parameters have already been estimated for Watson–Crick base pairs (SantaLucia 1998; Sugimoto et al. 1996; SantaLucia et al. 1996), internal mismatches (Aboul-ela et al. 1985), G·A mismatches (Li et al. 1991; Li and Agrawal 1995), G·T mismatches (Allawi and SantaLucia 1997), A·C mismatches (Allawi and SantaLucia 1998a), C·T mismatches (Allawi and SantaLucia 1998b), G·A mismatches (Allawi and SantaLucia 1998c), N·N mismatches (N represents A, T, C, or G) (Peyret et al. 1999), dangling ends (Senior et al. 1988; Bommarito et al. 2000), and bulge loops (Tanaka et al. 2004). The thermodynamic parameters for various loops are calculated approximately as a function of loop length; shorter loops have lower free energy while longer loops have higher free energy.

◻ **Table 1**

Nearest-neighbor thermodynamic parameters for Watson–Crick base pairs

Sequence	ΔH (Tanaka et al. 2004) (kcal/mol)	ΔS (Tanaka et al. 2004) (eu)	ΔG (Tanaka et al. 2004) (kcal/mol)	ΔG (Allawi and SantaLucia 1997) (kcal/mol)
AA/TT	-8.3 ± 1.1	-23.3 ± 3.1	-1.04 ± 0.07	-1.00 ± 0.01
AT/TA	-8.5 ± 0.8	-24.6 ± 2.4	-0.91 ± 0.04	-0.88 ± 0.04
AC/TG	-9.5 ± 1.1	-26.0 ± 2.7	-1.49 ± 0.09	-1.44 ± 0.04
AG/TC	-8.2 ± 0.9	-22.3 ± 2.7	-1.33 ± 0.06	-1.28 ± 0.03
TA/AT	-6.5 ± 0.6	-19.0 ± 1.7	-0.60 ± 0.04	-0.58 ± 0.06
TC/AG	-8.7 ± 0.8	-23.6 ± 2.4	-1.35 ± 0.03	-1.30 ± 0.03
TG/AC	-8.9 ± 0.4	-24.3 ± 1.3	-1.51 ± 0.05	-1.45 ± 0.06
CC/GG	-8.8 ± 0.6	-21.7 ± 1.7	-1.90 ± 0.04	-1.84 ± 0.04
CG/GC	-11.4 ± 0.9	-29.7 ± 2.5	-2.25 ± 0.06	-2.17 ± 0.05
GC/CG	-10.1 ± 1.0	-25.4 ± 2.6	-2.28 ± 0.06	-2.24 ± 0.03
Init_AT	4.4 ± 1.4	10.7 ± 3.9	1.19 ± 0.09	1.03 ± 0.05
Init_GC	1.7 ± 0.5	2.4 ± 1.3	1.12 ± 0.03	0.98 ± 0.05
Symmetry	$0 \pm -$	$-1.4 \pm -$	$0.4 \pm -$	$0.4 \pm -$

2.2 Design Constraint

2.2.1 Uniform Melting Temperature

DNA sequences used in DNA computing should hybridize with their complements at the same efficiency to prevent hybridization bias. Thus, making the melting temperature uniform among pairs of complementary sequences is important for obtaining uniform hybridization efficiency. The melting temperature is calculated using formula (1).

Another indicator for the melting temperature is the GC content, which is the ratio of the sum of the number of G and C nucleotides over the total number of nucleotides in the sequence. Double strands with a higher GC content tend to have a higher melting temperature because the hydrogen bond between G and C is more stable than that between A and T. Thus, the GC content can be used to approximately calculate the melting temperature. In DNA computing, sequences are often designed such that their GC content is 50%.

2.2.2 Specific Hybridization

Specific hybridization, which is also called intended/expected/planned hybridization/interaction, denotes the hybridization between an intended pair of sequences while nonspecific hybridization, or unintended/unexpected/unplanned hybridization/interaction, denotes the hybridization between an unintended pair of sequences. Nonspecific hybridization causes various problems in DNA computing, including errors in the computation, failure in the construction of intended DNA nanostructures, and misdiagnosis in DNA microarrays.

Therefore, in DNA computing, multiple sequences need to be designed, such that they do not hybridize nonspecifically with each other, while in RNA secondary structure design, a single sequence needs to be designed that folds into the desired secondary structure (Hofacker 2003; Andronescu et al. 2003, 2004; Dirks et al. 2004). Specific sequences are sequences between which only specific hybridizations can form. Since the number of combinations increases exponentially with the number of sequences, designing specific sequences is an NP-hard problem and is equivalent to the independent set problem (Frutos et al. 1997).

To design specific sequences, we must design sequences among which specific hybridizations are stable and nonspecific ones are unstable. Thus, specific-sequence design requires a metric for stability between two sequences. The most often used criterion for stability is the h-measure proposed by Garzon et al. (1997, 1998). The h-measure for sequences x_i, x_j is defined as

$$|x_i, x_j| := \min_{-n<k<n} H(x_i, \sigma^k(\overline{x_j})),$$ (2)

where $H(*, *)$ denotes the Hamming distance, σ^k denotes the right (left) shift for $k>0$ ($k<0$), k denotes the number of the shift, and $\overline{x_j}$ is the sequence which is Watson–Crick complementary to x_j. The h-measure is based on the fact that the stability depends on the number of base pairs: a double strand with a certain number of base pairs is more stable than one with fewer base pairs. Therefore, if the h-measure between two sequences is large, the double strand that they form tends to be unstable.

2.2.3 Preventing Secondary Structures

Secondary structures often prevent the sequence from hybridizing to the correct complementary sequence. Thus, we must avoid sequences that form stable secondary structures unless they are specifically designed to form them. The mfold program (Zuker 2003) is commonly used to calculate the stability of secondary structures.

2.2.4 Other Constraints

1. *Repeated Bases*
   Repeated bases have a harmful effect on specific hybridization. For example, guanine-rich motifs can form four-stranded complexes (Sen and Gilbert 1988, 1990). Thus, sequences are often designed not to have repeated bases.
2. *Forbidden Subsequences*
   When a restriction enzyme, which cuts off the sequence at specific sites, is used, the recognition sequence must appear only at an intended site. Otherwise, sequences will be cut off at some unintended site. Therefore, recognition sequences are forbidden except for intended sites.
3. *Three-Base Constraint*
   To prevent nonspecific hybridization, several strategies have been devised that use only three kinds of bases (A, T, and C) for sequence design (Cukras et al. 1999; Faulhammer et al. 2000; Braich et al. 2000, 2002). The reason for omitting base G is that it can form G·C and G·T base-pairings. In addition, Whiplash polymerase chain reaction (PCR) needs to use only three bases because the fourth base is used as stopper sequences to stop the

polymerase extension at the designated position (Hagiya et al. 1999). This constraint is effective to some extent for the specific hybridization. However, omitting G greatly reduces the variety of sequences.

2.3 Strategy for Sequence Design

This subsection summarizes strategies for producing sequences that satisfy the design criteria discussed in the previous section. Many strategies have been proposed for searching appropriate sequences.

In the coding theory research area, researchers have focused on a classical problem: maximize the size of a set consisting of equi-length code words so that the Hamming distances between them are above a certain threshold. This problem resembles the sequence design problem: design sequences so that the Hamming distances between them are above a certain threshold. Thus, the theoretical analyses and achievements in coding theory can be used for DNA sequence design.

On the basis of coding theory, sequences are produced by crossing two binary codes of length n, namely a template and map. First, the positions for [AT] and [GC] are determined using a template $t = t_1 t_2 \ldots t_n$ ($t_i \in \{0, 1\}$), where 0 indicates [GC] and 1 indicates [AT]. Since the number of occurrences of 0 in the template is equal to the number of G and C in a DNA sequence, the GC content of DNA sequences can be adjusted by controlling the occurrences of 0 in a template. Either A or T is chosen for template positions $t_i = 1$ and either G or C for positions $t_i = 0$ using a map $m = m_1 m_2 \ldots m_n$ ($m_i \in 0, 1$). The conversion rule is to convert 0 in a template and 0 in a map into C, 0 and 1 into G, 1 and 0 into A, and 1 and 1 into T. For example, the sequence "ATTGCTATGCTA" is produced by conversion of codes T="111001110011" and M="011101011010" (see ❷ *Fig. 2*).

The problem is to find as many templates and maps as possible such that sequences obtained from them satisfy certain criteria such as uniform GC content or maximal Hamming distance. Frutos et al. proposed a strategy for designing sequences that satisfy a Hamming distance constraint (Frutos et al. 1997). Although the design procedure does not consider frameshift hybridization, their study is a pioneering work in DNA sequence design using coding theory. Li et al. developed a similar design strategy based on the Hamming distance without the frameshift (Li et al. 2002).

❑ **Fig. 2**
Conversion from two binary codes into a DNA sequence. The bases of the sequence are determined on the basis of the conversion rule.

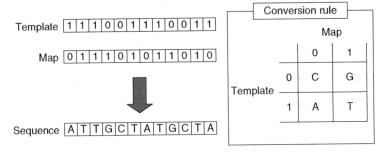

Marathe et al. clarified the lower and upper bounds on the maximum size of DNA sequences that satisfy the constraints on the basis of the Hamming distance without the frameshift (Marathe et al. 2001). Their study was based on several achievements in coding theory. As Brenneman and Condon pointed out (Brenneman and Condon 2002), however, the upper bound is probably still unknown when the Hamming distance is considered with the frameshift.

Liu et al. proposed a design strategy that considers frameshift hybridization in accordance with the h-measure (Liu et al. 2003). To prevent low h-measure values, they kept equal 01 content in the map. This strategy is based on the observation that deviation in the occurrence of 0 or 1 in the map tends to reduce the h-measure value. Although the heuristics may be partially effective, they are not sufficient because this strategy does not completely guarantee the exclusion of maps, which can cause a low h-measure value.

A more sophisticated method, called the "template method," was proposed by Arita and Kobayashi (2002). In this method, an error-correcting code is used as a map (see ❷ *Fig. 3*). The error-correcting code is the code for which codewords have at least k mismatches with each other. It has been already investigated in traditional coding theory. For example, a BCH code of length 15 provides at least 128 words with a minimum of five mismatches. Thus, the template method needs to design a template that satisfies a specified constraint. The constraint imposed by Arita and Kobayashi was that the number of mismatches must be larger than a threshold. The template method has two strengths: (i) frameshift hybridization is considered and (ii) mismatches between sequences and concatenations of two sequences are considered. Templates satisfying the specified constraint are found by an exhaustive search, although the number of candidates is dramatically reduced by theoretical consideration. As a result, sets of templates of length up to 30 can be designed.

This template method was later extended by using an AG-template, which determines the position of [AG] rather than [GC] (Kobayashi et al. 2003). The mismatches obtained by using AG-templates outnumber those by using GC-templates. However, using an AG-template does not guarantee uniform GC content. This problem was solved by using a constant weight code in which the codewords have the same number of occurrences of 1 (❷ *Fig. 3*). Thus, maps (i.e., constant weight codes) guarantee the uniform GC content instead of templates.

Another important strategy is a *de novo* design proposed by Seeman (1990). The *de novo* design generates sequences for which each subsequence of length l appears at most once. Let one consider a sequence CATGGGAGATGCTTAG as an example of $l = 6$. Any subsequence of length 6 such as CATGGG appears only once in this sequence. The first design step is to choose a sequence of length l that is not included in a given list of prohibited subsequences and to add this sequence to the prohibited list. This sequence is prolonged to the right by concatenating a letter in such a way that the resulting sequence of length $l+1$ does not have any sequence in the prohibited list as its suffix is of length l. Sequences generated by repeating this process are guaranteed not to have two identical subsequences of length l in it. This constraint is motivated by the fact that consecutive complementary bases are likely to form stable structures. However, there is no guarantee that the *de novo* design prevents nonspecific hybridization. For example, although sequences AAAAACAAAAA and AAAAAGAAAAA do not share any 6-mer subsequence, the Hamming distance between them is one, resulting in cross-hybridization between one sequence and the complement of the other. Therefore, the sequences produced by the *de novo* design must be checked using another program such as mfold (Zuker 2003) before their practical use.

◻ **Fig. 3**
Template method with GC template (*upper part*) and AG template (*lower part*).

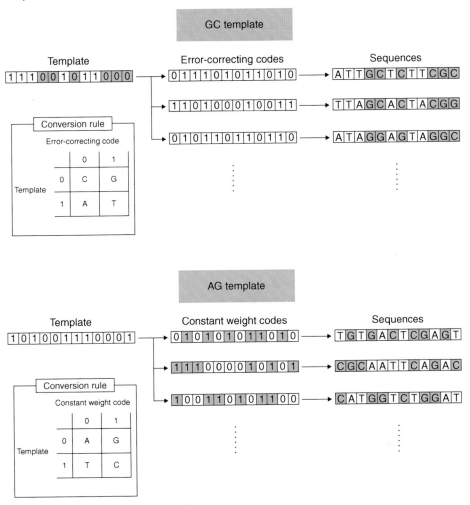

Feldkamp et al. developed a software program with a GUI called *DNASequenceGenerator* for the construction of sets of DNA sequences (Feldkamp et al. 2001). The program can design sequences that satisfy uniqueness (i.e., *de novo* design), uniform melting temperature, and GC content, and that prevent palindromes, start codons and restriction sites. It creates *de novo* sequences using a graph representing DNA sequences (see ❷ *Fig. 4*). Each path in the graph represents a DNA sequence. The program cuts off from the graph forbidden sequences defined by users and sequences already used. Consequently, *de novo* sequences can be designed.

Pancoska et al. developed a design algorithm for sequences with the same melting temperature that uses an Eulerian graph (Pancoska et al. 2004). The algorithm first constructs an Eulerian graph and an adjacency matrix from a given sequence (see ❷ *Fig. 5*). The key idea is that every Euler path has the same occurrence of each nearest-neighbor parameter. Thus, if we calculate the melting temperature using formula (1), all sequences represented by Euler

■ Fig. 4

Graph representation of DNA sequences. Each path represents a DNA sequence.

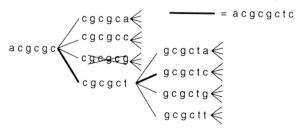

■ Fig. 5

Transformation from a DNA sequence into the Eulerian graph and the adjacency matrix.

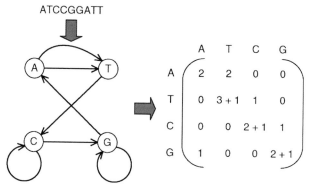

paths have the same melting temperature. Consequently, sequences with the same melting temperature can be designed by enumerating all Euler paths and converting them into the corresponding sequences.

With an appropriate evaluation function or set of constraints, the sequence design problem can be formulated as a search for sequences that maximize (minimize) the evaluation function or satisfy the constraints.

The simplest strategy is to generate sequences randomly, and then modify them until they satisfy the constraints. Faulhammer et al. developed a program called PERMUTE that designs sequences on the basis of this strategy (Faulhammer et al. 2000).

A similar strategy is to generate a sequence randomly and then add it to a sequence set if it satisfies all the constraints. This process is repeated until the sequence set contains a predetermined number of sequences. Suyama's group developed a procedure using this strategy for designing orthogonal DNA sequences (Yoshida and Suyama 2000). The sequences are designed to satisfy three constraints: (i) uniform melting temperature, (ii) no mishybridization, and (iii) no formation of stable intra-molecular structures. A disadvantage of this strategy is that it easily falls into a local optimum, resulting in the saturation of the number of sequences successfully designed. With the aim of overcoming this drawback, Kashiwamura et al. developed a modified algorithm called "two-step search" (Kashiwamura et al. 2004). They observed that critical sequences, which have higher scores of Hamming distance with sequences in a sequence set, prevent the design of more sequences. In line with this analysis, the two-step search generates sequences with Hamming distances as short as possible. This is analogous to packing particles into a limited space, for which an effective strategy is to pack them as tightly as possible.

Ruben et al. developed a program called "PUNCH" that designs sequences using a local search algorithm (Ruben et al. 2001). The strategy is to find sequences based on uniqueness, mishybridization, base equality, and folding scores. The program searches neighbors obtained by mutating previous best sequences randomly. Therefore, this search procedure cannot escape from local optima.

A more sophisticated algorithm is the stochastic local search (SLS) algorithm. The algorithm designs sequences by improving a given sequence set iteratively. First, it chooses a pair of sequences s_1 and s_2 that violate one of the given constraints. Then, it generates a set S_1 of sequences obtained by substituting some bases of s_1 and a set S_2 by substituting some bases of s_2. With a probability of θ, it selects a sequence s' from $S_1 \cup S_2$ at random. Otherwise, it selects a sequence s' from $S_1 \cup S_2$ such that the number of conflict violations is maximally decreased. If $s' \in S_1$, it replaces s_1 by s'. Otherwise, it replaces s_2 by s'. This procedure is iterated for a predefined number of steps. Using the SLS algorithm, Tulpan et al. found sequence sets that matched or exceeded the best previously known constructions though they did not take into account a frameshift for the Hamming distance (Tulpan et al. 2003; Tulpan and Hoos 2003). They determined that it was effective to use "hybrid randomized neighborhoods" obtained by adding random sequences to neighborhoods by substituting one base (Tulpan and Hoos 2003).

Sequences can also be designed using metaheuristics such as the genetic algorithm (GA). Deaton et al. used GA to design sequences that satisfied various constraints such as the Hamming distances between them being a certain length (Deaton et al. 1996, 1998). Although they considered only Hamming distances without a frameshift, they showed that the Hamming distances for some sequences were identical to the upper bound, called the "Hamming bound." Arita et al. proposed a design strategy in which an evaluation function calculated as the linear sum of multiple evaluation terms is optimized (Arita et al. 2000). The evaluation function considers restriction sites, GC content, Hamming distance, repetition of the same base, and complete hybridization at the 3′-end. They also utilized a GA for the optimization.

The sequence design methods described above approximate the stability of double strand on the basis of the number of base pairs. Some recently proposed design methods estimate the stability on the basis of the minimum free energy (ΔG_{min}). This thermodynamic approach approximates the stability with greater accuracy because it takes into account the stability of some loops and the stacking between base pairs. However, it takes longer to calculate the minimum free energy. The time complexity of the minimum free energy calculation by dynamic programming is $O(l^3)$, where l is the sequence length (Lyngso et al. 1999). Therefore, efforts have been made to reduce the computational time. Tanaka et al. proposed approximating the calculation of the minimum free energy in a greedy manner to reduce the time (Tanaka et al. 2005). They defined the problem of designing a pool S containing n sequences of length l for which the minimum free energy is greater than the threshold (ΔG_{min}^*) in any pairwise duplex of sequences in S and any concatenation of two sequences in S, plus their complements except for the complementary pairs. ❷ *Figure 6* shows a conceptual diagram of this greedy calculation. The approximated value for the minimum free energy, denoted by ΔG_{gre}, is calculated by searching for the most stable continuous base pairs and fixing them iteratively. The time complexity of ΔG_{gre} calculation is $O(l^2)$. Obviously, ΔG_{gre} is an upper bound of ΔG_{min} (i.e., $\Delta G_{min} \leq \Delta G_{gre}$). Thus, ΔG_{gre} can be used to check the sequence before calculating ΔG_{min}.

⬛ **Fig. 6**

Schematic diagram for sequence design using greedy calculation of ΔG_{min}. In this diagram, S and l denote a sequence set already designed and the sequence length, respectively. ΔG_{gre} denotes the approximate value of ΔG_{min} calculated in a greedy manner.

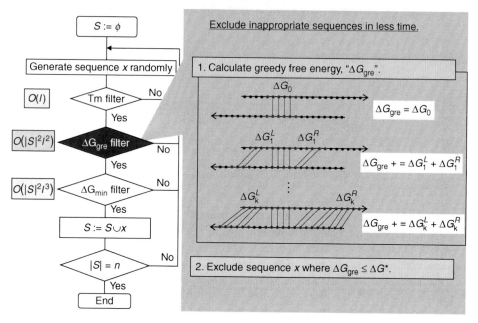

Kawashimo et al. also developed a design procedure based on the thermodynamic approach (Kawashimo et al. 2008, 2009). They reduced the computation time by introducing a sophisticated search order and by limiting the loop size.

2.4 Future Direction of Sequence Design

We have discussed sequence design to date, but research on sequence design continues. What are the future directions?

One will be the development of sequence design that takes the kinetics of DNAs into consideration. Kitajima et al. proposed using a design criterion for rapid hybridization between two complementary sequences (Kitajima et al. 2008). They demonstrated that the rapidity of hybridization depended on the stability of the self-folded secondary structures and on the nucleation capability at the tails of their self-folded secondary structures. Although this finding cannot be directly used for sequence design, it can be used to exclude the sequences for which the hybridization rate is low.

Another direction will be sequence design that considers the dynamics of DNA machinery, in which DNA ruled by the machinery has multiple stable states. In this case, sequences must be designed that can transit from one state to another. To achieve this, we should estimate the energy landscape produced by the DNA sequences. However, it is difficult to predict the energy landscape because the number of possible structures grows exponentially with the number of sequences. There have been several efforts to estimate the energy landscape efficiently (Uejima and Hagiya 2004; Kubota and Hagiya 2005).

Uejima and Hagiya developed a sequence design method for their multi-state molecular machine that is based on the minimum free energies, the structure transition paths, and the total frequency of the optimal and suboptimal structures (Uejima and Hagiya 2004). The transition paths were estimated using Morgan and Higgs' heuristics. This estimation was expanded by Kubota and Hagiya (2005). They proposed a technique for analyzing the energy landscapes given the minimax path, which is a path on the minimum spanning tree.

In short, sequence designs that consider the above constraints will be developed for more complex DNA machinery.

3 Memory Operations

As mentioned in the introduction of this chapter, one of the most crucial issues for constructing a molecular memory is how to access a specified word in its address space. This section summarizes various techniques that have been developed to implement memory operations on a molecular memory.

To compare various implementations, an associative molecular memory consisting of words that are divided into two portions is considered: the address part and the data part. The address part can be further divided into multiple address digits, comprising a hierarchical address space. For each memory operation, it is first assumed that the address part consists of a single address digit, and then it touches on the possibility of extending the memory operation to a hierarchical memory.

The data part of a memory word can be designed in various ways. It can represent a single bit, that is, 0 or 1. In some molecular memories, the base sequence of a data part is not relevant. For example, a long data part represents bit 1, and a short data part represents bit 0. Data parts can also be empty. In such a molecular memory, existence of a word having a specified address means bit 1 of the address and nonexistence of a word means bit 0. Each address (digit) can have its complementary address (digit). This is typical when variable assignments are represented by a DNA molecule. Each variable and its negation correspond to complementary address digits and they comprise a hierarchical molecular memory.

The data part can represent richer information. For example, if an artificial tag sequence is attached to a gene extracted from a chromosome or cDNA in a library of genes, then the tag sequence is regarded as the address part and the gene sequence as the data part.

This section is organized as follows. In the first subsection, various implementations of the access operation as mentioned above are summarized. In the next sections, the read and write operations are briefly explained. The final section touches on some other operations on a molecular memory.

3.1 Access Operation

This section explains how to access a memory word having a specified address in a molecular memory. If there is no data part in a memory word, success in accessing the word amounts to reading bit 1 of the address, and failure amounts to reading bit 0.

3.1.1 Extraction by Magnetic Beads

As in the seminal work of Adleman, extraction using magnetic beads is a typical access operation on a molecular memory. More generally, any kind of affinity separation based on the address part of a word can be used as an access operation. Affinity separation by magnetic beads is performed as follows.

The address part of a memory word should be represented by a single-stranded segment of a DNA molecule. Typically, a whole word takes the form of a single strand of DNA as in ❷ *Fig. 7*. As Baum pointed out (Baum 1995), in order to avoid unexpected interaction, the data part of a memory word can be double-stranded. Such words comprising a molecular memory are mixed and diluted in an aqueous solution.

DNA molecules containing the sequence complementary to the address part, called probes, are also single-stranded and attached on the surface of magnetic beads as in ❷ *Fig. 8*. The protein called biotin is covalently bonded to the probes. Such modified DNA can be ordered to a DNA synthesizing company as ordinary plain DNA with some additional cost. The surface of the magnetic beads is coated with another kind of protein called streptavidin. Strong interaction between streptavidin and biotin attaches the probes to the surface of the magnetic beads.

The magnetic beads are then put into the aqueous solution of the molecular memory in a test tube. The resulting test tube looks like muddy water, in which the molecular words and the probes attached on the surface of magnetic beads interact, and those words having the address part complementary to the probes are trapped by DNA hybridization. Those with different address parts are still in the solution.

A strong magnet is then placed at the side of the test tube. Not surprisingly, all the magnet beads are pulled together near the magnet (❷ *Fig. 9*), and the solution quickly becomes transparent, in which memory words not trapped in the beads are still diluted. The solution is then extracted by a pipet while the beads are fixed to the magnet. Pure water is then poured into the test tube and the magnet is removed. The water is then turned into mud. This process is repeated to separate all memory words that are not trapped in the beads.

◻ **Fig. 7**

A simple memory word. A single strand of DNA consists of the address part and the data part.

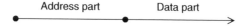

Address part   Data part

◻ **Fig. 8**

A strand trapped by a magnetic bead. Probes containing the sequence complementary to the target are attached on the surface of a magnetic bead. A single strand containing the target sequence is trapped by the bead.

◘ Fig. 9

Magnetic beads. In the left panel, magnetic beads are diluted in the tube to trap the target DNA. In the right panel, magnetic beads with the trapped DNA are collected by a magnet.

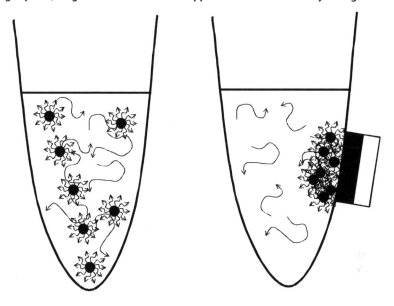

◘ Fig. 10

A hot tube connected to a cool tube. DNA molecules released from the hot tube are moved and trapped in the cool tube.

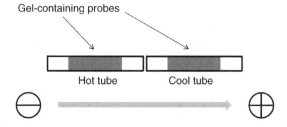

Finally, hot water is poured into the test tube. At this time, since hydrogen bonds in DNA hybridization are broken at high temperatures, the memory words trapped in the beads become free and diluted in the hot water. In this way, memory words having the specified address are separated from those having different addresses.

In the experiment conducted by Adleman et al., in which they solved a 20-variable SAT problem, probes are fixed in a gel filled inside a thin tube (Braich et al. 2002). Probes are chemically modified so that they are fixed in a gel and cannot move.

As in ❷ *Fig. 10*, two such tubes are connected and placed in a specially designed electrophoresis box, in which one tube is heated while the other is cooled down. In the hot tube, trapped DNA molecules are released and moved toward the cool tube according to the voltage

◻ **Fig. 11**

A word in a molecular memory with a hierarchical address space. Each word consists of multiple address digits (three digits in this figure) and a data part.

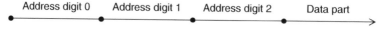

Address digit 0   Address digit 1   Address digit 2   Data part

of electrophoresis, and trapped in the cool tube if they are complementary to the probes fixed in the cool tube.

A hierarchical molecular memory consists of memory words whose address is comprised of multiple address digits as in ❯ *Fig. 11*. If a single digit has r different values and each address consists of n digits, the whole address space contains r^n different addresses.

It is easy to implement a hierarchical molecular memory whose access operation is implemented by affinity separation as explained so far. The access operation is iterated for each digit, that is, affinity separation by the ith digit is applied to the result for $(i-1)$th digit. A serious problem of this iterative approach is that at each round of affinity separation, some memory words are lost due to the limitation of the employed experimental technique. The yield of extraction thus exponentially decreases as n increases. This problem is solved by the approach using PCR, since copies of the extracted memory word are made.

3.1.2 Extraction by PCR

As is well known, PCR (polymerase chain reaction) is a method to amplify specified DNA molecules, that is, make their copies. Copied DNA molecules should have the specified sequences at their ends, with which short DNA molecules, called primers, hybridize.

To be more precise, copied DNA molecules are double-stranded, formed by a sequence x and its complement $\bar{x}$. The sequence x should end with a short sequence r, while the sequence $\bar{x}$ should also end with a short sequence l. Consequently, x should be of the form $\bar{l}yr$ (❯ *Fig. 12*). Primers are single-stranded DNA molecules whose sequence is $\bar{r}$ and $\bar{l}$.

PCR is conducted as follows (❯ *Fig. 12*). Double strands consisting of x and $\bar{x}$ are mixed with the pair of primers $\bar{r}$ and $\bar{l}$ in an appropriate buffer, containing polymerase, nucleotides, and adenosine triphosphate (ATP), which is called the polymerization buffer. The concentration of primers is much larger than that of double strands. The solution is then heated up to over 90°C, denaturing double strands into single strands x and $\bar{x}$. When the solution is cooled down, single strands hybridize with primers rather than their counterparts because they have more chances to contact the primers. During an appropriate duration for incubation, the 3′-end of $\bar{r}$ in the hybrid of x and $\bar{r}$ is extended by polymerase and the hybrid turns into a complete double strand consisting of x and $\bar{x}$. This process is called polymerization. The hybrid of $\bar{x}$ and $\bar{l}$ similarly turns into the same double strand. In this way, in a cycle of heating up, cooling down, and incubation, the concentration of x and $\bar{x}$ is doubled, and repeating the cycle amplifies the concentration of the double strands.

In order to use PCR for accessing memory words in a DNA memory, each word should have its data part surrounded by two primer sequences at its ends as in ❯ *Fig. 13*. These primer sequences comprise the address part, since a memory word can be amplified only if it has both

◘ Fig. 12

PCR (polymerase chain reaction). A double strand is first denatured to single strands at high temperature. If cooled down, they hybridize with short DNA molecules, called primers. Polymerase then extends the primers and two copies of the original double strand are made.

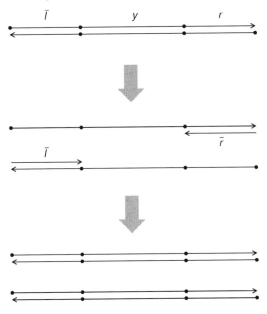

◘ Fig. 13

A memory word extractable by PCR. Each word is a double strand having primer sequences as its address.

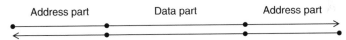

primer sequences. Therefore, in order to extract a memory word, the solution of the DNA memory is diluted and mixed with the primers in the polymerization buffer. Since PCR only amplifies the target memory word, if the resulting buffer is diluted, the concentration of memory words other than the target becomes less than detectable. This means that the target word is extracted by PCR.

A hierarchical address space can be obtained by introducing multiple primer pairs into a memory word. These primer pairs are nested as in ❷ *Fig. 14*. In order to extract a memory word, PCR is conducted with the outermost primer pair first. The obtained memory words are then processed by the second iteration of PCR with the next outer primer pair. Note that after PCR is conducted with an inner primer pair, the outer primer sequences are lost.

In ❷ Sect. 4.2, an actual PCR-based molecular memory is described in detail.

3.1.3 Hybridization and Conformational Change

Instead of extracting a memory word with a specified address, separating it from other words, it is also possible to access a memory word by changing its state, while words with different addresses are unchanged. In this kind of molecular memory, since memory words are not separated or mixed again, they can be fixed on a surface or in a space. In particular, it is possible to make a two-dimensional molecular memory by fixing DNA molecules on a surface of gold or other appropriate material.

If memory words are represented as single strands, simply adding a single-stranded probe whose sequence is complementary to the address of the target changes it into a double strand (❷ *Fig. 15*). In a simple setting, where a memory word does not have a data part, this process is considered as a write operation. The single-stranded form represents bit 0 and the double-stranded form represents bit 1.

As explained in detail in ❷ Sect. 5, Suyama's group employs hairpin molecules as memory words. The use of hairpins can introduce an additional condition for the access operation. Since hairpin formation is an intra-molecular reaction, it is more efficient than hybridization between a hairpin and its complement, which is an inter-molecular reaction. Therefore, it is necessary to denature a hairpin structure by raising temperature in order for its complement to hybridize with the hairpin (❷ *Fig. 16*).

They attach hairpins with different addresses to a gold surface, and employ a laser beam to locally raise temperature on the surface. Those hairpins on the laser spot are denatured and if their sequence is complementary to that of single strands in the surrounding solution, they can hybridize with the single strands.

It is not obvious to construct a hierarchical molecular memory based on hybridization, in which a memory word contains multiple address digits. For accessing a target memory word of the form in ❷ *Fig. 11*, single strands whose sequences are complementary to the address digits of the target are mixed with the whole molecular memory. Then the address part of the target

◘ Fig. 14

A memory word in a hierarchical molecular memory extractable by PCR. Each word contains nested primer pairs.

Address digit 1   Address digit 0   Data part   Address digit 0   Address digit 1

◘ Fig. 15

Hybridization of a memory word and a probe that is complementary to the address part of the word. The address part is changed from a single strand to a double strand.

Address part   Data part

Probe

word becomes fully double-stranded. It is therefore necessary to distinguish the target word from other words whose address part is not double-stranded or partially double-stranded.

Hagiya's group employed hairpins to construct a hierarchal molecular memory based on hybridization (Kameda et al. 2008b). A memory word consists of concatenated hairpins and a single-stranded end as in ❷ *Fig. 17*.

A memory word has a single-stranded part to the left of the first hairpin. Single strands called openers are used to open the hairpins successively. The first opener is complementary to the single-stranded part of a word and the stem of the first (leftmost) hairpin. It first hybridizes with the single-stranded part, and then invades into the stem of the first hairpin by branch migration. Eventually, it completely hybridizes with the single-stranded part and the adjacent part of the stem. Consequently the first hairpin is opened, and the other part of the stem is exposed as single-stranded. Therefore, if the second opener, which is complementary to the exposed stem and the adjacent part of the second hairpin, is present, it first

☐ **Fig. 16**

A hairpin molecule as a memory word. If the temperature is raised, the hairpin is denatured and it can hybridize with a probe that is complementary to the hairpin loop and one side of the stem.

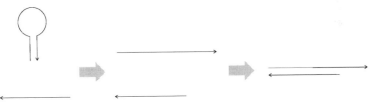

☐ **Fig. 17**

A memory word consisting of three consecutive hairpins. A single strand called an opener hybridizes with the single-stranded part of the memory word and opens the adjacent (first) hairpin. As a result, the next opener that opens the second hairpin can hybridize with the exposed stem of the first hairpin.

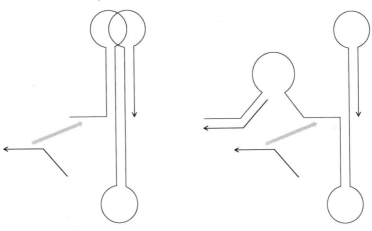

hybridizes with the exposed stem and then invades the stem of the second hairpin by branch migration. Note that without the first opener, the second cannot hybridize with the memory word and open the second hairpin. Therefore, only if all the openers are present, are all the hairpins opened. In particular, the final hairpin is opened only if all the openers are present. This means that a memory word is accessed if and only if the final hairpin is opened.

Hagiya's group actually constructed a memory word consisting of four hairpins and verified that the proposed scheme actually worked (Kameda et al. 2008b). They also constructed a molecular memory consisting of 3^3 addresses by employing the scheme. In this molecular memory, a memory word consists of three hairpins, and there are three variations for each opener that opens one of the hairpins. Consequently, there are 3^3 combinations of three openers. They verified that the final hairpin of each memory word can only be opened by the corresponding combination of three openers.

It is also possible to combine schemes described in this section. For example, in the scheme by Suyama's group a hairpin molecule can hybridize with its complement only if the local temperature is raised (Takinoue and Suyama 2004, 2006). If the stem of the hairpin is longer, raising the temperature is not sufficient to denature the hairpin. In this case, a single strand can be used to partially open the stem as in ❷ *Fig. 18*. This is a combination of the scheme of Hagiya's group (Kameda et al. 2008b) and that of Suyama's group (Takinoue and Suyama 2004, 2006).

3.2 Read Operation

In an extraction-based molecular memory, if a memory word consists of only the address part, and if existence and nonexistence of a word means bit 1 and bit 0, respectively, then extracting a memory word, whether by affinity separation or by PCR, amounts to reading the bit of the address of the word.

◻ **Fig. 18**
A hairpin memory word accessible by the combination of an opener and temperature. If an opener is present, it partially opens the hairpin. If the temperature is raised (from the upper right diagram to the lower left one), then the hairpin is completely opened and it can hybridize with the complementary strand.

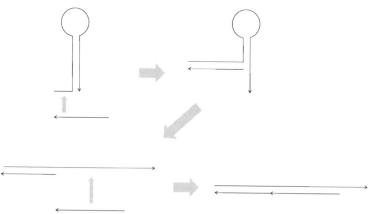

If a memory word contains a data part that encodes some additional information, then it is necessary to read the data part by a separate read operation. In an extraction-based molecular memory, the extracted word is given to a read operation such as gel electrophoresis, sequencing, or DNA chip detection. If the length of the data part encodes the information stored in the memory word, gel electrophoresis is the most efficient way to read the data. If the sequence of the data part encodes information, then sequencing is required to determine the bases in the data part.

As sequencing technology is progressing rapidly, it is becoming more and more efficient to read the base sequence of a given DNA molecule. However, it is still time consuming to read out individual bases in DNA. Moreover, sequencing requires a certain number of copies of the given DNA molecule, and additional amplification by PCR is usually necessary. Therefore, PCR-based memories are more suitable for sequencing.

DNA chips are also used if the target memory word is amplified with a fluorescent label. Single strands whose sequences are complementary to possible data parts are attached on a DNA chip. Single-stranded copies of the target word with a fluorescent label are then put on the DNA chip. The fluorescent label is observed at the spot having the sequence complementary to the data part.

Fluorescence resonance energy transfer (FRET) is more useful because the target need not be amplified. It can be used if the extracted target word (or its data part) is single-stranded. A molecular beacon is a hairpin structure having a fluorescent group at its end and a quencher group at the other end (❷ *Fig. 19*). If the data part is complementary to the loop sequence of the beacon, the hairpin is opened and the beacon hybridizes with the data part. Consequently, the fluorescent group and the quencher group part and fluorescence is observed.

In a hybridization-based molecular memory, since the target memory word is not extracted but is still mixed with other memory words, the methods mentioned above (except that of FRET) cannot be used in general.

In the simplest hybridization-based scheme in which each memory word consists of only a nonhierarchical address part, a molecular beacon corresponding to the address part of a memory word can be used to detect the existence of the memory word.

If each memory word contains a data part in addition to a nonhierarchical address part, a single-stranded probe whose sequence is complementary to the address part can be used together with another probe for the data part. Both probes are labeled with different

◻ **Fig. 19**

A molecular beacon. A fluorescent group and a quencher group are attached at the terminals of a small hairpin. Since the stem of the hairpin is short, it can hybridize with the strand whose sequence is complementary to the hairpin loop. After hybridization, the fluorescent group parts from the quencher and fluorescence is observed.

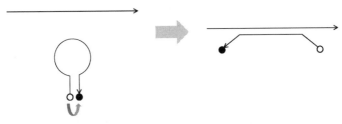

☐ Fig. 20

FRET between an address probe and a data probe. The two probes are complementary to the address part and the data part of a memory word. If they hybridize with the memory word, they come in close proximity and FRET occurs.

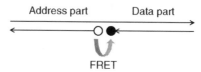

fluorescent groups. If both probes hybridize with the memory word, these groups come in close proximity to each other and energy transfer occurs (❷ *Fig. 20*).

Note that this method cannot be applied to a naïve hierarchical memory in which address digits are simply concatenated. The address digit next to the data part is only significant for energy transfer, and other digits are irrelevant.

The hairpin-based scheme proposed by Hagiya's group solves this problem (Kameda et al. 2008b). In this scheme, the last hairpin is regarded as a data part. Remember that the last hairpin can be opened only after all the preceding hairpins have been opened by their openers. If the last hairpin is opened, since it becomes single-stranded, FRET as in a molecular beacon, can be used to detect this single-stranded part.

So far, we assumed that the sequence (or length) of the data part of a memory word represents the content of the word. In general, however, the read operation of a molecular memory depends on the write operation. For example, in some hybridization-based memory schemes, each memory word consists of only an address part and does not have a data part. The single-stranded and double-stranded states of a memory word are regarded as bit 0 and 1, respectively. If a memory word hybridizes with the corresponding probe and becomes double-stranded, the word is considered to have bit 1, while if it is single-stranded, it is considered to have bit 0. In this kind of scheme, it is not obvious how to read out the single- or double-stranded state of the memory word.

A molecular beacon can be used here because it does not hybridize with the memory word having bit 1, since it is already double-stranded. However, if a beacon hybridizes with the memory word having bit 0, it becomes double-stranded. So this method is destructive. It is also possible to raise the temperature of the whole molecular memory, and collect all the probes attached to memory words. Each probe denotes that the corresponding memory word has had bit 1. This read operation is also destructive. Undestructive read operations require more sophisticated schemes in this kind of hybridization-based memory.

There are some other operations related to the read operation. For example, extraction with respect to the data part is a kind of read operation. They are explained in ❷ Sect. 3.3.

3.3 Write Operation

If existence and nonexistence of a memory word mean bit 1 and bit 0, respectively, then writing bit 1 to an address amounts to simply inserting the memory word of that address into the molecular memory. This operation is always easy. On the other hand, writing bit 0 to an address amounts to deleting the memory word of that address from the molecular memory.

This is easy in a molecular memory based on extraction by affinity separation. It is difficult or impossible to implement the delete operation in a molecular memory based on extraction by PCR or hybridization.

If the data part of a memory word encodes some information, then the write operation should change the data part according to the specified data.

In a molecular memory based on extraction by affinity separation, changing the data part of a memory word having a specified address amounts to deleting the memory word and adding a new memory word having the same address and the specified data. If a molecular memory is based on extraction by PCR, it is also possible to add a new memory word with the specified data, while it is difficult to delete the old memory word with the original data.

In a (nonhierarchical) hybridization-based molecular memory, it is possible to add a new data part to the accessed memory word or delete the original data part. For example, if the new data part is put together with the probe complementary to the address part and the new data part, they form a double strand with the target memory word (without a data part) as in ❷ *Fig. 21*. After ligation is performed, the target word is concatenated with the new data part. On the other hand, it is also possible to cut the target memory word with some restriction enzyme, if it hybridizes with a probe that is complementary to the address part and the adjacent portion of the data part. The double strand formed by the probe and the word is supposed to contain the recognition site of the restriction enzyme (❷ *Fig. 22*).

The hairpin-based scheme proposed by Hagiya's group also allows the write operation in a hierarchical molecular memory (Kameda et al. 2008b). After the last hairpin is opened, it is possible to add a specified data part with the help of an appropriate probe.

As explained in ❷ Sect. 3.2, if a memory word does not have a data part, hybridization between a memory word and its probe is considered as writing bit 1 onto the word. Although writing bit 1 is easy, reading the bit is difficult. Writing bit 0 is also nontrivial. Using branch migration as in molecular tweezers (Yurke et al. 2000), the strand complementary to the probe can remove it from the memory word (❷ *Fig. 23*).

◨ **Fig. 21**

Writing a new data part to a memory word. The new data part is put together with the probe complementary to the address part and the new data part. After ligation, the new data part is concatenated with the address part.

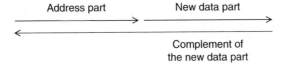

◨ **Fig. 22**

Deleting the data part of a memory word. The probe is complementary to the address part and the adjacent portion of the data part. The double strand formed by the word and the probe is cut by a restriction enzyme.

◘ Fig. 23

Removing a probe from a memory word. The strand complementary to the probe hybridizes with the single-stranded portion of the probe and eventually removes it from the memory word.

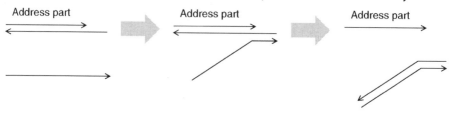

◘ Fig. 24

Writing on an address digit. A probe hybridizes with the address digit 1. This operation is regarded as writing bit 1 to the address digit.

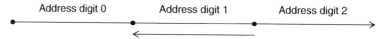

3.4   Other Operations

Due to the features of molecular memories, there are some more operations that can be implemented efficiently.

In a nonhierarchical molecular memory, the address part and the data part are symmetric. Therefore, it is also possible to extract memory words with respect to their data part. This kind of operation is called content-based addressing, and memories that allow such an operation are called associative memories.

In a hierarchical molecular memory, it is also possible to extract memory words based on a specified address digit or on a combination of address digits. For example, when an NP-complete problem like SAT is solved, each candidate of a solution is regarded as a memory word consisting of multiple address digits. Iteration of extraction results in a solution of such a problem. For solving a SAT problem, extraction is performed for each clause of a given Boolean formula. Each extraction operation selects those memory words that contain at least one digit corresponding to a literal in the corresponding clause.

Like extraction, it is also possible to write data onto a specified address digit. For example, in a hybridization-based hierarchical memory, a probe for a particular address digit hybridizes with those memory words having the address digit (❷ *Fig. 24*). In the simplest case of this scheme, each address digit has only one possible sequence, and whether or not a probe hybridizes with an address digit is based on whether the address digit is 1 or 0. In this case, the write operation is considered to change the address of a memory molecule from, for example, 000 to 010 (❷ *Fig. 24*).

As a result of this operation, each memory word contains a certain number of probes that hybridize with one of its address digits. Another possible operation is separation by counting. Using gel electrophoresis, for example, it is possible to separate memory words according to the number of probes they contain.

The two operations mentioned above, writing bit 1 to an address digit and separation by counting, compose the molecular computing paradigm called aqueous computing (Head and Gal 2001). Note that this scheme is different from the scheme described in ❷ Sect. 4.1, where each address digit has two possible sequences.

4 In Vitro DNA Memory

This section briefly touches on molecular memories in an aqueous solution, that is, in a test tube. They are manipulated using the operations mentioned in the previous section.

4.1 Solving NP-Complete Problems

Mainly for solving NP-complete problems, various memory schemes were proposed in the early days of DNA computing.

As mentioned in the previous section, Adleman's group solved a 20-variable SAT problem (Braich et al. 2002). For solving the problem, they constructed a pool consisting of 2^{20} different assignments to 20 Boolean variables. Each assignment can be regarded as a hierarchical memory word with 20 address digits, where each digit has two possible values, true or false. The size of this molecular memory is therefore 2^{20}.

The initial pool of assignments, that is, the whole address space, was constructed in two steps. In the first step, assignments to 10 variables were generated using a mix-and-split synthesis technique (Faulhammer et al. 2000). They used two tubes (or columns) beginning with synthesizing the two sequences for the tenth variable. The results were mixed and then separated into two again. The two sequences for the ninth variable were then attached. The assignments to the first 10 variables and those to the rest were combined using PCR (Stemmer et al. 1995).

The only memory operation employed for solving the problem is extraction with respect to a combination of address digits. For each clause in a given Boolean expression, those memory words that contain an address digit corresponding to a literal in the clause are extracted. If there remain memory words after all the extraction operations are performed, the formula is known to be satisfiable.

4.2 Nested Primer Molecular Memory (NPMM)

Yamamoto et al. constructed a molecular memory consisting of 16^6 addresses (Yamamoto et al. 2008). Their molecular memory is called NPMM, which stands for nested primer molecular memory. Each memory word consists of a data part and an address part. The data part is surrounded by the address part. Each address digit to the left of the data part is paired with a digit to the right, and they comprise a primer pair. In the 16^6-version of NPMM, there are six address digits, denoted by AL, AR, BL, BR, CL, and CR. The digits AL, BL, and CL are paired with AR, BR, and CR, respectively. The sequence CL is put to the leftmost end of a word and CR is put to the rightmost end. They comprise the first primer pair to extract a memory word. The sequence BL is put to the right of CL, and BR is put to the left of CR. The sequences AL and AR are put to the left and right of the data part. Therefore, each memory word having the data part, Data, is of the form

$$CL - BL - AL - Data - AR - BR - CR$$

These memory words are prepared as double strands of DNA. It suffices to repeat three PCR experiments to extract a single memory word.

In the 16^6-version of NPMM, 16 sequences were prepared for each address digit, denoted by ALi, ARi, BLi, etc. Consequently, the entire address space consists of $16^6 = 16,777,216$ addresses. In addition to the 16×6 sequences for the address digits, six linker sequences were inserted between adjacent address digits and between the data part and the address part. As for the data part, three sequences of different lengths were prepared. In total, 105 different sequences were designed, using a two-step search method, which was developed for this molecular memory.

The entire memory was constructed in a style of parallel overlap assembly (Stemmer et al. 1995) as in ❷ *Fig. 25*. The common linker sequences were used as primers in PCR.

In order to verify that the access operation was correctly implemented, Yamamoto et al. picked up one address (CL0, BL0, AL0, AR0, BR0, CR0) and inserted a memory word with a longer data part (data40) with the address. The initial memory consists of memory words having a fixed data part (data20). Similarly, they inserted a memory word with the longest

❑ **Fig. 25**

The construction of NPMM. PCR is iterated four times to construct a molecular memory with six digits.

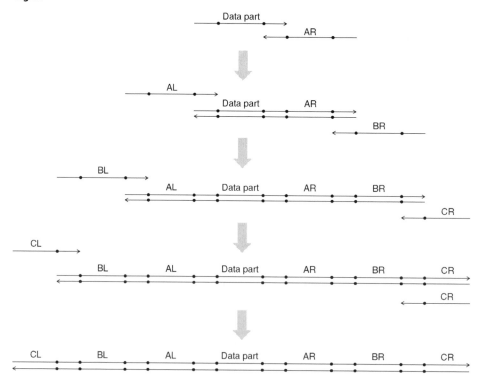

data part (data60) at another address (CL8, BL8, AL8, AR8, BR8, CR8). Then they accessed 16 different addresses (CLi, BLi, ALi, ARi, BRi, CRi) and verified by gel electrophoresis that the access operation was always correct.

One of the most important factors for implementing this kind of molecular memory is to reduce variation of concentrations of memory words. Using real-time PCR, Yamamoto et al. measured the concentrations of the first products in the construction of NPMM, that is, memory words of the form ALi − data20 − ARj. They observed that, compared to the least concentration of a memory word, the largest concentration was 2.36 times larger and the average was 1.4. In the 16^6-version of NPMM with six address digits, it is expected that compared to the least concentration of a memory word, the largest concentration is 173 and the average is 7.53. The molecular memory they actually constructed was supposed to contain 185 molecules for each memory address on average. Therefore, the least number of molecules for a word is expected to be 25.

Yamamoto et al. also estimated the limitation of their approach by analyzing a mathematical model based on the current experimental technologies including PCR. (The numbers in parentheses are their tentative estimations.)

- The total number of molecules should not exceed the number of primers (50 pmol).
- The number of PCR iterations is limited (30 cycles).
- After amplification, the copy number of a target word should be significantly larger than the total number of molecules (1,000 times larger).
- The total number of sequences for address digits is limited (200 sequences).

Using the model, they estimated that it was not possible to construct a molecular memory having more than 170M addresses. In this sense, the molecular memory they constructed was comparable to this limit.

As a possible application of a molecular memory with a huge address space, Hagiya's group proposed to use a molecular memory for DNA ink (Kameda et al. 2008a). By simply diluting a molecular memory, the total number of molecules becomes less than the total number of addresses. Consequently, each memory address may or may not exist in the diluted solution. This amounts to tossing a coin for each address. The diluted solution can be amplified with common primers, and if the primers are removed after amplification, it is practically impossible to completely determine the contents of the memory. Therefore, it can be used for authentication or as a seed for key generation. For example, it can be mixed with ink and used for printing passports or signing secret documents. It is possible to obtain more and more information from the memory by sampling more and more addresses by PCR.

5   DNA Memory on Surfaces

5.1   Introduction

A DNA molecule is a molecular memory stably storing genetic information. The information is described in a DNA base sequence and is accurately read out when DNA is replicated or RNA is transcribed. The specificity of the Watson–Crick base pairing between complementary DNA strands realizes high fidelity in replication and transcription. The specificity of the DNA base pairing is also utilized to construct structures, devices, and computers working in a nanometer scale (Bath and Turberfield 2007; Seeman and Lukeman 2005; Simmel and Dittmer

2005), which are actively developed today: DNA nanostructures (He et al. 2008; Rinker et al. 2008; Rothemund 2006), DNA nanomachines (Goodman et al. 2008; Venkataraman et al. 2007), DNA computers (Benenson et al. 2004; Komiya et al. 2006; Seelig et al. 2006; Takinoue et al. 2008; Zhang et al. 2007), etc.

The specificity of the DNA base pairing can achieve even rewritable molecular memories (Shin and Pierce 2004; Takinoue and Suyama 2006). This section introduces one of the rewritable DNA molecular memories. This memory is based on physicochemical reactions of hairpin-shaped DNA molecules; that is, a temperature-controlled conformational transition and DNA hybridization of the hairpin DNA molecules achieve memory addressing and data storing (Takinoue and Suyama 2004, 2006). Although, in life systems, a DNA molecule is a read-only and non-rewritable molecular memory storing information as a base sequence, the physicochemical reactions of the hairpin DNA can realize not only the readout but also the rewriting of the DNA molecular memory. By cooperation between the base sequence information and the physicochemical reactions, the hairpin DNA memory achieves massively parallel and highly selective addressing in a very large memory space.

5.2 Hairpin DNA Memory Dissolved in a Solution

First, the hairpin DNA memory dissolved in a solution is explained. A hairpin DNA memory has two states: a single-stranded hairpin structure (unwritten state, state I of ❷ Fig. 26a) and a double-stranded linear structure (written state, state II of ❷ Fig. 26a). A memory strand has a hairpin DNA sequence with a memory address in its loop region. A data strand is composed of an address part and a content part. The address part is a linear DNA sequence, whereas the data part can be composed of various substances, such as DNA/RNA sequences, proteins, metal nanoparticles, or quantum dots. The DNA sequence of the address part of the data strand is complementary to that of the hairpin loop of the memory strand, so that the memory and the corresponding data strands can hybridize with each other by base pairing between the

❑ **Fig. 26**
A hairpin DNA memory dissolved in a solution. (a) Structures of memory and data. State I: unwritten state (hairpin structure). State II: written state (linear structure). (b) Writing and erasing by temperature control. T_R, T_W, and T_E: temperatures of readout, writing, and erasing ($T_R < T_W < T_E$). State III: dissociated state.

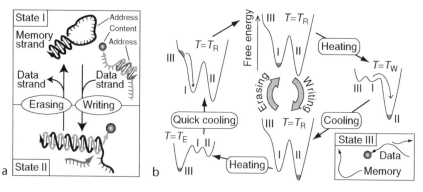

complementary sequences. The address part of the data strand partly forms a double helix with an auxiliary DNA strand to prevent the data strand from spontaneously hybridizing with the memory strand.

The writing and erasing of the hairpin-DNA memory are performed by temperature control of a solution containing memory and data strands. ❷ *Figure 26b* illustrates the mechanism of memory writing and erasing. Data is written to the memory through a state transition from state I to state II. When a solution containing the memory and the data strands is exposed to a writing temperature (T_W), the hairpin structure of the memory strand opens and the data strand is released from the auxiliary strand, and therefore the data strand hybridizes with the memory strand. Conversely, data is erased through a state transition from state II to state I. When the solution is quickly cooled down to a readout temperature (T_R) after being exposed to an erasing temperature (T_E), the hairpin structure of the memory strand closes before the data strand hybridizes with it again. The intramolecular hairpin formation of the memory strand is faster than the intermolecular hybridization between the memory and the data strands. Once the memory and the data strands are separated, they can hardly hybridize with each other until the solution temperature is raised to T_W.

The writing and erasing of the hairpin DNA memory can be performed even when multiple memory addresses coexist in a memory solution. ❷ *Figure 27* shows the experimental results of

◘ Fig. 27
Alternate repetition of writing and erasing of hairpin DNA memory. M1, M2: Memory 1, 2. D1, D2: Data 1, 2. (a) Selective writing and erasing of M1-D1, and the enlarged view. (b) Selective writing and erasing of M2-D2. (c) Parallel writing and erasing of M1-D1 and M2-D2. T_R=25°C, T_W = 50°C, T_E = 90°C. Experimental detail: reference Takinoue and Suyama (2006).

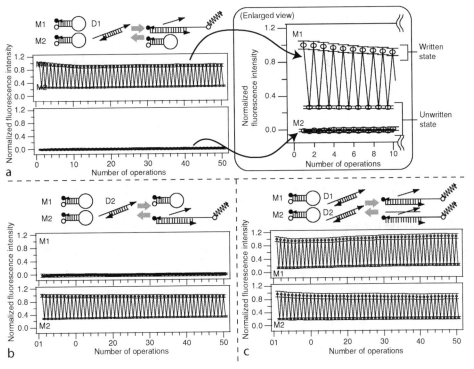

selective and parallel writing and erasing of the hairpin DNA memory in the case of the coexistence of two memory addresses. In ❯ *Fig. 27a* and ❯ *b*, Data 1 (D1) and Data 2 (D2) are selectively written to the memories with the corresponding address. ❯ *Figure 27c* shows that D1 and D2 are written to the memories with the corresponding address in parallel. The hairpin DNA memory works very stably as shown in the 50-times repeated writing and erasing operations in ❯ *Fig. 27*.

5.3 Hairpin DNA Memory Immobilized on a Surface

The functionality of the hairpin DNA memory can be extended by immobilizing hairpin DNA memory strands onto a solid surface. The immobilization allows spatial addressing of the memory by physically positioning memory locations on the surface, as well as molecular addressing of the memory by hybridization between memory and data strands (❯ *Fig. 28*). In the memory substrate of ❯ *Fig. 28a*, the memory strand is uniformly immobilized on the surface of a thin gold layer via a thiol-Au bond. The spatial addressing is achieved through local heating of the memory substrate by IR laser beam irradiation. No irradiation, weak irradiation, and strong irradiation of the IR laser beam correspond to the temperature controls T_R, T_W, and T_E in ❯ *Fig. 26b*. On the other hand, molecular addressing is performed through hybridization of a data strand with the immobilized memory strand. The reaction is the same as the reaction shown in ❯ *Fig. 26a*.

❯ *Figure 28b* shows a fluorescence microscope image of a hairpin DNA memory immobilized on a surface. The white and black areas indicate the written and unwritten states,

◘ **Fig. 28**
A hairpin DNA memory immobilized on a surface. (a) Hairpin DNA memories are immobilized on a gold surface. Writing and erasing are performed by IR laser beam irradiation.
(b) A fluorescence microscope image of the hairpin DNA memory. White area: written state. Black area: unwritten state.

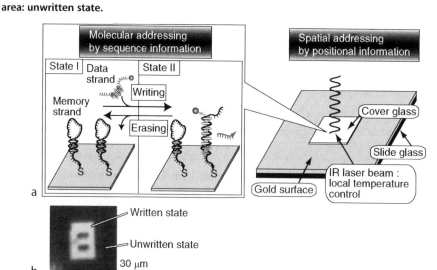

respectively. Local irradiation with the IR laser beam allows data storing with microscale spatial patterning. The memory addressing is not limited only to a single memory address. Selective and parallel writing and erasing for multiple memory addresses can also be realized. In addition, this system is stable for many repetitions of the writing and erasing.

5.4 Conclusion

This section explained the hairpin DNA memory and its immobilization on a solid surface. Hybridization of data strands with memory strands and the structural transition of hairpin-shaped memory strands realize molecular addressing and data storing to the memory. Immobilization makes possible spatial addressing to the memory using locations on the surface.

Hairpin DNA memory immobilized on a solid surface has a lot of possibilities for applications in micro and nanotechnology. In information technology, it will provide a large amount of memory space and massively parallel addressing to store plenty of information, especially molecular information such as molecular computing outputs. It will also provide a universal plate for assembling DNA nanostructures, nanoparticles, and other functional molecules by both molecular addressing and spatial addressing. The memory substrate can be extended to a fundamental technique for a molecular imaging plate to directly store the spatial distribution of mRNAs and proteins in tissues and cells. No silicon-based memory devices can succeed in these applications.

6 In Vivo DNA Memory

6.1 Introduction

The perfect design of in vivo memory requires substantial knowledge of cellular and molecular mechanisms. Ideally, the memory function should be isolated from the native physiological functions to avoid unexpected interactions among sequences. However, it is difficult to meet this requirement, especially if memory mechanisms are implemented with standard nucleic acids. Knowledge of the mechanisms of DNA transcription or translation remains elementary; previous attempts to computationally estimate transcribed regions (or gene structures) never matched the findings of tiling arrays, that as much as 70% of the human genome is transcribed (Kapranov et al. 2002). It is also unknown why bacterial genomes can detect and excise foreign DNA sequences with significantly biased GC contents. Due to these difficulties, in vivo memories have not been as widely investigated as in vitro approaches.

6.2 Encryption into Non-coding Regions

The simplest way to encode information into DNA is to apply a code table translating alphabet letters and digits to DNA triplets. ❷ *Table 2* was used by Clelland et al. to encode the message "June 6 Invasion: Normandy" (all in capitals) into a short DNA fragment in vitro (Clelland et al. 1999). A similar translation table was used in other approaches, such as encoding a song phrase of "It's a small world" into the genomes of *Escherichia coli* and *Deinococcus radiodurans* (Leier et al. 2000; Wong et al. 2003). This straightforward method has a fatal flaw; DNA

◘ **Table 2**

Direct encryption table used in Clelland et al. (1999)

A = CGA	B = CCA	C = GTT	D = TTG	E = GGT	G = TTT	H = CGC	I = ATG
J = AGT	K = AAG	L = TGC	M = TCC	N = TCT	O = GGC	Q = AAC	R = TCA
S = ACG	T = TTC	U = CTG	V = CCT	W = CCG	X = CTA	Y = AAA	_ = ATA
, = TCG	. = GAT	: = GCT	U^a = ACT	1 = ACC	2 = TAG	3 = GCA	4 = GAG
5 = AGA	7 = ACA	8 = AGG	9 = GCG				

[a] In the original paper, symbols F, P, Z, 6 are missing and U is duplicated.

◘ **Table 3**

An example of comma-free words of DNA triplets that also consider reverse strands. The maximum number of words is 10

ACA	AGA	AAU	ACC	CGA	ACU	GCA	AGG	AGU	CGG

sequences are retrieved by polymerase chain reaction (PCR) using short primer sequences, but such sequences may appear in the encoded message. It is an absolute requirement for primer sequences and their reverse complements to never occur in any combination of encrypted sequences irrespective of frame shifts. This condition is met by the design of comma-free codes.

A code is said to be comma-free if separator symbols are unnecessary to parse its words unambiguously. The concept of comma-freeness was first introduced by Francis Crick, as an elegant (but mistaken) association between codons and 20 amino acids (Crick et al. 1957). Indeed, the maximum number of comma-free DNA triplets is exactly 20 out of $4^3 = 64$. (It is easy to say that in this association there is no room for the "stop" sign. It is probably because we now know the correct codon table.)

How many comma-free words can be designed with DNA quadruplets or pentuplets? Obviously, any circular permutation of a comma-free word cannot be another word in the same code. Soon after Crick offered his beautiful hypothesis, Solomon Golomb proved that the maximum number of comma-free words of length k using n letters is $(n^k - n)/k$ (k: prime) (Golomb et al. 1958). When the comma-free condition for reverse complements is also considered, the number is halved (❷ *Table 3*). Since encryption of alphabet letters, digits, and primer sequences requires at least 40 letters, the necessary comma-free words can be no shorter than DNA pentuplets.

DNA pentuplets may raise different types of complication. The first is a larger bias for GC content, resulting in a broader range of melting temperatures. Since the melting temperature should be the same for all primer sequences, words for primers must be carefully chosen from comma-free DNAs. The second complication is vulnerability to mutations or sequencing errors. Enhancement of the error-correcting ability further lengthens the DNA words, and practical codes may become as long as 12 base-pairs (Arita and Kobayashi 2002). The third and the most serious complication is the possible interaction with native genome sequences. Especially in eukaryotic cells, any noncoding or intergenic region may be functional and it is impossible to guarantee the safety of inserting a long sequence into their genome.

6.3 Encryption into Coding Regions

A better way to achieve in vivo memories is to override artificial information on sequences whose biological information is well known. If we can superimpose a message in protein coding regions in a way that the message does not affect amino acid sequences to be translated, then the message can be used as a secret signature, or *DNA watermark*. This idea can be realized easily.

The translation table of amino acids specifies only 21 patterns (20 amino acids and the stop sign) using the 64 codons; the third position of the codons for frequently used amino acids may change without changing the translation result. This is called a *wobbled* codon (❿ *Table 4*).

Arita and Ohashi proposed association of a single bit for each codon where a value of 0 indicates the presence of the normal (wild-type) bases at the third position, and a value of one indicates altered bases at this position (wobbled codon by mutation) (Arita and Ohashi 2004).

In this scheme, the normal sequence of an organism is considered to carry information 000...0, and once some of its codons are wobbled, those positions are interpreted as one. The decoding of information requires the sequencing of the protein coding regions, for example, using PCR. Its obvious disadvantage is therefore little storage capacity; the sequence length that can be PCR-amplified is limited.

When wobbling between base U↔C and base A↔G is utilized in the standard codon table (❿ *Table 4*), a single bit can be overridden for 60 codons out of 64. Exceptions are UGA, UGG, AUA, and AUG, whose wobbling counterparts code different amino acids or a stop codon. Such codons are skipped in the encoding. For example, if alphabet letters and digits are encoded using the ASCII code (7 bits), one character can be watermarked in approximately

◘ Table 4

Codon usage chart of a standard eukaryotic genome. "Stop" indicates stop codons. Bold letters are codons that cannot be used for encoding artificial information (see main text)

UUU	UUA	UAU	UAA	GAA	GAU	GUG	GUU
Phe	Leu	Tyr	Stop	Glu	Asp	Val	Val
UUC	UUG	UAC	UAG	GAG	GAC	GUA	GUC
Phe	Leu	Tyr	Stop	Glu	Asp	Val	Val
UCU	UCA	UGU	**UGA**	GGA	GGU	GCA	GCU
Ser	Ser	Cys	**Stop**	Gly	Gly	Ala	Ala
UCC	UCG	UGC	**UGG**	GGG	GGC	GCG	GCC
Ser	Ser	Cys	**Trp**	Gly	Gly	Ala	Ala
CCC	CCG	CGC	CGG	AGG	AGC	ACG	ACC
Pro	Pro	Arg	Arg	Arg	Ser	Thr	Thr
CCU	CCA	CGU	CGA	AGA	AGU	ACA	ACU
Pro	Pro	Arg	Arg	Arg	Ser	Thr	Thr
CUC	CUA	CAC	CAG	AAG	AAC	**AUA**	AUC
Leu	Leu	His	Gln	Lys	Asn	**Ile**	Ile
CUU	CUG	CAU	CAA	AAA	AAU	**AUG**	AUU
Leu	Leu	His	Gln	Lys	Asn	**Met**	Ile

□ **Table 5**

Translation from binaries to DNA bases

00 → T	01 → G	10 → C	11 → A

every $7 \times 3 = 21$ base-pairs (this is an approximation because we may need to skip some codons). Using 6 bits for each letter, Arita and Ohashi demonstrated the encoding of "KEIO," the name of their university, into an essential gene of *Bacillus subtilis* (*ftzZ*). The watermarked bacterium did not exhibit any superficial anomaly with respect to colony morphology, growth rate, and the level of protein expression.

Their idea was later extended by Heider and Barnekow (2007). Instead of associating a single bit for all codons, they first translate binary information into DNA bases according to ❯ *Table 5*, and the DNA bases are embedded at the third position of codons one by one. If the substitution produces a wobbled codon, the embedding takes place. If it does not (i.e., if it alters the amino acid in ❯ *Table 4* or if it is identical to the native sequence), the codon is skipped and the next codon is tested. This method can assign 2 bits for each embedded letter as in ❯ *Table 5*, although the total length of the embedded sequence is information- and genome-dependent: every time a base to be embedded is identical to the third position of the target codon, or if the substitution does not produce a wobbled codon, then the encryption span becomes one codon longer. The actual encoding length is therefore unpredictable.

They also proposed the use of several cryptographic algorithms and error-correcting codes in the translation of the original information into the binary bits. Although they did not demonstrate this error-correcting mechanism, they tested the encoding method in the *Vam7* gene of *Saccharomyces cerevisiae*, a diploidal eukaryote (Heider and Barnekow 2008). The watermarked yeast again exhibited no superficial anomalies including sporulation. The issue of sexual reproduction with recombination, a notable characteristic of eukaryotic cells, was not addressed. These authors subsequently proposed the use of Y-chromosomal or mitochondrial sequences in future trials (Heider and Kessler 2008).

6.4   Conclusion

The use of error-correcting codes or cryptographic algorithms such as RSA is essentially irrelevant to the implementation of DNA memories because the technology under focus is to address how and where to encode information in genomic DNA. We must remember a simple way of implementing the error-correcting ability, that is, the encoding of any of the above code multiple times at different locations in a genome. In fact, using *Bacillus subtilis*, Yachie et al. demonstrated an application of a multiple sequence alignment for embedded memory fragments, each of which was retrieved by PCR experimentally (Yachie et al. 2007). They also proposed whole genome sequencing for retrieving memory sequences. Having seen the recent advances in high-speed genome sequencing, we no longer dismiss such a proposal as unrealistic.

Research on synthetic biology may lead to another type of information storage in vivo. Gardner et al. used two DNA-binding regulatory genes, each of which represses the other, to implement a bistable, genetic toggle switch (Gardner et al. 2000). This switch can be regarded as a single-bit memory although the reliability of keeping the memory bit under cell replication is not clear. In their system, toggling was performed by externally adding a chemical agent that blocks the DNA-binding ability of a protein used. Although their approach is obviously not scalable, it exemplifies the advantage of read-write, visual memory. Since it is not difficult to visualize the cellular state by using several different fluorescent proteins, one interesting application area would be scalable, multi-bit, visual in vivo memory.

7 Conclusion

We hope the reader now realizes that there exist a variety of models and methods to store information in terms of molecules. In particular, types of DNA memories range from those in solution or on surfaces to those stored in vivo. However, they all require careful design of base sequences of DNA. In particular, DNA memories in solution and on surfaces rely on selective hybridization between complementary sequences of DNA, and efforts to construct a reasonably large memory space in those types have led to the theory and practice of sequence design explained in ❷ Sect. 2. Such theory and practice can also be applied to molecular machines or nanostructures constructed from DNA.

Many of the achievements reported in this chapter, including NPMM and hairpin-based memories, were supported by the JST-CREST project on molecular memory.

References

Aboul-ela F, Koh D, Tinoco I, Martin F (1985) Base-base mismatches. Thermodynamics of double helix formation for dCA3XA3G + dCT3YT3G (X, Y = A,C, G,T). Nucleic Acids Res 13:4811–4824

Adleman L (1994) Molecular computation of solutions to combinatorial problems. Science 266:1021–1024

Allawi H, SantaLucia J (1997) Thermodynamics and NMR of internal G.T mismatches in DNA. Biochemistry 36:10581–10594

Allawi H, SantaLucia J (1998a) Nearest-neighbor thermodynamics of internal A.C mismatches in DNA: sequence dependence and pH effects. Biochemistry 37:9435–9444

Allawi H, SantaLucia J (1998b) Thermodynamics of internal C.T mismatches in DNA. Nucleic Acids Res 26:2694–2701

Allawi H, SantaLucia J (1998c) Nearest neighbor thermodynamic parameters for internal G.A mismatches in DNA. Biochemistry 37:2170–2179

Andronescu M, Aguirre-Hernandez R, Condon AE, Hoos HH (2003) RNAsoft: a suite of RNA secondary structure prediction and design software tools. Nucleic Acids Res 31:3416–3422

Andronescu M, Fejes AP, Hutter F, Hoos HH, Condon AE (2004) A new algorithm for RNA secondary structure design. J Mol Biol 336:607–624

Arita M, Kobayashi S (2002) DNA sequence design using templates. New Gen Comput 20:263–277

Arita M, Ohashi Y (2004) Secret signatures inside genomic DNA. Biotechnol Prog 20:1605–1607

Arita M, Nishikawa A, Hagiya M, Komiya K, Gouzu H, Sakamoto K (2000) Improving sequence design for DNA computing. In Proceedings of the genetic and evolutionary computation conference (GECCO-2000) Orlando, pp 875–882

Bath J, Turberfield AJ (2007) DNA nanomachines. Nat Nanotechnol 2:275–284

Baum EB (1995) Building an associative memory vastly larger than the brain. Science 268:583–585

Benenson Y, Gil B, Ben-Dor U, Adar R, Shapiro E (2004) An autonomous molecular computer for logical control of gene expression. Nature 429: 423–429

Bommarito S, Peyret N, SantaLucia J (2000) Thermodynamic parameters for DNA sequences with dangling ends. Nucleic Acids Res 28:1929–1934

Borer P, Dengler B, Tinoco I, Uhlenbeck O (1974) Stability of ribonucleic acid double-stranded helices. J Mol Biol 86:843–853

Braich RS, Chelyapov N, Johnson C, Rothemund PWK, Adleman L (2002) Solution of a 20-variable 3-SAT problem on a DNA computer. Science 296:499–502

Braich RS, Johnson C, Rothemund PWK, Hwang D, Chelyapov N, Adleman L (2000) Solution of a satisfiability problem on a gel-based DNA computer.

Proceedings of the 6th International Workshop on DNA-Based Computers, LNCS 2054: Berlin, pp 27–42

Brenneman A, Condon AE (2002) Strand design for bio-molecular computation. Theor Comput Sci 287: 39–58

Clelland CT, Risca V, Bancroft C (1999) Hiding messages in DNA microdots. Nature 399:533–534

Crick FH, Griffith JS, Orgel LE (1957) Codes without commas. Proc Natl Acad Sci USA 43:416–421

Cukras A, Faulhammer D, Lipton R, Landweber L (1999) Chess games: a model for RNA based computation. Biosystems 52:35–45

Deaton R, Murphy RC, Rose JA, Garzon M, Franceschetti DT, Stevens SEJ (1996) Genetic search for reliable encodings for DNA-based computation. In First genetic programming conference, Stanford, pp 9–15

Deaton R, Murphy RC, Garzon M, Franceschetti DR, Stevens SEJ (1998) Good encodings for DNA-based solutions to combinatorial problems. Proceedings of the second annual meeting on DNA based computers, DIMACS: series in discrete mathematics and theoretical computer science, vol 44, Princeton, pp 247–258

Dirks RM, Lin M, Winfree E, Pierce NA (2004) Paradigms for computational nucleic acid design. Nucleic Acids Res 32:1392–1403

Faulhammer D, Cukras A, Lipton R, Landweber L (2000) Molecular computation: RNA solutions to chess problems. Proceedings of the Natl Acad Sci USA 97:1385–1389

Feldkamp U, Saghafi S, Rauhe H (2001) DNASequence-Generator – a program for the construction of DNA sequences. Proceedings 7th International workshop on DNA-based computers, LNCS 2340, Tampa, pp 23–32

Frutos A, Liu Q, Thiel A, Sanner A, Condon AE, Smith L, Corn RM (1997) Demonstration of a word design strategy for DNA computing on surfaces. Nucleic Acids Res 25:4748–4757

Gardner TS, Cantor CR, Collins JJ (2000) Construction of a genetic toggle switch in *Escherichia coli*. Nature 403:339–342

Garzon M, Deaton R, Neather P, Franceschetti DR, Murphy RC (1997) A new metric for DNA computing. In Proceedings of the 2nd Genetic Programming Conference, San Mateo, pp 472–478

Garzon M, Deaton R, Nino L, Stevens SEJ, Wittner M (1998) Encoding genomes for DNA computing. In Proceedings of the 3rd Genetic Programming Conference, San Mateo, pp 684–690

Golomb S, Gordon B, Welch L (1958) Comma-free codes. Can J Math 10:202–209

Goodman RP, Heilemann M, Doose S, Erben CM, Kapanidis AN, Turberfield AJ (2008) Reconfigurable, braced, three-dimensional DNA nanostructures. Nat Nanotechnol 3:93–96

Hagiya M, Arita M, Kiga D, Sakamoto K, Yokoyama S (1999) Towards parallel evaluation and learning of boolean μ-formulas with molecules. DIMACS Series Discrete Math Theor Comput Sci 48:57–72

He Y, Ye T, Su M, Zhang C, Ribbe AE, Jiang W, Mao C (2008) Hierarchical self-assembly of DNA into symmetric supramolecular polyhedra. Nature 452:198–201

Head T, Gal S (2001) Aqueous computing: writing into fluid memory. Bull EATCS 75:190–198

Heider D, Barnekow A (2007) DNA-based watermarks using the DNA-crypt algorithm. BMC Bioinformatics 8:176

Heider D, Barnekow A (2008) DNA watermarks: a proof of concept. BMC Mol Biol 9:40

Heider D, Kessler D, Barnekow A (2008) Watermarking sexually reproducing diploid organisms. Bioinformatics 24:1961–1962

Hofacker IL (2003) Vienna RNA secondary structure server. Nucleic Acids Res 31:3429–3431

Kameda A, Kashiwamura S, Yamamoto M, Ohuchi A, Hagiya M (2008a) Combining randomness and a high-capacity DNA memory. 13th International meeting on DNA computing, DNA13, LNCS 4848, Berlin, pp 109–118

Kameda A, Yamamoto M, Ohuchi A, Yaegashi S, Hagiya M (2008b) Unravel four hairpins! Nat Comput 7:287–298

Kapranov P, Cawley SE, Drenkow J, Bekiranov S, Strausberg RL, Fodor SPA, Gingeras TR (2002) Large-scale transcriptional activity in chromosomes 21 and 22. Science 296:916–919

Kashiwamura S, Kameda A, Yamamoto M, Ohuchi A (2004) Two-step search for DNA sequence design. IEICE TRANSACTIONS Fundam Electron Commun Comput Sci E87-A:1446–1453

Kawashimo S, Ng YK, Ono H, Sadakane K, Yamashita M (2009) Speeding up local-search type algorithms for designing DNA sequences under thermodynamical constraints. Proceedings of the 14th International meeting on DNA computing, LNCS 5347, Prague, pp 168–179

Kawashimo S, Ono H, Sadakane K, Yamashita M (2008) Dynamic neighborhood searches for thermodynamically designing DNA sequence. 13th International meeting on DNA computing, DNA13, LNCS 4848, Memphis, pp 130–139

Kitajima T, Takinoue M, Shohda K, Suyama A (2008) Design of code words for DNA computers and nanostructures with consideration of hybridization kinetics. 13th International meeting on DNA computing, DNA13, LNCS 4848, Berlin, pp 119–129

Kobayashi S, Kondo T, Arita M (2003) On template method for DNA sequence design. Proceedings of the 8th International workshop on DNA-based computers, LNCS 2568, Berlin, pp 205–214

Komiya K, Sakamoto K, Kameda A, Yamamoto M, Ohuchi A, Kiga D, Yokoyama S, Hagiya M (2006) DNA polymerase programmed with a hairpin DNA incorporates a multiple-instruction architecture into molecular computing. Biosystems 83:18–25

Kubota M, Hagiya M (2005) Minimum basin algorithm: an effective analysis technique for DNA energy landscapes. Proceedings of the 10th International workshop on DNA computing, LNCS 3384, New York, pp 202–214

Leier A, Richter C, Banzhaf W, Rauhe H (2000) Cryptography with DNA binary strands. Biosystems 57:13–22

Li M, Lee HJ, Condon AE, Corn RM (2002) DNA word design strategy for creating sets of non-interacting oligonucleotides for DNA microarrays. Langmuir 18:805–812

Li Y, Agrawal S (1995) Oligonucleotides containing G.A pairs: effect of flanking sequences on structure and stability. Biochemistry 34:10056–10062

Li Y, Zon G, Wilson W (1991) Thermodynamics of DNA duplexes with adjacent G.A mismatches. Biochemistry 30:7566–7572

Lipton RJ (1995) DNA solution of hard computational problems. Science 268:542–545

Liu W, Wang S, Gao L, Zhang F, Xu J (2003) DNA sequence design based on template strategy. J Chem Inf Comput Sci 43:2014–2018

Lyngso R, Zuker M, Pedersen C (1999) Fast evaluation of internal loops in RNA secondary structure prediction. Bioinformatics 15:440–445

Marathe A, Condon AE, Corn RM (2001) On combinatorial DNA word design. J Comput Biol 8:201–219

Pancoska P, Moravek Z, Moll UM (2004) Rational design of DNA sequences for nanotechnology, microarrays and molecular computers using Eulerian graphs. Nucleic Acids Res 32:4630–4645

Peyret N, Seneviratne P, Allawi H, SantaLucia J (1999) Nearest-neighbor thermodynamics and NMR of DNA sequences with internal A.A, C.C, G.G, and T.T mismatches. Biochemistry 38:3468–3477

Rinker S, Ke Y, Liu Y, Chhabra R, Yan H (2008) Self-assembled DNA nanostructures for distance-dependent multivalent ligand-protein binding. Nat Nanotechnol 3:418–422

Rothemund PWK (2006) Folding DNA to create nanoscale shapes and patterns. Nature 440:297–302

Ruben AJ, Freeland SJ, Landweber LF (2001) PUNCH: An evolutionary algorithm for optimizing bit set selection. Proceedings of the 7th International workshop on DNA-based computers, LNCS 2340, Tampa, pp 150–160

SantaLucia J (1998) A unified view of polymer, dumbbell, and oligonucleotide DNA nearest-neighbor thermodynamics. Proc Natl Acad Sci USA 95:1460–1465

SantaLucia J, Allawi H, Seneviratne P (1996) Improved nearest-neighbor parameters for predicting DNA duplex stability. Biochemistry 35:3555–3562

Seelig G, Soloveichik D, Zhang DY, Winfree E (2006) Enzyme-free nucleic acid logic circuits. Science 314:1585–1588

Seeman N (1990) De novo design of sequences for nucleic acid structural engineering. J Biomol Struct Dyn 8:573–581

Seeman NC, Lukeman PS (2005) Nucleic acid nanostructures: bottom-up control of geometry on the nanoscale. Rep Prog Phys 68:237–270

Sen D, Gilbert W (1988) Formation of parallel four-stranded complexes by guanine-rich motifs in DNA and its implications for meiosis. Nature 334:364–366

Sen D, Gilbert W (1990) A sodium-potassium switch in the formation of four-stranded G4-DNA. Nature 344:410–414

Senior M, Jones R, Breslauer K (1988) Influence of dangling thymidine residues on the stability and structure of two DNA duplexes. Biochemistry 27:3879–3885

Shin JS, Pierce NA (2004) Rewritable memory by controllable nanopatterning of DNA. Nano Lett 4:905–909

Simmel FC, Dittmer WU (2005) DNA nanodevices. Small 1:284–299

Stemmer WP, Crameri A, Ha KD, Brennan TM, Heyneker HL (1995) Single-step assembly of a gene and entire plasmid from large numbers of oligodeoxyribonucleotides. Gene 164:49–53

Sugimoto N, Nakano S, Yoneyama M, Honda K (1996) Improved thermodynamic parameters and helix initiation factor to predict stability of DNA duplexes. Nucleic Acids Res 24:4501–4505

Takinoue M, Suyama A (2006) Hairpin-DNA memory using molecular addressing. Small 2:1244–1247

Takinoue M, Suyama A (2004) Molecular reactions for a molecular memory based on hairpin DNA. ChemBio Inform J 4:93–100

Takinoue M, Kiga D, Shohda K, Suyama A (2008) Experiments and simulation models of a basic computation element of an autonomous molecular computing system. Phys Rev E 78:041921

Tanaka F, Kameda A, Yamamoto M, Ohuchi A (2004) Thermodynamic parameters based on a nearest-neighbor model for DNA sequences with a single-bulge loop. Biochemistry 43:7143–7150

Tanaka F, Kameda A, Yamamoto M, Ohuchi A (2005) Design of nucleic acid sequences for DNA computing based on a thermodynamic approach. Nucleic Acids Res 33:903–911

Tulpan D, Hoos H (2003) Hybrid randomised neighbourhoods improve stochastic local search for DNA code design. In: Advances in artificial intelligence: 16th conference of the Canadian society for computational studies of intelligence, Berlin, vol 2671, pp 418–433

Tulpan D, Hoos H, Condon A (2003) Stochastic local search algorithms for DNA word design. Proceedings of the 8th International workshop on DNA-based computers, LNCS 2568, Berlin, pp 229–241

Uejima H, Hagiya M (2004) Secondary structure design of multi-state DNA machines based on sequential structure transitions. Proceedings of the 9th International workshop on DNA-based computers, LNCS 2943, Berlin, pp 74–85

Venkataraman S, Dirks RM, Rothemund PWK, Winfree E, Pierce NA (2007) An autonomous polymerization motor powered by DNA hybridization. Nat Nanotechnol 2:490–494

Watson J, Crick F (1953) Molecular structure of nucleic acids; a structure for deoxyribose nucleic acid. Nature 171:737–738

Wong P, Wong K-K, Foote H (2003) Organic data memory using the DNA approach. Commun ACM 46(1):95–98

Yachie N, Sekiyama K, Sugahara J, Ohashi Y, Tomita M (2007) Alignment-based approach for durable data storage into living organisms. Biotechnol Prog 23:501–505

Yamamoto M, Kashiwamura S, Ohuchi A, Furukawa M (2008) Large-scale DNA memory based on the nested PCR. Nat Comput 7:335–346

Yoshida H, Suyama A (2000) Solution to 3-SAT by breadth first search. DIMACS Ser Discrete Math Theor Comput Sci 54:9–22

Yurke B, Turberfield AJ, Mills AP, Simmel FC, Neumann JL (2000) A DNA-fuelled molecular machine made of DNA. Nature 406:605–608

Zhang DY, Turberfield AJ, Yurke B, Winfree E (2007) Engineering entropy-driven reactions and networks catalyzed by DNA. Science 318:1121–1125

Zuker M, Stiegler P (1981) Optimal computer folding of large RNA sequences using thermodynamics and auxiliary information. Nucleic Acids Res 9:133–148

Zuker M (2003) Mfold web server for nucleic acid folding and hybridization prediction. Nucleic Acids Res 31:3406–3415

39 Engineering Natural Computation by Autonomous DNA-Based Biomolecular Devices

John H. Reif[1] · *Thomas H. LaBean*[2]
[1]Department of Computer Science, Duke University, Durham, NC, USA
reif@cs.duke.edu
[2]Department of Computer Science and Department of Chemistry and
Department of Biomedical Engineering, Duke University, Durham,
NC, USA
thomas.labean@duke.edu

G. Rozenberg et al. (eds.), *Handbook of Natural Computing*, DOI 10.1007/978-3-540-92910-9_39,
© Springer-Verlag Berlin Heidelberg 2012

Abstract

This chapter overviews the past and current state of a selected part of the emerging research area of DNA-based biomolecular devices. We particularly emphasize molecular devices that are: *autonomous*, executing steps with no exterior mediation after starting; and *programmable*, the tasks executed can be modified without entirely redesigning the nanostructure.

We discuss work in this area that makes use of synthetic DNA to self-assemble into DNA nanostructure devices. Recently, there has been a series of impressive experimental results that have taken the technology from a state of intriguing possibilities into demonstrated capabilities of quickly increasing scale. We discuss various such programmable molecular-scale devices that achieve: computation, 2D patterning, amplified sensing, and molecular or nanoscale transport.

This article is written for a general audience, and particularly emphasizes the interdisciplinary aspects of this quickly evolving and exciting field.

1 Introduction

1.1 Autonomous Programmable Molecular Devices

This chapter introduces the reader to a rapidly evolving and very exciting topic of nanoscience, namely, self-assembled DNA nanostructures, and their use in molecular computation and molecular transport. To tighten the focus, and to emphasize techniques that seem most promising, the discussion will be limited to processes that are *autonomous* (executing steps with no exterior mediation after starting), and *programmable* (the tasks executed can be modified without entirely redesigning the nanostructure).

1.2 Questions About Biomolecular Devices

We will see that DNA self-assembly processes can be made computational-based and programmable, and it seems likely that computer science techniques will be essential to the further development of this emerging field of biomolecular computation. This chapter particularly illustrates the way in which computer science techniques and methods impact on this emerging field. Some of the key questions one might ask are given below.

- What is the theoretical basis for these devices?
- How will such devices be designed and implemented?
- How can we simulate them prior to manufacture?
- How can we optimize their performance?
- How will such devices be manufactured?
- How much do the devices cost?
- How scalable is the device design?
- How will efficient I/O be accomplished?
- How will they be programmed?

- What efficient algorithms can be programmed?
- What will be their applications? What can they control?
- How can we correct errors or repair them?

Note that these questions are exactly the sort of questions that computer scientists routinely ask about conventional computing devices. The discipline of computer science has developed a wide variety of techniques to address such basic questions, and we will later point out some which have an important impact on molecular-scale devices.

1.3   Why Bottom-Up Self-assembly?

Construction of molecular-scale structures and devices is one of the key challenges facing science and technology in the twenty-first century. This challenge is at the core of the emerging discipline of nanoscience. A key deficiency is the need for robust, error-free methods for self-assembly of complex devices out of large numbers of molecular components. This key challenge requires novel approaches.

For example, the macroelectronics industry is now reaching the limit of miniaturization possible by top-down lithographic fabrication techniques. New bottom-up methods are needed for self-assembling complex, aperiodic structures for nanofabrication of molecular electronic circuits that are significantly smaller than conventional electronics.

1.4   Why Use DNA for Assembly of Biomolecular Devices?

The particular molecular-scale devices that are the topic of this chapter are known as DNA nanostructures. As will be explained, DNA nanostructures have some unique advantages among nanostructures: they are relatively easy to design, fairly predictable in their geometric structures, and have been experimentally implemented in a growing number of labs around the world. They are primarily made up of synthetic DNA. A key principle in the study of DNA nanostructures is the use of self-assembly processes to actuate the molecular assembly. Conventional electronics fabrication is reaching the limit of miniaturization possible by top-down techniques. Since self-assembly operates naturally at the molecular scale, it does not suffer from the limitation in scale reduction that so restricts lithography or other more conventional top-down manufacturing techniques.

In attempting to understand the modern development of DNA self-assembly, it is interesting to recall that mechanical methods for computation date back to the very onset of computer science, for example, to the cog-based mechanical computing machine of Babbage. Lovelace stated in 1843 that Babbage's "Analytical Engine weaves algebraic patterns just as the Jacquard-loom weaves flowers and leaves." In some of the recently demonstrated methods for biomolecular computation described here, computational patterns were essentially woven into molecular fabric (DNA lattices) via carefully controlled and designed self-assembly processes.

1.5   The Dual Role of Theory and Experimental Practice

In many cases, self-assembly processes are programmable in ways analogous to more conventional computational processes. We will overview theoretical principles and techniques (such as tiling assemblies and molecular transducers) developed for a number of DNA self-assembly processes that have their roots in computer science theory (e.g., abstract tiling models and finite state transducers).

However, the area of DNA self-assembled nanostructures and molecular robotics is by no means simply a theoretical topic – many dramatic experimental demonstrations have already been made and a number of these will be discussed. The complexity of these demonstrations has increased at an impressive rate (even in comparison to the rate of improvement of silicon-based technologies).

1.6 Applications of Autonomous Programmable Molecular Devices

This chapter discusses the accelerating scale of complexity of DNA nanostructures (such as the number of addressable pixels of 2D patterned DNA nanostructures) and provides some predictions for the future. Molecular-scale devices using DNA nanostructures have been engineered to have various capabilities, ranging from (1) execution of molecular-scale computation, (2) use as scaffolds or templates for the further assembly of other materials (such as scaffolds for various hybrid molecular electronic architectures or perhaps high-efficiency solar cells), (3) robotic movement and molecular transport (akin to artificial, programmable versions of cellular transport mechanisms), (4) exquisitely sensitive molecular detection and amplification of single molecular events, and (5) transduction of molecular sensing to provide drug delivery. Error-correction techniques for correct assembly and repair of DNA self-assemblies have also been recently developed. Computer-based design and simulation are also essential to the development of many complex DNA self-assembled nanostructures and systems.

1.7 The Operation of Programmable Molecular Devices

While we will describe a wide variety of methods for executing molecular computation, these methods can be partitioned into two basic classes:

- *Distributed parallel molecular computations* execute a computation for which they require multiple distinct molecules that interact to execute steps of each computation. An example is the tiling assembly computations described in ❷ Sect. 5.
- *Local molecular computations* execute computations within a single molecule, possibly in parallel with other molecular computing devices. An example is Whiplash polymerase chain reaction (PCR) described in ❷ Sect. 9.1.

2 Introducing DNA, Its Structure and Its Manipulation

2.1 Introducing DNA

In general, nanoscience research is highly interdisciplinary. In particular, DNA self-assembly uses techniques from multiple disciplines such as biochemistry, physics, chemistry, and material science, as well as computer science and mathematics. While this makes the topic quite intellectually exciting, it also makes it challenging for a typical computer science reader. Having no training in biochemistry, one must obtain a coherent understanding of the topic from a short article. For this reason, this chapter was written with the expectation that the reader is a computer scientist with little background knowledge of chemistry or biochemistry. On the other hand, a reader with a basic knowledge of DNA, its structure and its enzymes can skip this section and proceed to the next.

2.2 DNA and Its Structure

A brief introduction to DNA is given here. *Single stranded DNA* (denoted ssDNA) is a linear polymer consisting of a sequence of nucleotide bases spaced along a backbone with chemical directionality (i.e., the 5-prime and 3-prime ends of the backbone are chemically distinct). By convention, the base sequence is listed starting from the 5-prime end of the polymer and ending at the 3-prime end (these names refer to particular carbon atoms in the deoxyribose sugar units of the sugar-phosphate backbone, the details of which are not critical to the present discussion). The consecutive monomer units (base + sugar + phosphate) of an ssDNA molecule are connected via covalent bonds. There are four types of *DNA* bases: adenine, thymine, guanine, and cytosine typically denoted by the symbols A, T, G, and C, respectively. These bases form the alphabet of DNA; the specific sequence comprises DNA's information content. The bases are grouped into *complementary pairs* (G, C) and (A, T).

The most basic DNA operation is *hybridization* where two ssDNA oriented in opposite directions can bind to form a double-stranded *DNA helix* (dsDNA) by pairing between complementary bases. DNA hybridization occurs in a physiologic-like buffer solution with appropriate temperature, pH, and salinity (❿ *Fig. 1*).

Since the binding energy of the pair (G, C) is approximately half-again the binding energy of the pair (A, T), the association strength of hybridization depends on the sequence of complementary bases, and can be approximated by known software packages. The *melting temperature* of a DNA helix is the temperature at which half of all the molecules are fully hybridized as double helix, while the other half are single-stranded. The kinetics of the DNA hybridization process is quite well understood; it often occurs in a (random) zipper-like manner, similar to a biased one-dimensional random walk.

❑ **Fig. 1**

Structure of a *DNA double helix*. (Image by Michael Ströck and released under the GNU Free Documentation License (GFDL).)

Whereas ssDNA is a relatively floppy molecule, dsDNA is quite stiff (over lengths of less than 150 or so bases) and has the well characterized double helix structure. The exact geometry (angles and positions) of each segment of a double helix depends slightly on the component bases of its strands and can be determined from known tables. There are about 10.5 bases per full rotation on this helical axis. A *DNA nanostructure* is a multimolecular complex consisting of a number of ssDNA that have partially hybridized along their subsegments.

2.3 Manipulation of DNA

Some techniques and known enzymes used for manipulation of DNA nanostructures are listed here.

Strand displacement, which is the displacement of a DNA strand from a prior hybridization with another complementary strand, is a key process in many of the DNA protocols for running DNA autonomous devices. ❯ *Figure 2a* illustrates the displacement of DNA strand via branch migration.

In addition to the hybridization reaction described above, there is a wide variety of known enzymes and other proteins used for manipulation of DNA nanostructures that have predictable effects. (Interestingly, these proteins were discovered in natural bacterial cells and tailored for laboratory use.) These include:

- *Restriction enzymes*, some of which can cut (or nick, which is to cut only one strand) strands of a DNA helix at locations determined by short specific DNA base sequences (❯ *Fig. 3*).
- *Ligase enzymes* that can heal nicks in a DNA helix (i.e., form covalent bonds to join two backbones).
- *Polymerase enzymes*, which, given an initial "primer" DNA strand hybridized onto a segment of a template DNA strand, can extend the primer strand in the 5' to 3' direction by appending free DNA nucleotides complementary to the template's nucleotides (see ❯ *Fig. 2b*). Certain polymerase enzymes (e.g., phi-29) can, as a side effect of their polymerization reaction, efficiently displace previously hybridized strands.
- In addition, *Deoxyribozymes* (*DNAzymes*) is a class of nucleic acid molecules that possess enzymatic activity – they can, for example, cleave specific target nucleic acids. Typically, they are discovered by in vivo evolution search. They have had some use in DNA computations (e.g., see Stojanovic and Stefanovic 2003).

Besides their extensive use in other biotechnology, the above reactions, together with hybridization, are often used to execute and control DNA computations and DNA robotic operations. The restriction enzyme reactions are programmable in the sense that they are site specific, only executed as determined by the appropriate DNA base sequence. The latter two reactions, using ligase and polymerase, require the expenditure of energy via consumption of adenosine triphosphate (ATP) molecules, and thus can be controlled by ATP concentration.

2.4 Why Use DNA to Assemble Molecular-Scale Devices?

Listed below are some reasons why DNA is uniquely suited for the assembly of molecular-scale devices.

◻ **Fig. 2**

(a) (*Left*) Strand displacement of dsDNA via a branch migration hybridization reaction: Figure illustrates DNA strand displacement of a DNA strand (indicated in red) induced by the hybridization of a longer strand (indicated in yellow), allowing the structure to reach a lower energy state. **(b)** (*Right*) Extension of primer strand bound to the template by DNA polymerase.

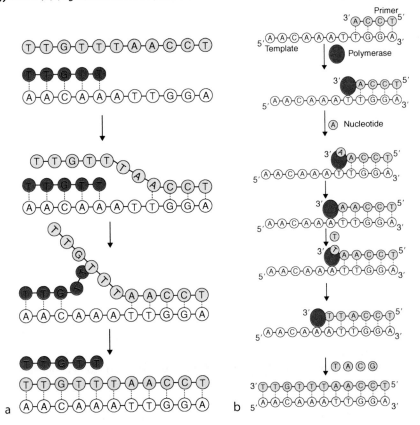

◻ **Fig. 3**

Example of restriction enzyme cuts of a doubly stranded DNA subsequence.

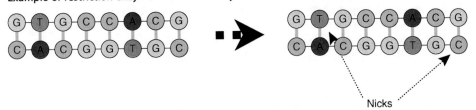

Nicks

There are many advantages of DNA as a material for building things at the molecular scale.

(a) From the perspective of design, the advantages are:

- The structure of most complex DNA nanostructures can be reduced to determining the structure of short segments of dsDNA. The basic geometric and thermodynamic properties of dsDNA are well understood and can be predicted by available software systems from key relevant parameters like sequence composition, temperature, and buffer conditions.
- The design of DNA nanostructures can be assisted by software. To design a DNA nanostructure or device, one needs to design a library of ssDNA strands with specific segments that hybridize to (and only to) specific complementary segments on other ssDNA. There are a number of software systems (developed at NYU, Caltech, Arizona State, and Duke University) for the design of the DNA sequences composing DNA tiles and for optimizing their stability, which employ heuristic optimization procedures for this combinatorial sequence design task (see ❯ Sect. 4.4 for more details).

(b) From the perspective of experiments, the advantages are:

- The solid-phase chemical synthesis of custom ssDNA is now routine and inexpensive; a test tube of ssDNA consisting of any specified short sequence of bases (<150) can be obtained from commercial sources for modest cost (about half a US dollar per base at this time); it will contain a very large number (typically at least 10^{12}) identical ssDNA molecules. The synthesized ssDNA can have errors (premature termination of the synthesis is the most frequent error), but can be easily purified by well-known techniques (e.g., electrophoresis as mentioned below).
- The assembly of DNA nanostructures is a very simple experimental process: in many cases, one simply combines the various component ssDNA into a single test tube with an appropriate buffer solution at an initial temperature above the melting temperature, and then slowly cools the test tube below the melting temperature.
- The assembled DNA nanostructures can be characterized by a variety of techniques. One such technique is electrophoresis. It can provide information about the relative molecular mass of DNA molecules, as well as some information regarding their assembled structures. Other techniques like atomic force microscopy (AFM) and transmission electron microscopy (TEM) provide images of the actual assembled DNA nanostructures on 2D surfaces.

3 Adleman's Initial Demonstration of a DNA-Based Computation

3.1 Adleman's Experiment

The field of DNA computing began in 1994 with a laboratory experiment (Adleman 1994, 1998). The goal of the experiment was to find a Hamiltonian path in a graph, which is a path that visits each node exactly once. To solve this problem, a set of ssDNA was designed based on the set of edges of the graph. When combined in a test tube and cooled, they self-assembled into dsDNA. Each of these DNA nanostructures was a linear DNA helix that corresponded to a path in the graph. If the graph had a Hamiltonian path, then one (or a subset) of these DNA nanostructures encoded the Hamiltonian path. By conventional biochemical extraction methods, Adleman was able to isolate only DNA nanostructures encoding Hamiltonian paths, and by determining their sequence, the explicit Hamiltonian path. It should be mentioned that this landmark experiment was designed and experimentally demonstrated by Adleman alone, a computer scientist with limited training in biochemistry.

3.2 The Non-scalability of Adleman's Experiment

While this experiment founded the field of DNA computing, it was not scalable in practice, since the number of different DNA strands needed increased exponentially with the number of nodes of the graph. Although there can be an enormous number of DNA strands in a test tube (10^{15} or more, depending on solution concentration), the size of the largest graph that could be solved by his method was limited to at most a few dozen nodes. This is not surprising, since finding the Hamiltonian path is an NP-complete problem, whose solution is likely to be intractable using conventional computers. Even though DNA computers operate at the molecular-scale, they are still equivalent to conventional computers (e.g., deterministic Turing machines) in computational power. This experiment taught a healthy lesson to the DNA computing community (which is now well recognized): to carefully examine scalability issues and to judge any proposed experimental methodology by its scalability.

3.3 Autonomous Biomolecular Computation

Shortly following Adleman's experiment, there was a burst of further experiments in DNA computing, many of which were quite ingenious. However, almost none of these DNA computing methods were autonomous, and instead required many tedious laboratory steps to execute. In retrospect, one of the most notable aspects of Adleman's experiment was that the self-assembly phase of the experiment was completely autonomous – it required no exterior mediation. This autonomous property makes an experimental laboratory demonstration much more feasible as the scale increases. The remaining article mostly discusses autonomous devices for biomolecular computation based on self-assembly.

4 Self-assembled DNA Tiles and Lattices

4.1 DNA Nanostructures

Recall that a DNA nanostructure is a multimolecular complex consisting of a number of ssDNA that have partially hybridized along their subsegments. The field of DNA nanostructures was pioneered by Seeman (Robinson and Seeman 1987; Sherman and Seeman 2004).

Particularly useful types of motifs often found in DNA nanostructures include:

A *stem-loop* (a), where ssDNA loops back to hybridize on itself (that is, one segment of the ssDNA (near the 5′ end) hybridizes with another segment further along (nearer the 3′ end) on the same ssDNA strand). The shown stem consists of the dsDNA region with sequence CACGGTGC on the bottom strand. The shown loop consists of the ssDNA region with sequence TTTT. Stem-loops are often used to form patterns on DNA nanostructures: a *sticky end* (b), where unhybridized ssDNA protrudes from the end of a double helix. The sticky end shown (ATCG) protrudes from dsDNA (CACG on the bottom strand). Sticky ends are often used to combine two DNA nanostructures together via hybridization of their complementary ssDNA. ❷ *Figure 4* shows the antiparallel nature of dsDNA with the 5′ end of each strand pointing toward the 3′ end of its partner strand.

A *Holliday junction*, as illustrated in ❷ *Fig. 5*, where two parallel DNA helices form a junction with one strand of each DNA helix (blue and red) crossing over to the other DNA helix. Holliday junctions are often used to tie together various parts of a DNA nanostructure.

◘ Fig. 4

DNA stem-loop and a DNA sticky end.

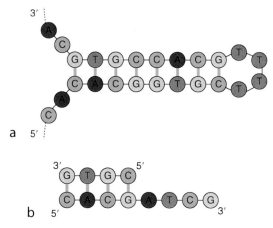

◘ Fig. 5

DNA Holliday junction. (Created by Miguel Ortiz-Lombardía, CNIO, Madrid, Spain.)

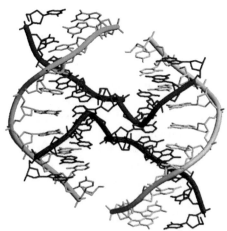

4.2 Computation By Self-assembly

The most basic way that computer science ideas have impacted DNA nanostructure design is via the pioneering work by theoretical computer scientists on a formal model of 2D tiling due to Wang in 1961, which culminated in a proof by Berger in 1966 that universal computation could be done via tiling assemblies. Winfree et al. (1998) was the first to propose applying the concepts of computational tiling assemblies to DNA molecular constructs. His core idea was to use tiles composed of DNA to perform computations during their self-assembly process. To understand this idea, we will need an overview of DNA nanostructures, as presented in ❷ Sect. 4.3.

4.3 DNA Tiles and Lattices

A *DNA tile* is a DNA nanostructure that has a number of sticky ends on its sides, which are termed *pads*. A DNA lattice is a DNA nanostructure composed of a group of DNA tiles that are assembled together via hybridization of their pads. Generally, the strands composing the DNA tiles are designed to have a melting temperature above those of the pads, ensuring that when the component DNA molecules are combined together in solution, first the DNA tiles assemble, and only then, as the solution is further cooled, do the tiles bind together via hybridization of their pads.

❷ *Figure 6* describes some principal DNA tiles (also see LaBean et al. 2006).

Seeman and Winfree, in 1998, developed a family of DNA tiles known collectively as DX tiles (see ❷ *Fig. 6a*) that consisted of two parallel DNA helices linked by immobile Holliday junctions. They demonstrated that these tiles formed large 2D lattices, as viewed by AFM.

Subsequently, other DNA tiles were developed to provide more complex strand topology and interconnections, including a family of DNA tiles known as *TX tiles* (see ❷ *Fig. 6b*) composed of three DNA helices. Both the DX tiles and the TX tiles are rectangular in shape, where two opposing edges of the tile have pads consisting of ssDNA sticky ends of the component strands. In addition, TX tiles have topological properties that allow for strands to propagate in useful ways through tile lattices (this property is often used for aid in patterning DNA lattices as described below).

◻ **Fig. 6**
DNA tiles. (a) DX tile. (b) TX tile. (c) Cross tile. (d) AFM image of 2D DNA lattice of cross tiles.

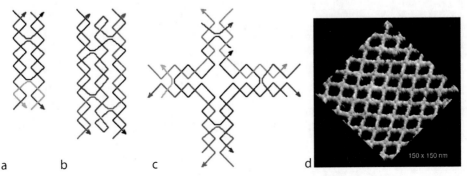

a b c d

Other DNA tiles known as *Cross tiles* (see ❷ *Fig. 6c*) (Yan et al. 2003b) are shaped roughly square (or more accurately, square cruciform), and have pads on all four sides, allowing for binding of the tile directly with neighbors in all four directions in the lattice plane. ❷ *Figure 6d* shows an AFM image of a 2D DNA lattice using Cross tiles.

To program a tiling assembly, the pads of tiles are designed so that tiles assemble together as intended. Proper designs ensure that only the adjacent pads (two pairs of sticky ends in the case of Cross tiles) of neighboring tiles are complementary, so only those pads hybridize together.

4.4 Software for the Design of DNA Tiles

A number of prototype computer software systems have been developed for the design of the DNA sequences composing DNA tiles, and for optimizing their stability. ❷ *Figure 7* gives a screen shot of a software system known as TileSoft (Yin et al. 2004a, b), developed jointly by Duke and Caltech, which provides a graphically interfaced sequence optimization system for designing DNA secondary structures. A more recent commercial product, NanoEngineer, with enhanced capabilities for DNA design and a more sophisticated graphic interface, was developed by Nanorex, Inc.

◼ Fig. 7
TileSoft: sequence optimization software for designing DNA secondary structures.

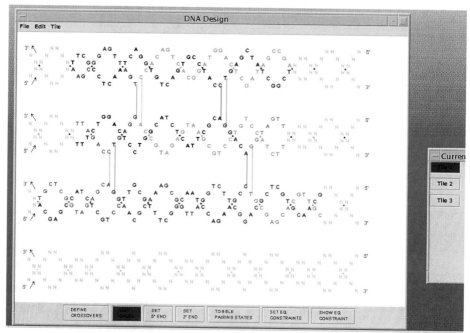

5 Autonomous Finite State Computation Using Linear DNA Nanostructures

5.1 The First Experimental Demonstration of Autonomous Computations Using Self-assembly of DNA Nanostructures

The first experimental demonstrations of computation using DNA tile assembly were done in 1999 (LaBean et al. 1999, 2000; Mao et al. 2000). Among the experiments demonstrated (Mao et al. 2000) was a two-layer, linear assembly of TX tiles that executed a bit-wise cumulative XOR computation. In this computation, n bits are input and n bits are output, where the ith output is the XOR of the first i input bits. This is the computation occurring when one determines the output bits of a full-carry binary adder circuit found on most computer processors. The experiment (Mao et al. 2000) is illustrated in ❷ *Fig. 8*.

These experiments (LaBean et al. 1999, 2000; Mao et al. 2000) provided initial answers to some of the most basic questions of how autonomous molecular computation might be done:

- *How can one provide data input to a molecular computation using DNA tiles?*

In this experiment the input sequence of n bits was defined using a specific series of "input" tiles with the input bits (1's and 0's) encoded by distinct short subsequences. Two different tile types (depending on whether the input bit was 0 or 1, these had specific sticky-ends and also specific subsequences at which restriction enzymes can cut the DNA backbone) were assembled according to specific sticky-end associations, forming the blue input layer illustrated in ❷ *Fig. 8*.

❷ *Figure 8* shows a unit TX tile (a) and the sets of input and output tiles (b) with geometric shapes conveying sticky-end complementary matching. The tiles of (b) execute binary computations depending on their pads, as indicated by the table in (b). The (blue) input layer and (green) corner condition tiles were designed to assemble first (see, e.g., computational assemblies (c) and (d)). The (red) output layer then assembles specifically starting from the bottom left using the inputs from the blue layer. (See Mao et al. (2000) for more details of this molecular computation.) The tiles were designed such that an output reporter strand ran through all the n tiles of the assembly by bridges across the adjoining pads in input, corner, and output tiles. This reporter strand was pasted together from the short ssDNA sequences within the tiles using the ligation enzyme mentioned previously. When the solution was warmed, this output strand was isolated and identified. The output data was read by experimentally determining the sequence of cut sites (see below). In principle, the output could be used for subsequent computations.

The next question of concern is:

- *How can one execute a step of computation using DNA tiles?*

To execute steps of computation, the TX tiles were designed to have pads at one end that encoded the cumulative XOR value. Also, since the reporter strand segments ran though each such tile, the appropriate input bit was also provided within its structure. These two values implied the opposing pad on the other side of the tile to be the XOR of these two bits.

A final question of concern is:

- *How can one determine and/or display the output values of a DNA tiling computation?*

◘ **Fig. 8**

Sequential Boolean computation via a linear DNA tiling assembly. (Adapted with permission from Mao et al. (2000).) (**a**) TX tile used in assembly. (**b**) Set of TX tiles providing logical programming for computation. (**c**) and (**d**) Example resulting computational tilings.

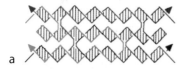

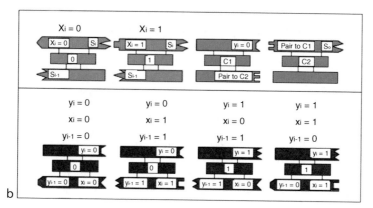

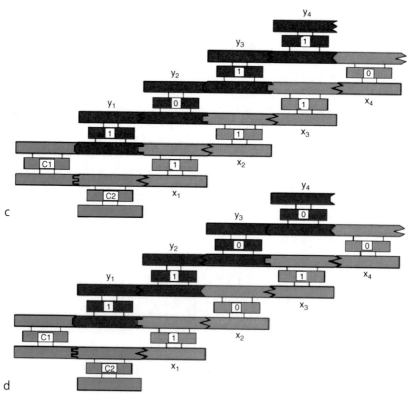

The output in this case was read by determining which of the two possible cut sites (endonuclease cleavage sites) were present at each position in the tile assembly. This was executed by first isolating the reporter strand, then digesting separate aliquots with each endonuclease separately and the two together, and finally these samples were examined by gel electrophoresis and the output values were displayed as banding patterns on the gel.

Another method for output (presented below) is the use of AFM observable patterning. The patterning was made by designing the tiles computing a bit 1 to have a stem loop protruding from the top of the tile. The sequence of this molecular patterning was clearly observable under appropriate AFM imaging conditions.

Although only very simple computations, the experiments of LaBean et al. 1999, 2000; Mao et al. 2000; and Yan et al. 2003b did demonstrate for the first time methods for autonomous execution of a sequence of finite-state operations via algorithmic self-assembly, as well as for providing inputs and for outputting the results. Further DNA tile assembly computations will be presented below.

5.2 Autonomous Finite-State Computations via Disassembly of DNA Nanostructures

An alternative method for autonomous execution of a sequence of finite-state transitions was subsequently developed by Shapiro and Benenson (2006). Their technique essentially operated in the reverse of the assembly methods described above, and instead can be thought of as disassembly. They began with a linear double-stranded DNA nanostructure whose sequence encoded the inputs, and then they executed a series of steps that digested the DNA nanostructure from one end. On each step, a sticky end at one end of the nanostructure encoded the current state, and the finite transition was determined by hybridization of the current sticky end with a small "rule" nanostructure encoding the finite-state transition rule. Then a restriction enzyme, which recognized the sequence encoding the current input as well as the current state, cut the appended end of the linear DNA nanostructure, to expose a new sticky end encoding the next state (❷ *Fig. 9*).

The hardware–software complex for this molecular device is composed of dsDNA with an ssDNA overhang (shown at the top left ready to bind with the input molecule) and a protein restriction enzyme (shown as gray pinchers).

This ingenious design is an excellent demonstration that there is often more than one way to do any task at the molecular scale. Adar et al. (2004) (see the conclusion, ❷ Sect. 11 for further discussion) demonstrated in the test tube a potential application of such a finite-state computing device to medical diagnosis and therapeutics.

6 Assembling Patterned and Addressable 2D DNA Lattices

One of the most appealing applications of tiling computations is their use to form patterned nanostructures to which other materials can be selectively bound.

An *addressable* 2D DNA lattice is one that has a number of sites with distinct ssDNA. This provides a superstructure for selectively attaching other molecules at addressable locations. Examples of addressable 2D DNA lattices will be given in ❷ Sect. 6.2.

☐ Fig. 9

Autonomous finite-state computations via disassembly of a double-stranded DNA nanostructure. (Figure adapted with permission from Shapiro and Benenson (2006).)

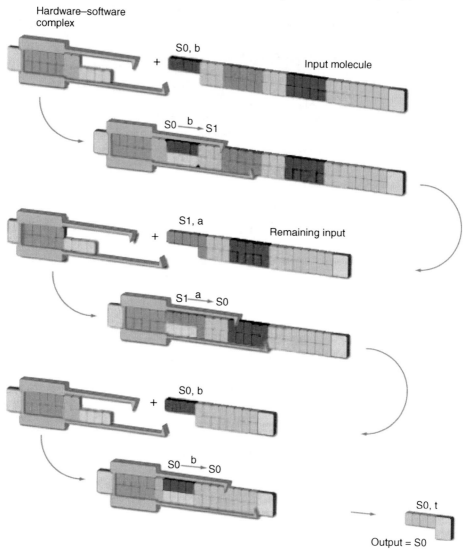

As discussed below, there are many types of molecules to which one can attach DNA. Known attachment chemistry allows them to be tagged with a given sequence of ssDNA. Each of these DNA-tagged molecules can then be assembled by hybridization of their DNA tags to a complementary sequence of ssDNA located within an addressable 2D DNA lattice. In this way, we can program the assembly of each DNA-tagged molecule onto a particular site of the addressable 2D DNA lattice.

6.1 Attaching Materials to DNA

There are many materials that can be made to directly or indirectly bind to specific segments of DNA using a variety of known attachment chemistries. Materials that can directly bind to specific segments of DNA include other (complementary) DNA, RNA, proteins, peptides, and various other materials. Materials that can be made to indirectly bind to DNA include a variety of metals (e.g., gold) that bind to sulfur compounds, carbon nanotubes (via various attachment chemistries), etc. These attachment technologies provide for a molecular-scale method of attaching heterogeneous materials to DNA nanostructures. For example, it can potentially be used for attaching molecular electronic devices to a 2D or 3D DNA nanostructure. Yan et al. (2003c) and Park et al. (2006b) describes conductive wires fabricated from self-assembled DNA tubes plated with gold or silver, as illustrated in ❷ *Fig. 10.*

6.2 Methods for Programmable Assembly of Patterned 2D DNA Lattices

The first experimental demonstration of 2D DNA lattices by Winfree and Seeman provided very simple patterning by repeated stripes determined by a stem loop projecting from every

❏ **Fig. 10**
Conductive wires fabricated from self-assembled DNA tubes plated with silver. (**a**) DNA tubes prior to plating. (**b**) DNA tubes after silver plating. (**c**) Illustration of conductivity test on silicon oxide substrate. (**d**) TEM image of conductivity test on silicon oxide substrate.

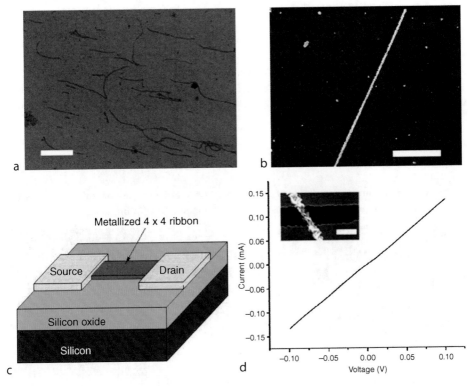

DNA tile on an odd column. This limited sort of patterning needed to be extended to large classes of patterns.

In particular, the key capability needed is a programmable method for forming distinct patterns on 2D DNA lattices, without having to completely redesign the lattice to achieve any given pattern. There are at least three methods for assembling patterned 2D DNA lattices that now have been experimentally demonstrated, as described in the next few sections.

6.2.1 Methods for Programmable Assembly of Patterned 2D DNA Lattices by Use of Scaffold Strands

A *scaffold strand* is a long ssDNA around which shorter ssDNA assemble to form structures larger than individual tiles. Scaffold strands were used to demonstrate programmable patterning of 2D DNA lattices in Yan et al. (2003a) by propagating 1D information from the scaffold into a second dimension to create AFM observable patterns. The scaffold strand weaves through the resulting DNA lattice to form the desired distinct sequence of 2D barcode patterns (left panel of ❿ *Fig. 11*). In this demonstration, identical scaffold strands ran through each row of the 2D lattices, using short stem loops extending above the lattice to form pixels. This determined a bar code sequence of stripes over the 2D lattice that was viewed by AFM. In principle, this method may be extended to allow for each row's patterning, to be determined by a distinct scaffold strand, defining an arbitrary 2D pixel image. A spectacular experimental demonstration of patterning via scaffold strand is also known as *DNA origami* (Rothemund 2006). This approach makes use of a long strand of "scaffold" ssDNA (such as from the sequence of a viral phage) that has only weak secondary structure and few long repeated or complementary subsequences. To this is added a large number of relatively short "staple" ssDNA sequences, with subsequences complementary to certain subsequences of the scaffold ssDNA. These staple sequences are chosen so that they bind to the scaffold ssDNA by hybridization and induce the scaffold ssDNA to fold together into a DNA nanostructure. A schematic trace of the scaffold strand is shown in the middle panel of ❿ *Fig. 11*, and an AFM image of the resulting assembled origami is shown in the right panel of ❿ *Fig. 11*. This landmark work of Rothemund (2006)

◻ Fig. 11

Methods for programmable assembly of patterned 2D DNA lattices by use of scaffold strands. (**a**) Barcode patterning. (**b**) DNA origami design. (**c**) AFM imaging of DNA origami.

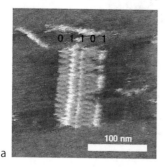
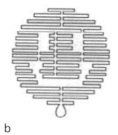

a   b   c

very substantially increases the scale of 2D patterned assemblies to hundreds of molecular pixels (that is, stem loops viewable via AFM) within a square area less than 100 nm on a side. In principle, this "molecular origami" method with staple strands can be used to form arbitrary complex 2D patterned nanostructures as defined. DNA origami has now been extended to simple 3D filaments that were used to partially orient membrane proteins in structural studies employing solution NMR (Douglas et al. 2007).

6.2.2 Programmable Assembly of Patterned 2D DNA Lattices by Computational Assembly

Another very promising method is to use the DNA tile's pads to program a 2D computational assembly. Recall that computer scientists in the 1970s showed that any computable 2D pattern can be so assembled. Winfree's group has experimentally demonstrated various 2D computational assemblies, and furthermore provided AFM images of the resulting nanostructures.

❷ *Figure 12* gives a modulo-2 version of Pascal's Triangle (known as the Sierpinski Triangle) (Rothemund et al. 2004), where each tile determines and outputs to neighborhood pads the XOR of two of the tile pads. Example AFM images (scale bars = 100 nm) of the assembled structures are shown in the three panels of ❷ *Fig. 12*.

❷ *Figure 13* gives Rothemund's and Winfree's design for a self-assembled binary counter, starting with 0 at the first row, and on each further row being the increment by 1 of the row below. The pads of the tiles of each row of this computational lattice were designed in a similar way to that of the linear XOR lattice assemblies described in the prior section. The resulting 2D counting lattice is found in MUX designs for address memory, and so this patterning may have major applications to patterning molecular electronic circuits (Barish et al. 2005).

☐ **Fig. 12**
Programmable assembly of Sierpinski triangle by use of computational assembly. (Figure adapted with permission from Rothemund et al. (2004).)

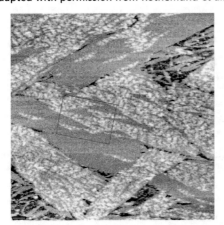

❏ **Fig. 13**

Rothemund's and Winfree's design for a self-assembled binary counter using tilings. (a) Tiles used and (b) binary counter assembly.

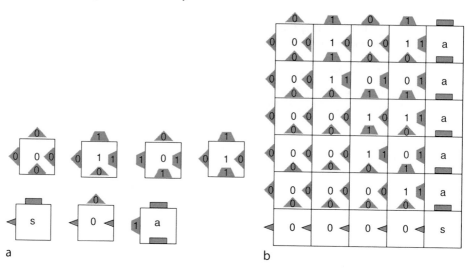

a

b

6.2.3 Programmable Assembly of Patterned 2D DNA Lattices by Hierarchical Assembly

A further approach, known as hierarchical assembly (Park et al. 2006a), is to assemble DNA lattices in multiple stages. ❏ *Figure 14* gives three examples of preprogrammed patterns displayed on addressable DNA tile lattices. Tiles are assembled prior to mixing with other preformed tiles. Unique ssDNA pads direct tiles to designed locations. White pixels are "turned on" by binding a protein (avidin) at programmed sites as determined in the tile assembly step by the presence or absence of a small molecule (biotin) appended to a DNA strand within the tile. Addressable, hierarchical assembly has been demonstrated for only modest size lattices to date, but has considerable potential particularly in conjunction with the above methods for patterned assembly.

7 Error Correction and Self-repair at the Molecular-Scale

7.1 The Need for Error Correction at the Molecular-Scale

In many of the self-assembled devices described here, there can be significant levels of error. These errors occur both in the synthesis of the component DNA, and in the basic molecular processes that are used to assemble and modify the DNA nanostructures, such as hybridization and the application of enzymes. There are various purification and optimization procedures developed in biochemistry for minimization of many of these types of errors. However, there remains a need for the development of methods for decreasing the errors of assembly and for self-repair of DNA tiling lattices comprising a large number of tiles. A number of

☐ **Fig. 14**
2D patterns by hierarchical assembly. (Figure adapted with permission from Park et al. (2006a).)
AFM images of characters D, N, and A.

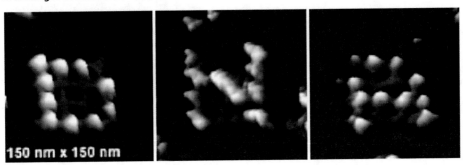

techniques have been proposed for decreasing the errors of a DNA tiling assembly, by providing increased redundancy, as described below.

7.2 Winfree's Proofreading Scheme for Error-Resilient Tilings

Winfree and Bekbolatov (2004) developed a "proofreading" method of replacing each tile with a subarray of tiles that provide sufficient redundancy to quadratically reduce errors, as illustrated in ❷ *Fig. 15*. This method, however, increased the area of the assembly by a factor of 4.

7.3 Reif's Compact Scheme for Error-Resilient Tilings

Reif et al. (2006) proposed a more compact method for decreasing assembly errors, as illustrated in ❷ *Fig. 16*. This method modifies the pads of each tile, so that essentially each tile executes both the original computation required at that location, as well as the computation of a particular neighbor, providing a quadratic reduction of errors without increasing the assembly size. The experimental testing of these and related error-reduction methods is ongoing. It seems possible that other error-correction techniques (such as error-correcting codes) developed in computer science may also be utilized.

7.4 Activatable Tiles for Reducing Errors

The uncontrolled assembly of tiling assemblies in reverse directions is potentially a major source of errors in computational tiling assemblies, and a roadblock in the development of applications of large patterned computational DNA lattices. Methods for controlled directional assembly of tiling assemblies would eliminate these errors. Majumder et al. (2007) have recently developed novel designs for an enhanced class of error-resilient DNA tiles (known as *activatable tiles*) for controlled directional assembly of tiles. While conventional DNA tiles store no state, the activatable tiling systems make use of a powerful DNA polymerase enzyme that allows the tiles to make a transition between active (allowing assembly) and inactive states. A *protection-deprotection* process strictly enforces the direction of tiling assembly growth, so that the assembly process is robust against entire classes of growth errors.

◻ **Fig. 15**
Winfree's proofreading scheme for error-resilient tilings. (a) Original tiles and (b) error-resilient tiles.

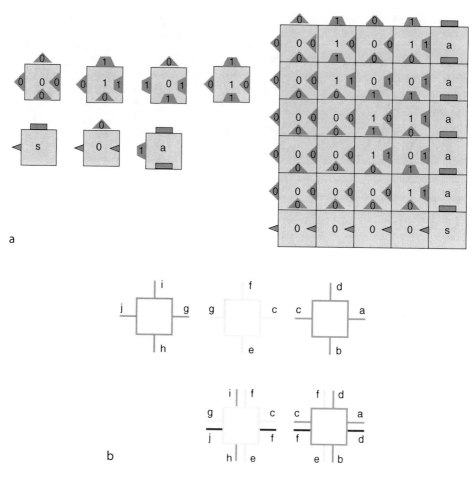

Initially, prior to binding with other tiles, the tile will be in an inactive state, where the tile is protected from binding with other tiles and thus prevents lattice growth in the (unwanted) reverse direction. After appropriate binding, other tiles bind to this tile, and then the tile makes a transition to an active state, allowing further growth.

8 Three-Dimensional DNA Lattices

8.1 Potential Applications of Three-Dimensional DNA Lattices

Most of the DNA lattices described in this article have been limited to 2D sheets. It appears to be much more challenging to assemble 3D DNA lattices of high regularity. There are some very important applications to nanoelectronics and biology if this can be done, as described below.

■ Fig. 16

A compact scheme for error-resilient tilings. (a) Original tile. (b) Error-resilient tiling. (c) A single pad mismatch causes another pad mismatch so destabilizing the assembly.

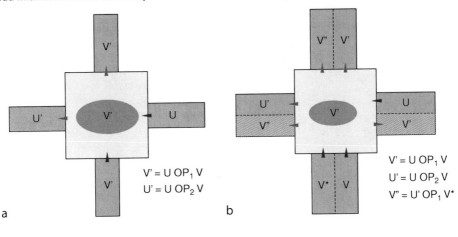

$$V' = U\ OP_1\ V$$
$$U' = U\ OP_2\ V$$

$$V' = U\ OP_1\ V$$
$$U' = U\ OP_2\ V$$
$$V'' = U'\ OP_1\ V^*$$

a   b

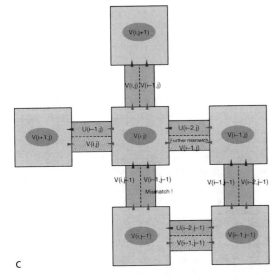

c

8.2 Scaffolding of 3D Nanoelectronic Architectures

The density of conventional nanoelectronics is limited by lithographic techniques to only a small number of layers. The assembly of even quite simple 3D nanoelectronic devices such as memory would provide much improvement in density. ❷ *Figure 17* shows DNA (cyan) and protein (red) organizing functional electronic structures.

It has been estimated that at least one half of all natural proteins cannot be readily crystallized, and have unknown structure, and determining these structures would have a major impact in the biological sciences. Suppose a 3D DNA lattice can be assembled with sufficient regularity and with regular interstices (say within each DNA tile comprising the

◻ Fig. 17
Scaffolding of 3D nanoelectronic architectures. (Adapted with permission from Robinson and Seeman (1987).)

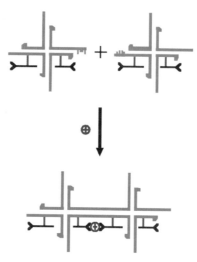

◻ Fig. 18
Scaffolding of proteins into regular 3D arrays. (Adapted with permission from Robinson and Seeman (1987).)

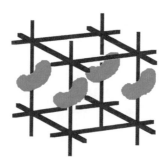

lattice). Then a given protein might be captured within each of the lattice's interstices, allowing it to be in a fixed orientation at each of its regularly spaced locations in 3D (see ❷ *Fig. 18*). This would allow the protein to be arranged in 3D in a regular way to allow for X-ray crystallography studies of its structure. This visionary idea is due to Seeman. So far, there has been only limited success in assembling 3D DNA lattices, and they do not yet have the degree of regularity (down to 2 or 3 Å) required for the envisioned X-ray crystallography studies. However, given the success up to now for 2D DNA lattices, this seems eventually achievable.

9 From Nucleic Detection Protocols to Autonomous Computation

9.1 The Detection Problem

A fundamental task of many biochemical protocols is to sense a particular molecule and then amplify the response. In particular, the detection of specific strands of RNA or DNA is an important problem for medicine. Typically, a protocol for nucleic detection is specialized to a subsequence of single-stranded nucleic acid (DNA or RNA oligonucleotide) to be detected. Given a sample containing a very small number of nucleic strand molecules to be detected, a detection protocol must amplify this to a much larger signal. Ideally, the detection protocol is exquisitely sensitive, providing a response from the presence of only a few of the target molecules.

There are a number of novel methods for doing DNA computation that can be viewed as being derived from protocols for detection of DNA. Therefore, understanding the variety of detection protocols can provide insight into these methods used for DNA computation.

9.2 Methods for Autonomous Molecular Computation Derived from PCR

9.2.1 The Polymerase Chain Reaction (PCR)

The original and still the most frequently used method for DNA detection is the *polymerase chain reaction* (*PCR*) (Saiki et al. 1985), which makes use of DNA polymerase to amplify a strand of DNA by repeated replication, using rounds of thermal cycling. (Recall that given an initial "primer" DNA strand hybridized onto a segment of a template DNA strand, polymerase enzyme can extend the primer strand by appending free DNA nucleotides complementary to the template's nucleotides.) In addition to DNA polymerase, the protocol requires a pair of "primer" DNA strands, which are extended by the DNA polymerase, each followed by heating and cooling, to allow displacement of the product strands.

9.2.2 Whiplash PCR: A Method for Local Molecular Computation

A method for DNA computation, known as Whiplash PCR (Sakamoto et al. 1999) makes use of a strand of DNA that essentially encodes a "program" describing state transition rules of a finite state computing machine; the strand is comprised of a sequence of "rule" subsequences (each encoding a state transition rule), and each separated by stopper sequences (which can stop the action of DNA polymerase). On each step of the computation, the 3' end of the DNA strand has a final sequence encoding a state of the computation. A computation step is executed when this 3' end hybridizes to a portion of a "rule" subsequence, and the action of DNA polymerase extends the 3' end to a further subsequence encoding a new state (❷ *Fig. 19*).

Whiplash PCR is interesting, since it executes a local molecular computation (recall that a molecular computation is local if the computation is within a single molecule, possibly in

☐ **Fig. 19**

Whiplash PCR: Repeated rounds of heating and cooling allow for the execution of a finite state transition machine with state transitions $a_i => b_i$ encoded by the DNA strand sequence. Each of the state transitions is executed in the cool stages, where if the current state is a_i, the 3' end of the DNA strand has the complement at its 3' end, which hybridizes to a sequence a_i and the state transition via polymerization extension step at the 3' end, allowing a transition to a state b_i. The heating stage allows for displacement of the extended 3' end.

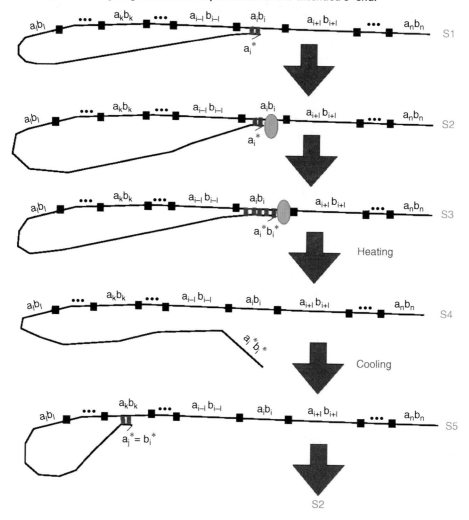

parallel with other molecular computing devices). In contrast, most methods for autonomous molecular computation (such as those based on the self-assembly of tiles) provide only a capability for distributed parallel molecular computation since to execute a computation they require multiple distinct molecules that interact to execute steps of each computation.

9.3 Isothermal and Autonomous PCR Detection and Whiplash PCR Computation Protocols

Neither the original PCR protocol nor the Whiplash PCR of Sakamoto et al. (1998) executes autonomously – they require thermal cycling for each step of their protocols. Walker et al. (1992a, b) developed isothermal (requiring no thermal cycling) methods for PCR known as strand displacement amplification (SDA) in which strands displaced from DNA polymerase are used for the further stages of the amplification reaction. Reif and Majumder recently developed (Reif and Majumder 2009) an autonomously executing version of Whiplash PCR (known as isothermal reactivating Whiplash PCR) that makes use of a strand-displacing polymerization enzyme (recall, however, that certain polymerase enzymes such as phi-29 can, as a side effect of their polymerization reaction, displace previously hybridized strands) with techniques to allow the reaction to process isothermally. In summary, an isothermal variant (strand-displacement PCR) of the basic PCR detection protocol provided insight on how to design an autonomous method for DNA computation. Like Whiplash PCR, this new isothermal reactivating Whiplash PCR provides for local molecular computation.

9.4 Autonomous Molecular Cascades and Related Hybridization Reactions for Autonomous DNA Computation

9.4.1 Autonomous Molecular Cascades for DNA Detection

Dirks and Pierce (2004) demonstrated an isothermal, enzyme-free (most known detection protocols require the use of protein enzymes) method for highly sensitive detection of a particular DNA strand. This protocol makes a triggered amplification by the hybridization chain reaction briefly illustrated in ❷ *Fig. 20*.

The protocol made use of multiple copies of two distinct DNA nanostructures T and T' that are initially added to a test tube. When ssDNA sequence S is added to the test tube, S initially has a hybridization reaction with a part of T, thus exposing a second ssDNA S' that had

❏ Fig. 20
Autonomous molecular cascade for signal amplification.

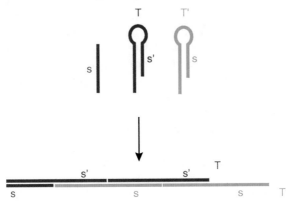

been previously hidden within the nanostructure of T. Next, S' has a hybridization reaction with a part of T', thus exposing a second copy of S that had been previously hidden within the nanostructure of T'. The second copy of S then repeats the process of other similar (but so far unaltered) copies of T and T', allowing a cascade effect to occur completely autonomously such autonomous molecular cascade devices have applications to a variety of medical applications, where a larger response (e.g., a cascade response) is required in response to one of the multiple molecular detection events.

9.4.2 Hybridization Reactions for Autonomous DNA Computation

Winfree (Zhang et al. 2007) developed a general methodology for designing systems of DNA molecules by the use of catalytic reactions that are driven by entropy. In particular, Zhang et al. (2007) demonstrates a general, powerful scheme for executing any Boolean circuit computation via cascades of DNA hybridization reactions. The unique common property of the above detection protocol (Dirks and Pierce 2004) and the molecular computations of Zhang et al. (2007) are their use of only hybridization, making no use of restriction enzyme or any other protein enzymes.

Following on this work, Yin et al. (2008) developed an elegant and highly descriptive labeled diagram scheme (with nodes indicating inputs, products, etc.) for illustrating the programming of biomolecular self-assembly and reaction pathways.

9.5 Autonomous Detection Protocols and Molecular Computations Using DNAzyme

In addition, Chen et al. (2004) demonstrated a novel method for DNA detection, which made use of a dual set of DNAzyme (recall a DNAzyme is a DNA molecule that possesses enzymatic activity, in particular cutting particular single-stranded DNA) that provided for amplified DNA detection. This led to the DNAzyme-based autonomous DNA walker (Tian et al. 2005) described in ❯ Sect. 10.4.2.

10 Autonomous Molecular Transport Devices Self-assembled from DNA

10.1 Molecular Transport

Many molecular-scale tasks may require the transport of molecules and there are a number of other tasks that can be done at the molecular scale that would be considerably aided by an ability to transport within and/or along nanostructures. For example, regarding the importance of molecular transport in nanoscale systems, consider the cell, which uses protein motors fueled by ATP to do this.

10.2 Nonautonomous DNA Motor Devices

In the early 2000s a number of researchers developed and demonstrated motors composed of DNA nanostructures; for example, Yurke et al. (2000) demonstrated a DNA actuator powered by DNA hybridization (complementary pairing between DNA strands). However, all of these

DNA motor devices required some sort of externally mediated changes (such as temperature-cycling, addition or elimination of a reagent, etc.) per work cycle of the device, and so did not operate autonomously.

10.3 The Need for Autonomous Molecular Transport

Almost all of the conventionally scaled motors used by mankind run without external mediation, and almost all natural systems for molecular motors are also autonomous (e.g., the cell's protein motors are all autonomous). The practical applications of molecular devices requiring externally mediated changes per work cycle are quite limited. So it is essential to develop autonomous DNA devices that do not require external mediation while executing movements.

10.4 Autonomous DNA Walkers

Reif (2003) first described the challenge of autonomous molecular transport devices which he called "DNA walkers" that traversed DNA nanostructures, and proposed two designs that gave bidirectional movement. In 2004, Sherman and Seeman demonstrated a DNA walker (Sherman and Seeman 2004), but it was nonautonomous since it required external mediation for every step it made.

10.4.1 Restriction Enzyme-Based Autonomous DNA Walkers

The first autonomous DNA walker was experimentally demonstrated in 2004 by Yin et al. (2004a, b). It employed restriction enzymes and ligase; see Yin et al. (2005) for its detailed general design (❷ *Fig. 21*).

The device is described in ❷ *Fig. 19*.

- Initially, a linear DNA nanostructure (the "road") with a series of attached ssDNA strands (the "steps") is self-assembled. Also, a fixed-length segment of DNA helix (the "walker") with short sticky ends (its "feet") are hybridized to the first two steps of the road.
- Then the walker proceeds to make a sequential movement along the road, where at the start of each step the feet of the walker are hybridized to two further consecutive steps of the road.
- Then a restriction enzyme cuts the DNA helix where the backward foot is attached, exposing a new sticky end forming a new replacement foot that can hybridize to the next step that is free, which can be the step just after the step where the other foot is currently attached. A somewhat complex combinatorial design for the sequences composing the steps and the walker ensures that there is unidirectional motion forward, along the road.

10.4.2 DNAzyme-Based Autonomous DNA Walkers

Subsequently in 2005 Mao's group (Tian et al. 2005) demonstrated an autonomous DNA walker that made use of a DNAzyme motor (Chen et al. 2004) that used the cuts provided by the enzymatic activity of DNAzyme to progress along a DNA nanostructure (❷ *Fig. 22*).

◘ Fig. 21

Autonomous molecular transport devices self-assembled from DNA.

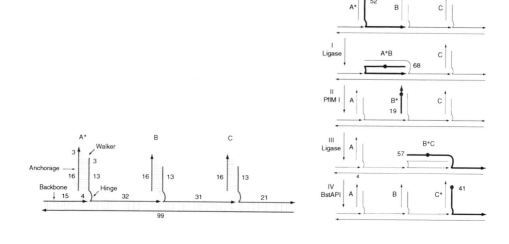

◘ Fig. 22

Mao's DNAzyme walker.

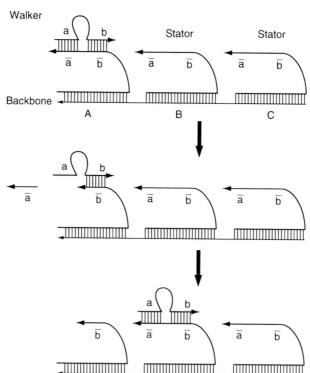

Bath and Turberfield (2007) also give an extensive survey of these and further recent DNA motor and walker devices.

10.5 Programmable Autonomous DNA Devices: Nanobots

There are some important applications of these autonomous DNA walkers including transport of molecules within large self-assembled DNA nanostructures. However, the potential applications may be vastly increased if they can be made to execute computations while moving along a DNA nanostructure. This would allow them, for example, to make programmable changes to their state and to make movements programmable. Such programmable autonomous DNA walker devices will be called "programmable DNA nanobots." Yin at al. (2006) describe an extension of the design of the restriction enzyme-based autonomous DNA walker (Yin et al. 2004a, b) described above in ❷ Sect. 10.4.1, to allow programmed computation while moving along a DNA nanostructure.

Another DNA nanobot design (see ❷ Fig. 23) for programmed computation while moving along a DNA nanostructure was developed by Sahu and Reif (2008) using in this case an extension of the design of the DNAzyme-based autonomous DNA walker (Tian et al. 2005) also described below (❷ Fig. 24).

It remains a challenge to experimentally demonstrate these.

11 Conclusions and Challenges

11.1 What Was Covered and What Was Missed: Further Reading

This chapter has covered most of the major known techniques and results for autonomous methods for DNA-based computation and transport.

However, there is a much larger literature of DNA-based computation that includes methods that are nonautonomous, but otherwise often ingenious and powerful. As just one notable example, Stojanovic demonstrated a deoxyribozyme-based molecular automaton (Stojanovic and Stefanovic 2003) and demonstrated its use to play the optimal strategy for a simple game.

◻ **Fig. 23**

Reif and Sahu's DNA nanobot: Figure illustrates the implementation of a state transition through DNAzymes. $D_{0,s1}$ in the transition machinery for state transition at 0 combines with input nanostructure when active input symbol encoded by the sticky end is 0. When the active input symbol encoded by the sticky end is 1, $D_{1,s1}$ in the transition machinery for state transition at 1 combines with the input nanostructure.

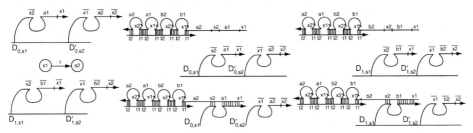

■ **Fig. 24**

Use of Reif and Sahu's DNA nanobot to execute a programmed traversal of a grid DNA nanostructure: **(a)** The DNAzyme implementation of the finite state machine shown on the left. **(b)** Illustration of programmable routing in two dimensions. **(a)** State transition diagram. **(b)** DNAzyme strands programming transitions. **(c)** Programmed traversal of DNA nanostructure grid.

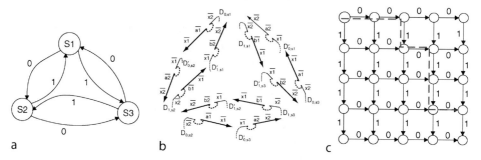

Other excellent surveys of DNA nanostructures and devices have been given by Seeman (Sherman and Seeman 2004; Sha et al. 2005), Mao (Deng et al. 2006), de Castro (2006), LaBean and Li (2007), Yan and Liu (2006), and Bath and Turberfield (2007).

11.2 Future Challenges for Self-assembled DNA Nanostructures

There are a number of key challenges still confronting this emerging field:
Experimentally demonstrate:

1. *Complex, error-free DNA patterning to the scale of, say, at least 10,000 pixels – as required, say, for a functional molecular electronic circuit for a simple processor.*
   Note: This would probably entail the use of a DNA tiling error correction method as well as a significant improvement over existing DNA patterning techniques.
2. *A programmable DNA nanobot autonomously executing a task critical to nano-assembly.*
   Note: The first stage might be a DNA walker that can be programmed to execute various distinct, complex traversals of a 2D DNA nanostructure, and to load and unload molecules at chosen sites on the nanostructure.
3. *An application of self-assembled DNA nanostructures to medical diagnosis.*

Adar et al. (2004) were the first to propose and to demonstrate in the test tube a finite-state computing DNA device for *medical diagnosis*: the device detects RNA levels (either over- or underexpression of particular RNA), computes a diagnosis based on a finite-state computation, and then provides an appropriate response (e.g., the controlled release of a single-stranded DNA that either promotes or interferes with expression). They demonstrated, in the test tube, a potential application of such a finite-state computing device to medical diagnosis and therapeutics. Sahu and Reif (2008) described a DNAzyme-based autonomous DNA nanobot (see ❷ Sect. 10.4) that can also function as a finite-state computing DNA device for medical diagnosis (❷ *Fig. 25*).

◻ Fig. 25

A finite-state computing DNA device for medical diagnosis based on Reif and Sahu's DNAzyme-based autonomous DNA nanobot. A state diagram for a DNAzyme doctor nanobot that controls the release of a "drug" RNA on the basis of the RNA expression tests for a disease. The figure shows the consequences of overexpression and underexpression of different RNAs on the concentrations of the respective characteristic sequences. The overexpression of $R1$ and $R2$ results in excess of $y1$ and $y2$, respectively, and they block the path of the input nanostructure by hybridizing with $D1$ and $D2$. Similarly, underexpression of $R3$ and $R4$ results in excess of $y3$ and $y4$, respectively, to block the path of input nanostructure.

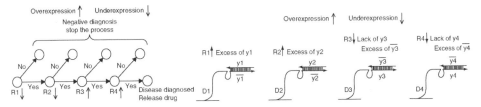

It remains a challenge to apply such a finite-state computing DNA device for medical diagnosis within the cell, rather than in the test tube.

Acknowledgments

Thanks to Nikhil Gopalkrishnan, Urmi Majumder, and Sudheer Sahu for their very useful comments on this article. Supported by NSF Grants CCF-0829797 and CCF-0829798.

References

Adar R, Benenson Y, Linshiz G, Rozner A, Tishby N, Shapiro E (2004) Stochastic computing with biomolecular automata. Proc Natl Acad Sci USA 101:9960–9965

Adleman LM (1994) Molecular computation of solutions to combinatorial problems. Science 266:1021–1024

Adleman L (Aug 1998) Computing with DNA. Sci Am 279(2):34–41

Barish RD, Rothemund PWK, Winfree E (2005) Two computational primitives for algorithmic self-assembly: copying and counting. NanoLetters 5(12):2586–2592

Bath J, Turberfield A (2007) DNA nanomachines. Nat Nanotechnol 2:275–284

Chen Y, Wang M, Mao C (2004) An autonomous DNA nanomotor powered by a DNA enzyme. Angew Chem Int Ed 43:3554–3557

de Castro LN (2006) Fundamentals of Natural Computing: Basic Concepts, Algorithms, and Applications.

Chapman & Hall/CRC Computer and Information Sciences

Deng Z, Chen Y, Tian Y, Mao C (2006) A fresh look at DNA nanotechnology, In: Chen J, Jonoska N, Rozenberg G (eds) Nanotechnology: science and computation, Springer series in Natural Computing, pp 23–34

Dirks RM, Pierce NA (2004) Triggered amplification by hybridization chain reaction. Proc Natl Acad Sci USA 101:15275–15278

Douglas SM, Chou JJ, Shih WM (2007) DNA-nano-tube-induced alignment of membrane proteins for NMR structure determination. PNAS 104:6644–6648

LaBean T, Li H (2007) Using DNA for construction of novel materials. Nano Today 2:26–35

LaBean TH, Gothelf KV, Reif JH (2006). Self-assembling DNA nanostructures for patterned molecular assembly, invited chapter in textbook. In: Mirkin CA,

Niemeyer CM (eds) Nanobiotechnology. Wiley, Weinheim, Germany

LaBean TH, Winfree E, Reif JH (1999) Experimental progress in computation by self-assembly of DNA tilings. In: Winfree E, Gifford DK (eds) DIMACS series in discrete mathematics and theoretical computer science, Proceedings of the 5th DIMACS workshop on DNA based computers, Vol 54. MIT, Cambridge, MA, pp 123–140

LaBean TH, Yan H, Kopatsch J, Liu F, Winfree E, Reif JH, Seeman NC (2000). The construction, analysis, ligation and self-assembly of DNA triple crossover complexes. J Am Chem Soc (JACS) 122:1848–1860

Majumder U, LaBean TH, Reif JH (2007) Activatable tiles: compact, robust programmable assembly and other applications, In: Garzon M, Yan H (eds) DNA computing: DNA13. Springer, Lecture notes in computer science (LNCS), Vol 4848. Springer, Berlin, Germany, pp 15–25

Mao C, LaBean, Reif TH, Seeman JH (Sept 28, 2000) Logical computation using algorithmic self-assembly of DNA triple-crossover molecules. Nature 407:493–495

Park SH, Pistol C, Ahn SJ, Reif JH, Lebeck AR, Dwyer C, LaBean TH (Jan 23, 2006a) Finite-size, fully addressable DNA tile lattices formed by hierarchical assembly procedures. Angew Chem [Int Ed] 45(5):735–739

Park S-H, Prior MW, LaBean TH, Finkelstein G (2006b) Optimized fabrication and electrical analysis of silver nanowires templated on DNA molecules. Appl Phys Lett 89:033901

Reif JH (2003) The design of autonomous DNA nanomechanical devices: walking and rolling DNA. In: Hagiya M, Ohuchi A (eds) DNA based computers (DNA8), Sapporo, Japan, June 10–13, 2002. Lecture notes in computer science, No. 2568, Springer, New York, pp 22–37. Published in Natural Computing, DNA8 special issue, Vol 2, pp 439–461

Reif JH, Majumder U (2009) Isothermal reactivating whiplash PCR for locally programmable molecular computation, Fourteenth international meeting on DNA based computers (DNA14). In: Goel A, Simmel FC (eds) Prague, Czech Republic (June, 2008). Lecture notes for computer science (LNCS), New York, NY, Springer, New York. Invited Paper, Special issue in Natural Computing

Reif JH, Sahu S, Yin P (2006) Compact error-resilient computational DNA tiling assemblies. In: Chen J, Jonoska N, Rozenberg G (eds) Nanotechnology: science and computation, Springer series in Natural Computing, New York, pp 79–104

Robinson BH, Seeman NC (1987) Protein Eng 1:295–300

Rothemund PWK (March 16, 2006) Folding DNA to create nanoscale shapes and patterns. Nature 440:297–302

Rothemund PWK, Papadakis N, Winfree E (Dec 2004). Algorithmic self-assembly of DNA Sierpinski triangles. PLoS Biol 2(12)

Sahu S, Reif JH (2008) DNA-based self-assembly and nanorobotics. VDM Verlag Dr. Mueller e. K, Saarbrücken, Germany, 128 p. ISBN-10: 363909770X, ISBN-13: 978-3639097702

Saiki RK, Scharf S, Faloona F, Mullis KB, Horn GT, Erlich HA, Arnheim N (Dec 20, 1985) Enzymatic amplification of beta-globin genomic sequences and restriction site analysis for diagnosis of sickle cell anemia. Science 230(4732):1350–1354

Sakamoto K, Kiga D, Momiya K, Gouzu H, Yokoyama S, Ikeda S, Sugiyama H, Hagiya M (1998) State transitions with molecules. In: Proceedings of the 4th DIMACS meeting on DNA based computers, held at the University of Pennsylvania, June 16–19, 1998. DIMACS series in discrete mathematics and theoretical computer science. American Mathematical Society, Providence, RI

Sha R, Zhang X, Liao S, Constantinou PE, Ding B, Wang T, Garibotti AV, Zhong H, Israel LB, Wang X, Wu G, Chakraborty B, Chen J, Zhang Y, Yan H, Shen Z, Shen W, Sa-Ardyen P, Kopatsch J, Zheng J, Lukeman PS, Sherman WB, Mao C, Jonoska N, Seeman NC (2005) Structural DNA nanotechnology: molecular construction and computation. In: Calude CS et al. (eds) UC 2005, LNCS 3699, Springer, Berlin, Germany, pp 20–31

Shapiro E, Benenson Y (May 2006) Bringing DNA computers to life. Sci Am 45–51

Sherman WB, Seeman NC (2004) A precisely controlled DNA biped walking device. Nano Lett Am Chem Soc 4:1203

Stojanovic MN, Stefanovic D (2003) A deoxyribozyme-based molecular automaton. Nat Biotechnol 21:1069–1074

Tian Y, He Y, Chen Y, Yin P, Mao C (2005) Molecular devices – a DNAzyme that walks processively and autonomously along a one-dimensional track. Angew Chem Int Ed 44:4355–4358

Walker GT, Fraiser MS, Schram JL, Little MC, Nadeau JG, Malinowski DP (Apr 1992a) Strand displacement amplification – an isothermal, in vitro DNA amplification technique. Nucleic Acids Res 20(7):1691–1696

Walker GT, Little MC, Nadeau JG, Shank DD (Jan 1992b) Isothermal in vitro amplification of DNA by a restriction enzyme/DNA polymerase system. Proc Natl Acad Sci USA 189(1):392–396

Winfree E, Bekbolatov R (2004) Proofreading tile sets: error correction for algorithmic self-assembly, in DNA computers 9. LNCS (2943):126–144

Winfree E, Yang X, Seeman NC (1998) Universal computation via self-assembly of DNA: some theory and experiments. In: DNA based computers II, American Mathematical Society, Providence, RI, pp 191–213

Yan H, Liu Y (2006) DNA nanotechnology: an evolving field, invited chapter in nanotechnology: science and computation. In: Chen J, Jonoska N, Rozenberg G (eds) Natural computing series. Springer, Berlin, Germany, pp 35–53

Yan H, Feng L, LaBean TH, Reif J (2003a) DNA nanotubes, parallel molecular computations of pairwise exclusive-or (XOR) using DNA "String Tile". Self-Assembly J Am Chem Soc (JACS) 125(47): 14246–14247

Yan H, LaBean TH, Feng L, Reif JH (2003b) Directed nucleation assembly of barcode patterned DNA lattices. PNAS 100(14):8103–8108

Yan H, Park SH, Finkelstein G, Reif JH, LaBean TH (Sept 26, 2003c) DNA-templated self-assembly of protein arrays and highly conductive nanowires. Science 301:1882–1884

Yin P, Choi HMT, Calvert CR, Pierce NA (2008) Programming biomolecular self-assembly pathways. Nature 451:318–322

Yin P, Guo B, Belmore C, Palmeri W, Winfree E, LaBean TH, Reif JH (June 7–10, 2004a) TileSoft: sequence optimization software for designing DNA secondary structures, abstract, preliminary proceedings, Tenth international meeting on DNA based computers (DNA10), Milan, Italy

Yin P, Turberfield AJ, Sahu S, Reif JH (2005) Designs for autonomous unidirectional walking DNA devices, tenth international meeting on DNA based computers (DNA10). In: Ferretti C, Mauri G, Zandron C (eds) Milan, Italy, June 7–10, 2004. Lecture notes in computer science, Vol 3384. Springer, New York, pp 410–425

Yin P, Turberfield AJ, Reif JH (2006) Design of autonomous DNA cellular automata, Eleventh international meeting on DNA based computers (DNA11). In: Carbone A, Pierce N (eds) London, UK (June 2005). Springer Lecture notes in computer science (LNCS), New York, NY, Vol 3892, Springer, New York, pp 399–416

Yin P, Yan H, Daniel XG, Turberfield AJ, Reif JH (Sept 20, 2004b) A unidirectional DNA walker moving autonomously along a linear track, Angew Chem [Int Ed], 43(37):4906–4911

Yurke B, Turberfield AJ, Mills AP Jr, Simmel FC, Neumann JL (2000) A DNA-fuelled molecular machine made of DNA. Nature 406:605–608

Zhang DY, Turberfield AJ, Yurke B, Winfree E (2007) Engineering entropy-driven reactions and networks catalyzed by DNA. Science 318:1121–1125

40 Membrane Computing

Gheorghe Păun
Institute of Mathematics of the Romanian Academy, Bucharest, Romania
Department of Computer Science and Artificial Intelligence, University of Seville, Spain
gpaun@us.es
george.paun@imar.ro

G. Rozenberg et al. (eds.), *Handbook of Natural Computing*, DOI 10.1007/978-3-540-92910-9_40,
© Springer-Verlag Berlin Heidelberg 2012

Abstract

This chapter introduces the basic ideas, results, and applications of membrane computing, a branch of natural computing inspired by the structure and the functioning of the living cell, as well as by the cooperation of cells in tissues, colonies of cells, and neural nets.

1 Introduction

Membrane computing is a branch of natural computing introduced by Păun (2000). Its main goal is to abstract computing models from the architecture and the functioning of living cells, as well as from the organization of cells in tissues, organs (brain included) or other higher-order structures such as colonies of cells (e.g., bacteria). The initial goal was to learn from cell biology something possibly useful to computer science, and the area quickly developed in this direction. Several classes of computing models, called P systems, were defined in this context and mainly investigated from mathematical and computer science points of view. Soon, other applications appeared, first in biology, then also in linguistics, computer science, economics, approximate optimization, etc. This chapter will introduce the main classes of P systems, with simple examples illustrating their structure and functioning, and then will briefly discuss some types of results and applications.

Consider that a cell is a hierarchical structure of compartments defined by membranes, with solutions of chemicals swimming in compartments and proteins bound on membranes, which also define selective channels among compartments. The computing model abstracted in this context is a distributed one, with (virtual) membranes defining its structure, with "reactions" taking place in compartments, and with a precise way of communicating among compartments. Thus, the main ingredients of a P system are (1) the membrane structure, (2) the multisets of objects placed in the compartments of the membrane structure, and (3) the rules for processing the objects and the membranes. The rules are used for changing the current configuration of the device and in this way we get computations, sequences of transitions among configurations starting with an initial configuration and, when halting, providing a result of the computation. Many classes of P systems have been defined, with biological, mathematical, or computer science motivation; these variants concern the arrangement of membranes, the form of rules, the way of using the rules, the way of defining the result of a computation, and so on. The main classes of P systems (cell-like, tissue-like, neural-like) will be discussed below.

Most of these classes of P systems are rather powerful from a computational point of view. They are Turing complete, even when using ingredients of a reduced complexity – a small number of membranes, rules of simple forms, ways of controlling the use of rules directly inspired from biology. Certain classes of P systems are also efficient. They can be used for solving computationally hard problems – typically **NP**-complete problems, in a feasible time, which are typically polynomial.

Membrane computing proved to be rather appealing as a modeling framework, especially for handling discrete (biological or biomedical, economic, etc.) processes. They have many attractive features: easy understandability, scalability and programmability, inherent compartmentalization, etc.

Membrane computing is a young research area, but already its bibliography is rather large. A comprehensive presentation can be found in the 2002 monograph (Păun 2002), with several applications presented in Ciobanu et al. (2006), where a friendly introduction to membrane computing is given in the first chapter. An up-to-date coverage of membrane computing can be found in a recent handbook (Păun et al. 2009). The complete bibliography of the domain, which at the end of 2008 totaled around 1,250 titles, can be found on the P Systems website at www.ppage.psystems.eu. In general, the reader can refer to these bibliographical sources, as this chapter only specifies a few references.

2 Cell-Like P Systems

In this section we discuss in some detail the first introduced and most investigated type of P system, inspired by the structure and functioning of the biological cell. Because the ingredients used in this model are also present, with variations, in other types of P systems, we discuss separately the main ingredients: membrane structure, types of rules, ways of using the rules and of defining the results.

2.1 Membrane Structure

One starts from the observation that the (eukaryotic) cell is compartmentalized by means of membranes, so that one considers hierarchical arrangements of membranes as suggested (in a 2D representation) in the left side of ❷ *Fig. 1*. Here we distinguish an external membrane (corresponding to the plasma membrane and usually called the *skin* membrane) as well as several membranes placed inside the skin membrane (corresponding to the membranes present in a cell, around the nucleus, in Golgi apparatus, vesicles, mitochondria, etc.). A membrane without any other membrane inside it is said to be *elementary*. Each membrane determines a compartment, called a *region*, which is the space delimited by it from above and

■ Fig. 1

A membrane structure and its tree representation.

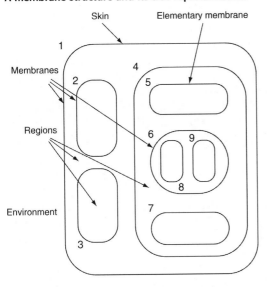
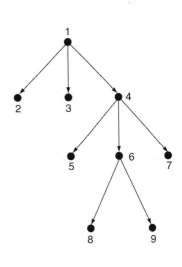

from below by the membranes placed directly inside, if any exist. Clearly, the correspondence between the membrane and region is one-to-one; that is why the terms are sometimes used interchangeably.

Usually, the membranes are identified by *labels* from a given set. In ❷ *Fig. 1*, we use numbers, starting with number 1 assigned to the skin membrane. (This is the standard labeling, but the labels can be more informative "names" associated with the membranes.) Also, in the figure the labels are assigned in a one-to-one manner to membranes, but this is possible only in the case of membrane structures that cannot grow (indefinitely); otherwise, several membranes would have the same label (such cases will be seen later). Due to the membrane–region correspondence, we identify by the same label a membrane and its associated region.

This hierarchical structure of membranes can be represented by a rooted unordered tree, with labeled nodes; the tree that describes the membrane structure in ❷ *Fig. 1* is given in the right-hand side of the same figure. The root of the tree is associated with the skin membrane and the leaves are associated with the elementary membranes.

Directly suggested by the tree representation is the symbolic representation of a membrane structure, by strings of labeled matching parentheses. For instance, a string corresponding to the structure from ❷ *Fig. 1* is the following:

$$[_1 \ [_2 \ ]_2 \ [_3 \ ]_3 \ [_4 \ [_5 \ ]_5 \ [_6 \ [_8 \ ]_8 \ [_9 \ ]_9 \ ]_6 \ [_7 \ ]_7 \ ]_4 \ ]_1 \qquad (*)$$

Because the tree is not ordered (and also corresponds to a biological reality), the membranes of the same level can float around. Therefore, by swapping two subexpressions placed at the same level, we get an expression that represents the same membrane structure. For instance, in the previous case, the expression

$$[_1 \ [_3 \ ]_3 \ [_4 \ [_6 \ [_8 \ ]_8 \ [_9 \ ]_9 \ ]_6 \ [_7 \ ]_7 \ [_5 \ ]_5 \ ]_4 \ [_2 \ ]_2 \ ]_1$$

is a representation of the same membrane structure, equivalent to (❷ *).

2.2 Multisets

In the compartments of a cell, there are various chemicals swimming in water (at this stage, we ignore the chemicals, mainly proteins, bound on membranes). Because in many cases for the biochemistry taking place in a cell "numbers matter", the natural data structure to use in this framework is the *multiset*, the set with multiplicities associated with its elements. For this reason, some basic elements related to multisets and to their representation by strings, which is standard in membrane computing, are discussed here.

A *multiset over* a given set U is a mapping $M : U \to \mathbf{N}$, where $\mathbf{N}$ is the set of nonnegative integers. For $a \in U$, $M(a)$ is the multiplicity of a in M. If the set U is finite, $U = \{a_1, \ldots, a_n\}$, then the multiset M can be explicitly given in the form $\{(a_1, M(a_1)), \ldots, (a_n, M(a_n))\}$, thus specifying for each element of U its multiplicity in M. In membrane computing, the usual way to represent a multiset $M = \{(a_1, M(a_1)), \ldots, (a_n, M(a_n))\}$ over a finite set $U = \{a_1, \ldots, a_n\}$ is by using strings: $w = a_1^{M(a_1)} a_2^{M(a_2)} \ldots a_n^{M(a_n)}$ and all permutations of w represent M; the empty multiset is represented by λ, the empty string. The total multiplicity of elements of a multiset (this is also called the *weight* of the multiset) clearly corresponds to the length of a string representing it. This compact representation is so frequent in membrane computing that we can usually say "the multiset w" instead of "the multiset represented by string w."

A few basic notions about multisets are useful in membrane computing, but they are not discussed here; the reader can refer to Calude et al. (2001) for details.

2.3 Evolution Rules

The main way the chemicals present in the compartments of a cell evolve is by means of reactions that consume certain chemicals and produce other chemicals. In what follows, the chemicals are considered to be unstructured and hence are described by symbols from a given alphabet; these symbols are called *objects*. Later, structured objects described by strings will also be considered.

Thus, in each compartment of a membrane structure, we consider a multiset of objects over a given alphabet. Corresponding to the biochemical reactions, the main type of rules for evolving these multisets are multiset rewriting rules (simply, evolution rules). Later, rules for moving objects across membranes, called communication rules, as well as rules for handling membranes will also be considered.

The multiset-rewriting rules are of the form $u \rightarrow v$, where u and v are multisets of objects. An example is $aab \rightarrow abcc$ (remember that the multisets are represented by strings). However, in order to make the compartments cooperate, objects have to be moved across membranes, and for this we add *target indications* to the objects produced by a rule as above (to the objects from multiset v). These indications are *here, in*, and *out*, with the meanings that an object associated with the indication *here* remains in the same region, one associated with the indication *in* goes immediately into an adjacent lower membrane that is nondeterministically chosen, and *out* indicates that the object has to exit the membrane, thus becoming an element of the region surrounding it. For instance, we can have $aab \rightarrow (a, here)(b, out)(c, here)(c, in)$. Using this rule in a given region of a membrane structure means to consume two copies of a and one of b (they are removed from the multiset of that region), and one copy of a, one of b, and two of c are produced; the resulting copy of a remains in the same region, and the same happens with one copy of c (indications *here*), while the new copy of b exits the membrane, going to the surrounding region (indication *out*), and one of the new copies of c enters one of the child membranes, nondeterministically chosen. If no such child membrane exists, that is, the membrane with which the rule is associated is elementary, then the indication *in* cannot be followed, and the rule cannot be applied. In turn, if the rule is applied in the skin region, then b will exit into the environment of the system (and it is "lost" there, since it can never come back). In general, the indication *here* is not specified when giving a rule.

The evolution rules are classified according to the complexity of their left-hand side. A rule with at least two objects in its left-hand side is said to be *cooperative*; a particular case is that of *catalytic* rules of the form $ca \rightarrow cv$, where c is an object (called catalyst), which assists the object a to evolve into the multiset v; rules of the form $a \rightarrow v$, where a is an object, are called *noncooperative*.

The rules can also contain other ingredients or indications about their use. For instance, a rule of the form $u \rightarrow v\delta$ entails the dissolution of the membrane where it is used: if the rule is applied, then the corresponding membrane disappears and its contents, object and membranes alike, are left free in the surrounding membrane; the rules of the dissolved membrane disappear with the membrane. The skin membrane is never dissolved.

The rules can also be of the form $u \rightarrow v|_p$, where p is an object; it is said that p is a *promoter* and the rule can be applied only if p is present in the region where the rule is applied.

Similarly, we have rules with *inhibitors*: $u \rightarrow v|_{\neg p}$ can be applied only if the object p is not present in the region.

Very important are also the communication rules and the rules for evolving the membrane structure, but, before introducing them we discuss another central feature, the way of using the rules.

2.4   Ways of Using the Rules

The most investigated way of applying the rules in the regions of a P system is described by the phrase: *in the maximally parallel manner, nondeterministically choosing the rules and the objects.*

This means the following: we assign objects to rules, nondeterministically choosing the objects and the rules until no further assignment is possible. Mathematically stated, we look to the *set* of rules and try to find a *multiset* of rules by assigning multiplicities to rules with two properties: (1) the multiset of rules is *applicable* to the multiset of objects available in the respective region; that is, there are enough objects to apply the rules a number of times as indicated by their multiplicities; and (2) the multiset is *maximal*, that is, no further rule can be added to it (no multiplicity of a rule can be increased) such that the obtained multiset is still applicable.

Thus, an evolutionary step in a given region consists of finding a maximal applicable multiset of rules, removing from the region all objects specified in the left-hand sides of the chosen rules (with multiplicities as indicated by the rules and by the number of times each rule is used), producing the objects from the right-hand sides of the rules, and then distributing these objects as indicated by the targets associated with them. If at least one of the rules introduces the dissolving action δ, then the membrane is dissolved and its contents become part of the parent membrane, provided that this membrane was not dissolved at the same time; otherwise we stop at the first upper membrane that was not dissolved (the skin membrane at least remains intact).

The maximal parallelism both corresponds to the apparent simultaneity of reactions taking place in a cell and it is also very useful from a mathematical point of view, but there are several other possibilities, in general closer to the biological reality. *Partial parallelism* is one of them: in each step, one uses a specified number of rules, or at least/at most a specified number of rules. In the limit, we have the *sequential* use of rules: only one rule is used in each region, nondeterministically chosen.

An interesting case is that of *minimal parallelism*. The intuition is that each region that *can* evolve (using at least one rule) *must* evolve (by means of at least one rule). More specifically, we start by nondeterministically choosing one rule from each region where a rule can be used; this combination of rules or any applicable supermultiset of it can be used in evolving the system. Note that in comparison with the maximal parallelism, a considerable degree of nondeterminism is possible, as many choices of multisets of rules can be applied (in particular, any maximal multiset).

2.5   Computations and Results of Computations

A membrane structure and the multisets of objects from its regions identify a *configuration* of a P system. Note that both objects and rules are associated with regions; they are *localized*.

By using the rules in one of the ways suggested above, we pass to another configuration; such a step is called a *transition*. Note that in all cases considered in the previous section, the evolution of the system is synchronized; a global clock exists that marks the time uniformly for all regions and all regions evolve in each time unit. A sequence of transitions constitutes a *computation*. A computation is successful if it halts – it reaches a configuration where no rule can be applied to the existing objects. With a halting computation we can associate a *result* in various ways. One possibility is to count the objects present in the halting configuration in a specified elementary membrane; this is called the *internal output*. We can also count the objects that leave the system during the computation; this is called the *external output*. In both cases, the result is a number. If we distinguish among different objects, then we can have as the result a vector of natural numbers. The objects that leave the system can also be arranged in a sequence according to the moments when they exit the skin membrane, and in this case the result is a string.

Total halting as mentioned above is only one of the ways of defining successful computations and the most investigated in membrane computing (also because it is similar to the way Turing machines provide a result). Another possibility is *partial halting*: The computation stops when at least one region of the system halts; it cannot use any of its rules. Then, we can also use *signals* of various kinds for marking the step when a result is provided by a computation; for instance, if a specified object exits the system, then we count the number of objects in a given membrane in order to get a result.

An important issue is the way we use the system. In the previous discussion, it was assumed that one can start from an initial configuration and proceed through transitions (according to a given way of using the rules). Because of the nondeterminism of the application of rules, one can get several successful computations, and hence several results. This is the *generative* way of using a P system, and in this way a P system *computes* (or *generates*) a set of numbers, or a set of vectors of numbers, or a language. Another important possibility is the *accepting* case: we start from the initial configuration, where some objects codify an input, and that input is accepted if and only if the computation halts (or a specified object, say yes, is sent to the environment). A more general case is that of *computing* a function: We start with the argument introduced in the initial configuration and the value of the function is obtained at the end of the computation.

2.6 A Formal Definition and an Example

We now have all the prerequisites for formally defining a cell-like P system. Such a system is a construct

$$\Pi = (O, \mu, w_1, \ldots, w_m, R_1, \ldots, R_m, i_{\text{in}}, i_{\text{out}})$$

where O is the alphabet of objects, μ is the membrane structure (with m membranes), given as an expression of labeled parentheses, $w_1, \ldots, w_m$ are (strings over O representing) multisets of objects present in the m regions of μ at the beginning of a computation, $R_1, \ldots, R_m$ are finite sets of evolution rules associated with the regions of μ, and $i_{\text{in}}, i_{\text{out}}$ are the labels of input and output membranes, respectively. If the system is used in the generative mode, then i_{in} is omitted; if the system is used in the accepting mode, then i_{out} is omitted. If the system is a catalytic one, then a subset C of O is specified, containing the catalysts. The number m of membranes in μ is called the *degree* of Π.

In the generative case, the set of numbers computed by Π when using the rules in mode *mode*, one of those mentioned in ❷ Sect. 2.4, is denoted by $N_{\text{mode}}(\Pi)$; in what follows, the rules are always used in the maximally parallel way, hence the subscript is omitted. The family of all sets $N(\Pi)$ computed by systems Π of degree at most $m \geq 1$ and using rules of *types-of-rules* forms is denoted by $NOP_m(\text{types-of-rules})$; if there is no bound on the degree of systems, then the subscript m is replaced with $*$.

The architecture and the functioning of a cell-like P system is illustrated with a simple example of a computing P system. Formally, it is given as follows:

$$\Pi = (O, C, \mu, w_1, w_2, R_1, R_2, i_o)$$

where

$O = \{a, b_1, b'_1, b_2, c, e\}$ (the set of objects)

$C = \{c\}$ (the set of catalysts)

$\mu = [_1 [_2]_2]_1$ (membrane structure)

$w_1 = c$ (initial objects in region 1)

$w_2 = \lambda$ (initial objects in region 2)

$R_1 = \{a \rightarrow b_1 b_2, \ cb_1 \rightarrow cb'_1, \ b_2 \rightarrow b_2 e_{in}|_{b_1}\}$ (rules in region 1)

$R_2 = \emptyset$ (rules in region 2)

$i_o = 2$ (the output region)

❷ *Figure 2* indicates the initial configuration (the rules included) of the system Π. It computes a function, namely $n \rightarrow n^2$, for any natural number $n \geq 1$. Besides catalytic and non-cooperating rules, the system also contains a rule with promoters, $b_2 \rightarrow b_2(e, in)|_{b_1}$: the object b_2 evolves to $b_2 e$ only if at least one copy of object b_1 is present in the same region.

We start with only one object in the system, the catalyst c. If we want to compute the square of a number n, then we have to input n copies of the object a in the skin region of the system. In that moment, the system starts working by using the rule $a \rightarrow b_1 b_2$, which has to be applied in parallel to all copies of a; hence, in one step, all objects a are replaced by n copies of b_1 and n copies of b_2. From now on, the other two rules from region 1 can be used. The catalytic rule $cb_1 \rightarrow cb'_1$ can be used only once in each step, because the catalyst is present in only one copy. This means that in each step one copy of b_1 gets primed. Simultaneously (because of the

◻ **Fig. 2**
A P system with catalysts and promoters.

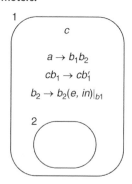

maximal parallelism), the rule $b_2 \rightarrow b_2(e, in)|_{b_1}$ should be applied as many times as possible — n times, because we have n copies of b_2. Note the important difference between the promoter b_1, which allows us to use the rule $b_2 \rightarrow b_2(e, in)|_{b_1}$, and the catalyst c: the catalyst is involved in the rule and is counted when applying the rule, while the promoter makes possible the use of the rule but it is not counted; the same (copy of an) object can promote any number of rules. Moreover, the promoter can evolve at the same time by means of another rule (the catalyst is never changed).

In this way, in each step we change one b_1 to b_1' and we produce n copies of e (one for each copy of b_2); the copies of e are sent to membrane 2 (the indication in from the rule $b_2 \rightarrow b_2(e, in)|_{b_1}$). The computation should continue as long as there are applicable rules. This means exactly n steps: in n steps, the rule $cb_1 \rightarrow cb_1'$ will exhaust the objects b_1 and in this way neither this rule can be applied, nor $b_2 \rightarrow b_2(e, in)|_{b_1}$, because its promoter no longer exists. Consequently, in membrane 2, considered as the output membrane, we get n^2 copies of object e.

Note that the computation is deterministic, always the next configuration of the system is unique, and that, changing the rule $b_2 \rightarrow b_2(e, in)|_{b_1}$ with $b_2 \rightarrow b_2(e, out)|_{b_1}$, the n^2 copies of e will be sent to the environment; hence we can read the result of the computation outside the system, and in this case membrane 2 is useless.

2.7 Symport and Antiport Rules

An important part of the cell activity is related to the passage of substances through membranes, and one of the most interesting ways to handle this transmembrane communication is by coupling molecules as suggested by cell biology. The process by which two molecules pass together across a membrane (through a specific protein channel) is called *symport*; when the two molecules pass simultaneously through a protein channel in opposite directions the process is called *antiport*.

These operations can be formalized by considering symport rules of the form (x, in) and (x, out), and antiport rules of the form $(z, out; w, in)$, where $x, z,$ and w are multisets of arbitrary size; one says that the length of x, denoted by $|x|$, is the *weight* of the symport rule, and $max(|z|, |w|)$ is the *weight* of the antiport rule. Thus, we obtain the following important class of P systems.

A *P system with symport/antiport rules* is a construct of the form
$$\Pi = (O, \mu, w_1, \ldots, w_m, E, R_1, \ldots, R_m, i_{in}, i_{out})$$

where all components $O, \mu, w_1, \ldots, w_m, i_{in}, i_{out}$ are as in a P system with multiset rewriting rules, $E \subseteq O$, and $R_1, \ldots, R_m$ are finite sets of symport/antiport rules associated with the m membranes of μ. The objects of E are supposed to be present in the environment of the system with an arbitrary multiplicity. (Note that the symport/antiport rules do not change the number of objects, but only their place, which is why we need a supply of objects in the environment; this supply is inexhaustible, that is, it does not matter how many objects are introduced in the system because arbitrarily many still remain in the environment.)

As above, the rules can be used in various ways, but here we consider again the nondeterministic maximally parallel manner. Transitions, computations, and halting computations will be defined in the usual way. The number of objects present in region i_{out} in the halting configuration is said to be computed by the system by means of that computation; the set of all numbers computed in this way by Π is denoted by $N(\Pi)$. The family of sets $N(\Pi)$ computed by systems Π of degree at most $m \geq 1$, using symport rules of degree at most $p \geq 0$ and antiport rules of degree at most $q \geq 0$ (the cases $p = 0$ or $q = 0$ correspond to not having symport or antiport rules, respectively), is denoted by $NOP_m(\text{sym}_p, \text{anti}_q)$.

The definition of P systems is illustrated with symport/antiport with an example of more general interest: a construction for simulating a *register machine* is given. In this way, one of the widely used proof techniques for the universality results in this area is also introduced.

A (nondeterministic) *register machine* is a device $M = (m, B, l_0, l_h, R)$, where $m \geq 1$ is the number of registers, B is the (finite) set of instruction labels, l_0 is the initial label, l_h is the halting label, and R is the finite set of instructions labeled (and hence uniquely identified) by elements from B. The labeled instructions are of the following forms:

- $l_1 : (\text{add}(r), l_2, l_3)$, $1 \leq r \leq m$ (add 1 to register r and go nondeterministically to one of the instructions with labels l_2, l_3),
- $l_1 : (\text{sub}(r), l_2, l_3)$, $1 \leq r \leq m$ (if register r is not empty, then subtract 1 from it and go to the instruction with label l_2, otherwise go to the instruction with label l_3).

A register machine generates a set of natural numbers in the following manner: we start computing with all m registers empty, with the instruction labeled l_0; if the label l_h is reached, then the computation *halts* and the value of register 1 is the number generated by the computation (all other registers can be assumed to be empty at that time). The set of all natural numbers generated in this way by M is denoted by $N(M)$. It is known that nondeterministic register machines (with three registers) can compute any set of Turing computable sets of natural numbers.

A register machine can be easily simulated by a P system with symport/antiport rules. The idea is illustrated in ❷ *Fig. 3*, where we have represented the initial configuration of the system, the rules associated with the unique membrane, and the set E of objects present in the environment.

The value of each register r is represented by the multiplicity of object $a_r, 1 \leq r \leq m$, in the unique membrane of the system. The labels from B, as well as their primed versions, are also objects of our system. We start with the unique object l_0 present in the system. In the presence of a label object l_1 we can simulate the corresponding instruction $l_1 : (\text{add}(r), l_2, l_3)$ or $l_1 : (\text{sub}(r), l_2, l_3)$.

◼ **Fig. 3**

An example of symport/antiport P system.

The simulation of an add instruction is clear, so we discuss only a sub instruction. The object l_1 exits the system in exchange for the two objects $l_1' l_1''$ (rule $(l_1, out; l_1' l_1'', in)$). In the next step, if any copy of a_r is present in the system, then l_1' has to exit (rule $(l_1' a_r, out; l_1''', in)$), thus diminishing the number of copies of a_r by one, and bringing inside the object l_1'''; if no copy of a_r is present, which corresponds to the case when the register r is empty, then the object l_1' remains inside. Simultaneously, rule $(l_1'', out; l_1^{iv}, in)$ is used, bringing inside the "checker" l_1^{iv}. Depending on what this object finds in the system, either l_1''' or l_1', it introduces the label l_2 or l_3, respectively, which corresponds to the correct continuation of the computation of the register machine.

When the object l_h is introduced, it is expelled into the environment and the computation stops. Clearly, the (halting) computations in Π directly correspond to (halting) computations in M; hence $N(M) = N(\Pi)$.

2.8 P Systems with Active Membranes

In the systems discussed above, the membrane structure is static, the membranes can be at most dissolved, which is not quite similar to what happens in biology, where membranes can be created or divided, and operations of exocytosis, endocytosis, phagocytosis change the membrane structure. In this section only the operation of membrane division is considered – actually, a case when the membranes play a direct part in the rules is considered.

This leads to a very important class of P systems, the main one used in devising polynomial solutions to computationally hard problems, namely the class of *P systems with active membranes*, which are constructs of the form

$$\Pi = (O, H, \mu, w_1, \ldots, w_m, R)$$

where $O, w_1, \ldots, w_m$ are as in a P system with multiset rewriting rules or with symport/antiport rules, H is a finite set of *labels* for membranes, μ is a membrane structure of degree $m \geq 1$, with polarizations associated with membranes, that is "electrical charges" $\{+, -, 0\}$, and R is a finite set of rules, of the following forms:

(a) $[_h a \to v]_h^e$, for $h \in H, e \in \{+, -, 0\}, a \in O, v \in O^*$ (object evolution rules);
(b) $a[_h]_h^{e_1} \to [_h b]_h^{e_2}$, for $h \in H, e_1, e_2 \in \{+, -, 0\}, a, b \in O$ (in communication rules);
(c) $[_h a]_h^{e_1} \to [_h]_h^{e_2} b$, for $h \in H, e_1, e_2 \in \{+, -, 0\}, a, b \in O$ (out communication rules);
(d) $[_h a]_h^e \to b$, for $h \in H, e \in \{+, -, 0\}, a, b \in O$ (dissolving rules);
(e) $[_h a]_h^{e_1} \to [_h b]_h^{e_2} [_h c]_h^{e_3}$, for $h \in H, e_1, e_2, e_3 \in \{+, -, 0\}, a, b, c \in O$
    (division rules for elementary membranes; in reaction with an object, the membrane is divided into two membranes with the same label, and possibly of different polarizations; the object specified in the rule is replaced in the two new membranes possibly by new objects; the remaining objects are duplicated and may evolve in the same step by rules of type (a)).

Note that each rule has specified the membrane where it is applied; that is why a global set of rules R is considered. The rules are applied in a maximally parallel manner, with the following details: a membrane can be subject to only one rule of types (b)–(e) (the rules of type (a) are not considered to involve the membrane where they are applied); inside each membrane, the rules of type (a) are applied in parallel; each copy of an object is used by only one rule of any type. The rules are used in a bottom-up manner: we use first the rules of type (a), and then the rules of other types; in this way, in the case of dividing membranes, the result of using first the rules of type (a) is duplicated in the newly obtained membranes). As usual, only halting computations give a result in the form of the number (or the vector) of objects expelled into the environment during the computation.

The set H of labels has been specified because it is possible to allow the change of membrane labels. For instance, a division rule can be of the more general form

$$(e') \; [_{h_1} a \,]_{h_1}^{e_1} \rightarrow [_{h_2} b \,]_{h_2}^{e_2} [_{h_3} c \,]_{h_3}^{e_3} \quad \text{for } h_1, h_2, h_3 \in H, e_1, e_2, e_3 \in \{+, -, 0\}, a, b, c \in O.$$

The change of labels can also be considered for rules of types (b) and (c). Also, we consider the possibility of dividing membranes into more than two copies, or even of dividing nonelementary membranes (in such a case, all inner membranes are duplicated in the new copies of the membrane).

P systems with active membranes can be used for computing numbers, but their main usefulness is in devising polynomial-time solutions to computationally hard problems, by a time–space trade-off. The space is created both by duplicating objects and by dividing membranes. This possibility is illustrated by the way of using membrane division for the generation of all 2^n truth assignments possible for n propositional variables; this is the basic step in solving SAT (satisfiability of propositional formulas in the conjunctive normal form) in polynomial time in this framework.

Assume that we have the variables $x_1, x_2, \ldots, x_n$; we construct the following system (of degree 2):

$$\Pi = (O, \{1, 2\}, [_1 [_2 \,]_2^0]_1^0, \lambda, a_1 a_2 \ldots a_n c_1, R)$$
$$O = \{a_i, c_i, t_i, f_i \mid 1 \leq i \leq n\} \cup \{\text{check}\}$$
$$R = \{[_2 a_i]_2^0 \rightarrow [_2 t_i]_2^0 [_2 f_i]_2^0 \mid 1 \leq i \leq n\}$$
$$\cup \{[_2 c_i \rightarrow c_{i+1}]_2^0 \mid 1 \leq i \leq n - 1\}$$
$$\cup \{[_2 c_n \rightarrow \text{check}]_2^0, \; [_2 \text{check}]_2^0 \rightarrow \text{check}[_2 \,]_2^+\}$$

We start with the objects $a_1, \ldots, a_n$ in the inner membrane and we divide this membrane repeatedly by means of the rules $[_2 a_i]_2^0 \rightarrow [_2 t_i]_2^0 [_2 f_i]_2^0$; note that the object a_i used in each step is nondeterministically chosen, but each division replaces that object by t_i (for *true*) in one membrane and with f_i (for *false*) in the other membrane; hence after n steps the configuration obtained is the same regardless of the order of expanding the objects. Specifically, we get 2^n membranes with label 2, with each one containing a truth assignment for the n variables. Simultaneously with the division, we have to use the rules of type (a) that update the "counter" c; hence at each step we increase by one the subscript of c. Therefore, when all variables have been expanded, we get the object check in all membranes (the rule of type (a) is used first, and after that the result is duplicated in the newly obtained membranes). In step $n + 1$, this object exits each copy of membrane 2, changing its polarization to positive; this is meant to signal the fact that the generation of all truth assignments is completed, and we can start checking the truth values of (the clauses of) the propositional formula.

The previous example was also chosen to show that the polarizations of membranes are not used while generating the truth assignments, though they might be useful after that; until now, this has been the case in all polynomial-time solutions to **NP**-complete problems obtained in this framework. An important *open problem* in this area is whether or not the polarizations can be avoided. This can be done if other ingredients are considered, such as label changing or division of nonelementary membranes, but without adding such features the best result obtained so far is that the number of polarizations can be reduced to two.

2.9 Structuring the Objects

In the previous classes of P systems, the objects were considered atomic, identified only by their name, but in a cell many chemicals are complex molecules (e.g., proteins, DNA molecules, other large macromolecules), whose structures can be described by strings or more complex data, such as trees, arrays, etc. Also from a mathematical point of view, it is natural to consider P systems with string objects.

Such a system has the form

$$\Pi = (V, T, \mu, M_1, \ldots, M_m, R_1, \ldots, R_m)$$

where V is the alphabet of the system, $T \subseteq V$ is the terminal alphabet, μ is the membrane structure (of degree $m \geq 1$), $M_1, \ldots, M_m$ are finite sets of strings present in the m regions of the membrane structure, and $R_1, \ldots, R_m$ are finite sets of string-processing rules associated with the m regions of μ.

We have given the system in general form, with a specified terminal alphabet (we say that the system is *extended*; if $V = T$, then the system is said to be *non-extended*), and without specifying the type of rules. These rules can be of various forms, but we consider here only two cases: rewriting and splicing.

In a *rewriting P system*, the string objects are processed by rules of the form $a \rightarrow u(\text{tar})$, where $a \rightarrow u$ is a context-free rule over the alphabet V and tar is one of the target indications here, in, and out. When such a rule is applied to a string $x_1 a x_2$ in a region i, we obtain the string $x_1 u x_2$, which is placed in region i, in any inner region, or in the surrounding region, depending on whether tar is here, in, or out, respectively. The strings that leave the system do not come back; if they are composed only of symbols from T, then they are considered as generated by the system. The language of all strings generated in this way is denoted by $L(\Pi)$.

In a *splicing P system*, we use splicing rules like those in DNA computing, that is, of the form $u_1 \# u_2 \$ u_3 \# u_4$, where $u_1, u_2, u_3,$ and u_4 are strings over V. For four strings $x, y, z, w \in V^*$ and a rule $r : u_1 \# u_2 \$ u_3 \# u_4$, we write

$$(x, y) \vdash_r (z, w) \text{ if and only if } x = x_1 u_1 u_2 x_2, \; y = y_1 u_3 u_4 y_2,$$
$$z = x_1 u_1 u_4 y_2, \; w = y_1 u_3 u_2 x_2,$$
$$\text{for some } x_1, x_2, y_1, y_2 \in V^*$$

We say that x and y are spliced at the sites $u_1 u_2$ and $u_3 u_4$, respectively, and the result of the splicing (obtained by recombining the fragments obtained by cutting the strings as indicated by the sites) is the strings z and w.

In our case, we add target indications to the two resulting strings, that is, we consider rules of the form $r : u_1 \# u_2 \$ u_3 \# u_4 (\text{tar}_1, \text{tar}_2)$, with tar_1 and tar_2 one of here, in, and out. The meaning is as standard: After splicing the strings x, y from a given region, the resulting strings z, w are moved to the regions indicated by $\text{tar}_1, \text{tar}_2$, respectively. The language generated by such a system consists again of all strings over T sent into the environment during the computation, without considering only halting computations.

An example of a rewriting or a splicing P system is not given here, but an important extension of rewriting rules — namely, *rewriting with replication* — is introduced. In such systems, the rules are of the form $a \rightarrow (u_1, \text{tar}_1) \| (u_2, \text{tar}_2) \| \ldots \| (u_n, \text{tar}_n)$, with the meaning that by rewriting a string $x_1 a x_2$ we get n strings, $x_1 u_1 x_2, x_1 u_2 x_2, \ldots, x_1 u_n x_2$, which have to be moved in the regions indicated by targets $\text{tar}_1, \text{tar}_2, \ldots, \text{tar}_n$, respectively.

The replicated rewriting is important because it provides the possibility to replicate strings, thus enlarging the workspace. Indeed, this is one of the frequently used ways to generate an exponential workspace in linear time, used then for solving computationally hard problems in polynomial time.

2.10 P Systems with Objects on Membranes

Consider that objects placed only in membranes cover only part of the biological reality, where most of the reactions taking place in a cell are controlled by proteins bound on membranes. Thus, it is a natural challenge for membrane computing to take care of this situation and, indeed, several types of P systems were investigated where membranes carry objects.

Three main approaches are considered in the literature. The first one is inspired from Cardelli (2005), where six basic operations were introduced. Counterparts of biological operations with membranes and two calculi were defined on this basis: pino/exo/phago calculus and mate/drip/bud calculus. Part of these operations were rephrased in terms of membrane computing and used in defining classes of P systems where objects are placed only on membranes. We call *membrane systems* the obtained P systems. A few ideas from Cardelli and Păun (2006) will be recalled below.

In the above-mentioned systems, we do not have objects also in compartments, as in usual P systems. Two models taking care of this aspect are what we call *the Ruston model* and *the Trento model*. In both cases, objects are placed both on the membranes and in their compartments, with the difference that in the former case the objects do not change their places, those bound on membranes remain there, while in the latter model the objects can move from compartments to membranes and back. The first type of systems were introduced in Păun and Popa (2006) and the second one in Brijder (2007). Several further papers were devoted to these systems, but the reader can refer to Păun et al. (2009) and www.ppage.psystems.eu for details.

Just to have an idea about this direction of research, here the P systems based on mate/drip operations are introduced.

One can consider an alphabet A of special objects called "proteins." A membrane that has associated a multiset $u \in A^*$ is represented by $[\,]_u$; we also say that u *marks* the membrane. When necessary, we can also use labels for identifying membranes, but in most cases the objects marking them are sufficient. The operations considered here are the following:

$$\text{mate: } [\,]_{ua}[\,]_v \rightarrow [\,]_{uxv}$$
$$\text{drip: } [\,]_{uav} \rightarrow [\,]_{ux}[\,]_v$$

where $a \in V$ and $u, v, x \in A^*$. The length of the string uxv (i.e., the total multiplicity of the multiset represented by this string) from each rule is called the *weight* of the rule.

In each case, multisets of objects are transferred from membranes appearing in the left-hand side of rules to membranes appearing in the right-hand side of rules, with protein a evolved into the multiset x. It is important to note that in *drip* rules the multiset uxv is precisely split into ux and v, with these two multisets assigned to the two new membranes.

The rules are applied to given membranes if they are marked with multisets of objects, which include the multisets mentioned in the left-hand side of rules. All objects placed on membranes but not involved in the rules are not affected by the use of rules; in the case of *drip*, these objects are randomly distributed to the two resulting membranes. In the case of *mate*, the result is uniquely determined.

The contents of membranes involved in these operations (i.e., the membranes possibly placed inside) is transferred from the input membranes to the output membranes in the same way as in brane calculi. Denoting these contents (empty or consisting of other membranes) by **P**, **Q**, we can indicate the effect of the two operations as follows:

$$\text{mate: } [\mathbf{P}]_{ua}[\mathbf{Q}]_v \rightarrow [\mathbf{P}\ \mathbf{Q}]_{uxv}$$
$$\text{drip: } [\mathbf{P}]_{uav} \rightarrow [\ ]_{ux}[\mathbf{P}]_v$$

Now, a P system of the following form can be defined

$$\Pi = (A, \mu, u_0, u_1, \dots, u_m, R)$$

where A is an alphabet (its elements are called *objects* or *proteins*), μ is a membrane structure with at least two membranes (hence $m \geq 1$), labeled with $0, 1, \dots, m$, where 0 is the skin membrane, $u_0, u_1, \dots, u_m \in A^*$ are multisets of objects bound to the membranes of μ at the beginning of the computation, with $u_0 = \lambda$, R is a finite set of mate, drip rules, of the forms specified above, using objects from the set A.

As usual in membrane computing, the evolution of the system proceeds through transitions among configurations, based on the nondeterministic maximally parallel use of rules. In each step, each membrane and each object can be involved in only one rule. A membrane remains unchanged if no rule is applied to it. The skin membrane never evolves.

A computation that starts from the initial configuration (the one described by μ and multisets $u_0, u_1, \dots, u_m$) is successful only if (1) it halts, that is, it reaches a configuration where no rule can be applied, and (2) in the halting configuration there are only two membranes, the skin (marked with the empty multiset) and an inner one. The result of a successful computation is the number of proteins that mark the inner membrane in the halting configuration.

The set of all numbers computed in this way by Π is denoted by $N(\Pi)$. The family of all sets $N(\Pi)$ computed by P systems Π using at any moment during a halting computation at most m membranes, and mate, drip rules of weight at most p, q, respectively, is denoted by $NOP_m(\text{mate}_p, \text{drip}_q)$.

3 Tissue-Like P Systems

We pass now to consider a very important generalization of the membrane structure, passing from the cell-like structure, described by a tree, to a tissue-like structure, with the membranes placed in the nodes of an arbitrary graph (which corresponds to the complex communication networks established among adjacent cells by making their protein channels cooperate, moving molecules directly from one cell to another). Actually, in the basic variant of tissue-like P systems, this graph is virtually a total one; what matters is the communication graph, dynamically defined during computations. In short, several (elementary) membranes – also called cells – are freely placed in a common environment; they can communicate either with each other or with the environment by symport/antiport rules. Specifically, antiport rules of the form $(i, x/y, j)$ are considered, where i, j are labels of cells or at most one is zero, identifying the environment, and x, y are multisets of objects. This means that the multiset x is moved from i to j at the same time as the multiset y is moved from j to i. If one of the multisets x, y is empty, then we have, in fact, a symport rule. Therefore, the communication among cells is done either directly, in one step, or indirectly, through the environment: One cell throws some objects out and other cells can take these objects in the next step or later. As in symport/antiport P systems, the environment contains a specified set of objects in arbitrarily many copies. A computation develops as standard, starting from the initial configuration and using the rules in the nondeterministic maximally parallel manner. When halting, we count the objects from a specified cell, and this is the result of the computation.

The graph plays a more important role in so-called *tissue-like P systems with channel-states*, whose rules are associated with synapses and are of the form $(s, x/y, s')$, where s, s' are states from a given finite set and x, y are multisets of objects. Such a rule of the form $(s, x/y, s') \in R_{(i,j)}$ is interpreted as an antiport rule $(i, x/y, j)$ as above, acting only if the synapse (i, j) has the state s; the application of the rule means (1) moving the objects specified by x from cell i (from the environment, if $i = 0$) to cell j, at the same time with the move of the objects specified by y in the opposite direction, and (2) changing the state of the synapse from s to s'. The use of states provides a powerful way to control the application of rules.

A still more elaborated class of systems, called *population P systems*, was introduced in Bernardini and Gheorghe (2004) with motivations related to the dynamics of cells in skin-like tissues, populations of bacteria, and colonies of ants. These systems are highly dynamical: not only the links between cells, corresponding to the channels from the previous model with states assigned to the channels, can change during the evolution of the system, but also the cells can change their names, can disappear (get dissolved), and can divide, thus producing new cells; these new cells inherit, in a well-specified sense, links with the neighboring cells of the parent cell. The generality of this model makes it rather attractive for applications in areas such as those mentioned above, related to tissues, populations of bacteria, etc.

A further interesting version of tissue-like P systems is that of *neural-like P systems*, where we again use a population of cells (each one identified by its label) linked by a specified set of synapses. This time, each cell has at every moment a state from a given finite set of states, its contents in the form of a multiset of objects from a given alphabet of objects, and a set of rules for processing these objects.

The rules are of the form $sw \rightarrow s'(x, here)(y, go)(z, out)$, where s, s' are states and w, x, y, z are multisets of objects; in state s, the cell consumes the multiset w and produces the multisets x, y, z; the objects from multiset x remain in the cell, those of multiset y have to be communicated to the cells toward which there are synapses starting in the current cell; a multiset z, with the indication *out*, is allowed to appear only in a special cell, designated as the output cell, and for this cell the use of the previous rule entails sending the objects of z to the environment.

The computation starts with all cells in specified initial states, with initially given contents, and proceeds by processing the multisets from all cells, simultaneously, according to the local rules. Then the obtained objects are redistributed along synapses and send a result into the environment through the output cell; a result is accepted only when the computation halts.

Because of the use of states, there are several possibilities for processing the multisets of objects from each cell. In the *minimal* mode, a rule is chosen and applied once to the current pair (state, multiset). In the *parallel* mode, a rule is chosen, for example, $sw \rightarrow s'w'$, and used in the maximally parallel manner: The multiset w is identified in the cell contents in the maximal manner, and the rule is used for processing all these instances of w. Finally, in the *maximal* mode, we apply in the maximally parallel manner all rules of the form $sw \rightarrow s'w'$, that is, with the same states s and s' (note the difference with the parallel mode, where in each step a rule is chosen and only that rule is used as many times as possible).

There are also three ways to move the objects between cells (of course, we only move objects produced by rules in multisets with the indication *go*). Assume that a rule $sw \rightarrow s'(x, here)(y, go)$ is applied in a given cell i. In the *spread* mode, the objects from y are nondeterministically distributed to all cells j such that (i,j) is a synapse of the system. In the *one* mode, all the objects from y are sent to one cell j, provided that the synapse (i,j) exists. Finally, we can also *replicate* the objects of y, and each object from y is sent to all cells j such that (i,j) is an available synapse.

We do not enter into more details; on the one hand, this class of P systems still waits for a systematic investigation; on the other hand, there is another class of P systems inspired from the brain's organization, whose study is much more developed, the *spiking neural P systems*.

4   Spiking Neural P Systems

Spiking neural P systems (SN P systems) were introduced in Ionescu et al. (2006) with the aim of defining P systems based on ideas specific to spiking neurons, recently much investigated in neural computing.

Very shortly, an SN P system consists of a set of *neurons* (cells consisting of only one membrane) placed in the nodes of a directed graph and sending signals (*spikes*, denoted in what follows by the symbol a) along *synapses* (arcs of the graph). Thus, the architecture is that of a tissue-like P system, with only one kind of object present in the cells. The objects evolve by means of *spiking rules*, which are of the form $E/a^c \rightarrow a; d$, where E is a regular expression over $\{a\}$ and c, d are natural numbers, $c \geq 1$, $d \geq 0$. The meaning is that a neuron containing k spikes such that $a^k \in L(E)$, $k \geq c$, can consume c spikes and produce one spike, after a delay of d steps. This spike is sent to all neurons to which a synapse exists outgoing from the neuron where the rule was applied. There also are *forgetting rules*, of the form $a^s \rightarrow \lambda$, with the meaning that $s \geq 1$ spikes are forgotten, provided that the neuron contains exactly s spikes. We say that the rules "cover" the neuron: all spikes are taken into consideration when using a rule. The system works in a synchronized manner; that is, in each time unit, each neuron that can use a rule should do it, but the work of the system is sequential in each neuron: Only (at most) one rule is used in each neuron. One of the neurons is considered to be the *output neuron*, and its spikes are also sent to the environment. The moments of time when a spike is emitted by the output neuron are marked with 1, the other moments are marked with 0. This binary sequence is called the *spike train* of the system – it might be infinite if the computation does not stop.

In the spirit of spiking neurons, the result of a computation is encoded in the distance between consecutive spikes sent into the environment by the (output neuron of the) system. For example, we can consider only the distance between the first two spikes of a spike train, or the distances between the first k spikes, the distances between all consecutive spikes, taking into account all intervals or only intervals that alternate, all computations or only halting computations, etc.

An SN P system can also be used in the accepting mode: A neuron is designated as the *input neuron* and two spikes are introduced in it, at an interval of n steps; the number n is accepted if the computation halts.

Another possibility is to consider the spike train itself as the result of a computation, and then we obtain a (binary) language generating device. We can also consider both input and output neurons and then an SN P system can work as a transducer. Languages on arbitrary alphabets can be obtained by generalizing the form of rules: Take rules of the form $E/a^c \rightarrow a^p; d$, with the meaning, provided that the neuron is covered by E, c spikes are consumed and p spikes are produced, and sent to all connected neurons after d steps (such rules are called *extended*). Then, with a step when the system sends out i spikes, we associate a symbol b_i, and thus we get a language over an alphabet with as many symbols as the number of spikes simultaneously produced. Another natural extension is to consider several output neurons, thus producing vectors of numbers, not only single numbers.

The technical details are skipped, but a simple example is considered (if a rule $E/a^c \rightarrow a; d$ has $L(E) = \{a^c\}$, then it can be written in the simplified form $a^c \rightarrow a; d$):

$$\Pi_1 = (O, \sigma_1, \sigma_2, \sigma_s, syn, out), \text{ with}$$
$$O = \{a\} \text{ (alphabet, with only one object, the spike)}$$
$$\sigma_1 = (2, \{a^2/a \rightarrow a; 0, \quad a \rightarrow \lambda\})$$
$$\text{(first neuron: initial number of spikes, rules)}$$
$$\sigma_2 = (1, \{a \rightarrow a; 0, \quad a \rightarrow a; 1\})$$
$$\text{(second neuron: initial number of spikes, rules)}$$
$$\sigma_3 = (3, \{a^3 \rightarrow a; 0, \quad a \rightarrow a; 1, \quad a^2 \rightarrow \lambda\})$$
$$\text{(third neuron: initial number of spikes, rules)}$$
$$syn = \{(1, 2), (2, 1), (1, 3), (2, 3)\} \text{ (synapses)}$$
$$out = 3 \text{ (output neuron).}$$

This system functions as follows. All neurons can fire in the first step, with neuron σ_2 choosing nondeterministically between its two rules. Note that neuron σ_1 can fire only if it contains two spikes; one spike is consumed, the other remains available for the next step.

Both neurons σ_1 and σ_2 send a spike to the output neuron, σ_3; these two spikes are forgotten in the next step. Neurons σ_1 and σ_2 also exchange their spikes; thus, as long as neuron σ_2 uses the rule $a \rightarrow a; 0$, the first neuron receives one spike, thus completing the needed two spikes for firing again.

However, at any moment, starting with the first step of the computation, neuron σ_2 can choose to use the rule $a \rightarrow a; 1$. On the one hand, this means that the spike of neuron σ_1 cannot enter neuron σ_2 and it only goes to neuron σ_3; in this way, neuron σ_2 will never work again because it remains empty. On the other hand, in the next step neuron σ_1 has to use its forgetting rule $a \rightarrow \lambda$, while neuron σ_3 fires, using the rule $a \rightarrow a; 1$. Simultaneously, neuron σ_2 emits its spike, but it cannot enter neuron σ_3 (it is closed at this moment); the spike enters neuron σ_1, but it is forgotten in the next step. In this way, no spike remains in the system. The computation ends with the expelling of the spike from neuron σ_3. Because of the waiting moment imposed by the rule $a \rightarrow a; 1$ from neuron σ_3, the two spikes of this neuron cannot be consecutive, but at least two steps must exist in between.

Thus, it can be concluded that Π computes/generates all natural numbers greater than or equal to 2.

There are several classes of SN P systems, for instance, asynchronous SN P systems (no clock is considered, any neuron may use or not a rule), with exhaustive use of rules (when enabled, a rule is used as many times as made possible by the spikes present in a neuron), with certain further conditions imposed on the halting configuration, etc.

5 Computing Power

As previously mentioned, many classes of P systems that combine various ingredients (as described above or similar ones) are able to simulate Turing machines; hence they are *computationally complete*. Always, the proofs of results of this type are constructive, and this has an important consequence from the computability point of view: in this way we obtain *universal* (hence *programmable*) P systems. In short, starting from a universal Turing machine (or an equivalent universal device), we get an equivalent universal P system. Among others, this implies that in the case of Turing complete classes of P systems, the hierarchy on the number of membranes always collapses — at most at the level of the universal P systems. There are, however, classes of nonuniversal P systems for which "the number of membranes matters," to cite the title of Ibarra (2005), where two classes of P systems were defined for which the hierarchies on the number of membranes are infinite. Actually, the number of membranes in a P system sufficient to characterize the power of Turing machines is always rather small.

Here only a few of the most interesting types of universality results for cell-like P systems are mentioned:

1. P systems with symbol-objects with catalytic rules, using only two catalysts and two membranes, are universal.
2. P systems with symport/antiport rules of a restricted size (e.g., three membranes, symport rules of weight 2, and no antiport rules, or three membranes and minimal symport and antiport rules) are universal.
3. P systems with symport/antiport rules (of arbitrary size), using only three membranes and only three objects, are universal.
4. P systems with active membranes, using rules of types (a), (b), (c) and two polarizations are universal.
5. P systems with string objects and rewriting rules usually generate only matrix languages, but adding priorities, ways to control the permeability of membranes, inhibitors, or using the rules in the leftmost way lead to universality.
6. P systems with *mate, drip* rules controlled by at most four objects and using at most five membranes are universal.

Universality can also be obtained for other types of P systems, such as tissue-like systems, SN P systems (provided that the number of spikes present in the neurons is not bounded, otherwise only semilinear sets of numbers can be computed), accepting P systems, etc. However, we do not recall here such results; one can refer to Păun et al. (2009) and www.ppage.psystems. eu for details. In general, around these results there are a series of open problems, mainly related to their optimality. Some of these results are continuously improved, so the interested reader is advised to consult the membrane computing current bibliography for the most recent results.

In the accepting mode, an interesting issue appears, that of using deterministic systems. Most universality results were obtained in the deterministic case, but there also are situations where the deterministic systems are strictly less powerful than the nondeterministic ones. This is proven in Ibarra and Yen (2006) for the accepting catalytic P systems.

It is worth noting that the proofs of computational completeness are based on simulating various types of grammars with restricted derivation (mainly matrix grammars with appearance checking) or on simulating register machines. In the case of SN P systems, this has an interesting consequence: Starting the proofs from small universal register machines, one can find small universal SN P systems. For instance, as shown in Păun and Păun (2007), there are universal computing SN P systems with 84 neurons using standard rules and with only 49 neurons using extended rules. In the generative case, the best results are 79 and 50 neurons, respectively.

6 Computational Efficiency

The computational power (the "competence") is only one of the important questions to be dealt with when defining a new (bio-inspired) computing model. The other fundamental question concerns the computing *efficiency*. Because P systems are parallel computing devices, it is expected that they can solve hard problems in an efficient manner – and this expectation is confirmed for systems provided with ways for producing an exponential workspace in a linear time. Mainly, three such biologically inspired possibilities have been considered so far in the literature, and all of them were proven to lead to polynomial solutions to **NP**-complete problems, by a time–space trade-off.

These three ideas are membrane division, membrane creation, and string replication. The standard problems addressed in this framework were decidability problems, starting with SAT, the Hamiltonian path problem, and the node covering problem. Other types of problems were considered, such as the problem of inverting one-way functions, or the subset-sum and the knapsack problems (note that the last two are numerical problems, where the answer is not of the yes/no type, as in decidability problems). In general, the approach is of a brute force type: All candidate solutions are generated, making use of the massive parallelism (e.g., all truth assignments of the variables appearing in a SAT problem); then they are checked, again in parallel, in search of a solution. The computation may proceed nondeterministically provided that it is *confluent*: Either eventually we reach the same configuration and then we continue deterministically (strong confluence), or all computations halt and each of them provides the same result (weak confluence).

The formal framework for dealing with complexity matters in terms of P systems is provided by the *accepting P systems with input*: A family of P systems of a given type is constructed starting from a given problem, and an instance of the problem is introduced as an input in one such systems; working in a deterministic (or a confluent) mode, the system sends to the environment the answer to the respective instance. The family of systems should be constructed in a uniform mode by a Turing machine, working a polynomial time.

This direction of research has been much investigated. A large number of problems were considered, the membrane computing complexity classes were refined, characterizations of the **P** $\neq$ **NP** conjecture were obtained in this framework, several characterizations of the class **P**, even problems that are **PSPACE**-complete, were proven to be solvable in polynomial time by means of membrane systems provided with membrane division or membrane creation. An important (and difficult) problem is that of finding the border between efficiency and non-efficiency: which ingredients should be used in order to be able to solve hard problems in a polynomial time? The reader is referred to the complexity chapter from Păun et al. (2006) and to the literature available at www.ppage.psystems.eu for details – including many problems that are still open in this area.

7 Applications

There are many features of membrane computing that make it attractive for applications in several disciplines, especially for biology and biomedicine: *distribution* (with the important system–part interaction, emergent behavior, nonlinearly resulting from the composition of local behaviors), easy programmability, scalability/extensibility, transparency (multiset rewriting rules are nothing other than reaction equations such as those customarily used in chemistry and biochemistry), parallelism, nondeterminism, communication, and so on and so forth.

As expected, most of the applications have been carried out in biology. These applications are usually based on experiments using programs for simulating/implementing P systems on usual computers, and there are already several such programs, more and more elaborated (e.g., with better and better interfaces, which allow for the friendly interaction with the program). An overview of membrane computing software reported in the literature (some programs are available at the webpage www.ppage.psystems.eu together with sample applications) can be found in the volume Ciobanu et al. (2006).

Of course, when using a P system for simulating a biological process, we are no longer interested in its computing behavior (power, efficiency, etc.), but in its evolution in time; the P system is then interpreted as a dynamical system, and its trajectories are of interest, its "life." Moreover, the ingredients we use are different from those considered in theoretical investigations. For instance, in mathematical terms, we are interested in results obtained with a minimum of premises and with weak prerequisites, while the rules are used in ways inspired from automata and language theory (e.g., in a maximally or minimally parallel way), but when dealing with applications the systems are constructed in such a way that they capture the features of reality (for instance, the rules are of a general form, they are applied according to probabilistic strategies, based on stoichiometric calculations, the systems are not necessarily synchronized, and so on).

The typical applications run as follows. One starts from a biological process described in general in graphical terms (chemicals are related by reactions represented in a graph-like manner, with special conventions for capturing the context-sensitivity of reactions, the existence of promoters or inhibitors, etc.) or already available in data bases in SBML (system biology mark-up language) form; these data are converted into a P system that is introduced in a simulator. The way the evolution rules (reactions) are applied is the key point in constructing this simulator (often, the classical Gillespie algorithm is used in compartments, or multi-compartmental variants of it are considered). As a result, the evolution in time of the multiplicity of certain chemicals is displayed, thus obtaining a graphical representation of the interplay in time of certain chemicals, their growth and decay, and so on. Many illustrations of this scenario can be found in the literature, many times dealing with rather complex processes.

Besides applications in biology, applications were reported in computer graphics, linguistics (both as a representation language for various concepts related to language evolution, dialogue, semantics, and making use of the parallelism, in solving parsing problems in an efficient way), economics (where many biochemical metaphors find a natural counterpart, with the mentioning that the "reactions" that take place in economics, for instance, in market-like frameworks, are not driven only by probabilities/stoichiometric calculations, but also by psychological influences, which makes the modeling still more difficult than in biology), computer science (in devising sorting and ranking algorithms), cryptography, etc.

A very promising direction of research, namely, applying membrane computing in devising approximate algorithms for hard optimization problems, was initiated in Nishida (2004), who proposed *membrane algorithms* as a new class of distributed evolutionary algorithms. These algorithms can be considered as high-level (distributed and dynamically evolving their structure during the computation) evolutionary algorithms. In short, candidate solutions evolve in compartments of a (dynamical) membrane structure according to local algorithms, with better solutions migrating down in the membrane structure; after a specified halting condition is met, the current best solution is extracted as the result of the algorithm.

This strategy was checked for several optimization problems, and the results were more than encouraging for a series of benchmark problems: The convergence is very fast, the number of membranes is rather influential on the quality of the solution, the method is reliable, both the average quality and the worst solutions were good enough and always better than the average and the worst solutions given by existing optimization procedures.

8 Closing Remarks

This chapter aimed to provide only a quick presentation of membrane computing, of basic ideas and of types of results and applications. Many more things are available in the literature: further classes of P systems (e.g., we said nothing about conformon-P systems, discussed in detail in Frisco (2008)), other problems to investigate (e.g., normal forms, semantics), links with other areas (e.g., X-machines, Petri nets), software (many programs are available, most of them also on www.ppage.psystems.eu, together with examples of applications), and so on. The area also has many unsolved problems and directions for research that wait for further efforts.

We mentioned before the borderline between universality and non-universality and between efficiency and non-efficiency concerning the succinctness of universal P systems or P systems able to solve hard problems in polynomial time. Then, because universality implies undecidability of all nontrivial questions, an important issue is that of finding classes of P systems with decidable properties.

This is also related to the use of membrane computing as a modeling framework: If no insights can be obtained in an analytical manner algorithmically, then what remains is to simulate the system on a computer. To this aim, better programs are still needed, maybe parallel implementations able to handle real-life questions.

Several research topics concern complexity investigations: uniform versus semiuniform solutions, the role of unary encoding, confluent versus deterministic solutions, the possibility of using precomputed resources, activated during the computation (as done for SN P systems in Chen et al. (2006)), etc.

We conclude by stating that membrane computing is a young and well-developed branch of natural computing, rather promising for applications, which still needs considerable research efforts – which might be a challenge for the reader.

References

Bernardini F, Gheorghe M (2004) Population P systems. J Univ Comput Sci 10(5):509–539

Brijder R, Cavaliere M, Riscos-Núñez A, Rozenberg G, Sburlan D (2007) Membrane systems with marked membranes. Electron Notes Theor Comput Sci 171:25–36

Calude CS, Păun Gh, Rozenberg G, Salomaa A (eds) (2001) Multiset processing. Mathematical, computer science, and molecular computing points of view. Lecture notes in computer science, vol. 2235, Springer, Berlin

Cardelli L (2005) Brane calculi – interactions of biological membranes. Lecture notes in computer science vol. 3082, pp 257–280

Cardelli L, Păun Gh (2006) An universality result for a (mem)brane calculus based on mate/drip operations. Intern J Found Comput Sci 17:49–68

Chen H, Ionescu M, Ishdorj T-O (2006) On the efficiency of spiking neural P systems. In: Proceedings of 8th international conference on electronics, information, and communication, Ulan Bator, Mongolia, June 2006, pp 49–52

Ciobanu G, Păun Gh, Pérez-Jiménez MJ (eds) (2006) Applications of membrane computing. Springer, Berlin

Frisco P (2008) Computing with cells. Advances in membrane computing. Oxford University Press, Oxford, UK

Ibarra OH (2005) On membrane hierarchy in P systems. Theor Comput Sci 334(1–2):115–129

Ibarra OH, Yen HC (2006) Deterministic catalytic systems are not universal. Theor Comput Sci 363(2): 149–161

Ionescu M, Păun Gh, Yokomori T (2006) Spiking neural P systems. Fundamenta Informaticae 71(2–3): 279–308

Nishida TY (2004) An application of P systems: a new algorithm for NP-complete optimization problems. In: Callaos N et al. (eds) Proceedings of the 8th world multi-conference on systems, cybernetics and informatics, vol. V, pp 109–112

Păun A, Păun Gh (2007) Small universal spiking neural P systems. BioSystems 10(1):48–60

Păun A, Popa B (2006) P systems with proteins on membranes. Fundamenta Informaticae 72:467–483

Păun Gh (2001) Computing with membranes. J Comput Syst Sci 61(1):108–143 (and Turku Center for Computer Science-TUCS Report 208, November 1998, http://www.tucs.fi)

Păun Gh (2002) Membrane computing. An introduction. Springer, Berlin

Păun Gh, Rozenberg G, Salomaa A (eds) (2009) Handbook of membrane computing. Oxford University Press, Oxford, UK

The P Systems Website: ppage.psystems.eu. Accessed Apr 2010

Quantum Computation

Mika Hirvensalo

41 Mathematics for Quantum Information Processing

Mika Hirvensalo
Department of Mathematics, University of Turku, Finland
mikhirve@utu.fi

G. Rozenberg et al. (eds.), *Handbook of Natural Computing*, DOI 10.1007/978-3-540-92910-9_41,
© Springer-Verlag Berlin Heidelberg 2012

Abstract

Information processing in the physical world must be based on a physical system representing the information. If such a system has a quantum physical description, then one talks about *quantum information* processing. The mathematics needed to handle quantum information is somewhat more complicated than that used for classical information. The purpose of this chapter is to introduce the basic mathematical machinery used to describe quantum systems with finitely many potential observable values. A typical example of such a system is the *quantum bit* (or *qubit*, for short), which has at most two potential values for any physical observable. The mathematics for quantum information processing is based on vector spaces over complex numbers, and can be seen as a straightforward generalization of linear algebra of real vector spaces.

1   History

The earliest ideas on *quantum information* theory were introduced by von Neumann very soon after the initial mathematical machinery of quantum mechanics was developed (von Neumann 1927). On the other hand, the upcoming ideas of *quantum information processing* were still decades away. In the 1980s Paul Benioff presented the first ideas of computing in microsystems obeying quantum mechanics (Benioff 1980, 1982).

From the point of view of modern quantum information processing, the most important early idea was that of Richard Feynman. In his seminal article (Feynman 1982), he introduced a most interesting idea: that it may be impossible to simulate quantum mechanical systems with classical computers without an exponential slowdown (when the simulation efficiency is measured with respect to the number of particles in the system to be simulated).

The next important step toward quantum computing was taken by David Deutsch. In (Deutsch 1985), the aim was to address the Church–Turing thesis from the physical point of view. The Church–Turing thesis (see, e.g., Papadimitriou (1994)) says roughly that a Turing machine is the mathematical counterpart of the intuitive notion of an algorithm: For any intuitive algorithm, there exists a Turing machine performing the same task. Briefly, Deutsch's argument is as follows: Computation of any form is a physical process anyway, and to prove the Church–Turing thesis it is therefore enough to show that there is a physical system capable of simulating all other physical systems. Deutsch did not pay attention to the complexity of computation, and the pillars of quantum complexity theory were later established by Bernstein and Vazirani (1997).

Interesting quantum algorithms were already introduced by Deutsch and subsequently by Simon (1994), but the most famous quantum algorithms, polynomial-time factoring and the discrete logarithm, were designed by Shor (1994).

2   Hilbert Space Formalism

The mathematical objects that establish the basis for Newtonian mechanics are not very complicated: everything can be laid on the notion of vectors. To describe a point-sized body in a three-dimensional space, one can give the *position* $\mathbf{x} \in \mathbb{R}^3$ and *momentum* $\mathbf{p} \in \mathbb{R}^3$, related via $\mathbf{p} = m\frac{d}{dt}\mathbf{x}$, where m is the mass of the body. The dynamics of this very simple system is described by the Newtonian equation of motion

$$\mathbf{F} = \frac{d}{dt}\mathbf{p}$$

where $\mathbf{F}$ is the force affecting the body.

Toward a more holistic approach to mechanics, consider first a one-dimensional motion of a particle. The *total energy* of the system in this simple case is $H = \frac{1}{2}mv^2 + V(x) = \frac{p^2}{2m} + V(x)$, where $\frac{1}{2}mv^2$ is the *kinetic energy* and $V(x) = -\int_{x_0}^{x} F(s)ds$ is the *potential energy* of the particle depending only on the position x. Hence

$$\frac{\partial}{\partial x}H = V'(x) = -F(x) = -\frac{d}{dt}p$$

and

$$\frac{\partial}{\partial p}H = \frac{p}{m} = v = \frac{d}{dt}x$$

Differential equations

$$\frac{d}{dt}x = \frac{\partial}{\partial p}H, \quad \frac{d}{dt}p = -\frac{\partial}{\partial x}H$$

thus obtained are an example of *Hamiltonian reformulation* of classical mechanics. A general mechanical system can consist of many bodies, and the bodies themselves need not be point-like, but they may have internal degrees of freedom (such as capability of rotating).

▶ It is possible to a formulate a general mechanical system in terms of Hamiltonian mechanics, and the equation of motion becomes

$$\frac{d}{dt}x_i = \frac{\partial}{\partial p_i}H$$
$$\frac{d}{dt}p_i = -\frac{\partial}{\partial x_i}H$$

where $H = H(x_1, \ldots, x_n, p_1, \ldots, p_n, t)$ is the *Hamiltonian function* of the system, describing the total energy.

Variables $x_1, \ldots, x_n$ are called *generalized coordinates* and may involve, for example, the usual spatial coordinates and angles of rotation, whereas $p_1, \ldots, p_n$ are called *generalized momenta* (may include the usual momenta and the angular momenta, for instance).

The general form of Hamiltonian mechanics thus contains $\mathbf{x} = (x_1, \ldots, x_n)$, and $\mathbf{p} = (p_1, \ldots, p_n)$ (both vectors in $\mathbb{R}^n$) as a description of the *system state* at a specified moment of time. Collecting these two, one can say that the pair $(\mathbf{x}, \mathbf{p}) \in \mathbb{R}^{2n}$ is the *state vector* of the system. One can also say the system has n *degrees of freedom*.

It should not then come as a surprise that the description of quantum mechanics also needs vector spaces. However, it is not evident that quantum mechanics should be based on *complex* vector spaces. In fact, to introduce complex numbers is not unavoidable (this is trivial, as a complex number can be presented as a pair of real numbers), but mathematically very helpful. In what follows we will introduce the basic ingredients needed for the Hilbert space formalism of quantum mechanics.

The formalism of quantum mechanics was developed from *wave mechanics* (due to Erwin Schrödinger) and *matrix mechanics* (credited to Werner Heisenberg, Max Born, and Pascual Jordan). These mathematically equivalent formalisms were developed most notably by von Neumann (1932) into a Hilbert space formalism, the major topic of this chapter. It is noteworthy that, in general, the most interesting features emerge in *infinite-dimensional* Hilbert spaces, whereas from the quantum computational point of view, usually finite-dimensional Hilbert spaces are satisfactory enough.

To follow this representation, it is assumed that the reader has a basic knowledge of finite-dimensional *real* vector spaces $\mathbb{R}^n$ strong enough to cover objects such as vectors, linear mappings, matrices, bases, eigenvalues, eigenvectors, inner products, orthogonality, and subspaces. These notions can be found in many basic linear algebra textbooks, for instance in Cohn (1994). In the following, these notions are generalized to *complex* vector spaces. The resulting mathematical objects obtained are then used to describe quantum mechanics.

Definition 1 A *Hilbert space* is a vector space over complex numbers, equipped with a Hermitian inner product, complete with respect to the metric induced by the inner product.

To explain the terminology of the above definition, we choose the standpoint of quantum computing, and then the most important examples of Hilbert spaces are the finite-dimensional ones. As a prototype of such, one can introduce the following: Let

$$\mathbb{C}^n = \{(x_1, \ldots, x_n) \mid x_i \in \mathbb{C}\}$$

be the set of all n-tuples of complex numbers. When equipped with pointwise scalar multiplication

$$a(x_1, \ldots, x_n) = (ax_1, \ldots, ax_n)$$

and pointwise vector addition

$$(x_1, \ldots, x_n) + (y_1, \ldots, y_n) = (x_1 + y_1, \ldots, x_n + y_n)$$

set $\mathbb{C}^n$ evidently becomes a vector space. The (Hermitian) inner product in $\mathbb{C}^n$ is defined by

$$\langle \mathbf{x} \mid \mathbf{y} \rangle = x_1^* y_1 + \ldots + x_n^* y_n$$

where x^* means the complex conjugate of x. Defined as above, the inner product obviously satisfies the following properties.

▶ **Hermitian Inner Product**

- $\langle \mathbf{x} \mid \mathbf{y} \rangle = \langle \mathbf{y} \mid \mathbf{x} \rangle^*$
- $\langle \mathbf{x} \mid \mathbf{x} \rangle \geq 0$ and $\langle \mathbf{x} \mid \mathbf{x} \rangle = 0$ if and only if $\mathbf{x} = \mathbf{0}$
- $\langle \mathbf{x} \mid a\mathbf{y} + b\mathbf{z} \rangle = a\langle \mathbf{x} \mid \mathbf{y} \rangle + b\langle \mathbf{x} \mid \mathbf{z} \rangle$

According to our definition, the Hermitian inner product is linear with respect to the second argument, and *antilinear* with respect to the first one (meaning that $\langle a\mathbf{x} + b\mathbf{y} \mid \mathbf{z} \rangle = a^* \langle \mathbf{x} \mid \mathbf{z} \rangle + b^* \langle \mathbf{y} \mid \mathbf{z} \rangle$). In the mathematics literature, the Hermitian inner product is usually linear with respect to the first argument, but the convention chosen here is more familiar to quantum physics.

The inner product induces a *norm* by $\|\mathbf{x}\| = \sqrt{\langle \mathbf{x} \mid \mathbf{x} \rangle}$, and the completeness of Definition 1 means completeness with respect to this norm. That is, for each sequence $\mathbf{x}_n$ satisfying

$\lim_{n,m \to \infty} ||\mathbf{x}_n - \mathbf{x}_m|| = 0$, there is a limit $\mathbf{x} = \lim_{n \to \infty} \mathbf{x}_n$ also belonging to the space. It should be noted that the completeness is self-evident in the finite-dimensional case. In fact, the following theorem is not very difficult to prove.

Theorem 1 *An n-dimensional Hilbert space H_n (n finite) is isomorphic to $\mathbb{C}^n$.*

Hence the basic structure of quantum computing, a finite-dimensional Hilbert space, can in practice be treated as $\mathbb{C}^n$, the set of n-tuples of complex numbers. However, it is not only the space H_n on which the formalism of quantum mechanics will be based, but space $L(H_n)$, the set of all linear mappings $H_n \to H_n$, will be needed for the description. But when restricting to so-called *pure states* (defined later), space H_n is (almost) sufficient to accommodate the mathematical description of quantum mechanics. Especially in the theory of quantum computing, this is frequently the case. In any case, H_n is called the *state space* of an n-level quantum system. Here and hereafter, the dimension n of H_n equals the number of the *perfectly distinguishable states* of the system to be described. This means that there are n states of the system that can be mutually separated by observations with 100% accuracy.

In quantum computing, the spatial location of the system under investigation is not usually the most essential one; more frequently, the system's internal degrees of freedom are those of interest. This can be compared to a classical system, which, in addition to its position, has also its dynamics of rotation. Analogously, in quantum systems there can be internal degrees of freedom, and in current physical realizations of quantum computing, one of the most noteworthy is the *spin* of a particle. (As the name suggests, spin is closely related to the rotation of the particle around its axis, but spin has also nonclassical features.) It should be remembered that the Hamiltonian reformulation of classical mechanics can treat *all degrees of freedom* in an equal way, and the same is true also in quantum mechanics: the *Schrödinger equation of motion* treats all the degrees of freedom in an analogous way; the position of a particle would involve potentially an infinite number of potential values, but there are also internal properties (such as the spin) that can assume only a finite number of measurable values. It is also noteworthy that obviously one can concentrate only on one degree of freedom of a given system, leaving a lot of system information out. This means, for example, that it is evidently possible to create a mathematical description of a particle spin only, leaving out the particle position and other properties.

Before going into the formalism, it is useful to present some notation. So-called Dirac notation will be useful: for a vector $\mathbf{x} = (x_1, \ldots, x_n) \in \mathbb{C}^n$, notation

$$| \mathbf{x} \rangle = \begin{pmatrix} x_1 \\ \vdots \\ x_n \end{pmatrix} \tag{1}$$

stands for a column vector, and notation

$$\langle \mathbf{x} | = (x_1^*, \ldots, x_n^*) \tag{2}$$

for the row vector, the *adjoint* version of ❷ Eq. 1. Adjointness for matrices will be treated in more detail later.

For $\mathbf{x}$ and $\mathbf{y} \in \mathbb{C}^n$, notations $|\mathbf{x}\rangle\langle\mathbf{y}|$ and $\langle \mathbf{x} || \mathbf{y} \rangle$ are both interpreted as *tensor products* (alternative term: *Kronecker product*).

Definition 2 The tensor product of matrices A $(k \times l)$ and B $(m \times n)$ is a $km \times ln$-matrix defined as (in block form).

$$A \otimes B = \begin{pmatrix} a_{11}B & a_{12}B & \dots & a_{1l}B \\ a_{21}B & a_{22}B & \dots & a_{2l}B \\ \vdots & \vdots & \ddots & \vdots \\ a_{k1}B & a_{k2}B & \dots & a_{kl}B \end{pmatrix}$$

For vectors $\mathbf{x}=(x_1,\dots,x_n)$ and $\mathbf{y}=(y_1,\dots,y_n)$, we can hence write (identifying 1×1 matrices and real numbers)

$$\langle \mathbf{x} \,\|\, \mathbf{y} \rangle = \left(x_1^*, x_2^*, \dots, x_n^* \right) \otimes \begin{pmatrix} y_1 \\ y_2 \\ \vdots \\ y_n \end{pmatrix} = x_1^* y_1 + x_2^* y_2 + \dots + x_n^* y_n$$

which is exactly the same as given by the definition of inner product $\langle \mathbf{x} \mid \mathbf{y} \rangle$.

The tensor product, on the other hand,

$$| \mathbf{x} \rangle\langle \mathbf{y} | = \begin{pmatrix} x_1 \\ x_2 \\ \vdots \\ x_n \end{pmatrix} \otimes \left(y_1^*, y_2^*, \dots, y_n^* \right) = \begin{pmatrix} x_1 y_1^* & x_1 y_2^* & \dots & x_1 y_n^* \\ x_2 y_1^* & x_2 y_2^* & \dots & x_2 y_n^* \\ \vdots & \vdots & \ddots & \vdots \\ x_n y_1^* & x_n y_2^* & \dots & x_n y_n^* \end{pmatrix} \tag{3}$$

is not so evident, but also this matrix has a clear interpretation. Indeed, ❷ Definition 3 speaks about the *linear mapping*, whose matrix is in line (❷ Eq. 3).

Definition 3 The set of linear mappings $H_n \to H_n$ is denoted by $L(H_n)$. For any pair of vectors $\mathbf{x}, \mathbf{y} \in H_n$ we define $|\mathbf{x}\rangle\langle\mathbf{y}| \in L(H_n)$ by

$$|\mathbf{x}\rangle\langle\mathbf{y}|\mathbf{z} = \langle \mathbf{y} \mid \mathbf{z} \rangle \mathbf{x}$$

As mentioned, the matrix (❷ Eq. 3) is the matrix of linear mapping $|\mathbf{x}\rangle\langle\mathbf{y}|$, with respect to the natural basis (i.e., the basis $\mathbf{e}_1 = (1,0, \dots, 0)$, $\dots$, $\mathbf{e}_n = (0,0, \dots, 1)$). Recalling that $|\mathbf{z}\rangle$ actually represents the column vector consisting of the coordinates of $\mathbf{z}$ (with respect to the natural basis), we can also write

$$|\mathbf{x}\rangle\langle\mathbf{y}||\mathbf{z}\rangle = \langle \mathbf{y} \mid \mathbf{z} \rangle |\mathbf{x}\rangle \tag{4}$$

which differs from Definition 3 only in the following way: in Definition 3, $|\mathbf{x}\rangle\langle\mathbf{y}|$ stands for the *linear mapping*, whereas in ❷ Eq. 4, the same notation stands for a matrix. While mathematically the linear mapping and its matrix are certainly distinct objects, it is customary to identify them when the basis is fixed and clear from the context. Unless otherwise stated, this identification will be used in what follows.

For later purposes, special attention should be paid to mapping $|\mathbf{x}\rangle\langle\mathbf{x}|$ (assuming $||\mathbf{x}||=1$). As $|\mathbf{x}\rangle\langle\mathbf{x}|\mathbf{y} = \langle \mathbf{x} \mid \mathbf{y} \rangle \mathbf{x}$, it is obvious that $|\mathbf{x}\rangle\langle\mathbf{x}|$ is the projection onto the one-dimensional space spanned by $\mathbf{x}$.

When given a fixed orthonormal basis $\{\mathbf{e}_1, \dots, \mathbf{e}_n\}$ of H_n, one can represent any mapping $A \in L(H_n)$ uniquely as

$$A = \sum_{i=1}^{n} \sum_{j=1}^{n} A_{ij} |\mathbf{e}_i\rangle\langle\mathbf{e}_j| \tag{5}$$

where $A_{ij} = \langle e_i | A e_j \rangle$. Equation (❯ 5) is hence nothing more than the matrix representation of mapping A (with respect to basis $\{e_1, \ldots, e_n\}$). It is evident that also $L(H_n)$ is a vector space, and from representation (❯ 5) one can directly read its dimension, which is n^2. The advantage of representation (❯ 5) lies in the fact that, unlike in the ordinary matrix representation, here we have also the basis e_i explicitly shown.

Definition 4 The adjoint mapping A^* of A is defined by $\langle x \mid Ay \rangle = \langle A^* x \mid y \rangle$ for any vectors $x, y \in L(H_n)$.

In the matrix form, the adjoint mapping is easy to recover

$$\left(\sum_{i=1}^{n} \sum_{j=1}^{n} A_{ij} |e_i\rangle \langle e_j| \right)^* = \sum_{i=1}^{n} \sum_{j=1}^{n} A_{ji}^* |e_i\rangle \langle e_j|$$

which means that the matrix of an adjoint mapping is obtained by transposing and conjugating the original matrix. It is an easy exercise to verify that taking the adjoint mapping is a linear mapping $L(H_n) \rightarrow L(H_n)$, that is, an element of $L(L(H_n))$. This means that $(A+B)^* = A^* + B^*$, and $(cA)^* = c^* A^*$. In addition to that, it is easy to see that $(AB)^* = B^* A^*$. Moreover, the definitions presented above imply directly that $A|x\rangle\langle x|B^* = |Ax\rangle\langle By|$, no matter which particular interpretation (matrices of mappings) the mentioned objects possess.

Definition 5 Mapping $A \in L(H_n)$ is self-adjoint, if $A^* = A$.

The following analogy may be useful sometimes: Self-adjoint mappings relate to $L(H_n)$ as real numbers relate to complex numbers. To be more precise, we have the following representation theorem.

Theorem 2 For any $A \in L(H_n)$, there are unique self-adjoint mappings B and C so that

$$A = B + iC.$$

Proof If such a presentation exists, then $A^* = B - iC$, and addition gives $B = \frac{1}{2}(A + A^*)$, whereas subtraction shows $C = \frac{1}{2i}(A - A^*)$. This proves the existence and uniqueness of self-adjoint mappings B and C.

A polar representation analogous to $z = |z| e^{i\theta}$ also exists, but will be represented later. At this moment, it is worth mentioning that the analogy between real numbers and self-adjoint mappings becomes more concrete in the following proposition.

Proposition 1 All eigenvalues of a self-adjoint mapping $A \in L(H_n)$ are real.

Proof Let λ be an eigenvalue of A and x an eigenvector belonging to λ. Then

$$\lambda \langle x|x \rangle = \langle x|\lambda x \rangle = \langle x|Ax \rangle = \langle A^* x|x \rangle = \langle Ax|x \rangle = \langle \lambda x|x \rangle = \lambda^* \langle x|x \rangle$$

which implies $\lambda = \lambda^*$.

Definition 6 A mapping $A \in L(H_n)$ is *positive*, if $\langle x \mid Ax \rangle \geq 0$ for each $x \in L(H_n)$.

In the terminology more familiar to mathematicians, "positive" is replaced with "positive semidefinite." It is easy to see that the converse to the above theorem also holds: if $\langle x \mid Ax \rangle \in \mathbb{R}$

for each $\mathbf{x} \in H_n$, then A is necessarily self-adjoint. Hence Definition 6 necessarily speaks only about self-adjoint mappings. The previously mentioned analogy has its extension here: Positive mappings relate to self-adjoint mappings as the nonnegative real numbers relate to all real numbers, since it easy to see that a mapping is positive if and only if all of its eigenvalues are nonnegative.

One more definition is needed before introducing the quantum mechanical counterpart of the state vector $(\mathbf{x}, \mathbf{p})$ of classical mechanics.

Definition 7   For any mapping $A \in L(H_n)$, the *trace* is defined as

$$\mathrm{Tr}(A) = \sum_{i=1}^{n} \langle \mathbf{e}_i \mid A\mathbf{e}_i \rangle$$

where $\{\mathbf{e}_1, \dots, \mathbf{e}_n\}$ is an orthonormal basis of H_n.

It is possible to show that the trace is independent of the orthonormal basis chosen, and by the definition the trace equals the sum of the diagonal elements of the matrix. The trace also equals the sum of the eigenvalues (see Axler (1997), for example).

Definition 8   A *state* of an n-level quantum system is a self-adjoint, unit-trace, positive mapping $A \in L(H_n)$. A *density matrix* is a matrix representation of a state.

Recall that in classical mechanics one can represent the state of a physical system as $(\mathbf{x}, \mathbf{p}) \in \mathbb{R}^{2n}$, where $\mathbf{x}$ stands for the (generalized) coordinates and $\mathbf{p}$ for the (generalized) momentum. In quantum mechanics, a state is represented as linear mapping $A \in L(H_n)$ satisfying the conditions of Definition 8.

However, it should be noted carefully that n in the description of classical mechanics has a very different role than n in quantum mechanics. Recall that in classical descriptions n (the number of coordinates of $\mathbf{x}$ and $\mathbf{p}$) stands for the degrees of freedom of the system under investigation, and in quantum mechanics n, the dimension of H_n, stands for the maximal number of perfectly distinguishable measurement results that can be extracted from the system, for a suitably chosen measurement. The Hilbert space H_n can stand for a *single* degree of freedom: It accommodates the mathematical structure describing how the value of anything observable on this specific degree of freedom is produced.

To describe *various* degrees of freedom in quantum mechanics, one should set up a Hilbert space individually for any such degree. The total description is then in the Hilbert space obtained as a *tensor product* of all "component spaces." It should be noted that usually observables with an infinite number of potential values (and consequently infinite-dimensional Hilbert spaces) are needed, but in the context of quantum computing physical objects with a finite number of potential values are satisfactory: to describe a *bit*, one needs only a physical system capable of existing in two perfectly distinguishable states.

In the classical description $(\mathbf{x}, \mathbf{p})$ the value of any observable can calculated directly from the coordinates of $\mathbf{x}$ and $\mathbf{p}$. In quantum mechanics, the setup is dramatically different: quantum mechanics is a fundamentally stochastic theory, and for any state of the system there are always observables with values not definitely prescribed, but obtained stochastically. Conversely, for any observable, there are even *pure states* (introduced later) of the system for which the measurement result is not determined uniquely.

A very important structural property of the state set is given in the following proposition, whose proof is actually straightforward from the definitions.

Proposition 2 *All states of an n-level quantum system form a convex set, meaning that if S_1 and S_2 are states then their convex combination*

$$pS_1 + (1 - p)S_2$$

with $p \in [0,1]$ also is a state.

The above proposition actually expresses the principle that it is possible to obtain new states from previous ones as a *stochastic mixture*: the new state $pS_1 + (1-p)S_2$ can be seen as a description of the stochastic process, where S_1 occurs with probability p, and S_2 with a complementary probability $1-p$.

The *extreme points* of the state set deserve special attention.

Definition 9 State S is *pure* if it cannot be presented as a convex combination $S = pS_1 + (1 - p)S_2$, where $S_1 \neq S_2$. A state that is not pure is *mixed*.

There exists a well-known and simple mathematical characterization for pure quantum states, but to introduce that we need to define one more notion.

Definition 10 Mapping $A \in L(H_n)$ is an *(orthogonal) projection* if $A^* = A$ and $A^2 = A$.

Hereafter, the attribute "orthogonal" will be ignored as all projections treated in this chapter will be indeed orthogonal. It can be shown that there is one-to-one correspondence between projections and subspaces of H_n. This connection can be briefly explained as follows: for any subspace $W \subseteq H_n$ there exists an *orthogonal complement* $W^\perp$, also a subspace of H_n, which has the property that any $\mathbf{x} \in W$ and any $\mathbf{y} \in W^\perp$ are orthogonal: $\langle \mathbf{x} \mid \mathbf{y} \rangle = 0$. Moreover, H_n can be represented as a *direct sum* of the subspaces, $H_n = W \oplus W^\perp$, which means that each $\mathbf{x} \in H_n$ can be uniquely represented as $\mathbf{x} = \mathbf{x}_W + \mathbf{x}_{W^\perp}$, where $\mathbf{x}_W \in W$ and $\mathbf{x}_{W^\perp} \in W^\perp$. Now if A is a projection, then there exists a unique subspace W so that $A\mathbf{x} = \mathbf{x}_W$. Conversely, for any subspace $W \subseteq H_n$ there exists a unique projection A so that $A\mathbf{x} = \mathbf{x}_W$.

We conclude this section with the well-known characterization of pure states.

Proposition 3 *A state $A \in L(H_n)$ of a quantum system is pure if $A = |\mathbf{x}\rangle\langle \mathbf{x}|$ is a projection onto a one-dimensional subspace (here $||\mathbf{x}|| = 1$).*

3 Spectral Representation

Spectral representation, also known as *spectral decomposition*, is a very powerful tool of matrix theory. It also provides a representation for an arbitrary quantum state as a convex combination of pure states, as will be shortly seen.

Definition 11 A matrix A is *normal* if it commutes with its adjoint, that is $AA^* = A^*A$.

Especially, self-adjoint mappings are normal, since then $AA^* = A^2 = A^*A$. Normal matrices have the property that they can be diagonalized unitarily, meaning that there exists a possibility to rotate the natural basis of H_n so that the matrix in the new basis is diagonal. Indeed, we understand "rotate" here a bit more generally, meaning that for any normal matrix A, there is a unitary matrix U (defined later in ❷ Section 7.1.1) so that UAU^* is diagonal. This can also be expressed in the following theorem.

Theorem 3 (Spectral representation) *Let $A \in L(H_n)$ be normal. Then there exists an orthonormal basis $\{x_1, \ldots, x_n\}$ of H_n so that*

$$A = \lambda_1 |x_1\rangle\langle x_1| + \ldots + \lambda_n |x_n\rangle\langle x_n| \tag{6}$$

It is noteworthy that in the spectral representation numbers λ_i are the eigenvalues of A and x_i is an eigenvector belonging to λ_i (the set of the eigenvalues is also called the *spectrum* of the matrix or a mapping). Equation (❷ 6) hence simply means that the spectral representation corresponds to choosing an orthonormal basis $x_1, \ldots, x_n$ in which the matrix A is diagonal. The above theorem just states that for normal matrices (or mappings as well), there always exists an orthonormal basis allowing a diagonal presentation. The proof of the spectral representation theorem for *self-adjoint* mappings can be found in Axler (1997), for instance, and its extension to normal mappings is quite an easy exercise when we know that all mappings can be represented as $C = A + iB$, where A and B are self-adjoint.

Definition 12 Let $f : \mathbb{C} \to \mathbb{C}$ be an arbitrary function. Then f is defined on normal mappings by

$$f(A) = f(\lambda_1)|x_1\rangle\langle x_1| + \ldots + f(\lambda_n)|x_n\rangle\langle x_n| \tag{7}$$

if A has spectral representation (❷ 6).

It is an easy exercise to show that the above definition is independent of the choice of the spectral representation.

Example 1 Matrix $\begin{pmatrix} \frac{\pi}{2} & 0 \\ 0 & -\frac{\pi}{2} \end{pmatrix}$ defines a normal mapping. By fixing the basis of H_2 as $|0\rangle = \begin{pmatrix} 1 \\ 0 \end{pmatrix}$ and $|1\rangle = \begin{pmatrix} 0 \\ 1 \end{pmatrix}$ we can represent this mapping as $A = \frac{\pi}{2} \cdot |0\rangle\langle 0| - \frac{\pi}{2} \cdot |1\rangle\langle 1|$. Now

$$e^{iA} = e^{i\frac{\pi}{2}}|0\rangle\langle 0| + e^{i \cdot (-\frac{\pi}{2})}|1\rangle\langle 1| = i|0\rangle\langle 0| - i|1\rangle\langle 1|$$

which can be represented as a matrix $\begin{pmatrix} i & 0 \\ 0 & -i \end{pmatrix}$ as well.

As the states of quantum systems are depicted by self-adjoint, positive, unit-trace operators, we will be interested especially in their representations. Since self-adjoint mappings are normal, they admit a spectral representation. Moreover, it should be noted that the eigenvalues λ of self-adjoint operators in presentation (❷ 6) are real.

The positivity $\langle x \,|\, Ax \rangle \geq 0$ always implies that for each i, $\lambda_i \geq 0$. As representation (❷ 6) corresponds to a diagonal matrix, it is also easy to verify that $\mathrm{Tr}(A) = \lambda_1 + \ldots + \lambda_n$. We can conclude this as follows:

▶ A state A of an n-level quantum system can be represented as

$$A = \lambda_1 |x_1\rangle\langle x_1| + \ldots + \lambda_n |x_n\rangle\langle x_n| \tag{8}$$

where $\lambda_i \geq 0$, and $\lambda_1 + \ldots + \lambda_n = 1$. This is to say that each state $A \in L(H_n)$ can be represented as a convex combination of at most n pure states, and, moreover, the pure states can be found by searching the eigenvectors of A.

Representation (❷ 8) may recommend that we interpret a general quantum state A as a probability distribution over pure states $|\mathbf{x}_1\rangle\langle\mathbf{x}_1|, \ldots, |\mathbf{x}_n\rangle\langle\mathbf{x}_n|$. Unfortunately, representation (❷ 8) is not always unique.

Example 2   Let $|0'\rangle = \frac{1}{\sqrt{2}}|0\rangle + \frac{1}{\sqrt{2}}|1\rangle$ and $|1'\rangle = \frac{1}{\sqrt{2}}|0\rangle - \frac{1}{\sqrt{2}}|1\rangle$. Then

$$|0'\rangle\langle 0'| + |1'\rangle\langle 1'| = |\frac{1}{\sqrt{2}}|0\rangle + \frac{1}{\sqrt{2}}|1\rangle\rangle\langle\frac{1}{\sqrt{2}}|0\rangle + \frac{1}{\sqrt{2}}|1\rangle|$$

$$+ |\frac{1}{\sqrt{2}}|0\rangle - \frac{1}{\sqrt{2}}|1\rangle\rangle\langle\frac{1}{\sqrt{2}}|0\rangle - \frac{1}{\sqrt{2}}|1\rangle|$$

$$= \frac{1}{2}|0\rangle\langle 0| + \frac{1}{2}|0\rangle\langle 1| + \frac{1}{2}|1\rangle\langle 0| + \frac{1}{2}|1\rangle\langle 1|$$

$$+ \frac{1}{2}|0\rangle\langle 0| - \frac{1}{2}|0\rangle\langle 1| - \frac{1}{2}|1\rangle\langle 0| + \frac{1}{2}|1\rangle\langle 1|$$

$$= |0\rangle\langle 0| + |1\rangle\langle 1|$$

Let $||\mathbf{x}|| = 1$. A simple calculation shows that $|a\mathbf{x}\rangle\langle a\mathbf{x}| = |a|^2|\mathbf{x}\rangle\langle\mathbf{x}|$, which implies that vectors $\mathbf{x}$ and $e^{i\theta}\mathbf{x}$ (with $\theta \in \mathbb{R}$) specify exactly the same projection onto a one-dimensional vector space. Recall that those projections are special cases of quantum states (pure states), and they play an important role in quantum computing and in general quantum theory: most quantum computing models are based on pure states only. Quantum information processing in general will of course require more than pure states.

A pure state $A = |\mathbf{x}\rangle\langle\mathbf{x}|$ is often identified with a unit-length *state vector* or *vector state* $\mathbf{x}$, but it should be noted that this identification is not unambiguous: Vectors $e^{i\theta}\mathbf{x}$ with $\theta \in \mathbb{R}$ are all vector representations of the same pure state.

A notion constantly present in quantum physics (as well as in classical wave theory) is that of *superposition*.

Definition 13   Let $\{\mathbf{x}_1, \ldots, \mathbf{x}_n\}$ be an orthonormal basis of H_n and

$$\mathbf{x} = \alpha_1\mathbf{x}_1 + \ldots + \alpha_n\mathbf{x}_n$$

a unit-length vector describing a pure state. We say then that $\mathbf{x}$ is a *superposition* of vector states $\mathbf{x}_1, \ldots, \mathbf{x}_n$ with *amplitudes* $\alpha_1, \ldots, \alpha_n$.

In classical mechanics, one can always introduce a mixed state when two pure states T_1 and T_2 are given: The new mixed state is a statistical combination $pT_1 + (1-p)T_2$ of the pure ones, meaning that T_1 is present with probability p, and T_2 with the complementary probability $1-p$.

In quantum mechanics, one can also create mixtures from the pure state analogously to the classical case, but a very important nonclassical feature is that from given pure states one can form another pure state by superposition: If $\mathbf{x}_1$ and $\mathbf{x}_2$ are perpendicular unit-length vectors representing pure states $|\mathbf{x}_1\rangle\langle\mathbf{x}_1|$ and $|\mathbf{x}_2\rangle\langle\mathbf{x}_2|$, then their linear combination $\alpha_1\mathbf{x}_1 + \alpha_2\mathbf{x}_2$ (where $|\alpha_1|^2 + |\alpha_2|^2 = 1$) is again a pure (vector) state, the superposition of vector states $\mathbf{x}_1$ and $\mathbf{x}_2$.

A simple calculation shows that

$$|\alpha_1\mathbf{x}_1 + \alpha_2\mathbf{x}_2\rangle\langle\alpha_1\mathbf{x}_1 + \alpha_2\mathbf{x}_2|$$

$$= |\alpha_1|^2|\mathbf{x}_1\rangle\langle\mathbf{x}_1| + \alpha_1\alpha_2^*|\mathbf{x}_1\rangle\langle\mathbf{x}_2| + \alpha_2\alpha_1^*|\mathbf{x}_2\rangle\langle\mathbf{x}_1| + |\alpha_2|^2|\mathbf{x}_2\rangle\langle\mathbf{x}_2|$$

Term $\alpha_1\alpha_2^*|\mathbf{x}_1\rangle\langle\mathbf{x}_2| + \alpha_2\alpha_1^*|\mathbf{x}_2\rangle\langle\mathbf{x}_1|$ in the above superposition is called then an *interference term*, whereas the remaining term

$$|\alpha_1|^2|\mathbf{x}_1\rangle\langle\mathbf{x}_1| + |\alpha_2|^2|\mathbf{x}_2\rangle\langle\mathbf{x}_2|.$$

is indeed a statistical mixture of the original states $|\mathbf{x}_1\rangle\langle\mathbf{x}_1|$ and $|\mathbf{x}_2\rangle\langle\mathbf{x}_2|$.

To create a superposition state, unfortunately the vector representatives of pure states $|\mathbf{x}_1\rangle\langle\mathbf{x}_1|$ and $|\mathbf{x}_2\rangle\langle\mathbf{x}_2|$ are needed. Recall that for example $\mathbf{x}_2$ and $-\mathbf{x}_2$ both define the same pure state, meaning that $|\mathbf{x}_2\rangle\langle\mathbf{x}_2| = |-\mathbf{x}_2\rangle\langle-\mathbf{x}_2|$, but nevertheless the state vectors $\frac{1}{\sqrt{2}}\mathbf{x}_1 + \frac{1}{\sqrt{2}}\mathbf{x}_2$ and $\frac{1}{\sqrt{2}}\mathbf{x}_1 - \frac{1}{\sqrt{2}}\mathbf{x}_2$ obtained from the different representations for $|\mathbf{x}_2\rangle\langle\mathbf{x}_2|$ are completely different objects.

4 The Meaning of the Quantum States

Quantum physics is structurally a stochastic theory. The stochastic nature comes into the picture when observing a quantum system in a fixed state; there is not usually a single outcome, but only a probability distribution of potential outcomes. By saying that the stochastic nature is structurally built in quantum theory, we mean that the outcome distribution instead of a single outcome does not emerge from the ignorance: even the most accurate knowledge of the system does not admit more than a distribution of the outcomes. The pure states correspond, in a sense, to maximal information about the quantum system, but even in a pure state one must accept a probability distribution of the potential outcomes.

In this section, the *minimal interpretation* of quantum mechanics will be presented, which actually is an axiom describing how the probabilities will be calculated from the mathematical objects describing quantum systems. For that purpose, we need to describe how *observables* are specified in this formalism. This is of course similar to classical physics: For instance, pair $(\mathbf{x}, \mathbf{p})$ describes the state of a classical system, but the value of the position is revealed from the state by projection $(\mathbf{x}, \mathbf{p}) \mapsto \mathbf{x}$.

Definition 14 An *observable* of a quantum system H_n is a collection $P_1, \ldots, P_k$ of projections onto mutually orthogonal subspaces so that $P_1 + \ldots + P_k = I$ (the identity mapping), equipped with a collection of real numbers $\lambda_1, \ldots, \lambda_k$.

Another equivalent viewpoint is offered in the following definition.

Definition 15 An *observable* of a quantum system H_n is a self-adjoint mapping $A \in L(H_n)$.

It is not difficult to see that the above two definitions speak about the same object. Indeed, if A is a self-adjoint mapping, then it admits a spectral representation

$$A = \sum_{k=1}^{n} \lambda_k |\mathbf{x}_k\rangle\langle\mathbf{x}_k|$$

which can also be written in the form

$$A = \sum_{k=1}^{n'} \lambda_k' P_k$$

where $\lambda'_1, \ldots, \lambda'_{n'}$ are the distinct eigenvalues of A, and

$$P_k = \sum_{l=1}^{d_l} |\mathbf{x}_{k_l}\rangle\langle\mathbf{x}_{k_l}|$$

is the projection onto the eigenspace V_k of λ'_k. As eigenspaces belonging to distinct eigenvalues of a self-adjoint operator A are orthogonal (Axler 1997), the projections satisfy the orthogonality required in the first definition.

On the other hand, a collection $P_1, \ldots, P_k$ of projections of H_n onto orthonormal subspaces of H_n together with real numbers $\lambda_1, \ldots, \lambda_k$ evidently define a self-adjoint mapping

$$A = \lambda_1 P_1 + \ldots + \lambda_k P_k$$

If $P_1 + \ldots + P_k \neq I$, then another projection $P_{k+1} = I - (P_1 + \ldots + P_k)$ with $\lambda_{k+1} = 0$ is introduced to satisfy the requirement $P_1 + \ldots + P_{k+1} = I$.

Remark 1 The notion of "a collection of projections equipped with real numbers" can be expressed by using the notion of *projection-valued measure*. A projection-valued measure on the real line is a mapping $E : \mathbb{R} \to L(H_n)$ which, in the case of Definition 14, satisfies $E(\lambda_i) = P_i$, and $E(\lambda) = 0$ for $\lambda \notin \{\lambda_1, \ldots, \lambda_k\}$. In general, a projection-valued measure must satisfy conditions similar to an ordinary measure on a real line (see Hirvensalo (2004), for instance).

Whichever definition is used for an observable, the intuitive meaning will be that the values $\lambda_1, \ldots, \lambda_k$ are the *potential* values of observable A, and probability distributions of the values are expressed in the following definition.

Definition 16 (Minimal interpretation of quantum mechanics)

Let $T \in L(H_n)$ be a state of a quantum system and $A = \lambda_1 P_1 + \ldots + \lambda_k P_k$ be an observable, where P_i are mutually orthogonal projections summing up to the identity mapping. Then the probability that λ_i will be observed when the system is in state T is

$$\mathbb{P}_T(\lambda_i) = \text{Tr}(P_i T)$$

The minimal interpretation has its mathematical basis in Gleason's theorem (Gleason 1957), which shows that for dimension $n \geq 3$ all probability measures in $L(H_n)$ are of the above form.

Example 3 The presentation of an observable as a self-adjoint mapping A is useful when talking about the *expected value* of the observable. Indeed, the expectation (in state T) is given by

$$\mathbb{E}_T(A) = \sum_{i=1}^{k} \mathbb{P}_T(\lambda_i)\lambda_i = \sum_{i=1}^{k} \lambda_i \text{Tr}(P_i T) = \text{Tr}(AT)$$

since $A = \sum_{i=l}^{k} \lambda_i P_i$.

The mathematical formalism presented so far implies rather easily quite deep consequences. One such is the famous *Heisenberg uncertainty relation* (see Hirvensalo (2004) for the proof of the next proposition).

Proposition 4 (Uncertainty relation) *Let A and B be any observables. Then the product of variances* $\mathbb{V}_T(A)$ *and* $\mathbb{V}_T(B)$ *in a pure state* $T = |\mathbf{x}\rangle\langle\mathbf{x}|$ *can be bounded below as*

$$\mathbb{V}_T(A)\mathbb{V}_T(B) \geq \frac{1}{4}|\langle\mathbf{x}|[A, B]\mathbf{x}\rangle|^2$$

where $[A, B] = AB - BA$ *is the* commutator *of A and B.*

The above uncertainty relation, however, has its greatest importance in infinite-dimensional vector spaces: in such cases it is possible that $[A, B] = cI$, a multiple of the identity mapping, and then the lower bound would be state-independent. A traditional example is the commutator of the position and momentum $[X, P] = i\hbar I$.

Let $\mathbf{x} \in H_n$ be a unit-length vector and $T = |\mathbf{x}\rangle\langle\mathbf{x}|$ a pure state determined by $\mathbf{x}$. Fix also an observable $A = \lambda_1 |\mathbf{x}_1\rangle\langle\mathbf{x}_1| + \ldots + \lambda_n |\mathbf{x}_n\rangle\langle\mathbf{x}_n|$ with n distinct eigenvalues (potential values of observable A). As $\{\mathbf{x}_1, \ldots, \mathbf{x}_n\}$ is an orthonormal basis of H_n, we can always write

$$\mathbf{x} = \alpha_1\mathbf{x}_1 + \ldots + \alpha_n\mathbf{x}_n$$

and the probability to observe λ_i is given by

$$P(\lambda_i) = \mathrm{Tr}(|\mathbf{x}_i\rangle\langle\mathbf{x}_i||\mathbf{x}\rangle\langle\mathbf{x}|) = |\alpha_i|^2$$

as easily verified. To express this in other words: assume that a quantum system is in a vector state

$$\mathbf{x} = \alpha_1\mathbf{x}_1 + \ldots + \alpha_n\mathbf{x}_n$$

and the observable studied is

$$A = \lambda_1|\mathbf{x}_1\rangle\langle\mathbf{x}_1| + \ldots + \lambda_n|\mathbf{x}_n\rangle\langle\mathbf{x}_n|$$

(distinct eigenvalues λ_i). Then the probability that value λ_i is observed is $|\alpha_i|^2$. For such an observable, it is also typical to omit also the *value* of the observable and merely to refer to the vector $\mathbf{x}_i$ (intuitively one could think about observing $\mathbf{x}_i$ instead of value λ_i) generating a one-dimensional subspace of H_n. The following version of the minimal interpretation is written for pure states, the aforementioned special type of observables, and for ignoring the (eigen)values of the observables.

Definition 17 (Minimal interpretation, special version) Let

$$\mathbf{x} = \alpha_1\mathbf{x}_1 + \ldots + \alpha_n\mathbf{x}_n$$

be a vector representation of a pure quantum state. Then the probability of observing $\mathbf{x}_i$ is $|\alpha_i|^2$.

5 Quantum Bits

When speaking about computation, bits are a most interesting object. Therefore it is worth devoting a section to their quantum physical representations, *quantum bits* or *qubits* for short.

A physical system capable of representing a bit must have two distinguishable states, but there are no further restrictions on the nature of the system. The mathematical description of the system therefore lives in H_2, two-dimensional Hilbert space.

Definition 18 A *computational basis* of H_2 refers to a fixed orthonormal basis $\{|0\rangle, |1\rangle\}$ of H_2.

The above definition refers rather to a physical nature and interpretation of the quantum system than to any mathematical properties of H_2. Indeed, in a physical realization there may be some preferred states that are more natural to interpret as 0 and 1 than any other states. For instance, if representing quantum bits with electron energy states, one may want to associate logical zero with the ground state, and logical one with an exited state.

Mathematically, it is always possible to construct H_2 as the Cartesian product $\mathbb{C} \times \mathbb{C}$ and choose $|0\rangle = \begin{pmatrix} 1 \\ 0 \end{pmatrix}$ and $|1\rangle = \begin{pmatrix} 0 \\ 1 \end{pmatrix}$. Of course, any other choice for the computational basis is equally good, and hence the notion "computational basis" should be understood only contextually. However, in this chapter the aforementioned choice is used for the computational basis.

A *state* of the qubit is, according to the previous definition, a self-adjoint, positive, unit-trace mapping in $L(H_2)$. When thinking about the matrix representation

$$S = \begin{pmatrix} a & b \\ c & d \end{pmatrix} = a|0\rangle\langle 0| + b|0\rangle\langle 1| + c|1\rangle\langle 0| + d|1\rangle\langle 1|$$

one sees that the self-adjointness immediately imposes restrictions such as $a, d \in \mathbb{R}$, and that $c = b^*$. Furthermore, from the unit trace condition we see that $a + d = 1$, so a general form of a self-adjoint mapping with unit trace must be of the form

$$S = \begin{pmatrix} a & b \\ b^* & 1 - a \end{pmatrix}$$

where $a \in \mathbb{R}$. Hence there are three real numbers (a and the real and imaginary parts of b) that specify a unit-trace self-adjoint mapping. The positivity obviously imposes extra conditions, and for finding them we choose a slightly different approach.

Definition 19 The *Pauli spin matrices* are defined as follows:

$$\sigma_x = \begin{pmatrix} 0 & 1 \\ 1 & 0 \end{pmatrix}, \sigma_y = \begin{pmatrix} 0 & -i \\ i & 0 \end{pmatrix}, \sigma_z = \begin{pmatrix} 1 & 0 \\ 0 & -1 \end{pmatrix}$$

Yet adding the identity matrix $I = \begin{pmatrix} 1 & 0 \\ 0 & 1 \end{pmatrix}$ we get an interesting set: it is easy to prove that $\{I, \sigma_x, \sigma_y, \sigma_z\}$ is a basis of $L(H_2)$, when regarding $L(H_2)$ as a vector space over $\mathbb{C}$. Especially this means that any 2×2 complex matrix S can be uniquely represented as

$$S = wI + x\sigma_x + y\sigma_y + z\sigma_z \tag{9}$$

where $w, x, y,$ and $z \in \mathbb{C}$. Furthermore, all matrices in $\{I, \sigma_x, \sigma_y, \sigma_z\}$ are self-adjoint, which implies that

$$S^* = w^*I + x^*\sigma_x + y^*\sigma_y + z^*\sigma_z$$

Now because the representation is unique, we conclude that if S is self-adjoint, then necessarily coefficients $w, x, y,$ and z in ❷ Eq. 9 are real. A further observation shows that $\sigma_x, \sigma_y,$ and σ_z have zero trace, meaning that the trace of ❷ Eq. 9 is $Tr(wI) = 2w$. This shows that a unit-trace, self-adjoint mapping in $L(H_2)$ has a unique representation

$$S = \frac{1}{2}(I + x\sigma_x + y\sigma_y + z\sigma_z) \tag{10}$$

where x, y, and $z \in \mathbb{R}$. This is perfectly compatible with the aforementioned fact that unit-trace self-adjoint mappings in $L(H_2)$ have three real degrees of freedom: via representation (➲ 10) we can identify such a mapping with a point in $\mathbb{R}^3$, but the restrictions imposed by the positivity condition are still to be discovered.

For that, we notice first that the spectral representation directly implies that a matrix is positive if and only if its eigenvalues are nonnegative. To discover the characteristic polynomial of ➲ Eq. 10, we write explicitly

$$S = \frac{1}{2}(I + x\sigma_x + y\sigma_y + z\sigma_z) = \frac{1}{2}\begin{pmatrix} 1+z & x-iy \\ x+iy & 1-z \end{pmatrix}$$

and find the roots of the characteristic polynomial

$$\lambda = \frac{1}{2}\left(1 \pm \sqrt{x^2 + y^2 + z^2}\right) \tag{11}$$

Both roots are nonnegative if and only if $x^2 + y^2 + z^2 \leq 1$, and hence we see that there is one-to-one correspondence between the points of the unit sphere in $\mathbb{R}^3$ and pure states of a qubit.

Definition 20 The *Bloch sphere* (also known as the *Poincaré sphere*) is the image of the sphere

$$\{(x, y, z) \mid x^2 + y^2 + z^2 \leq 1\} \subseteq \mathbb{R}^3$$

in $L(H_2)$ under mapping

$$(x, y, z) \mapsto \frac{1}{2}(I + x\sigma_x + y\sigma_y + z\sigma_z) \tag{12}$$

Even more information is easily available: A pure state $|\mathbf{x}\rangle\langle\mathbf{x}|$ is a projection onto a one-dimensional subspace, and the sole eigenvalues of such a mapping are hence 1 and 0. By ➲ Eq. 11 this is possible only if $x^2 + y^2 + z^2 = 1$. This fact, which would as well follow from the convexity-preserving nature of mapping (➲ Eq. 12), shows that the *pure states of a quantum bit lie exactly on the surface of the Bloch sphere*. By using the standard spherical coordinate representation, we can say that the pure states of a quantum bit have two real degrees of freedom; indeed the use of spherical coordinates result in a vector state $\cos\theta\,|0\rangle + e^{i\phi}\sin\theta\,|1\rangle$.

Example 4 Pure state $|0\rangle$ corresponds to the projection $|0\rangle\langle0|$, whose matrix representation is

$$(1,0) \otimes \begin{pmatrix} 1 \\ 0 \end{pmatrix} = \begin{pmatrix} 1 & 0 \\ 0 & 0 \end{pmatrix} = \frac{1}{2}(I + \sigma_z) = \frac{1}{2}(I + 0\cdot\sigma_x + 0\cdot\sigma_y + 1\cdot\sigma_z)$$

so the Bloch sphere point $(0,0,1)$ corresponds to the pure state $|0\rangle$. This point is referred to as the *north pole* of the Bloch sphere. Analogously one can see that the *south pole* of the Bloch sphere corresponds to the pure state $|1\rangle$.

Let us find the point corresponding to the pure state $\frac{1}{\sqrt{2}}(|0\rangle + |1\rangle)$. The density matrix of this state is

$$\left(\frac{1}{\sqrt{2}}, \frac{1}{\sqrt{2}}\right) \otimes \begin{pmatrix} \frac{1}{\sqrt{2}} \\ \frac{1}{\sqrt{2}} \end{pmatrix} = \frac{1}{2}\begin{pmatrix} 1 & 1 \\ 1 & 1 \end{pmatrix} = \frac{1}{2}(I + \sigma_x) = \frac{1}{2}(I + 1\cdot\sigma_x + 0\cdot\sigma_y + 0\cdot\sigma_z)$$

so the point $(1,0,0)$ corresponding to this state is found on the equator of the Bloch sphere.

For finding the point corresponding to $\frac{1}{2}|0\rangle\langle0| + \frac{1}{2}|1\rangle\langle1|$, we first find the matrix

$$\frac{1}{2}\begin{pmatrix} 1 & 0 \\ 0 & 0 \end{pmatrix} + \frac{1}{2}\begin{pmatrix} 0 & 0 \\ 0 & 1 \end{pmatrix} = \begin{pmatrix} \frac{1}{2} & 0 \\ 0 & \frac{1}{2} \end{pmatrix} = \frac{1}{2}I + 0 \cdot \sigma_x + 0 \cdot \sigma_y + 0 \cdot \sigma_z$$

which means that the center $(0,0,0)$ corresponds to the mixed state $\frac{1}{2}|0\rangle\langle0| + \frac{1}{2}|1\rangle\langle1|$.

What is then, for example, the pure state corresponding to the equator point $(0,1,0)$? This is easily found, as

$$\frac{1}{2}(I + 0 \cdot \sigma_x + 1 \cdot \sigma_y + 0 \cdot \sigma_z) = \frac{1}{2}\begin{pmatrix} 1 & -i \\ i & 1 \end{pmatrix}$$

This matrix is self-adjoint, so it admits a spectral representation, which can be found by finding its eigenvectors: Vector $\frac{1}{\sqrt{2}}\begin{pmatrix} i \\ 1 \end{pmatrix}$ belongs to the eigenvalue 0, and $\frac{1}{\sqrt{2}}\begin{pmatrix} -i \\ 1 \end{pmatrix}$ to eigenvalue 1. Hence (note that the factor $\frac{1}{2}$ is omitted now)

$$\begin{pmatrix} 1 & -i \\ i & 1 \end{pmatrix} = 0 \cdot \begin{pmatrix} i \\ 1 \end{pmatrix}^* \otimes \begin{pmatrix} i \\ 1 \end{pmatrix} + 1 \cdot \begin{pmatrix} -i \\ 1 \end{pmatrix}^* \otimes \begin{pmatrix} -i \\ 1 \end{pmatrix}$$

$$= 0 \cdot (-i, 1) \otimes \begin{pmatrix} i \\ 1 \end{pmatrix} + 1 \cdot (i, 1) \otimes \begin{pmatrix} -i \\ 1 \end{pmatrix}$$

and we can see that the state vector corresponding to point $(0,1,0)$ can be chosen as

$$\frac{1}{\sqrt{2}}\begin{pmatrix} -i \\ 1 \end{pmatrix} = \frac{-i}{\sqrt{2}}|0\rangle + \frac{1}{\sqrt{2}}|1\rangle.$$

6 Compound Systems

The notion of a *joint quantum system* or *compound quantum system* is established via tensor product construction.

Definition 21 Let $\mathcal{S}_1$ and $\mathcal{S}_2$ be quantum systems with state spaces H_n and H_m, respectively. Then the state space of the compound system $\mathcal{S}_{12}$ is the *mn*-dimensional tensor product $H_n \otimes H_m$.

The tensor product is actually a very clever construction in algebra allowing us to represent bilinear mappings as linear ones preceded by a very natural bilinear embedding. However, this sophisticated algebraic definition is not needed to perform the necessary calculations. Hence we take a constructive approach and merely mention the following facts: if $\mathbf{x}_1, \ldots, \mathbf{x}_n$ and $\mathbf{y}_1, \ldots, \mathbf{y}_m$ are the orthonormal bases of H_n and H_m, respectively, then $\{\mathbf{x}_i \otimes \mathbf{y}_j \mid i \in \{1,\ldots, n\}, j \in \{1,\ldots, m\}\}$ is an orthonormal basis of $H_n \otimes H_m$, as the inner product takes the form

$$\langle \mathbf{x}_1 \otimes \mathbf{y}_1 \mid \mathbf{x}_2 \otimes \mathbf{y}_2 \rangle = \langle \mathbf{x}_1 \mid \mathbf{x}_2 \rangle \langle \mathbf{y}_1 \mid \mathbf{y}_2 \rangle$$

Any pair of linear mappings $A \in L(H_n)$, $B \in L(H_m)$ together define a linear mapping $A \otimes B \in L(H_n \otimes H_m)$ via

$$(A \otimes B)(\mathbf{x}_i \otimes \mathbf{y}_j) = A\mathbf{x}_i \otimes B\mathbf{y}_j$$

for basis vectors $\mathbf{x}_i \otimes \mathbf{y}_j$, and the extension to the whole space is straightforwardly done by linearity. It should already here be carefully noted that as there are vectors other than $\mathbf{x} \otimes \mathbf{y} \in H_n \otimes H_m$ (so-called *decomposable vectors*), there are also linear mappings other than those of the form $A \otimes B$ in $L(H_n \otimes H_m)$. In the next section, this very evident fact will be examined again, but another identity very useful for the continuation follows easily from the previous ones:

$$\mathrm{Tr}(A \otimes B) = \mathrm{Tr}(A)\mathrm{Tr}(B)$$

It must of course be mentioned here that in the field of quantum computing the tensor sign is very generally omitted. For example, let $\{|0\rangle, |1\rangle\}$ be an orthonormal basis of H_2. Then the orthonormal basis of $H_4 \simeq H_2 \otimes H_2$ can be constructed as $\{|0\rangle \otimes |0\rangle, |0\rangle \otimes |1\rangle, |1\rangle \otimes |0\rangle, |1\rangle \otimes |1\rangle\}$, but the abbreviations $\{|0\rangle|0\rangle, |0\rangle|1\rangle, |1\rangle|0\rangle, |1\rangle|1\rangle\}$ and $\{|00\rangle, |01\rangle, |10\rangle, |11\rangle\}$ are very frequent.

Example 5 Using the notations $|0\rangle = \begin{pmatrix} 1 \\ 0 \end{pmatrix}$ and $|1\rangle = \begin{pmatrix} 0 \\ 1 \end{pmatrix}$ one can get concrete representations for the aforementioned vectors:

$$|00\rangle = |0\rangle \otimes |0\rangle = \begin{pmatrix} 1 \\ 0 \end{pmatrix} \otimes \begin{pmatrix} 1 \\ 0 \end{pmatrix} = \begin{pmatrix} 1 \\ 0 \\ 0 \\ 0 \end{pmatrix}, |01\rangle = |0\rangle \otimes |1\rangle = \begin{pmatrix} 1 \\ 0 \end{pmatrix} \otimes \begin{pmatrix} 0 \\ 1 \end{pmatrix} = \begin{pmatrix} 0 \\ 1 \\ 0 \\ 0 \end{pmatrix},$$

$$|10\rangle = |1\rangle \otimes |0\rangle = \begin{pmatrix} 0 \\ 1 \end{pmatrix} \otimes \begin{pmatrix} 1 \\ 0 \end{pmatrix} = \begin{pmatrix} 0 \\ 0 \\ 1 \\ 0 \end{pmatrix}, |11\rangle = |1\rangle \otimes |1\rangle = \begin{pmatrix} 0 \\ 1 \end{pmatrix} \otimes \begin{pmatrix} 0 \\ 1 \end{pmatrix} = \begin{pmatrix} 0 \\ 0 \\ 0 \\ 1 \end{pmatrix}.$$

The tensor product of two vector spaces generalizes to that of three or more spaces by defining $H_n \otimes H_m \otimes H_l = H_n \otimes (H_m \otimes H_l)$, which can be shown to be isomorphic to $(H_n \otimes H_m) \otimes H_l$. Subsequently, the notations such as $|0\rangle \otimes |1\rangle \otimes |0\rangle$ are abbreviated in quantum computing as $|0\rangle|1\rangle|0\rangle$, or as $|010\rangle$.

7 Subsystem States

In order to call a system $\mathscr{S}_1$ a *subsystem* of a quantum system $\mathscr{S}$, it is necessary that $\mathscr{S}_1$ must be *identifiable* as a quantum system itself.

Let $\mathscr{S}_1$ and $\mathscr{S}_2$ be identifiable quantum systems, and S_1 and S_2 their states, respectively. From the previous section it is known that for example $S_1 \otimes S_2$ is an eligible state of the compound system $\mathscr{S}_{12}$. State $S_1 \otimes S_2$ is an example of a *decomposable state*.

Example 6 Let H_2 be equipped with an orthonormal basis $(|0\rangle, |1\rangle)$. Then the basis vectors $|0\rangle$ and $|1\rangle$ determine pure states of the system, and states other than these can be obtained for example by superposition: any vector $\alpha|0\rangle + \beta|1\rangle$ with $|\alpha|^2 + |\beta|^2 = 1$ determines a pure state as well.

In the tensor product $H_2 \otimes H_2$, states $|0\rangle \otimes |0\rangle$ (abbreviated as $|00\rangle$) and $|11\rangle$ are pure states, and so is their superposition $\frac{1}{\sqrt{2}}|00\rangle + \frac{1}{\sqrt{2}}|11\rangle$. But unlike original states $|00\rangle$ and

$|11\rangle$, (pure) state $\frac{1}{\sqrt{2}}|00\rangle + \frac{1}{\sqrt{2}}|11\rangle$ cannot be represented as a tensor product of pure states in H_2 (which can be verified easily). It even turns out that the pure state $\frac{1}{\sqrt{2}}|00\rangle + \frac{1}{\sqrt{2}}|11\rangle$ cannot be represented in the form

$$\sum_i p_i S_i \otimes T_i$$

where S_i and T_i are states of H_2 (pure or mixed) and $\sum_i p_i = 1$. Pure state

$$\frac{1}{\sqrt{2}}|00\rangle + \frac{1}{\sqrt{2}}|11\rangle$$

is hence called *entangled*.

▶ A state $V \in L(H_n \otimes H_m)$ is said to be *decomposable* if V can be expressed as a finite sum

$$V = \sum_i p_i S_i \otimes T_i$$

where $S_i \in L(H_n)$ and $T_i \in L(H_m)$. A state that is not decomposable is *entangled*.

The fact that the trace of a state must be 1 implies that $\sum_i p_i$ must equal to 1 in the above representation, if such a representation is possible. For pure states, the entanglement appears to be a bit simpler than for general states: in fact, a pure state $\mathbf{z} \in H_n \otimes H_m$ is decomposable if and only if $\mathbf{z}$ is of the form $\mathbf{x} \otimes \mathbf{y}$ ($\mathbf{x} \in H_n$, $\mathbf{y} \in H_m$, and entangled otherwise).

Decomposable pure states of compound quantum systems are the easiest ones when determining the subsystem states: if $\mathbf{z} = \mathbf{x} \otimes \mathbf{y}$ is a state of a compound system, then the (pure) subsystem states are (trivially) $\mathbf{x}$ and $\mathbf{y}$. On the other hand, for an entangled state it is not clear, from an algebraic viewpoint, how to fix the subsystem states, even when the compound system state is pure.

Example 7   Let $\frac{1}{\sqrt{2}}|00\rangle + \frac{1}{\sqrt{2}}|11\rangle$ be a (pure) state of $H_2 \otimes H_2$. What should the subsystem states look like? As mentioned previously, it is not possible to find pure states $\mathbf{x}, \mathbf{y} \in H_2$ to satisfy $\mathbf{x} \otimes \mathbf{y} = \frac{1}{\sqrt{2}}|00\rangle + \frac{1}{\sqrt{2}}|11\rangle$.

The way to extract the subsystems states from the compound state is based on the statistical interpretation of quantum states. To begin with, consider any observable A on the subsystem $\mathscr{S}_1$. Clearly A can be extended into an observable on the whole system $\mathscr{S}_{12}$ as $A \otimes B$, whatever the observable B of the subsystem $\mathscr{S}_2$, but choice $B = I$ serves a very special purpose. Indeed, the observable I on the subsystem $\mathscr{S}_2$ has always value 1 in any state, which means that statistically observable $A \otimes I$ of the compound system does not differ from the observable A of the subsystem $\mathscr{S}_1$. With the minimal interpretation of quantum mechanics (Definition 16), this is enough to give the firm mathematical basis for the following definition.

Definition 22   Let $\mathscr{S}_1$ and $\mathscr{S}_2$ be quantum systems with state spaces H_n and H_m, respectively. Let also $S \in L(H_n \otimes H_m)$ be the state of a compound quantum system. Then the subsystem $\mathscr{S}_1$ state is $S_1 \in L(H_n)$, which satisfies

$$\mathrm{Tr}(S_1 A) = \mathrm{Tr}(S(A \otimes I))$$

for all observables $A \in L(H_n)$. We say that the subsystem state S_1 is obtained by *tracing over* H_m and denote

$$S_1 = \mathrm{Tr}_{H_m}(S)$$

The states of the subsystem $\mathscr{S}_2$ have an analogous definition and notation. The result of tracing over is also known as a *partial trace*.

For the proof of the following proposition, see Hirvensalo (2004), for instance.

Proposition 5 *State S_1 of the previous definition exists and is unique. It can be expressed as*

$$\mathrm{Tr}_{H_m}(S) = \sum_{i=1}^{n}\sum_{j=1}^{n}\sum_{k=1}^{m}\langle \mathbf{x}_i \otimes \mathbf{y}_k \mid T(\mathbf{x}_j \otimes \mathbf{y}_k)\rangle |\mathbf{x}_i\rangle\langle\mathbf{x}_j|$$

where $\{\mathbf{x}_1, \ldots, \mathbf{x}_n\}$ is an orthonormal basis of H_n and $\{\mathbf{y}_1, \ldots, \mathbf{y}_m\}$ an orthonormal basis of H_n.

Example 8 It follows directly from the above proposition that $\mathrm{Tr}_{H_m}(S_1 \otimes S_2) = S_1$. Another easy corollary is the linearity of tracing over: $\mathrm{Tr}_{H_m}(p_1 S + p_2 T) = p_1\mathrm{Tr}_{H_m}(S) + p_2\mathrm{Tr}_{H_m}(T)$, which subsequently implies that $\mathrm{Tr}_{H_m}(p_1 S_1 \otimes S_2 + p_2 T_1 \otimes S_2) = p_1 S_1 + p_2 T_1$.

By the above proposition, one can compute the subsystem states for a pure state

$$\frac{1}{\sqrt{2}}|00\rangle + \frac{1}{\sqrt{2}}|11\rangle \tag{13}$$

They both are

$$\frac{1}{2}|0\rangle\langle 0| + \frac{1}{2}|1\rangle\langle 1| = \frac{1}{2}I_2 \tag{14}$$

a mixed state. On the other hand, a reconstruction of ❷ Eq. 13 from subsystem states (❷ Eq. 14) is doomed to fail:

$$\left(\frac{1}{2}|0\rangle\langle 0| + \frac{1}{2}|1\rangle\langle 1|\right) \otimes \left(\frac{1}{2}|0\rangle\langle 0| + \frac{1}{2}|1\rangle\langle 1|\right)$$

$$= \frac{1}{4}|00\rangle\langle 00| + \frac{1}{4}|01\rangle\langle 01| + \frac{1}{4}|10\rangle\langle 10| + \frac{1}{4}|11\rangle\langle 11| = \frac{1}{4}I_4$$

which is different from the pure state determined by $\frac{1}{\sqrt{2}}|00\rangle + \frac{1}{\sqrt{2}}|11\rangle$:

$$\left|\frac{1}{\sqrt{2}}|00\rangle + \frac{1}{\sqrt{2}}|11\rangle\right\rangle\left\langle\frac{1}{\sqrt{2}}|00\rangle + \frac{1}{\sqrt{2}}|11\rangle\right|$$

$$= \frac{1}{2}|00\rangle\langle 00| + \frac{1}{2}|00\rangle\langle 11| + \frac{1}{2}|11\rangle\langle 00| + \frac{1}{2}|11\rangle\langle 11|$$

The above observations can be expressed as follows: A compound quantum system $\mathscr{S}_{12}$ always determines its subsystem states, but if states of the subsystems $\mathscr{S}_1$ and $\mathscr{S}_2$ are known, it is not possible to determine the state of the compound system uniquely without further information. A further analysis in fact shows that the case where both subsystem states are pure is the only one when the (decomposable) compound system state can be determined uniquely.

7.1 Transformations of Quantum States

Since unitary mappings play an important role in quantum systems, we will first devote a section to them.

7.1.1 Unitary Mappings

Definition 23   Mapping $U \in L(H_n)$ is *unitary* if $U^* = U^{-1}$.

Another equivalent form of the definition is as follows.

Definition 24   Mapping U is *unitary* if it preserves the inner products, meaning that $\langle U\mathbf{x} \mid U\mathbf{y} \rangle = \langle \mathbf{x} \mid \mathbf{y} \rangle$ for all $\mathbf{x}, \mathbf{y} \in H_n$.

The equivalence of the above two definitions is obvious, since

$$\langle U\mathbf{x} \mid U\mathbf{y} \rangle = \langle U^* U\mathbf{x} \mid \mathbf{y} \rangle$$

The latter form of the definition directly implies that unitary mappings also preserve the norms of the vectors: $\|U\mathbf{x}\|^2 = \langle U\mathbf{x} \mid U\mathbf{x} \rangle = \langle \mathbf{x} \mid \mathbf{x} \rangle = \|\mathbf{x}\|^2$, hence $\|U\mathbf{x}\| = \|\mathbf{x}\|$. The latter form can also be interpreted so that the unitary mappings preserve the angles between vectors; hence it is justified to say that unitary mappings are *rigid motions* or *rotations* of H_n. What is not so straightforward is that the condition $\|U\mathbf{x}\| = \|\mathbf{x}\|$ for each $\mathbf{x} \in H_n$ is also enough to guarantee the unitarity of mapping U. This follows from the so-called *polarization identity*; for the proof, see Hirvensalo (2004) for instance.

Recall from ❷ Sect. 2 that the self-adjoint mappings of H_n are, in a sense, extensions of real numbers. We will shortly see that the unitary mappings relate to the unit circle of complex numbers in the same way as self-adjoint mappings relate to real numbers. The following simple proposition draws the first connection.

Proposition 6   *The eigenvalues of unitary mappings lie in the unit circle.*

Proof   If $U\mathbf{x} = \lambda \mathbf{x}$, then by a direct consequence of ❷ Definition 24,

$$|\lambda| \|\mathbf{x}\| = \|\lambda \mathbf{x}\| = \|U\mathbf{x}\| = \|\mathbf{x}\|$$

hence $|\lambda| = 1$.

Like self-adjoint operators, also the unitary ones are normal: $U^* U = I = U U^*$, hence unitary mappings admit a spectral representation

$$U = \lambda_1 |\mathbf{x}_1\rangle\langle\mathbf{x}_1| + \ldots + \lambda_n |\mathbf{x}_n\rangle\langle\mathbf{x}_n| \tag{15}$$

and from Proposition 6 we know that each eigenvalue λ_k lies in the unit circle. Conversely, since any λ in the unit circle satisfies $\lambda^* = \lambda^{-1}$, we see that any mapping of form ❷ Eq. 15 with $|\lambda_i| = 1$ is indeed unitary. Since any unit circle number λ_k can be represented as $\lambda_k = e^{i\theta_k}$, where $\theta_k \in \mathbb{R}$, we get the following proposition.

Proposition 7   *If H is a self-adjoint mapping, then e^{iH} is unitary. Conversely, for any unitary mapping $U \in L(H_n)$ there exists a self-adjoint mapping H so that $U = e^{iH}$.*

We will present another analogy between unitary mappings and unit circle complex numbers. For that purpose, the notion of the absolute value defined for the operators will be needed.

Definition 25 Let $A \in L(H_n)$. Then the *absolute value* of A is defined as

$$|A| = \sqrt{A^*A}$$

Notice that this definition makes perfect sense: As $(A^*A)^* = A^*(A^*)^* = A^*A$, operator A^*A is self-adjoint and hence normal. Moreover $\langle \mathbf{x} \mid A^*A\mathbf{x} \rangle = \langle A\mathbf{x} \mid A\mathbf{x} \rangle = \|A\mathbf{x}\|^2 \geq 0$, which means that A^*A is a positive mapping; hence its eigenvalues are nonnegative, and the square root can be defined via spectral decomposition: if

$$A^*A = \sum_{k=1}^{n} \lambda_k |\mathbf{x}_k\rangle \langle \mathbf{x}_k|$$

then

$$\sqrt{A^*A} = \sum_{k=1}^{n} \sqrt{\lambda_k} |\mathbf{x}_k\rangle \langle \mathbf{x}_k|$$

Numbers $\sqrt{\lambda_1}, \ldots, \sqrt{\lambda_k}$ (the eigenvalues of $|A|$) are called the *singular values* of matrix A and are frequently denoted by symbols $s_1, \ldots, s_k$.

The next proposition, whose proof will be omitted, establishes another analogue between unitary mappings and unit circle numbers.

Proposition 8 (Polar decomposition) *Each $A \in L(H_n)$ can be represented as $A = U|A|$, where U is a unitary mapping. However, U is unique only if A is invertible.*

In an infinite-dimensional vector space, the above proposition does not necessarily hold, but "unitary" must be replaced with "partial isometry."

7.1.2 State Transformations

As the state of a quantum system $\mathscr{S}$ with state space H_n corresponds to a linear mapping in $L(H_n)$, it is natural to understand a state transformation (also called *discrete time evolution*) as a mapping $V: L(H_n) \rightarrow L(H_n)$, which takes the state S_1 of system $\mathscr{S}$ at time t_1 into the state S_2 at time t_2: $V(S_1) = S_2$.

To discover the nature of mapping V, we can think about the obvious constraints. As the most obvious one, $S_2 = V(S_1)$ should be a quantum state if S_1 is. This means that V should at least preserve the positivity and the unit trace. Already these requirements impose some restrictions on the form of mapping V, but the heaviest constraint comes from the linearity assumption. Typically, mapping $V: L(H_n) \rightarrow L(H_n)$ is assumed to be linear, and this assumption is supported, among others, by the idea of mixing states: the statistical mixture $T = pT_1 + (1-p)T_2$ should evolve so that T_1 and T_2 are kept independent, meaning that $V(T) = pV(T_1) + (1-p)V(T_2)$. By questioning the very ontology of the mixed quantum states, one can obviously present arguments against the linearity, too, but such a discussion is not the purpose of this chapter. Instead, this presentation follows the mainstream, accepting the linearity of state transformations. Already these ideas give rise to the following definition.

Definition 26 A linear mapping $V: L(H_n) \rightarrow L(H_n)$ is *positive* if $V(A)$ is a positive mapping whenever A is.

When accepting the aforementioned conditions (positivity and preserving the unit trace, plus the linearity) on the state transformations, we could imagine that no other conditions are needed. However, this is not the case. Indeed, it is possible to imagine a compound system $\mathscr{S}_{12}$ with quantum state transformation $V \otimes I$ that leaves the subsystem $\mathscr{S}_2$ virtually untouched.

Example 9 Mapping $V : L(H_n) \rightarrow L(H_n)$ defined by $V(A) = A^*$ preserves positivity: Indeed, if A is positive, then $\langle \mathbf{x} \mid A\mathbf{x} \rangle \geq 0$ for any $\mathbf{x}$. Then also $\langle \mathbf{x} \mid V(A)\mathbf{x} \rangle = \langle \mathbf{x} \mid A^*\mathbf{x} \rangle = \langle A\mathbf{x} \mid \mathbf{x} \rangle = \langle \mathbf{x} \mid A\mathbf{x} \rangle^* = \langle \mathbf{x} \mid A\mathbf{x} \rangle \geq 0$.

Now in the case $n = 2$ V takes the form $V(|0\rangle\langle0|) = |0\rangle\langle0|$, $V(|0\rangle\langle1|) = |1\rangle\langle0|$, $V(|1\rangle\langle0|) = |0\rangle\langle1|$, and $V(|1\rangle\langle1|) = |1\rangle\langle1|$. Mapping

$$A = |00\rangle\langle00| + |00\rangle\langle11| + |11\rangle\langle00| + |11\rangle\langle11| \in L(H_2 \otimes H_2)$$

is positive, as easily seen from the spectral representation $A = |\,|00\rangle + |11\rangle\rangle\langle\langle00| + |11\rangle|$. But $V \otimes I_2$ takes $|00\rangle\langle11| = |0\rangle\langle1| \otimes |0\rangle\langle1|$ into $|1\rangle\langle0| \otimes |0\rangle\langle1| = |10\rangle\langle01|$ and $|11\rangle\langle00| = |1\rangle\langle0| \otimes |1\rangle\langle0|$ into $|0\rangle\langle1| \otimes |1\rangle\langle0| = |01\rangle\langle10|$, hence

$$(V \otimes I_2)(A) = |00\rangle\langle00| + |10\rangle\langle01| + |01\rangle\langle10| + |11\rangle\langle11|$$

but the mapping thus obtained *is not positive*, as it has -1 as an eigenvalue:

$$(V \otimes I_2)(A)(|10\rangle - |01\rangle) = |01\rangle - |10\rangle = -(|10\rangle - |01\rangle)$$

The above example shows that we should impose a heavier restriction on mapping V to be an eligible quantum state transform. This stronger condition is expressed in the definition below.

Definition 27 Linear mapping $V : L(H_n) \rightarrow L(H_n)$ is a *completely positive mapping* if $V \otimes I_k$ is a positive mapping for any k (here I_k stands for the identity mapping on k-dimensional Hilbert space).

The above definition completes the characterization of discrete quantum time evolutions (quantum state transformations).

Definition 28 The discrete time evolution operator of a quantum system with state space H_n is the completely positive, trace-preserving linear mapping $L(H_n) \rightarrow L(H_n)$.

The discrete quantum time evolution operators are also known as *superoperators*. There are two well-known characterizations for them, both of which follow from the *Stinespring dilation theorem* (Stinespring 1955). The first characterization is known as the *Kraus representation*.

Theorem 4 *A completely positive mapping $V : L(H_n) \rightarrow LL(H_n)$ can be represented as*

$$V(A) = \sum_{k=1}^{m} V_k A V_k^* \tag{16}$$

where each $V_k \in L(H_n)$ and $m \leq n^2$. V is trace-preserving if and only if

$$\sum_{k=1}^{m} V_k^* V_k = I$$

Even though the second characterization also follows from the Stinespring dilation theorem as does the first one, this second one is more frequently known as the *Stinespring representation*. Another name for the representation below is the *Ozawa representation* (Ozawa 1984).

Theorem 5   *For any completely positive, trace-preserving mapping $V : L(H_n) \rightarrow L(H_n)$ there exists a Hilbert space H_m with $m \leq n^2$, pure state $|0\rangle\langle 0| \in H_m$, and a unitary mapping $U \in L(H_n \otimes H_m)$ so that*

$$V(A) = \mathrm{Tr}_{H_m}(U(A \otimes |0\rangle\langle 0|)U^*) \tag{17}$$

Representation (❯ 17) offers an interesting interpretation, which will be discussed after introducing a special case in the next section.

7.1.3   Closed Systems

The state transformation as introduced in the previous chapter can be irreversible in general. For instance, $V : L(H_2) \rightarrow L(H_2)$ defined via Kraus representation

$$V(A) = |0\rangle\langle 0|A|0\rangle\langle 0| + |1\rangle\langle 1|A|1\rangle\langle 1|$$

is clearly a completely positive, trace-preserving mapping. The images of the basis elements of $L(H_2)$ under V can be easily computed:

$$V(|0\rangle\langle 0|) = |0\rangle\langle 0||0\rangle\langle 0||0\rangle\langle 0| + |1\rangle\langle 1||0\rangle\langle 0||1\rangle\langle 1| = |0\rangle\langle 0|$$

and similarly $V(|0\rangle\langle 1|) = 0$, $V(|1\rangle\langle 0|) = 0$, and $V(|1\rangle\langle 1|) = |1\rangle\langle 1|$. We can readily observe that V is not injective, hence V^{-1} does not exist. It is also noteworthy that the action of V on a pure (vector) state $\frac{1}{\sqrt{2}}|0\rangle + \frac{1}{\sqrt{2}}|1\rangle$ yields

$$V(|\frac{1}{\sqrt{2}}|0\rangle + \frac{1}{\sqrt{2}}|1\rangle\rangle\langle\frac{1}{\sqrt{2}}|0\rangle + \frac{1}{\sqrt{2}}|1\rangle|)$$

$$= |0\rangle\langle 0||\frac{1}{\sqrt{2}}|0\rangle + \frac{1}{\sqrt{2}}|1\rangle\rangle\langle\frac{1}{\sqrt{2}}|0\rangle + \frac{1}{\sqrt{2}}|1\rangle||0\rangle\langle 0|$$

$$+ |1\rangle\langle 1||\frac{1}{\sqrt{2}}|0\rangle + \frac{1}{\sqrt{2}}|1\rangle\rangle\langle\frac{1}{\sqrt{2}}|0\rangle + \frac{1}{\sqrt{2}}|1\rangle||1\rangle\langle 1|$$

$$= \frac{1}{2}|0\rangle\langle 0| + \frac{1}{2}|1\rangle\langle 1|$$

a mixed state (analogously one can see that V turns vector state $\frac{1}{\sqrt{2}}|0\rangle - \frac{1}{\sqrt{2}}|1\rangle$ into a mixed state $\frac{1}{2}|0\rangle\langle 0| + \frac{1}{2}|1\rangle\langle 1|$).

In general, a quantum system is called *closed* if it does not interact with any other system and if it is not entangled with another system. In this chapter, however, the following definition will be taken for a closed system.

Definition 29   Discrete time evolution in $L(H_n)$ is called *closed* if it is of the form

$$V(A) = UAU^* \tag{18}$$

where $U \in L(H_n)$ is unitary. A quantum system with closed (discrete) time evolution is called closed. A system that is not closed is called *open*.

Notice that both the Kraus representation and the Stinespring representation are extensions of ❷ Eq. 18. Indeed, ❷ Eq. 18 is a Kraus representation with $m=1$ and a Stinespring representation with $m=0$.

Closed discrete time evolution $V(A) = UAU^*$ on a pure (vector) state $\mathbf{x}$ especially means that $|\mathbf{x}\rangle\langle\mathbf{x}|$ is converted into $U|\mathbf{x}\rangle\langle\mathbf{x}|U^* = |U\mathbf{x}\rangle\langle U\mathbf{x}|$, which is another pure state. Moreover, a closed discrete evolution is always invertible: A can be recovered as $A = U^*V(A)U$, which even reveals that the inverse of a closed state transformation is obtained by replacing U with $U^* = U^{-1}$. A closed state transformation means that the state vector turns into another state vector $U\mathbf{x}$.

▶ In the theory of quantum computing, it frequently assumed that the computational device has only closed state transformations and handles only pure states. In particular, this implies that one can then remain in the vector state formalism, and any state transformation takes the form

$$\mathbf{x} \mapsto U\mathbf{x}$$

where U is a unitary mapping. This is usually referred to by saying that (discrete) quantum evolution is *unitary*.

It is now interesting to see that the Stinespring representation (❷ 17) can be interpreted as follows: a quantum system with an open state transformation can be augmented with an auxiliary system, and the state transformation in the compound system is closed. Open state transformations of the subsystem hence result from watching the system only partially; the state transformation is closed in a larger system.

7.1.4 Schrödinger Equation

Previously, only the state transformations of quantum systems were considered. In other words, the evolution studied so far has been discrete, meaning that in a fixed time interval state S turns into $V(S)$ when the time has elapsed. There is nothing wrong with such an approach, but evidently it gives only part of the picture. For a refinement, one should learn how to approach *continuous state transformations*. Continuous state transformations are referred to generally as the *time evolution*.

To establish the mathematical machinery for that purpose, we restrict ourselves, for the sake of simplicity, to the closed systems beginning on a pure state. In particular this means that the images of the initial state vector $\mathbf{x}_0$ are studied under unitary mappings U_t for $t \in \mathbb{R}$; the state at time t is $\mathbf{x}_t = U_t\mathbf{x}_0$. How to continue from this is briefly, but not very thoroughly, described as follows: It is natural to require that $U_{t_1+t_2}\mathbf{x} = U_{t_1}U_{t_2}\mathbf{x}$, which is not far from saying that there is a morphism $t \mapsto U_t$ from real numbers to unitary mappings. Together with a continuity assumption, such a morphism can in fact be represented as $U_t = e^{itH}$, where H is a fixed self-adjoint mapping (Hamiltonian of the quantum system) (Stone 1932). Consequently,

$$\mathbf{x}_t = U_t\mathbf{x}_0 = e^{itH}\mathbf{x}_0$$

and componentwise differentiation gives

$$\frac{d}{dt}\mathbf{x}_t = iHe^{itH}\mathbf{x}_0 = iHU_t\mathbf{x}_0 = iH\mathbf{x}_t$$

▶ The equation

$$\frac{d}{dt}\mathbf{x}_t = iH\mathbf{x}_t$$

is called the *Schrödinger* equation.

The Schrödinger equation is a description of how the pure state of a closed system evolves continuously in time. The above form is frequently called the *abstract Schrödinger equation*, but the "concrete" ones can be obtained by substituting "concrete" Hamiltonians for H.

7.2 Models for Quantum Computing

Any classical model of computing also has a quantum version. There is in fact almost a "canonical" way of converting a classical model into a quantum version, namely that of replacing the classical description of the physical objects by the quantum description. In this section, the quantum time evolution means *discrete* time evolution, that is, the state transformation.

7.2.1 Finite Automata

Finite automata are models for real-time computation with a finite memory. The automaton reads the input word one letter at a time. Reading of a letter imposes a change of the internal state of the automaton (there are only finitely many of them), and having read the whole input word, the state of the automaton is observed. If the automaton has then reached a *final state* (can also be called *accepting state*), then the input word is accepted, otherwise rejected. Hence finite automata classify all potential input words into two piles: accepted and rejected. This can be expressed in other words by saying that the finite automata solve computational problems with yes/no answers: The problem instance is encoded into the input word, and the accept/reject behavior of the automaton is interpreted as the yes/no answer.

It turns out that finite automata can solve only a very restricted class of computational problems. Nevertheless, the research community has shown (and still shows) enormous interest in them. This is due to the following facts: the finite automata offer a very simple computational model with a huge number of connections to algebra, logic, combinatorics, logic, etc., not forgetting that the basic model of the finite automaton has beautiful closure properties; for example, constructing new automata of known ones for simultaneous computation or sequential computation is a straightforward task. Moreover, it is easy to present the generalizations and variants of finite automata: nondeterministic, probabilistic, and even quantum, as will be shortly seen. In addition, comparison between the different variants has been relatively successful, a natural measure for the complexity is the number of the states, and, for instance, a classical result shows that nondeterministic finite automata recognizing a language can have exponentially fewer states than the minimal deterministic one (Yu 1997). Even further, almost all classical models of computing (including the Turing machines in the next section) can be obtained by starting with finite automata, then augmenting them with an extra part. A wider description of finite automata can be found in Yu (1997), here we present only the definition.

A finite automaton A is a fivetuple $A = (Q, \Sigma, \delta, q_0, F)$, where Σ is a finite alphabet, Q is the state set, $\delta : Q \times \Sigma \rightarrow Q$ is the transition function, $q_0 \in Q$ the initial state, and $F \subseteq Q$ the set of final (accepting) states.

A *probabilistic finite automaton* (see Paz (1971) for more details) is obtained by replacing the transition function $\delta : Q \times \Sigma \to Q$ by the transition function $\delta : Q \times \Sigma \times Q \to [0,1]$ whose value at triplet (q, σ, p) is interpreted as the probability of entering state p when being in state q and reading σ as an input letter. Equivalently, one can describe a probabilistic finite automaton as a collection of $|Q| \times |Q|$ probabilistic matrices $(M_a)_{ij} = \delta(q_j, a, q_i)$ together with an initial probability distribution.

To get a *quantum finite automaton*, we assume that the set of states is described by a quantum mechanical systems. Consequently, we can fix a Hilbert space H_n ($n = |Q|$) together with an orthonormal basis $\{|q\rangle \,|\, q \in Q\}$. For each $a \in \Sigma$, we define a set of completely positive mappings $V_a : L(H_n) \to L(H_n)$, fix an *initial state* $S \in L(H_n)$, and a *final projection* $P \in L(H_n)$. The *acceptance probability* $p(w)$ for a word $w = a_1 \ldots a_k \in \Sigma^*$ (note that in the previous sections notion A^* was used for the adjoint mapping and here, as in theoretical computer science generally, Σ^* stands for the free monoid generated by Σ) is then computed as

$$p(w) = \mathrm{Tr}(P V_n \ldots V_1 S)$$

A *measure once* quantum automaton (introduced in Moore and Crutchfield 2000) is obtained from the above description by assuming that the initial state is pure, and that each mapping V_a is indeed closed time evolution $V_a(T) = U_a T U_a^*$ (unitary time evolution).

In the model of the *measure many* quantum automaton (introduced in Kondacs and Watrous (1997)), one also assumes a pure initial state and unitary time evolution. The state set is divided into accepting, neutral, and rejecting states, and the subspaces generated by them are denoted by H_A, H_N, and H_R, respectively. For an input letter a, the model includes unitary time evolution V_a followed by a measurement of the observable defined by mutually orthogonal subspaces H_A, H_N, and H_R. If the state observed was accepting (resp. rejecting), then the input word is accepted (resp. rejected), but if the observed state was neutral, then the computation is continued with the next input letter.

It is trivial to note that measure-once quantum automata are subcases of the measure many automata, but it also turns out that both quantum automata models together with the model of probabilistic automata are subcases of the general model with an open time evolution (Hirvensalo 2008).

7.2.2 Quantum Turing Machines

While finite automata are very restricted in their computational power, *Turing machines* can compute anything one can even imagine to be computed. Or at least this is a very sincere hope expressed in the famous *Church–Turing thesis* (see Papadimitriou (1994) for discussion), which essentially says that for any intuitive algorithm there is a Turing machine doing the same job. Turing machines were introduced by Turing (1936) to satisfy the need for a formal definition of *algorithm*.

Turing machines extend the notion of finite automata by an *infinite memory*, which, in the basic model, is a linear array of memory cells, each capable of containing a symbol from a finite alphabet. This infinite memory of the Turing machine is called the *tape*, and it is accessed via a *read–write head* that can point only at one particular cell at a time, and on a single computational step can move to point at most to the adjacent cells.

Besides the tape, the Turing machines have, as finite automata do, their finite set of internal states. The computation of a Turing machine runs as follows: The input word is initially written on the tape, but all other tape cells are empty. At the initial stage of computation, the read–write head is set to point at the cell containing the first letter of the input word. A computational step of a Turing machine then can change the internal state of the machine, the content of the tape cell being scanned, and the position of the read–write head. The computational steps follow each other until the machine reaches an internal final state, which can be either of accepting or rejecting type. It may also happen that the final state is never reached, but the computation continues forever.

It can be understood that Turing machines, unlike finite automata, classify the input words into rejected, accepted, and neutral (for which the computation never ends), but also other classifications are possible: we could for example group together rejecting and unending computations to get a yes/no answer when solving computational problems. On the other hand, there is no way to tell in general whether a Turing machine computation is going to halt or not (Papadimitriou 1994).

As a theoretical model of computing, Turing machines do not have intuitively good programming properties. In fact, almost the converse holds true: The program of a Turing machine, which is encoded in the transition function, is usually the most unintuitive one could ever imagine. But, on the other hand, the Turing machine formalism offers the clearest picture on the computational resources one could ever imagine: The time needed for the computation is just the number of the computational steps from the beginning to the end, and the space devoured by the computation is the number of tape cells used during it.

Just as with finite automata, Turing machines also bear notable closure properties: For example, given two Turing machines, it is possible to construct a new machine for parallel computations, as well as for consecutive computation. Moreover, it is possible to construct a single Turing machine acting as a *clock* in the following sense: For an input of length n, the machine performs computations for exactly $f(n)$ steps, and then halts (the construction can be made for a large class of functions f, but of course not for all functions). This offers a potential way of handling nonfinishing computations: Given a Turing machine M, one can construct a time-counting machine T, and join them into a new machine M', which halts either when the input word is accepted or when the time limit is reached. This construction is particularly good for algorithms with a known time complexity function, as the termination of the computation can then be always guaranteed.

For an extensive exposure on Turing machines, we refer to Papadimitriou (1994), but a brief formal definition follows immediately.

Definition 30 A Turing machine is a sixtuple $T = (Q, \Gamma, \delta, q_0, q_a, q_r)$, where Q is the set of the machine's internal states, $\Gamma = \Sigma \cup \{\square\}$ is the input alphabet Σ augmented with the blank symbol $\square$, and q_0, q_a, and q_r are the initial, accepting, and rejecting states, respectively.

The transition function $\delta : Q \times \Sigma \mapsto Q \times \Sigma \times \{L, S, R\}$ specifies the action of the machine: if $\delta(q_1, \sigma_1) = (q_2, \sigma_2, L)$, then the machine being in state q_1 and reading σ_1 writes σ_2 in the memory cell being scanned, enters state q_2, and moves the read–write head to the left. Correspondingly, values S and R would order the read–write head to stay, and move to the right, respectively.

Now the memory of a Turing machine consists of the finite set of internal states together with the tape content. As the tape is initially assumed to contain only the (finite) input word, there are at any moment of the computation only finitely many non-blank symbols on the tape. This implies that the number of potential *configurations* of the machine is countable. To specify what a configuration of the Turing machine means, there are two traditional ways.

The first one is a triplet (q, i, w), where $q \in Q$ is the state of the machine, $i \in \mathbb{Z}$ tells the number of the memory cell being scanned by the read–write head (initially assumed 0), and $w \in \Gamma^*$ is the nonblank content of the tape presented in the shortest way. In the second representation, the position of the read–write head is indicated by splitting the content of the tape into two sections: configuration (q, w_1, w_2) then means that the state of the machine is q, the non-blank content of the tape is $w_1 w_2$, and that the read–write head is scanning the first symbol of w_2.

The quantum version of Turing machines, first introduced by David Deutsch and subsequently analyzed in Bernstein and Vazirani (1997) is obtained by assuming that both the tape and the set of internal states are quantum physical. The internal state set will be obviously described as in the case of finite automata: Hilbert space $H_{|Q|}$ with an orthonormal basis $\{q | q \in Q\}$ will serve as a quantum mechanical description of the state set. The tape itself, being an infinite object, is a bit more complicated. A single tape cell can of course be described by using Hilbert space $H_{|\Gamma|}$ with an orthonormal basis $\{|\gamma\rangle | \gamma \in \Gamma\}$, two tape cells as $H_{|\Gamma|} \otimes H_{|\Gamma|}$, three cells as $H_{|\Gamma|} \otimes H_{|\Gamma|} \otimes H_{|\Gamma|}$, etc. The quantum physical description of the tape is hence given by an infinite tensor product of finite dimensional Hilbert spaces $H_{|\Gamma|}$ called T, but the action of the machine is nevertheless defined by a local rule.

Deutsch's model presented the configuration of the machine as (q, i, w), where i presents the number of the cell being scanned. To get a quantum version of this, one needs to introduce an infinite-dimensional Hilbert space $H_\mathbb{Z}$ with orthonormal basis $\{|i\rangle \mid i \in \mathbb{Z}\}$, and the quantum version of the configuration is simply a state in space $H_{|Q|} \otimes H_\mathbb{Z} \otimes T$.

The other model, see Hirvensalo (2004) for instance, uses configurations of the form (q, w_1, w_2). There are only numerable many of those configurations, and we set up a Hilbert space H_C (the configuration space) with an orthonormal basis $\{(q, w_1, w_2) | q \in Q, w_1, w_2 \in \Gamma^*\}$.

If open time evolution is allowed, then the relationship between classical and quantum Turing machines is not a very problematic one: it is possible to interpret an ordinary Turing machine as a special case of quantum Turing machines merely by defining the time evolution by $V | c \rangle = | c' \rangle$, where c' is the configuration obtained from c by transition function δ.

On the other hand, most papers on quantum computing, including the pioneering articles, assume that the quantum Turing machine model operates with *closed (unitary) time evolution*, V which imposes a heavy restriction on δ. In general it is possible that δ does not specify any reversible time evolution. Fortunately it has been thoroughly examined to what extent the computation allows reversibility. A classical result (Bennett 1973) showed that *all computation* whatsoever can be performed also in a reversible manner, if extra workspace is available. In particular, the result of Bennett (1973) implies that an arbitrary computation can be simulated by a reversible one with the cost of a polynomial space increase.

7.2.3 Quantum Circuits

Quantum Turing machines offer, as their classical counterparts do, a good theoretical model for computation, as they provide a clear idea of the space and time required by computation. However, as in the classical case, it is very difficult to express even simple algorithms by means of a quantum Turing machine. For that purpose, *quantum circuits* are much better. They were first introduced by Deutsch (1989) and provide an easy way of expressing the technical details of quantum algorithms.

Boolean circuits (see Papadimitriou (1994) or Hirvensalo (2004) for instance) can be seen as acyclic graphs with nodes labeled by logical gates. A typical selection of gates includes AND,

NOT, and OR. The idea of the circuit computation is to encode the input into a binary string, and then let the circuit decide whether or not the input is accepted. Hence one needs actually a family C_1, C_2, C_3, ..., of circuits so that circuit C_n, which takes the inputs of length n, has n input wires and one output wire. C_n then computes the output value 1 or 0 (yes/no) plainly obeying the structure of the Boolean circuit, using the gates as elementary computational devices.

Computation by circuits differs essentially from finite automata and Turing machines. As the latter devices can handle inputs of all lengths by a single machine, the circuit model assumes a particular circuit C_n for each input word length n. This is usually expressed by saying that finite automata and Turing machines are *uniform* computational devices, and the set of circuits is not.

Furthermore, it turns out that for *any* subset $L \subseteq \{0,1\}^*$ there is a family C_1, C_2, C_3, ... of circuits accepting L, meaning that for a string w of length n, C_n outputs 1 if and only if $w \in L$ (Papadimitriou 1994; Hirvensalo 2004). On the contrary, it is clear that for an arbitrary subset $L \subseteq \{0,1\}^*$ there does not exist a Turing machine accepting L. This does not contradict the Church–Turing thesis claiming that the Turing machines possess the ultimate computational power, as only the *existence* of the circuit family C_1, C_2, C_3, ... is guaranteed no construction. Without a construction, a circuit family cannot be regarded as a computational device at all, but when the construction is included, circuit formalism can have great advantage over the sole Turing machine formalism, especially in the theory of quantum computing.

The construction of circuit C_n requires an algorithm: given n, the task is to produce the graph of C_n, hopefully in a very efficient manner. As an algorithm is called for, one immediately realizes that we are again speaking of Turing machines (or of some equivalent uniform computational model). One could hence again question the meaningfulness of the circuit model, but it has turned out that splitting the algorithm into two major parts (construction of the circuit and the computation performed by the circuit) can offer understanding, which outperforms that given by a direct uniform computational model.

The aforementioned division (construction of the circuit followed by circuit computation) seems to work especially well in quantum computing. The famous quantum algorithms for integer factoring by Peter W. Shor (Shor 1994) and quadratic speedup for search problems by Lov Grover (Grover 1996) can be very well explained using the quantum circuit model (see Hirvensalo (2004)). Indeed, the construction of the quantum circuit for both tasks turns out to be very easy, and the quantum effects (interference and entanglement) can be explored in the circuit computation.

By allowing open time evolution, one gets the notion of *quantum circuits with mixed states* introduced in Aharonov et al. (1998). Quantum circuits with open time evolution can simulate effectively ordinary Boolean circuits, but when restricted to closed time evolution, the computational power seemingly diminishes. However, such a great variety of research papers use quantum circuits with closed (unitary) time evolution that it is worth concentrating on them in more detail.

A *quantum circuit* with unitary time evolution is a sequence U_1, ..., U_k of unitary mappings acting on $H_{2^n} = H_2 \otimes \ldots H_2$ (n terms), where n is the number of input qubits. Mappings U_1, ..., U_k have the following restrictions: there is a finite set $\{G_1, \ldots, G_m\}$ of *quantum gates* (unitary mappings), from which mappings U_i are constructed. This means each U_i is one of the mappings G_j operating on some of the n qubits and leaving all others untouched.

The computation of the circuit is then very straightforward: Initially, the qubits contain the input word $\mathbf{x} \in \{0, 1\}^n$, so that the initial state of the circuit is $|\mathbf{x}\rangle$. Then mappings $U_1, \ldots, U_k$ are applied to get the final state

$$U_k \ldots U_1 |\mathbf{x}\rangle$$

and the output qubits can be observed. Number k is called the *depth* of the circuit.

Note that as the time evolution is reversible, one must have equally many input and output qubits, but of course one can restrict this to read only one of the output bits. In the second place, the finite set $G = \{G_1, \ldots, G_m\}$ is the counterpart of classical AND, NOT, and OR gates, and it is natural to assume that G is *universal* in the sense that all unitary mappings $U \in L(H_{2^n})$ can be approximated by gates in set G. The actual choice of G does not matter so much. One can also note that in the classical case one of the gates AND and OR could be omitted, or it could be possible to use only a single gate NAND. In the quantum context, universality must mean approximability: set $L(H_{2^n})$ has continuum cardinality, but mappings obtained from a finite set G in the above fashion form only a numerable set, implying that *most* unitary mappings cannot be represented exactly.

The question of universality has been studied very thoroughly. Already in his seminal paper (Deutsch 1989) Deutsch proved that there is a single three-qubit universal quantum gate. Later, it was shown in Barenco et al. (1995) that unary gates (those acting only on one qubit) together with controlled NOT gate form a universal set even in the strict sense (exact representation). A very simple set of universal gates was given in Shi (2003): the Toffoli gate together with the Hadamard–Walsh gate is a universal set. Another very important point is the approximation efficiency: how many gates from a universal set are needed in general to approximate a given gate with accuracy ε? The famous *Solovay–Kitaev* theorem (see Nielsen and Chuang (2000)) implies that only $O\left(n \log^c \frac{n}{\varepsilon}\right)$ gates from *any* universal gate set are needed to simulate (within accuracy ε) a quantum circuit with n arbitrary gates. (In the Solovay–Kitaev theorem, $c \in [1, 2]$ is a constant whose optimal value is not known.)

To conclude this section, we discuss briefly the relationship between quantum circuits with mixed states (open time evolution) and quantum circuits with closed (unitary) time evolution. As mentioned previously, quantum circuits with unitary time evolution are apparently more restricted than those with open time evolution, but there is a standard way of circumventing this restriction, namely the *ancilla bits*. This means that in addition to the n input qubits, there will be a number, say $a(n)$, of ancilla qubits initially all set to state $|0\rangle$, so that the initial state of the circuit is $|\mathbf{x}\rangle |\mathbf{0}\rangle$. Then an extended circuit operates on $n + a(n)$ qubits to get the desired result, and it can be shown that an arbitrary quantum circuit with open time evolution can be simulated in this manner, using $a(n)$ (where a is a polynomial) ancilla qubits (Aharonov et al. 1998).

References

Aharonov A, Kitaev A, Nisan N (1998) Quantum circuits with mixed states. In: proceedings of the 30th annual ACM symposium on theory of computation, Dallas, TX, May 1998, pp 20–30

Axler S (1997) Linear algebra done right. Springer, New York, http://www.springer.com/mathematics/algebra/book/978-0-387-98259-5

Barenco A, Bennett CH, Cleve R et al. (1995) Elementary gates for quantum computation. Phys Rev A 52(5): 3457–3467

Benioff PA (1980) The computer as a physical system: a microscopic quantum mechanical Hamiltonian model of computers as represented by Turing machines. J Stat Phys 22(5):563–591

Benioff PA (1982) Quantum mechanical Hamiltonian models of discrete processes that erase their own histories: application to Turing machines. Int J Theor Phys 21(3/4):177–202

Bennett CH (1973) Logical reversibility of computation. IBM J Res Dev 17:525–532

Bernstein E, Vazirani U (1997) Quantum complexity theory. SIAM J Comput 26(5):1411–1473

Cohn PM (1994) Elements of linear algebra. Chapman & Hall, Boca Raton, FL. CRC Press reprint (1999). http://www.amazon.com/Elements-Linear-Algebra-Chapman-Mathematics/dp/0412552809

Feynman RP (1982) Simulating physics with computers. Int J Theor Phys 21(6/7):467–488

Deutsch D (1985) Quantum theory, the Church-Turing principle and the universal quantum computer. Proc R Soc Lond Ser A Math Phys Sci 400:97–117

Deutsch D (1989) Quantum computational networks. Proc R Soc Lond A 425:73–90

Gleason AM (1957) Measures on the closed subspaces of a Hilbert space. J Math Mech 6:885–893

Grover LK (1996) A fast quantum-mechanical algorithm for database search. In: Proceedings of the 28th annual ACM symposium on the theory of computing, Philadelphia, PA, May 1996, pp 212–219

Hirvensalo M (2004) Quantum computing, 2nd edn. Springer, Heidelberg

Hirvensalo M (2008) Various aspects of finite quantum automata. In: Proceedings of the 12th international conference on developments in language theory, Kyoto, Japan, September 2008. Lecture notes in computer science, vol 5257. Springer, Berlin, pp 21–33

Kondacs A, Watrous J (1997) On the power of quantum finite state automata. In: Proceedings of the 38th annual symposium on foundations of computer science, Miami Beach, FL, October 1997, pp 66–75

Moore C, Crutchfield JP (2000) Quantum automata and quantum grammars. Theor Comput Sci 237(1–2):275–306

Nielsen MA, Chuang IL (2000) Quantum computation and quantum information. Cambridge University Press

Ozawa M (1984) Quantum measuring processes of continuous observables. J Math Phys 25:79–87

Papadimitriou CH (1994) Computational complexity. Addison-Wesley, Reading, MA

Paz A (1971) Introduction to probabilistic automata. Academic, New York

Shi Y (2003) Both Toffoli and controlled-NOT need little help to do universal quantum computation. Quantum Info Comput 3(1):84–92

Shor PW (1994) Algorithms for quantum computation: discrete log and factoring. In: Proceedings of the 35th annual IEEE symposium on foundations of computer science, Santa Fe, NM, November 1994, pp 20–22

Simon DR (1994) On the power of quantum computation. In: Proceedings of the 35th annual IEEE symposium on foundations of computer science, Santa Fe, NM, November 1994, pp 116–123

Stinespring WF (1955) Positive functions on C^*-algebras. Proc Am Math Soc 6:211–216

Stone MH (1932) On one-parameter unitary groups in Hilbert space. Ann Math 33:643–648

Turing AM (1936) On computable numbers, with an application to the Entscheidungsproblem. Proc Lond Math Soc 2(42):230–265

von Neumann J (1927) Thermodynamik quantummechanischer Gesamheiten. Nachrichten von der Gesellschaft der Wissenschaften zu Göttingen 1:273–291

von Neumann J (1932) Mathematische Grundlagen der Quantenmechanik. Springer, Berlin

Yu S (1997) Regular languages. In: Rozenberg G, Salomaa A (eds) Handbook of formal languages. Springer, Berlin

42 Bell's Inequalities — Foundations and Quantum Communication

Časlav Brukner[1] · *Marek Żukowski*[2]
[1]Faculty of Physics, University of Vienna, Vienna, Austria
caslav.brukner@univie.ac.at
[2]Institute of Theoretical Physics and Astrophysics, University of Gdansk,
Poland
marek.zukowski@univie.ac.at
fizmz@univ.gda.pl

G. Rozenberg et al. (eds.), *Handbook of Natural Computing*, DOI 10.1007/978-3-540-92910-9_42,
© Springer-Verlag Berlin Heidelberg 2012

Abstract

For individual events, quantum mechanics makes only probabilistic predictions. Can one go beyond quantum mechanics in this respect? This question has been a subject of debate and research since the early days of the theory. Efforts to construct a deeper, realistic level of physical description, in which individual systems have, like in classical physics, preexisting properties revealed by measurements are known as hidden-variable programs. Demonstrations that a hidden-variable program necessarily requires outcomes of certain experiments to disagree with the predictions of quantum theory are called "no-go theorems." The Bell theorem excludes local hidden variable theories. The Kochen–Specker theorem excludes non-contextual hidden variable theories. In local hidden-variable theories faster-than-light-influences are forbidden, thus the results for a given measurement (actual, or just potentially possible) are independent of the settings of other measurement devices which are at space-like separation. In non-contextual hidden-variable theories, the predetermined results of a (degenerate) observable are independent of any other observables that are measured jointly with it.

It is a fundamental doctrine of quantum information science that quantum communication and quantum computation outperform their classical counterparts. If this is to be true, some fundamental quantum characteristics must be behind the better-than-classical performance of information-processing tasks. This chapter aims at establishing connections between certain quantum information protocols and foundational issues in quantum theory. After a brief discussion of the most common misinterpretations of Bell's theorem and a discussion of what its real meaning is, we will demonstrate *how quantum contextuality and violation of local realism can be used as useful resources* in quantum information applications. In any case, the readers should bear in mind that this chapter is not a review of the literature of the subject, but rather a quick introduction to it.

1 Introduction

Which quantum states are useful for quantum information processing? All non-separable states? Only distillable non-separable states? Only those which violate constraints imposed by local realism? Entanglement is the most distinct feature of quantum physics with respect to the classical world (Schrödinger 1935). On the one hand, entangled states violate Bell's inequalities, and thus rule out a local realistic explanation of quantum mechanics. On the other hand, they enable certain communication and computation tasks to have an efficiency not achievable by the laws of classical physics. Intuition suggests that these two aspects, the fundamental one, and the one associated with applications, are intimately linked. It is natural to assume that the quantum states, which allow the no-go theorems of quantum theory, such as Kochen–Specker, Bell's or the Greenberger–Horne–Zeilinger theorem should also be useful for quantum information processing. If this were not true, one might expect that the efficiency of quantum information protocols could be simulatable by classical, local realistic or non-contextual models, and thus achievable already via classical means. This intuitive reasoning is supported by the results of, for example, Ekert (1991), Scarani and Gisin (2001) and Acin et al. (2006): violation of a Bell's inequality is a criterion for the security of quantum key distribution protocols. Also, it was shown that violation of Bell's inequalities by a quantum state implies that pure-state entanglement can be distilled from it (Acin et al. 2002) and that Bell's inequalities are related to optimal solutions of quantum state targeting (Bechmann-Pasquinucci 2005). This overview will give other examples that demonstrate the strong link between fundamental features of quantum states and their applicability in quantum information protocols, such as in quantum communication complexity problems, quantum random access coding, or certain quantum games.

2 Quantum Predictions for Two-Qubit Systems

To set the stage for the story first, a description of two-qubit systems is given in full detail.

Predictions shall be presented for all possible local yes–no experiments on two spin-1/2 systems (in modern terminology, qubits) for all possible quantum states, that is, from the pure maximally entangled singlet state (or the Bohm-EPR state), via factorizable (i.e., non-entangled) states, up to any mixed state. This will enable us to reveal the distinguishing traits of the quantum predictions for entangled states of the simplest possible compound quantum system. The formalism can be applied to any system consisting of two subsystems, such that each of them is described by a two-dimensional Hilbert space. The spin-1/2 convention is chosen to simplify the description.

2.1 Pure States

An important tool simplifying the analysis of the pure states of two subsystems is the so-called Schmidt decomposition.

2.1.1 Schmidt Decomposition

For any nonfactorizable (i.e., entangled) pure state, $|\psi\rangle$ of a *pair* of quantum subsystems, one described by a Hilbert space of dimension N, the other by space of dimension M, $N \leq M$, it is always possible to find preferred bases, one basis for the first system, another one for the second, such that the state becomes a sum of bi-orthogonal terms, that is

$$|\psi\rangle = \sum_{i=1}^{N} c_i |a_i\rangle_1 |b_i\rangle_2 \tag{1}$$

with $_n\langle x_i | x_j\rangle_n = \delta_{ij}$, for $x = a, b$ and $n = 1, 2$. It is important to stress that the appropriate single subsystem bases, here $|a_i\rangle_1$ and $|b_j\rangle_2$, depend upon the state that we want to Schmidt decompose.

The ability to Schmidt decompose the state is equivalent to a well-known fact from matrix algebra, that any $N \times M$ matrix $\hat{A}$ can be always put into a diagonal form $\hat{D}$, by applying a pair of unitary transformations, U and V such that: $\sum_{j=1}^{N} \sum_{k=1}^{M} U_{ij} A_{jk} V_{kl} = D_l \delta_{il}$.

The interpretation of the above formula could be put as follows. If the quantum pure state of two systems is non-factorizable, then there exists a pair of local observables (for system 1 with eigenstates $|a_i\rangle$, and for system 2 with eigenstates $|b_i\rangle$) such that the results of their measurements are perfectly correlated.

The method of the Schmidt decomposition allows us to put every pure normalized state of two spins into

$$|\psi\rangle = \cos\alpha/2|+\rangle_1|+\rangle_2 + \sin\alpha/2|-\rangle_1|-\rangle_2 \tag{2}$$

for some angle α and local bases spent by states $|+\rangle_i$ and $|-\rangle_i$, i=1,2. The Schmidt decomposition generally allows the coefficients to be real. This is achievable via trivial phase transformations of the preferred bases.

2.2   Arbitrary States

Systems can be in mixed states. Such states describe situations in which there does not exist any *nondegenerate* observable for which the measurement result is deterministic. This is the case when the system can be with various probabilities $P(x) \geq 0$ in some nonequivalent states $|\psi(x)\rangle$, with $\sum_x P(x) = 1$. Mixed states are represented by self-adjoint nonnegative (density) operators $\rho = \sum_x P(x)|\psi(x)\rangle\langle\psi(x)|$. As $\text{Tr}|\psi(x)\rangle\langle\psi(x)| = 1$ one has $\text{Tr}\rho = 1$.

The properties of mixed states of the two spin-1/2 systems will now be presented in detail. Any self-adjoint operator for one spin-1/2 particle is a linear combination of the Pauli matrices σ_i, i = 1, 2, 3, and the identity operator, $\sigma_0 = \mathbf{1}$, with *real* coefficients. Thus, any self-adjoint operator in the tensor product of the two spin-1/2 Hilbert spaces must be a real linear combination of all possible products of the operators $\sigma_\mu^1 \sigma_\nu^2$, where the Greek indices run from 0 to 3, and the superscripts denote the particle. As the trace of σ_i is zero, we arrive at the following form of the general density operator for two spin 1/2 systems:

$$\rho = \frac{1}{4}\left(\sigma_0^{(1)}\sigma_0^{(2)} + \mathbf{r}\cdot\boldsymbol{\sigma}^{(1)}\sigma_0^{(2)} + \sigma_0^{(1)}\mathbf{s}\cdot\boldsymbol{\sigma}^{(2)} + \sum_{m,n=1}^{3} T_{nm}\sigma_n^{(1)}\sigma_m^{(2)}\right) \tag{3}$$

where $\mathbf{r}$ and $\mathbf{s}$ are real three-dimensional (Bloch) vectors and $\mathbf{r}\cdot\boldsymbol{\sigma} \equiv \sum_{i=1}^{3} r_i\sigma_i$ is the scalar product between vectors $\mathbf{r}$ and $\boldsymbol{\sigma}$. The tensor product symbol $\otimes$ shall be used only sparingly, only whenever it is deemed necessary. The condition $\text{Tr}\rho = 1$ is satisfied, thanks to the first term.

Since the average of any real variable which can have only two values +1 and −1 cannot be larger than 1 and less than −1, the real coefficients T_{mn} satisfy relations

$$-1 \leq T_{mn} = \text{Tr}\rho\sigma_n^{(1)}\sigma_m^{(2)} \leq 1 \tag{4}$$

and they form a matrix (also called a correlation tensor), which will be denoted by $\hat{T}$. One also has

$$-1 \leq r_n = \text{Tr}\rho\sigma_n^{(1)} \leq 1 \tag{5}$$

and

$$-1 \leq s_m = \text{Tr}\rho\sigma_m^{(2)} \leq 1 \tag{6}$$

2.2.1   Reduced Density Matrices for Subsystems

A reduced density matrix represents the local state of a compound system. If there are two subsystems, then the average of any observable, which pertains to the first system only, that is, of the form $A \otimes \mathbf{1}$, where $\mathbf{1}$ is the identity operation for system 2, can be expressed as follows

$\mathrm{Tr}_{12}(A \otimes \mathbf{1}\rho) = \mathrm{Tr}_1[A(\mathrm{Tr}_2\rho)]$. Here, Tr_i represents a trace with respect to system i. As trace is a basis independent notion, one can always choose a factorizable basis, and therefore split the trace calculation into two stages.

The reduced one particle matrices for spins 1/2, are of the following form:

$$\rho_1 \equiv \mathrm{Tr}_2\rho = \frac{1}{2}(\mathbf{1} + \mathbf{r} \cdot \boldsymbol{\sigma}^{(1)}) \tag{7}$$

$$\rho_2 \equiv \mathrm{Tr}_1\rho = \frac{1}{2}(\mathbf{1} + \mathbf{s} \cdot \boldsymbol{\sigma}^{(2)}) \tag{8}$$

with $\mathbf{r}$ and $\mathbf{s}$ the two local Bloch vectors of the spins.

Let us denote the eigenvectors of the spin projection along direction $\mathbf{a}$ of the first spin as: $|\psi(\pm 1, \mathbf{a})\rangle_1$. They are defined by the relation

$$\mathbf{a} \cdot \boldsymbol{\sigma}^{(1)}|\psi(\pm 1, \mathbf{a})\rangle_1 = \pm 1|\psi(\pm 1, \mathbf{a})\rangle_1 \tag{9}$$

where $\mathbf{a}$ is a real vector of unit length (i.e., $\mathbf{a} \cdot \boldsymbol{\sigma}^{(1)}$ is a Pauli operator in the direction of $\mathbf{a}$). The probability of a measurement of this Pauli observable to give a result ± 1 is given by

$$P(\pm 1|\mathbf{a})_1 = \mathrm{Tr}_1\rho_1\pi_{(\mathbf{a},\pm 1)}^{(1)} = \frac{1}{2}(1 \pm \mathbf{a} \cdot \mathbf{r}) \tag{10}$$

and it is positive for arbitrary $\mathbf{a}$, if and only if, the norm of $\mathbf{r}$ satisfies

$$|\mathbf{r}| \le 1 \tag{11}$$

Here $\pi_{(\mathbf{a},\pm 1)}^{(1)}$ is the projector $|\psi(\pm 1, \mathbf{a})\rangle_{11}\langle\psi(\pm 1, \mathbf{a})|$

2.3 Local Measurements on Two Qubits

The probabilities for local measurements to give the result $l = \pm 1$ for particle 1 and the result $m = \pm 1$ for particle 2, under specified local settings, $\mathbf{a}$ and $\mathbf{b}$ respectively, are given by:

$$P(l, m|\mathbf{a}, \mathbf{b})_{1,2} = \mathrm{Tr}\rho\pi_{(\mathbf{a},l)}^{(1)}\pi_{(\mathbf{b},m)}^{(2)} = \frac{1}{4}(1 + l\mathbf{a} \cdot \mathbf{r} + m\mathbf{b} \cdot \mathbf{s} + lm\mathbf{a} \cdot \hat{\mathrm{T}}\mathbf{b}) \tag{12}$$

where $\hat{\mathrm{T}}\mathbf{b}$ denotes the transformation of the column vector $\mathbf{b}$ by the matrix $\hat{\mathrm{T}}$ (Euclidean vectors are treated here as column matrices).

One can simplify all these relations by performing suitable local unitary transformations upon each of the subsystems, that is, via factorizable unitary operators $U^{(1)}U^{(2)}$. It is well known that any unitary operation upon a spin 1/2 is equivalent to a three-dimensional rotation in the space of Bloch vectors. In other words, for any real vector $\mathbf{w}$

$$U(\hat{\mathrm{O}})\mathbf{w} \cdot \boldsymbol{\sigma}U(\hat{\mathrm{O}})^\dagger = (\hat{\mathrm{O}}\mathbf{w}) \cdot \boldsymbol{\sigma} \tag{13}$$

where $\hat{\mathrm{O}}$ is the orthogonal matrix of the rotation. If the density matrix is subjected to such a transformation on either spins subsystem, that is, to the $U^1(\hat{\mathrm{O}}_1)U^2(\hat{\mathrm{O}}_2)$ transformation, the parameters $\mathbf{r}$, $\mathbf{s}$, and $\hat{\mathrm{T}}$ transform themselves as follows

$$\mathbf{r}' = \hat{\mathrm{O}}_1\mathbf{r}$$

$$\mathbf{s}' = \hat{\mathrm{O}}_2\mathbf{s} \tag{14}$$

$$\hat{\mathrm{T}}' = \hat{\mathrm{O}}_1\hat{\mathrm{T}}\hat{\mathrm{O}}_2^\mathrm{T}$$

Thus, for an arbitrary state, we can always choose such factorizable unitary transformation that the corresponding rotations (i.e., orthogonal transformations) will diagonalize the correlation tensor (matrix) $\hat{T}$. This can be seen as another application of Schmidt's decomposition, this time in the case of the second rank tensors.

The physical interpretation of the above is that one can always choose two (local) systems of coordinates, one for the first particle, the other for the second particle, in such a way that the $\hat{T}$ matrix will be diagonal.

Let us note that one can decompose the two spin density matrices into:

$$\rho = \rho_1 \otimes \rho_2 + \frac{1}{4} \sum_{m,n=1}^{3} C_{nm} \sigma_n^1 \otimes \sigma_m^2 \tag{15}$$

that is, it is a sum of the product of the two reduced density matrices and a term $\hat{C} = \hat{T} - \mathbf{r}\mathbf{s}^T$, which is responsible for the correlation effects.

Any density operator satisfies the inequality $\frac{1}{d} < \mathrm{Tr}\rho^2 \leq 1$, where d is the dimension of the Hilbert space in which it acts, that is, of the system it describes. The value of $\mathrm{Tr}\rho^2$ is a measure of the purity of the quantum state. It is equal to 1 only for single dimensional projectors, that is, the pure states. In the studied case of two qubits one must have

$$|\mathbf{r}|^2 + |\mathbf{s}|^2 + ||\hat{T}||^2 \leq 3 \tag{16}$$

For pure states, represented by Schmidt decomposition (❷ 2), $\hat{T}$ is diagonal with entries $T_{xx} = -\sin\alpha$, $T_{yy} = \sin\alpha$, and $T_{zz} = 1$, whereas $\mathbf{r} = \mathbf{s}$, and their z component is nonzero: $s_z = m_z = \cos\alpha$. Thus, in the case of a maximally entangled state, $\hat{T}$ has only diagonal entries equal to +1 and −1. In the case of the singlet state,

$$|\psi\rangle = \frac{1}{\sqrt{2}} \left(|+\rangle_1|-\rangle_2 - |-\rangle_1|+\rangle_2 \right) \tag{17}$$

which can be obtained from ❷ Eq. 2, by putting $\alpha = -\frac{\pi}{2}$ and rotating one of the subsystems such that $|+\rangle$ and $|-\rangle$ interchange (this is equivalent to a 180° rotation with respect to the axis x where $|+\rangle$ and $|-\rangle$ are eigenstates of spin along z; see above (❷ Eq. 14)), the diagonal elements of the correlation tensor are all −1.

3 Einstein–Podolsky–Rosen Experiment

In their seminal 1935 paper (Einstein et al. 1935) entitled *"Can quantum-mechanical description of physical reality be considered complete?"* Einstein, Podolsky, and Rosen (EPR) consider quantum systems consisting of two particles such that, while neither position nor momentum of either particle is well-defined, the difference of their positions and the sum of their momenta are both precisely defined. It then follows that measurement of either position or momentum performed on, say, particle 1 immediately implies, for particle 2, a precise position or momentum, respectively, even when the two particles are separated by arbitrary distances without any actual interaction between them.

We shall present the EPR argumentation for incompleteness of quantum mechanics in the language of spins 1/2. This was done by Bohm (1952) in 1952. A two-qubit example of an EPR state is the singlet state (❷ Eq. 17). The properties of a singlet can be inferred without mathematical considerations given above. This is a state of zero total spin. Thus, measurements of the same component of the two spins must always give opposite values – this is simply the conservation of angular momentum at work. In terms of the language of Pauli matrices the product of the local results is then always -1. We have (infinitely many) *perfect (anti-)correlations* for every possible choice of spin component. We assume that the two spins are very far away, but nevertheless in the singlet state.

After the translation into Bohm's example, the EPR argument runs as follows. Here are their premises:

1. *Perfect correlations*: If whatever spin components of particles 1 and 2 are measured, and the settings are identical for both particles, then with certainty the outcomes will be found to be perfectly anti-correlated.
2. *Locality*: "Since at the time of measurements the two systems no longer interact, no real change can take place in the second system in consequence of anything that may be done to the first system."
3. *Reality*: "If, without in any way disturbing a system, we can predict with certainty (i.e., with probability equal to unity) the value of a physical quantity, then there exists an element of physical reality corresponding to this physical quantity."
4. *Completeness*: "Every element of the physical reality must have a counterpart in the [complete] physical theory."

In contrast to the last three premises, which though they are quite plausible are still indications of a certain philosophical viewpoint, the first premise is a statement about a well-established property of a singlet state.

The EPR argument is as follows. Because of the perfect anti-correlations (1), we can predict with certainty the result of measuring either the x component or the y component of the spin of particle 2 by previously choosing to measure the same quantity of particle 1. By locality (2), the measurement of particle 1 cannot cause any real change in particle 2. This implies that by the premise (3) both the x *and* the y components of the spin of particle 2 are elements of reality. This is also the case for particle 1 by a parallel argument where particle 1 and 2 interchange their roles. Yet, (according to Heisenberg's uncertainty principle) there is no quantum state of a single spin in which both x and y spin components have definite values. Therefore, by premise (4), quantum mechanics cannot be a complete theory.

In his answer (Bohr 1935), published in the same year and under the same title as the EPR paper, Bohr criticized the EPR concept of "reality" as assuming the systems have intrinsic properties independently of whether they are observed or not and he argued for "the necessity of a final renunciation of the classical ideal of causality and a radical revision of one's attitude toward the problem of physical reality." Bohr pointed out that the wording of the criterion of physical reality (3) proposed by EPR contains an ambiguity with respect to the expression, "without in any way disturbing the system." And, while, as Bohr wrote, there is "no question of mechanical disturbance of the system," there is "the question of *an influence on the very conditions, which define the possible types of predictions regarding the future behavior of the system.*" Bohr thus pointed out that the results of quantum measurements, in contrast to those of classical measurements, depend on the complete experimental arrangement (*context*), which can even be nonlocal as in the EPR case. Before any measurement is performed, only the correlations between the spin components of two particles, but not spin components of individual particles, are defined. The x or y component (but never both) of an individual particle becomes defined only when the respective observable of the distant particle is measured.

Perhaps the clearest way to see how strongly the philosophical viewpoints of EPR and Bohr differ is in their visions of the future development of quantum physics. While EPR wrote: "We believe that such a [complete] theory is possible," Bohr's opinion is that (his) complementarity "provides room for new physical law, the coexistence of which might at first sight appear irreconcilable with the basic principles of science."

4 Bell's Theorem

Bell's theorem can be thought of as a disproof of the validity of the EPR ideas. The elements of physical reality cannot be an internally consistent notion. A broader interpretation of this result is that a local and realistic description of nature, at the fundamental level, is untenable. Further consequences are that there exist quantum processes, which cannot be modeled by any classical ones, not necessarily physical processes, but also some classical computer simulations with a communication constraint. This opened up the possibility of the development of quantum communication.

We shall now present a derivation of Bell's inequalities. The stress will be put on clarification of the underlying assumptions. These will be presented in the most reduced form.

4.1 Thought Experiment

At two measuring stations A and B, which are far away from each other, two characters Alice and Bob observe simultaneous flashy appearances of numbers $+1$ or -1 at the displays of their local devices (or the monitoring computers). The flashes appear in perfect coincidence (with respect to a certain reference frame). In the middle, between the stations, is something that they call "source." When it is absent, or switched off, the numbers ± 1's do not appear at the displays. The activated source always causes two flashes, one at A, and one at B. They appear slightly after a relativistic retardation time with respect to the activation of the source, never before. Thus there is enough "evidence" for Alice and Bob that the source causes the flashes. The devices at the stations have a knob, which can be put in two positions: $m = 1$ or 2 at A station, and $n = 1$ or 2 at B. Local procedures used to generate random choices of local knob positions are equivalent to *independent, fair coin tosses*. Thus, each of the four possible values of the pair n, m are equally likely, that is, the probability $P(n, m) = P(n)P(m) = \frac{1}{4}$. The "tosses," and knob settings, are made at random times, and often enough, so that the information on these is never available at the source during its activation periods (the tosses and settings cannot have a causal influence on the workings of source). The local measurement data (setting, result, and moment of measurement) are stored and very many runs of the experiment are performed (❯ *Fig. 1*).

4.1.1 Assumptions Leading to Bell's Inequalities

A concise *local realistic* description of such an experiment would use the following assumptions (Gill et al. 2002, 2003):

1. We assume *realism*, which is any logically self-consistent model that allows one to use *eight* variables in the theoretical description of the experiment: $A_{m,n}$, $B_{n,m}$ where $n, m = 1, 2$. The variable $A_{m,n}$ gives the value, ± 1, which could be obtained at station A, if the knob settings, at A and B, were at positions n, m, respectively. Similarly, $B_{n,m}$ plays the same role for station B, under the same settings. This is equivalent to the assumption that a joint (nonnegative, properly normalized) probability distribution of these variables,

□ **Fig. 1**

Test of Bell's inequalities. Alice and Bob are two separated parties who share entangled particles. Each of them is free to choose two measurement settings 1 and 2 and they observe flashes in their detection stations, which indicate one of the two possible measurement outcomes +1 or −1.

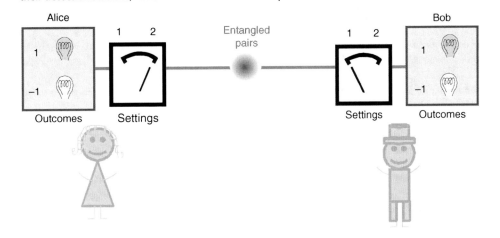

$P(A_{1,1}, A_{1,2}, A_{2,1}, A_{2,2}; B_{1,1}, B_{1,2}, B_{2,1}, B_{2,2}, n, m)$, is always allowed to exist. (Note, that no hidden variables appear, beyond these ten. However, given a (possibly stochastic) hidden variables theory, one will be able to define the ten variables as (possibly random) functions of the variables in that theory.)

2. The assumption of *locality* does not allow influences to go outside the light cone.

3. Alice and Bob are free to choose their settings "at the whim." This is the *freedom, or "free will,"* often only a tacit assumption (Bell 1985, Kofler et al. 2006). A less provocative version of this assumption: *There exist stochastic processes, which could be used to choose the values of the local settings of the devices which are independent of the workings of the source, that is, they neither influence it nor are influenced by it.* By the previous assumptions, the events of activation of the source and of the choice and fixing of the local settings must be space-like separated and they furthermore cannot have common cause in the past.

Note that when setting labels m, n are sent to the measurement devices, they are likely to cause some unintended disturbance: by these assumptions, *any disturbance at A, as far as it influences the outcome at A, is not related to the coin toss nor to the potential outcomes at B, and vice versa.*

Note further, that $A_{n,m}$ and $B_{n,m}$ are not necessarily actual properties of the systems. The only thing that is assumed is that there is a theoretical description, which allows one to use all these *eight* values.

4.1.2 First Consequences

Let us write down the immediate consequences of these assumptions:

- By *locality*: for all n, m:

$$A_{m,n} = A_m, \quad B_{n,m} = B_n \tag{18}$$

That is, the outcome which would appear at A does not depend on which setting might be chosen at B, and vice versa. *Thus $P(A_{1,1}, \ldots, B_{2,2}, n, m)$ can be reduced to $P(A_1, A_2, B_1, B_2, n, m)$.*

- By *freedom*

$$(n, m) \text{ is statistically independent of } (A_1, A_2, B_1, B_2) \tag{19}$$

Thus, the *overall* probability distributions for potential settings and potential outcomes satisfy

$$P(n, m, A_1, A_2, B_1, B_2) = P(n, m)P(A_1, A_2, B_1, B_2) \tag{20}$$

The choice of settings in the two randomizes, A and B, is causally separated from the local realistic mechanism, which produces the potential outcomes.

4.1.3 Lemma: Bell's Inequality

The probabilities, Pr, of the four logical propositions, $A_n = B_m$, satisfy

$$\Pr\{A_1 = B_2\} - \Pr\{A_1 = B_1\} - \Pr\{A_2 = B_1\} - \Pr\{A_2 = B_2\} \leq 0 \tag{21}$$

Proof: Only four, or two, or none of the propositions in the left-hand side of the inequality can be true, thus (21). QED.

Now, if the observation settings are totally random (dictated by "coin tosses"), $P(n, m) = \frac{1}{4}$. Then, according to all the assumptions

$$P(A_n = B_m | n, m) = P(n, m) \Pr\{A_n = B_n\} = \frac{1}{4} \Pr\{A_n = B_m\} \tag{22}$$

Therefore, we have a Bell's inequality: under the *conjunction* of the assumptions for the *experimentally accessible* probabilities one has

$$P(A_1 = B_2 \mid 1, 2) - P(A_1 = B_1 \mid 1, 1) - P(A_2 = B_1 \mid 2, 1) - P(A_2 = B_2 \mid 2, 2) \leq 0 \tag{23}$$

This is the well-known Clauser–Horne–Shimony–Holt (CHSH) inequality (Clauser et al. 1969).

4.2 Bell's Theorem

Quantum mechanics predicts for experiments satisfying all the features of the thought experiment the left-hand side of inequality (❷ 23) to be as high as $\sqrt{2} - 1$, which is larger than the local realistic bound 0. *Hence, one has Bell's theorem* (Bell 1964): *if quantum mechanics holds, local realism, defined by the full set of the above assumptions, is untenable.* But, how does nature behave – according to local realism or quantum mechanics? It seems that we are approaching the moment in which one could have, as perfect as possible, a laboratory realization of the thought experiment (the locality loophole was closed in Weihs et al. (1998) and Aspect et al. (1982), the detection loophole was closed in Rowe et al. (2001), and in recent experiments measurement settings were space-like separated from the photon pair emission (Scheidl et al. 2008)). Hence, the local realistic approach to the description of physical phenomena is close to being shown untenable, too.

4.2.1 The Assumptions as a Communication Complexity Problem

Assume that we have two programmers P_k, where $k = 1, 2$, each possessing an enormously powerful computer. They share certain joint classically correlated strings of arbitrary lengths and/or some computer programs. All these will be collectively denoted as λ. But, once they both possess λ, no communication whatsoever between them is allowed. After this initial stage, each one of them gets from a referee a one-bit random number $x_k \in \{0, 1\}$, known only to him/her (P_1 knows only x_1, P_2 knows only x_2). The *individual* task of each of them is to produce, via whatever computational program, a one-bit number $I_k(x_k, \lambda)$, and communicate only this one bit to a Referee, who just compares the received bits. There is no restriction on the form and complication of the *possibly stochastic* functions I_k, or any actions taken to define the values, but any communication between the partners is absolutely not allowed. The *joint* task of the partners is to devise a computer code which, under the constraints listed above and without any cheating, allows them to have, after very many repetitions of the procedures (each starting with establishing a new shared λ), the following functional dependence of the probability that their bits sent back to the Referee are equal:

$$P\{I_1(x_1) = I_2(x_2)\} = \frac{1}{2} + \frac{1}{2}\cos\left[-\pi/4 + (\pi/2)(x_1 + x_2)\right] \tag{24}$$

This is a variant of communication complexity problems. The current task is absolutely impossible to achieve with the classical means at their disposal, and without communication. Simply because whatever the protocol

$$\Pr\{I_1(1) = I_2(1)\} - \Pr\{I_1(0) = I_2(0)\} - \Pr\{I_1(1) = I_2(0)\} - \Pr\{I_1(0) = I_2(1)\} \leq 0 \tag{25}$$

whereas the value of this expression in the quantum strategy can be as high as $\sqrt{2} - 1$. This value can be obtained on average if the programmers use entanglement as a resource and receive their respective qubits from an entangled pair (e.g., singlet) during the communication stages (when λ is established). Instead of computing, the partners make a local measurement on their qubits. They measure Pauli observables $\mathbf{n} \cdot \boldsymbol{\sigma}$, where $\|\mathbf{n}\| = 1$. Since the probability for them to get identical results, r_1, r_2, for observation directions $\mathbf{n}_1, \mathbf{n}_2$ is

$$P_Q\{r_1 = r_2 | \mathbf{n}_1, \mathbf{n}_2\} = \frac{1}{2} - \frac{1}{2}\mathbf{n}_1 \cdot \mathbf{n}_2 \tag{26}$$

for suitably chosen $\mathbf{n}_1(x_1), \mathbf{n}_2(x_2)$, they get values of P_Q equal to those in (❯ Eq. 24). The messages sent back to the Referee encode the local results of measurements of $\mathbf{n}_1 \cdot \boldsymbol{\sigma} \otimes \mathbf{n}_2 \cdot \boldsymbol{\sigma}$, and the local measurement directions are suitably chosen as functions of x_1 and x_2. The relation between Bell's inequalities and quantum communication complexity problems will be reviewed in more detail in ❯ Sect. 6.

4.2.2 Philosophy or Physics? Which Assumptions?

The assumptions behind Bell's inequalities are often criticized as being "philosophical." If we remind ourselves about Mach's influence on Einstein, philosophical discussions related to physics may be very fruitful.

For those who are, however, still skeptical one can argue as follows. The whole (relativistic) classical theory of physics is realistic (and local). Thus, we have an important exemplary realization of the postulates of local realism. Philosophical propositions could be defined as those which *are not* observationally or experimentally falsifiable at the given moment of the development of human knowledge or, in pure mathematical theory, are not logically derivable. Therefore, the *conjunction* of all assumptions of Bell's inequalities is not a philosophical statement, as it is *testable* both experimentally and logically (within the known current mathematical formulation of the fundamental laws of physics). Thus, Bell's theorem removed the question of the possibility of a local realistic description from the realm of philosophy. Now, this is just a question of a good experiment.

The other criticism is formulated in the following way. Bell's inequalities can be derived using a single assumption of the existence of a joint probability distribution for the observables involved in them, or that the probability calculus of the experimental propositions involved in the inequalities is of Kolmogorovian nature, and nothing more. But if we want to understand the violation of Bell's inequalities, we stumble on the following question: *Does the joint probability take into account the full experimental context or not?* The experimental context is in the present case (at least) the full state of the settings (m, n). Thus, if we use the same notation as above for the realistic values, this time applied to the possible results of measurements of observables, initially we can assume the existence of only $p(A_{1,1}, A_{1,2}, A_{2,1}, A_{2,2}; B_{1,1}, B_{1,2}, B_{2,1}, B_{2,2})$. Note that such a probability could be, for example, factorizable into $\prod_{n,m} P(A_{n,m}, B_{n,m})$. That is, one could, in such a case, have different probability distributions pertaining to different experimental contexts (which can even be defined through the choice of measurement settings in space-like separated laboratories!).

Let the discussion of this be from the quantum mechanical point of view, only because such considerations have a nice formal description within this theory, familiar to all physicists. Two observables, say $\hat{A}_1 \otimes \hat{B}_1$ and $\hat{A}_2 \otimes \hat{B}_2$, as well as other possible pairs are functions of two different *maximal* observables for the whole system (which are nondegenerate by definition). If one denotes such a maximal observable linked with $\hat{A}_m \otimes \hat{B}_n$ by $\hat{M}_{m,n}$ and its eigenvalues by $M_{m,n}$ the existence of the aforementioned joint probability is equivalent to the existence of a $p(M_{1,1}, M_{1,2}, M_{2,1}, M_{2,2})$ in the form of a proper probability distribution. Only if one assumes, additionally, context independence, can this be reduced to the question of the existence of (nonnegative) probabilities $P(A_1, A_2, B_1, B_2)$, where A_m and B_n are eigenvalues of $\hat{A}_m \otimes \mathbf{1}$ and $\mathbf{1} \otimes \hat{B}_n$, where, in turn, $\mathbf{1}$ is the unit operator for the given subsystem. While context independence is physically doubtful when the measurements are not spatially separated, and thus one can have mutual causal dependence, it is well justified for spatially separated measurements. That is, *locality* enters one's reasoning, whether we like it or not. Of course one cannot derive any Bell's inequality of the usual type if the random choice of settings is not independent of the distribution of A_1, A_2, B_1, B_2, that is without (❷ Eq. 20).

There is yet another challenge to the set of assumptions presented above. It is often claimed that realism can be derived, once one considers the fact that maximally entangled quantum systems reveal perfect correlations, and one additionally assumes locality. Therefore, it would seem that the only basic assumption behind Bell's inequalities is locality, with the other auxiliary one of freedom. Such a claim is based on the ideas of EPR, who conjectured that one can introduce "elements of reality" of a remote system, provided this system is perfectly correlated with another system. To show the fallacy of such a hope, three particle correlations are now discussed, in the case of which consideration of just a few "elements of reality" reveals that they are a logically inconsistent notion. Therefore, they cannot be a starting point for deriving a self-consistent realistic theory. The three-particle reasoning is used here because of its beauty and simplicity, not because one cannot reach a similar conclusion for two-particle correlations.

4.3 Bell's Theorem Without Inequalities: Three Entangled Particles or More

As the simplest example, take a Greenberger–Horne–Zeilinger (Greenberger et al. 1989, 1990) (GHZ) state of $N = 3$ particles (❷ Fig. 2):

$$|\text{GHZ}\rangle = \frac{1}{\sqrt{2}}(|a\rangle|b\rangle|c\rangle + |a'\rangle|b'\rangle|c'\rangle) \tag{27}$$

where $\langle x|x'\rangle = 0$ ($x = a$, b, c, and kets denoted by one letter pertain to one of the particles). The observers, Alice, Bob, and Cecil measure the observables: $\hat{A}(\phi_A)$, $\hat{B}(\phi_B)$, $\hat{C}(\phi_C)$, defined by

$$\hat{X}(\phi_X) = |+, \phi_X\rangle\langle+, \phi_X| - |-, \phi_X\rangle\langle-, \phi_X| \tag{28}$$

◘ Fig. 2
Test of the GHZ theorem. Alice, Bob, and Cecil are three separated parties who share three entangled particles in the GHZ state. Each of them is free to choose between two measurement settings, 1 and 2, and they observe flashes in their detection stations, which indicate one of the two possible measurement outcomes +1 or −1.

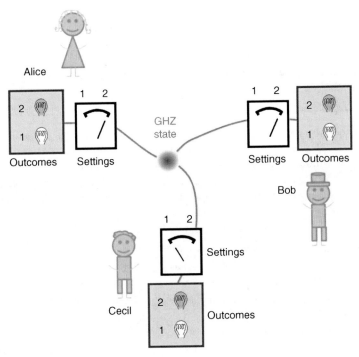

where

$$|\pm, \phi_X\rangle = \frac{1}{\sqrt{2}}(\pm i|x'\rangle + \exp(i\phi_X)|x\rangle) \tag{29}$$

and $\hat{X} = \hat{A}, \hat{B}, \hat{C}$. The quantum prediction for the expectation value of the product of the three local observables is given by

$$E(\phi_A, \phi_B, \phi_C) = \langle\text{GHZ}|\hat{A}(\phi_A)\,\hat{B}(\phi_B)\hat{C}(\phi_C)|\text{GHZ}\rangle = \sin(\phi_A + \phi_B + \phi_C) \tag{30}$$

Therefore, if $\phi_A + \phi_B + \phi_C = \pi/2 + k\pi$, quantum mechanics predicts perfect (anti)correlations. For example, for $\phi_A = \pi/2$, $\phi_B = 0$ and $\phi_C = 0$, whatever the results of local measurements of the observables, for, say, the particles belonging to the ith triple of the measured ensemble represented by the quantum state $|\text{GHZ}\rangle$, their product must be unity. In a local realistic theory, one would have

$$A^i(\pi/2)B^i(0)C^i(0) = 1 \tag{31}$$

where $X^i(\phi)$, $X = A$, B, or C is the local realistic value of a local measurement of the observable $\hat{X}(\phi)$ that *would have been* obtained for the ith particle triple, if the setting of the measuring device were ϕ. By locality $X^i(\phi)$ depends solely on the local parameter. The ❷ Eq. 31 indicates that we can predict, with certainty, the result of measuring the observable, pertaining to one of the particles (say c), by choosing to measure suitable observables for the other two. Hence, the values $X^i(\phi)$ are EPR elements of reality.

However, if the local apparatus settings are different, one *would have had*, for example,

$$A^i(0)B^i(0)C^i(\pi/2) = 1 \tag{32}$$

$$A^i(0)B^i(\pi/2)C^i(0) = 1 \tag{33}$$

$$A^i(\pi/2)B^i(\pi/2)C^i(\pi/2) = -1 \tag{34}$$

Yet, the four statements (31–34) are inconsistent within local realism. Since $X^i(\phi) = \pm 1$, if one multiplies side by side ❷ Eqs. 31–33, the result is

$$1 = -1 \tag{35}$$

This shows that the mere concept of the existence of "elements of physical reality" as introduced by EPR is in contradiction with quantum mechanical predictions. We have a "Bell's theorem without inequalities" (Greenberger et al. 1989, 1990).

Some people still claim that EPR correlations together with the assumption of locality allow us to derive realism. The above example clearly shows that such a realism would allow us to infer that $1 = -1$.

4.4 Implications of Bell's Theorem

Violations of Bell's inequalities imply that the underlying *conjunction of assumptions of realism, locality, and "free will"* is not valid, and *nothing more*.

It is often said that the violations indicate "(quantum) nonlocality." However, if one wants *nonlocality* to be *the* implication, one has to assume "free will" and realism. But this is only at the moment a philosophical choice (it seems that there is no way to falsify it). *It is not a necessary condition for violations of Bell's inequalities.*

Bell's theorem shows that even a local inherently probabilistic hidden-variable theory cannot agree with all the predictions of the quantum theory (considerations are based on $p(A_1, A_2, B_1, B_2)$ without assuming its actual structure, or whether the distribution for a single run is essentially deterministic, all we require is a joint "coexistence" of the variables $A_1, \ldots, B_2$ in a *theoretical* description). Therefore, the above statements cover theories that treat probabilities as irreducible, and for which one can define $p(A_1, A_2, B_1, B_2)$. Such theories contradict quantum predictions. This, for some authors, indicates that nature is nonlocal. While the mere existence of Bohm's model (Bohm 1952) demonstrates that nonlocal hidden-variables are a logically valid option, it is now known that there are plausible models, such as Leggett's crypto-nonlocal hidden-variable model (Leggett 2003), that are in disagreement with both quantum predictions and experiment (Gröblacher et al. 2007; Branciard et al. 2007). But, perhaps more importantly, if one is ready to consider inherently probabilistic theories, then there is no immediate reason to require the existence of (nonnegative and normalized) probabilities $p(A_{1,1}, \ldots, B_{2,2})$. Violation of this condition on realism, together with locality, which allows us to reduce the distribution to $p(A_1, \ldots, B_2)$, is not in a *direct* conflict with the theory of relativity, as it does not necessarily imply the possibility of signaling superluminally. To the contrary, quantum correlations cannot be used for direct communication between Alice to Bob, but still violate Bell's inequalities. It is therefore legitimate to consider quantum theory as a probability theory subject to, or even derivable from, more general principles, such as the non-signaling condition (Popescu and Rohrlich 1994; Barrett 2007) or information-theoretical principles (von Weizsäcker 1985; Zeilinger 1999; Pawłowski et al. 2009; Dakic and Brukner 2009).

Note that complementarity, inherent in the quantum formalism (which can be mathematically expressed as the nonexistence of joint probabilities for non-commuting, i.e., non-commeasurable, observables), completely contradicts the form of realism defined above. So why quantum-nonlocality?

To put it short, Bell's theorem does not imply *any* property of quantum mechanics. It just tells us what it is not.

5 All Bell's Inequalities for Two Possible Settings on Each Side

A general method of deriving *all* standard Bell's inequalities shall now be presented (that is, Bell's inequalities involving two-outcome measurements and with two settings per observer). Although these will not be spelt out explicitly, all the assumptions discussed above are behind the algebraic manipulations leading to the inequalities. A derivation for the two-observer problem is presented in detail, because the generalization to more observers is, surprisingly, obvious.

Consider pairs of particles (say, photons) simultaneously emitted in well-defined opposite directions. After some time, the photons arrive at two very distant measuring devices A and B operated by Alice and Bob. Alice, chooses to measure either observable $\hat{A}_1$ or $\hat{A}_2$, and Bob either $\hat{B}_1$ or $\hat{B}_2$. The hypothetical results that they may get for the jth pair of photons are A_1^j and A_2^j, for Alice's two possible choices, and B_1^j and B_1^j, for Bob's. The numerical values of these results ($+1$ or -1) are defined by the two eigenvalues of the observables.

Since always either $|B_1^j - B_2^j| = 2$ and $|B_1^j + B_2^j| = 0$, or $|B_1^j - B_2^j| = 0$ and $|B_1^j + B_2^j| = 2$, with a similar property of Alice's hypothetical results, the following relation holds

$$|A_1^j \pm A_2^j| \cdot |B_1^j \pm B_2^j| = 0 \tag{36}$$

for all possible sign choices within ❯ Eq. 36 except one, for which we have 4. Therefore

$$\sum_{k,l=0}^{1} |(A_1^j + (-1)^k A_2^j)(B_1^j + (-1)^l B_2^j)| = 4 \tag{37}$$

or equivalently we have the set of identities

$$\sum_{s_1,s_2=-1}^{1} S(s_1, s_2)[(A_1^j + s_1 A_2^j)(B_1^j + s_2 B_2^j)] = \pm 4 \tag{38}$$

with any $S(s_1, s_2) = \pm 1$. There are $2^{2^2} = 16$ such S functions.

Imagine now that N pairs of photons are emitted, pair by pair (N is sufficiently large, such that $\sqrt{1/N} \ll 1$). The average value of the products of the local values is given by

$$E(A_n, B_m) = \frac{1}{N} \sum_{j=1}^{N} A_n^j B_m^j \tag{39}$$

where $n, m = 1, 2$.

Therefore, after averaging, the following single Bell-type inequality emerges:

$$\sum_{k,l=0}^{1} |E(A_1, B_1) + (-1)^l E(A_1, B_2) + (-1)^k E(A_2, B_1) + (-1)^{k+l} E(A_2, B_2)| \le 4 \tag{40}$$

or, equivalently, a series of inequalities:

$$-4 \le \sum_{s_1,s_2=-1}^{1} S(s_1, s_2)[E(A_1, B_1) + s_2 E(A_1, B_2) + s_1 E(A_2, B_1) + s_1 s_2 E(A_2, B_2)] \le 4 \tag{41}$$

As the choice of the measurement settings is assumed to be statistically independent of the working of the source, that is, of the distribution of A_1's, A_2's, B_1's, and B_2's, the averages $E(A_n, B_m)$ cannot differ much, for high N, from the *actually observed* ones in the subsets of runs for which the given pair of settings was selected.

5.1 Completeness of the Inequalities

The inequalities form a complete set. That is, they define the faces of the convex polytope formed out of all possible local realistic models for the given set of measurements. Whenever the local realistic model exists, inequality (❯ 40) is satisfied by its predictions. To prove the sufficiency of condition (❯ 40), we construct a local realistic model for any correlation functions that satisfy it, that is, we are interested in the local realistic models for $E_{k_1 k_2}^{LR}$, such that they fully agree with the measured correlations $E(k_1, k_2)$ for all possible observables $k_1, k_2 = 1, 2$.

One can introduce $\hat{E}$ which is a "tensor" or matrix built out of E_{ij}, with $i, j = 1, 2$. If all its components can be derived from local realism, one must have

$$\hat{E}_{LR} = \sum_{A,B} P(\mathbf{A}, \mathbf{B}) \mathbf{A} \otimes \mathbf{B} \tag{42}$$

with $\mathbf{A} = (A_1, A_2)$, $\mathbf{B} = (B_1, B_2)$ and nonnegative probabilities $P(\mathbf{A}, \mathbf{B})$. The summation is over all possible values of $\mathbf{A}$ and $\mathbf{B}$.

Let us ascribe for fixed s_1, s_2, a hidden probability that $A_1 = s_1 A_2$ and $B_1 = s_2 B_2$ in the form familiar from ❯ Eq. 40:

$$P(s_1, s_2) = \frac{1}{4} \left| \sum_{k_2, k_2 = 1}^{2} s_1^{k_1 - 1} s_2^{k_2 - 1} E(k_1, k_2) \right| \tag{43}$$

Obviously these probabilities are positive. However, they sum up to identity only if inequality (❯ 40) is saturated, otherwise there is a "probability deficit," ΔP. This deficit can be compensated without affecting the correlation functions, see below.

First we construct the following structure, which is indeed the local realistic model of the set of correlation functions if the inequality is saturated:

$$\sum_{s_1, s_2 = -1}^{1} \Sigma(s_1, s_2) P(s_1, s_2)(1, s_1) \otimes (1, s_2) \tag{44}$$

where $\Sigma(s_1, s_2)$ is the sign of the expression within the modulus in ❯ Eq. 43.

Now, if $\Delta P > 0$, a "tail" is added to this expression given by:

$$\frac{\Delta P}{16} \sum_{A_1 = -1}^{1} \sum_{A_2 = -1}^{1} \sum_{B_1 = -1}^{1} \sum_{B_2 = -1}^{1} (A_1, A_2) \otimes (B_1, B_2) \tag{45}$$

This "tail" does not contribute to the values of the correlation functions, because it represents the fully random noise. The sum of ❯ Eq. 44 is a valid local realistic model for $\hat{E} = (E(1, 1), E(1, 2), E(2, 1), E(2, 2))$. The sole role of the "tail" is to make all hidden probabilities add up to 1.

To give the reader some intuitive grounds for the actual form and the completeness of the derived inequalities, some remarks shall now be given. The gist is that the consecutive terms in the inequalities are just expansion coefficients of the tensor $\hat{E}$ in terms of a complete orthogonal sequence of basis tensors. Thus, the expansion coefficients represent the tensors in a one-to-one way.

In the four-dimensional real space, where both $\hat{E}_{LR}$ and $\hat{E}$ are defined, one can find an orthonormal basis set $\hat{S}_{s_1 s_2} = \frac{1}{2}(1, s_1) \otimes (1, s_2)$. Within these definitions the hidden probabilities acquire a simple form:

$$P(s_1, s_2) = \frac{1}{2} |\hat{S}_{s_1 s_2} \cdot \hat{E}| \tag{46}$$

where the dot denotes the scalar product in $\mathbf{R}^4$. Now the local realistic correlations, $\hat{E}_{LR}$, can be expressed as:

$$\hat{E}_{LR} = \sum_{s_1, s_2 = -1}^{1} |\hat{S}_{s_1 s_2} \cdot \hat{E}| \Sigma(s_1, s_2) \hat{S}_{s_1 s_2} \tag{47}$$

The modulus of any number $|x|$ can be split into $|x| = x \, \text{sign}(x)$, and we can always demand that the product $A_1 B_1$ has the same sign as the expression inside the modulus. Thus, we have

$$\hat{E} = \sum_{s_1, s_2 = -1}^{1} (\hat{S}_{s_1 s_2} \cdot \hat{E}) \hat{S}_{s_1 s_2} \tag{48}$$

The expression in the bracket is the coefficient of tensor $\hat{E}$ in the basis $\hat{S}_{s_1 s_2}$. These coefficients are then summed over the same basis vectors, therefore the last equality appears.

5.2   Two-Qubit States That Violate the Inequalities

A general two-qubit state can be expressed in the following concise form

$$\hat{\rho} = \frac{1}{4} \sum_{\mu,\nu=0}^{3} T_{\mu\nu}(\hat{\sigma}_\mu^1 \otimes \hat{\sigma}_\nu^2) \tag{49}$$

The two-qubit correlation function for measurements of spin 1 along direction $\mathbf{n}(1)$ and of spin 2 along $\mathbf{n}(2)$ is given by

$$E_{QM}(\mathbf{n}(1), \mathbf{n}(2)) = \mathrm{Tr}\left[\hat{\rho}\left(\mathbf{n}(1) \cdot \hat{\boldsymbol{\sigma}}^1 \otimes \mathbf{n}(2) \cdot \hat{\boldsymbol{\sigma}}^2\right)\right] \tag{50}$$

and it reads

$$E_{QM}(\mathbf{n}(1), \mathbf{n}(2)) = \sum_{i,j=1}^{3} T_{ij} n(1)_i n(2)_j \tag{51}$$

Two-particle correlations are fully defined once we know the components of T_{ij}, $i, j = 1, 2, 3$, of the tensor $\hat{T}$. ❷ Equation 51 can be put into a more convenient form:

$$E_{QM}(\mathbf{n}(1), \mathbf{n}(2)) = \hat{T} \cdot \mathbf{n}(1) \otimes \mathbf{n}(2) \tag{52}$$

where "$\cdot$" is the scalar product in the space of tensors, which in turn is isomorphic with $\mathbf{R}^3 \otimes \mathbf{R}^3$.

According to (40) the quantum correlation $E_{QM}(\mathbf{n}(1), \mathbf{n}(2))$ can be described by a local realistic model if and only if, for *any* choice of the settings $\mathbf{n}(1)^{k_1}$ and $\mathbf{n}(2)^{k_2}$, where k_1, $k_2 = 1, 2$, we have

$$\frac{1}{4} \sum_{k,l=1}^{2} \left| \hat{T} \cdot [\mathbf{n}(1)^1 + (-1)^k \mathbf{n}(1)^2] \otimes [\mathbf{n}(2)^1 + (-1)^l \mathbf{n}(2)^2)] \right| \leq 1 \tag{53}$$

Since there always exist two mutually orthogonal unit vectors $\mathbf{a}(x)^1$ and $\mathbf{a}(x)^2$ such that

$$\mathbf{n}(x)^1 + (-1)^k \mathbf{n}(x)^2 = 2\alpha(x)_k \mathbf{a}(x)^k \text{ with } k = 1, 2 \tag{54}$$

and with $\alpha(x)_1 = \cos\theta(x)$, $\alpha(x)_2 = \sin\theta(x)$, we obtain

$$\sum_{k,l=1}^{2} \left| \alpha(1)_k \alpha(2)_l \hat{T} \cdot \mathbf{a}(1)^k \otimes \mathbf{a}(2)^l \right| \leq 1 \tag{55}$$

Note that $\hat{T} \cdot \mathbf{a}(1)^k \otimes \mathbf{a}(2)^l$ is a component of the tensor $\hat{T}$ after a transformation of the local coordinate systems of each of the particles into those where the two first basis vectors are $\mathbf{a}(x)^1$ and $\mathbf{a}(x)^2$. We shall denote such transformed components again by T_{kl}.

The necessary and sufficient condition for a two-qubit correlation to be described, within a local realistic model, is that in any plane of observation for each particle (defined by the two observation directions) we must have

$$\sum_{k,l=1}^{2} |\alpha(1)_k \alpha(2)_l T_{kl}| \leq 1 \tag{56}$$

for arbitrary $\alpha(1)_k$, $\alpha(2)_l$.

Using the Cauchy inequality, we obtain

$$\sum_{k,l=1}^{2} |\alpha(1)_k \alpha(2)_l T_{kl}| \leq \sqrt{\sum_{k,l=1}^{2} T_{kl}^2} \tag{57}$$

Therefore, if

$$\sum_{k,l=1}^{2} T_{kl}^2 \leq 1 \tag{58}$$

for any set of local coordinate systems, the two-particle correlation functions of the form of ❷ Eq. 51 can be understood within the local realism (in a two-settings-per-observer experiment).

As will be shown below, this condition is both necessary and sufficient.

5.2.1 Sufficient Condition for Violation of the Inequality

The full set of inequalities is derivable from the identity (❷ 38) where we put non-factorable sign function $S(s_1, s_2) = \frac{1}{2}(1 + s_1) + (1 - s_1)s_2$. In this case, we obtain the CHSH inequality in its standard form

$$\left| \langle (A_1 + A_2)B_1 + (A_1 - A_2)B_2 \rangle_{\text{avg}} \right| \leq 2 \tag{59}$$

where $\langle \ldots \rangle_{\text{avg}}$ denotes the average over an ensemble of particles. All other nontrivial inequalities are obtainable by all possible sign changes $X_k \to -X_k$ (with $k = 1, 2$ and $X = A, B$). It is easy to see that factorizable sign functions, such as, for example, $S(s_1, s_2) = s_1 s_2$, lead to trivial inequalities $|E(A_n, B_m)| \leq 1$. As noted above, the quantum correlation function, $E_Q(\mathbf{a}_k, \mathbf{b}_l)$, is given by the scalar product of the correlation tensor $\hat{T}$ with the tensor product of the local measurement settings represented by unit vectors $\mathbf{a}_k \otimes \mathbf{b}_l$, that is, $E_Q(\mathbf{a}_k, \mathbf{b}_l) = (\mathbf{a}_k \otimes \mathbf{b}_l) \cdot \hat{T}$. Thus, the condition for a quantum state endowed with the correlation tensor $\hat{T}$ to satisfy the inequality (❷ 59), is that for all directions $\mathbf{a}_1, \mathbf{a}_2, \mathbf{b}_1, \mathbf{b}_2$ we have

$$\left| \left[\left(\frac{\mathbf{a}_1 + \mathbf{a}_2}{2} \right) \otimes \mathbf{b}_1 + \left(\frac{\mathbf{a}_1 - \mathbf{a}_2}{2} \right) \otimes \mathbf{b}_2 \right] \cdot \hat{T} \right| \leq 1 \tag{60}$$

where both sides of ❷ Eq. 59 were divided by 2.

Next notice that $\mathbf{A}_\pm = \frac{1}{2}(\mathbf{a}_1 \pm \mathbf{a}_2)$ satisfy the following relations: $\mathbf{A}_+ \cdot \mathbf{A}_- = 0$ and $||\mathbf{A}_+||^2 + ||\mathbf{A}_-||^2 = 1$. Thus $\mathbf{A}_+ + \mathbf{A}_-$ is a unit vector, and $\mathbf{A}_\pm$ represent its decomposition into two orthogonal vectors. If one introduces unit vectors $\mathbf{a}_\pm$ such that $\mathbf{A}_\pm = a_\pm \mathbf{a}_\pm$, we have $a_+^2 + a_-^2 = 1$. Thus, we can put inequality (❷ 60) into the following form:

$$|\hat{S} \cdot \hat{T}| \leq 1 \tag{61}$$

where $\hat{S} = a_+ \mathbf{a}_+ \otimes \mathbf{b}_1 + a_- \mathbf{a}_- \otimes \mathbf{b}_2$. Note that since $\mathbf{a}_+ \cdot \mathbf{a}_- = 0$, we have $\hat{S} \cdot \hat{S} = 1$, that is, $\hat{S}$ is a tensor of unit norm. Any tensor of unit norm, $\hat{U}$, has the following Schmidt decomposition $\hat{U} = \lambda_1 \mathbf{v}_1 \otimes \mathbf{w}_1 + \lambda_2 \mathbf{v}_2 \otimes \mathbf{w}_2$, where $\mathbf{v}_i \cdot \mathbf{v}_j = \delta_{ij}, \mathbf{w}_i \cdot \mathbf{w}_j = \delta_{ij}$ and $\lambda_1^2 + \lambda_2^2 = 1$. The (complete) freedom of the choice of the measurement directions $\mathbf{b}_1$ and $\mathbf{b}_2$ allows us by choosing $\mathbf{b}_2$ orthogonal to $\mathbf{b}_1$ to put $\hat{S}$ in the form isomorphic with $\hat{U}$, and the freedom of choice of $\mathbf{a}_1$ and $\mathbf{a}_2$ allows $\mathbf{A}_+$ and $\mathbf{A}_-$ to be arbitrary orthogonal unit vectors, and $\mathbf{a}_+$ and $\mathbf{a}_-$ to also be arbitrary. Thus $\hat{S}$ can be equal to any unit tensor. To get the maximum of the left-hand side of ❷ Eq. 60, we Schmidt decompose the correlation tensor and take two terms of the decomposition which have the largest coefficients. In this way, we get a tensor $\hat{T}'$, of Schmidt rank two. We put $\hat{S} = \frac{1}{||\hat{T}'||} \hat{T}'$, and the maximum is $||\hat{T}'|| = \sqrt{\hat{T}' \cdot \hat{T}'}$. Thus, in other words,

$$\max \left[\sum_{k,l=1}^{2} T_{kl}^2 \right] \leq 1 \tag{62}$$

is the necessary and sufficient condition for the inequality (➍ 40) to hold, provided the maximization is taken over all local coordinate systems of two observers. The condition is equivalent to the necessary and sufficient condition of the Horodecki family (Horodecki et al. 1995) for violation of the CHSH inequality.

5.3 Bell's Inequalities for N Particles

Let us consider a Bell's inequality test with N observers. Each of them chooses between two possible observables, determined by local parameters $\mathbf{n}_1(j)$ and $\mathbf{n}_2(j)$, where $j = 1, \ldots, N$. Local realism implies the existence of two numbers A_1^j and A_2^j, each taking values $+1$ or -1, which describe the predetermined result of a measurement by the jth observer for the two observables. The following algebraic identity holds:

$$\sum_{s_1,\ldots,s_N=-1}^{1} S(s_1,\ldots,s_N) \prod_{j=1}^{N} \left[A_1^j + s_j A_2^j\right] = \pm 2^N \tag{63}$$

where $S(s_1, \ldots, s_N)$ is an arbitrary "sign" function, that is, $S(s_1, \ldots, s_N) = \pm 1$. It is a straightforward generalization of that for two observers as given in ➍ Eq. 38. The correlation function is the average over many runs of the experiment $E_{k_1,\ldots,k_N} = \left\langle \prod_{j=1}^{N} A_{k_j}^j \right\rangle_{avg}$ with $k_1, \ldots, k_N \in \{1, 2\}$. After averaging ➍ Eq. 63 over the ensemble of the runs, we obtain the Bell's inequalities. (This set of inequalities is a sufficient and necessary condition for the correlation functions entering them to have a local realistic model; compare it to the two-particle case in 5.1.)

$$\left| \sum_{s_1,\ldots,s_N=-1}^{1} S(s_1,\ldots,s_N) \sum_{k_1,\ldots,k_N=1}^{2} s_1^{k_1-1} \ldots s_N^{k_N-1} E_{k_1,\ldots,k_N} \right| \leq 2^N \tag{64}$$

Since there are 2^{2^N} different functions S, the above inequality represents a set of 2^{2^N} Bell's inequalities.

All these boil down to just one inequality (!):

$$\sum_{s_1,\ldots,s_N=-1}^{1} \left| \sum_{k_1,\ldots,k_N=1}^{2} s_1^{k_1-1} \ldots s_N^{k_N-1} E_{k_1,\ldots,k_N} \right| \leq 2^N \tag{65}$$

The proof of this fact is a trivial exercise with the use of the property that either $|X| = X$ or $|X| = -X$, where X is a real number. An inequality equivalent to (65) was derived independently in Weinfurter and Żukowski (2001) and Werner and Wolf (2001). The presented derivation follows Żukowski and Brukner (2002) where the necessary and sufficient condition for the violation of the inequality was obtained.

5.4 N-Qubit Correlations

A general N-qubit state can be put in the form

$$\hat{\rho} = \frac{1}{2^N} \sum_{\mu_1,\cdots,\mu_N=0}^{3} T_{\mu_1\cdots\mu_N} (\otimes_{k=1}^{N} \hat{\sigma}_{\mu_K}^k) \tag{66}$$

Thus, the N qubit correlation function has the following structure

$$E_{QM}(\mathbf{n}(1), \mathbf{n}(2), \dots, \mathbf{n}(N)) = \hat{\mathbf{T}} \cdot \mathbf{n}(1) \otimes \mathbf{n}(2) \dots \otimes \mathbf{n}(N) \qquad (67)$$

This expression when inserted into the Bell inequality gives

$$1/2^N \sum_{k_1, \dots, k_N=1} |\hat{\mathbf{T}} \cdot \prod_{x=1}^{N} (\mathbf{n}^1(x) + (-1)^x \mathbf{n}^2(x)| \leq 1$$

By introducing for every observer $x = 1, \dots, N$ two mutually orthogonal unit vectors $\mathbf{a}^1(x)$ and $\mathbf{a}^2(x)$, just as in (54), and denoting $\hat{\mathbf{T}} \cdot \mathbf{a}^{k_1}(1) \otimes \dots \otimes \mathbf{a}^{k_2}(N) = T_{k_1 \dots k_N}$, where $k_i = 1, 2, 3$, we obtain the necessary and sufficient condition for a description of the correlation function within local realism:

$$\sum_{k_1, k_2 \dots, k_N=1}^{2} \left| \alpha(1)_{k_1} \alpha(2)_{k_2} \dots \alpha(N)_{k_N} T_{k_1 k_2 \dots k_N} \right| \leq 1 \qquad (68)$$

for any possible choice of local coordinate systems for individual particles. Again if

$$\sum_{k_1, \dots, k_N=1}^{2} T_{k_1 \dots k_N}^2 \leq 1 \qquad (69)$$

for any set of local coordinate systems, the N-qubit correlation function can be described by a local realistic model. The proofs of these facts are generalizations of those presented earlier, pertaining to two particles. The sufficient condition for violation of the general Bell's inequality for N particles by a general state of N qubits can be found in Żukowski and Brukner (2002).

5.5 Concluding Remarks

The inequalities presented above represent the full set of standard "tight" Bell's inequalities for an arbitrary number of parties. Any non-tight inequality is weaker than tight ones. Such Bell's inequalities can be used to detect entanglement, however not as efficiently as entanglement witnesses. Nevertheless, they have the advantage over the witnesses, as they are system independent. They detect entanglement no matter the actual Hilbert space which describes the subsystems.

As will be shown below, the entanglement violating Bell's inequalities analyzed above is directly applicable in some quantum informational protocols that beat any classical ones of the same kind. This will be shown via an explicit construction of such protocols.

6 Quantum Reduction of Communication Complexity

In his review paper entitled "Quantum Communication Complexity (A Survey)" Brassard (2003) posed a question: "*Can entanglement be used to save on classical communication?*" He continued that there are good reasons to believe at first that the answer to the question is negative. Holevo's theorem (Holevo 1973) states that no more than n bits of classical information can be communicated between parties by the transmission of n qubits regardless of the coding scheme as long as no entanglement is shared between parties. If the communicating parties share prior entanglement, twice as much classical information can be transmitted (this is the so-called "superdense coding" (Bennett and Wiesner 1992)), but no more. It is thus reasonable to expect that even if the parties share entanglement, no savings in communication can be achieved beyond that of the superdense coding ($2n$ bits per n qubits transmitted).

It is also well known that entanglement alone cannot be used for communication. Local operations performed on any subsystem of an entangled composite system cannot have any observable effect on any other subsystem; otherwise it could be exploited to communicate faster than light. One would thus intuitively conclude that entanglement is useless for saving communication. Brassard, however, concluded "... *all the intuition in this paragraph is wrong.*"

The topic of classical communication complexity was introduced and first studied by Yao (1979). A typical communication complexity problem can be formulated as follows. Let Alice and Bob be two separated parties who receive some input data of which they know only their own data and not the data of the partner. Alice receives an input string x and Bob an input string y and the goal for both of them is to determine the value of a certain function $f(x, y)$. Before they start the protocol, Alice and Bob are even *allowed to share* (*classically correlated*) *random strings* or any other data, which might improve the success of the protocols. They are allowed to process their data locally in whatever way. The obvious method to achieve the goal is for Alice to communicate x to Bob, which allows him to compute $f(x, y)$. Once obtained, Bob can then communicate the value $f(x, y)$ back to Alice. It is the topic of communication complexity to address the questions: *Could there be more efficient solutions for some functions f(x, y)? What are these functions?*

A trivial example that there could be more efficient solutions than the obvious one given above is a constant function $f(x, y) = c$, where c is a constant. Obviously here Alice and Bob do not need to communicate at all, as they can simply take c for the value of the function. However, there are functions for which the only obvious solution is optimal, that is, only the transmission of x to Bob warrants that he reaches the correct result. For instance, it is shown that n bits of communication are necessary and sufficient for Bob to decide whether or not Alice's n-bit input is the same as his (Brassard 2003; Kushilevitz and Nisan 1997).

Generally, one might distinguish the following two types of communication complexity problems:

1. What is the minimal amount of communication (minimal number of bits) required for the parties to determine the value of the function with certainty?
2. What is the highest possible probability for the parties to arrive at the correct value for the function if only a *restricted* amount of communication is allowed?

Here, only the second class of problems will be considered. Note that in this case one does not insist on the correct value of the function to be obtained with certainty. While an error in computing the function is allowed, the parties try to compute it correctly with as high probability as possible.

From the perspective of the physics of quantum information processing, the natural questions is: *Are there communication complexity tasks for which the parties could increase the success in solving the problem if they share prior entanglement?* In their original paper Cleve and Buhrman (1997) showed that entanglement can indeed be used to save classical communication. They showed that, to solve a certain three-party problem with certainty, the parties need to broadcast at least 4 bits of information in a classical protocol, whereas in the quantum protocol (with entanglement shared), it is sufficient for them to broadcast only 3 bits of information. This was the first example of a communication complexity problem that could be solved with higher success than would be possible with any classical protocol. Subsequently, Buhrman et al. (1997) found a two-party problem that could be solved with a probability of success exceeding 85% and 2 bits of information communicated if prior shared entanglement is available, whereas the probability of success in a classical protocol could not exceed 75% with the same amount of communication.

The first problem whose quantum solution requires a significantly smaller amount of communication compared to classical solutions was discovered by Buhrman et al. (1999). They considered a k-party task, which requires roughly $k \ln k$ bits of communication in a classical protocol, and exactly k bits of classical communication if the parties are allowed to share prior entanglement. The quantum protocol of Buhrman et al. (1997) is based on the violation of the CHSH inequality by a two-qubit maximally entangled state. Similarly, the quantum protocols for multiparty problems (Buhrman et al. 1997, 1999; Cleve and Buhrman 1997) are based on an application of the GHZ-type argument against local realism for multi-qubit maximally entangled states. Galvao (2001) has shown an equivalence between the CHSH and GHZ tests for three particles and the two- and three-party quantum protocols of Buhrman et al. (1997), respectively. In a series of papers (Brukner et al. 2002, 2003, 2004b; Trojek et al. 2005), it was shown that entanglement violating Bell's inequality can always be exploited to find a better-than-any-classical solution to some communication complexity problems. In this brief overview, the approach mainly followed is the one introduced in these papers. This approach was further developed and applied in Augusiak and Horodecki (2006) and Tamir (2007) (see also Marcovitch and Reznik 2008).

6.1 The Problem and Its Optimal Classical Solution

Imagine several spatially separated partners, P_1 to P_N, each of which has some data known to him/her only, denoted here as X_i, with $i = 1, \ldots, N$. They face a joint task: to compute the value of a function $T(X_1, \ldots, X_N)$. This function depends on all data. Obviously, they can get the value of T by sending all their data to partner P_N, who does the calculation and announces the result. But are there ways to reduce the amount of communicated bits, that is, to reduce the communication complexity of the problem?

Assume that every partner P_k receives a two-bit string $X_k = (z_k, x_k)$ where $z_k, x_k \in \{0, 1\}$. We shall consider specific binary task functions that have the following form

$$T = f(x_1, \ldots, x_N)(-1)^{\sum_{k=1}^{N} z_k}$$

where $f \in \{-1, 1\}$. The partners also know the probability distribution ("promise") of the bit strings ("inputs"). There are two constraints on the problem. First, only distributions shall be considered that are completely random with respect to z_k's, that is a class of the form $p(X_1, \ldots, X_N) = 2^{-N} p'(x_1, \ldots, x_N)$. Second, communication between the partners is restricted to $N - 1$ bits. Assume that we ask the last partner to give his/her answer $A(X_1, \ldots, X_N)$, equal to ± 1, to the question of what is the functional value of $T(X_1, \ldots, X_N)$ in each run for the given set of inputs $X_1, \ldots, X_N$.

For simplicity, we introduce $y_k = (-1)^{z_k}$, $y_k \in \{-1, 1\}$. We shall use y_k as a synonym of z_k. Since T is proportional to $\prod_k y_k$, the final answer A is completely random if it does not depend on *every* y_k. Thus, information on z_k's from all $N - 1$ partners must somehow reach P_N.

Therefore, the only communication "trees" which might lead to success are those in which each P_k sends only a one-bit message $m_k \in \{0, 1\}$. Again we introduce $e_k = (-1)^{m_k}$ and e_k will be treated as a synonym of m_k.

The average success of a communication protocol can be measured with the following fidelity function

$$F = \sum_{X_1,\ldots,X_N} p(X_1,\ldots,X_N) T(X_1,\ldots,X_N) A(X_1,\ldots,X_N) \tag{70}$$

or equivalently

$$F = \frac{1}{2^N} \sum_{x_1,\ldots,x_N=0}^{1} p'(x_1,\ldots,x_N) f(x_1,\ldots,x_N) \sum_{y_1,\ldots,y_N=-1}^{1} \prod_{k=1}^{N} y_k A(x_1,\ldots,x_N; y_1,\ldots,y_N) \tag{71}$$

The probability of success is $P = (1 + F)/2$.

The first steps of a derivation of the reduced form of the fidelity function for an optimal classical protocol will now be presented (the reader may reconstruct the other steps or consult references Brukner et al. (2004b) and Trojek et al. (2005)). In a classical protocol, the answer A of the partner P_N can depend on the local input y_N, x_N, and messages, $e_{i_1},\ldots,e_{i_l}$, received *directly* from a subset of l partners $P_{i_1},\ldots,P_{i_l}$:

$$A = A(x_N, y_N, e_{i_1},\ldots,e_{i_l}) \tag{72}$$

Let us fix x_N, and treat A as a function A_{x_N} of the remaining $l + 1$ dichotomic variables

$$y_N, e_{i_1},\ldots,e_{i_l}$$

That is, we now treat x_N as a fixed index. All such functions can be thought of as 2^{l+1}-dimensional vectors, because the values of each such function form a sequence of length equal to the number of elements in the domain. In the 2^{l+1}-dimensional space containing such functions, we have *an orthogonal basis* given by

$$V_{jj_1\ldots j_l}(y_N, e_{i_1},\ldots,e_{i_l}) = y_N^j \prod_{k=1}^{l} e_{i_k}^{j_k} \tag{73}$$

where $j, j_1,\ldots,j_l \in \{0, 1\}$. Thus, we can expand $A(x_N, y_N, e_{i_1},\ldots,e_{i_l})$ with respect to this basis and the expansion coefficients read

$$c_{jj_1\ldots j_l}(x_N) = \frac{1}{2^{l+1}} \sum_{y_N, e_{i_1},\ldots,e_{i_l}=-1}^{1} A(x_N, y_N, e_{i_1},\ldots,e_{i_l}) V_{jj_1\ldots j_l}(y_N, e_{i_1},\ldots,e_{i_l}) \tag{74}$$

Since $|A| = |V_{jj_1,\ldots,j_l}| = 1$, we have $|c_{jj_1\ldots j_l}(x_N)| \leq 1$. The expansion is put into the expression (71) for F and

$$F = \frac{1}{2^N} \sum_{x_1,\ldots,x_N=0}^{1} g(x_1,\ldots,x_N) \sum_{y_1,\ldots,y_N=-1}^{1} \prod_{h=1}^{N} y_h \left[\sum_{j,j_1,\ldots,j_l=0}^{1} c_{jj_1\ldots j_l}(x_N) y_N^j \prod_{k=1}^{l} e_{i_k}^{j_k} \right] \tag{75}$$

is obtained, where $g(x_1,\ldots,x_N) \equiv f(x_1,\ldots,x_N) p'(x_1,\ldots,x_N)$. Because $\sum_{y_N=-1}^{1} y_N y_N^0 = 0$, and $\sum_{y_k=-1}^{1} y_k e_k^0 = 0$, only the term with all $j, j_1,\ldots,j_l$ equal to unity can give a nonzero contribution to F. Thus, A in F can be replaced by

$$A' = y_N c_N(x_N) \prod_{k=1}^{l} e_{i_k} \tag{76}$$

where $c_N(x_N)$ stands for $c_{11...1}(x_N)$. Next, notice that, for example, e_{i_1}, can depend only on local data x_{i_1}, y_{i_1} and the messages obtained by P_{i_1} from a subset of partners: $e_{p_1}, \ldots, e_{p_m}$. This set does not contain any of the e_{i_k}'s of the formula (❯ 76) above. In analogy with A, the function e_{i_1}, for a fixed x_{i_1}, can be treated as a vector, and thus can be expanded in terms of orthogonal basis functions (of a similar nature as ❯ Eq. 73), etc. Again, the expansion coefficients satisfy $|c'_{j_1 \ldots j_m}(x_{i_1})| \leq 1$. If we put this into A', we obtain a new form of F, which after a trivial summation over y_N and y_{i_1} depends on $c_N(x_N)c_{i_1}(x_i)\prod_{k=2}^{l} e_{i_k}$, where $c_{i_1}(x_i)$ stands for $c'_{11...1}(x_{i_1})$, and its modulus is again bounded by 1. Note that y_N and y_{i_1} disappear, as $y_k^2 = 1$.

As each message appears in the product only once, we continue this procedure of expanding those messages which depend on earlier messages, till it halts. The final reduced form of the formula for the fidelity of an optimal protocol reads

$$F = \sum_{x_1,\ldots,x_N=0}^{1} g(x_1,\ldots,x_N) \prod_{n=1}^{N} c_n(x_n) \tag{77}$$

with $|c_n(x_n)| \leq 1$. Since F in ❯ Eq. 77 is linear in every $c_n(x_n)$, its extrema are at the limiting values $c_n(x_n) = \pm 1$. *In other words, a Bell-like inequality $|F| \leq \text{Max}(F) \equiv B(N)$ gives the upper fidelity bound.* Note that the above derivation shows that optimal classical protocols include one in which partners P_1 to P_{N-1} send to P_N one-bit messages which encode the value of $e_k = y_k c(x_k)$, where $k = 1, 2, \ldots, N-1$.

6.2 Quantum Solutions

The inequality for F suggests that some problems may have quantum solutions, which surpass any classical ones in their fidelity. Simply, one may use an entangled state $|\psi\rangle$ of N qubits that violates the inequality. Send to each of the partners one of the qubits. In a protocol run all N partners make measurements on the local qubits, the settings of which are determined by x_k. They measure a certain qubit observable $\mathbf{n}_k(x_k) \cdot \boldsymbol{\sigma}$. The measurement results $\gamma_k = \pm 1$ are multiplied by y_k, and the partner P_k, for $1 \leq k \leq N-1$, sends a bit message to P_N encoding the value of $m_k = y_k \gamma_k$. The last partner calculates $y_N \gamma_N \prod_{k=1}^{N-1} m_k$, and announces this as A. The average fidelity of such a process is

$$F = \sum_{x_1,\ldots,x_N=0}^{1} g(x_1,\ldots,x_N) \langle \psi | \otimes_{k=1}^{N} (\mathbf{n}_k(x_k) \cdot \boldsymbol{\sigma}_k) | \psi \rangle \tag{78}$$

and in certain problems can even reach *unity*.

For some tasks, the quantum versus classical fidelity ratio grows *exponentially* with N. This is the case, for example, for the so-called *modulo-4 sum* problem. Each partner receives a two-bit input string ($X_k = 0, 1, 2, 3$; $k = 1, \ldots, N$). The promise is that X_ks are distributed such that $\left(\sum_{k=1}^{N} X_k\right) \text{mod} 2 = 0$. The task is: P_N must tell whether the sum modulo-4 of all inputs is 0 or 2. (It can be formulated in terms of a task function $T = 1 - \left(\sum_{k=1}^{N} X_k\right) \text{mod} 4$. An alternative formulation of the problem reads $f = \cos\left(\frac{\pi}{2}\sum_{k=1}^{N} X_k\right)$ with $p' = 2^{-N+1}\left|\cos\left(\frac{\pi}{2}\sum_{k=1}^{N} X_k\right)\right|$.)

For this problem, the classical fidelity bounds decrease exponentially with N, that is $B(F) \leq 2^{-K+1}$, where $K = N/2$ for even and $K = (N+1)/2$ for odd number of parties.

If we use the N qubit GHZ states: $|GHZ\rangle = \frac{1}{\sqrt{2}}(|z+,\ldots,z+\rangle + |z-,\ldots,z-\rangle)$, where $|z\pm\rangle$ is the state of spin ± 1 along the z-axis, and suitable pairs of local settings, the associated Bell's inequality can be violated maximally. Thus, we have a quantum protocol that always gives the correct answer.

In all quantum protocols considered here, entanglement that leads to a violation of Bell's inequality is a resource that allows for better-than-classical efficiency of the protocol. Surprisingly, we can also show a version of a quantum protocol without entanglement (Galvão 2001; Trojek et al. 2005). The partners exchange a single qubit, P_k to P_{k+1} and so on, and each of them makes a suitable unitary transformation on it (which depends on z_k and x_k). The partner P_N, who receives the qubit as the last one, additionally performs a dichotomic measurement. The result we get is equal to T. For details, including an experimental realization, see Trojek et al. (2005). The obvious conceptual advantage of such a procedure is that the partners exchange a single qubit, from which, due to the Holevo bound (Holevo 1973), we can read out at most one bit of information. In contrast with the protocol involving entanglement, no classical transfer of any information is required, except from the announcement by P_N of his measurement result!

In summary, if we have a pure entangled state of many qubits (this can be generalized to higher-dimensional systems and Bell's inequalities involving more than two measurement settings per observer), there exists a Bell's inequality that is violated by this state. This inequality has some coefficients $g(x_1, \ldots, x_n)$, in front of correlation functions, that can always be renormalized in such a way that

$$\sum_{x_1,\ldots,x_n=0}^{1} |g(x_1,\ldots,x_n)| = 1$$

The function g can always be interpreted as a product of the dichotomic function $f(x_1, \ldots, x_n) = \frac{g(x_1,\ldots,x_N)}{|g(x_1,\ldots,x_N)|} = \pm 1$ and a probability distribution $p'(x_1, \ldots, x_n) = |g(x_1, \ldots, x_n)|$. Thus, we can construct a communication complexity problem that is tailored to a given Bell's inequality, with task function $T = \prod_i^N y_i f$. All this can be extended beyond qubits, see Brukner et al. (2002, 2003).

As was shown, for three or more parties, $N \geq 3$, quantum solutions for certain communication complexity problems can achieve probabilities of success of unity. This is not the case for $N = 2$ and the problem based on the CHSH inequality. The maximum quantum value for the left-hand side of the CHSH inequality (❷ 25) is just $\sqrt{2} - 1$. This is much bigger than the Bell bound of 0, but still not the largest possible value, for an arbitrary theory that is not following local realism, which equals 1. Because the maximum possible violation of the inequality is not attainable by quantum mechanics, several questions arise. Is this limit forced by the theory of probability, or by physical laws? We will address this question in the next section and we will look at the consequences of the maximal logically possible violation of the CHSH inequality.

6.3 Stronger-Than-Quantum Correlations

The Clauser–Horne–Shimony–Holt (CHSH) inequality (Clauser et al. 1969) for local realistic theories gives the upper bound on a certain combination of correlations between two space-like separated experiments. Consider Alice and Bob who independently perform one out of two measurements on their part of the system, such that in total there are four experimental setups: $(x, y) = (0, 0), (0, 1), (1, 0),$ or $(1, 1)$. For any local hidden variable theory, the CHSH inequality must hold. We put it in the following form:

$$p(a = b|x = 0, y = 0) + p(a = b|x = 0, y = 1) + p(a = b|x = 1, y = 0)$$
$$+ p(a = -b|x = 1, y = 1) \leq 3 \tag{79}$$

or equivalently,

$$\sum_{x,y=0,1} p(a \oplus b = x \cdot y) \leq 3 \tag{80}$$

In the latter form, we interpret the dichotomic measurement results as binary values, 0 or 1, and their relations are "modulo 2 sums," denoted here by $\oplus$. We have $0 \oplus 0 = 1 \oplus 1 = 0$ and $0 \oplus 1 = 1$. For example, $p(a = b|x = 0, y = 0)$ is the probability that Alice's and Bob's outcomes are the same when she chooses setting $x = 0$ and he setting $y = 0$.

As discussed in previous sections, quantum mechanical correlations can violate the local realistic bound of inequality (❷ 80) and the limit was proven by Cirel'son (1980) to be $2 + \sqrt{2}$. Popescu and Rohrlich (1994) asked why quantum mechanics allows a violation of the CHSH inequality with a value of $2 + \sqrt{2}$, but not more, though the maximal logically possible value is 4. Would a violation with a value larger than $2 + \sqrt{2}$ lead to (superluminal) signaling? If this were true, then quantum correlations could be understood as maximal allowed correlations respecting non-signaling requirement. This could give us an insight into the origin of quantum correlations, without any use of the Hilbert space formalism.

The non-signaling condition is equivalent to the requirement that the marginals are independent of the partner's choice of setting

$$p(a|x, y) \equiv \sum_{b=0,1} p(a, b|x, y) = p(a|x) \tag{81}$$

$$p(b|x, y) \equiv \sum_{b=0,1} p(a, b|x, y) = p(b|y) \tag{82}$$

where $p(a, b|x, y)$ is the joint probability for outcomes a and b to occur, given x and y are the choices of measurement settings respectively, $p(a|x)$ is the probability for outcome a, given x is the choice of measurement setting and $p(b|y)$ similarly. Popescu and Rohrlich constructed a toy-theory where the correlations reach the maximal algebraic value of 4 for left-hand expression of the inequality (❷ 79), but are nevertheless not in contradiction with no-signaling. The probabilities in the toy model are given by

$$\left. \begin{array}{l} p(a = 0, b = 0|x, y) = \frac{1}{2} \\ p(a = 1, b = 1|x, y) = \frac{1}{2} \end{array} \right\} \text{ if } xy \in \{00, 01, 10\}$$
$$\left. \begin{array}{l} p(a = 1, b = 0|x, y) = \frac{1}{2} \\ p(a = 0, b = 1|x, y) = \frac{1}{2} \end{array} \right\} \text{ if } xy = 11 \tag{83}$$

Indeed we have

$$\sum_{x,y=0,1} p(a \oplus b = x \cdot y) = 4 \tag{84}$$

Van Dam (2000) and, independently, Cleve considered how plausible are stronger-than-quantum correlations from the point of view of communication complexity, which describes how much communication is needed to evaluate a function with distributed inputs. It was shown that the existence of correlations that maximally violate the CHSH inequality would allow us to perform all distributed computations (between two parties) of dichotomic functions with a communication constraint of just one bit. If one is ready to believe that nature should not allow for "easy life" concerning communication problems, this could be a reason why superstrong correlations are indeed not possible.

Instead of superstrong correlations, one usually speaks about a "nonlocal box" (NLB), an imaginary device that takes as inputs x at Alice's and y at Bob's side, and outputs a and b at respective sides, such that $a \oplus b = x \cdot y$. Quantum mechanical measurements on a maximally entangled state allow for a success probability of $p = \cos^2 \frac{\pi}{8} = \frac{2+\sqrt{2}}{4} \approx 0.854$ at the game of simulating NLBs. Recently, it was shown that in any "world" in which it is possible to implement an approximation to the NLB, that works correctly with probability greater than $\frac{3+\sqrt{6}}{6} = 90.8\%$, for all distributed computations of dichotomic functions with a one-bit communication constraint, one can find a protocol that always gives the correct values (Brassard et al. 2006). This bound is an improvement over van Dam's one, but still has a gap with respect to the bound imposed by quantum mechanics.

6.3.1 Superstrong Correlations Trivialize Communication Complexity

We shall present a proof that availability of a perfect NLB would allow for a solution of a general communication complexity problem for a binary function, with an exchange of a single bit of information. The proof is due to van Dam (2000).

Consider a Boolean function $f \colon \{0, 1\}^n \to \{0, 1\}$, which has as inputs two n-bit strings $\mathbf{x} = (x_1, \ldots, x_n)$ and $\mathbf{y} = (y_1, \ldots, y_n)$. Suppose that Alice receives the $\mathbf{x}$ string and Bob, who is separated from Alice, the $\mathbf{y}$-string, and they are to determine the function value $f(\mathbf{x}, \mathbf{y})$ by communicating as little as possible. They have, however, NLBs as resources.

First, let us notice that any dichotomic function $f(\mathbf{x}, \mathbf{y})$ can be rewritten as a finite summation:

$$f(\mathbf{x}, \mathbf{y}) = \sum_{i=1}^{2^n} P_i(\mathbf{x}) Q_i(\mathbf{y}) \tag{85}$$

where $P(\mathbf{x})$ are polynomials in $\mathbf{x} \in \{0, 1\}$ and $Q_i(\mathbf{y}) = y_1^{i_1} \cdot \ldots \cdot y_n^{i_n}$ are monomials in $y_i \in \{0, 1\}$ with $i_1, \ldots, i_n \in \{0, 1\}$. Note that the latter ones constitute an orthogonal basis in a 2^n-dimensional space of polynomials with no higher power of any variable than 1. The decomposed function f is treated as a function of y's, while the inputs $x_1, \ldots, x_n$ are considered as indices numbering functions f. Note that there are 2^n different monomials. Alice can locally compute all the P_i values by herself and likewise Bob can compute all Q_i by himself. These values determine the settings of Alice and Bob that will be chosen in the ith run of the experiment. Note that to this end they need, in general, exponentially many NLBs. Alice and Bob perform for every $i \in \{1, \ldots, 2^n\}$ a measurement on the ith NLB in order to obtain without any communication a collection of bit values a_i and b_i, with the property $a_i \oplus b_i = P_i(\mathbf{x}) Q_i(\mathbf{y})$. Bob can add all his b_i to $\sum_{i=1}^{2^n} b_i$ values without requiring any information from Alice, and he can broadcast this single bit to Alice. She, on her part, computes the sum of her a_i to $\sum_{i=1}^{2^n} a_i$ and adds Bob's bit to it. The final result

$$\sum_{i=1}^{2^n}(a_i \oplus b_i) = \sum_{i=1}^{2^n} P_i(\mathbf{x})Q_i(\mathbf{y}) = f(\mathbf{x},\mathbf{y}) \tag{86}$$

is the function value. Thus, *superstrong correlations trivialize every communication complexity problem.*

7 The Kochen–Specker Theorem

In the previous sections, we saw that tests of Bell's inequalities are not only theory-independent tests of nonclassicality, but also have applications in quantum information protocols. Examples are communication complexity problems (Brassard 2003, Brukner et al. 2004b), entanglement detection (Hyllus et al. 2005), security of key distribution (Ekert 1991; Scarani and Gisin 2001; Acín et al. 2006), and quantum state discrimination (Schmid et al. 2008). Thus, entanglement which violates local realism can be seen as a resource for efficient information processing. Can quantum contextuality – the fact that quantum predictions disagree from those of non-contextual hidden-variable theories – also be seen as such a resource? An affirmative answer shall be given to this question by considering an explicit example of a quantum game.

The Kochen–Specker theorem is a "no go" theorem that proves a contradiction between predictions of quantum theory and those of *non-contextual* hidden variable theories. It was proved by Bell (1966), and independently by Kochen and Specker (1967). The non-contextual hidden-variable theories are based on the conjecture of the following three assumptions:

1. *Realism*: It is a model that allows us to use all variables $A_m(n)$ in the theoretical description of the experiment, where $A_m(n)$ gives the value of some observable A_m which *could* be obtained if the knob setting were at positions m. The index n describes the entire experimental "context" in which A_m is measured and is operationally defined through the positions of all other knob settings in the experiment, which are used to measure other observables jointly with A_m. All $A_m(n)$'s are treated as, perhaps, unknown, but still fixed, (real) numbers, or variables for which a proper joint probability distribution can be defined.
2. *Non-contextuality*: The value assigned to an observable $A_m(n)$ of an individual system is independent of the experimental context n in which it is measured, in particular of any properties that are measured *jointly* with that property. This implies that $A_m(n) = A_m$ for all contexts n.
3. "Free will." The experimenter is free to choose the observable and its context. The choices are independent of the actual hidden values of A_m's, etc.

Note that "non-contextuality" implies locality (i.e., non-contextually with respect to a remote context), but there is no implication the other way round. One might have theories which are local, but locally contextual.

It should be stressed that the local realistic and non-contextual theories provide us with predictions, which can be tested experimentally, and which can be derived *without making any reference to quantum mechanics* (though many derivations in the literature give exactly the opposite impression). In order to achieve this, it is important to realize that predictions for non-contextual realistic theories can be derived in a completely operational way (Simon et al. 2001). For concreteness, imagine that an observer wants to perform a measurement of an observable, say the square, S_n^2, of a spin component of a spin-1 particle along a certain direction **n**. There will be an experimental procedure for trying to do this as accurately as possible. This procedure will be referred to by saying that one sets the "control switch" of one's apparatus to the position **n**. In all the experiments that will be discussed, only a finite number of different switch positions is required. By definition, different switch positions are clearly distinguishable for the observer, and the switch position is all he knows about. Therefore, in an operational sense, the measured physical observable is entirely defined by the switch position. From the above definition, it is clear that the same switch position can be chosen again and again in the course of an experiment. Notice that in such an approach as described above, it does not matter which observable is "really" measured and to what precision. One just derives general predictions, provided that certain switch positions are chosen.

In the original Kochen–Specker proof (Kochen and Specker 1968), the observables that are considered are squares of components of spin 1 along various directions. Such observables have values 1 or 0, as the components themselves have values 1, 0, or −1. The squares of spin components $\hat{S}_{n_1}^2$, $\hat{S}_{n_2}^2$, and $\hat{S}_{n_3}^2$ along any three orthogonal directions $\mathbf{n}_1$, $\mathbf{n}_2$, and $\mathbf{n}_3$ can be measured jointly. Simply, the corresponding quantum operators commute with each other. In the framework of a hidden-variable theory, one assigns to an individual system a set of numerical values, say +1, 0, +1,... for the square of spin component along each direction $S_{n_1}^2$, $S_{n_2}^2$, $S_{n_3}^2$,... that can be measured on the system. If any of the observables is chosen to be measured on the individual system, the result of the measurement would be the corresponding value. In a non-contextual hidden variable theory, one has to assign to an observable, say $S_{n_1}^2$, the *same* value independently of whether it is measured in an experimental procedure jointly as a part of some set $\{S_{n_1}^2, S_{n_2}^2, S_{n_3}^2\}$ or of some other set $\{S_{n_1}^2, S_{n_4}^2, S_{n_5}^2\}$ of physical observables, where $\{\mathbf{n}_1, \mathbf{n}_2, \mathbf{n}_3\}$ and $\{\mathbf{n}_1, \mathbf{n}_4, \mathbf{n}_5\}$ are triads of orthogonal directions. Note that within quantum theory, some of the operators corresponding to the observables from the first set may *not commute* with some corresponding to the observables from the second set.

The squares of spin components along orthogonal directions satisfy

$$\hat{S}_{n_1}^2 + \hat{S}_{n_2}^2 + \hat{S}_{n_3}^2 = s(s+1) = 2 \tag{87}$$

This is *always* so for a particle of spin 1 ($s = 1$). This implies that for every measurement of three squares of mutually orthogonal spin components, two of the results will be equal to one, and one of them will be equal to zero. The Kochen–Specker theorem considers a set of triads of orthogonal directions $\{\mathbf{n}_1, \mathbf{n}_2, \mathbf{n}_3\}$, $\{\mathbf{n}_1, \mathbf{n}_4, \mathbf{n}_5\}$,..., for which at least some of the directions have to appear in several of the triads. The statement of the theorem is that there are sets of directions for which it is not possible to give any assignment of 1's and 0's to the directions consistent with the constraint (❷ 87). The original theorem in Kochen and Specker (1968) used 117 vectors, but this was subsequently reduced to 33 vectors (Peres 1991) and 18 vectors (Cabello et al. 1996). Mathematically, the contradiction with quantum predictions has its origin in the fact that the classical structure of non-contextual hidden variable theories is represented by commutative algebra, whereas quantum mechanical observables need not be commutative, making it impossible to embed the algebra of these observables in a commutative algebra.

The disproof of non-contextually relies on the assumption that the same value is assigned to a given physical observable, $\hat{S}_n^2$, regardless with which two other observables the experimenter chooses to measure it. In quantum theory, the additional observables from one of those sets correspond to operators that do not commute with the operators corresponding to additional observables from the other set. As was stressed in a masterly review on hidden variable theories by Mermin (1993), Bell wrote (Bell 1966): "These different possibilities require different experimental arrangements; there is no *a priori* reason to believe that the results ... should be the same. The result of observation may reasonably depend not only on the state of the system (including hidden variables) but also on the complete disposition apparatus." Nevertheless, as Bell himself showed, the disagreement between predictions of quantum mechanics and of the hidden-variables theories can be strengthened if non-contextuality is replaced by a much more compelling assumption of locality. Note that in Bohr's doctrine of the inseparability of the object and the measuring instrument, an observable *is* defined through the entire measurement procedure applied to measure it. Within this doctrine one would not speak about measuring the same observable in different contexts, but rather about measuring entirely different maximal observables, and deriving from it the value of a degenerate observable. Note that the Kochen–Specker argument necessarily involves degenerate observables. This is why it does not apply to single qubits.

7.1   A Kochen–Specker Game

A quantum game will now be considered which is based on the Kochen–Specker argument strengthened by the locality condition (see Svozil (2004)). A pair of entangled spin 1 particles is considered which forms a singlet state with total spin 0. A formal description of this state is given by

$$|\Psi\rangle = \frac{1}{\sqrt{3}}\left(|1\rangle_n|-1\rangle_n + |-1\rangle_n|1\rangle_n - |0\rangle_n|0\rangle_n\right) \tag{88}$$

where, for example, $|1\rangle_n|-1\rangle_n$ is the state of the two particles with spin projection $+1$ for the first particle and spin projection -1 for the second particle 1 along the same direction $\mathbf{n}$. It is important to note that this state is invariant under a change of the direction $\mathbf{n}$. This implies that if the spin components for the two particles are measured along an arbitrary direction, however the same on both sides, the sum of the two local results is always zero. This is a direct consequence of the conservation of angular momentum.

Let us present the quantum game introduced in Cleve et al. (2004). The requirement in the proof of the Kochen–Specker theorem can be formulated as the following problem in geometry. There exists an explicit set of vectors $\{\mathbf{n}_1, \ldots, \mathbf{n}_m\}$ in $\mathbf{R}^3$ that cannot be colored in red (i.e., assign the value 1 to the spin squared component along that direction) or blue (i.e., assign the value 0) such that both of the following conditions hold:

1. For every orthogonal pair of vectors $\mathbf{n}_i$ and $\mathbf{n}_j$, they are not both colored blue.
2. For every mutually orthogonal triple of vectors $\mathbf{n}_i$, $\mathbf{n}_j$, and $\mathbf{n}_k$, at least one of them is colored blue.

For example, the set of vectors can consist of 117 vectors from the original Kochen–Specker proof (Kochen and Specker 1968), 33 vectors from Peres' proof or 18 vectors from Cabello's proof (Cabello et al. 1996).

The Kochen–Specker game employs the above sets of vectors. Consider two separated parties, Alice and Bob. Alice receives a random triple of orthogonal vectors as her input and Bob receives a single vector randomly chosen from the triple as his input. Alice is asked to give a trit indicating which of her three vectors is assigned color red or 1 (implicitly, the other two vectors are assigned color blue or 0). Bob outputs a bit assigning a color to his vector. The requirement in the game is that Alice and Bob assign the same color to the vector that they receive in common. It is straightforward to show that the existence of a perfect classical strategy in which Alice and Bob can share classically correlated strings for this game would violate the reasoning used in the Kochen–Specker theorem. On the other hand, there is a perfect quantum strategy using the entangled state (❯ Eq. 88). If Alice and Bob share two particles in this state, Alice can perform a measurement of squared spin components pertaining to directions $\{\mathbf{n}_i, \mathbf{n}_j, \mathbf{n}_k\}$, which are equal to those of the three input vectors, and Bob measures squared spin component in direction $\mathbf{n}_l$ for his input. Then Bob's measurement will necessarily yield the same answer as the measurement by Alice along the same direction.

Concluding this section, we note that quantum contextuality is also closely related to quantum error correction (DiVincenzo and Peres 1997), quantum key distribution (Nagata 2005), one-location quantum games (Aharon and Vaidman 2008), and entanglement detection between internal degrees of freedom.

8 Temporal Bell's Inequalities (Leggett–Garg Inequalities)

In the last section, one more basic information-processing task, the random access code problem, will be considered. It can be solved with a quantum setup with a higher efficiency than is classically possible. It will be shown that the resource for better-than-classical efficiency is a violation of "temporal Bell's inequalities" – the inequalities that are satisfied by temporal correlations of certain classes of hidden-variable theories. Instead of considering correlations between measurement results on distantly located physical systems, here the focus is on one and the same physical system and there is an analysis of the correlations between measurement outcomes at different times. The inequalities were first introduced by Leggett and Garg (1985), Leggett (2002) in the context of testing superpositions of macroscopically distinct quantum states. Since the aim here is different, we shall discuss general assumptions which allow us to derive temporal Bell's inequalities irrespective of whether the object under consideration is macroscopic or not. This is why the assumptions differ from the original ones of Leggett and Garg (1985), and Leggett (2002). Compare also to Paz (1993), Shafiee and Golshani (2003), and Brukner et al. (2004a).

Theories which are based on the conjunction of the following four assumptions are now considered. (There is one more difference between the present approach and that of Leggett and Garg (1985) and Leggett (2002). While there the observer measures a single observable having a choice between different times of measurement, here, at any given time, the observer has a choice between two (or more) different measurement settings. One can use both approaches to derive temporal Bell's inequalities.)

1. *Realism*: It is a model that allows us to use all variables $A_m(t)$ $m = 1, 2, \ldots$ in the theoretical description of the experiment performed at time t, where $A_m(t)$ gives the value of some observable, which *could* be obtained if it were measured at time t. All $A_m(t)$'s are treated as perhaps unknown, but still fixed numbers, or variables for which a proper joint probability distribution can be defined.

2. *Non-invasiveness*: The value assigned to an observable $A_m(t_1)$ at time t_1 is independent whether or not a measurement was performed at some earlier time t_0 or which observable $A_n(t_0)$ $n = 1, 2, \ldots$ at that time was measured. In other words, (actual or potential) measurement values $A_m(t_1)$ at time t_1 are *independent* of the measurement settings chosen at earlier times t_0.

3. *Induction*: The standard arrow of time is assumed. In particular, the values $A_m(t_0)$ at earlier times t_0 do not depend on the choices of measurement settings at later times t_1. (Note that this already follows from the "noninvasiveness" when applied symmetrically to both arrows of time.)

4. *"Free will"*: The experimenter is free to choose the observable. The choices are independent of the actual hidden values of A_m's, etc.

Consider an observer and allow her to choose at time t_0 and at some later time t_1 to measure one of two dichotomic observables $A_1(t_i)$ and $A_2(t_i)$, $i \in \{0, 1\}$. The assumptions given above imply the existence of numbers for $A_1(t_i)$ and $A_2(t_i)$, each taking values either $+1$ or -1, which describe the (potential or actual) predetermined result of the measurement. For the temporal correlations in an individual experimental run, the following identity holds: $A_1(t_0)[A_1(t_1) - A_2(t_1)] + A_2(t_0)[A_1(t_1) + A_2(t_1)] = \pm 2$. With similar steps as in the derivation of the standard Bell's inequalities, one easily obtains:

$$p(A_1 A_1 = 1) + p(A_1 A_2 = -1) + p(A_2 A_1 = 1) + p(A_2 A_2 = 1) \leq 3 \qquad (89)$$

where we omit the dependence on time.

An important difference between tests of non-contextuality and temporal Bell's inequalities is that the latter are also applicable to single qubits or *two-dimensional* quantum systems. The temporal correlation function will now be calculated for consecutive measurements of a single qubit. Take an arbitrary mixed state of a qubit, written as $\rho = \frac{1}{2}(1 + \mathbf{r} \cdot \boldsymbol{\sigma})$, where 1 is the identity operator, $\hat{\boldsymbol{\sigma}} \equiv (\sigma_x, \sigma_y, \sigma_z)$ are the Pauli operators for three orthogonal directions x, y, and z, and $\mathbf{r} \equiv (r_x, r_y, r_z)$ is the Bloch vector with the components $r_i = \text{Tr}(\rho \sigma_i)$.

Suppose that the measurement of the observable $\boldsymbol{\sigma} \cdot \mathbf{a}$ is performed at time t_0, followed by the measurement of $\boldsymbol{\sigma} \cdot \mathbf{b}$ at t_1, where $\mathbf{a}$ and $\mathbf{b}$ are directions at which the spin is measured. The quantum correlation function is given by $E_{QM}(\mathbf{a}, \mathbf{b}) = \sum_{k,l=\pm 1} k \, l \, \text{Tr}(\rho \pi_{\mathbf{a},k}) \, \text{Tr}(\pi_{\mathbf{a},k} \pi_{\mathbf{b},l})$, where, for example, $\pi_{\mathbf{a},k}$ is the projector onto the subspace corresponding to the eigenvalue $k = \pm 1$ of the spin along $\mathbf{a}$. Here, we use the fact that after the first measurement the state is projected on the new state $\pi_{\mathbf{a},k}$. Therefore, the probability to obtain the result k in the first measurement and l in the second one is given by $\text{Tr}(\rho \pi_{\mathbf{a},k}) \text{Tr}(\pi_{\mathbf{a},k} \pi_{\mathbf{b},l})$. Using $\pi_{\mathbf{a},k} = \frac{1}{2}(1 + k \boldsymbol{\sigma} \cdot \mathbf{a})$ and $\frac{1}{2} \text{Tr}[(\boldsymbol{\sigma} \cdot \mathbf{a})(\boldsymbol{\sigma} \cdot \mathbf{b})] = \mathbf{a} \cdot \mathbf{b}$ one can easily show that the quantum correlation function can simply be written as

$$E_{QM}(\mathbf{a}, \mathbf{b}) = \mathbf{a} \cdot \mathbf{b} \qquad (90)$$

Note that, in contrast to the usual (spatial) correlation function, the temporal one (❯ Eq. 90) does not depend on the initial state ρ. Note also that a slight modification of the derivation of ❯ Eq. 90 can also apply to the cases in which the system evolves between the two measurements following an arbitrary unitary transformation.

The scalar product form of quantum correlations (❯ 90) allows for the violation of the temporal Bell's inequality and the maximal value of the left-hand side of (❯ Eq. 89) is achieved for the choice of the measurement settings: $\mathbf{a}_1 = \frac{1}{\sqrt{2}}(\mathbf{b}_1 - \mathbf{b}_2)$, $\mathbf{a}_2 = \frac{1}{\sqrt{2}}(\mathbf{b}_1 + \mathbf{b}_2)$ and is equal to $2 + \sqrt{2}$.

8.1 Quantum Random Access Codes

Random access code is a communication task for two parties, say again Alice and Bob. Alice receives some classical *n*-bit string known only to her (her local input). She is allowed to send just a one-bit message, m, to Bob. Bob is asked to tell the value of the bth bit of Alice, $b = 1, 2\ldots$, n. However b is known only to him (this is his local input data). The goal is to construct a protocol enabling Bob to tell the value of the bth bit of Alice, with as high average probability of success as possible, for a uniformly random distribution of Alice's bit-strings, and a uniform distribution of b's. Note that Alice does not know in advance which bit Bob is to recover. Thus, she has no option to send just this required bit.

If they share a quantum channel, then one speaks about a quantum version of the previous problem. Alice is asked to encode her classical n-bit message into 1 qubit (quantum bit) and send it to Bob. He performs some measurement on the received qubit to extract the required bit. In general, the measurement that he uses will depend on which bit he wants to reveal. The idea behind these so-called quantum random access codes already appeared in a paper written circa 1970, and published in 1983, by Wiesner (1983).

We illustrate the concept of random access code with the simplest scheme, in which in a classical framework Alice needs to encode a two-bit string b_0b_1 into a single bit, or into a single qubit in a quantum framework.

In the classical case, Alice and Bob need to decide on a protocol defining which bit-valued message is to be sent by Alice, for each of the four possible values of her two-bit string b_0b_1. There are only $2^4 = 16$ different deterministic protocols, thus the probability of success can be evaluated in a straightforward way. The optimal deterministic classical protocols can then be shown to have a probability of success $P_C = 3/4$. For example, if Alice sends one of the two bits, then Bob will reveal this bit with certainty and have probability of 1/2 to reveal the other one. Since any probabilistic protocol can be represented as a convex combination of the 16 deterministic protocols, the corresponding probability of success for any such probabilistic protocol will be given by the weighted sum of the probabilities of success of the individual deterministic protocols. This implies that the optimal probabilistic protocols can at best be as efficient as the optimal deterministic protocol, which is 3/4.

Ambainis et al. (1999) showed that there is a quantum solution of the random access code with probability of success $P_Q = \cos^2(\pi/8) \approx 0.85$. It is realized as follows: depending on her two-bit string b_0b_1, Alice prepares one of the four states $|\psi_{b_0b_1}\rangle$. These states are chosen to be on the equator of the Bloch sphere, separated by equal angles of $\pi/2$ radians (see ❷ *Fig. 3*). Using the Bloch sphere parametrization $|\psi(\theta, \phi)\rangle = \cos(\theta/2)|0\rangle + \exp(i\phi)\sin(\theta/2)|1\rangle$, the four encoding states are represented as:

$$\begin{aligned}
|\psi_{00}\rangle &= |\psi(\pi/2, \pi/4)\rangle \\
|\psi_{01}\rangle &= |\psi(\pi/2, 7\pi/4)\rangle \\
|\psi_{10}\rangle &= |\psi(\pi/2, 3\pi/4)\rangle \\
|\psi_{11}\rangle &= |\psi(\pi/2, 5\pi/4)\rangle
\end{aligned} \tag{91}$$

Bob's measurements, which he uses to guess the bits, will depend on which bit he wants to obtain. To guess b_0, he projects the qubit along the x-axis in the Bloch sphere, and to decode b_1 he projects it along the y-axis. He then estimates the bit value to be 0 if the measurement outcome was along the positive direction of the axis and 1 if it was along the negative axis. It can easily be calculated that the probability of successful retrieving of the correct bit value is the same in all cases: $P_Q = \cos^2(\pi/8) \approx 0.85$, which is higher than the optimal probability of success $P_C = 0.75$ of the classical random access code using one bit of communication.

We will now introduce a hidden variable model of the quantum solution to see that the key resource in its efficiency lies in violation of temporal Bell's inequalities. Galvao (2002) was

◻ **Fig. 3**
The set of encoding states and decoding measurements in quantum random access code represented in the $x - y$ plane of the Bloch sphere. Alice prepares one of the four quantum states $\psi_{b_0 b_1}$ to encode two bits b_0, $b_1 \in \{0, 1\}$. Depending on which bit Bob wants to reveal he performs either a measurement along the x (to reveal b_0) or along the y axis (to reveal b_1).

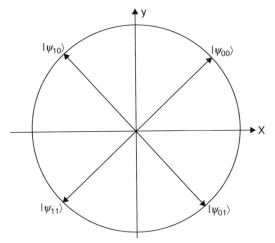

the first to point to the relation between violation of Bell's type inequalities and quantum random access codes. See also Spekkens et al. (2009) for a relation with the parity-oblivious multiplexing.

A hidden-variable model equivalent to the quantum protocol, which best fits the temporal Bell's inequalities can be considered as a description of the following modification of the original quantum protocol. Alice prepares the initial state of her qubit as a completely random state, described by a density matrix proportional to the unit operator, σ_0. Her parity of bit values $b_0 \oplus b_1$ defines a measurement basis, which is used by her to prepare the state to be sent to Bob. Note that the result of the dichotomic measurement in the basis defined by $b_0 \oplus b_1$ is due to the nature of the initial state, completely random and totally uncontrollable by Alice. To fix the bit value b_1 (and thus also the value b_0, since the parity is defined by the choice of the measurement basis) on her wish, Alice either leaves the state unchanged, if the result of measurement corresponds to her wish of b_1, or she rotates the state in the $x - y$ plane at $180°$ to obtain the orthogonal state, if the result corresponds to $b_1 \oplus 1$. Just a glance at the states involved in the standard quantum protocol shows the two complementary (unbiased) bases which define her measurement settings, and which resulting states are linked with which values of $b_0 b_1$. After the measurement, the resulting state is sent to Bob, while Alice is in possession of a bit pair $b_0 b_1$, which is perfectly correlated with the qubit state on the way to Bob. That is, we have exactly the same starting point as in the original quantum protocol.

Now, it is obvious that the quantum protocol violates the temporal inequalities, while any hidden variable model of the above procedure using the four assumptions (1–4) behind the temporal inequalities is not violating them. What is important is the saturation of the temporal inequalities which is equivalent to a probability of success of 3/4.

The link with temporal Bell's inequalities points to another advantage of quantum over classical random access codes. Usually, one considers the advantage to be *resource dependent*.

With this, we mean that there is an advantage as far as one compares one classical bit with one qubit. Yet the proof given above shows that the quantum strategy has an advantage over *all* hidden variable models respecting (1–4), that is, also those where Alice and Bob use temporal classical correlations between systems of arbitrarily large number of degrees of freedom.

Let us conclude this chapter by pointing out an interesting research avenue. Here, a brief review was given on the results demonstrating that "no go theorems" for various hidden variable classes of theories are behind better-than-classical efficiency in many quantum communication protocols. It would be interesting to investigate the link between fundamental features of quantum mechanics and the power of quantum computation. We showed that temporal Bell's inequalities distinguish between classical and quantum search (Grover) algorithms (Morikoshi 2006). Also cluster states – entangled states on which the logic gates are applied as sequences of only single-qubit measurements – are known to violate Bell's inequalities (Scarani et al. 2005; Tóth et al. 2006). These results point to the aforementioned link but we are still far away from understanding what the key non-classical ingredients are that give rise to the enhanced quantum computational power. The question gets even more fascinating after realizing that not only too little (Van den Nest et al. 2006; Brennen and Miyake 2008; Anders and Browne 2009; Vidal 2007; Markov and Shi 2008; Fannes et al. 1992) but also too much entanglement does not allow powerful quantum computation (Gross et al. 2008; Bremner et al. 2008).

Acknowledgments

Support from the Austrian Science Foundation FWF within Project No. P19570-N16, SFB FoQuS and CoQuS No. W1210-N16, and the European Commission, Projects QAP and QESSENCE, is acknowledged. MZ is supported by MNiSW grant N202 208538. The collaboration is a part of an ÖAD/MNiSW program.

References

Acín A, Scarani V, Wolf MM (2002) Bell inequalities and distillability in *N*-quantum-bit systems. Phys Rev A 66:042323

Acín A, Gisin N, Masanes L (2006) From Bell's theorem to secure quantum key distribution. Phys Rev Lett 97:120405

Aharon N, Vaidman L (2008) Quantum advantages in classically defined tasks. Phys Rev A 77:052310

Ambainis A, Nayak A, Ta-Shma A, Vazirani U (1999) Dense quantum coding and a lower bound for 1-way quantum automata. Proceedings of the 31st annual ACM symposium on the theory of computing, New York, pp 376–383

Anders J, Browne DE (2009) Computational power of correlations. Phys Rev Lett 102:050502

Aspect A, Dalibard J, Roger G (1982) Experimental test of Bell's inequalities using time-varying analyzers. Phys Rev Lett 49:1804–1807

Augusiak R, Horodecki P (2006) Bound entanglement maximally violating Bell inequalities: quantum entanglement is not fully equivalent to cryptographic security. Phys Rev A 74:010305

Barrett J (2007) Information processing in generalized probabilistic theories. Phys Rev A 75:032304

Bechmann-Pasquinucci H (2005) From quantum state targeting to Bell inequalities. Found Phys 35:1787–1804

Bell JS (1964) On the Einstein-Podolsky-Rosen paradox. Physics 1:195–200; reprinted in Bell JS (1987) Speakable and unspeakable in quantum mechanics. Cambridge University Press, Cambridge

Bell JS (1966) On the problem of hidden variables in quantum mechanics. Rev Mod Phys 38:447–452

Bell JS (1985) Free variables and local causality. Dialectica 39:103–106

Bennett CH, Wiesner SJ (1992) Communication via one- and two-particle operators on Einstein-Podolsky-Rosen states. Phys Rev Lett 69:2881–2884

Bohm D (1952) A suggested interpretation of the quantum theory in terms of "hidden" variables, I and II. Phys Rev 85:166–193

Bohr N (1935) Can quantum-mechanical description of physical reality be considered complete? Phys Rev 48:696–702

Branciard C, Ling A, Gisin N et al. (2007) Experimental falsification of Leggett's nonlocal variable model. Phys Rev Lett 99:210407

Brassard G (2003) Quantum communication complexity (a survey). Found Phys 33:1593–1616. Available via http://arxiv.org/abs/quant-ph/0101005

Brassard G, Buhrman H, Linden N et al. (2006) Limit on nonlocality in any world in which communication complexity is not trivial. Phys Rev Lett 96:250401

Bremner M J, Mora C, Winter A (2009) Are random pure states useful for quantum computation? Phys Rev Lett 102:190502

Brennen GK, Miyake A (2008) Measurement-based quantum computer in the gapped ground state of a two-body Hamiltonian. Phys Rev Lett 101:010502

Brukner Č, Paterek T, Żukowski M (2003) Quantum communication complexity protocols based on higher-dimensional entangled systems. Int J Quantum Inf 1:519–525

Brukner Č, Żukowski M, Zeilinger A (2002) Quantum communication complexity protocol with two entangled qutrits. Phys Rev Lett 89:197901

Brukner Č, Taylor S, Cheung S, Vedral V (2004a) Quantum entanglement in time. Available via arXiv: quant-ph/0402127

Brukner Č, Żukowski M, Pan J-W, Zeilinger A (2004b) Bell's inequality and quantum communication complexity. Phys Rev Lett 92:127901

Buhrman H, Cleve R, van Dam W (1997) Quantum entanglement and communication complexity. Available via http://arxiv.org/abs/quant-ph/9705033

Buhrman H, van Dam W, Høyer P, Tapp A (1999) Multiparty quantum communication complexity. Phys Rev A 60:2737–2741

Cabello A, Estebaranz JM, García-Alcaine G (1996) Bell-Kochen-Specker theorem: a proof with 18 vectors. Phys Lett A 212:183–187

Cirel'son BS (1980) Quantum generalizations of Bell's inequality. Lett Math Phys 4:93–100

Clauser JF, Horne MA, Shimony A, Holt RA (1969) Proposed experiment to test local hidden-variable theories. Phys Rev Lett 23:880–884

Cleve R, Buhrman H (1997) Substituting quantum entanglement for communication. Phys Rev A 56:1201–1204

Cleve R, Høyer P, Toner B, Watrous J (2004) Consequences and limits of nonlocal strategies. Proceedings of the 19th IEEE annual conference on computational complexity, Amherst, MA, pp 236–249

Dakic B, Brukner C (2009) Quantum theory and beyond: is entanglement special? arXiv:0911.0695

DiVincenzo DP, Peres A (1997) Quantum code words contradict local realism. Phys Rev A 55: 4089–4092

Einstein A, Podolsky B, Rosen N (1935) Can quantum-mechanical description of physical reality be considered complete? Phys Rev 47:777–780

Ekert A (1991) Quantum cryptography based on Bell's theorem. Phys Rev Lett 67:661–663

Fannes M, Nachtergaele B, Werner RF (1992) Finitely correlated states on quantum spin chains. Commun Math Phys 144:443–490

Galvão EF (2001) Feasible quantum communication complexity protocol. Phys Rev A 65:012318

Galvão EF (2002) Foundations of quantum theory and quantum information applications. Ph.D. thesis, University of Oxford, Oxford. Available via arXiv: quant-ph/0212124

Gill RD, Weihs G, Zeilinger A, Żukowski M (2002) No time loophole in Bell's theorem: the Hess-Philipp model is nonlocal. Proc Nat Acad Sci USA 99:14632–14635

Gill RD, Weihs G, Zeilinger A, Żukowski M (2003) Comment on "Exclusion of time in the theorem of Bell" by Hess K and Philipp W. Europhy Lett 61:282–283

Greenberger DM, Horne MA, Zeilinger A (1989) Going beyond Bell's theorem. In: Kafatos M (ed) Bell's theorem, quantum theory, and conceptions of the universe. Kluwer, Dordrecht, pp 73–76

Greenberger DM, Horne MA, Shimony A, Zeilinger A (1990) Bell's theorem without inequalities. Am J Phys 58:1131–1143

Gröblacher S, Paterek T, Kaltenbaek R et al. (2007) An experimental test of non-local realism. Nature 446:871–875

Gross D, Flammia S, Eisert J (2009) Most quantum states are too entangled to be useful as computational resources. Phys Rev Lett 102:190501

Holevo AS (1973) Bounds for the quantity of information transmitted by a quantum communication channel. Problemy Peredachi Informatsii 9:3–11. English translation in Probl Inf Transm 9:177–183

Horodecki R, Horodecki P, Horodecki M (1995) Violating Bell inequality by mixed spin-$\frac{1}{2}$ states: necessary and sufficient condition. Phys Lett A 200:340–344

Hyllus P, Gühne O, Bruß D, Lewenstein M (2005) Relations between entanglement witnesses and Bell inequalities. Phys Rev A 72:012321

Kochen S, Specker E (1968) The problem of hidden variables in quantum mechanics. J Math Mech 17:59–87

Kofler J, Paterek T, Brukner C (2006) Experimenter's freedom in Bell's theorem and quantum cryptography, Phys Rev A 73:022104

Kushilevitz E, Nisan N (1997) Communication complexity. Cambridge University Press, Cambridge

Leggett AJ (2002) Testing the limits of quantum mechanics: motivation, state of play, prospects. J Phys Condensed Matter 14:15 R415

Leggett AJ (2003) Nonlocal hidden-variable theories and quantum mechanics: an incompatibility theorem. Found Phys 33:1469–1493

Leggett AJ, Garg A (1985) Quantum mechanics versus macroscopic realism: is the flux there when nobody looks? Phys Rev Lett 54:857–860

Marcovitch S, Reznik B (2008) Implications of communication complexity in multipartite systems. Phys Rev A 77:032120

Markov IL, Shi Y (2008) Simulating quantum computation by contracting tensor networks. SIAM J Comput 38(3):963–981

Mermin ND (1993) Hidden variables and the two theorems of John Bell. Rev Mod Phys 65:803–815

Morikoshi F (2006) Information-theoretic temporal Bell inequality and quantum computation. Phys Rev A 73:052308

Nagata K (2005) Kochen-Specker theorem as a precondition for secure quantum key distribution. Phys Rev A 72:012325

Pawłowski M, Paterek T, Kaszlikowski D, Scarani V, Winter A and Zukowski M (2009) Information causality as a physical principle. Nature (London) 461(7267):1101–1104

Paz JP, Mahler G (1993) Proposed test for temporal Bell inequalities. Phys Rev Lett 71:3235–3239

Peres A (1991) Two simple proofs of the Kochen-Specker theorem. J Phys A 24:L175–L178

Peres A (1994) Quantum theory: concepts and methods. Kluwer, Boston

Popescu S, Rohrlich D (1994) Quantum nonlocality as an axiom. Found Phys 24:379–385

Rowe MA, Kielpinski D, Meyer V, Sackett CA, Itano WM, Monroe C, Wineland DJ (2001) Experimental violation of a Bell's inequality with efficient detection. Nature 409:791–794

Scarani V, Gisin N (2001) Quantum communication between N partners and Bell's inequalities. Phys Rev Lett 87:117901

Scarani V, Acín A, Schenck E, Aspelmeyer M (2005) Nonlocality of cluster states of qubits. Phys Rev A 71:042325

Schmid C, Kiesel N, Laskowski W et al. (2008) Discriminating multipartite entangled states. Phys Rev Lett 100:200407

Scheidl T, Ursin R, Kofler J et al. (2008) Violation of local realism with freedom of choice. Available via http://arxiv.org/abs/0811.3129

Schrödinger E (1935) Die gegenwärtige Situation in der Quantenmechanik. Naturwissenschaften 23:807–812; 823–828; 844–849. Translation published in Proc Am Philos Soc 124:323–338 and in Wheeler JA, Zurek WH (eds) (1983) Quantum

theory and measurement. Princeton University Press, Princeton, NJ, pp 152–167

Shafiee A, Golshani M (2003) Single-particle Bell-type inequality. Annales de la Fondation de Broglie 28:105–118

Simon C, Brukner Č, Zeilinger A (2001) Hidden-variable theorems for real experiments. Phys Rev Lett 86:4427–4430

Spekkens RW, Buzacott DH, Keehn AJ et al. (2009) Preparation contextuality powers parity-oblivious multiplexing. Phys Rev Lett 102:010401

Svozil K (2004) Quantum mechanics is noncontextual. Available via arXiv:quant-ph/0401112

Tamir B (2007) Communication complexity protocols for qutrits. Phys Rev A 75:032344

Tóth G, Gühne O, Briegel HJ (2006) Two-setting Bell inequalities for graph states. Phys Rev A 73:022303

Trojek P, Schmid C, Bourennane M et al. (2005) Experimental quantum communication complexity. Phys Rev A 72:050305

van Dam W (2000) Implausible consequences of superstrong nonlocality. Chapter 9 in van Dam W, Nonlocality & communication complexity. Ph.D. Thesis, University of Oxford, Department of Physics. Available via arXiv:quant-ph/0501159

Van den Nest M, Miyake A, Dür W, Briegel HJ (2006) Universal resources for measurement-based quantum computation. Phys Rev Lett 97:150504

Vidal G (2007) Classical simulation of infinite-size quantum lattice systems in one spatial dimension. Phys Rev Lett 98:070201

Weihs G, Jennewein T, Simon C, Weinfurter H, Zeilinger A (1998) Violation of Bell's inequality under strict Einstein locality conditions. Phys Rev Lett 81:5039–5043

Weinfurter H, Żukowski M (2001) Four-photon entanglement from down-conversion. Phys Rev A 64:010102

Werner RF, Wolf MM (2001) All multipartite Bell correlation inequalities for two dichotomic observables per site. Phys Rev A 64:032112

Wiesner S (1983) Conjugate coding, ACM Sigact News 15:78–88

von Weizsäcker CF (1985) Aufbau der Physik. Hanser, Munich

Yao AC (1979) Some complexity questions related to distributed computing. Proceedings of the 11th Annual ACM Symposium on Theory of Computing, pp 209–213

Zeilinger A (1999) A foundational principle for quantum mechanics. Found Phys 29:631–643

Żukowski M, Brukner Č (2002) Bell's theorem for general N-qubit states. Phys Rev Lett 88:210401

43 Algorithms for Quantum Computers

Jamie Smith[1] · *Michele Mosca*[2]
[1]Institute for Quantum Computing and Department of Combinatorics & Optimization, University of Waterloo, Canada
ja5smith@iqc.ca
[2]Institute for Quantum Computing and Department of Combinatorics & Optimization, University of Waterloo and St. Jerome's University and Perimeter Institute for Theoretical Physics, Waterloo, Canada
mmosca@iqc.ca

G. Rozenberg et al. (eds.), *Handbook of Natural Computing*, DOI 10.1007/978-3-540-92910-9_43,
© Springer-Verlag Berlin Heidelberg 2012

Abstract

This chapter surveys the field of quantum computer algorithms. It gives a taste of both the breadth and the depth of the known algorithms for quantum computers, focusing on some of the more recent results. It begins with a brief review of quantum Fourier transform-based algorithms, followed by quantum searching and some of its early generalizations. It continues with a more in-depth description of two more recent developments: algorithms developed in the quantum walk paradigm, followed by tensor network evaluation algorithms (which include approximating the Tutte polynomial).

1   Introduction

Quantum computing is a new computational paradigm created by reformulating information and computation in a quantum mechanical framework (Feynman 1982; Deutsch 1985). Since the laws of physics appear to be quantum mechanical, this is the most relevant framework to consider when considering the fundamental limitations of information processing. Furthermore, in recent decades a major shift from just observing quantum phenomena to actually controlling quantum mechanical systems has been seen. The communication of quantum information over long distances, the "teleportation" of quantum information, and the encoding and manipulation of quantum information in many different physical media were also observed. We still appear to be a long way from the implementation of a large-scale quantum computer; however, it is a serious goal of many of the world's leading physicists, and progress continues at a fast pace.

In parallel with the broad and aggressive program to control quantum mechanical systems with increased precision, and to control and interact a larger number of subsystems, researchers have also been aggressively pushing the boundaries of what useful tasks one could perform with quantum mechanical devices. These include improved metrology, quantum communication and cryptography, and the implementation of large-scale quantum algorithms.

It was known very early on (Deutsch 1985) that quantum algorithms cannot compute functions that are not computable by classical computers, however they might be able to efficiently compute functions that are not efficiently computable on a classical computer. Or, at the very least, quantum algorithms might be able to provide some sort of speed-up over the best possible or best-known classical algorithms for a specific problem.

The purpose of this chapter is to survey the field of quantum algorithms, which has grown tremendously since Shor's breakthrough algorithms (Shor 1994, 1997) over 15 years ago. Much of the work in quantum algorithms is now textbook material (e.g., Nielsen and Chuang (2000); Hirvensalo (2001); Kitaev et al. (2002); Kaye et al. (2007); Mermin (2007)), and these examples will only be briefly mentioned in order to provide a broad overview. Other parts of this chapter, in particular, ❷ Sects. 4 and ❷ 5, give a more detailed description of some more recent work.

We organized this chapter according to underlying tools or approaches taken, and we include some basic applications and specific examples, and relevant comparisons with classical algorithms.

In ❷ Sect. 2 we begin with algorithms one can naturally consider to be based on a quantum Fourier transform, which includes the famous factoring and discrete logarithm algorithms of Peter Shor (1994, 1997). Since this topic is covered in several textbooks and recent surveys, this topic will only be briefly surveyed. Several other sections could have been added on algorithms for generalizations of these problems, including several cases of the non-Abelian hidden subgroup problem, and hidden lattice problems over the reals (which have important applications in number theory); however, these are covered in a recent survey (Mosca 2009) and in substantial detail in Childs and van Dam (2010).

❯ Section 3 continues with a brief review of classic results on quantum searching and counting, and more generally amplitude amplification and amplitude estimation.

In ❯ Sect. 4, algorithms based on quantum walks are discussed. The related topic of adiabatic algorithms, which was briefly summarized in Mosca (2009) will not be covered here; a broader survey of this and related techniques ("quantum annealing") can be found in Das and Chakrabarti (2008).

❯ Section 5 concludes with algorithms based on the evaluation of the trace of an operator, also referred to as the evaluation of a tensor network, and which has applications such as the approximation of the Tutte polynomial.

The field of quantum algorithms has grown tremendously since the seminal work in the mid-1990s, and a full detailed survey would simply be infeasible for one chapter. Several other sections could have been added. One major omission is the development of algorithms for simulating quantum mechanical systems, which was Feynman's original motivation for proposing a quantum computer. This field was briefly surveyed in Mosca (2009), with emphasis on the recent results in Berry et al. (2007) (more recent developments can be found in Wiebe et al. (2008)). This remains an area worthy of a comprehensive survey; like many other areas that researchers have tried to survey, it is difficult because it is still an active area of research. It is also an especially important area because these algorithms tend to offer a fully exponential speed-up over classical algorithms, and thus are likely to be among the first quantum algorithms to be implemented that will offer a speed-up over the fastest-available classical computers.

Finally, one could also write a survey of quantum algorithms for intrinsically quantum information problems, such as entanglement concentration, or quantum data compression. This topic is not covered in this chapter, though there is a very brief survey in Mosca (2009).

One can find a rather comprehensive list of the known quantum algorithms (up to mid-2008) in Stephen Jordan's PhD thesis (Jordan 2008). It is hoped that this chapter complements some of the other recent surveys in providing a reasonably detailed overview of the state of the art in quantum algorithms.

2 Algorithms Based on the Quantum Fourier Transform

The early line of quantum algorithms was developed in the "black-box" or "oracle" framework. In this framework, part of the input is a black box that implements a function $f(x)$, and the only way to extract information about f is to evaluate it on inputs x. These early algorithms used a special case of quantum Fourier transform, the Hadamard gate, in order to solve the given problem with fewer black-box evaluations of f than a classical algorithm would require.

Deutsch (1985) formulated the problem of deciding whether a function $f : \{0, 1\} \to \{0, 1\}$ was constant or not. Suppose one has access to a black box that implements f reversibly by mapping $x, 0 \mapsto x, f(x)$; one can further assume that the black box in fact implements a unitary transformation U_f that maps $|x\rangle |0\rangle \mapsto |x\rangle |f(x)\rangle$. Deutsch's problem is to output "constant" if $f(0) = f(1)$ and to output "balanced" if $f(0) \neq f(1)$, given a black box for evaluating f. In other words determine $f(0) \oplus f(1)$ (where $\oplus$ denotes addition modulo 2). Outcome "0" means f is constant and "1" means f is not constant.

A classical algorithm would need to evaluate f twice in order to solve this problem. A quantum algorithm can apply U_f only once to create

$$\frac{1}{\sqrt{2}}|0\rangle|f(0)\rangle + \frac{1}{\sqrt{2}}|1\rangle|f(1)\rangle$$

Note that if $f(0) = f(1)$, then applying the Hadamard gate to the first register yields $|0\rangle$ with probability 1, and if $f(0) \neq f(1)$, then applying the Hadamard gate to the first register and ignoring the second register leaves the first register in the state $|1\rangle$ with probability $\frac{1}{2}$; thus a result of $|1\rangle$ could only occur if $f(0) \neq f(1)$.

As an aside, one can note that in general, given

$$\frac{1}{\sqrt{2}}|0\rangle|\psi_0\rangle + \frac{1}{\sqrt{2}}|1\rangle|\psi_1\rangle$$

applying the Hadamard gate to the first qubit and measuring it will yield "0" with probability $\frac{1}{2} + \mathrm{Re}(\langle\psi_0|\psi_1\rangle)$; this "Hadamard test" is discussed in more detail in ❿ Sect. 5.

In Deutsch's case, measuring a "1" meant $f(0) \neq f(1)$ with certainty, and a "0" was an inconclusive result. Even though it was not perfect, it still was something that could not be done with a classical algorithm. The algorithm can be made exact (Cleve et al. 1998) (i.e., one that outputs the correct answer with probability 1) if one assumes further that U_f maps $|x\rangle|b\rangle \mapsto |x\rangle|b \oplus f(x)\rangle$, for $b \in \{0, 1\}$, and one sets the second qubit to $\frac{1}{\sqrt{2}}|0\rangle - \frac{1}{\sqrt{2}}|1\rangle$. Then U_f maps

$$\left(\frac{|0\rangle + |1\rangle}{\sqrt{2}}\right)\left(\frac{|0\rangle - |1\rangle}{\sqrt{2}}\right) \mapsto (-1)^{f(0)}\left(\frac{|0\rangle + (-1)^{f(0)\oplus f(1)}|1\rangle}{\sqrt{2}}\right)\left(\frac{|0\rangle - |1\rangle}{\sqrt{2}}\right)$$

Thus, a Hadamard gate on the first qubit yields the result

$$(-1)^{f(0)}|f(0) \oplus f(1)\rangle\left(\frac{|0\rangle - |1\rangle}{\sqrt{2}}\right)$$

and measuring the first register yields the correct answer with certainty.

The general idea behind the early quantum algorithms was to compute a black-box function f on a superposition of inputs, and then extract a global property of f by applying a quantum transformation to the input register before measuring it. It is usually assumed that we have access to a black box that implements

$$U_f : |\mathbf{x}\rangle|\mathbf{b}\rangle \mapsto |\mathbf{x}\rangle|\mathbf{b} \oplus f(\mathbf{x})\rangle$$

or in some other form where the input value $\mathbf{x}$ is kept intact and the second register is shifted by $f(\mathbf{x})$ in some reversible way.

Deutsch and Jozsa (1992) used this approach to get an exact algorithm that decides whether $f : \{0, 1\}^n \mapsto \{0, 1\}$ is constant or "balanced" (i.e., $|f^{-1}(0)| = |f^{-1}(1)|$), with a promise that one of these two cases holds. Their algorithm evaluated f only twice, while classically any exact algorithm would require $2^{n-1} + 1$ queries in the worst case. Bernstein and Vazirani (1997) defined a specific class of such functions $f_{\mathbf{a}}: \mathbf{x} \mapsto \mathbf{a} \cdot \mathbf{x}$, for any $\mathbf{a} \in \{0, 1\}^n$, and showed how the same algorithm that solves the Deutsch–Jozsa problem allows one to determine $\mathbf{a}$ with two evaluations of $f_{\mathbf{a}}$ while a classical algorithm requires n evaluations. (Both of these algorithms can be done with one query if we have $|\mathbf{x}\rangle|b\rangle \mapsto |\mathbf{x}\rangle|b \oplus f(\mathbf{x})\rangle$.) They further showed how a related "recursive Fourier sampling" problem could be solved super-polynomially faster on a quantum computer. Simon (1994) later built on these tools to develop a black-box quantum algorithm that was exponentially faster than any classical algorithm for finding a hidden string $s \in \{0, 1\}^n$ that is encoded in a function $f: \{0, 1\}^n \rightarrow \{0, 1\}^n$ with the property that $f(\mathbf{x}) = f(\mathbf{y})$ if and only if $\mathbf{x} = \mathbf{y} \oplus \mathbf{s}$.

Shor (1994, 1997) built on these black-box results to find an efficient algorithm for finding the order of an element in the multiplicative group of integers modulo N (which implies an efficient classical algorithm for factoring N) and for solving the discrete logarithm problem in the multiplicative group of integers modulo a large prime p. Since the most widely used public key cryptography schemes at the time relied on the difficulty of integer factorization, and others relied on the difficulty of the discrete logarithm problem, these results had very serious practical implications. Shor's algorithms straightforwardly apply to black-box groups, and thus permit finding orders and discrete logarithms in any group that is reasonably presented, including the additive group of points on elliptic curves, which is currently one of the most widely used public key cryptography schemes (see e.g., Menezes et al. (1996)).

Researchers tried to understand the full implications and applications of Shor's technique, and a number of generalizations were soon formulated (e.g., Boneh and Lipton 1995; Grigoriev 1997). One can phrase Simon's algorithm, Shor's algorithm, and the various generalizations that soon followed as special cases of the *hidden subgroup problem*. Consider a finitely generated Abelian group G, and a hidden subgroup K that is defined by a function $f: G \rightarrow X$ (for some finite set X) with the property that $f(x) = f(y)$ if and only if $x - y \in K$ (we use additive notation, without loss of generality). In other words, f is constant on cosets of K and distinct on different cosets of G. In the case of Simon's algorithm, $G = \mathbb{Z}_2^n$ and $K = \{0, s\}$. In the case of Shor's order-finding algorithm, $G = \mathbb{Z}$ and $K = r\mathbb{Z}$ where r is the unknown order of the element. Other examples and how they fit in the hidden subgroup paradigm are given in Mosca (2008).

Soon after, Kitaev (Kitaev 1995) solved a problem he called the *Abelian stabilizer problem* using an approach that seemed different from Shor's algorithm, one based in eigenvalue estimation. Eigenvalue estimation is in fact an algorithm of independent interest for the purpose of studying quantum mechanical systems. The Abelian stabilizer problem is also a special case of the hidden subgroup problem. Kitaev's idea was to turn the problem into one of estimating eigenvalues of unitary operators. In the language of the hidden subgroup problem, the unitary operators were shift operators of the form $f(x) \mapsto f(x + y)$. By encoding the eigenvalues as relative phase shifts, he turned the problem into a phase estimation problem.

The Simon/Shor approach for solving the hidden subgroup problem is to first compute $\sum_x |x\rangle |f(x)\rangle$. In the case of finite groups G, one can sum over all the elements of G, otherwise one can sum over a sufficiently large subset of G. For example, if $G = \mathbb{Z}$, and $f(x) = a^x \mod N$ for some large integer N, we first compute $\sum_{x=0}^{2^n-1} |x\rangle |a^x\rangle$, where $2^n > N^2$ (we omit the "mod N" for simplicity). If r is the order of a (i.e., r is the smallest positive integer such that $a^r \equiv 1$) then every value x of the form $x = y + kr$ gets mapped to a^y. Thus, we can rewrite the above state as

$$\sum_{x=0}^{2^n-1} |x\rangle |a^x\rangle = \sum_{y=0}^{r-1} \left(\sum_j |y + jr\rangle \right) |a^y\rangle \qquad (1)$$

where each value of a^y in this range is distinct. Tracing out the second register one is thus left with a state of the form

$$\sum_j |y + jr\rangle$$

for a random y and where j goes from 0 to $\lfloor (2^n - 1)/r \rfloor$. We loosely refer to this state as a "periodic" state with period r. We can use the inverse of the quantum Fourier transform (or the quantum Fourier transform) to map this state to a state of the form $\sum_x \alpha_x |x\rangle$, where the amplitudes are biased toward values of x such that $x/2^n \approx k/r$. With probability at least $4/\pi^2$ we obtain an x such that $|x/2^n - k/r| \leq 1/2^{n+1}$. One can then use the continued fractions algorithm to find k/r (in lowest terms) and thus find r with high probability. It is important to note that the continued fractions algorithm is not needed for many of the other cases of the Abelian hidden subgroup considered, such as Simon's algorithm or the discrete logarithm algorithm when the order of the group is already known.

In contrast, Kitaev's approach for this special case was to consider the map $U_a: |b\rangle \mapsto |ba^x\rangle$. It has eigenvalues of the form $e^{2\pi i k/r}$, and the state $|1\rangle$ satisfies

$$|1\rangle = \frac{1}{\sqrt{r}} \sum_{k=0}^{r-1} |\psi_k\rangle$$

where $|\psi_k\rangle$ is an eigenvector with eigenvalue $e^{2\pi i k/r}$:

$$U_a: |\psi_k\rangle \mapsto e^{2\pi i k x/r} |\psi_k\rangle$$

If one considers the controlled-U_a, denoted $c - U_a$, which maps $|0\rangle |b\rangle \mapsto |0\rangle |b\rangle$ and $|1\rangle |b\rangle \mapsto |1\rangle |ba\rangle$, and if one applies it to the state $(|0\rangle + |1\rangle)|\psi_k\rangle$ one gets

$$(|0\rangle + |1\rangle)|\psi_k\rangle \mapsto (|0\rangle + e^{2\pi i k/r}|1\rangle)|\psi_k\rangle$$

In other words, the eigenvalue becomes a relative phase, and thus one can reduce eigenvalue estimation to phase estimation. Furthermore, since one can efficiently compute a^{2^j} by performing j multiplications modulo N, one can also efficiently implement $c - U_{a^{2^j}}$ and thus easily obtain the qubit $|0\rangle + e^{2\pi i 2^j (k/r)}|1\rangle$ for integer values of j without performing $c - U_a$ a total of 2^j times. Kitaev developed an efficient ad hoc phase estimation scheme in order to estimate k/r to high precision, and this phase estimation scheme could be optimized further (Cleve et al. 1998). In particular, one can create the state

$$(|0\rangle + |1\rangle)^n |1\rangle = \sum_{x=0}^{2^n-1} |x\rangle |1\rangle = \sum_{k=0}^{r-1} \sum_{x=0}^{2^n-1} |x\rangle |\psi_k\rangle$$

(we use the standard binary encoding of the integers $x \in \{0, 1, \ldots, 2^n - 1\}$ as bit strings of length n) to apply the $c - U_{a^{2^j}}$ using the $(n - j)$th bit as the control bit, and using the second register (initialized in $|1\rangle = \sum_k |\psi_k\rangle$) as the target register, for $j = 0, 1, 2, \ldots, n - 1$, to create

$$\sum_k (|0\rangle + e^{2\pi i 2^{n-1}\frac{k}{r}}|1\rangle) \cdots (|0\rangle + e^{2\pi i 2\frac{k}{r}}|1\rangle)(|0\rangle + e^{2\pi i \frac{k}{r}}|1\rangle)|\psi_k\rangle \qquad (2)$$

$$= \sum_{x=0}^{2^n-1} |x\rangle |a^x\rangle = \sum_{k=0}^{r-1} \sum_{x=0}^{2^n-1} e^{2\pi i x \frac{k}{r}} |x\rangle |\psi_k\rangle \qquad (3)$$

If the second register is ignored or discarded, one is left with a state of the form $\sum_{x=0}^{2^n-1} e^{2\pi i x k/r} |x\rangle$ for a random value of $k \in \{0, 1, \ldots, r - 1\}$. The inverse quantum Fourier transformation maps this state to a state $\sum_y \alpha_y |y\rangle$ where most of the weight of the amplitudes is near values of y such that $y/2^n \approx k/r$ for some integer k. More specifically $|y/2^n - k/r| \leq 1/2^{n+1}$ with probability at least $4/\pi^2$; furthermore $|y/2^n - k/r| \leq 1/2^n$ with probability at least $8/\pi^2$. As in the case of Shor's algorithm, one can use the continued fractions algorithm to determine k/r (in lowest terms) and thus determine r with high probability.

It was noted (Cleve et al. 1998) that this modified eigenvalue estimation algorithm for order-finding was essentially equivalent to Shor's period-finding algorithm for order-finding. This can be seen by noting that we have the same state in ❯ Eqs. 1 and ❯ 2, and in both cases, we discard the second register and apply an inverse Fourier transform to the first register. The only difference is the basis in which the second register is mathematically analyzed.

The most obvious direction in which to try to generalized the Abelian hidden subgroup algorithm is to solve instances of the hidden subgroup problem for non-Abelian groups. This includes, for example, the graph automorphism problem (which corresponds to finding a hidden subgroup of the symmetric group). There has been nontrivial, but limited, progress in this direction, using a variety of algorithmic tools, such as sieving, "pretty good measurements," and other group theoretic approaches. Other generalizations include the hidden shift problem and its generalizations, hidden lattice problems on real lattices (which has important applications in computational number theory and computationally secure cryptography), and hidden nonlinear structures. These topics and techniques would take several dozen pages just to summarize, so the reader can refer to Mosca (2009) or Childs and van Dam (2010) so that more room is left to summarize other important topics.

3 Amplitude Amplification and Estimation

A very general problem for which quantum algorithms offer an improvement is that of searching for a solution to a computational problem in the case that a solution x can be easily verified. One can phrase such a general problem as finding a solution x to the equation $f(x) = 1$, given a means for evaluating the function f. One can further assume that $f: \{0, 1\}^n \mapsto \{0, 1\}$.

The problems from ❯ Sect. 2 can be rephrased in this form (or a small number of instances of problems of this form), and the quantum algorithms for these problems exploit some nontrivial algebraic structure in the function f in order to solve the problems superpolynomially faster than the best possible or best-known classical algorithms. Quantum computers also allow a more modest speed-up (up to quadratic) for searching for a solution to a function f without any particular structure. This includes, for example, searching for solutions to *NP*-complete problems.

Note that classically, given a means for guessing a solution $\mathbf{x}$ with probability p, one could amplify this success probability by repeating many times, and after a number of guesses in $O(1/p)$, the probability of finding a solution is in $\Omega(1)$. Note that the quantum implementation of an algorithm that produces a solution with probability p will produce a solution with probability amplitude $\sqrt{p}$. The idea behind quantum searching is to somehow amplify this probability to be close to 1 using only $O(1/\sqrt{p})$ guesses and other steps.

Lov Grover (1996) found precisely such an algorithm, and this algorithm was analyzed in detail and generalized (Boyer et al. 1998; Brassard and Høyer 1997; Grover 1998; Brassard et al. 2002) to what is known as "amplitude amplification." Any procedure for guessing a solution with probability p can be (with modest overhead) turned into a unitary operator A that maps $|0\rangle \mapsto \sqrt{p}|\psi_1\rangle + \sqrt{1-p}|\psi_0\rangle$, where $\mathbf{0} = 00\ldots0$, $|\psi_1\rangle$ is a superposition of states encoding solutions $\mathbf{x}$ to $f(\mathbf{x}) = 1$ (the states could in general encode $\mathbf{x}$ followed by other "junk" information) and $|\psi_0\rangle$ is a superposition of states encoding values of $\mathbf{x}$ that are not solutions.

One can then define the quantum search iterate (or "Grover iterate") to be

$$Q = -AU_0A^{-1}U_f$$

where $U_f: |\mathbf{x}\rangle \mapsto (-1)^{f(\mathbf{x})}|\mathbf{x}\rangle$, and $U_0 = I - 2|\mathbf{0}\rangle\langle\mathbf{0}|$ (in other words, maps $|\mathbf{0}\rangle \mapsto -|\mathbf{0}\rangle$ and $|\mathbf{x}\rangle \mapsto |\mathbf{x}\rangle$ for any $\mathbf{x} \neq \mathbf{0}$). Here, for simplicity, it is assumed that there are no "junk" bits in the unitary computation of f by A. Any such junk information can either be "uncomputed" and reset to all 0s, or even ignored (letting U_f act only on the bits encoding $\mathbf{x}$ and applying the identity to the junk bits).

This algorithm is analyzed thoroughly in the literature and in textbooks, so the main results and ideas are only summarized here to help the reader understand the later sections on searching via quantum walk.

If one applies Q a total of k times to the input state $A|00\ldots0\rangle = \sin(\theta)|\psi_1\rangle + \cos(\theta)|\psi_0\rangle$, where $\sqrt{p} = \sin(\theta)$, then one obtains

$$Q^k A|00\ldots0\rangle = \sin((2k+1)\theta)|\psi_1\rangle + \cos((2k+1)\theta)|\psi_0\rangle$$

This implies that with $k \approx \frac{\pi}{4\sqrt{p}}$ one obtains $|\psi_1\rangle$ with probability amplitude close to 1, and thus measuring the register will yield a solution to $f(x) = 1$ with high probability. This is quadratically better than what could be achieved by preparing and measuring $A|0\rangle$ until a solution is found.

One application of such a generic amplitude amplification method is for searching. One can also apply this technique to approximately count (Brassard et al. 1997) the number of solutions to $f(x) = 1$, and more generally to estimate the amplitude (Brassard et al. 2002) with which a general operator A produces a solution to $f(x) = 1$ (in other words, the transition amplitude from one recognizable subspace to another).

There are a variety of other applications of amplitude amplification that cleverly incorporate amplitude amplification in an unobvious way into a larger algorithm. Some of these examples are discussed in Mosca (2009) and most are summarized in Jordan (2008).

Since there are some connections to some of the more recent tools developed in quantum algorithms, we will briefly explain how amplitude estimation works.

Consider any unitary operator A that maps some known input state, say $|\mathbf{0}\rangle = |00\ldots0\rangle$, to a superposition $\sin(\theta)|\psi_1\rangle + \cos(\theta)|\psi_0\rangle$, $0 \leq \theta \leq \pi/2$, where $|\psi_1\rangle$ is a normalized superposition of "good" states $\mathbf{x}$ satisfying $f(\mathbf{x}) = 1$ and $|\psi_0\rangle$ is a normalized superposition of "bad" states $\mathbf{x}$ satisfying $f(\mathbf{x}) = 0$ (again, for simplicity, one can ignore extra junk bits). If we measure $A|\mathbf{0}\rangle$, we would measure a "good" $\mathbf{x}$ with probability $\sin^2(\theta)$. The goal of amplitude estimation is to approximate the amplitude $\sin(\theta)$. One can assume for convenience that there are $t > 0$ good states and $n - t > 0$ bad states, and that $0 < \sin(\theta) < \pi/2$.

It can be noted that the quantum search iterate $Q = -AU_0A^{-1}U_f$ has eigenvalues $|\psi_+\rangle = \frac{1}{\sqrt{2}}(|\psi_0\rangle + i|\psi_1\rangle)$ and $|\psi_-\rangle = \frac{1}{\sqrt{2}}(|\psi_0\rangle - i|\psi_1\rangle)$ with respective eigenvalues $e^{i2\theta}$ and $e^{-i2\theta}$. It also has $2^n - 2$ other eigenvectors; $t - 1$ of them have eigenvalue $+1$ and are the states orthogonal to $|\psi_1\rangle$ that have support on the states $|\mathbf{x}\rangle$ where $f(\mathbf{x}) = 1$, and $n - t - 1$ of them have eigenvalue -1 and are the states orthogonal to $|\psi_0\rangle$ that have support on the states $|\mathbf{x}\rangle$ where $f(\mathbf{x}) = 0$. It is important to note that $A|\psi\rangle = \frac{e^{i\theta}}{\sqrt{2}}|\psi_+\rangle + \frac{e^{-i\theta}}{\sqrt{2}}|\psi_-\rangle$ has its full support on the two-dimensional subspace spanned by $|\psi_+\rangle$ and $|\psi_-\rangle$.

It is worth noting that the quantum search iterate $Q = AU_0A^{-1}U_f$ can also be thought of as two reflections

$$-U_{A|0\rangle}U_f = (2A|0\rangle\langle0|A^\dagger - I)(I - 2\sum_{\mathbf{x}|f(\mathbf{x})=1}|\mathbf{x}\rangle\langle\mathbf{x}|)$$

one that flags the "good" subspace with a -1 phase shift, and then one that flags the subspace orthogonal to $A|0\rangle$ with a -1 phase shift. In the two-dimensional subspace spanned by $A|0\rangle$ and $U_f A|0\rangle$, these two reflections correspond to a rotation by angle 2θ. Thus, it should not be surprising to find eigenvalues $e^{\pm 2\pi i \theta}$ for states in this two-dimensional subspace. (In ❯ Sect. 4, a more general situation where we have an operator Q with many nontrivial eigenspaces and eigenvalues will be considered.)

Thus, one can approximate $\sin(\theta)$ by estimating the eigenvalue of either $|\psi_+\rangle$ or $|\psi_-\rangle$. Performing the standard eigenvalue estimation algorithm on Q with input $A|0\rangle$ (as illustrated in ❯ Fig. 1) gives a quadratic speed-up for estimating $\sin^2(\theta)$ versus simply repeatedly measuring $A|0\rangle$ and counting the frequency of 1s. In particular, we can obtain an estimate $\tilde{\theta}$ of θ such that $|\tilde{\theta} - \theta| < \varepsilon$ with high (constant) probability using $O(1/\varepsilon)$ repetitions of Q, and thus $O(1/\varepsilon)$ repetitions of U_f, A and A^{-1}. For fixed p, this implies that $\tilde{p} = \sin^2(\tilde{\theta})$ satisfies $|p - \tilde{p}| \in O(\varepsilon)$ with high probability. Classical sampling requires $O(1/\varepsilon^2)$ samples for the same precision. One application is speeding up the efficiency of the "Hadamard" test mentioned in ❯ Sect. 5.2.3.

Another interesting observation is that increasingly precise eigenvalue estimation of Q on input $A|0\rangle$ leaves the eigenvector register in a state that gets closer and closer to the mixture

◻ **Fig. 1**
The circuit estimates the amplitude with which the unitary operator A maps $|0\rangle$ to the subspace of solutions to $f(x) = 1$. One uses $m \in \Omega(\log(1/\varepsilon))$ qubits in the top register, and prepares it in a uniform superposition of the strings y representing the integers 0, 1, 2, ..., $2^m - 1$ (one can in fact optimize the amplitudes of each y to achieve a slightly better estimate (D'Ariano et al. 2007)). The controlled-Q^y circuit applies Q^y to the bottom register when the value y is encoded in the top qubits. If $A|0\rangle = \sin(\theta)|\psi_1\rangle + \cos(\theta)|\psi_0\rangle$, where $|\psi_1\rangle$ is a superposition of solutions x to $f(x) = 1$, and $|\psi_0\rangle$ is a superposition of values x with $f(x) = 0$, then the value $y = y_1 y_2 \ldots y_m$ measured in the top register corresponds to the phase estimate $2\pi y/2^m$, which is likely to be within $\frac{2\pi}{2^m}$ (modulo 2π) of either 2θ or -2θ. Thus, the value of $\sin^2 \pi y / 2^m$ is likely to satisfy $|\sin^2 \pi y / 2^m - \sin^2 \theta| \in O(\varepsilon)$.

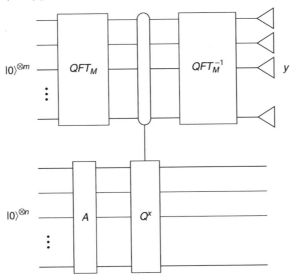

$\frac{1}{2}|\psi_+\rangle\langle\psi_+| + \frac{1}{2}|\psi_-\rangle\langle\psi_-|$ that equals $\frac{1}{2}|\psi_1\rangle\langle\psi_1| + \frac{1}{2}|\psi_0\rangle\langle\psi_0|$. Thus, eigenvalue estimation will leave the eigenvector register in a state that contains a solution to $f(\mathbf{x}) = 1$ with probability approaching $\frac{1}{2}$. One can in fact, using this entire algorithm as subroutine in another quantum search algorithm, obtain an algorithm with success probability approaching 1 (Mosca 2001; Kaye et al. 2007), and the convergence rate can be improved further (Tulsi et al. 2006). Another important observation is that for the purpose of searching, the eigenvalue estimate register is never used except in order to determine a random number of times in which to apply Q to the second register. This in fact gives the quantum searching algorithm of Boyer et al. (1998). In other words, applying Q a random number of times decoheres or approximately projects the eigenvector register in the eigenbasis, which gives a solution with high probability (❷ *Fig. 2*). The method of approximating projections by using randomized evolutions was recently refined, generalized, and applied in Boixo et al. (2009).

Quantum searching as discussed in this section has been further generalized in the quantum walk paradigm, which is the topic of the next section.

❏ Fig. 2
The amplitude estimation circuit can also be used for searching for a solution to $f(\mathbf{x}) = 1$, as illustrated in the figure on the left. The second register starts off in the state $A|0\rangle$, and after a sufficiently precise eigenvalue estimation of the quantum search iterate Q, the second register is approximately projected in the eigenbasis. The idea projection would yield the mixed state $\frac{1}{2}|\psi_+\rangle\langle\psi_+| + \frac{1}{2}|\psi_-\rangle\langle\psi_-| = \frac{1}{2}|\psi_1\rangle\langle\psi_1| + \frac{1}{2}|\psi_0\rangle\langle\psi_0|$, and thus yields a solution to $f(\mathbf{x}) = 1$ with probability approaching $\frac{1}{2}$ as m gets large (the probability is in $\Omega(1)$ once $m \in \Omega(1/\sqrt{p})$). The same algorithm can in fact be achieved by discarding the top register and instead randomly picking a value $y \in \{0, 1, \ldots, 2^m - 1\}$ and applying Q^y to $A|0\rangle$, as illustrated on the right.

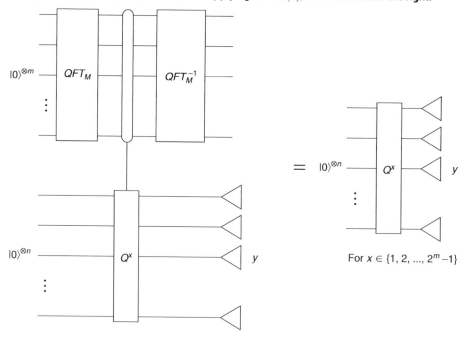

For $x \in \{1, 2, \ldots, 2^m - 1\}$

4 Quantum Walks

Random walks are a common tool throughout classical computer science. Their applications include the simulation of biological, physical, and social systems, as well as probabilistic algorithms such as Monte Carlo methods and the Metropolis algorithm. A classical random walk is described by a $n \times n$ matrix P, where the entry $P_{u,v}$ is the probability of a transition from a vertex v to an adjacent vertex u in a graph G. In order to preserve normalization, it is required that P is *stochastic* – that is, the entries in each column must sum to 1. We denote the initial probability distribution on the vertices of G by the column vector q. After n steps of the random walk, the distribution is given by $P^n q$.

Quantum walks were developed in analogy to classical random walks, but it was not initially obvious how to do this. Most generally, a quantum walk could begin in an initial state $\rho_0 = |\psi_0\rangle\langle\psi_0|$ and evolve according to any completely positive map $\mathscr{E}$ such that, after t time steps, the system is in state $\rho(t) = \mathscr{E}^t(\rho_0)$. Such a quantum walk is simultaneously using classical randomness and quantum superposition. We can focus on the power of quantum mechanics by restricting to unitary walk operations, which maintain the system in a coherent quantum state. So, the state at time t can be described by $|\psi(t)\rangle = U^t|\psi_0\rangle$, for some unitary operator U. However, it is not initially obvious how to define such a unitary operation. A natural idea is to define the state space A with basis $\{|v\rangle : v \in V(G)\}$ and walk operator $\tilde{P}$ defined by $\tilde{P}_{u,v} = \sqrt{P_{u,v}}$. However, this will not generally yield a unitary operator $\tilde{P}$, and a more complex approach is required. Some of the earliest formulations of unitary quantum walks appear in papers by Nayak and Vishwanath (2000), Ambainis et al. (2001), Kempe (2003), and Aharonov et al. (2001). These early works focused mainly on quantum walks on the line or a cycle. In order to allow unitary evolution, the state space consisted of the vertex set of the graph, along with an extra "coin register." The state of the coin register is a superposition of $|\text{LEFT}\rangle$ and $|\text{RIGHT}\rangle$. The walk then proceeds by alternately taking a step in the direction dictated by the coin register and applying a unitary "coin tossing operator" to the coin register. The coin tossing operator is often chosen to be the Hadamard gate. It was shown in Aharonov et al. (2001), Ambainis (2003), Ambainis et al. (2001), Kempe (2003), and Nayak and Vishwanath (2000) that the mixing and propagation behavior of these quantum walks was significantly different from their classical counterparts. These early constructions developed into the more general concept of a discrete time quantum walk, which will be defined in detail.

Two methods will be described for defining a unitary walk operator. In a *discrete time* quantum walk, the state space has basis vectors $\{|u\rangle \otimes |v\rangle : u, v \in V\}$. Roughly speaking, the walk operator alternately takes steps in the first and second registers. This is often described as a walk on the edges of the graph. In a *continuous time* quantum walk, the attention will be restricted to symmetric transition matrices P. We take P to be the Hamiltonian for the system. Applying Schrödinger's equation, this will define continuous time unitary evolution in the state space spanned by $\{|u\rangle : u \in V\}$. Interestingly, these two types of walk are not known to be equivalent. An overview of both types of walk as well as some of the algorithms that apply them will be given.

4.1 Discrete Time Quantum Walks

Let P be a stochastic matrix describing a classical random walk on a graph G. We would like the quantum walk to respect the structure of the graph G, and take into account the transition probabilities $P_{u,v}$. The quantum walk should be governed by a unitary operation, and is therefore reversible. However, a classical random walk is not, in general, a reversible process. Therefore, the quantum walk will necessarily behave differently than the classical walk. While the state space of the classical walk is V, the state quantum walk takes place in the space spanned by $\{|u, v\rangle : u, v \in V\}$. One can think of the first register as the current location of the walk, and the second register as a record of the previous location. To facilitate the quantum walk from a state $|u, v\rangle$, one can first mix the second register over the neighbors of u, and then swap the two registers. The method by which one mixes over the neighbors of u must be chosen carefully to ensure that it is unitary. To describe this formally, one can define the following states for each $u \in V$:

$$|\psi_u\rangle = |u\rangle \otimes \sum_{v \in V} \sqrt{P_{vu}}|v\rangle \tag{4}$$

$$|\psi_u^*\rangle = \sum_{v \in V} \sqrt{P_{vu}}|v\rangle \otimes |u\rangle \tag{5}$$

Furthermore, define the projections onto the space spanned by these states:

$$\Pi = \sum_{u \in V} |\psi_u\rangle\langle\psi_u| \tag{6}$$

$$\Pi^* = \sum_{u \in V} |\psi_u^*\rangle\langle\psi_u^*| \tag{7}$$

In order to mix the second register, we perform the reflection $(2\Pi - I)$. Letting S denote the swap operation, this process can be written as

$$S(2\Pi - I) \tag{8}$$

It turns out that we will get a more elegant expression for a single step of the quantum walk if we define the walk operator W to be two iterations of this process:

$$W = S(2\Pi - I)S(2\Pi - I) \tag{9}$$

$$= (2(S\Pi S) - I)(2\Pi - I) \tag{10}$$

$$= (2\Pi^* - I)(2\Pi - I) \tag{11}$$

So, the walk operator W is equivalent to performing two reflections.

Many of the useful properties of quantum walks can be understood in terms of the spectrum of the operator W. We define D, the $n \times n$ matrix with entries $D_{u,v} = \sqrt{P_{u,v}P_{v,u}}$. This is called the *discriminant matrix*, and has eigenvalues in the interval $[0, 1]$. In the theorem that follows, the eigenvalues of D that lie in the interval $(0, 1)$ will be expressed as $\cos(\theta_1), \ldots, \cos(\theta_k)$. Let $|\theta_1\rangle, \ldots, |\theta_k\rangle$ be the corresponding eigenvectors of D. Now, define the subspaces

$$A = \text{span}\{|\psi_u\rangle\} \tag{12}$$

$$B = \text{span}\{|\psi_u^*\rangle\} \tag{13}$$

Finally, define the operator

$$Q = \sum_{v \in V} |\psi_v\rangle\langle v| \tag{14}$$

and

$$\left|\phi_j\right\rangle = Q\left|\theta_j\right\rangle \tag{15}$$

The following spectral theorem for quantum walks can be now stated:

Theorem 1 (Szegedy 2004) *The eigenvalues of W acting on the space A + B can be described as follows:*

1. *The eigenvalues of W with non-zero imaginary part are $e^{\pm 2i\theta_1}, e^{\pm 2i\theta_2}, \ldots, e^{\pm 2i\theta_k}$ where $\cos(\theta_1), \ldots, \cos(\theta_k)$ are the eigenvalues of D in the interval (0, 1). The corresponding (un-normalized) eigenvectors of W(P) can be written as $\left|\phi_j\right\rangle - e^{\pm 2i\theta_j} S\left|\phi_j\right\rangle$ for j = 1, ..., k.*
2. *$A \cap B$ and $A^\perp \cap B^\perp$ span the +1 eigenspace of W. There is a direct correspondence between this space and the +1 eigenspace of D. In particular, the +1 eigenspace of W has the same degeneracy as the +1 eigenspace of D.*
3. *$A \cap B^\perp$ and $A^\perp \cap B$ span the −1 eigenspace of W.*

We say that P is *symmetric* if $P^T = P$ and *ergodic* if it is aperiodic. Note that if P is symmetric, then the eigenvalues of D are just the absolute values of the eigenvalues of P. It is well-known that if P is ergodic, then it has exactly one stationary distribution (i.e., a unique + 1 eigenvalue). Combining this fact with theorem (1) gives the following corollary.

Corollary 1 *If P is ergodic and symmetric, then the corresponding walk operator W has a unique +1 eigenvector in span(A,B):*

$$|\psi\rangle = \frac{1}{\sqrt{n}} \sum_{v \in V} |\psi_v\rangle = \frac{1}{\sqrt{n}} \sum_{v \in V} |\psi_v^*\rangle \tag{16}$$

Moreover, if we measure the first register of $|\psi\rangle$, we get a state corresponding to vertex u with probability

$$\Pr(u) = \frac{1}{n} \sum_{v \in V} P_{u,v} = \frac{1}{n} \tag{17}$$

This is the uniform distribution, which is the unique stationary distribution for the classical random walk.

4.1.1 The Phase Gap and the Detection Problem

In this section, an example of a quadratic speedup will be given for the problem of detecting whether there are any "marked" vertices in the graph G. First, the following are defined:

Definition 1 The *phase gap* of a quantum walk is defined as the smallest positive value 2θ such that $e^{\pm 2i\theta}$ are eigenvalues of the quantum walk operator. It is denoted by $\Delta(P)$.

Definition 2 Let $M \subseteq V$ be a set of marked vertices. In the *detection problem*, one is asked to decide whether M is empty.

Let P be ergodic and reversible, with positive eigenvalues. We define the following modified walk P':

$$P'_{uv} = \begin{cases} P_{uv} & v \notin M \\ 0 & u \neq v, v \in M \\ 1 & u = v, v \in M \end{cases} \tag{18}$$

This walk resembles P, except that it acts as the identity on the set M. That is, if the walk reaches a marked vertex, it stays there. Let P_M denote the operator P' restricted to $V \backslash M$. Then, arranging the rows and columns of P', we can write

$$P' = \begin{pmatrix} P_M & 0 \\ P'' & I \end{pmatrix} \tag{19}$$

By Theorem 1, if $M = \emptyset$, then $P_M = P' = P$ and $\|P_M\| = 1$. Otherwise, we have the strict inequality $\|P_M\| < 1$. The following theorem bounds $\|P_M\|$ away from 1:

Theorem 2 *If $(1 - \delta)$ is the absolute value of the eigenvalue of P with second largest magnitude, and $|M| \geq |\varepsilon|V|$, then $\|P_M\| \leq 1 - \frac{\delta\varepsilon}{2}$.*

It will be now shown that the detection problem can be solved using eigenvalue estimation. ❷ Theorem 2 will allow us to bound the running time of this method. First, we describe the discriminant matrix for P':

$$D(P')_{uv} = \begin{cases} P_{uv} & u, v \notin M \\ 1 & u = v, v \in M \\ 0 & \text{otherwise} \end{cases} \tag{20}$$

Now, beginning with the state

$$|\psi\rangle = \frac{1}{\sqrt{n}} \sum_{v \in V} |\psi_v\rangle = \frac{1}{\sqrt{n}} \sum_{v \in V} \sqrt{P_{v,u}} |u\rangle |v\rangle \tag{21}$$

we measure whether or not we have a marked vertex; if so, we are done. Otherwise, we have the state

$$|\psi_M\rangle = \frac{1}{\sqrt{|V \backslash M|}} \sum_{u, v \in V \backslash M} \sqrt{P_{vu}} |u\rangle |v\rangle \tag{22}$$

If $M = \emptyset$, then this is the state $|\psi\rangle$ defined in ❷ Eq. 16, and is the +1 eigenvector of $W(P)$. Otherwise, by Theorem 1, this state lies entirely in the space spanned by eigenvectors with values of the form $e^{\pm 2i\theta_j}$, where θ_j is an eigenvalue of P_M. Applying ❷ Theorem 2, we know that

$$\theta \geq \cos^{-1}\left(1 - \frac{\delta\varepsilon}{2}\right) \geq \sqrt{\frac{\delta\varepsilon}{2}} \tag{23}$$

So, the task of distinguishing between M being empty or nonempty is equivalent to that of distinguishing between a phase parameter of 0 and a phase parameter of at least $\sqrt{\frac{\delta\varepsilon}{2}}$. Therefore, applying phase estimation to $W(P')$ on state $|\psi_M\rangle$ with precision $O(\sqrt{\delta\varepsilon})$ will decide whether M is empty with constant probability. This requires time $O\left(\frac{1}{\sqrt{\delta\varepsilon}}\right)$.

By considering the modified walk operator P', it can be shown that the detection problem requires $O(\frac{1}{\delta\varepsilon})$ time in the classical setting. Therefore, the quantum algorithm provides a quadratic speedup over the classical one for the detection problem.

4.1.2 Quantum Hitting Time

Classically, the first hitting time is denoted $H(\rho, M)$. For a walk defined by P, starting from the probability distribution ρ on V, $H(\rho, M)$ is the smallest value n such that the walk reaches a marked vertex $v \in M$ at some time $t \in \{0, \ldots, n\}$ with constant probability. This idea is captured by applying the modified operator P' and some n times, and then considering the probability that the walk is in some marked state $v \in M$. Let ρ_M be any initial distribution restricted to the vertices $V \backslash M$. Then, at time t, the probability that the walk is in an unmarked state is $\|P_M^t \rho_M\|_1$, where $\| \cdot \|_1$ denotes the L_1 norm. Assuming that M is nonempty, we can see that $\|P_M\| < 1$. So, as $t \to \infty$, we have $\|P_M^t \rho_M\|_1 \to 0$. So, as $t \to \infty$, the walk defined by P' is in a marked state with probability 1. As a result, if we begin in the uniform distribution π on V, and run the walk for some time t, we will "skew" the distribution toward M, and thus away from the unmarked vertices. So, we define the classical hitting time to be the minimum t such that

$$\|P_M^t \rho_M\|_1 < \varepsilon \tag{24}$$

for any constant ε of our choosing. Since the quantum walk is governed by a unitary operator, it does not converge to a particular distribution the way that the classical walk does. We cannot simply wait an adequate number of time steps and then measure the state of the walk; the walk might have already been in a state with high overlap on the marked vertices and then evolved away from this state! Quantum searching has the same problem when the number of solutions is unknown. We can get around this in a similar way by considering an expected value over a sufficiently long period of time. This will form the basis of the definition of hitting time. Define

$$|\pi\rangle = \frac{1}{\sqrt{|V \backslash M|}} \sum_{v \in V \backslash M} |\psi_v\rangle \tag{25}$$

Then, if M is empty, $|\pi\rangle$ is a $+1$ eigenvector of W, and $W^t |\pi\rangle = |\pi\rangle$ for all t. However, if M is nonempty, then the spectral theorem tells that $|\pi\rangle$ lies in the space spanned by eigenvectors with eigenvalues $e^{\pm 2i\theta_j}$ for nonzero θ_j. As a result, it can be shown that, for some values of t, the state $W^t |\pi\rangle$ is "far" from the initial distribution $|\pi\rangle$. The quantum hitting is defined in the same way as Szegedy (2004). The hitting time $H_Q(W)$ as the minimum value T such that

$$\frac{1}{T+1} \sum_{t=0}^{T} \||W^t |\pi\rangle - |\pi\rangle\|^2 \geq 1 - \frac{|M|}{|V|} \tag{26}$$

This leads to Szegedy's hitting time theorem (2004).

Theorem 3 *The quantum hitting time $H_Q(W)$ is*

$$O\left(\frac{1}{\sqrt{1 - \|P_M\|}} \right)$$

Corollary 2 *Applying* ❷ *Theorem 2, if the second largest eigenvalue of P has magnitude $(1 - \delta)$ and $|M|/|V| \geq \varepsilon$, then $H_Q(W) \in O\left(\frac{1}{\sqrt{\delta\varepsilon}}\right)$.*

Notice that this corresponds to the running time for the algorithm for the detection problem, as described above. Similarly, the classical hitting time is in $O\left(\frac{1}{\delta\varepsilon}\right)$, corresponding to the best classical algorithm for the detection problem.

It is also worth noting that, if there are no marked elements, then the system remains in the state $|\pi\rangle$, and the algorithm never "hits." This gives an alternative way to approach the detection problem. We run the algorithm for a randomly selected number of steps $t \in \{0, \ldots, T\}$ with T of size $O(\sqrt{1/\delta\varepsilon})$, and then measure whether the system is still in the state $|\pi\rangle$; if there are any marked elements, then we can expect to find some other state with constant probability.

4.1.3 The Element Distinctness Problem

In the element distinctness problem, a black box is given that computes the function

$$f : \{1, \ldots, n\} \rightarrow S \qquad (27)$$

and we are asked to determine whether there exist $x, y \in \{1, \ldots, n\}$ with $x \neq y$ and $f(x) = f(y)$. We would like to minimize the number of queries made to the black box. There is a lower bound of $\Omega(n^{2/3})$ on the number of queries, due indirectly to Aaronson and Shi (2004). The algorithm of Ambainis (2004) proves that this bound is tight. The algorithm uses a quantum walk on the Johnson graph $J(n, p)$ has vertex set consisting of all subsets of $\{1, \ldots, n\}$ of size p. Let S_1 and S_2 be p-subsets of $\{1, \ldots, n\}$. Then, S_1 is adjacent to S_2 if and only if $|S_1 \cap S_2| = p - 1$. The Johnson graph $J(n, p)$ therefore has $\binom{n}{p}$ vertices, each with degree $p(n - p)$. The state corresponding to a vertex S of the Johnson graph will not only record which subset $\{s_1, \ldots, s_p\} \subseteq \{1, \ldots, n\}$ it represents, but the function values of those elements. That is,

$$|V\rangle = |s_1, s_2 \ldots, s_p; f(s_1), f(s_2), \ldots, f(s_p)\rangle \qquad (28)$$

Setting up such a state requires p queries to the black box.

The walk then proceeds for t iteration, where t is chosen from the uniform distribution on $\{0, \ldots, T\}$ and $T \in O(1/\sqrt{\delta\varepsilon})$. Each iteration has two parts to it. First, one needs to check if there are distinct s_i, s_j with $f(s_i) = f(s_j)$ – that is, whether the vertex S is marked. This requires no calls to the black box, since the function values are stored in the state itself. Second, if the state is unmarked, one needs to take a step of the walk. This involves replacing, say, s_i with s_i', requiring one query to erase $f(s_i)$ and another to insert the value $f(s_i')$. So, each iteration requires a total of 2 queries. ε and δ will be bounded now. If only one pair x, y exists with $f(x) = f(y)$, then there are $\binom{n-2}{p-2}$ marked vertices. This tells that, if there are any such pairs at all, epsilon is $\Omega(p^2/n^2)$. Johnson graphs are very well understood in graph theory. It is a well-known result that the eigenvalues for the associated walk operator are given by

$$\lambda_j = 1 - \frac{j(n+1-j)}{p(n-p)} \qquad (29)$$

for $0 \leq j \leq p$. For a proof, see Brouwer (1989). This gives $\delta = \frac{n}{p(n-p)}$. Putting this together, we find that $\sqrt{1/\delta\varepsilon}$ is $O\left(\frac{n}{\sqrt{p}}\right)$. So, the number of queries required is $O\left(p + \frac{n}{\sqrt{p}}\right)$. In order to minimize this quantity, we choose p to be of size $\Theta(n^{2/3})$. The query complexity of this algorithm is $O(n^{2/3})$, matching the lower bound of Aaronson and Shi (2004).

4.1.4 Unstructured Search as a Discrete Time Quantum Walk

Unstructured search in terms of quantum walks will be considered now. For unstructured search, we are required to identify a marked element from a set of size n. Let M denote the set of marked elements, and U denote the set of unmarked elements. Furthermore, let $m = |M|$ and $q = |U|$. It is assumed that the number of marked elements, m, is very small in relation to the total number of elements n. If this were not the case, we could efficiently find a marked element with high probability by simply checking a randomly chosen set of vertices. Since the set lacks any structure or ordering, the corresponding walk takes place on the complete graph on n vertices. One can define the following three states:

$$|UU\rangle = \frac{1}{q} \sum_{u,v \in U} |u, v\rangle \tag{30}$$

$$|UM\rangle = \frac{1}{\sqrt{nq}} \sum_{\substack{u \in U \\ v \in M}} |u, v\rangle \tag{31}$$

$$|MU\rangle = \frac{1}{\sqrt{nq}} \sum_{\substack{u \in M \\ v \in U}} |u, v\rangle \tag{32}$$

Noting that $\{|UU\rangle, |UM\rangle, |MU\rangle\}$ is an orthonormal set, the action of the walk operator on the three-dimensional space will be considered

$$\Gamma = \mathrm{span}\{|UU\rangle, |UM\rangle, |MU\rangle\} \tag{33}$$

In order to do this, we will express the spaces A and B, defined in ❷ Eqs. 12 and ❷ 13 in terms of a different basis. First we can label the unmarked vertices:

$$U = \{u_0, ..., u_{q-1}\} \tag{34}$$

Then, we can define

$$|\gamma_k\rangle = \frac{1}{\sqrt{q}} \sum_{j=0}^{q-1} e^{\frac{2\pi i j k}{q}} |\psi_{u_j}\rangle \tag{35}$$

and

$$|\gamma_k^*\rangle = \frac{1}{\sqrt{q}} \sum_{j=0}^{q-1} e^{\frac{2\pi i j k}{q}} |\psi_{u_j}^*\rangle \tag{36}$$

with k ranging from 0 to $q - 1$. Note that $|\gamma_0\rangle$ corresponds to the definition of $|\pi\rangle$ in ❷ Eq. 25. We can then rewrite A and B:

$$A = \mathrm{span}\ \{|\gamma_k\rangle\} \tag{37}$$

$$B = \mathrm{span}\ \{|\gamma_k^*\rangle\} \tag{38}$$

Now, note that for $k \neq 0$, the space Γ is orthogonal to $|\gamma_k\rangle$ and $|\gamma_k^*\rangle$ Furthermore,

$$|\gamma_0\rangle = \frac{1}{\sqrt{n}} \left(\sqrt{m}|UM\rangle + \sqrt{q}|UU\rangle\right) \tag{39}$$

and

$$|\gamma_0^*\rangle = \frac{1}{\sqrt{n}} \left(\sqrt{m}|MU\rangle + \sqrt{q}|UU\rangle\right) \tag{40}$$

Therefore, the walk operator

$$W = (2\Pi^* - I)(2\Pi - I) \tag{41}$$

when restricted to Γ is simply

$$W' = \left(2|\gamma_0^*\rangle\langle\gamma_0^*| - I\right)\left(2|\gamma_0\rangle\langle\gamma_0| - I\right) \tag{42}$$

❷ *Figure 3* illustrates the space Γ that contains the vectors $|\gamma_0\rangle$ and $|\gamma_k^*\rangle$.

At this point, it is interesting to compare this algorithm to Grover's search algorithm. It can be defined that

$$|\rho_t\rangle = W^t|UU\rangle. \tag{43}$$

Now, define $|\rho_t^1\rangle$ and $|\rho_t^2\rangle$ to be the projection of $|\rho_t\rangle$ onto span$\{|UU\rangle,|UM\rangle\}$ and span$\{|UU\rangle,|MU\rangle\}$, respectively. We can think of these as the "shadow" cast by the vector $|\rho_t\rangle$ on two-dimensional planes within the space Γ. Note that $|\gamma_0\rangle$ lies in the span$\{|UU\rangle,|UM\rangle\}$ and its projection onto span$\{|UU\rangle,|MU\rangle\}$ is $\sqrt{q/n}|UU\rangle$. So, the walk operator acts on $|\rho_t\rangle$ by reflecting it around the vector γ_0 and then around $\sqrt{q/n}|UU\rangle$. This is very similar to Grover search, except for the fact that the walk operator in this case does not preserve the magnitude of $|\rho_t^1\rangle$. So, with each application of the walk operator, $|\rho_t^1\rangle$ is rotated by 2θ, where $\theta = \tan^{-1}(\sqrt{m/q})$ is the angle between $|UU\rangle$ and $|\gamma_0\rangle$. The case for $|\rho_t^2\rangle$ is exactly analogous, with the rotation taking place in the plane span$\{|UU\rangle,|MU\rangle\}$. So, we can think of the quantum walk as Grover search taking place simultaneously in two intersecting planes. It should not come as a surprise, then, that we can achieve a quadratic speedup similar to that of Grover search.

A slightly modified definition of hitting time will be now used to show that the walk can indeed be used to find a marked vertex in $O(\sqrt{n})$ time. Rather than using the hitting time as defined in ❷ Sect. 4.1.2, the state $|\pi\rangle = |\gamma_0\rangle$ will be replaced with $|UU\rangle$. Note that we can create $|UU\rangle$ from $|\pi\rangle$ by simply measuring whether the second register contains a marked vertex. This is only a small adjustment, since $|\gamma_0\rangle$ is close to $|UU\rangle$. Furthermore, both lie in the space Γ, and the action of the walk operator is essentially identical for both starting states.

◻ **Fig. 3**

The three-dimensional space Γ. The walk operator acts on this space by alternately performing reflections in the vector $|\gamma_0\rangle$ and $|\gamma_k^*\rangle$.

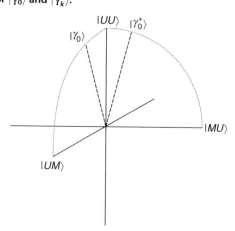

It should not be surprising to the reader that the results of ❷ Sect. 4.1.2 apply to this modified definition as well.

In this case, the operator P is defined by

$$P_{uv} = \begin{cases} 0 & \text{if } u = v \\ \frac{1}{n-1} & \text{if } u \neq v \end{cases} \qquad (44)$$

Let $v_0, \ldots, v_{n-1}$ be a labeling of the vertices of G. Let $x^0, \ldots, x^{n-1}$ denote the eigenvectors of P, with $x^k_{v_j}$ denoting the amplitude on v_j in x^k. Then, the eigenvalues of P are as follows:

$$x^k_{v_j} = \frac{1}{n} \cdot e^{\frac{2\pi i j k}{n}} \qquad (45)$$

Then, x^0 has eigenvalue 1, and is the stationary distribution. All the other x^k have eigenvalue $-1/(n-1)$, giving a spectral gap of $\delta = \frac{n-2}{n-1}$. Applying corollary 2 gives the quantum hitting time for the corresponding quantum walk operator W

$$H_Q(W) \leq \sqrt{\frac{(n-1)n}{n-2}} = O(\sqrt{n}) \qquad (46)$$

So, we run the walk for some randomly selected time $t \in \{0, \ldots, T-1\}$ with T of size $O(\sqrt{n})$, then measure whether either the first or second register contains a marked vertex. Applying Theorem 3, the probability that neither contains a marked vertex is

$$\frac{1}{T} \sum_{t=0}^{T-1} |\langle \rho_t | UU \rangle|^2 \leq \frac{1}{T} \sum_{t=0}^{T-1} |\langle \rho_t | UU \rangle|$$

$$= 1 - \frac{1}{2T} \sum_{t=0}^{T-1} \| |\rho_t\rangle - |UU\rangle \|^2$$

$$\leq 1 - \frac{|M|}{2|V|}$$

Assuming that $|M|$ is small compared to $|V|$, we find a marked vertex in either the first or second register with high probability. Repeating this procedure a constant number of times, the success probability can be forced arbitrarily close to 1.

4.1.5 The MNRS Algorithm

Magniez et al. (2007) developed an algorithm that generalizes the search algorithm described above to any graph. A brief overview of this algorithm and others is also given in the survey paper by Santha (2008). The MNRS algorithm employs similar principles to Grover's algorithm; we apply a reflection in M, the space of unmarked states, followed by a reflection in $|\pi\rangle$, a superposition of marked and unmarked states. This facilitates a rotation through an angle related to the number of marked states. In the general case considered by Magniez et al., $|\pi\rangle$ is the stationary distribution of the walk operator. It turns out to be quite difficult to implement a reflection in $|\pi\rangle$ exactly. Rather, the MNRS algorithm employs an approximate version of this reflection. This algorithm requires $O\left(\frac{1}{\sqrt{\delta\varepsilon}}\right)$ applications of the walk operator, where δ is the eigenvalue gap of the operator P, and ε is the proportion vertices that are marked. In his survey paper, Santha (2008) outlines some applications of the MNRS algorithm, including a version of the element distinctness problem where we are asked to find elements x and y such that $f(x) = f(y)$.

4.2 Continuous Time Quantum Walks

To define a classical continuous time random walk for a graph with no loops, we define a matrix similar to the adjacency matrix of G, called the *Laplacian*:

$$L_{uv} = \begin{cases} 0 & u \neq v, \ uv \notin E \\ 1 & u \neq v, \ uv \in E \\ -\deg(v) & u = v \end{cases} \tag{47}$$

Then, given a probability distribution $p(t)$ on the vertices of G, the walk is defined by

$$\frac{\mathrm{d}}{\mathrm{d}t} p(t) = L p(t) \tag{48}$$

Using the Laplacian rather than the adjacency matrix ensures that $p(t)$ remains normalized. A continuous time quantum walk is defined in a similar way. For simplicity, it will be assumed that the Laplacian is symmetric, although it is still possible to define the walk in the asymmetric case. Then, since the Laplacian is Hermitian, we can simply take it to be the Hamiltonian of our system. Letting $|\rho(t)\rangle$ be a normalized vector in $\mathbf{C}^V$, Schrödinger's equation gives

$$i \frac{\mathrm{d}}{\mathrm{d}t} |\rho(t)\rangle = L |\rho(t)\rangle \tag{49}$$

Solving this equation, we get an explicit expression for $\rho(t)$:

$$|\rho(t)\rangle = e^{-iLt} |\rho(0)\rangle \tag{50}$$

Let $\{|\lambda_1\rangle \ldots |\lambda_n\rangle\}$ be the eigenvectors of L with corresponding values $\{\lambda_1, \ldots, \lambda_n\}$. We can rewrite the expression for $\rho(t)$ in terms of this basis as follows:

$$|\rho(t)\rangle = \sum_{j=1}^{n} e^{-i\lambda_j t} \langle \lambda_j | \rho_0 \rangle |\lambda_j\rangle \tag{51}$$

Clearly, the behavior of a continuous time quantum walk is very closely related to the eigenvectors and spectrum of the Laplacian.

Note that we are not required to take the Laplacian as the Hamiltonian for the walk. We could take any Hermitian matrix we like including the adjacency matrix, or the transition matrix of a (symmetric) Markov chain.

4.2.1 A Continuous Time Walk for Unstructured Search

For unstructured search, the walk takes place on a complete graph. First, the following two states are defined:

$$|V\rangle = \frac{1}{\sqrt{|V|}} \sum_{v \in V} |v\rangle \tag{52}$$

$$|M\rangle = \frac{1}{\sqrt{|M|}} \sum_{v \in M} |v\rangle \tag{53}$$

The Hamiltonian that will be used is a slightly modified version of the Laplacian for the complete graph, with an extra "marking term" added:

$$H = |V\rangle\langle V| + |M\rangle\langle M| \tag{54}$$

It is convenient to consider the action of this Hamiltonian in terms of the vectors $|M\rangle$ and $|M^\perp\rangle$, where

$$|M^\perp\rangle = \frac{1}{\sqrt{|V\setminus M|}}\sum_{v\in V\setminus M}|v\rangle = \frac{1}{\sqrt{1-\langle M|V\rangle^2}}(|V\rangle - |M\rangle\langle M|V\rangle) \tag{55}$$

As outlined in Childs (2008), we let $\alpha = \langle M|V\rangle$. We can then rewrite the Hamiltonian in terms of the basis $\{|M\rangle, |M^\perp\rangle\}$:

$$H = \begin{pmatrix} \alpha^2 & \alpha\sqrt{1-\alpha^2} \\ \alpha\sqrt{1-\alpha^2} & 1-\alpha^2 \end{pmatrix} + \begin{pmatrix} 1 & 0 \\ 0 & 0 \end{pmatrix} \tag{56}$$

$$= \begin{pmatrix} 1 & 0 \\ 0 & 1 \end{pmatrix} + \alpha^2\begin{pmatrix} 1 & 0 \\ 0 & -1 \end{pmatrix} + \alpha\sqrt{1-\alpha^2}\begin{pmatrix} 0 & 1 \\ 1 & 0 \end{pmatrix} \tag{57}$$

$$= I + \alpha^2\sigma_Z + \alpha\sqrt{1-\alpha^2}\sigma_X \tag{58}$$

$$= I + \alpha(\alpha\sigma_Z + \sqrt{1-\alpha^2}\sigma_X) \tag{59}$$

where σ_X and σ_Z are the Pauli X and Z operators. Note that the identity term in the sum simply introduces a global phase, and can be ignored. Note that the operator

$$A = \alpha\sigma_Z + \sqrt{1-\alpha^2}\sigma_X \tag{60}$$

has eigenvalues ± 1, and is therefore Hermitian and unitary. Therefore, we can write

$$e^{-iHt} = e^{-iA\alpha t} = \cos(\alpha t)I - i\sin(\alpha t)A \tag{61}$$

Also note that

$$A|V\rangle = (\alpha\sigma_Z + \sqrt{1-\alpha^2}\sigma_X)(\alpha|M\rangle + \sqrt{1-\alpha^2}|M^\perp\rangle) \tag{62}$$

$$= |M\rangle \tag{63}$$

If we start with the state $|V\rangle$, we can now calculate the state of the system at time t:

$$|\rho(t)\rangle = \cos(\alpha t)|V\rangle - i\sin(\alpha t)A|V\rangle \tag{64}$$

$$= \cos(\alpha t)|V\rangle - i\sin(\alpha t)|M\rangle \tag{65}$$

So, at time t, the probability of finding the system in a state corresponding to a marked vertex is

$$\sum_{v\in M}|\langle v|\rho(t)\rangle|^2 = |M|\left(\frac{\cos^2(\alpha t)}{|V|} + \frac{\sin^2(\alpha t)}{|M|}\right) \tag{66}$$

$$= \alpha^2\cos^2(\alpha t) + \sin^2(\alpha t) \tag{67}$$

At time $t = 0$, this is the same as sampling from the uniform distribution, as we would expect. However, at time $t = \pi/2\alpha = \pi/2 \cdot \sqrt{N/M}$, we observe a marked vertex with probability 1. Therefore, this search algorithm runs in time $O(\sqrt{N/M})$, which coincides with the running time of Grover's search algorithm and the related discrete time quantum walk algorithm.

4.2.2 Mixing in Quantum Walks and the Glued Tree Traversal Problem

While continuous time quantum walks give a generic quadratic speedup over their classical counterparts, there are some graphs on which the quantum walk gives exponential speedup. One such example is the "glued trees" of Childs et al. (2003). In this example, the walk takes place on the following graph, obtained by taking two identical binary trees of depth d and joining their leaves using a random cycle, as illustrated in ❷ *Fig. 4*. Beginning at the vertex ENTRANCE, we would like to know how long we need to run the algorithm to reach EXIT. It is straightforward to show that classically this will take time exponential in d, the depth of the tree. Intuitively, this is because there are so many vertices and edges in the "middle" of the graph, that there is small probability of a classical walk "escaping" to the exit. It will then be proved that a continuous time quantum walk can achieve an overlap of $\Omega(P(d))$ with the EXIT vertex in time $O\left(\frac{1}{Q(d)}\right)$, where Q and P are polynomials.

Let V_s denote the set of vertices at depth s, so that $V_0 = \{\text{ENTRANCE}\}$ and $V_{2d+1} = \{\text{EXIT}\}$. Taking the adjacency matrix to be the Hamiltonian, the operator $U(t) = e^{-iHt}$ acts identically on the vertices in V_s for any s. Therefore, the states

$$|s\rangle = \frac{1}{\sqrt{|V_s|}} \sum_{v \in V_s} |v\rangle \qquad (68)$$

◘ Fig. 4
A small glued tree graph.

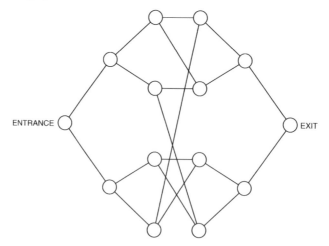

ENTRANCE

EXIT

form a convenient basis. We can therefore think of the walk on G as a walk on the states $\{|s\rangle : 0 \leq s \leq 2d + 1\}$. We also note that, if A is the adjacency matrix of G,

$$A|s\rangle = \begin{cases} \sqrt{2}|s+1\rangle & s = 0 \\ \sqrt{2}|s-1\rangle & s = 2d+1 \\ \sqrt{2}|s-1\rangle + 2|s+1\rangle & s = d \\ \sqrt{2}|s+1\rangle + 2|s-1\rangle & s = d+1 \\ \sqrt{2}|s+1\rangle + \sqrt{2}|s-1\rangle & \text{otherwise} \end{cases} \tag{69}$$

Aside from the exceptions at ENTRANCE, EXIT, and the vertices in the center, this walk looks exactly like the walk on a line with uniform transition probabilities.

Continuous time classical random walks will eventually converge to a limiting distribution. The time that it takes for the classical random walk to get "close" to this distribution is called the *mixing time*. More formally, we can take some small constant γ, and take the mixing time to be the amount of time it takes to come within γ of the stationary distribution, according to some metric. In general, we express the mixing time in terms of $1/\gamma$ and n, the number of vertices in the graph.

Since quantum walks are governed by a unitary operator, the same convergent behavior cannot be expected. However, the limiting behavior of quantum walks can be defined by taking an average over time. In order to do this, we define $\Pr(u,v,T)$. If we select $t \in [0,T]$ uniformly at random and run the walk for time t, beginning at vertex u, then $\Pr(u,v,T)$ is the probability that we find the system in state v. Formally, this can be written as

$$\Pr(u,v,T) = \frac{1}{T} \int_0^T |\langle v|e^{-iAt}|u\rangle|^2 dt \tag{70}$$

where A is the adjacency matrix of the graph. Now, if $\{|\lambda\rangle\}$ is taken to be the set of eigenvectors of A with corresponding eigenvalues $\{\lambda\}$, then $\Pr(u,v,T)$ can be rewritten as follows:

$$\Pr(u,v,T) = \frac{1}{T} \int_0^T \left| \sum_\lambda e^{-i\lambda t} \langle v|\lambda\rangle\langle\lambda|u\rangle \right|^2 dt \tag{71}$$

$$= \frac{1}{T} \int_0^T \sum_{\lambda,\lambda'} \left(e^{-i(\lambda-\lambda')t} \langle v|\lambda\rangle\langle\lambda|u\rangle\langle v|\lambda'\rangle\langle\lambda'|u\rangle \right) dt \tag{72}$$

$$= \sum_\lambda |\langle v|\lambda\rangle\langle\lambda|u\rangle|^2 \tag{73}$$

$$+ \frac{1}{T} \sum_{\lambda \neq \lambda'} \langle v|\lambda\rangle\langle\lambda|u\rangle\langle v|\lambda'\rangle\langle\lambda'|u\rangle \int_0^T e^{-i(\lambda-\lambda')t} dt \tag{74}$$

$$= \sum_\lambda |\langle v|\lambda\rangle\langle\lambda|u\rangle|^2 \tag{75}$$

$$+ \sum_{\lambda \neq \lambda'} \langle v|\lambda\rangle\langle\lambda|u\rangle\langle v|\lambda'\rangle\langle\lambda'|u\rangle \frac{1 - e^{-i(\lambda-\lambda')T}}{i(\lambda-\lambda')T} \tag{76}$$

In particular, we have

$$\lim_{T \to \infty} \mathrm{Pr}(u, v, T) = \sum_{\lambda} |\langle v|\lambda\rangle\langle\lambda|u\rangle|^2 \tag{77}$$

This value will be denoted by $\mathrm{Pr}(u,v,\infty)$. This is the quantum analogue of the limiting distribution for a classical random walk. We would now like to apply this to the specific case of the glued tree traversal problem. First, we will lower bound $\mathrm{Pr}(\text{ENTRANCE, EXIT}, \infty)$. It will be then shown that $\mathrm{Pr}(\text{ENTRANCE, EXIT}, T)$ approaches this value rapidly as T is increased, implying that we can traverse the glued tree structure efficiently using a quantum walk.

Define the reflection operator Θ by

$$\Theta|j\rangle = |2d - 1 - j\rangle \tag{78}$$

This operator commutes with the adjacency matrix, and hence the walk operator because of the symmetry of the glued trees. This implies that Θ can be diagonalized in the eigenbasis of the walk operator e^{-iAt} for any t. What is more, the eigenvalues of Θ are ± 1. As a result, if $|\lambda\rangle$ is an eigenvalue of e^{-iAt}, then

$$\langle\lambda|\text{ENTRANCE}\rangle = \pm\langle\lambda|\text{EXIT}\rangle \tag{79}$$

This can be applied to ❯ Eq. 77, yielding

$$\mathrm{Pr}(\text{ENTRANCE, EXIT}, \infty) = \sum_{\lambda} |\langle\text{ENTRANCE}|\lambda\rangle|^4 \tag{80}$$

$$\geq \frac{1}{2d + 2} \sum_{\lambda} |\langle\text{ENTRANCE}|\lambda\rangle|^2 \tag{81}$$

$$= \frac{1}{2d + 2} \tag{82}$$

Now, we need to determine how quickly $\mathrm{Pr}(\text{ENTRANCE, EXIT}, T)$ approaches $\mathrm{Pr}(\text{ENTRANCE, EXIT}, \infty)$ as we increase T:

$$|\mathrm{Pr}(\text{ENTRANCE, EXIT}, T) - \mathrm{Pr}(\text{ENTRANCE, EXIT}, \infty)| \tag{83}$$

$$= \left|\sum_{\lambda\neq\lambda'} \langle\text{EXIT}|\lambda\rangle\langle\lambda|\text{ENTRANCE}\rangle\langle\text{EXIT}|\lambda'\rangle\langle\lambda'|\text{ENTRANCE}\rangle \frac{1 - e^{-i(\lambda-\lambda')T}}{i(\lambda - \lambda')T}\right| \tag{84}$$

$$\leq \sum_{\lambda\neq\lambda'} |\langle\lambda|\text{ENTRANCE}\rangle|^2 |\langle\lambda'|\text{ENTRANCE}\rangle|^2 \frac{1 - e^{-i(\lambda-\lambda')T}}{i(\lambda - \lambda')T} \tag{85}$$

$$\leq \frac{2}{T\delta} \tag{86}$$

where δ is the difference between the smallest gap between any two distinct eigenvalues of A. As a result, we get

$$\mathrm{Pr}(\text{ENTRANCE, EXIT}, T) \geq \frac{1}{2d - 1} - \frac{2}{T\delta} \tag{87}$$

Childs et al. (2003) show that δ is $\Omega(1/d^3)$. Therefore, if we take T of size $O(d^4)$, we get success probability $O(1/d)$. Repeating this process, we can achieve an arbitrarily high probability of success in time polynomial in d – an exponential speedup over the classical random walk.

4.2.3 AND-OR **Tree Evaluation**

AND-OR trees arise naturally when evaluating the value of a two player combinatorial game. The players will be called P_0 and P_1. The positions in this game are represented by nodes on a tree. The game begins at the root, and players alternate, moving from the root toward the leaves of the tree. For simplicity, we assume that a binary tree is being dealt with; that is, for each move, the players have exactly two moves. While the algorithm can be generalized to any approximately balanced tree, but the binary case is considered for simplicity. We also assume that every game lasts some fixed number of turns d, where a turn consists of one player making a move. The total number of leaf nodes is denoted by $n = 2^d$. We can label the leaf nodes according to which player wins if the game reaches each node; they are labeled with a 0 if P_0 wins, and a 1 if P_1 wins. We can then label each node in the graph by considering its children. If a node corresponds to P_0's turn, we take the AND of the children. For P_1's move, we take the OR of the children. The value at the root node tells which player has a winning strategy, assuming perfect play.

Now, since

$$\text{AND}(x_1, \ldots, x_k) = \text{NAND}(\text{NAND}(x_1, \ldots, x_k)) \tag{88}$$

$$\text{OR}(x_1, \ldots, x_k) = \text{NAND}(\text{NAND}(x_1), \ldots, \text{NAND}(x_k)) \tag{89}$$

we can rewrite the AND-OR tree using only NAND operations. Furthermore, since

$$\text{NOT}(x_1) = \text{NAND}(x_1) \tag{90}$$

rather than label leaves with a 1, we can insert an extra node below the leaf and label it 0. Also, consecutive NAND operations cancel out, and will be removed. This is illustrated in ❷ *Fig. 5*. Note that the top-most node will be omitted from now on, and the shaded node will be referred to as the ROOT node. Now, the entire AND-OR tree is encoded in the structure of the NAND tree. We will write NAND(v) to denote the NAND of the value obtained by evaluating the tree up to a vertex v. The value of the tree is therefore NAND(ROOT).

The idea of the algorithm is to use the adjacency matrix of the NAND tree as the Hamiltonian for a continuous time quantum walk. It will be shown that the eigenspectrum of this walk operator is related to the value of NAND(ROOT). This idea first appeared in a paper of Farhi et al. (2007), and was later refined by Ambainis et al. (2007). We will begin with the following lemma:

Lemma 1 *Let $|\lambda\rangle$ be an eigenvector of A with value λ. Then, if v is a node in the NAND tree with parent node p and children C,*

$$\langle p|\lambda\rangle = -\lambda\langle v|\lambda\rangle + \sum_{c \in C}\langle c|\lambda\rangle \tag{91}$$

Therefore, if A has an eigenvector with value 0, then

$$\langle p|\lambda_0\rangle = \sum_{c \in C}\langle c|\lambda_0\rangle \tag{92}$$

□ Fig. 5

An AND-OR tree and the equivalent NAND tree, with example values. Note that the top-most node will be omitted from now on, and the shaded node in the NAND tree will be referred to as the ROOT node.

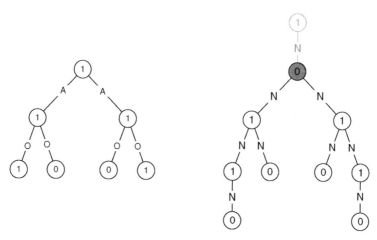

Using this fact and some inductive arguments, Ambainis et al. (2007) prove the following theorem:

Theorem 4 *If* NAND(ROOT) = 0, *then there exists an eigenvector* $|\lambda_0\rangle$ *of A with eigenvalue 0 such that* $|\langle \text{ROOT}|\lambda_0\rangle| \geq \frac{1}{\sqrt{2}}$. *Otherwise, if* NAND(ROOT) = 1, *then for any eigenvector* $|\lambda\rangle$ *of A with* $\langle\lambda|A|\lambda\rangle < \frac{1}{2\sqrt{n}}$, *we have* $\langle \text{ROOT}|\lambda_0\rangle = 0$.

This result immediately leads to an algorithm. We perform phase estimation with precision $O(1/\sqrt{n})$ on the quantum walk, beginning in the state $|\text{ROOT}\rangle$. If NAND(ROOT) = 0, get a phase of 0 with probability $\geq \frac{1}{2}$. If NAND(ROOT) = 1, we will never get a phase of 0. This gives a running time of $O(\sqrt{n})$. While attention has been restricted to binary trees here, this result can be generalized to any m-ary tree.

It is worth noting that unstructured search is equivalent to taking the OR of n variables. This is an AND-OR tree of depth 1, and n leaves. The $O(\sqrt{n})$ running time of Grover's algorithm corresponds to the running time for the quantum walk algorithm.

Classically, the running time depends not just on the number of leaves, but on the structure of the tree. If it is a balanced m-ary tree of depth d, then the running time is

$$O\left(\left(\frac{m-1+\sqrt{m^2+14m+1}}{4}\right)^d\right) \tag{93}$$

In fact, the quantum speedup is maximal when we have an n-ary tree of depth 1. This is just unstructured search, and requires $\Omega(n)$ time classically.

Reichardt and Spalek (2008) generalize the AND-OR tree problem to the evaluation of a broader class of logical formulas. Their approach uses *span programs*. A span program P consists of a set of target vector t, and a set of input vectors $\{v_1^0, v_1^1, \ldots, v_n^0, v_n^1,\}$ corresponding to logical literals $\{x_1, \overline{x_1}, \ldots, x_n, \overline{x_n}\}$. The program corresponds to a Boolean function $f_P: \{0, 1\}^n \to \{0, 1\}$ such that, for $\sigma \in \{0, 1\}^n$, $f(\sigma) = 1$ if and only if (Karchmer and Wigderson 1993)

$$\sum_{j=1}^{n} v_j^{\sigma_j} = t$$

Reichardt and Spalek outline the connection between span programs and the evaluation of logical formulas. They show that finding $\sigma \in f^{-1}(1)$ is equivalent to finding a zero eigenvector for a graph G_P corresponding to the span program P. In this sense, the span program approach is similar to the quantum walk approach of Childs et al. – both methods evaluate a formula by finding a zero eigenvector of a corresponding graph.

5 Tensor Networks and Their Applications

A tensor network consists of an underlying graph G, with an algebraic object called a *tensor* assigned to each vertex of G. The value of the tensor network is calculated by performing a series of operations on the associated tensors. The nature of these operations is dictated by the structure of G. At their simplest, tensor networks capture basic algebraic operations, such as matrix multiplication and the scalar product of vectors. However, their underlying graph structure makes them powerful tools for describing combinatorial problems as well. Two such examples will be explored – the Tutte polynomial of a planar graph, and the partition function of a statistical mechanical model defined on a graph. As a result, the approximation algorithm that will be described below for the value of a tensor network is implicitly an algorithm for approximating the Tutte polynomial, as well as the partition function for these statistical mechanics models. We begin by defining the notion of a tensor. We then outline how these tensors and tensor operations are associated with an underlying graph structure. For a more detailed account of this algorithm, the reader is referred to Arad and Landau (2008). Finally, we will describe the quantum approximation algorithm for the value of a tensor network, as well as the applications mentioned above.

5.1 Tensors: Basic Definitions

Tensors are formally defined as follows:

Definition 3 A tensor M of rank m and dimension q is an element of $\mathbf{C}^{q^m}$. Its entries are denoted by $M_{j_1, j_2, \ldots, j_m}$, where $0 \leq j_k \leq q - 1$ for all j_k.

Based on this definition, a vector is simply a tensor of rank 1, while a square matrix is a tensor of rank 2. Several operations on tensors, which will generalize many familiar operations from linear algebra, will be now defined.

Definition 4 Let M and N be two tensors of dimension q and rank m and n, respectively. Then, their product, denoted $M \otimes N$, is a rank $m + n$ tensor with entries

$$(M \otimes N)_{j_1, \ldots, j_m, k_1, \ldots, k_n} = M_{j_1, \ldots, j_m} \cdot N_{k_1, \ldots, k_n} \tag{94}$$

This operation is simply the familiar tensor product. While the way that the entries are indexed is different, the resulting entries are the same.

Definition 5 Let M be a tensor of rank m and dimension q. Now, take a and b with $1 \le a < b \le m$. The contraction of M with respect to a and b is a rank $m - 2$ tensor N defined as follows:

$$N_{j_1,\dots,j_{a-1},j_{a+1},\dots,j_{b-1},j_{b+1},\dots,j_m} = \sum_{k=0}^{q-1} M_{j_1,\dots,j_{a-1},k,j_{a+1},\dots,j_{b-1},k,j_{b+1},\dots,j_m} \qquad (95)$$

One way of describing this operation is that each entry in the contracted tensor is given by summing along the "diagonal" defined by a and b. This operation generalizes the partial trace of a density operator. The density operator of two-qubit system can be thought of as a rank 4 tensor of dimension two. Tracing out the second qubit is then just taking a contraction with respect to 3 and 4. It is also useful to consider the combination of these two operations.

Definition 6 If M and N are two tensors of dimension q and rank m and n, then for a and b with $1 \le a \le m$ and $1 \le b \le n$, the contraction of M and N is the result of contracting the product $M \otimes N$ with respect to a and $m + b$.

We now have the tools to describe a number of familiar operations in terms of tensor operations. For example, the inner product of two vectors can be expressed as the contraction of two rank 1 tensors. Matrix multiplication is just the contraction of 2 rank 2 tensors M and N with respect to the second index of M and the first index of N. Finally, if a Hilbert space $H = \mathbf{C}^q$ is taken, then we can identify a tensor M of dimension q and rank m with a linear operator $M^{s,t} : H^{\otimes t} \to H^{\otimes s}$ where $s + t = m$:

$$M^{s,t} = \sum_{j_1,\dots,j_m} M_{j_1,\dots,j_m} |j_1\rangle \otimes \dots \otimes |j_s\rangle \langle j_{s+1}| \otimes \dots \otimes \langle j_m| \qquad (96)$$

This correspondence with linear operators is essential to understanding tensor networks and their evaluation.

5.2 The Tensor Network Representation

A tensor network $T(G, \mathbf{M})$ consists of a graph $G = (V, E)$ and a set of tensors $\mathbf{M} = \{M[v] : v \in V\}$. A tensor is assigned to each vertex, and the structure of the graph G encodes the operations to be performed on the tensors. It can be said that the *value* of a tensor network is the tensor that results from applying the operations encoded by G to the set of tensors $\mathbf{M}$. When the context is clear, we will let $T(G, \mathbf{M})$ denote the value of the network. In addition to the typical features of a graph, G is allowed to have *free edges* – edges that are incident with a vertex on one end, but are unattached at the other. For simplicity, it will be assumed that all of the tensors in $\mathbf{M}$ have the same dimension q. It is also required that deg (v), the degree of vertex v, is equal to the rank of the associated tensor $M[v]$. If G consists of a single vertex v with m free edges, then $T(G, \mathbf{M})$ represents a single tensor of rank m. Each index of M is associated with a particular edge. It will often be convenient to refer to the edge associated with an index a as e_a. A set of rules that will allow us to construct a tensor network corresponding to any sequence of tensor operations will be defined now:

- If M and N are the value of two tensor networks $T(G, \mathbf{M})$ and $T(H, \mathbf{N})$, then taking the product $M \otimes N$ is equivalent to taking the disjoint union of the two tensor networks, $T(G \cup H, \mathbf{M} \cup \mathbf{N})$.
- If M is the value of a tensor network $T(G, \mathbf{M})$, then contracting M with respect to a and b is equivalent to joining the free edges e_a and e_b in G.
- As a result, taking the contraction of M and N with respect to a and b corresponds to joining the associated edge e_a from $T(G, \mathbf{M})$ and e_b from $T(H, \mathbf{N})$.

Some examples are illustrated in ❯ *Fig. 6*. Applying these simple operations iteratively allows one to construct a network corresponding to any series of products and contractions of tensors. Note that the number of free edges in $T(G, \mathbf{M})$ is always equal to the rank of the corresponding tensor M. Tensor networks that have no free edges, and whose value is therefore a scalar, will be of particular interest.

The tensor network interpretation of a linear operator $M^{s,t}$ defined in Eq. 96 can now be considered. $M^{s,t}$ is associated with a rank $m = s + t$ tensor M. An equivalent tensor network could consist of a single vertex with m free edges $(e_1, \ldots, e_m)$. The operator $M^{s,t}$ acts on elements of $(\mathbf{C}^q)^{\otimes t}$, which are rank t tensors, and can therefore be represented by a single vertex with t free edges $(e'_1, \ldots, e'_t)$. The action of $M^{s,t}$ on an element $N \in (\mathbf{C}^q)^{\otimes t}$ is therefore represented by connecting e'_k to e_{s+k} for $k = 1, \ldots, t$. ❯ *Figure 7* illustrates the operator $M^{2,2}$ acting on a rank 2 tensor. Note that the resulting network has two free edges, corresponding to the rank of the resulting tensor.

◻ **Fig. 6**
The network representation of a tensor M of rank 4, and the contraction of two rank 4 tensors, M and N.

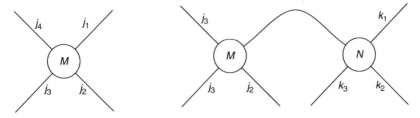

◻ **Fig. 7**
The network corresponding to the operator $M^{2,2}$ acting on a rank 2 tensor N.

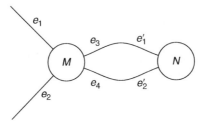

Now, the tensor networks that are of most interest – those with no free edges – will be reconsidered. We first consider the example of an inner product of two rank 1 tensors. This corresponds to a graph consisting of two vertices u and v joined by a single edge. The value of the tensor is then

$$T(G, \mathbf{M}) = \sum_{j=0}^{q} (M[u])_j \cdot (M[v])_j \qquad (97)$$

That is, it is a sum of q terms, each of which corresponds to the assignment of an element of $\{0, \ldots, q - 1\}$ to the edge (u, v). We can refer to this assignment as an q-edge coloring of the graph. In the same way, the value of a more complex tensor network is given by a sum whose terms correspond to q-edge colorings of G. Given a coloring of G, let $M[v]^\gamma = (M[v])_{i_1,\ldots,i_m}$ where $i_1, \ldots, i_m$ are the values assigned by γ to the edges incident at v. That is, a q-edge coloring specifies a particular entry of each tensor in the network. Then, we can rewrite the value of $T(G, \mathbf{M})$ as follows:

$$T(G, \mathbf{M}) = \sum_{\gamma} \left(\prod_{v \in V} M[v]^\gamma \right) \qquad (98)$$

where γ runs over all q-edge colorings of G. Thinking of the value of a tensor network as a sum over all q-edge colorings of G will prove useful when we consider applications of tensor networks to statistical mechanics models.

5.3 Evaluating Tensor Networks: the Algorithm

In this section, we will first show how to interpret a tensor network as a series of linear operators. We will then show how to apply these operators using a quantum circuit. Finally, we will see how to approximate the value of the tensor network using the Hadamard test.

5.3.1 Overview

In order to evaluate a tensor network $T(G, \mathbf{M})$, some structure is first given to it, by imposing an ordering on the vertices of G,

$$V(G) = \{v_1, \ldots, v_n\}$$

The sets $S_j = \{v_1, \ldots, v_j\}$ are further defined. This gives a natural way to depict the tensor network, which is shown in ❯ *Fig. 8*. The evaluation algorithm will "move" from left to right, applying linear operators corresponding to the tensors M_{v_i}. One can think of the edges in

■ **Fig. 8**
Part of the graph G, with an ordering on V (G).

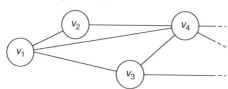

analogy to the "wires" of a quantum circuit diagram. To make this more precise, let v_i be a vertex of degree d. Let t be the number of vertices connecting v_i to S_{i-1} and s be the number of edges connecting v_i to $V(G) \backslash S_i$. Then, the operator that is applied at v_i is $M_{v_i}^{s,t}$. Letting j be the number of edges from S_{i-1} to $V(G) \backslash S_i$, we apply the identity operator $\mathbf{I}_j$ on the j indices that correspond to edges not incident at v_i. Combining these, we get an operator

$$N_i = M_{v_i}^{s,t} \otimes \mathbf{I}_j \tag{99}$$

Note that, taking the product of these operators, $\prod_{i=1}^{n} N_i$, gives the value of $T(G, \mathbf{M})$. However, there are a few significant details remaining. Most notably, the operators N_i are not, in general, unitary. In fact, they may not even correspond to square matrices. A method for converting the N_i into equivalent unitary operators will be now outlined.

5.3.2 Creating Unitary Operators

The first task is to convert the operators N_i into "square" operators – that is, operators whose domain and range have the same dimension. Referring to ❯ Eq. 99, the identity I_j is already square, so we need only modify $M_{v_i}^{s,t}$. In order to do this, some extra qubits will be added in the $|0\rangle$ state. We consider three cases:

1. If $s = t$, we define $\widetilde{M_{v_i}^{s,t}} = M_{v_i}^{s,t}$
2. If $s > t$, then we add $s - t$ extra qubits in the $|0\rangle$ state, and define $\widetilde{M_{v_i}^{s,t}}$ such that

$$\widetilde{M_{v_i}^{s,t}}\left(|\psi\rangle \otimes |0\rangle^{\otimes(s-t)}\right) = M_{v_i}^{s,t}|\psi\rangle \tag{100}$$

For completeness, we say that if the last $s - t$ qubits are not in the $|0\rangle$ state, $\widetilde{M_{v_i}^{s,t}}$ takes them to the 0 vector.

3. If $s < t$, then we define

$$\widetilde{M_{v_i}^{s,t}}|\psi\rangle = \left(M_{v_i}^{s,t}|\psi\rangle\right) \otimes |0\rangle^{\otimes(t-s)} \tag{101}$$

Finally, we define

$$\widetilde{N_i} = \widetilde{M_{v_i}^{s,t}} \otimes \mathbf{I}_j \tag{102}$$

which is a square operator. Now, to derive corresponding unitary operators, we make use of the following lemma:

Lemma 2 *Let M be a linear map $A: H^{\otimes t} \to H^{\otimes t}$, where H is a Hilbert space $H = \mathbf{C}^q$. Furthermore, let $G = \mathbf{C}^2$ be a spanned by $\{|0\rangle, |1\rangle\}$. Then, there exists a unitary operator U_M: $(H^{\otimes t} \otimes G) \to (H^{\otimes t} \otimes G)$ such that*

$$U_M(|\psi\rangle \otimes |0\rangle) = \frac{1}{\|M\|} M|\psi\rangle \otimes |0\rangle + |\phi\rangle \otimes |1\rangle \tag{103}$$

U_M can be implemented on a quantum computer in $\text{poly}(q^t)$ time.

A proof can be found in Arad and Landau (2008) as well as Aharonov et al. (2007). Applying this lemma, we can create n unitary operators $U_{\widetilde{N_j}}$ for $1 \le j \le n$, and define

$$U = \prod_{j=1}^{n} U_{\widetilde{N_j}} \tag{104}$$

It is easily verified that

$$\langle 0|^{\otimes r} U|0\rangle^{\otimes r} = \frac{T(G, \mathbf{M})}{\prod_j \|M_{v_j}^{s,t}\|} \tag{105}$$

where r is dependent on the number of vertices n as well as the structure of G and the ordering of the vertices $v_1, \ldots, v_n$.

5.3.3 The Hadamard Test

In order to approximate $\langle 0|^{\otimes r} U|0\rangle^{\otimes r}$, most authors suggest the *Hadamard test* – a well-known method for approximating the weighted trace of a unitary operator. First, we add an ancillary qubit that will act as a control register. We then apply the circuit outlined below, and measure the ancillary qubit in the computational basis.

It is not difficult to show that we measure

$$|0\rangle \text{ with probability } \frac{1}{2}(1 + \text{Re}\langle \psi|U|\psi\rangle)$$

$$|1\rangle \text{ with probability } \frac{1}{2}(1 - \text{Re}\langle \psi|U|\psi\rangle)$$

So, if one assigns a random variable X such that $X = 1$ when one measures $|0\rangle$ and $X = -1$ when one measures $|1\rangle$, then X has expected value $\text{Re}\langle \psi|U|\psi\rangle$. So, in order to approximate $\text{Re}\langle \psi|U|\psi\rangle$ to a precision of ε with constant probability $p > 1/2$, one requires $O(\varepsilon^{-2})$ repetitions of this protocol. In order to calculate the imaginary portion, we apply the gate

$$R = \begin{pmatrix} 1 & 0 \\ 0 & -i \end{pmatrix}$$

to the ancillary qubit, right after the first Hadamard gate. It is important to note that the Hadamard test gives an *additive approximation* of $\langle 0|^{\otimes r} U|0\rangle^{\otimes r}$. Additive approximations will be examined in ❯ Sect. 5.3.5.

5.3.4 Approximating $\langle 0|^{\otimes r} U|0\rangle^{\otimes r}$ Using Amplitude Estimation

In order to estimate $\langle 0|^{\otimes r} U|0\rangle^{\otimes r}$, most of the literature applies the Hadamard test (❯ *Fig. 9*). However, it is worth noting that we can improve the running time by using the process of *amplitude estimation*, as outlined in ❯ Sect. 3. Using the notation from ❯ Sect. 3, we have $U_f = U_0$ and $A = U$, the unitary corresponding to the tensor network $T(G, \mathbf{M})$. We begin in the state $U|0\rangle^{\otimes r}$, and the search iterate is

$$Q = -UU_0 U^{-1} U_0 \tag{106}$$

◻ Fig. 9

The quantum circuit for the Hadamard test.

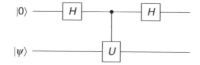

In order to approximate $\langle 0|^{\otimes r} U|0\rangle^{\otimes r}$ to a precision of ε, the running time is in $O(1/\varepsilon)$, a quadratic improvement over the Hadamard test.

5.3.5 Additive Approximation

Let $f: S \to \mathbf{C}$ be a function that we would like to evaluate. Then, an *additive approximation* A with approximation scale $\Delta: S \to \mathbf{R}$ is an algorithm that, on input $x \in S$ and parameter $\varepsilon > 0$, outputs $A(x)$ such that

$$\Pr(|A(x) - f(x)| \geq \varepsilon \Delta(x)) \leq c \tag{107}$$

for some constant c with $0 \leq c < 1/2$. If $\Delta(x)$ is $O(f(x))$, then A is a *fully polynomial randomized approximation scheme*, or FPRAS.

The approximation scale for the algorithm outlined above will be now determined. Using amplitude estimation, we estimate

$$\langle 0|^{\otimes r} U|0\rangle^{\otimes r} = \frac{T(G, \mathbf{M})}{\prod_j \|M_{v_j}^{s,t}\|} \tag{108}$$

to within ε with time requirement in $O(1/\varepsilon)$. However, the quantity we actually want to evaluate is $T(G, \mathbf{M})$, so we must multiply by $\prod_j \|M_{v_j}^{s,t}\|$. Therefore, the algorithm has approximation scale

$$\Delta(G, \mathbf{M}) = \prod_j \|M_{v_j}^{s,t}\| \tag{109}$$

We apply ❷ Lemma 2 to each vertex in G, and require $O(1/\varepsilon)$ repetitions of the algorithm to approximate $T(G, \mathbf{M})$ with the desired accuracy using amplitude estimation. This gives an overall running time

$$O\left(\text{poly}(q^{D(G)}) \cdot |V(G)| \cdot \varepsilon^{-1}\right) \tag{110}$$

where $D(G)$ is the maximum degree of any vertex in G. The algorithm is, therefore, an additive approximation of $T(G, \mathbf{M})$.

5.4 Approximating the Tutte Polynomial for Planar Graphs Using Tensor Networks

In 2000, Kitaev, Freedman, Larsen, and Wang (Freedman et al. 2000, 2001) demonstrated an efficient quantum simulation for topological quantum field theories. In doing so, they implied that the Jones polynomial can be efficiently approximated at $e^{2\pi i/5}$. This suggests that efficient quantum approximation algorithms might exist for a wider range of knot invariants and values. Aharonov et al. (2008) developed such a Jones polynomial for any complex value. Yard and Wocjan (2006) developed a related algorithm for approximating the HOMFLYPT polynomial of a braid closure. Both of these knot invariants are special cases of the Tutte polynomial. While the tensor network algorithm in ❷ Sect. 3.2 follows directly from a later paper of Aharonov et al. (2007), it owes a good deal to these earlier results for knot invariants.

A definition of the Tutte polynomial, as well as an overview of its relationship to tensor networks and the resulting approximation algorithm, will be given.

5.4.1 The Tutte Polynomial

The multivariate Tutte Polynomial is defined for a graph $G = (V, E)$ with edge weights $\mathbf{w} = \{w_e\}$ and variable q as follows:

$$Z_G(q, \mathbf{w}) = \sum_{A \subseteq E} q^{k(A)} \prod_{e \in A} w_e \tag{111}$$

where $k(A)$ denotes the number of connected components in the graph induced by A. The power of the Tutte polynomial arises from the fact that it captures nearly all functions on graphs defined by a *skein relation*. A skein relation is of the form

$$f(G) = x \cdot f(G/e) + y \cdot f(G \backslash e) \tag{112}$$

where G/e is created by contracting an edge e and $G \backslash e$ is created by deleting e. Oxley and Welsh (Welsh 1993) show that, with a few additional restrictions on f, if f is defined by a skein relation, then computing f can be reduced to computing Z_G. It turns out that many functions can be defined in terms of a skein relation, including

1. The partition functions of the Ising and Potts models from statistical physics
2. The chromatic and flow polynomials of a graph G
3. The Jones Polynomial of an alternating link

The exact evaluation of the Tutte polynomial, even when restricted to planar graphs, turns out to be #P-hard for all but a handful of values of q and $\mathbf{w}$. time FPRAS seems very unlikely for the Tutte polynomial of a general graph and any parameters q and $\mathbf{w}$. The interesting question seems to be *which* graphs and parameters admit an efficient and accurate approximation. By describing the Tutte polynomial as the evaluation of a Tensor network, an additive approximation algorithm for the Tutte polynomial is immediately produced. However, we do not have a complete characterization of the graphs and parameters for which this approximation is nontrivial.

5.4.2 The Tutte Polynomial as a Tensor Network

Given a planar graph G, and an embedding of G in the plane, we define the *medial graph* L_G. The vertices of L_G are placed on the edges of G. Two vertices of L_G are joined by an edge if they are adjacent on the boundary of a face of G. If G contains vertices of degree 2, then L_G will have multiple edges between some pair of vertices. Also note that L_G is a regular graph with valency 4. An example of a graph and its associated medial graph is depicted in ❷ *Fig. 10*.

In order to describe the Tutte polynomial in terms of a tensor network, we make use of the *generalized Temperley–Lieb algebra*, as defined in Aharonov et al. (2007). The basis elements of the Temperley–Lieb algebra $GTL(d)$ can be thought of as diagrams in which m upper pegs are joined to n lower pegs by a series of strands. The diagram must contain no crossings or loops. See ❷ *Fig. 11* for an example of such an element. Two basis elements are considered equivalent if they are isotopic to each other or if one can be obtained from the other by "padding" it on the right with a series of vertical strands. The algebra consists of all complex weighted sums of these basis elements. Given two elements of the algebra T_1 and T_2, we can take their product $T_2 \cdot T_1$ by simply placing T_2 on top of T_1 and joining the strands. Note that if the number of strands do not match, we can simply pad the appropriate element

☐ **Fig. 10**
A graph *G* and the resulting medial graph L_G.

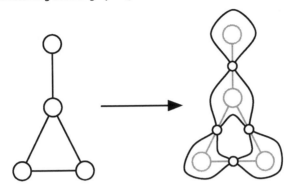

☐ **Fig. 11**
An element of the Temperley–Lieb algebra with six upper pegs and four lower pegs.

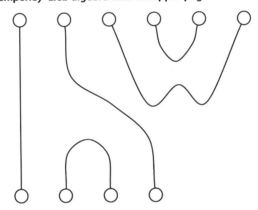

with some number of vertical strands. As a consequence, the identity element consists of any number of vertical strands. In composing two elements in this way, a loop may be created. Let us say that $T_1 \cdot T_2$ contains one loop. Then, if T_3 is the element created by removing the loop, we define $T_1 \cdot T_2 = dT_3$, where d is the complex parameter of $GTL(d)$.

We would also like this algebra to accommodate drawings with crossings. Let T_1 be such a diagram, and T_2 and T_3 be the diagrams resulting from "opening" the crossing in the two ways indicated in ❷ *Fig. 12*. Then, we define $T_1 = aT_2 + bT_3$, for appropriately defined a and b.

If we think of L_G as the projection of a knot onto the plane, where the vertices of L_G are the crossings of the knot, we see that L_G can be expressed in terms of the generalized Temperley–Lieb algebra. We would like to take advantage of this in order to map L_G to a series of tensors that will allow us to approximate the Tutte polynomial of G. Aharonov et al. define such a representation ρ of the Temperley–Lieb algebra. If $T \in GTL(d)$ is a basis element with m lower pegs and n upper pegs, then ρ is a linear operator such that

$$\rho(T) \colon H^{\otimes m+1} \to H^{\otimes n+1} \tag{113}$$

◘ **Fig. 12**

An element that contains a crossing is equated to a weighted sum of the elements obtained by "opening" the crossing.

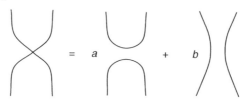

where $H = \mathbf{C}^k$ for some k, which depends on the particular embedding of G, such that

1. ρ preserves multiplicative structure. That is, if T_1 has m upper pegs and T_2 has m lower pegs, then

$$\rho(T_2 \cdot T_1) = \rho(T_2) \cdot \rho(T_1) \tag{114}$$

2. ρ is linear. That is,

$$\rho(\alpha T_1 + \beta T_2) = \alpha \rho(T_1) + \beta \rho(T_2) \tag{115}$$

For these purposes, k can be upper bounded by the number of edges $|E(L_G)| = 2|E(G)|$, and corresponds to the value r in ❷ Eq. 105. To represent L_G, each crossing (vertex) of L_G is associated with a weighted sum of two basis elements, and therefore is represented by a weighted sum of the corresponding linear operators, $\rho(T_1) = a\rho(T_2) + b\rho(T_3)$. The minima and maxima of L_G are also basis elements of $GTL(d)$, and can be represented as well. Finally, since L_G is a closed loop, and has no "loose ends", $\rho(L_G)$ must be a scalar multiple of the identity operator on $H = \mathbf{C}^k$. This relates well to our notion that the value of a tensor network with no loose ends is just a scalar.

We would like this scalar to be the Tutte polynomial of the graph G. In order to do this, we need to choose the correct values for a, b, and d. Aharonov et al. show how to choose these values so that the scalar is a graph invariant called the Kauffman bracket, from which we can calculate the Tutte polynomial for planar graphs. So, we can construct a tensor network whose value is the Kauffman bracket of L_G, and this gives the Tutte polynomial of G by the following procedure:

1. Construct the medial graph L_G and embed it in the plane.
2. Let L_G' be constructed by adding a vertex at each local minimum and maximum of L_G. These vertices are required in order to assign the linear operator associated with each minimum/maximum with a vertex in the tensor network.
3. Each vertex v of L_G' has a Temperley–Lieb element T_v associated with it. For some vertices, this is a crossing; for others, it is a local minimum or maximum. Assign the tensor $M[v] = \rho(T_v)$ to the vertex v.

The tensor network we are interested in is then $T(L_G', \mathbf{M})$, where $\mathbf{M} = \{M[v] : v \in V(L_G')\}$. Approximating the value of this tensor network gives an approximation of the Kauffman bracket of L_G, and therefore of the Tutte polynomial of G. Furthermore, it is known that L_G' has maximum degree 4. Finally, we will assume that $|E(G)| \geq |V(G)|$ giving a running time of

$$O\left(\text{poly}(q^4) \cdot |E(G)| \cdot \varepsilon^{-1}\right) \tag{116}$$

In this case, q depends on the particular embedding f L_G, but can be upper bounded by $|E|$. For more details on the representation ρ and quantum approximations of the Tutte polynomial as well as the related Jones polynomial, the reader is referred to the work of Aharonov et al., in particular Aharonov and Arad (2006) and Aharonov et al. (2007).

5.5 Tensor Networks and Statistical Mechanics Models

Many models from statistical physics describe how simple local interactions between microscopic particles give rise to macroscopic behavior. See Beaudin et al. (2008) for an excellent introduction to the combinatorial aspects of these models. The models that we are concerned with here are described by a weighted graph $G = (V, E)$. In this graph, the vertices represent particles, while edges represent an interaction between particles. A configuration σ of a q-state model is an assignment of a value from the set $\{0, \ldots, q - 1\}$ to each vertex of G. We denote the value assigned to v by σ_v. For each edge $e = (u, v)$, we define a local Hamiltonian $h_e(\sigma_u, \sigma_v) \in \mathbb{C}$. The Hamiltonian for the entire system is then given taking the sum:

$$H(\sigma) = \sum_{e=(u,v)} h_e(\sigma_u, \sigma_v) \tag{117}$$

The sum runs over all the edges of G. The *partition function* is then defined as

$$Z_G(\beta) = \sum_{\sigma} e^{-\beta H(\sigma)} \tag{118}$$

where $\beta = 1/kT$ is referred to as the *inverse temperature* and k is Boltzmann's constant. The partition function is critical to understanding the behavior of the system. First, the partition function allows one to determine the probability of finding the system in a particular configuration, given an inverse temperature β:

$$\Pr(\sigma') = \frac{e^{-\beta H(\sigma')}}{Z_G(\beta)} \tag{119}$$

This probability distribution is known as the Boltzmann distribution. Calculating the partition function also allows us to derive other properties of the system, such as entropy and free energy. An excellent discussion of the partition function from a combinatorial perspective can be found in Welsh (1993).

A tensor network whose value is the partition function Z_G will be now constructed. Given the graph G, we define the vertices of $G' = (V', E')$ as follows:

$$V' = V \cup \{v_e : e \in E\} \tag{120}$$

We are simply adding a vertex for each edge e of G, and identifying it by v_e. We then define the edge set of G':

$$E' = \{(x, v_e), (v_e, y) : (x, y) = e \in E\} \tag{121}$$

So, the end product G' resembles the original graph; vertices in the middle of each edge of the original graph have been simply added. The tensors that will be identified with each vertex will be defined separately for the vertex set V of the original graph and the set $V_E = \{v_e\}$ of vertices that are added to define G'. In each case, they will be dimension q tensors for a q-state model. For $v \in V$, $M[v]$ is an "identity" operator; that is, it takes on the value 1 when all indices are equal, and 0 otherwise:

$$(M[v])_{i_1,\dots,i_m} = \begin{cases} 1 & \text{if } i_1 = i_2 = \dots = i_m \\ 0 & \text{otherwise} \end{cases} \tag{122}$$

The vertices in V_E encode the actual interactions between neighboring particles. The tensor associated with v_e is

$$(M_{v_e})_{s,t} = e^{-\beta h_e(s,t)} \tag{123}$$

Now, one can consider the value of the tensor network in terms of q-edge colorings of G:

$$T(G', \mathbf{M}) = \sum_{\gamma} \left(\prod_{v \in V'} M[v]^{\gamma} \right) \tag{124}$$

where γ runs over all q-edge colorings of G'. Based on the definitions of the tensors $M[v]$ for $v \in V$, we see that $\prod_{v \in V'} M[v]^{\gamma}$ is nonzero only when the edges incident at each vertex are all colored identically. This restriction ensures that each nonzero term of the sum (❷ Eq. 124) corresponds to a configuration σ of the q-state model. That is, a configuration σ corresponds to a q-edge coloring γ of G' where each $\gamma(e) = \sigma_v$ whenever e is incident with $v \in V$. This gives the following equality:

$$\begin{aligned} T(G', \mathbf{M}) &= \sum_{\gamma} \left(\prod_{v \in V'} M[v]^{\gamma} \right) \\ &= \sum_{\sigma} \prod_{e=(u,v)} e^{-\beta h_e(\sigma_u, \sigma_v)} \\ &= \sum_{\sigma} e^{-\beta H(\sigma)} \\ &= Z_G(\beta) \end{aligned}$$

The time required to approximate the value of this tensor network will be now considered. Recall that the running time is given by

$$O\left(\text{poly}(q^{D(G')}) \cdot |V(G')| \cdot \varepsilon^{-1} \right) \tag{125}$$

First, it is observed that, if one assumes that G is connected, then $|V(G')|$ is $O(|E|)$. Restrictions are not placed on the maximum degree $D(G')$. The vertices of G' of high degree must be in V, since all vertices in V_E have degree 2. Now, the tensors assigned to vertices of V are just identity operators; they ensure that the terms of the sum from ❷ Eq. 124 are nonzero only when all the edges incident at v are colored identically. As a result, one can replace each vertex $v \in V$ with $\deg(v) > 3$ by $(\deg(v) - 2)$ vertices, each of degree 3. See ❷ Fig. 13 for an example. The tensor assigned to each of these new vertices is the identity operator of rank 3. It is not difficult to show that this replacement does not affect the value of the tensor network. It does, however, affect the number of vertices in the graph, adding a multiplicative factor to the running time, which can now be written

$$O\left(\text{poly}(q) \cdot \text{d}(G) \cdot |E| \cdot \varepsilon^{-1} \right) \tag{126}$$

For more details on the approximation scale of this algorithm, the reader is referred to Arad and Landau (2008).

□ **Fig. 13**
Replacing a vertex *v* by (deg(*v*) − 2) vertices.

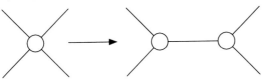

Arad and Landau also discuss a useful restriction of q-state models called *difference models*. In these models, the local Hamiltonians h_e depend only on the difference between the states of the vertices incident with e. That is, $h_e(\sigma_u, \sigma_v)$ is replaced by $h_e(|\sigma_u - \sigma_v|)$, where $|\sigma_u - \sigma_v|$ is calculated modulo q. Difference models include the well-known Potts, Ising and Clock models. Arad and Landau show that the approximation scale can be improved when attention is restricted to difference models.

Van den Nest et al. (Cuevas et al. 2008; Hübener et al. 2008) show that the partition function for difference models can be described as the overlap between two quantum states

$$Z_G = \langle \psi_G | \left(\bigotimes_{e \in E(G)} |\alpha_e\rangle \right) \tag{127}$$

Van den Nest uses the state $|\psi_G\rangle$ to encode the structure of the graph, while each state $|\alpha_e\rangle$ encodes the strength of the local interaction at e. The actual computation of this overlap is best described as a tensor network. While this description does not yield a computational speedup over the direct method described above, it is an instructive way to deconstruct the problem. In the case of planar graphs, it also relates the partition function of a graph G and its planar dual.

Geraci and Lidar (Geraci 2008; Geraci and Lidar 2008) show that the Potts partition may be efficiently and exactly evaluated for a class of graphs related to irreducible cyclic cocycle codes. While their algorithm does not employ tensor networks, it is interesting to consider the implications of their results in the context of tensor networks. Is there a class of tensor networks whose value can be efficiently and exactly calculated using similar techniques?

6 Conclusion

In this chapter, we reviewed quantum algorithms of several types: those based on the quantum Fourier transform, amplitude amplification, quantum walks, and evaluating tensor networks. This is by no means a complete survey of quantum algorithms; most notably absent are algorithms for simulating quantum systems. This is a particularly natural application of quantum computers, and these algorithms often achieve an exponential speedup over their classical counterparts.

Algorithms based on the quantum Fourier transform, as reviewed in ❷ Sect. 2, include some of the earliest quantum algorithms that give a demonstrable speedup over classical algorithms, such as the algorithms for Deutsch's problem and Simon's problem. Many problems in this area can be described as hidden subgroup problems. Although problems of this type have been well-studied, many questions remain open. For example, there is no efficient quantum algorithm known for most examples of non-Abelian groups. In particular, the quantum complexity of the graph isomorphism problem remains unknown.

Similarly, the family of algorithms based on amplitude amplification and estimation has a relatively long history. While Grover's original searching algorithm is the best-known example, this approach has been greatly generalized. For example, in Brassard et al. (2002) it is shown how the same principles applied by Grover to searching can be used to perform amplitude estimation and quantum counting. Amplitude amplification and estimation have become ubiquitous tools in quantum information processing, and are often employed as critical subroutines in other quantum algorithms.

Quantum walks provide an interesting analogue to classical random walks. While they are inspired by classical walks, quantum walks have many properties that set them apart. Some applications of quantum walks are natural and somewhat unsurprising; using a walk to search for a marked element is an intuitive idea. Others, such as evaluating AND-OR trees are much more surprising. Quantum walks remain an active area of research, with many open questions. For example, the relative computational capabilities of discrete and continuous time quantum walks are still not fully understood.

The approximation algorithm for tensor networks as described in this chapter (Arad and Landau 2008) is relatively new. However, it is inspired by the earlier work of Aharonov et al. (2008), as well as Wocjan and Yard (2006). The tensor network framework allows us to capture a wide range of problems, particularly problems of a combinatorial nature. Indeed, many #P-hard problems, such as evaluating the Tutte polynomial, can be captured as tensor networks. The difficulty arises from the additive nature of the approximation. One of the most important questions in this area is when this approximation is useful, and when the size of the approximation window renders the approximation trivial.

Numerous other families of new quantum algorithms that are not elaborated in this chapter have also been briefly mentioned, and many others unfortunately go unmentioned. We tried to balance an overview of some of the classic techniques with a more in-depth treatment of some of the more recent and novel approaches to finding new quantum algorithms.

We expect and hope to see many more exciting developments in the coming years: novel applications of existing tools and techniques, new tools and techniques in the current paradigms, as well as new algorithmic paradigms.

References

Aaronson S, Shi Y (2004) Quantum lower bounds for the collision and the element distinctness problems. J ACM 51(4):595–605. doi: http://doi.acm.org/10.1145/1008731.1008735

Aharonov D, Ambainis A, Kempe J, Vazirani U (2001) Quantum walks on graphs. In: STOC'01: proceedings of the 33rd annual ACM symposium on theory of computing. ACM Press, New York, pp 50–59. doi: http://doi.acm.org/10.1145/380752.380758

Aharonov D, Arad I (2006) The BQP-hardness of approximating the Jones polynomial. http://arxiv.org/abs/quant-ph/0605181

Aharonov D, Arad I, Eban E, Landau Z (2007) Polynomial quantum algorithms for additive approximations of the Potts model and other points of the Tutte plane. http://arxiv.org/abs/quant-ph/0702008

Aharonov D, Jones V, Landau Z (2008) A polynomial quantum algorithm for approximating the Jones polynomial. Algorithmica 55(3):395–421

Ambainis A (2003) Quantum walks and their algorithmic applications. Int J Quantum Inform 1:507–518

Ambainis A (2004) Quantum walk algorithm for element distinctness. In: Proceedings of the 45th annual IEEE symposium on foundations of computer science, pp 22–31. doi: 10.1109/FOCS.2004.54

Ambainis A, Bach E, Nayak A, Vishwanath A, Watrous J (2001) One-dimensional quantum walks. In: STOC' 01: proceedings of the 33rd annual ACM symposium on theory of computing. ACM Press, New York, pp 37–49. doi: http://doi.acm.org/10.1145/380752.380757

Ambainis A, Childs A, Reichardt B, Spalek R, Zhang S (2007) Any and-or formula of size n can be evaluated in time $n^{1/2+o(1)}$ on a quantum computer. In: Proceedings of the 48th annual IEEE symposium on foundations of computer science, pp 363–372. doi: 10.1109/FOCS.2007.57

Arad I, Landau Z (2008) Quantum computation and the evaluation of tensor networks. http://arxiv.org/abs/0805.0040

Beaudin L, Ellis-Monaghan J, Pangborn G, Shrock R (2008) A little statistical mechanics for the graph theorist. http://arxiv.org/abs/0804.2468

Bernstein BK, Vazirani U (1997) Quantum complexity theory. SIAM J Comput 26:1411–1473

Berry DW, Ahokas G, Cleve R, Sanders BC (2007) Efficient quantum algorithms for simulating sparse Hamiltonians. Commun Math Phys 270:359

Boixo S, Knill E, Somma R (2009) Quantum state preparation by phase randomization. http://arxiv.org/abs/0903.1652

Boneh D, Lipton R (1995) Quantum cryptanalysis of hidden linear functions (extended abstract). In: Proceedings of the 15th annual international cryptology conference on advances in cryptology. Lecture notes in computer science, vol. 963. Springer, London, UK, pp 424–437

Boyer M, Brassard G, Høyer P, Tapp A (1998) Tight bounds on quantum searching. Fortschritte der Physik 56(5–5):493–505

Brassard G, Høyer P (1997) An exact quantum polynomial-time algorithm for Simon's problem. In: Proceedings of the fifth Israeli symposium on theory of computing and systems (ISTCS'97). IEEE Press, Piscataway, pp 12–23

Brassard G, Høyer P, Mosca M, Tapp A (2002) Quantum amplitude amplification and estimation. Quantum Computation & Information, AMS Contemporary Math Series 305:53–74

Brassard G, Høyer P, Tapp A (1997) Cryptology column — quantum algorithm for the collision problem. ACM SIGACT News 28:14–19

Brouwer AE (1989) Distance-regular graphs. Springer, New York

Childs A (2008) CO781 Topics in quantum information: quantum algorithms. Lecture notes on quantum algorithms. http://www.math.uwaterloo.ca/~amchilds/teaching/w08/co781.html

Childs AM, Cleve R, Deotto E, Farhi E, Gutmann S, Spielman DA (2003) Exponential algorithmic speedup by a quantum walk. In: STOC '03: proceedings of the 35th annual ACM symposium on theory of computing. ACM Press, New York, pp 59–68. doi: http://doi.acm.org/10.1145/780542.780552

Childs A, van Dam W (2010) Quantum algorithms for algebraic problems. Rev Mod Phys 82(1):1–52

Cleve R, Ekert A, Macchiavello C, Mosca M (1998) Quantum algorithms revisited. Proc Roy Soc Lond A 454:339–354

D'Ariano GM, van Dam W, Ekert E, Macchiavello C, Mosca M (2007) General optimized schemes for phase estimation. Phys Rev Lett 98(9):090,501

Das A, Chakrabarti BK (2008) Quantum annealing and analog quantum computation. Rev Mod Phys 80:1061

De las Cuevas G, Dür W, Van den Nest M, Briegel HJ (2008) Completeness of classical spin models and universal quantum computation. http://arxiv.org/abs/0812.2368

Deutsch D (1985) Quantum theory, the Church-Turing principle and the universal quantum computer. Proc Roy Soc Lond A 400:97–117

Deutsch D, Jozsa R (1992) Rapid solutions of problems by quantum computation. Proc Roy Soc Lond, A 439:553–558

Farhi E, Goldstone J, Gutmann S (2007) A quantum algorithm for the Hamiltonian NAND tree. http://arxiv.org/abs/quant-ph/0702144

Feynman R (1982) Simulating physics with computers. Int J Theor Phys 21(6,7):467–488

Freedman MH, Kitaev A, Larsen MJ, Wang Z (2001) Topological quantum computation. http://arxiv.org/abs/quant-ph/0101025

Freedman MH, Kitaev A, Wang Z (2000) Simulation of topological field theories by quantum computers. http://arxiv.org/abs/quant-ph/0001071

Geraci J (2008) A BQP-complete problem related to the Ising model partition function via a new connection between quantum circuits and graphs. http://arxiv.org/abs/0801.4833

Geraci J, Lidar DA (2008) On the exact evaluation of certain instances of the Potts partition function by quantum computers. Commun Math Phys 279(3):735–768

Grigoriev D (1997) Testing shift-equivalence of polynomials by deterministic, probabilistic and quantum machines. Theor Comput Sci 180:217–228

Grover L (1996) A fast quantum mechanical algorithm for database search. In: Proceedings of the 28th annual ACM symposium on the theory of computing (STOC, 96). ACM Press, New York, pp 212–219

Grover L (1998) A framework for fast quantum mechanical algorithms. In: Proceedings of the 13th annual ACM symposium on theory of computing (STOC' 98). ACM Press, New York, pp 53–62

Hirvensalo M (2001) Quantum computing. Series: Natural Computing Series. Springer

Hübener R, Van den Nest M, Dür W, Briegel HJ (2008) Classical spin systems and the quantum stabilizer formalism: general mappings and applications. http://arxiv.org/abs/0812.2127

Jordan S (2008) Quantum computation beyond the circuit model. PhD thesis, MIT University, Cambridge

Karchmer M, Wigderson A (1993) On span programs. In: Proceedings of the 8th IEEE structures in complexity conference. IEEE Press, Piscataway, pp 102–111

Kaye P, Laflamme R, Mosca M (2007) An introduction to quantum computation. Oxford University Press, Oxford, UK

Kempe J (2003) Quantum random walks - an introductory overview. Contemp Phys 44(4):307–327

Kitaev AY (1995) Quantum measurements and the Abelian stabilizer problem. http://arxiv.org/abs/quant-ph/9511026

Kitaev A, Shen A, Vyalvi M (2002) Classical and quantum computation. American Mathematical Society, Providence, RI

Magniez F, Nayak A, Roland J, Santha M (2007) Search via quantum walk. In: STOC '07: proceedings of the 39th annual ACM symposium on theory of computing. ACM, New York, pp 575–584. doi: http://doi.acm.org/10.1145/1250790.1250874

Menezes A, van Oorschot P, Vanstone S (1996) Handbook of applied cryptography. CRC Press, Boca Raton

Mermin ND (2007) Quantum computer science: an introduction. Cambridge University Press, Cambridge

Mosca M (2001) Counting by quantum eigenvalue estimation. Theor Comput Sci 264:139–153

Mosca M (2008) Abelian hidden subgroup problem. In: Kao M-Y (ed) Encyclopedia of algorithms. Springer, Berlin

Mosca M (2009) Quantum algorithms. In: Meyers R (ed) Encyclopedia of complexity and systems science. Springer

Nayak A, Vishwanath A (2000) Quantum walk on the line. http://arxiv.org/abs/quant-ph/0010117

Nielsen M, Chuang I (2000) Quantum computation and quantum information. Cambridge University Press, Cambridge, UK

Reichardt BW, Spalek R (2008) Span-program-based quantum algorithm for evaluating formulas. In: STOC '08: proceedings of the 40th annual ACM symposium on theory of computing. ACM Press, New York, pp 103–112. doi: http://doi.acm.org/10.1145/1374376.1374394

Santha M (2008) Quantum walk based search algorithms. In: Agrawal M, Du D-Z, Duan Z, Li A (eds) Theory and applications of models of computation. Lecture notes in computer science, vol 4978. Springer, Berlin, Heidelberg, pp 31–46. doi: 10.1007/978-3-540-79228-4_3

Shor P (1994) Algorithms for quantum computation: Discrete logarithms and factoring. In: Proceedings of the 35th annual symposium on foundations of computer science. IEEE Computer Society, Washington, DC, pp 124–134

Shor P (1997) Polynomial-time algorithms for prime factorization and discrete logarithms on a quantum computer. SIAM J Comput 26:1484–1509

Simon D (1994) On the power of quantum computation. In: Proceedings of the 35th IEEE symposium on the foundations of computer science (FOCS). IEEE Computer Society, Washington, DC, pp 116–123

Szegedy M (2004) Quantum speed-up of Markov chain based algorithms. In: Proceedings of the 45th annual IEEE symposium on foundations of computer science (FOCS). IEEE Computer Society, Washington, DC, pp 32–41. doi: http://dx.doi.org/10.1109/FOCS.2004.53

Tulsi T, Grover L, Patel A (2006) A new algorithm for fixed point quantum search. Quant Inform Comput 6(6):483–494

Welsh D (1993) Complexity: knots, colourings and countings. Cambridge University Press, Cambridge, UK

Wiebe N, Berry DW, Høyer P, Sanders BC (2008) Higher order decompositions of ordered operator exponentials. http://arxiv.org/abs/0812.0562

Wocjan P, Yard J (2006) The Jones polynomial: quantum algorithms and applications in quantum complexity theory. http://arxiv.org/abs/quant-ph/0603069

44 Physical Implementation of Large-Scale Quantum Computation

Kalle-Antti Suominen
Department of Physics and Astronomy, University of Turku, Finland
kalle-antti.suominen@utu.fi

G. Rozenberg et al. (eds.), *Handbook of Natural Computing*, DOI 10.1007/978-3-540-92910-9_44,
© Springer-Verlag Berlin Heidelberg 2012

Abstract

The development of large-scale quantum computing started rapidly in 1994 after the presentation of the factoring algorithm by Peter Shor. In this review, the basic requirements for the successful implementation of quantum algorithms on physical systems are first discussed and then a few basic concepts in actual information processing are presented. After that, the current situation is evaluated, concentrating on the most promising methods for which actual experimental progress has taken place. Among these are trapped ions, nuclear spins, and various solid-state structures such as quantum dots and Josephson junctions.

1   Introduction

The term "quantum computing" was perhaps first expressed by Richard Feynman (1982), when he suggested that one might optimize the simulation of quantum systems with computers that are based on the principles of quantum mechanics. Such quantum simulators are now studied both theoretically and experimentally. A good example is provided by cold neutral atoms that are trapped into optical lattices formed by laser beams (Bloch 2005). As the atomic interactions can be tuned with magnetic fields, and quantum statistics plays a definite role in the system dynamics, one can hope to simulate nontrivial models for many-body dynamics in lattices, including quantum phase transitions (Sachdev 1999).

The modern view on quantum computing, however, leans toward an analogy with digital computers, in the sense that the framework consists of binary numbers (registers), logical operations (gates), and networks (circuits) (Steane 1998; Nielsen and Chuang 2000; Stenholm and Suominen 2005; Mermin 2007). Rather than solving problems in physics, the aim is to solve mathematical problems. Appropriate examples are the famous factoring algorithm of Shor (1994), or the search algorithm of Grover (1997). These algorithms use some of the special properties of quantum mechanics such as superpositions and entanglement, to achieve a computational speedup that has a more fortuitous scaling with the proportion of the problem than the corresponding classical approaches. Part of the wide attention given to quantum computing is based on the special role that the factoring of large integers has in the security of the public key cryptosystems (Mollin 2002). Apart from these two algorithms, the apparent benefits of quantum computers are still unclear.

The basic element in quantum computing is the physical representation of the quantum bit, also called a qubit. For quantum communication and quantum measurement purposes, it is quite meaningful to study systems consisting only of a few qubits; this is the limit that some of the experimentally studied systems have already reached. The number of qubits needed for solving nontrivial mathematical problems, however, is on the order of a hundred in the ideal case, closer to a thousand or more with error correction included (currently the best figure is on the order of ten qubits, with liquid nuclear magnetic resonance (NMR) systems (Negrevergne et al. 2006)). This defines the concept of large-scale quantum computing, which in practice usually means that one needs to construct a mesoscopic or even a macroscopic quantum system. There are many detailed studies of physical implementation of quantum computing, such as Chen et al. (2007); Nakahara and Ohmi (2008); Stolze et al. (2008). The purpose of this review is to provide a short overview of the topic.

2   DiVincenzo Criteria

The general conditions for building a successfully working quantum computer are rather obvious, and the best known formulation for them was given by DiVincenzo (2000) (he was also the first to make careful comparison of various physical systems suitable for quantum computing; see, for example, DiVincenzo (1995)). The five items in his famous "checklist" can be formulated roughly as

1. *A scalable set of well-defined and individually identifiable two-state quantum systems that can act as qubits and which relate as a whole to a well-defined and finite state space.*

   To represent binary numbers, we need to give each qubit a label. Scalability means that we should be able to increase the number of qubits without limit (at least in theory). Combined with the other items below, this can prove to be a hard demand.

2. *Physical preparation of single qubits or groups of qubits into pure states in the state space.*

   This often reduces to the possibility of preparing the fiducial state where all qubits are firmly set to the state representing the number 0. Any other state can be then obtained if the next item on the checklist is available.

3. *Implementation of unitary operations that involve one or two qubits.*

   Since it is expected that all algorithms can be mapped into networks of single qubit operations and two-qubit conditional logic, this is sufficient. Furthermore, any network can be realized with single qubit operations and control-NOT operations, and the latter in turn can be constructed from simple conditional phase flips. This last stage is important because it opens the possibility for adiabatic operations and the use of geometric phases.

   Physically, the requirement for conditional logic maps usually into controlling the interaction between two arbitrary qubits. In experimental realizations we can use the fact that as the computational power scales exponentially with the number of qubits, we nevertheless need to implement only a polynomially increasing number of SWAP gates to reduce the condition of interactions between arbitrary two qubits into nearest-neighbor operations only. This greatly helps in finding eligible physical implementations.

   It should be noted that in optimizing networks, it may be helpful to allow other operations such as the three-qubit control-NOT operation either with two control bits and one target bit (Toffoli gate) or one control bit and two target bits with conditional mutual swapping (Fredkin gate).

4. *Performance of all required unitary operations at a time scale that is faster than any physical mechanism that causes the quantum system to decohere.*

   Unfortunately, this requirement clashes often with the scalability requirement in item 1, and it is sometimes considered as a fundamental obstacle to quantum computing (Haroche and Raimond 1996). At worst, the decoherence times for quantum registers scale exponentially with the number of qubits (Unruh 1995). In fact, decoherence times for registers made of qubits can be much faster than the decoherence time for a single qubit times the number of qubits (Palma et al. 1996). In addition, for typical algorithms, the number of necessary basic operations can also increase rapidly with the number of qubits (at least polynomially as in the quantum Fourier transform (Coppersmith 1994; Barenco et al. 1996) needed in Shor's factoring algorithm).

   Although this requirement sets in principle the upper bound to the number of available qubits, it does not necessarily prevent one from reaching computationally useful numbers of qubits. Also, error correction (usually by redundancy) or the use of decoherence-free subspaces are among the tools used to further evade the effects of decoherence.

5. *The possibility to physically read the state of each qubit once the computation has been performed.*

This requirement is often fulfilled if one can satisfy items 1–3 as well and it could be combined with item 2. In actual proof-of-principle demonstrations, one often performs full quantum tomography on the final state, and uses the fidelity of the experimental state with the expected ideal result as the figure of merit.

For quantum communication purposes, these five requirements have been augmented by two additional ones (DiVincenzo 2000):

6. *The ability to interconvert stationary and flying qubits.*
7. *The ability to faithfully transmit qubits between specified locations.*

These additions also apply to quantum computing if we consider grid-like quantum computing, in which single qubits or collections of qubits form the nodes of a metacomputer; see, for example, the ion trap proposal in Kielpinski et al. (2000). If realizable, such an approach provides an option to circumvent the scalability requirement in item 1. Recently quantum teleportation has become a viable tool for satisfying items 6 and 7, and a suitable method to access quantum memories, for example.

These stringent requirements usually mean that we need to have a mesoscopic quantum system and the ability to control its dynamics with high accuracy. Several suggestions for physical implementation of quantum computing have appeared since 1994. Although many of them might have a chance of experimental realization at least on the level of a few qubits, very few have been selected as targets for serious studies. For few-qubit applications, their intended use in, for example, quantum communication sets other requirements such as the possibility of transmission or interaction with information carriers such as photons(items 6 and 7 in the list above). The purpose of this review is to discuss those realizations that are now studied by several groups of experimentalists, with some initial success, for the purpose of large-scale quantum computing (at least in principle). Before that, it is advantageous to provide a short representation of some of the concepts that were briefly mentioned above.

3 Elements of Quantum Computation

The purpose of a quantum computer is to solve problems following an algorithm. The implementation of an algorithm must be mapped into a unitary operation, and for practical purposes this operation must be factored into a series of few-qubit operations known as gates. In this sense, the quantum computer resembles its classical counterpart. For the more mathematical aspects of quantum algorithms and quantum computing, the book by Hirvensalo (2004) is suggested.

3.1 Quantum Algorithms

As mentioned above, the main existing algorithms aim at either factoring large integers (Shor) or searching databases (Grover). The factoring algorithm relies on calculating a modular function $f(x) = a^x \bmod M$, where a is an almost freely chosen integer and M is the number that we want to factor. The periodicity of this function is the key single number that one seeks.

The quantum parallelism means that for a combined set of input and output registers (both with N qubits and $2^N \simeq M^2$), the initial state can be written as

$$|\Psi_0\rangle_{2N} = |\Psi_0^{\text{input}}\rangle_N \otimes |\Psi_0^{\text{output}}\rangle_N = \left(\frac{1}{2^{N/2}} \sum_{x=0}^{2^N-1} |x\rangle_N\right) \otimes |0\rangle_N \qquad (1)$$

where the input state is an equal superposition of all possible numbers and in the output register all qubits are in the 0 state.

At this point, it is convenient to introduce the standard single-qubit Hadamard operation H_{Had}, that is of the form

$$U_{\text{Had}} = \frac{1}{\sqrt{2}}[(|0\rangle + |1\rangle)\langle 0| + (|0\rangle - |1\rangle)\langle 1|] \qquad (2)$$

and which clearly puts any qubit from a state 0 or 1 to their equal superposition. By applying U_{Had} to each input qubit, the above initial state for the input register is easily produced (assuming all qubits can be set first to state 0; even this is not always trivial in actual physical implementations because such resetting is not necessarily a unitary operation). An alternative representation of operations is the matrix form, where the qubit states are given as

$$|0\rangle = \begin{pmatrix} 1 \\ 0 \end{pmatrix} \text{ and } |1\rangle = \begin{pmatrix} 0 \\ 1 \end{pmatrix} \qquad (3)$$

The Hadamard operation is a 2×2 matrix

$$U_{\text{Had}} = \frac{1}{\sqrt{2}} \begin{pmatrix} 1 & 1 \\ 1 & -1 \end{pmatrix} \qquad (4)$$

and by applying it twice we get the identity $U_{\text{Had}}^2 = \mathbf{1}$.

Next, one applies the unitary transformation U_f corresponding to calculating $f(x)$ so that the result is stored into the output register while the input is left intact. Due to the linearity of quantum mechanics, this leads to the new two-register state

$$|\Psi_f\rangle_{2N} = \frac{1}{2^{N/2}} \sum_{x=0}^{2^N-1} |x\rangle_N |f(x)\rangle_N \qquad (5)$$

This is a remarkable state as it represents entanglement between the register states so that each outcome can exist only with its corresponding input, in a total quantum correlation. As one obtains the result by applying U_f only once, this is often called quantum parallelism. As one can extract by measurements only one $(x, f(x))$ pair at a time, this approach is not useful for just calculating values of $f(x)$.

Since $f(x)$ is periodic, the amount of possible numbers in $|f(x)\rangle$ is limited, and for each result there is a periodic subset of numbers in the input register. If one performs a quantum Fourier transform (QFT) on the input register, one obtains an integer multiple of the inverse of the period. This, together with inaccuracies due to the finiteness of the registers, means that the total running time for the algorithm with the standard QFT network scales as N^3, where N^2 is for the QFT, and additional N repetitions are needed to fix the period with sufficient accuracy. As the best known classical approach scales as $\exp(N^{1/3})$, the advantage increases nearly exponentially with increasing N.

The Shor algorithm is genuinely quantum in the sense that it utilizes the entanglement between the registers. Whether its speed is due to its quantum nature, or due to the inherent efficiency of QFT, is an open question. Another quantum property is the superposition and, together with infinitesimal projections, it is the basic ingredient in the Grover algorithm.

For the database search, we need only one quantum register. Assuming that there is a mechanism that allows the computer to identify the desired item with a number stored in a register of N qubits, one can take as the initial state

$$|\Psi_0\rangle_N = \frac{1}{2^{N/2}} \sum_{x=0}^{2^N-1} |x\rangle_N \tag{6}$$

The search operation then recognizes a special value x_0 related to the wanted item. As a unitary operation, it changes the superposition into

$$|\Psi_1\rangle_N = \frac{1}{2^{N/2}} \left(\sum_{x=0, x \neq x_0}^{2^N-1} |x\rangle_N - |x_0\rangle \right) \tag{7}$$

The next step is to perform a specific single-qubit operation on each qubit, and then repeat the above. It can be shown that after $N_0 \simeq (\pi/4)\sqrt{N}$ repetitions of this two-step process, one gets

$$|\Psi_{N_0}\rangle_N = |x_0\rangle \tag{8}$$

and the item is then identified through x_0. Compared to the Shor algorithm, the speed-up is not huge (classical search scales as $N/2$). For a nice description of the Grover algorithm, the book by Vedral (2006) is referred to.

A third algorithm that is often mentioned especially in few-qubit demonstrations is the Deutsch–Jozsa algorithm (Cleve et al. 1998) although in practice it serves only demonstration purposes. We assume an a priori unknown one-qubit function $f(x)$ that maps a one-digit binary number to a one-digit binary number. The question is whether the function is constant ($f(0) = f(1)$) or balanced ($f(0) \neq f(1)$). This is the Deutsch problem, and classically one solves it by calculating the function twice with different arguments (0 and 1), and comparing the result. The Deutsch–Jozsa algorithm is a generalization of this, so that the argument can be a number between 0 and $2^N - 1$, while the values of the function are still limited to 0 or 1.

In brief, the operation starts by setting the state (input register with N qubits and output with 1 qubit)

$$|\Psi_0\rangle_{N+1} = \frac{1}{2^{(N+1)/2}} \sum_{x=0}^{2^N-1} |x\rangle_N \otimes (|0\rangle_1 - |1\rangle_1) \tag{9}$$

and then calculates $f(x)$ by a unitary transformation U_f, so that $f(x)$ is added to the existing value of the output bit (binary addition, so $0 + 0 = 0$, $0 + 1 = 1 + 0 = 1$, $1 + 1 = 0$):

$$|\Psi_f\rangle_{N+1} = \frac{1}{2^{(N+1)/2}} \sum_{x=0}^{2^N-1} |x\rangle_N \otimes (|f(x)\rangle_1 - |f(x) + 1\rangle_1) \tag{10}$$

This is equivalent to

$$|\Psi_f\rangle_{N+1} = \frac{1}{2^{(N+1)/2}} \sum_{x=0}^{2^N-1} (-1)^{f(x)} |x\rangle_N \otimes (|0\rangle_1 - |1\rangle_1) \tag{11}$$

Now the output register can be discarded, and one performs a Hadamard operation to each qubit of the input register. If projected to the all-zero state $|0\rangle_N$, one gets that the projection is successful with probability 1 for constant case, and with probability 0 for the balanced case.

3.2   Quantum Gates and Quantum Networks

It is tempting to consider performing simple unitary operations on large registers without problems, but in practice that involves too many degrees of freedom for simultaneous control. A general operation on an N-qubit register would correspond to a $2^N \times 2^N$ matrix. Clearly, one loses the quantum advantage if one thinks of the register as a 2^N-dimensional quantum state instead of working with N two-state systems. In practice, all computations are made in networks of quantum gates that act typically on single qubits, or on two qubits (for conditional logic). Such operations are unitary and thus also reversible. The Hadamard gate above is a typical single-qubit operation; most of these can be described as rotations in the single-qubit Hilbert space. These operations set qubits to specific superposition states and can also affect the phase relation between the probability amplitudes of single qubit states, and they were studied extensively in the early days of quantum computing, and reviewed in many textbooks, see, for example, Nielsen and Chuang (2000); Stenholm and Suominen (2005); Mermin (2007).

Of the two-qubit gates, we concentrate here only on a few essential ones, and their relationship to each other. The controlled-NOT (CNOT) gate is a standard element in conditional logic. In short, it flips the state of the target qubit if the control qubit is in state 1. Specifically, if the control qubit is in a superposition state, CNOT entangles it with the target qubit (in case they were initially in uncorrelated states). An alternative gate is the controlled phase gate, which alters the global phase of the two qubits by ϕ if they are both in state 1. Especially, if one can effect a gate with conditional phase change of π, that is, $|11\rangle \rightarrow -|11\rangle$, this conditional phase flip gate can be used to construct the CNOT gate. This is important for geometric quantum computing discussed later. CNOT gates are also needed in error correction methods for quantum computers. For demonstration purposes and for quantum communication one wants to either create highly entangled states of qubits, or to detect entangled states such as the Bell states. The CNOT gate is a very convenient tool to produce such states, or in mapping them to easily distinguishable qubit states.

The two-qubit gates are usually described as 4×4 matrices in the joint basis $\{|00\rangle, |01\rangle, |10\rangle, |11\rangle\}$. The representations for the CNOT gate, phase gate, and the phase flip gate are, respectively,

$$
U_{\mathrm{CNOT}} = \begin{pmatrix} 1 & 0 & 0 & 0 \\ 0 & 1 & 0 & 0 \\ 0 & 0 & 0 & 1 \\ 0 & 0 & 1 & 0 \end{pmatrix}, \quad
U_\phi = \begin{pmatrix} 1 & 0 & 0 & 0 \\ 0 & 1 & 0 & 0 \\ 0 & 0 & 1 & 0 \\ 0 & 0 & 0 & e^{i\phi} \end{pmatrix}, \quad
U_{\mathrm{ph-flip}} = \begin{pmatrix} 1 & 0 & 0 & 0 \\ 0 & 1 & 0 & 0 \\ 0 & 0 & 1 & 0 \\ 0 & 0 & 0 & -1 \end{pmatrix} \tag{12}
$$

Another two-qubit gate is the SWAP gate that interchanges the state of two qubits:

$$
U_{\mathrm{SWAP}} = \begin{pmatrix} 1 & 0 & 0 & 0 \\ 0 & 0 & 1 & 0 \\ 0 & 1 & 0 & 0 \\ 0 & 0 & 0 & 1 \end{pmatrix}, \quad
U_{\mathrm{iSWAP}} = \begin{pmatrix} 1 & 0 & 0 & 0 \\ 0 & 0 & i & 0 \\ 0 & i & 0 & 0 \\ 0 & 0 & 0 & 1 \end{pmatrix} \tag{13}
$$

Here, we introduce also the iSWAP gate discussed later. Swapping is not conditional logic, but it is important because instead of moving qubits physically, one can alter their labeling instead. Thus, any two qubits can be made nearest neighbors, which enlarges the number of those physical systems that are realistic for quantum computing purposes. The swapping can be realized with a simple combination of three CNOT gates.

In general, the CNOT gate combined with a set of single-qubit operations form a universal set: all possible two-qubit operations can be realized with them. In the same manner as the CNOT can be constructed with conditional phase flips, we can find other forms of conditional operations that can be taken as a starting point for realizing actual computation (Bremner et al. 2002). Often the experimental realization of two-qubit operations determines which kind of conditional logic gate is the basic element. Earlier, the demonstration of the CNOT gate was considered as a necessary step for successful computation, but lately it has been replaced by many other alternatives such as CROT, $\sqrt{\text{SWAP}}$, and iSWAP (given above). The first two are represented by the following unitary transformations:

$$U_{\text{CROT}} = \begin{pmatrix} 1 & 0 & 0 & 0 \\ 0 & 1 & 0 & 0 \\ 0 & 0 & 0 & -1 \\ 0 & 0 & 1 & 0 \end{pmatrix}, \quad U_{\sqrt{\text{SWAP}}} = \frac{1}{\sqrt{2}} \begin{pmatrix} 1 & 0 & 0 & 0 \\ 0 & e^{i\pi/4} & ie^{i\pi/4} & 0 \\ 0 & ie^{i\pi/4} & e^{i\pi/4} & 0 \\ 0 & 0 & 0 & 1 \end{pmatrix} \quad (14)$$

The second definition is obviously not a unique one for $\sqrt{\text{SWAP}}$ (e.g., clearly $\pi/4$ can be replaced with $3\pi/4$). The variation $\sqrt{\text{iSWAP}}$ has also been used. We can see that the variations of the SWAP gate are suitable for situations where only the states $|01\rangle$ and $|10\rangle$ can be coupled.

A quantum network consists of the gates and the information to which qubits and in which order one applies the gates. The quantum Fourier transform network for N qubits, for example, consists of N single-qubit Hadamard gates, and $N(N-1)/2$ conditional phase shifts (with qubit index-dependent values of phase shifts) (Coppersmith 2002; Barenco et al. 1996). In some special cases, one can utilize gates that operate on three qubits or more. The Toffoli gate swaps the state of the target qubit, if the two control qubits are both in state 1. This can be generalized into n control qubits, naturally. The Fredkin gate performs a SWAP gate on two target qubits if the control qubit is in state 1.

3.3   Error Correction

The sensitivity of quantum mechanical superpositions to disturbances is perhaps the most serious problem for actually realizing quantum computing. In addition to decoherence and dissipation, due to the coupling of quantum systems to their environment, small errors due to imperfections in gate operations interfere with the computing process. Although small, the cumulative effect of small errors can be devastating if they are allowed to propagate freely in the system through conditional logic. The errors can be suppressed by applying error correction. However, it requires operations that are typically subjected to the same inaccuracies as the actual computational operations. If the error correction during computation is effective, that is, it reduces errors more than it adds them, computation is regarded as *fault-tolerant*.

The standard quantum error correction relies on redundancy. For example, for one qubit in state

$$\alpha|0\rangle + \beta|1\rangle \quad (15)$$

we write the highly entangled three-qubit state

$$\alpha|000\rangle + \beta|111\rangle \quad (16)$$

This is achieved by starting with $(\alpha|0\rangle + \beta|1\rangle) \otimes |0\rangle \otimes |0\rangle$ and applying the CNOT on the last two zero qubits with the first one as the control. The use of three qubits increases the size of the qubit Hilbert space from 2 to 2^3. If, at most, one of the spins flips, the location of the flip can be identified by projecting the state on four orthogonal states, which are

$$
\begin{aligned}
P_0 &= |000\rangle\langle000| + |111\rangle\langle111| \\
P_1 &= |100\rangle\langle100| + |011\rangle\langle011| \\
P_2 &= |010\rangle\langle010| + |101\rangle\langle101| \\
P_3 &= |001\rangle\langle001| + |110\rangle\langle110|
\end{aligned}
\tag{17}
$$

The result identifies which qubit flipped (or that none flipped), and one can perform a correcting flip on that qubit. This does not perturb the actual superposition since we do not know whether 0 became 1 or 1 became 0, but only that a flip occurred.

With a slightly more complicated operation (and more qubits) one can identify and correct phase flips as well. In 1995 Peter Shor showed how such a general error correction scheme can be constructed with nine qubits (Shor 1995), but the number was reduced later to seven (Steane 1996; Calderbank and Shor 1996). Finally the sufficient number of qubits needed was found to be five (Bennett et al. 1996; Laflamme et al. 1996), leading to fault-tolerant error correction (Knill 2001). Systematic studies in the field have produced more general methods such as stabilizer codes (Gottesman 1996). For a recent detailed description of quantum error correction and fault-tolerant computation see Gaitan (2008); see also the related chapter and references therein in Nielsen and Chuang (2000).

For noisy computation and specific cases, one can also utilize optimization of algorithms. For instance, the quantum Fourier transform relies on conditional phase shifts. It can be shown that since the noise per operation severely perturbs the smallest phase shifts, it becomes advantageous to reduce the overall noise by not performing these phase shifts at all (Barenco et al. 1996). The improvement is due to the decreased total number of operations, and additional benefits can come for nearest-neighbor realizations since the size of the implemented phase shift is inversely proportional to the difference in qubit labels (typically equivalent to the physical distance between qubits as well).

3.4 Decoherence-Free Subspaces

Error correction can target both the technical problems as well as the environmental effects. Even for technically perfect qubit operations, one is still left with the coupling of a quantum system with its environment. The thermal heat bath is a typical example of a source for dissipation and decoherence. The first leads to qubit state flips, and the latter erodes the phase relations between amplitudes either within the qubits or (and more rapidly) between qubits. A prime example is the transversal and longitudinal relaxation times in NMR; it also illustrates the general feature that decoherence time scales can be much faster than dissipation time scales, and, thus, in most cases the decoherence time scale compared with the gate operation time is the key figure of merit.

The standard approach to decoherence is to suppress the interaction between the system and its environment. Cooling the computing system to low temperatures is a basic solution; in some cases, such as flux qubits (❷ Sect. 8.2), the cooling is essential for the existence of the qubit system in the first place. Occasionally, the decoherence can actually be of advantage, as we shall see in connection with the liquid NMR quantum computation in ❯ Sect. 6.1, but such situations are exceptions with their own special difficulties.

For some systems, one can find, for some combinations of qubits, that they are more robust to decoherence than other combinations. Thus, one can select a decoherence-free subspace in this extended Hilbert space, and limit the computation there (Palma et al. 1996; Duan and Guo 1997; Zanardi and Rasetti 1997; Lidar et al. 1998).

3.5 Adiabatic and Geometric Quantum Computing

One of the very robust states, when it comes to decoherence and dissipation, is the ground state of a quantum system. In adiabatic evolution, the system remains always in the instantaneous eigenstate of the system. If the computing starts with the ground state of the computer, and one can perform the computation adiabatically (Farhi et al. 2000; Childs et al. 2001), it means that the system remains on the ground state all the time, and it is only the nature of the ground state that evolves into the final result of the computation.

One can describe the adiabatic process as a passage related to a changing parameter. If one performs a closed loop in the parameter space, one can affiliate a global geometric phase to the evolution. This global phase is the famous Berry phase, and it depends on the path that one takes in the parameter space. Later, the approach was generalized to non-closing loops as well. This provides the basis for geometrical quantum computation. An interacting combination of two quantum bits can be made to evolve along its changing ground state so that the global phase change is conditional to the values of the two qubits. It means that in practice one can realize, for example, the phase-flip operation. Another name for computations using geometric phases and closed loops in the parameter space is holonomic quantum computation. There is a nice recent review on the geometric quantum computing and its basic concepts (Sjöqvist 2008).

Recently, some interest was created by the quantum processor developed by the D-Wave company, based on superconducting flux qubits. The actual computation is taking place, assumably, through adiabatic evolution. This is still subject to much debate. One of the problems with adiabatic evolution is that although the lowest eigenstate of the system remains well-defined, it may get very close to some higher state, and nonadiabatic transitions between them can be suppressed only by an extremely slow evolution, which in turn may lead to other problems.

3.6 Cluster State Quantum Computing and Anyonic Quantum Computing

Normally, we assume that the quantum computation follows a very clear cycle. First the input is prepared, then processed, and after that the result is read out by measurement. This is "Hamiltonian computing." In quantum mechanics, one can challenge this view. One approach is to make the whole computation proceed by measurements instead of unitary evolution. If a two- or three-dimensional lattice of qubits interacts suitably (e.g., Ising-type interaction), it can form a large cluster of entangled states (Briegel and Raussendorf 2001). Computation can then proceed in this cluster by targeted measurements on individual particles (Raussendorf and Briegel 2001). The reliance on irreversible measurements, instead of unitary operations only, has given the cluster approach the name *one-way quantum computing*.

Another idea based on a lattice of qubits with nearest neighbor interactions was proposed by Kitaev (1997). The approach is expected to be fault-tolerant by construction. Computation is performed by using an anyonic excitation of the system, where an exchange of two particles changes the global phase of their wavefunction. The approach relies especially on having non-Abelian anyons, which might be found in two-dimensional degenerate quantum gases (fractional quantum Hall effect). This topic has been recently reviewed thoroughly, see Nayak et al. (2008).

4 Cavity QED

In a confined space, the electromagnetic field modes become quantized into standing waves and form a discrete set of energy states for the field. In the quantum electrodynamical (QED) description, the energy of each mode with frequency v is quantized into photon states, integer multiples of hv including the vacuum state of each mode (h is the Planck constant). Such a cavity with well-defined modes in one direction can be built by opposing mirrors. The photon states of the modes can occur in quantum superpositions of photon numbers n. However, the temperature of the mirrors defines the temperature of the cavity, so to avoid thermal occupancy of the photon states, one must go to very low temperatures. Also, the quality of the mirrors must be high, otherwise the loss of photons broadens the energy spectrum of each mode.

The key issue is the controlled interaction between the cavity modes and the atoms that fly through the cavity or are placed there by means of some electromagnetic trapping. In an ideal case, one can describe the atoms as two-level systems, and limit the interaction to a single mode. One can then explore some key phenomena such as Rabi oscillations between atomic states (Jaynes–Cummings model) but, more importantly, with the atoms, one can manipulate the photonic state of the cavity. States with definite photon number n are actually highly nonclassical states of light, and they can be created with the help of the atoms. Alternatively, the photonic state of the cavity can be explored in a nondestructive way (QND, quantum non-demolishing) with atoms that are highly off-resonant but still coupled with the relevant cavity mode.

Such cavities have been prepared both in the optical and microwave region. It is easier to obtain high quality (high-Q) cavities in the microwave range than in the optical range. For the microwave range, the atoms are usually pumped into a high-energy Rydberg state (main quantum number $\simeq 60$), so that the transition energy falls into the microwave regime, and in the QND photonic state measurements one applies atom interferometry. The best demonstrations of controlled and coherent dynamics of two-state quantum systems and interactions with a quantized field have been achieved with such systems. These are essential factors in single-qubit operations, and the cavity QED terminology has been adapted for many other physical implementations when demonstrating that the state of a qubit is indeed a quantum superposition and coherence can be established between the qubit states. For example, by implementing Rabi oscillations in a qubit, one can estimate single-qubit decoherence through the damping of these oscillations (and by observing them to prove that there is some coherence to begin with; see, for example, the flux qubit example of Chiorescu et al. (2003)). The concepts of cavity QED were reviewed in detail in the textbook by Haroche and Raimond (2006). In fact, one of the first demonstrations of conditional phase shifts was performed in 1995 with an optical cavity, where photon qubits (0 or 1 photon per polarization mode) experienced an interaction via off-resonant Cs atoms flying through the cavity (Turchette et al. 1995). See also the 2004 review in van Enk et al. (2004).

5 Physical Realization with Electromagnetically Trapped Particles

5.1 Trapped Ions

The trapping of ions with electromagnetic fields has provided much insight into quantum mechanics since the early 1980s (Horvath et al. 1997). Just the actual observation of the single ions, by detecting the light scattered by them, showed that single atoms are a meaningful concept in physics, not just an abstraction. The observed change of the quantum state of a single ion through a quantum jump in the mid-1980s demonstrated clearly the limits of ensemble quantum mechanics, and also the existence of the until-then hypothetic jumps. In 20 years, these jumps have evolved from a mystery into a practical tool in detecting the state of single ions (see item 5 on the DiVincenzo checklist); already in the first experiments, they were used to show that the scattered light actually originated from a specific integer number of ions.

Laser cooling of these ions was developed alongside the trapping methods, and gave control over the motional states of ions. Especially, it made it possible to reach situations where the motion itself becomes quantized, and in the case of more than one ion, one gets phonons. Thus, it is not perhaps surprising that very soon after the presentation of the Shor algorithm in 1994, Ignacio Cirac and Peter Zoller published their proposal for quantum computing with chains of ultracold ions (Cirac and Zoller 1995). Even now, this scheme provides the basic approach to quantum computing with trapped ions.

Trapping of charged particles by static electromagnetic fields is in principle not possible (Earnshaw theorem), but the problem can be averted by adding either a fast-oscillating electric field (Paul trap) or a static magnetic field (Penning trap). For quantum computing purposes, to obtain individual addressing and controllable interactions, the so far most suitable situation for many ions is provided by the linear ion trap, in which the trapping potential is an elongated cigar, into which the atoms at low temperatures form a chain as they are pushed toward each other by the trapping potential, but this is balanced by the Coulomb repulsion between charged ions. The quantized motion of ions is now a collective effect, and for sufficiently low temperatures any motion perpendicular to the cigar is suppressed, and along the cigar the lowest phonon modes correspond to the oscillation of the whole chain in unison, and to the breathing-like mode where each atom oscillated with a π phase shift compared to its two neighbors.

The cooling of such a chain is based on ions absorbing photons at lower energies, and then radiating them at higher energies, with the energy difference arising from a change of the motional state from a higher one to a lower one (sideband cooling). As the collective motion affects all ions of the chain, Cirac and Zoller proposed that if one codes one qubit per ion, using its internal electronic states, conditional logic can be mediated with the phonons in the chain. The scenario is simple: a change of the state of each qubit can be realized by addressing the particular qubit by laser light. Assuming that the chain is in its lowest motional state, one can address qubit 1 so that if it is in state 1, then a phonon appears in the chain (excitation to the next state of collective motion). When addressing qubit 2, one can make it conditional so that one has four possible outcomes, depending on the original state of the qubit 2, and the phonon state of the chain (and thus the state of qubit 1).

The lifetimes for excited atomic and ionic states are usually short, on the order of nanoseconds, so they are not a priori optimal states for qubit coding. The problem can be circumvented by using either the ground state and a long-living excited state that exists in most ions of alkaline-earth atoms (group II), or two Zeeman substates of the ground state, coupled then with a Raman setup instead of a single laser field. The shortlived electronic states have their purpose, however, as one can detect the state of qubits using the quantum jump approach.

Several groups have adopted the Cirac–Zoller scheme as a basis for demonstrating quantum computing. Basic problems on the road have been the elimination of fluctuations in the electromagnetic trapping fields and actual cooling of the chain of ions (Nägerl et al. 1998; Schmidt et al. 2003a). But successful demonstrations have been achieved. They can be coarsely divided into two groups: demonstration of gate operations, and preparation of specific entangled states of many qubits. Although the latter are often obtained as a result of logical operations and quantum computing, here we concentrate on the gate operations. The success of logical operations is verified by quantum tomography of the final state, and quantified by fidelity with the ideal result. In Innsbruck, in the group of Rainer Blatt, the control-NOT gate between two qubits was demonstrated in 2003 (Schmidt-Kaler et al. 2003b). Already in 1995, such a gate was demonstrated in Boulder (Monroe et al. 1995), but this was a nonscalable demonstration where the two qubits were coded in the same ion, and the Cirac–Zoller scheme was not used. Later in Innsbruck, the three-qubit Toffoli gate was demonstrated in 2009 (Monz et al. 2009). The three-qubit operation was achieved by taking advantage of the three lowest motional states, as an extension of the Cirac–Zoller scheme.

As the ionic chains become longer, their cooling is an increasing challenge. This has led to the proposals involving hot ions. Typically, the phonons are then replaced by photons for the conditional logic. Perhaps the most successful of the proposals is the Mølmer–Sørensen bichromatic gate (Sørensen and Mølmer 1999a, b). It relies on the idea of the conditional phase gate as the basic element. When two ions are simultaneously interacting with two light fields, the phase of the final state will depend on the internal states (acquires -1 if both qubits are in state 1). This scheme works seemingly independent of the motional states of the ions, but their presence is required nevertheless as they are absent only due to the use of the perturbation approach (which means that the gates are not necessarily very fast). Such a gate action was demonstrated recently in Innsbruck (Benhelm et al. 2008).

Despite the success with either cold or hot ions one can ask how scalable the ion trap computers are. Clearly, the lengthening of the chain will pose many problems, from cooling to individual addressing. The possibility to go for 2D structures (Wigner crystals) would allow better scaling, but also more complicated motional states and cooling of the structure. Another possibility is to distribute the computing between several small scale ion trap computers, using, for example, photons as mediators between the computers (and quantum teleportation as a tool for exchanging the quantum information between the ions and mediating photons). Perhaps a simpler alternative is to shuffle ions around, dividing the trap into computing and memory sections (Kielpinski et al. 2002; Schaetz et al. 2004). For qubit states involving the two Zeeman states of the ionic ground state one can utilize microwave fields. This opens the possibility to use high-quality cavities in the control of qubits. Cavity QED is a well-established field and can provide useful tools, although it applies best to transitions between atomic Rydberg states.

For further details, the reader is referred to an excellent review on ion trap quantum computing by Häffner et al. (2008).

5.2 Trapped Neutral Atoms

Trapping of neutral atoms into structures where they can be addressed one-by-one is much harder than for charged particles. There are basically two avenues for this: single atom traps or optical lattices. Light from a highly detuned laser field is unlikely to be absorbed by atoms, but they still interact in a manner that allows one to control the external degrees of freedom of the atoms. Basically, the atoms feel the spatial changes in the field as an external potential. If the light is tuned above the atomic transition (blue detuning) the field intensity maxima repel atoms, and for tuning below the transition (red detuning) the maxima are points of attraction. This is either called dipole or gradient force. In the simplest formulation, atoms are attracted by, for example, the focal point of a laser beam. If one adds some dissipative mechanism, one can confine atoms into such dipole traps (also called FORTs, i.e., far-off-resonant traps). This is one of the key points for laser cooling and trapping of atoms (Metcalf and Van der Straten 1999). If one adds the spin-induced degeneracy of the atomic states and the polarization of the light, one can additionally control the system with external magnetic fields, or use the laser field superpositions to create specific structures such as lattices with potential minima and maxima at regular intervals.

The trapping of atoms into microtraps is a challenging task. For quantum computing purposes, one would prefer a set of such traps with one atom (and only one) in each, which is not easy. The advantage of microtraps compared to optical lattices is that it is easier to address atoms individually, when the separation is in the micrometer range. Several possible setups are currently studied, including atom chip systems, where the atoms are localized very close to a solid surface, and wires on the surface provide the electromagnetic fields that confine the atoms. As the confining forces are weak, low temperatures and elimination of collisions with background atoms are required, not to mention an accurate trap-loading process. Conditional logic is also a challenge. If sufficiently close, the atoms can interact by a state-selective dipole–dipole interaction, which provides one possibility, but there are no proof-of-principle experimental demonstrations for quantum computation in such systems yet.

For optical lattices, the lattice site occupancy is low and not very well controlled with basic laser-cooling and trapping methods. However, the Bose–Einstein condensation of atoms provides a useful tool for preparing lattice occupation. The atoms are trapped and cooled so that they begin to occupy the lowest quantum state of the trap potential. If they are bosons, they undergo eventually a quantum phase transition to a condensed state, where the atoms are correlated completely and the whole sample is described by a single collective quantum wave function (and with a nonlinear Schrödinger equation where the atomic interaction appears as a mean-field potential) (Pethick and Smith 2008). In an optical lattice one can, for sufficiently high barriers between lattice sites, observe a transition to a Mott insulator, where (a) each lattice site is occupied and (b) with exactly the same number of atoms. This has already been achieved experimentally (Bloch 2005), and it opens possibilities for quantum computing, especially if single-atom addressability can be reached. But this is a future method and no demonstrations of quantum logic have been performed so far.

5.3 Trapped Single Electrons

Experimentally, the trapping of single electrons with electromagnetic fields preceded the trapping of ions. In general, many solid-state structures such as quantum dots or superconducting islands rely on confining potentials for electrons or equivalent quasiparticles. In principle, even atoms and molecules are just traps for electrons. A common theme for many systems is that the dynamics of a complicated system depends only on a few electrons while the nuclei and the majority of the electrons only form a background. Not surprisingly, the popular alkaline atoms for neutral atom trapping and ionized alkaline-earth atoms for ion trapping are effectively hydrogen-like single-electron atoms. In fact, even the nuclei for atoms such as C, N, and F are made of several protons and neutrons, but as composite particles they only appear to one as spin-1/2 systems.

If one nevertheless persists with the concept of an electron trapped by externally generated electromagnetic fields, possibilities for quantum computing have been considered in such systems as well. An example is the proposal where the electrons are confined on the surface of liquid helium and controlled by electromagnetic fields (Platzman and Dykman 1999; Dykman and Platzman 2000). The surface potential for electrons stops them from penetrating the liquid but it is weakly attractive at short distances, so that a few bound states appear, confining the electrons as a two-dimensional gas at the surface. The bound states can be addressed with microwave fields. Due to the Coulomb interaction, the electrons repulse each other (in analogy to trapped ions) and, as their temperature is lowered, one observes a phase transition to a Wigner crystal (Grimes and Adams 1979) (see also Rousseau et al. (2009) and references therein; the Wigner crystal has been observed for trapped ions as well (see, e.g., Birkl et al. 1992)).

However, one does not rely on the crystallization in qubit preparation but instead, by embedding electrodes into the liquid, one can provide the localization in the direction of the surface as well, and in addition one can induce individual (local) Stark shifts to resonance energies so that one can select the particular qubit, which is subsequently manipulated with the microwave field pulse (other qubits are off-resonant with the field).

The dipole–dipole interaction between adjacent electrons depends on the bound state because it shapes the electron wavefunction and thus the charge distribution (in very much the same manner as in the first quantum dot proposals that also used the dipole–dipole interaction for conditional logic (Barenco et al. 1995)). The read-out would be obtained by external field-assisted tunneling of the higher state electrons out of the surface potential; they would be individually detected by channel plates. It has been suggested also that the electron spin could be used to represent a qubit (Lyon 2006); the degeneracy of spin states is lifted with an external magnetic field, and the spin-dependence of the dipole–dipole interaction can be used. There is some experimental activity (see, e.g., Sabouret and Lyon 2006) in this field and also some theoretical studies have been done (see, e.g., Dykman and Platzman 2003; Dahm et al. 2003). The main decay channel would be the coupling of electrons to the liquid surface excitations (ripplons), but one of the advantages of the system is that this coupling is weak, and compared to other suggestions such as nuclear spins in solid-state structures (see, e.g., ❯ Sect. 6.2) the advantage is that here the electrons are trapped practically in vacuum.

6 Nuclear Spin Computers

An obvious candidate for a quantum bit is the spin-1/2 particle, which is the standard example of many quantum mechanics textbooks and often provides for physics students the first encounter with group theory through SU(2) representations. As discussed in the previous section, although the electron is a typical example, many of the atomic nuclei are spin-1/2 systems as well. As they form the key ingredients in many molecules or solid structures, and can be manipulated and studied with radio-frequency fields, it is natural to consider them as possible basic elements for quantum computing. The nuclear systems, however, behave very differently depending on their surroundings, and this leads to the fact that the quantum computing aspects are best discussed separately for liquid and solid systems.

6.1 Molecular Spins in Liquids

Nuclear magnetic resonance (NMR) is a well-established research topic, and an important method in many fields such as chemistry and medical sciences. The broad success also brings problems: the sophisticated concepts and terminology in modern NMR are not anymore so accessible to physicists, especially to those working on quantum systems such as trapped ions and atoms or solid-state electron systems.

At first, the idea of doing quantum computing with large numbers of molecules in liquid at room temperature seems counterintuitive, especially since it seems to be completely complementary to other approaches, such as trapped ions that rely on single isolated systems at extremely low temperatures. This kind of approach is generally called ensemble quantum computing.

The basic element is a molecule consisting of several atoms, many of which have spin-1/2 nuclei. In a magnetic field, the spin-up and spin-down states obtain different energies, and the spins oscillate with their characteristic frequencies given by the energy difference. This Larmor frequency is typically on the order of 100 MHz. The local environment in the molecule for each nucleus is usually unique and gives a chemical shift to this energy/oscillation frequency, on the order of ~ 1 MHz or less. It means that in principle each spin can be detected and manipulated with a field of specific frequency. The magnetic moment of a single spin is too weak for practical use, and thus one needs macroscopic numbers.

In liquid state, these molecules can move and rotate freely. This has the additional advantage that interactions between the molecules simply average to zero. Furthermore, the same applies to the anisotropic dipolar magnetic spin–spin interaction within each molecule. Eventually, one is left only with the isotropic interaction between nuclear spins mediated by the atomic valence electrons. For two nearby spins, this gets down to a simple Hamiltonian

$$H = \hbar\omega_1 I_1^z + \hbar\omega_2 I_2^z + 2J_{12}I_1^z I_2^z \tag{18}$$

where $\omega_{1,2}$ are the characteristic frequencies, $I_{1,2}^z$ are the spin projection operators in the direction of the external field, and J_{12} quantifies the spin–spin interaction. Thus, one can address spins with different RF-pulses, and the third term in the Hamiltonian allows for conditional logic in the form of controlled phase slips.

Although the large numbers of molecules and the averaging characteristics of the thermal liquid seem to be useful for some aspects, one would also assume that they wash out any quantum coherences as well. In fact, since the thermal energy exceeds well the energy differences of spin states, the total state of the system is almost equal to a statistical mixture of equally populated spin states. The key issue is the small deviation from the mixture, which has the characteristics of a pure quantum state, with 2^N states for N spin-1/2 nuclei of a single molecule. This pseudo-pure state is often written in the form

$$\rho \simeq \frac{1}{2^N}\mathbf{1} + \alpha|\Psi\rangle\langle\Psi| \tag{19}$$

Clearly, since the spin operators are traceless, the thermal part does not contribute to any expectation values. Also, any unitary operation on this density matrix will only reproduce the thermal part, and operate on the pure part in the normal quantum fashion.

Of course, the above picture is highly simplified and in practice one performs complicated sequences of pulses, and uses spin-echo methods to control phase evolution. The advantage is that the decoherence times for nuclear spins are extremely long compared to the pulse durations. However, ensemble quantum computing is beset with several problems.

The factor α is small, typically 10^{-4} to 10^{-5} for a few spins at room temperature, and it scales in the high temperature limit as

$$\alpha \simeq \frac{N\hbar\Delta\omega}{k_B T 2^N} \tag{20}$$

where $\Delta\omega$ is the energy difference between spin-up and spin-down states, T is the temperature, and k_B is the Boltzmann factor. To maintain the signal size means exponentially increasing the number of molecules or repeated measurements, as pointed out by Jones (2000).

The increase in the number of qubits means that one needs larger molecules with identifiable spins, and this is difficult to achieve. For instance, one of the largest systems is an *l*-histidine molecule with 14 spin-1/2 nuclei, of which two hydrogen nuclei are indistinguishable (forming a spin-0 and spin-1 system), providing at maximum a 12-qubit + 1 qutrit system (Negrevergne et al. 2006). It may be extremely difficult to go beyond this.

The pulse sequences needed for computation become exceedingly numerous and also complicated to design, tending to scale exponentially with the number of qubits. Thus, the duration of computation is reaching the decoherence limits. In addition, the small errors in each pulse seem to accumulate quickly as the number of pulses increases. This brings forward the apparently still unresolved question about the "quantumness" of ensemble quantum computing: Is it basically a simulation of actual quantum computing rather than the real thing?

To initiate a computation, one needs to set the pure component to a specific state, typically a state with all qubits in zero state. This is not easy to do and several methods have been applied in practice. Similarly, one can detect the state of the system at the end of a computation, but it is not possible to make projective measurements on selected spins during the computation process (which is a basic ingredient in many error correcting schemes). All this leads to the problem that ensemble quantum computing is not a very suitable candidate, for example, for grid-like quantum computing.

Despite the problems listed above, one should not neglect the many successes of liquid NMR quantum computing in demonstrating or benchmarking the basic operations and algorithms:

- Four-item search (Grover) with chloroform (decoherence time$\simeq 0.3$ s, computation time $\simeq 0.035$ s), in 1998 (Chuang et al. 1998). The two qubits were H and C nuclei. Another demonstration at the same time used partially deuterated cytosine, with two H nuclei as qubits (Jones et al. 1998). In 2000, a three-qubit version with ^{13}C-labeled $CHFBr_2$ with H, C, and F nuclei as qubits was demonstrated (Vandersypen et al. 2000), and in a more flexible way in 2002 with ^{13}C-labeled alanine, with three C nuclei as spins (Kim et al. 2002).
- Factoring of the number 15 in 2001 with 7-qubit perfluorobutadienyl iron complex (2 C and 5 F nuclei) (Vandersypen et al. 2001).
- The largest known qubit number: 12, with *l*-histidine (Negrevergne et al. 2006).

In addition, the NMR approach has also given special impetus to developing the concept of adiabatic quantum computing. Many reviews on liquid NMR quantum computing have been published during the last 10 years. Among the most comprehensive is the 2004 review article by Vandersypen and Chuang (2004), who have been involved closely with the theoretical and experimental development of the field. Another source is the recent multiauthored book (Oliveira et al. 2007).

6.2 Nuclear and Electron Spins in Silicon

One of the characteristics of solid-state systems is that the electron spin can also play a role. While the nuclear spin can be addressed with the NMR methods, the electron spin-resonance (ESR) method becomes an equivalent tool for control (and not forgetting ENDOR, electron-nuclear spin double resonance). Computing with quantum dots is based only on electrons and will be discussed in ❷ Sect. 7.

A much noted suggestion for using spins in solids was made by Kane (1998). His vision was a purified silicon lattice made of zero-spin ^{28}P atoms, with embedded spin 1/2 ^{31}P atoms (donors) at regular nanoscale intervals (10–20 nm), and at an equivalent distance from the silicon surface. The hyperfine interaction couples the donor nuclear spin strongly with the donor electron. The advantage is that these spins have very long dephasing times that could, at very low temperatures, reach 10^6 s. Low temperatures are also needed to inhibit the actual ionization of the donors. This must be compared with the expected gate operation time, for which Kane estimated 10^{-5} s. Both figures are probably optimistic but clearly the numbers look a priori promising.

The surface of the silicon lattice would be covered with some barrier material such as SiO_2 (to isolate the donor electrons from the leads), and on top of that one would place the metallic leads (A), positioned above each qubit. With the localized constant magnetic field ($B \sim 2$ T) one can split the degeneracy of the hyperfine states, and then further control the hyperfine resonance frequencies of each qubit independently with a current in the A lead (Stark shift induced by the current-generated electric field). Thus, an applied oscillating field ($B_{AC} \sim 10^{-3}$ T) would create transitions only at the selected qubit since others would be off-resonant.

The large distance between the donors ensures that they can be controlled independently, and electron-mediated interaction between the donor nuclear spins does not play a role. Conditional logic is, however, achieved through this interaction. Additional leads (J) will be placed between the A leads. When charged positively, they pull donor electrons from adjacent qubits and the electron-enhanced interaction, dependent on the spin states (exchange interaction between the electron spins), is sufficiently strong for gate operations, and its duration is controlled by the J lead. This effects conditional logic in very much the same manner as with liquid NMR systems.

The method is in principle scalable, but it is also quite challenging technologically. Further problems are produced by the readout, since now one deals with single spins, so the NMR signal is not sufficiently large for detection (in liquid NMR, this was circumvented by large numbers). The original scheme by Kane was to combine tunneling with single-electron transistor detectors, but this is considered to be problematic both technologically as well as in the light of decoherence times.

An alternative approach that uses the donor electrons was proposed in 2000 (Vrijen et al. 2000). The approach relies on a layered Si–Ge structure. The important gain is the increase in the distance between qubits by an order of magnitude, to make lithographic implementation easier. The role of the ^{31}P nuclei is now only to provide the location for the donor electrons. Si and Ge have different electronic g factors, so by moving electrons vertically between two Si–Ge layers with different Si/Ge proportion, one can modify the electron spin resonance for control with ESR (instead of NMR). Operation with electron spins alone speeds up the gate rate (75 MHz vs. 1 GHz). Conditional logic is still achieved with the increase of the electron wavefunction overlap but now this is achieved with the A gates (the possibility to achieve conditional logic with exchange interaction in general was suggested in Loss and DiVincenzo (1998) in the context of quantum dot computers; see also DiVincenzo et al. (2000).

The measuring scheme (conversion of spin into charge) is also simplified. But many of the technological challenges remain, such as the ^{31}P implantation in regular lattices. Although some progress has been made in the fabrication of the Si:P materials (see, e.g., O'Brien et al. 2001; Jamieson et al. 2005), experimental advances in silicon-based quantum computing are still in the future. Laboratory research on the Si:P computers is nevertheless carried out vigorously; see, for example, Andresen et al. (2007), but clearly there are still many problems remaining even for a demonstration of a two-qubit action. Recently, coherent qubit state transfer between nuclear and electron spins was reported (Morton et al. 2008), though under the name of quantum memory rather than quantum computer studies.

6.3 Nitrogen Vacancies in Diamond

Although single-crystal diamond is nowadays considered an old-fashioned form of carbon, compared to fullerenes and graphene, for quantum computing purposes it has recently emerged as a lucrative system (Prawer and Greentree 2008). Impurities in diamond appear as color centers, giving the specific colors for emeralds, rubies, and sapphires. Proper diamonds are not usually completely transparent but have a yellowish shade, for which there is a rigorous classification scheme. This shade is due to nitrogen impurities in diamond.

The basic ingredient for diamond qubits is a nitrogen vacancy (NV). It means that the nitrogen atom (N) replaces one of the carbon atoms, and in addition next to it there is a vacancy (V) in the carbon lattice. Typically, vacancies move in the lattice, until they find a nitrogen impurity, with which they then form a highly stable local structure. For a production of these NV centers, one can use chemical vapor-deposition or ion-implantation. The electron structure of the NV center is a spin triplet ground state that can be optically coupled with the excited triplet state (giving the yellow color, of course). Thus, one can induce and observe quantum coherences on the level of a single system in the same manner as one would observe single ions in electromagnetic traps (Jelezko et al. 2004). Even quantum jumps have been observed with such a transition.

The triplet ground state is split by the dipolar interaction so that the $m_s = 0$ state is the lowest state, and $m_s = \pm 1$ states are degenerate. This degeneracy of the $m_s = \pm 1$ states can be lifted with an external magnetic field. The splitting of the Zeeman substates is about 2.9 GHz, and can be accessed with microwave fields. The interesting point is that optical excitation is only allowed from the $m_s = 0$ state so that it can be used to detect the occupancy on that state.

A nearby ^{13}C nuclear spin can couple with one of the NV electrons through hyperfine interaction, giving an extra 130 MHz splitting to the electronic states. Thus, one has four states involving the nondegenerate combinations of the electron and nuclear spin. Then one can use optical pumping for the preparation of the initial state, and manipulation with microwave and radio-frequency fields, and make the read-out with a laser again. With this, one can effect conditional logic between the qubits in nuclear and electron spins (Jelezko et al. 2004). This is reminiscent of the first trapped ion two-qubit gate, where the qubits were stored in a single ion. The coherence properties of the electron spin have been determined experimentally (Childress et al. 2006), and recently a second ^{13}C atom was added to demonstrate the quantumness of the system by creation and detection of tripartite quantum entanglement (Neumann et al. 2008). In this room-temperature experiment, the weak point is the electron spin decoherence, which is a few microseconds. If one limits to entangling only the two nuclear spins, the time scale is pushed into the millisecond region.

The main issue with NV centers is obviously scalability, which cannot be achieved with the present setup. One solution could be an optical bus between, for example, different small crystals that host small numbers of qubits. But for the time being the NV centers are mainly suitable for demonstration purposes.

7 Solid-State Qubits with Quantum Dots

Quantum dots emerged as suitable candidates for quantum computing very early (Barenco et al. 1995; Loss and DiVincenzo 1998; Sherwin et al. 1999). The discrete energy states of the dots provide a tempting qubit representation, and the semiconductor technology is highly developed at nanoscale due to the continuous effort to follow Moore's law. The possibility for optical manipulation and the use of electron spin (spintronics) extend the possibilities further (see, e.g., the book Awschalom et al. 2002).

A quantum dot is usually obtained by preparing layered heterostructures, with electrons on the conduction band trapped into one of the layers (quantum wells), and the 2D confinement is augmented with trapping of electrons with fields provided by the electrodes that are etched on the surface of the structure (lateral quantum dot structure). One can obtain single-electron trapping by using the Coulomb blockade: a charge can only tunnel into a dot from the adjacent layer if no other charges are there, as they would block the tunneling by electric repulsion. This aspect can be further controlled by external electric fields that adjust the energy states of the dot in relation to the levels at adjacent layers. This is the basis for single-electron transistors and other such devices. The above is an example of laterally positioned and coupled dots, but vertical, multilayered structures are also possible (Burkard et al. 1999).

For scalable qubits, one could use the quantized motional states of the electrons, and for conditional logic the dipole–dipole interaction, which can be made to depend on the electron state by external fields that affect the electronic charge distributions (as proposed by Barenco et al. 1995). Alternatively, one can use the electron spin as a qubit, control the spin states by ESR fields, and use the spin-dependent exchange interaction for conditional logic, as proposed by Loss and DiVincenzo (1998).

The spin-based systems have been popular and extensive studies have been performed, not only on quantum computation but also on the dynamics and observation of spins in few electron quantum dots (see the recent review by Hanson et al. (2007)). Having more than one electron in a dot creates a singlet-triplet electron-spin structure that can be addressed with magnetic fields. An alternative system with a singlet-triplet structure is a double quantum dot, with a controllable interdot tunneling, see, for example, (Koppens et al. 2007). The tunneling is also controlled by a spin blockade, that is, to conserve the total spin a change between singlet and triplet states (and thus a specific tunneling event between dots) can only occur if a spin-flip takes place as well. The observed coherence times (Koppens et al. 2006) for a single qubit are promising on the order of microseconds (at temperatures below 1 K), but further work is required to estimate the possibilities of spin-based quantum dot qubits.

Finally, electrons can be transferred from the valence band to the conduction band by optical transition; thus one obtains an exciton, that is, an electron–hole pair, that can act as a qubit (Troiani et al. 2000). One dot can possess more than one exciton. In fact, one of the first demonstrations of a quantum dot quantum gate (Li et al. 2003) took place (in analogy to the trapped ion demonstration (Monroe et al. 1995)) in a single quantum dot with such a biexciton as the two-qubit system. The gate operation relied on different light polarizations since for the qubits the value "1" corresponded to different electron spins in the two electron–hole pairs. The gate that was performed was actually the CROT gate. Earlier, entanglement between the excitons was demonstrated in such a system (Chen et al. 2000). To make the excitonic computers scalable is a challenge, though.

In general, the semiconductors provide a challenging environment for quantum phenomena. In addition to coupling with phonons, the quantum dot electrons feel the other electrons through spin–orbit coupling, and the nuclear spins of the background material via hyperfine interaction. Apparently the latter mechanism is found to be the more relevant mechanism in experiments (Koppens et al. 2005, 2006). For excitons, the recombination of electrons and holes by spontaneous emission (coupling to the electromagnetic degrees of freedom) is a serious obstacle. It is possible to reach gate times on the order of 1 ps, and recombination times on the order of hundreds of picoseconds have been obtained (see discussion in Li et al. 2003), so more than a few gate operations could be achieved. For scalability, qubits in different dots could be coupled by setting them into an optical microcavity, so that excitons at different dots couple to the same cavity mode, and even interactions mediated by intermediate excitons have been proposed; see Chapter 7 in Chen et al. (2007) and references therein.

8   Superconducting Qubits

Superconductivity is a quantum phenomenon that takes place for certain materials at low temperatures (Tinkham 2004). The electric current runs then without resistance in the material, and such currents are quite unaffected by external conditions, except for sufficiently strong magnetic fields. In the standard case of superconductivity, the electrons in the material form Cooper pairs that are the unstoppable current carriers, and the resistance can occur only by a mechanism that provides sufficient energy that breaks the pair. However, the Cooper pairs and the single-electron states are separated by a distinct pairing gap in energy, which at low temperatures exceeds the thermal energy in the system.

Thus superconductivity is a prime example of a macroscopic quantum coherent system, and one can attach to the degenerate electron gas a wavefunction-like order parameter that has a specific phase. The spatial gradient of that phase gives the superconducting current. Superfluidity of liquids and Bose–Einstein condensation of atomic or molecular gases are similar low-temperature examples with overall phase coherence. The connection between the phase and current means that since the phase change in a closed loop can only be a multiple of 2π, the flux related to it must be quantized. If the loop can shrink without limit in the material, it leads to quantized vortices, which are observed especially with rotated or stirred liquids and gases (Tilley and Tilley 1990). For superconductors, one can build circuits so that the loops cannot shrink to zero size, and thus one can observe the quantization of the persistent superconducting current.

A superconducting current can pass through small regions of insulating material by quantum tunneling. One mechanism is by breaking the Cooper pair during the process (quasiparticle tunneling), or the Josephson tunneling of Cooper pairs. Typically, the latter effect is the more interesting one, and these junctions are often called Josephson or tunneling junctions. It is a quantum phenomena, since it is controlled by the phase difference ϕ over the junction, which means that the tunneling is a coherent process. The time derivative of ϕ gives the voltage over the junction, and the current through it is $I = I_c \sin(\phi)$, where I_c is the critical current that sets the upper limit. Interestingly, a current can exist across the junction even if the phase difference is constant and there is no voltage difference (DC Josephson effect). An externally created electric potential V across the junction gives a phase difference $\phi = 2eVt/h$ and thus an alternating current (AC Josephson effect). The maximum of the current depends on the energy gap of the superconducting material and the temperature. SQUIDs (superconducting quantum interference devices) are based on the sensitivity of this current to external magnetic fields through the field-dependence of the energy gap. For actual insulating material, the barrier thickness should not exceed a few hundred nanometers, but for normal state metal one can go to the micrometer scale.

8.1 Charge Qubits

If one prepares a simple setup with one Josephson junction and two superconducting regions, coupled into a loop by a voltage source, one gets the so called Cooper pair box. Typically, one of the superconducting regions is small and is called an island. The fascinating aspect is that although we have countless Cooper pairs on the island, its electrostatic properties can be governed by a single Cooper pair. We can tune the system so that, initially, there is no net charge on the island (zero excess Cooper pairs). The junction has capacitance C, and there is another one (C_e) between the island and the external circuit. The Hamiltonian for the system is then (Nakahara and Ohmi 2008)

$$H = \frac{1}{2}E_c(n - n_e)^2 - E_J \cos \phi, \quad E_c = \frac{(2e)^2}{C + C_e}, \quad n_e = \frac{C_e V_e}{2e}, \quad E_J = \frac{I_c \hbar}{2e} \quad (21)$$

where V_e is the applied external voltage across the system and n is the Cooper pair number of the island. Note that n_e is a continuous quantity whereas n is limited to integer values.

It shows that one can have a field-induced degeneracy between n and $n + 1$ Cooper pairs on an island. At this degeneracy, tunneling is possible, and the two charge states are in fact coupled by the tunneling coupling E_J (this is not a true coupling; it appears as a frame of reference term since we write the system in terms of Cooper pairs localized on either side of the junction, which is not true because of tunneling). The charge number n and the junction phase ϕ form a conjugate pair of quantities (in analogy to x and p), and thus $\cos \phi$ can be interpreted as a term that induces $n \to n \pm 1$ transitions between the charge parabola.

One can also have a superposition of charge states, and this forms the charge qubit (Shnirman et al. 1997; Makhlin et al. 1999, 2000). By driving the system into the resonance point (changing n_e) one can observe coherent oscillations (Rabi oscillations) driven by E_J, or by traversing the degeneracy with fixed speed one obtains Landau–Zener transitions. Coherent oscillations and their decay were observed already in 1999 (Nakamura et al. 1999). Such charge qubits can be coupled, for example, capacitatively, as demonstrated in 2003 (Pashkin et al. 2003), followed closely by a demonstration of conditional logic (Yamamoto et al. 2003). The superconducting systems are very robust to perturbations, but the electronics needed for manipulating and detecting the qubit states seem to be a sufficient source for non-negligible noise.

8.2   Flux Qubits

Parallel to the development of superconducting charge qubits, another idea emerged, namely the use of the current itself (Mooij et al. 1999; Friedman et al. 2000). The idea is to prefabricate the junction as a part of a small ring of superconducting material, so that the tunneling coupling E_J dominates over the charge state separation E_C. This is the standard RF-SQUID setup, and such a ring can then have a persistent current in it, controlled by the external magnetic field flux Φ_e. One can show that, due to the existence of quantum phase coherence in the ring, the total magnetic flux Φ through the ring must obey a quantization rule, so that $\Phi = (\phi + 2\pi n)\Phi_0/(2\pi)$, where the flux quantum is $\Phi_0 = h/(2e)$ and ϕ is the phase change over the junction. It means that to satisfy this relation, a current I arises in the ring, so that $\Phi = \Phi_e + LI$, where L is the self-inductance of the ring. As described, for example, in Friedman et al. (2000), the flux can have the appearance of a position-like quantity, with kinetic energy $C\dot{\Phi}^2/2$ and potential

$$U = U_0 \left[\frac{1}{2} \left(\frac{2\pi(\Phi - \Phi_e)}{\Phi_0} \right)^2 - \beta \cos(2\pi\Phi/\Phi_0) \right] \tag{22}$$

where C is the junction capacitance, $U_0 = \Phi_0^2/(4\pi^2 L)$ and $\beta = 2\pi L I_c/\Phi_0$, with I_c again as the critical current of the junction.

The above potential is for practical purposes a double-well structure with two minima. The β term arises from the phase change over the junction. For $\Phi_e = \Phi_0/2$ one obtains a symmetric structure, and let one consider it for simplicity. The two wells correspond to flux quanta $n = 0$ and $n = 1$, and to equal currents flowing in opposite directions. These fluxes and accompanying supercurrents form the flux qubit. At the bottom of the wells and for large E_J they are quite stable and robust to perturbations. For smaller E_J, the true lowest quantum states spread over the whole double-well and are roughly the symmetric and antisymmetric superpositions of the localized states, and their degeneracy is split by the tunnel coupling. Variations in Φ_e make the potential antisymmetric, separating the left- and right-localized states in energy, and thus reducing the effect of the tunnel coupling. Such coherent superpositions were observed experimentally in 2000 (Friedman et al. 2000; van der Wal et al. 2000).

The system can be further refined by adding more junctions, such as three or four (Mooij et al. 1999); this makes the size of the setup smaller and reduces possible disturbances. This is because with one junction a large inductance L and, thus, a large size is needed to allow for the generation of the sufficiently large compensation flux; in the multijunction setup the compensation flux can be as small as $10^{-3}\Psi_0$. The double-well description is nevertheless still fine (van der Wal et al. 2000). The energy separation of the flux states is in the microwave region and one can thus perform single-qubit operations; here the disturbances in the external flux cause decoherence, and its time scale has been estimated to be around 20 ns by considering Ramsay interferometry with two $\pi/2$ pulses (Chiorescu et al. 2003).

The state of a flux qubit can be read out with various schemes, see van der Wal et al. (2000) and Chiorescu et al. (2003). The conditional logic can be obtained, for example, by circling the qubits with another superconducting ring so that a four-state system emerges, with two-qubit states that can be manipulated with microwave fields (Mooij et al. 1999; Hime et al. 2006; van der Ploeg et al. 2007). Inductive coupling can be achieved by making the superconducting boxes for two qubits share one edge (Majer et al. 2005). A promising scheme is to use the shared edge method but having an auxiliary qubit between the actual qubits with a large tunneling energy, so that it can be adiabatically eliminated. With this method, conditional logic has been demonstrated with four-junction qubits experimentally (Niskanen et al. 2007).

8.3 Phase Qubits and Circuit QED

One can also place a strong bias current $I_e \simeq I_c$ on the Josephson junction when $E_J \gg E_c$. The potential for the phase of the current becomes a washboard potential that has shallow local minima with a few bound states. A similar situation can be obtained with the RF-SQUID by setting a large Φ_e that tilts the potential strongly. The two lowest states form a phase qubit. Transitions between them can be driven with microwave fields, and detection can happen by state-dependent tunneling that can be observed as a direct current. The qubit circuits are coupled with a capacitor, and one can excite entangled states with microwave radiation; this has been demonstrated experimentally (Berkley et al. 2003; Steffen et al. 2006), but an actual demonstration of conditional logic is still missing. Decoherence times are on the order of 100 ns, compared to the duration of single qubit operations that is an order of magnitude less (Steffen et al. 2006).

In above examples, the superconducting system states were usually controlled with microwave radiation. This has also led to the idea of using microwave cavities for transferring quantum information (Blais et al. 2004) and, hence, to the concept of circuit QED. A coupling of a single charge qubit to a cavity mode has been demonstrated (Shuster et al. 2004). Recently, such a coupling of two phase qubits was demonstrated (Majer et al. 2007). The qubits consisted of two parallel junctions so that the tunneling coupling and, thus, the energy separation for the qubit states were controllable with external magnetic flux through the qubit system. Both qubits were coupled to a finite-length coplanar waveguide and separated by a distance of about 4 mm. They did not exchange actual photons with the cavity because they were off-resonant with it, but their presence induced a dispersive qubit–qubit coupling that shifted resonances for joint states, which was spectroscopically detected. Also, resonant coupling of a phase qubit to the cavity mode has been demonstrated, including the transfer of quantum information from the phase qubit to the photon (Houck et al. 2007). The actual exchange of a photon between two qubits through a resonant cavity coupling has been demonstrated as well (Sillanpää et al. 2007). Although still at a very preliminary phase, this approach is especially interesting as it opens up the possibility of well-separated qubits.

9 All-Optical Quantum Computing

Photons are the ultimate tools for quantum communication, so it is natural to ask if they can be applied in large-scale quantum computing as well. The problem with photons is that they are generated or destroyed quite easily, and are thus very different from more physical forms of qubit systems. Also, they do not interact but require some medium for the process, so conditional logic with linear optical elements is not possible (starting with Bell state detection that would be simple if a linear optics CNOT gate was available (Lütkenhaus et al. 1999)). However, it is possible to circumvent this problem by using photodetectors and feedback loops as was shown in 2001 (Knill et al. 2001).

Demonstrations for simple operations in all-optical quantum computing have been performed (see, e.g., O'Brien et al. 2003), but it remains to be seen whether the actual large scale realization is a feasible goal. For a nice recent review, O'Brien (2007) is suggested.

10 Closing Remarks

The large-scale implementation of quantum computing is still clearly far beyond current experimental capabilities, as the examples given in this review demonstrate. Many ideas are only at a preliminary stage although vigorous work is going on. At least the large number of Science and Nature articles show that many of the proposed systems have broad interest, hopefully because they either hold great promise or are based on ingenious theoretical ideas and innovative experimental work. Both are definitely needed. It has been 15 years since the appearance of the Shor algorithm, but we are not yet even close to a working quantum computer. The future will show how well the attractive character of quantum information persists. Since all these attempts improve our understanding of quantum mechanics and its manifestations, and are usually linked with nanophysics research, it is quite possible that the most valuable outcome will eventually be something other than a working, large-scale quantum computer.

Acknowledgments

The author acknowledges the financial support by the Academy of Finland, the Väisälä Foundation, and the Magnus Ehrnrooth Foundation.

References

Andresen SES, Brenner R, Wellard CJ et al. (2007) Charge state control and relaxation in an atomically doped silicon device. Nano Lett 7:2000–2003

Awschalom DD, Loss D, Samarth N (eds) (2002) Semiconductor spintronics and quantum computation. Springer, Berlin

Barenco A, Deutsch D, Ekert A (1995) Conditional quantum dynamics and logic gates. Phys Rev Lett 74:4083–4086

Barenco A, Ekert A, Suominen K-A, Törmä P (1996) Approximate quantum Fourier transform and decoherence. Phys Rev A 54:139–146

Benhelm J, Kirchmair G, Roos CF, Blatt R (2008) Towards fault-tolerant quantum computing with trapped ions. Nat Phys 4:463–466

Bennett CH, DiVincenzo DP, Smolin JA, Wootters WK (1996) Mixed state entanglement and quantum error correction. Phys Rev A 54:3824–3851

Berkley AJ, Xu H, Ramos RC et al. (2003) Entangled macroscopic quantum states in two superconducting qubits. Science 300:1548–1550

Birkl G, Kassner S, Walther H (1992) Multiple-shell structures of laser-cooled ^{24}Mg$^+$ ions in a quadrupole storage ring. Nature 357:310–313

Briegel HJ, Raussendorf R (2001) Persistent entanglement in arrays of interacting particles. Phys Rev Lett 86:910–913

Blais A, Huang R-S, Wallraff A, Girvin SM, Schoelkopf RJ (2004) Cavity quantum electrodynamics for superconducting electrical circuits: an architecture for quantum computation. Phys Rev A 69:062320

Bloch I (2005) Ultracold quantum gases in optical lattices. Nat Phys 1:23–30

Bremner MJ, Dawson CM, Dodd JL et al. (2002) Practical scheme for quantum computation with any two-qubit entangling gate. Phys Rev Lett 89:247902

Burkard G, Loss D, DiVincenzo DP (1999) Coupled quantum dots as quantum gates. Phys Rev B 59:2070–2078

Calderbank AR, Shor PW (1996) Good quantum error-correcting codes exist. Phys Rev A 54:1098–1105

Chen G, Bonadeo NH, Steel DG et al. (2000) Optically induced entanglement of excitons in a single quantum dot. Science 289:1906–1909

Chen G, Church DA, Englert B-G et al. (2007) Quantum computing devices. Chapman & Hall/CRC, Boca Raton, FL

Childress L, Gurudev Dutt MV, Taylor JM et al. (2006) Coherent dynamics of coupled electron and nuclear spin qubits in diamond. Science 314:281–285

Childs AM, Farhi E, Preskill J (2001) Robustness of adiabatic quantum computation. Phys Rev A 65:012322

Chiorescu I, Nakamura Y, Harmans CJPM, Mooij JE (2003) Coherent quantum dynamics of a superconducting flux qubit. Science 299:1869–1871

Chuang IL, Gershenfeld N, Kubinec M (1998) Experimental implementation of fast quantum searching. Phys Rev Lett 80:3408–3411

Chuang IL, Vandersypen LMK, Zhou X et al. (1998) Experimental realization of a quantum algorithm. Nature 393:143–146

Cirac JI, Zoller P (1995) Quantum computations with cold trapped ions. Phys Rev Lett 74:4091–4094

Cleve R, Ekert A, Macchiavello C, Mosca M (1998) Quantum algorithms revisited. Proc R Soc Lond A 454:339–354

Coppersmith D (1994) An approximate Fourier transform useful in quantum factoring. IBM Research Report RC19642; see also arXiv quant-ph/0201067 (2002)

Dahm AJ, Heilman JA, Karakurt I, Peshek TJ (2003) Quantum computing with electrons on helium. Physica E 18:169–172

DiVincenzo DP (1995) Quantum computation. Science 270:255–261

DiVincenzo DP (2000) The physical implementation of quantum computation. Fortschritte der Physik 48:771–783

DiVincenzo DP, Bacon D, Kempe J et al. (2000) Universal quantum computation with the exchange interaction. Nature 408:339–342

Duan L-M, Guo G-C (1997) Preserving coherence in quantum computation by pairing quantum bits. Phys Rev Lett 79:1953–1956

Dykman MI, Platzman PM (2000) Quantum computing using electrons floating on liquid helium. Fortschritte der Physik 48:1095–1108

Dykman MI, Platzman PM, Seddighrad P (2003) Qubits with electrons on liquid helium. Phys Rev B 67:155402

van Enk SJ, Kimble HJ, Mabuchi H (2004) Quantum information processing in cavity-QED. Quantum Inf Process 3:75–90

Farhi E, Goldstone J, Gutmann S, Sipser M (2000) Quantum computation by adiabatic evolution. arXiv quant-ph/0001106

Feynman RP (1982) Simulating physics with computers. Int J Theor Phys 21:467–488

Friedman JR, Patel V, Chen W et al. (2000) Quantum superposition of distinct macroscopic states. Nature 406:43–46

Gaitan F (2008) Quantum error correction and fault tolerant quantum computing. CRC Press, Boca Raton, FL

Gershenfeld NA, Chuang IL (1997) Bulk spin-resonance quantum computation. Science 275:350–356

Gottesman D (1996) Class of quantum error-correcting codes saturating the quantum Hamming bound. Phys Rev A 54:1862–1868

Grimes CC, Adams G (1979) Evidence for a liquid-to-crystal phase transition in a classical, two-dimensional sheet of electrons. Phys Rev Lett 42:795–798

Grover LK (1997) Quantum mechanics helps in searching for a needle in a haystack. Phys Rev Lett 79:325–328

Häffner H, Roos CF, Blatt R (2008) Quantum computing with trapped ions. Phys Rep 469:155–203

Hanson R, Kouwenhoven LP, Petta JR et al. (2007) Spins in few-electron quantum dots. Rev Mod Phys 79:1217

Haroche S, Raimond J-M (1996) Quantum computing: dream or nightmare? Phys Today 49(8):51–52

Haroche S, Raimond J-M (2006) Exploring the quantum: atoms, cavities and photons. Oxford University Press, Oxford

Hime T, Reichardt PA, Plourde BLT et al. (2006) Solid-state qubits with current-controlled coupling. Science 314:1427–1429

Hirvensalo M (2004) Quantum computing, 2nd edn. Springer, Berlin

Horvath GZsK, Thompson RC, Knight PL (1997) Fundamental physics with trapped ions. Contemp Phys 38:25–48

Houck AA, Schuster DI, Gambetta JM et al. (2007) Generating single microwave photons in a circuit. Nature 449:328–331

Jamieson DN, Yang C, Hopf T et al. (2005) Controlled shallow single-ion implantation in silicon using an active substrate for sub-20-keV ions. Appl Phys Lett 86:202101

Jelezko F, Gaebel T, Popa I et al. (2004) Observation of coherent oscillations in a single electron spin. Phys Rev Lett 92:076401

Jelezko F, Gaebel T, Popa I et al. (2004) Observation of coherent oscillation of a single nuclear spin and realization of a two-qubit conditional quantum gate. Phys Rev Lett 93:130501

Jones JA (2000) NMR quantum computation: a critical evaluation. Fortschritte der Physik 48:909–924

Jones JA, Mosca M, Hansen RH (1998) Implementation of a quantum search algorithm on a quantum computer. Nature 393:344

Kane BE (1998) A silicon-based nuclear spin quantum computer. Science 393:133–137

Kielpinski D, Monroe C, Wineland DJ (2002) Architecture for a large-scale ion-trap quantum computer. Nature 417:709

Kim J, Lee J-S, Lee S (2002) Experimental realization of a target-accepting quantum search by NMR. Phys Rev A 65:054301

Kitaev A (1997) Fault-tolerant quantum computation by anyons. Ann Phys 303:2–30. see also arXiv quant-ph/9707021

Knill E (2005) Quantum computing with realistically noisy devices. Nature 434:39–44

Knill E, Laflamme R, Milburn GJ (2001) A scheme for efficient quantum computation with linear optics. Nature 409:46–52

Koppens FHL, Buizert C, Tielrooij KJ et al. (2006) Driven coherent oscillations of a single electron spin in a quantum dot. Nature 442:766–771

Koppens FHL, Folk JA, Elzerman JM et al. (2005) Control and detection of singlet-triplet mixing in a random nuclear field. Science 309:1346–1350

Laflamme R, Miquel C, Paz JP, Zurek WH (1996) Perfect quantum error correcting code. Phys Rev Lett 77:198–201

Li X, Wu Y, Steel D et al. (2003) An all-optical quantum gate in a semiconductor quantum dot. Science 301:809–811

Lidar DA, Chuang IL, Whaley KB (1998) Decoherence-free subspaces for quantum computation. Phys Rev Lett 81:2594–2597

Loss D, DiVincenzo DP (1998) Quantum computation with quantum dots. Phys Rev A 57:120–126

Lütkenhaus N, Calsamiglia J, Suominen K-A (1999) Bell measurements for teleportation. Phys Rev A 59:3295–3300

Lyon SA (2006) Spin-based quantum computing using electrons on liquid helium. Phys Rev A 74:052338

Majer J, Chow JM, Gambetta JM et al. (2007) Coupling superconducting qubits via a cavity bus. Nature 449:443–447

Majer JB, Paauw FG, ter Haar ACJ et al. (2005) Spectroscopy on two coupled superconducting flux qubits. Phys Rev Lett 94:090501

Makhlin Y, Schön G, Shnirman A (1999) Josephson-junction qubits with controlled couplings. Nature 398:305–307

Makhlin Y, Schön G, Shnirman A (2000) Quantum-state engineering with Josephson-junction devices. Rev Mod Phys 73:357–400

Mermin ND (2007) Quantum computer science: an introduction. Cambridge University Press, Cambridge

Metcalf HJ, Van der Straten P (1999) Laser cooling and trapping. Springer, Berlin

Mollin RA (2002) RSA and public-key cryptography. CRC Press, Boca Raton, FL

Monroe C, Meekhof DM, King BE et al. (1995) Demonstration of a fundamental quantum logic gate. Phys Rev Lett 75:4714–4717

Monz T, Kim K, Hänsel W et al. (2009) Realization of the quantum Toffoli gate with trapped ions. Phys Rev Lett 102:040501

Mooij JE, Orlando TP, Levitov L et al. (1999) Josephson persistent-current qubit. Science 285:1036–1039

Morton JJL, Tyryshkin AM, Brown RM et al. (2008) Solid-state quantum memory using the ^{31}P nuclear spin. Nature 455:1085–1088

Nägerl HC, Bechter W, Eschner J et al. (1998) Ion strings for quantum gates. Appl Phys B 66:603–608

Nakahara M, Ohmi T (2008) Quantum computing: from linear algebra to physical realizations. CRC Press, Boca Raton, FL

Nakamura Y, Pashkin YA, Tsai JS (1999) Coherent control of macroscopic quantum states in a single-Cooper-pair box. Nature 398:786–788

Nayak C, Simon SH, Stern A et al. (2008) Non-Abelian anyons and topological quantum computation. Rev Mod Phys 80:1083

Negrevergne C, Mahesh TS, Ryan CA et al. (2006) Benchmarking quantum control methods on a 12-qubit system. Phys Rev Lett 96:170501

Neumann P, Mizuochi N, Rempp F et al. (2008) Multipartite entanglement among single spins in diamond. Science 320:1326–1329

Nielsen MA, Chuang IL (2000) Quantum computation and quantum information. Cambridge University Press, Cambridge

Niskanen AO, Harrabi K, Yoshihara F et al. (2007) Quantum coherent tunable coupling of superconducting qubits. Science 316:723–726

O'Brien JL (2007) Optical quantum computing. Science 318:1567–1570

O'Brien JL, Pryde GJ, White AG et al. (2003) Demonstration of an all-optical quantum controlled-NOT gate. Nature 426:264–267

O'Brien JL, Schofield SR, Simmons MY et al. (2001) Towards the fabrication of phosphorus qubits for a silicon quantum computer. Phys Rev B 64:161401

Oliveira I, Sarthour R Jr, Bonagamba T et al. (2007) NMR quantum information processing, Elsevier, Amsterdam

Palma GM, Suominen K-A, Ekert AK (1996) Quantum computers and dissipation. Proc R Soc Lond A 452:567–584

Pashkin YA, Yamamoto T, Astafiev O et al. (2003) Quantum oscillations in two coupled charge qubits. Nature 421:823–826

Pethick CJ, Smith H (2008) Bose-Einstein condensation in dilute gases, 2nd edn. Cambridge University Press, Cambridge

Platzman PM, Dykman MI (1999) Quantum computing with electrons floating on liquid helium. Science 284:1967–1969

van der Ploeg SHW, Izmalkov A, van den Brink AM et al. (2007) Controllable coupling of superconducting flux qubits. Phys Rev Lett 98:057004

Prawer S, Greentree AD (2008) Diamond for quantum computing. Science 320:1601–1602

Raussendorf R, Briegel HJ (2001) A one-way quantum computer. Phys Rev Lett 86:5188–5191

Rousseau E, Ponarin D, Hristakos L et al. (2009) Addition spectra of Wigner islands of electrons on superfluid helium. Phys Rev B 79:045406

Sabouret G, Lyon SA (2006) Measurement of the charge transfer efficiency of electrons clocked on superfluid helium. App Phys Lett 88:254105

Sachdev S (1999) Quantum phase transitions. Cambridge University Press, Cambridge

Schaetz T, Leibfried D, Chiaverini J et al. (2004) Towards a scalable quantum computer/simulator based on trapped ions. Appl Phys B 79:979–986

Schmidt-Kaler F, Häffner H, Gulde S et al. (2003a) How to realize a universal quantum gate with trapped ions. Appl Phys B 77:789–796

Schmidt-Kaler F, Häffner H, Riebe M et al. (2003b) Realization of the Cirac-Zoller controlled-NOT quantum gate. Nature 422:408–411

Schuster DI, Blais A, Frunzio L et al. (2004) Strong coupling of a single photon to a superconducting qubit using circuit quantum electrodynamics. Nature 431:162–167

Sherwin MS, Imamoglu A, Montroy T (1999) Quantum computation with quantum dots and terahertz cavity quantum electrodynamics. Phys Rev A 60:3508–3514

Shnirman A, Schön G, Hermon Z (1997) Quantum manipulations of small Josephson junctions. Phys Rev Lett 79:2371–2374

Shor PW (1994) Algorithms for quantum computation: discrete log and factoring. Proceedings of the 35th annual IEEE symposium on foundations of computer science, FOCS, Paris, pp 20–22

Shor PW (1995) Scheme for reducing decoherence in quantum computer memory. Phys Rev A 52: R2493–R2496

Sillanpää MA, Park JI, Simmonds RW (2007) Coherent quantum state storage and transfer between two phase qubits via a resonant cavity. Nature 449: 438–442

Sjöqvist E (2008) A new phase in quantum computation. Physics 1:35

Sørensen A, Mølmer K (1999a) Quantum computation with ions in thermal motion. Phys Rev Lett 82: 1971–1974

Sørensen A, Mølmer K (1999b) Entanglement and quantum computation with ions in thermal motion. Phys Rev A 62:022311

Steane AM (1996) Error correcting codes in quantum theory. Phys Rev Lett 77:793–797

Steane AM (1998) Quantum computing. Rep Prog Phys 61:117–173

Steffen M, Ansmann M, Bialczak RC et al. (2006) Measurement of the entanglement of two superconducting qubits via state tomography. Science 313: 1423–1425

Stenholm S, Suominen K-A (2005) Quantum approach to informatics. Wiley, Hoboken, NJ

Stolze J, Suter D (2008) Quantum computing, revised and enlarged: a short course from theory to experiment, 2nd edn., Wiley-VCH, Weinheim

Tilley DR, Tilley J (1990) Superfluidity and superconductivity. Taylor & Francis, London

Tinkham M (2004) Introduction to superconductivity, 2nd edn. Dover, New York

Troiani F, Hohenester U, Molinari E (2000) Exploiting exciton-exciton interactions in semiconductor quantum dots for quantum-information processing. Phys Rev B 62:R2263–R2266

Turchette QA, Hood CJ, Lange W, Mabuchi H, Kimble HJ (1995) Measurement of conditional phase shifts for quantum logic. Phys Rev Lett 75:4710–4713

Unruh WG (1995) Maintaining coherence in quantum computers. Phys Rev A 51:992–997

Vandersypen LMK, Chuang IL (2004) NMR techniques for quantum control and computation. Rev Mod Phys 76:1037–1069

Vandersypen LMK, Steffen M, Sherwood MH et al. (2000) Implementation of a three-quantum-bit search algorithm. Appl Phys Lett 76:646–649

Vandersypen LMK, Steffen M, Breyta G et al. (2001) Experimental realization of Shor's quantum factoring algorithm using nuclear magnetic resonance. Nature 414:883–887

Vedral V (2006) Introduction to quantum information science. Oxford University Press, Oxford

Vrijen R, Yablonovitch E, Wang K et al. (2000) Electron-spin-resonance transistors for quantum computing in silicon-germanium heterostructures. Phys Rev A 62:012306

van der Wal CH, ter Haar ACJ, Wilhelm FK et al. (2000) Quantum superposition of macroscopic persistent-current states. Science 290:773–777

Yamamoto T, Pashkin YA, Astafiev O et al. (2003) Demonstration of conditional gate operation using superconducting charge qubits. Nature 425:941–944

Zanardi P, Rasetti M (1997) Noiseless quantum codes. Phys Rev Lett 79:3306–3309

45 Quantum Cryptography

Takeshi Koshiba
Graduate School of Science and Engineering, Saitama University, Japan
koshiba@mail.saitama-u.ac.jp

G. Rozenberg et al. (eds.), *Handbook of Natural Computing*, DOI 10.1007/978-3-540-92910-9_45,
© Springer-Verlag Berlin Heidelberg 2012

Abstract

Several results in quantum cryptography will be surveyed in this chapter. After a brief introduction to classical cryptography, we provide some cryptographic primitives from the viewpoint of quantum computational complexity theory, which are helpful to get an idea of quantum cryptographic protocols. We then examine cryptographic protocols of quantum key distribution, quantum bit commitment, quantum oblivious transfer, quantum zero-knowledge, quantum public-key encryption, quantum digital signature, and their security issues.

1 Introduction

Due to the rapid growth of electronic communication means, information security has become a crucial issue in the real world. Modern cryptography provides fundamental techniques for securing communication and information. While modern cryptography varies from encryption and digital signatures to cryptographic protocols, we can partition modern cryptography into public-key cryptography and secret-key cryptography. From the theoretical point of view, public-key cryptography has a computational complexity-theoretic flavor and secret-key cryptography has an information-theoretic flavor. Though the two disciplines of cryptography have different flavors, they are not separate from each other but rather complement each other.

The principal and classical task of cryptography is to provide confidentiality. Besides confidentiality, modern cryptography provides authentication, data integrity, and so on. Since Diffie and Hellman (1976) devised the notion of public-key cryptosystems, computational complexity theoretic approaches to cryptology have succeeded in theory and practice. The fundamental study of one-way functions and pseudorandom generators has developed computational complexity theory. Cryptographic protocols such as digital signatures, commitment schemes, oblivious transfer schemes, and zero-knowledge proof systems have contributed to building various information security systems. For each objective mentioned above, we should make models of adversaries so as to enable discussion of whether some cryptographic protocols or methods fulfill the objective. For public-key cryptography, it is typically thought that the adversary should be a probabilistic polynomial-time Turing machine or a polynomial-size circuit family. For secret-key cryptography, the adversary might be the Almighty or those who have some specific power.

It is when the adversary is physically realized that it can become a real threat. It is known that the real world behaves quantum mechanically. Thus, we may suppose that the adversary should run quantum mechanically. Shor's algorithm on quantum computers for the integer factorization problem (Shor 1997) illustrates that quantum adversaries would spoil the RSA cryptosystem (Rivest et al. 1978), which is widely used for secure communications. Shor also proposed an efficient quantum algorithm for the discrete logarithm problem. Moreover, since the security of many cryptographic protocols relies on the computational hardness of these two problems, cryptographic protocols might have to be reconstructed to maintain information security technologies in the future if the quantum adversary has a physical implementation.

The quantum mechanism also has an impact on the other discipline of cryptography, i.e., secret-key cryptography. In 1984, Bennett and Brassard (1984) proposed a quantum key distribution scheme, which is a key agreement protocol using quantum communication. For the last two decades, so-called quantum cryptography has dramatically developed. For example, its unconditional security proofs were provided, some alternatives were proposed, and so on. We should especially mention that Mayers (1997) and Lo and Chau (1997) independently demonstrated that quantum mechanics cannot necessarily make all cryptographic schemes unconditionally secure.

As mentioned, secret-key cryptography enjoys the benefits of the quantum mechanism. On the other hand, Shor's algorithms can be regarded as a negative effect of the quantum mechanism on public-key cryptography. From the computational point of view, his algorithms illustrate that the quantum mechanism could make the computation more powerful. It is natural to consider that the power of the quantum mechanism could open up an era of neo-modern cryptography. Over the last decade, new possibilities (such as the unconditional security of the BB84 protocol) of quantum cryptography have been explored. Moreover, since the negative impact on public-key cryptography due to Shor, quantum cryptography has been also discussed and developed from the complexity-theoretic point of view. These investigations have promoted the development of quantum information theory and quantum computational complexity theory. In this chapter, we will survey what has been studied in quantum cryptography.

2 Classical Cryptography: Overview

2.1 One-Time Pad

The invention of Shannon's information theory initiated the mathematical study of communication. He also showed that the *one-time pad* has perfect secrecy and it is the only way to achieve perfect secrecy.

Here, a basic definition of classical cryptosystems will be given. A *symmetric-key cryptosystem* is a quintuple $(\mathcal{K}, \mathcal{P}, \mathcal{C}, \mathcal{E}, \mathcal{D})$, where

- $\mathcal{K}$ is a finite set of possible *keys*, which must be shared with legitimate users.
- $\mathcal{P}$ is a finite set of possible *plaintexts*.
- $\mathcal{C}$ is a finite set of possible *ciphertexts*.
- For each k, there is an *encryption rule* $e_k \in \mathcal{E}$ and the corresponding *decryption rule* $d_k \in \mathcal{D}$, where $e_k : \mathcal{P} \to \mathcal{C}$ and $d_k : \mathcal{C} \to \mathcal{P}$ must satisfy $d_k(e_k(m)) = m$ for each $m \in P$. Typically, $\mathcal{E}$ and $\mathcal{D}$ can be considered as the *encryption algorithm* and *decryption algorithm*, respectively. Namely, $\mathcal{E}$ takes a key k and a plaintext m as input and computes $e_k(m)$. Similarly, $\mathcal{D}$ takes a key k and a ciphertext c as input and computes $d_k(c)$.

In the basic scenario, there are two parties who wish to communicate with each other over an insecure channel. These parties are usually referred to as Alice and Bob. The insecurity of the channel means that an eavesdropper, called Eve, may wiretap the messages that are sent over this channel. Alice and Bob can make use of a cryptosystem to keep their information secret in spite of the existence of the eavesdropper, by agreeing with each other on a secret key k via another secure communication method.

In order for Alice to send her secret message m, she uses the key k to compute the ciphertext $c = e_k(m)$, and she sends Bob the ciphertext c over the insecure channel. Since Bob also knows the key k, he can decrypt the ciphertext by using the decryption algorithm and obtain $d_k(c) = d_k(e_k(m)) = m$. On the other hand, since Eve wiretaps the ciphertext c but she does not know the key k, there is no easy way to obtain the original message m.

The *one-time pad* is the symmetric-key cryptosystem that achieves perfect secrecy. Suppose that $\mathcal{P} = \mathcal{C} = \mathcal{K} = \{0, 1\}^n$, $e_k(m) = m \oplus k$ and $d_k(c) = c \oplus k$, where $\oplus$ is a bitwise XOR operation.

Let $\mathbf{P}$ be a random variable over $\mathcal{P}$ and $\mathbf{K}$ a random variable over $\mathcal{K}$. A cryptosystem is said to have *perfect secrecy* if and only if for every $m \in \mathcal{P}$ and for every $c \in \mathcal{C}$, $\Pr[\mathbf{P} = m \mid \mathbf{C} = c] = \Pr[\mathbf{P} = m]$. That is, knowing the ciphertext c is not advantageous to the prediction of the original plaintext m. Shannon showed that for every perfectly secure cryptosystem the entropy of the key is at least as large as the entropy of the plaintext. This says that Alice cannot securely send her secret message of n-bits by using a key of length less than n bits.

Moreover, we have a characterization of perfect secrecy that is referred to as Shannon's Theorem. It states that the cryptosystem with $\mathcal{P} = \mathcal{C} = \mathcal{K}$ and $supp(\mathbf{P}) = \mathcal{P}$ has perfect secrecy if and only if $\mathbf{K}$ is the uniform distribution over $\mathcal{K}$ and there exists a unique key k such that $e_k(m) = c$ for every $m \in \mathcal{P}$ and for every $c \in \mathcal{C}$. By using Shannon's Theorem, it is not hard to see that the one-time pad has perfect secrecy.

2.2 Message Authentication

Confidentiality may be the prime function of cryptographic systems. Besides confidentiality, *authenticity* is another function of cryptographic systems. As a tool to provide message authentication, the universal hash function family (due to Carter and Wegman (1979) and Wegman and Carter (1981)) is well known. $\mathcal{H}$ is a *universal hash function family* if the following holds.

- Each $h \in \mathcal{H}$ is a function from X to Y. Namely, the domain X and the range Y depend only on $\mathcal{H}$.
- For distinct $x, x' \in X$,

$$\Pr[h(x) = h(x')] = \frac{1}{|Y|}$$

where the probability is taken over the uniform distribution on $\mathcal{H}$.

If for distinct $x_1, x_2 \in X$ and for $y_1, y_2 \in Y$

$$\Pr[h(x_1) = y_1 \wedge h(x_2) = y_2] = \frac{1}{|Y|^2}$$

then $\mathcal{H}$ is a pairwise independent (aka strongly universal) hash function family. It is easy to see that pairwise independent hash functions imply universal hash functions. For some domain X and range Y, there exists an efficient implementation of $\mathcal{H}$. Namely, there is an efficient algorithm that computes $h(x)$ on input x and the description of h. For example, we have an efficient implementation of the pairwise independent hash function family for $X = \{0, 1\}^n$ and $Y = \{0, 1\}^n$. Let us consider the following:

$$\mathcal{H} = \{h_{a,b}(x) = ax + b \mid a, b \in \mathbb{F}_{2^n}\}$$

where we identify the finite field $\mathbb{F}_{2^n}$ (of order 2^n) with $\{0, 1\}^n$ by the standard encoding and the arithmetic computation is carried on $\mathbb{F}_{2^n}$. Since the computation over $\mathbb{F}_{2^n}$ can be done efficiently, $\mathcal{H}$ has an efficient implementation. Also note that there exist efficient implementations for the pairwise independent hash function families whose domain and range are natural.

A typical application of the pairwise independent hash function is message authentication. Suppose that Alice and Bob share the hash key k and agree with the corresponding hash function h_k. The case where Alice sends a message m with $h_k(m)$ can be considered. If Bob receives (m, z) and checks if $z = h_k(m)$ and whenever the check fails, he can judge the message to be forged. On the other hand, since Eve does not know the hash key, the authentication part (i.e., z) seems to be perfectly random.

2.3 Block Ciphers and Cryptographic Hash Functions

Though the one-time pad has perfect secrecy, it is not practical since the key length must be at least as large as the message length. As a practical symmetric cryptosystem, blockwise encryption (called the *block cipher*) has been considered. Some block ciphers (e.g., data encryption standard (DES) and advanced encryption standard (AES)) can withstand some specific attacks. While several design paradigms have been developed so as to withstand known attacks, block ciphers cannot have perfect secrecy.

Universal hashing also has a desirable property. Because of its nature, the number of uses per key is limited. Instead, heuristic cryptographic hash functions such as SHA-1 are practically used.

2.4 Cryptographic Protocols

As mentioned, the era of modern cryptography started in the 1970s. The invention of the notion of one-way functions introduced a computational complexity viewpoint into cryptology. One-way functions are functions $f: \{0, 1\}^* \rightarrow \{0, 1\}^*$ such that, for each $x \in \{0, 1\}^*$, $f(x)$ is efficiently computable but $f^{-1}(y)$ is computationally tractable only for a negligible fraction of all y's. The notion of one-way functions is one of the most fundamental in cryptology. Though the existence of one-way functions (which is related to the **P** versus **NP** problem) is unproven, modern cryptography has been established based on it. Actually, computational assumptions such as the existence of one-way functions have brought about many cryptographic protocols, such as public-key encryption schemes, digital signature schemes, commitment schemes, oblivious transfer schemes, zero-knowledge proof/argument systems, and so on.

2.4.1 Public-Key Encryption Scheme

A *public-key encryption scheme* is different to a symmetric-key encryption scheme. While the encryption key and the decryption key are the same in the case of symmetric-key encryption, the encryption key is different from the decryption key in the case of public-key encryption. This means that in cases of public-key encryption the encryption key can be publicly issued. The decryption key must be still secret.

The public-key encryption scheme is usually defined by a triple $(\mathcal{G}, \mathcal{E}, \mathcal{D})$:

- $\mathcal{G}$ is a key generation algorithm that generates a pair of the secret key k_s and the public key k_p on input 1^n (the unary representation of the security parameter n). Here, the plaintext space $\mathcal{P}$ and the ciphertext space $\mathcal{C}$ are implicitly defined. $\mathcal{P}$ and $\mathcal{C}$ basically depend on the security parameter n. (They sometimes depend on the key pair (k_s, k_p).)
- $\mathcal{E}$ is an encryption algorithm that, given the public key k_p and the message m, computes the ciphertext $c = \mathcal{E}(k_p, m)$.
- $\mathcal{D}$ is a decryption algorithm that, given the secret key k_s and the ciphertext c, computes the plaintext $m = \mathcal{D}(k_s, c)$.
- For every possible key pair (k_s, k_s) and for every message $m \in \mathcal{P}$, $\mathcal{D}(k_s, \mathcal{E}(k_p, m)) = m$.

As in the case of the one-time pad, there are two parties, Alice and Bob. In the case of public-key encryption, the roles of Alice and Bob are asymmetric and all parties including Eve are modeled by probabilistic polynomial-time algorithms. First, Bob or the trusted third party runs the key generation algorithm to obtain the key pair (k_s, k_p). The public key k_p is published. If the trusted third party runs the algorithm, the secret key k_s must be transferred to Bob via a secure communication method. In order for Alice to send her secret message m to Bob, she obtains Bob's public key k_p, computes the ciphertext $c = \mathcal{E}(k_p, m)$ and sends Bob the ciphertext c. After Bob obtains the ciphertext c, he can compute $m = \mathcal{D}(k_s, c)$. Eve wiretaps the ciphertext c and she can also obtain the public key k_p. This means that she can choose an arbitrary message m' on her own and obtain the corresponding ciphertext $c' = \mathcal{E}(k_p, m')$. This ability is called the *chosen plaintext attack*. Thus, the security of public-key encryption must be guaranteed against the chosen plaintext attack at least. In recent theory, a more powerful attack model called the *chosen ciphertext attack* has been considered. Here, the details are not mentioned.

Besides the attack model, it is important to define the security goal. *One-wayness, semantic security* (Goldwasser and Micali 1984) and *non-malleability* (Dolev et al. 2000) are typical goals. One-wayness means that it is infeasible for Eve to recover the whole message from the target ciphertext. Semantic security means that it is infeasible for Eve to extract any partial information of the plaintext from the ciphertext. Non-malleability means that it is infeasible for Eve to obtain another ciphertext whose plaintext is related to the original plaintext of the target ciphertext. The bare RSA public-key encryption scheme (Rivest et al. 1978) cannot have semantic security. Thus, the RSA scheme should be used in a more sophisticated way, such as optimal asymmetric encryption padding (OAEP) (Bellare and Rogaway 1995). Furthermore, hybrid encryptions (e.g., Cramer and Shoup (2003)), among them public-key encryption, symmetric encryption and message authentication techniques, they have been extensively studied, because they easily provide a more practical and more secure public-key encryption scheme.

2.4.2 Digital Signature Scheme

A *digital signature scheme* is one of the most important protocols in cryptology. It is usually defined by a triple $(\mathcal{G}, \mathcal{S}, \mathcal{V})$:

- $\mathcal{G}$ is a key generation algorithm that generates a pair of the secret key k_s and the public key k_p on input 1^n.
- $\mathcal{S}$ is a signature algorithm that, given the secret key k_s and the message m, computes a signature $\sigma = \mathcal{S}(k_s, m)$.

- $\mathcal{V}$ is a verification algorithm that, given the public key k_p and the signed message (m, σ), checks the validity of σ for the message m.

In the basic scenario, there is a signer Alice who wishes to prove the authenticity of her message and a verifier Bob who verifies the validity of the signature. First, Alice or the trusted third party runs the key generation algorithm to obtain the key pair (k_s, k_p). The public key k_p is published. If the trusted third party runs the algorithm, the secret key k_s must be transferred to Alice via a secure communication method. In order for Alice to sign her message m, she obtains her secret key k_s, computes the signature $\sigma = \mathcal{S}(k_s, m)$, and issues the signed message (m, σ). If Bob wishes to verify the validity of the signature σ for the message m, he runs the verification algorithm $\mathcal{V}$ with the public key k_p and checks the validity. Bob rejects the signed message if the check fails and accepts it otherwise.

The most secure notion of digital signature schemes is *existential unforgeability* against the *adaptive chosen message attack*. It means that for any message m even if Eve can adaptively obtain signed messages (m', σ') for her own choice $m' \neq m$ then she cannot forge the signature σ valid for the message m. It is known that digital signature schemes satisfying existential unforgeability against the adaptive chosen message attack are constructible from any one-way function (Naor and Yung 1989).

2.4.3 Bit Commitment Scheme

A *bit commitment* is a fundamental cryptographic protocol between two parties. The protocol consists of two phases: commit phase and reveal phase. In the commit phase, the sender Alice has a bit b in her private space and she wants to commit b to the receiver Bob. They exchange messages and at the end of the commit phase Bob gets some information that represents b. In the reveal phase, Alice confides b to Bob by exchanging messages. At the end of the reveal phase, Bob judges whether the information from the reveal phase really represents b or not. Basically, there are three requirements for secure bit commitment: correctness, the hiding property, and the binding property. Correctness guarantees that if both parties are honest then for any bit $b \in \{0, 1\}$ Alice has, Bob accepts it with certainty. The hiding property guarantees that (cheating) Bob cannot reveal the committed bit during the commit phase. The binding property guarantees that (cheating) Alice cannot commit her bit b such that Alice maliciously reveals $b \oplus 1$ as her committed bit but Bob accepts it.

In the classical case, a simple argument shows the impossibility of bit commitment with the hiding and the binding properties both statistical. Thus, either hiding or binding must be computational. A construction of the statistically binding scheme (of the round complexity $O(1)$) from any pseudorandom generator was given by Naor (1991). Since the existence of one-way functions is equivalent to that of pseudorandom generators (Håstad 1999), a statistically binding scheme can be based on any one-way function. A construction of a statistically hiding scheme from one-way permutation was given by Naor, Ostrovsky, Venkatesan and Yung (Naor et al. 1998). After that, the assumption of the existence of one-way permutation was relaxed to that of a approximable-preimage-size one-way function (Haitner et al. 2005). Finally, Haitner and Reingold (2007) showed that a statistically hiding scheme can be based on any one-way function by using the excellent techniques in Nguyen et al. (2006).

Since a statistically binding (resp., statistically hiding) bit commitment scheme is a building block for zero-knowledge proof (resp., zero-knowledge argument) systems (Goldreich et al. 1991;

Brassard et al. 1988), it is desirable to be efficient from several viewpoints (e.g., the total size of messages exchanged during the protocol, or the round number of communications in the protocol).

2.4.4 Oblivious Transfer

Oblivious transfer (OT) is also an important two-party cryptographic protocol. The first known OT system was introduced by Rabin (1981) where a message is received with probability 1/2 and the sender cannot know whether his message reaches the receiver. Prior to this, Wiesner (1983) introduced a primitive called multiplexing, which is equivalent to the 1-out-of-2 OT (Even et al. 1985) known today, but it was then not seen as a tool in cryptography. Even et al. defined the 1-out-of-2 OT (Even et al. 1985), where the sender Alice has two secrets σ_0 and σ_1 and the receiver Bob can choose one of them in an oblivious manner. That is, Alice cannot know Bob's choice $i \in \{0, 1\}$ and Bob cannot know any information on σ_{1-i}. The former property is called *receiver's privacy* and the latter *sender's privacy*. Later, Crépeau (1988) showed that Rabin's OT and the 1-out-of-2 OT are equivalent. Furthermore, the more general 1-out-of-N OT (where the sender has N secrets) and the more specific 1-out-of-2 *bit* OT (where the secrets are one bit long) are similarly defined and the reductions among the variants of OT have been discussed in the literature, for example Brassard et al. (1996, 2003) and Crépeau and Savvides (2006).

OT protocols are fundamental building blocks of modern cryptography. Most notably, it is known that any secure function evaluation can be based on OT (Kilian 1988; Goldreich et al. 1987).

2.4.5 Zero-Knowledge Proof/Argument Systems

The notion of *zero-knowledge* was introduced by Goldwasser, Micali and Rackoff (Goldwasser et al. 1989). Roughly speaking, an interactive proof system has the zero-knowledge property if any (possibly cheating) verifier that communicates with the honest prover learns nothing through the interaction except the validity of the claimed statement. In the classical case, there are several variants of zero-knowledge corresponding to how to formally define the notion that the verifier "learns nothing." Anyway, the verifier "learns nothing" if there exists a polynomial-time simulator whose output is indistinguishable in some sense from the information the verifier would have after the interaction with the honest prover.

Goldreich, Micali and Wigderson (Goldreich et al. 1991) showed that any **NP** language has zero-knowledge proof by using a (nonuniform) computationally concealing statistically binding bit commitment scheme, that is, any (nonuniform) one-way function. This result can be extended for **PSPACE** languages. Moreover, a statistical zero-knowledge argument for any **NP** language is constructible from any one-way function (Nguyen et al. 2006).

3 Quantum Computational Cryptographic Primitive

Before showing several quantum cryptographic protocols, some computational primitive for quantum cryptography will be mentioned.

3.1   Quantum One-Way Permutations

In classical cryptography, some cryptographic protocols assume the existence of one-way permutation. Thus, some quantum cryptographic primitive might assume the existence of quantum one-way permutations. As will be seen, Dumais, Mayers and Salvail (Dumais et al. 2000) proposed a quantum bit commitment scheme based on the existence of quantum one-way permutations.

In the classical case, the RSA function and the discrete logarithm-based function could be candidates for one-way permutations. Unfortunately, such candidates are no longer one-way in the quantum setting because Shor's quantum algorithm (Shor 1997) solves the factorization problem and the discrete logarithm problem efficiently. Thus, it is important to consider the existence of quantum one-way permutations.

Kashefi, Nishimura and Vedral (Kashefi et al. 2002) gave a necessary and sufficient condition for the existence of *worst-case* quantum one-way permutations as follows.

Let $f: \{0, 1\}^n \to \{0, 1\}^n$ be a permutation. Then f is worst-case quantum one-way if and only if there exists a unitary operator in $Q = \{Q_j(f) \mid j = 0, 1, \ldots, n/2 - 1\}$ that is not efficiently computable. The reflection operators $Q_j(f)$ are defined as

$$Q_j(f) = \sum_{x \in \{0,1\}^n} |x\rangle\langle x| \otimes (2|\psi_{j,x}\rangle\langle\psi_{j,x}| - I)$$

where

$$|\psi_{j,x}\rangle = \frac{1}{\sqrt{2^{n-2j}}} \sum_{y:\text{pref}(f(y),2j) = \text{pref}(x,2j)} |y\rangle$$

and $\text{pref}(s, i)$ denotes the i bits-long prefix of a string s.

They also considered quantum weakly one-way permutations (i.e., in the cryptographic sense) and gave a sufficient condition on the existence. After that, Kawachi, Kobayashi, Koshiba and Putra (Kawachi et al. 2005b) gave a necessary condition and completed an algorithmic characterization of quantum weakly one-way permutations. The characterization is almost similar to the case of worst-case quantum one-way permutation except that the unitary operators in Q permit exponentially small errors. While Kawachi et al. (2005b) mentioned a characterization of quantum weakly one-way permutations only, a similar characterization holds for quantum *strongly* one-way permutations. These characterizations might be helpful either to search for candidates of quantum one-way permutations or to disprove their existence.

3.2   Quantum Hard-Core Predicates

Hard-core predicates for one-way functions are also important in computational cryptography. Goldreich and Levin (1989) showed that a hard-core predicate is constructible from any one-way function f. Let $f'(x, r) = (f(x), r)$ be a function where $x, r \in \{0, 1\}^n$. Here, it is easy to see that f' is also one-way. The predicate

$$GL(x, r) = \langle x, r \rangle = \sum_{i=1}^{n} x_i r_i \bmod 2$$

is a hard-core predicate for the one-way function f'. Moreover, they gave a way to construct a hard-core function of output length $O(\log n)$ for any one-way function. Adcock and

Cleve (2002) considered quantum hard-core predicates (i.e., hard-core predicates against quantum adversaries) for quantum one-way functions. They showed that for any quantum one-way function f, $GL(x, r)$ is also a quantum hard-core predicate for the quantum one-way function $f'(x, r) = (f(x), r)$. Furthermore, they proved that the reduction between quantum hard-core predicates and quantum one-way functions is simpler and tighter than the classical reduction. Actually, they showed that a lower bound on the number of oracle calls in the classical reduction is properly larger than an upper bound in the quantum reduction.

It is widely known that list decodable codes have many computational complexity theoretical applications including hard-core functions (Sudan 2000). Intuitively speaking, if we are given a corrupted codeword of a list decodable code, we may output a short list containing messages close to the correct one rather than uniquely recover it. For example, the Goldreich–Levin Theorem (Goldreich and Levin 1989) mentioned above can be regarded as an efficient list-decoding algorithm for the binary Hadamard code. Actually, for a message x, $GL(x, 0), \ldots$, $GL(x, 2^n - 1)$ corresponds to a codeword of the binary Hadamard code. The prediction algorithm for the hard-core predicate GL also corresponds to an access to a corrupted codeword. Suppose that we could obtain a polynomially long list that contains x with high probability by accessing the corrupted codeword. It implies that inverting the one-way function f is efficiently computable. Thus, we can say that there is no efficient algorithm to predict the hard-core value of GL.

From the quantum computational point of view, a general construction of quantum hard-core functions from quantum one-way functions was obtained due by Kawachi and Yamakami (2006), utilizing a classical code that has a quantum list-decoding algorithm. Roughly speaking, they showed that any code that has an almost orthogonal structure in a sense has a quantum list-decoding algorithm. Consequently, it enables one to discover new (quantum) hard-core functions. Especially, the quantum list-decoding technique affirmatively but quantumly settled an open problem whether Damgård's generator (Damgård 1988) is cryptographically secure or not.

4 Quantum Key Distribution

Quantum key distribution (QKD) is a quantum technique to share a random string between two parties Alice and Bob. It is the most successful protocol in quantum cryptography, which strengthens symmetric-key techniques. Usually, QKD protocols assume the authentication channel. It means that Alice and Bob beforehand have to agree on the hash key for universal hash functions. Thus, QKD can be regarded as a technique for stretching shared randomness between Alice and Bob.

Known QKD protocols can be partitioned into two types. One is based on Heisenberg's uncertainty principle and the other is based on Bell's inequality.

4.1 Based on Heisenberg's Uncertainty Principle

Quantum key distribution in the early stages, such as the BB84 protocol (Bennett and Brassard 1984) and the B92 protocol (Bennett 1992), were based on Heisenberg's uncertainty principle. The principle says that two complementary quantum states cannot be simultaneously measured. If the eavesdropper Eve wiretaps the quantum channel to obtain the secret information, then she has to measure the quantum states. However, Heisenberg's uncertainty principle says that the measurement of quantum states affects the system. Thus, we can check the presence of an eavesdropper by observing the system change.

4.1.1 The BB84 Protocol

In this protocol, we use two orthogonal bases. One is the computational basis $\{|0\rangle, |1\rangle\}$ and the other is the diagonal basis $\{|0\rangle_\times, |1\rangle_\times\}$, where

$$|0\rangle_\times = \frac{1}{\sqrt{2}}(|0\rangle + |1\rangle)$$

$$|1\rangle_\times = \frac{1}{\sqrt{2}}(|0\rangle - |1\rangle)$$

We also use $\{|0\rangle_+, |1\rangle_+\}$ to denote the computational basis $\{|0\rangle, |1\rangle\}$. The protocol is as follows:

1. Alice randomly chooses $4n$ qubits, each in one of the four states $|0\rangle_+, |0\rangle_\times, |1\rangle_+$ or $|1\rangle_\times$, and sends them to Bob.
2. For each qubit that Bob receives, he chooses either the computational basis $(+)$ or the diagonal basis $(\times)$ and measures the qubit with respect to the basis. If the channel is not wiretapped and Alice's basis coincides with Bob's, then his is the same as the bit of Alice. If the bases differ, his measurement result is randomly distributed.
3. Alice tells Bob the basis she used for each qubit via the classical authentication channel. Bob keeps the bits where Bob's basis coincides with Alice's. This happens in about half the cases, so Bob will have about $2n$ bits left.
4. Bob randomly partitions about $2n$ bits into two groups of about n bits. They use one group to estimate the error rate. Bob tells Alice all the bits in one group. If Alice's bits are too different from Bob's bits, they abort the protocol.
5. Alice and Bob obtain a common secret key from the remaining about n bits by performing error correction and privacy amplification.

Firstly, Mayers (1996) gave a (somewhat complex) proof of the unconditional security of the BB84 protocol. After that Shor and Preskill (2000) gave a simple proof by showing that the entanglement-based version of the BB84 protocol is equivalent for Eve to the original BB84 protocol. Note that the security proof in Mayers' fashion has been simplified (see, e.g., Koashi and Preskill (2003)).

4.1.2 The B92 Protocol

This protocol is similar to the BB84 protocol. While the BB84 protocol uses four types of quantum states, the B92 protocol instead uses two non-orthogonal quantum states.

Let $|\psi_0\rangle$ and $|\psi_1\rangle$ be two non-orthogonal quantum states. Then, two projective measurements P_0 and P_1 can be defined as follows:

$$P_0 = I - |\psi_1\rangle\langle\psi_1|$$
$$P_1 = I - |\psi_0\rangle\langle\psi_0|$$

The point is that $P_0|\psi_1\rangle = 0$ and $P_1|\psi_0\rangle = 0$.

The sketch of the B92 protocol is as follows: (1) Alice chooses qubits, each in one of the two quantum states $|\psi_0\rangle$ and $|\psi_1\rangle$ and sends them to Bob. (2) For each qubit that Bob receives, he chooses at random one of the two measurements P_0 and P_1 and measures the qubit. He keeps it secret whether he measures something or not. (3) Bob tells Alice the measurement he uses. (4) Alice can extract bits where $|\psi_i\rangle$ matches to P_i.

4.2   Based on Bell's Inequality

Experiments in the 1970s indicated that the correlation function of the entangled quantum states cannot be expressed by classical probability and these entangled quantum states are nonlocal. Ekert (1991) exploited these properties in the QKD protocol (E91 protocol). Namely, he used the property that the entangled photons have strong correlation to yield the secret key. On the other hand, he also used the property that the entangled photons do not satisfy Bell's inequality to detect the presence of the eavesdropper. Since the E91 protocol has a bit complex structure that relates to Bell's inequality, we give the description of a similar but simpler protocol. That is the protocol of Bennett, Brassard and Mermin (Bennett et al. 1992) based on the entanglement.

4.2.1   The BBM92 Protocol

The protocol is as follows:

1. A source S emits a pair of qubits such that

$$|\psi_{AB}\rangle = \frac{1}{\sqrt{2}}(|0\rangle_+|0\rangle_+ + |1\rangle_+|1\rangle_+)$$
$$= \frac{1}{\sqrt{2}}(|0\rangle_\times|0\rangle_\times + |1\rangle_\times|1\rangle_\times)$$

The first qubit is sent in opposite directions. The first qubit is received by Alice and the second by Bob.
2. Both Alice and Bob randomly and independently choose either the computational basis ($+$) or the diagonal basis ($\times$) and measure the qubit with respect to the basis.
3. Alice and Bob announce their bases via the classical channel and keep the bits where the bases coincide.

The principle is based on the fact that Alice's bits and Bob's are exactly the same only when the pure state $|\psi_{AB}\rangle$ is prepared. If Eve wants to yield the situation where Alice's bits and Bob's are exactly the same, the only thing she can do is to prepare a pure state on the system AB and send it to Alice and Bob. Thus, Eve cannot leave any correlation between her system and the measurement results by Alice and Bob.

5   Quantum Public-Key Encryption

As mentioned, public-key cryptosystems are indispensable, even in the future if quantum computers are physically realized. Unfortunately, almost all practical public-key cryptosystems are vulnerable to Shor's algorithm. On the other hand, there are public-key cryptosystems that have not been shown to be vulnerable to quantum adversaries. Lattice-based cryptography (Micciancio and Regev 2009) is one of them. In this section, we focus on public-key cryptosystems that use quantum computation.

5.1 The OTU00 Scheme

Okamoto, Tanaka and Uchiyama (Okamoto et al. 2000) first proposed a knapsack-based cryptosystem as a quantum public-key cryptosystem. Since knapsack-based cryptosystems in the early stages used linearity in the public key, attack algorithms were developed. To avoid attack algorithms, Chor and Rivest (1988) introduced nonlinearity into the public-key generation by using an easy discrete logarithm problem and proposed yet another knapsack-based cryptosystem. The OTU00 scheme can be regarded as an extension of the Chor–Rivest cryptosystem. This is because they use general discrete logarithm problems to generate public keys with the help of quantum computation.

The original OTU00 scheme uses the algebraic number field. Here, we give a simpler description of the OTU00 scheme based on the finite field.

[Key Generation Algorithm $\mathcal{G}$]
1. For the security parameter n, determine the parameters k, ℓ bounded by a polynomial in n.
2. Randomly generate a prime p and a generator g of $\mathbb{F}_p^*$.
3. Randomly choose k relatively prime integers $p_1, \ldots, p_k \in \mathbb{F}_p^*$ such that the product of any ℓ elements out of the k elements is bounded by p.
4. Run Shor's algorithm to compute a_i such that $p_i = g^{a_i} \bmod p$ for each i $(1 \leq i \leq k)$.
5. Randomly choose $d \in \mathbb{F}_{p-1}$.
6. Compute $b_i = (a_i + d) \bmod (p - 1)$ for each i.
7. Return the public key $(\ell, b_1, \ldots, b_k)$ and the secret key $(g, d, p, p_1, \ldots, p_k)$.

[Encryption Algorithm $\mathcal{E}$]
1. Encode the message $m \in \{0, 1\}^v$ $(v = \lfloor \log_2 \binom{k}{\ell} \rfloor)$ into a k-bit string $(s_1, \ldots, s_k)$ of the Hamming weight ℓ as follows:
   (a) Initialize the variable j to ℓ.
   (b) Repeat the following from $i = 1$ to $i = k$. For convenience, we let $\binom{a}{0} = \binom{0}{q} = 1$ for $a \geq 1$.
     - If $m \geq \binom{k-i}{j}$ then set $s_i \leftarrow 1$ and $m \leftarrow m - \binom{k-i}{j}$ and decrease j by 1.
     - If $m < \binom{k-i}{j}$ then set $s_i = 0$.
2. Compute the ciphertext $c = \sum_{i=1}^{k} b_i s_i$ and return it.

[Decryption Algorithm $\mathcal{D}$]
1. Compute $r = c - \ell d \bmod p - 1$.
2. Compute $u = g^r \bmod p$.
3. Compute $(s_1, \ldots, s_k)$ as follows:
   - If $p_i | u$ then set $s_i = 1$.
   - If $p_i \nmid u$ then set $s_i = 0$.

4. Recover the plaintext m from $(s_1, \ldots, s_k)$ as follows:
   (a) Initialize m to 0 and the variable j to ℓ.
   (b) Repeat the following from $i = 1$ to $i = k$.
   If $s_i = 1$ then set $m \leftarrow m + \binom{k-i}{j}$ and decrease j by 1.

5. Return m.

The correctness of the decryption comes from the following fact:

$$u \equiv g^r \equiv g^{c - \ell d} \equiv g^{\sum_i s_i b_i - \ell d} \equiv g^{\sum_i s_i a_i} \equiv \prod_i (g^{a_i})^{s_i} \equiv \prod_i p_i^{s_i} \pmod{p}$$

Nguyen and Stern (2005) show that breaking the OTU00 scheme is reducible to the shortest vector problem or the closest vector problem. However, since these problems are **NP**-hard, the reductions do not provide an effective attack.

5.2 The KKNY05 Scheme

Kawachi, Koshiba, Nishimura and Yamakami (Kawachi et al. 2005a) proposed a semantically secure (in a weak sense) quantum public-key cryptosystem based on the worst-case hardness of the graph automorphism problem. Actually, they introduced a computational problem of distinguishing between two specific quantum states as a new cryptographic problem to design the quantum public-key cryptosystem. The computational indistinguishability between quantum states is a generalization of the classical indistinguishability between two probability distributions, which plays an important role in computational cryptography. Their problem QSCD$_{ff}$ asks whether we can distinguish between two sequences of identical samples of $\rho_\pi^+(n)$ and of $\rho_\pi^-(n)$ for each fixed hidden permutation π for each length parameter n of a certain form. Let S_n be the *symmetric group* of degree n and let

$$\mathcal{K}_n = \{\pi \in S_n : \pi^2 = id \text{ and } \forall i \in \{1, \ldots, n\}[\pi(i) \neq i]\}$$

for $n \in N$, where id stands for the identity permutation and $N = \{2(2n' + 1) : n' \in \mathbb{N}\}$. For each $\pi \in \mathcal{K}_n$, let $\rho_\pi^+(n)$ and $\rho_\pi^-(n)$ be two quantum states defined by

$$\rho_\pi^+(n) = \frac{1}{2n!} \sum_{\sigma \in S_n} (|\sigma\rangle + |\sigma\pi\rangle)(\langle\sigma| + \langle\sigma\pi|) \text{ and}$$

$$\rho_\pi^-(n) = \frac{1}{2n!} \sum_{\sigma \in S_n} (|\sigma\rangle - |\sigma\pi\rangle)(\langle\sigma| - \langle\sigma\pi|)$$

The cryptographic properties of QSCD$_{ff}$ follow mainly from the definition of the set $\mathcal{K}_n$ of the hidden permutations. Although the definition seems somewhat artificial, the following properties of $\mathcal{K}_n$ lead to cryptographic and complexity-theoretic properties of QSCD$_{ff}$:

(i) $\pi \in \mathcal{K}_n$ is of order 2, which provides the trapdoor property of QSCD$_{ff}$.
(ii) For any $\pi \in \mathcal{K}_n$, the conjugacy class of π is equal to $\mathcal{K}_n$, which enables one to prove the equivalence between the worst-case/average-case hardness of QSCD$_{ff}$.
(iii) The graph automorphism problem is (polynomial-time Turing) equivalent to its subproblem with the promise that a given graph has a unique non-trivial automorphism in $\mathcal{K}_n$ or none at all. This equivalence is exploited to give a complexity-theoretic lower bound of QSCD$_{ff}$, that is, the worst-case hardness of the graph automorphism problem.

Their problem QSCD$_{ff}$ is closely related to a much harder problem: the hidden subgroup problem on the *symmetric groups* (SHSP). Note that no known subexponential-time quantum algorithm exists for SHSP. Hallgren et al. (2003) introduced a distinction problem between certain two quantum states, similar to $QSCD_{ff}$ to discuss the computational intractability of SHSP by a "natural" extension of Shor's algorithm (1997) with the quantum Fourier transformation. An efficient solution to this distinction problem gives an answer to a pending question on a certain special case of SHSP. To solve this distinction problem, as they showed, the so-called *weak Fourier sampling* on a single sample should require an exponential number of samples. This result was improved by Grigni et al. (2004) and Kempe et al. (2006). On the contrary, Hallgren et al. (2006) and Hayashi et al. (2008) proved that no time-unbounded quantum algorithm solves the distinction problem even from $o(n \log n)$ samples. Kawachi et al. (2005a) showed that the above distinction problem is polynomial-time reducible to $QSCD_{ff}$. This immediately implies that we have no time-unbounded quantum algorithm for $QSCD_{ff}$ from $o(n \log n)$ samples. Even with sufficiently many samples for $QSCD_{ff}$, there is no known subexponential-time quantum algorithm for $QSCD_{ff}$ and thus finding such an algorithm seems a daunting task.

One can move to the description of the quantum public-key cryptosystem in Kawachi et al. (2005a). Usually public-key cryptosystems consist of three algorithms for key generation, encryption, and decryption.

[Public-Key Generation Algorithm]
Input: $\pi \in \mathcal{K}_n$
Procedure:

Step 1. Choose a permutation σ from S_n uniformly at random and store it in the second register. Then, the entire system is in the state $|0\rangle|\sigma\rangle$.

Step 2. Apply the Hadamard transformation to the first register.

Step 3. Apply the Controlled-π to both registers.

Step 4. Apply the Hadamard transformation to the first register again.

Step 5. Measure the first register in the computational basis. If 0 is observed, then the quantum state in the second register is ρ_π^+. Otherwise, the state of the second register is ρ_π^-. Now, apply the conversion algorithm to ρ_π^-.

The encryption algorithm consists of two parts: one is the conversion algorithm below and the other is so simple that we describe it in the description of the total system.

[Conversion Algorithm]
The following transformation inverts, given ρ_π^+, its phase according to the sign of the permutation with certainty.

$$|\sigma\rangle + |\sigma\pi\rangle \longmapsto (-1)^{\mathrm{sgn}(\sigma)}|\sigma\rangle + (-1)^{\mathrm{sgn}(\sigma\pi)}|\sigma\pi\rangle$$

Since π is odd, the above algorithm converts ρ_π^+ into ρ_π^-.

Finally, we give a description of the decryption algorithm.

[Decryption Algorithm]
Input: unknown state χ which is either ρ_π^+ or ρ_π^-.
Procedure:

Step 1. Prepare two quantum registers: The first register holds a control bit and the second one holds χ. Apply the Hadamard transformation H to the first register. The state of the system now becomes

$$H|0\rangle\langle 0|H \otimes \chi$$

Step 2. Apply the Controlled-π operator C_π to the two registers, where $C_\pi |0\rangle|\sigma\rangle = |0\rangle|\sigma\rangle$ and $C_\pi |1\rangle|\sigma\rangle = |1\rangle|\sigma\pi\rangle$ for any $\sigma \in S_n$.

Step 3. Apply the Hadamard transformation to the first register.

Step 4. Measure the first register in the computational basis. Output the observed result.

The following is the description of the total cryptosystem consisting of two phases: key transmission phase and message transmission phase.

[Key Transmission Phase]

1. Bob chooses a decryption key π uniformly at random from $\mathcal{K}_n$.
2. Bob generates sufficiently many copies of the encryption key ρ_π^+ by using the public-key generation algorithm.
3. Alice obtains encryption keys from Bob.

[Message Transmission Phase]

1. Alice encrypts 0 or 1 into ρ_π^+ or ρ_π^-, respectively, by using the conversion algorithm, and sends it to Bob.
2. Bob decrypts Alice's message using the decryption algorithm.

6   Quantum Digital Signature

Gottesman and Chuang (2001) proposed a quantum digital signature scheme. In the classical case, the Lamport signature (Lamport 1979) is based on one-way functions. The GC01 scheme is a quantum counterpart of the Lamport signature scheme. In the classical case, the existence of one-way functions is unproven. They allowed one-way functions to have the image of quantum states and showed the existence of (extended) one-way functions in the information-theoretic sense. Note that if the one-way function in the GC01 scheme is replaced with any one-way function in the computational sense then the security (the existential unforgeability against the chosen message attack) of the GC01 scheme holds in the computational sense. The original GC01 scheme has the non-repudiation property based on our choice of parameters. (The non-repudiation property means that once the signer creates a signed message he cannot repudiate the validity of the signed message.)

6.1   The GC01 scheme

For simplicity, we give the description of a version of the GC01 scheme without the non-repudiation property.

[Key Generation Algorithm $\mathcal{G}$]

1. Let f be an extended quantum one-way function.
2. Randomly choose $x_0^{(i)}$ and $x_1^{(i)}$ from $\{0, 1\}^n$ for $i = 1, \ldots, s$ and return them as the secret key. (How to determine the value of s will be mentioned later.)
3. For $i = 1, \ldots, s$, make t copies of $(|\psi_0^{(i)}\rangle, |\psi_1^{(i)}\rangle) = (|f(x_0^{(i)})\rangle, |f(x_1^{(i)})\rangle)$ and distribute them to at most t verifiers.

[Signature Algorithm $\mathcal{S}$]

For a target bit $b \in \{0, 1\}$ to be signed, produce its signature σ as follows:

$$\sigma = (\sigma_1, \ldots, \sigma_s) = (x_b^{(1)}, \ldots, x_b^{(s)})$$

[Verification Algorithm $\mathcal{V}$]

1. Check the validity of (b, σ) by using the public key $(|\psi_0^{(i)}\rangle, |\psi_1^{(i)}\rangle)$.
2. Compute $(f(\sigma_1), \ldots, f(\sigma_s))$ from the signature $(\sigma_1, \ldots, \sigma_s)$.
3. Check the equivalence between $f(\sigma_i)$ and $\psi_b^{(i)}$ for $i = 1, \ldots, s$ by using the swap-test (Buhrman 2001). If all the swap-tests pass, then accept (b, σ) as a valid signed message. Otherwise, reject (b, σ).

6.2 Swap Test

As seen, the swap-test is used to check the equivalence between $|f(\sigma_i)\rangle$ and $|\psi_b^{(i)}\rangle$ in the verification algorithm $\mathcal{V}$. In general, the swap-test checks the equivalence between $|\phi\rangle$ and $|\psi\rangle$. Exactly speaking, the swap-test performs the following:

$$(H \otimes I)(\text{c-SWAP})(H \otimes I)|0\rangle|\phi\rangle|\psi\rangle \qquad (1)$$

where c-SWAP exchanges the last two registers if the first qubit is 1. Then ❷ Eq. 1 is written as

$$\frac{1}{2}|0\rangle(|\phi\rangle|\psi\rangle + |\psi\rangle|\phi\rangle) + \frac{1}{2}|1\rangle(|\phi\rangle|\psi\rangle - |\psi\rangle|\phi\rangle)$$

If $|\phi\rangle = |\psi\rangle$ then the measurements on the first register always give the value 0. Let $\delta = \langle\phi|\psi\rangle$. If $|\phi\rangle \neq |\psi\rangle$ then the probability that the measurements on the first register give the value 0 is $\frac{1}{2}(1 + \delta^2)$. Since the swap-test is a one-sided error algorithm, the error probability is at most $\frac{1}{2}(1 + \delta^2)$.

Now we assume that $|\langle f(\sigma_i)|\psi_b^{(i)}\rangle| < \delta$ for every $i = 1, \ldots, s$. Then, the probability that $\mathcal{V}$ accepts an invalid signed message is the same as the probability of all the swap-tests in the case that $\sigma_i \neq x_b^{(i)}$ for all $i = 1, \ldots, s$. Since it is less than $\left(\frac{1+\delta^2}{2}\right)^s$, we have to set s so that $\left(\frac{1+\delta^2}{2}\right)^s$ becomes negligible.

7 Quantum Commitment and Oblivious Transfer

Since the BB84 protocol (Bennett and Brassard 1984) was shown to be secure unconditionally, some other cryptographic protocols such as bit commitment and oblivious transfer schemes were expected to achieve security without any computational assumption.

The security of bit commitment schemes consists of a concealing property and a binding property. A concealing property satisfies even if a cheating receiver cannot predict the committed value before the reveal phase. A binding property satisfies even if a cheating sender cannot reveal two different values in the reveal phase. In the classical setting, either the concealing or the binding property must be computational. By introducing quantum states into the bit commitment scheme, we expected that unconditionally secure bit commitment schemes would be realized.

First, Crépeau and Kilian (1988) showed a construction from unconditionally secure quantum commitment schemes to 1-out-of 2 quantum oblivious transfer schemes. Its information-theoretic security was proved in Crépeau (1994), Mayers and Salvail (1994), and Yao (1995). (Note that such a reduction in the classical setting is unknown.) Unfortunately, it was later shown that commitment schemes with unconditional concealing and binding properties are impossible even if quantum states are applicable (Lo and Chau 1997; Mayers 1997). Thus, alternative approaches such as quantum string commitment (Kent 2003) or cheat-sensitive quantum bit commitment (Aharonov et al. 2000; Hardy and Kent 2004; Buhrman et al. 2008) have been studied. Since those approaches do not reach a satisfactory level, yet another approach will be considered. If either the concealing or the binding condition of the bit commitment scheme is allowed to be computational, interesting commitment schemes utilizing quantum information are still possible. Dumais, Mayers and Salvail (Dumais et al. 2000) proposed a perfectly concealing and computationally binding noninteractive quantum bit commitment scheme under the assumption of the existence of quantum one-way permutations. Koshiba and Odaira (2009) generalized the Dumais–Mayers–Salvail (DMS) scheme to obtain a statistically concealing and computationally binding noninteractive quantum bit commitment scheme under the assumption of the existence of a quantum one-way function of the special form. The use of quantum information enables us to transform from a computationally concealing and computationally binding bit commitment scheme to a statistically concealing and computationally binding (interactive) scheme (Crépeau et al. 2001). Since the former is constructible from any quantum one-way function, so is the latter.

Here, let us see some issues in defining quantum bit commitment. Though we can define the concealing condition of quantum bit commitments as in the classical case, some care must be taken to define the binding condition in the quantum case. In the classical case, the binding property means that the probability that the bit value committed in the commit phase can be revealed as both 0 and 1 is negligibly small. However, this definition is too strong in the quantum case. Suppose that a sender in some quantum bit commitment scheme could generate the following state:

$$\frac{1}{\sqrt{2}}|0\rangle|\phi_0\rangle + \frac{1}{\sqrt{2}}|1\rangle|\phi_1\rangle$$

where $|\phi_0\rangle$ (resp., $|\phi_1\rangle$) is a quantum state to be sent when 0 (resp., 1) is honestly committed in the commit phase. Then the sender Alice sends the quantum state only in the second register to the receiver Bob and keeps the quantum state in the first register at her side. Alice can change the committed value with probability 1/2 by measuring the quantum state left at her side just before the reveal phase. Since the above situation is essentially inevitable, the straightforward extension of the binding condition in the classical case is not satisfied in the quantum case. Thus, we have to weaken the definition of the binding condition in the quantum case. Actually, Dumais, Mayers and Salvail (Dumais et al. 2000) considered that the binding condition is satisfied if $p_0 + p_1 - 1$ is negligibly small, where p_0 (resp., p_1) is the probability that the committed value is revealed as 0 (resp., 1).

The following is the description of the quantum bit commitment scheme in Dumais et al. (2000). Let $f: \{0, 1\}^n \rightarrow \{0, 1\}^n$ be any quantum one-way permutation.

[Commit Phase]

1. Alice decides $w \in \{0, 1\}$ as her committing bit. She also chooses $x \in \{0, 1\}^n$ uniformly at random, and then computes $f(x)$.
2. Alice sends $|f(x)\rangle$ to Bob if $w = 0$ and $H^{\otimes n} |f(x)\rangle$ otherwise.

[Reveal Phase]

1. Alice announces x and w to Bob.
2. If $w = 1$, Bob applies $H^{\otimes n}$ to the state sent from Alice. Otherwise, he does nothing to the state.
3. Bob measures the resulting state in the computational basis. Let y be the outcome. He accepts if and only if $y = f(x)$.

Finally in this section, we mention a conversion from quantum bit commitment scheme with computational assumption to quantum oblivious transfer with computational assumption. In Crépeau and Kilian's conversion from quantum bit commitment to 1-out-of 2 quantum oblivious transfer, the commitment scheme must be statistically secure. Thus, since the security proof (Yao 1995) is information-theoretic, his proof is not applicable to the computational security. Crépeau et al. have shown a way to convert a statistically conceal-ing and computationally binding string commitment scheme of some non-standard type to 1-out-of-2 quantum oblivious transfer (Crépeau et al. 2004).

8 Quantum Zero-Knowledge Proof

At any round of interaction to simulate, a simulator typically generates a pair of a question from the verifier and a response from the honest prover. If this produces a pair that is inconsistent in itself (or with the other parts of the transcript of the interaction simulated so far), the simulator "rewinds" the process to simulate this round again. We now mention why the "rewinding" technique is not generally applicable to quantum verifiers. First, quantum information cannot be copied and the fact is also known as the no-cloning theorem. Second, measurements are generally irreversible processes. This difficulty was explicitly point-ed out by van de Graaf (1997).

Since then, the study of quantum zero-knowledge has developed from the computational point of view. For example, Watrous (2002) defined the quantum counterpart of honest verifier statistical zero-knowledge and studied its properties. We denote by **HVQSZK** the class of languages that have honest verifier quantum statistical zero-knowledge proof systems. In Watrous (2002), the following statements are shown:

1. There exists a complete promise problem, which is a natural generalization of a complete promise problem in statistical zero-knowledge (**SZK**) due to Sahai and Vadhan (2003).
2. **HVQSZK** is contained by **PSPACE**.
3. **HVQSZK** is closed under the complement.
4. Any HVQSZK protocol can be parallelized to a one-round HVQSZK protocol.

Kobayashi (2003) defined noninteractive quantum perfect zero-knowledge (**NIQPZK**) and noninteractive quantum statistical zero-knowledge (**NIQSZK**) and studied their properties. Specifically:

1. If the prover and the verifier do not beforehand have any shared randomness and any shared entanglement, languages that have noninteractive quantum zero-knowledge proof systems are in **BQP**.
2. Assuming that the verifier and the prover share polynomially many Einstein–Podolsky–Rosen (EPR) pairs a priori, **NIQPZK** with the one-sided error has a natural complete promise problem, which is a generalization of a complete promise problem in noninteractive statistical zero-knowledge (**NISZK**) due to Goldreich, Sahai and Vadhan (Goldreich et al. 1999).
3. The graph non-automorphism (GNA) problem has a **NIQPZK** proof system with prior EPR pairs. (Since it is unknown whether GNA is in **BQP** or not, **NIQPZK** with prior EPR pairs includes a nontrivial language.)

In order to circumvent problematic issues caused by the rewinding technique, Damgård, Fehr and Salvail (Damgård et al. 2004) showed a way to construct a computational QZK proof and perfect QZK argument against quantum adversaries in the *common reference string* model, wherein it is assumed that an honest third party samples a string from some specified distribution and provides both the prover and verifier with this string at the start of the interaction. They have also given a construction of unconditionally concealing and computationally binding string commitment protocols against quantum adversaries.

Watrous (2006) resolved in many cases the problematic issues caused by the rewinding technique in quantum zero-knowledge by introducing a success probability amplifying technique, which is successfully utilized to amplify the success probability in the Quantum Merlin–Arthur protocol without increasing witness sizes (Marriott and Watrous 2004). Specifically he showed:

1. The graph isomorphism problem is in perfect zero-knowledge (**PZK**) even against the adversarial verifier who uses quantum computation to cheat.
2. If quantum one-way permutations exist, every problem in **NP** has ZK proof systems even against the adversarial verifier who uses quantum computation to cheat.
3. **HVQSZK=QSZK**, where **QSZK** denotes the class of problems having quantum statistical zero-knowledge proof systems.

Together with his proof construction, this implies that all the properties proved for **HVQSZK** in Watrous (2002) are inherited to **QSZK** (except for those related to the round complexity).

Kobayashi (2008) proves a number of general properties on quantum zero-knowledge proofs, not restricted to the statistical zero-knowledge case. Specifically, for quantum computational zero-knowledge proofs, letting **QZK** and **HVQZK** denote the classes of problems having quantum computational zero-knowledge proofs and honest-verifier quantum computational zero-knowledge proof systems, respectively, he showed:

1. **HVQZK = QZK**.
2. Any problem in **QZK** has a public-coin quantum computational zero-knowledge proof system.
3. Any problem in **QZK** has a quantum computational zero-knowledge proof systems of perfect completeness.

References

Adcock M, Cleve R (2002) A quantum Goldreich-Levin theorem with cryptographic applications. In: Proceedings of the 19th annual symposium on theoretical aspects of computer science, Antibes-Juan les Pins, France, March 2002. Lecture notes in computer science, vol 2285. Springer, Berlin, pp 323–334

Aharonov D, Ta-Shma A, Vazirani UV, Yao AC-C (2000) Quantum bit escrow. In: Proceedings of the 32nd ACM symposium on theory of computing, ACM, Portland, OR, May 2000, pp 705–714

Bellare M, Rogaway P (1995) Optimal asymmetric encryption. In: EUROCRYPT '94: Advances in cryptology, Perugia, Italy, May 1994. Lecture notes in computer science, vol 950. Springer, Berlin, pp 92–111

Bennett CH (1992) Quantum cryptography using any two nonorthogonal states. Phys Rev Lett 68:3121–3124

Bennett CH, Brassard G (1984) Quantum cryptography: public key distribution and coin tossing. In: Proceeding of the IEEE international conference on computers, systems, and signal processing, Bangalore, India, December 1984. IEEE, New York, pp 175–179

Bennett CH, Brassard G, Mermin ND (1992) Quantum cryptography without Bell's theorem. Phys Rev Lett 68:557–559

Brassard G, Chaum D, Crépeau C (1988) Minimum disclosure proofs of knowledge. J Comput Syst Sci 37(2):156–189

Brassard G, Crépeau C, Santha M (1996) Oblivious transfers and intersecting codes. IEEE Trans Info Theory 42(6):1769–1780

Brassard G, Crépeau C, Wolf S (2003) Oblivious transfers and privacy amplification. J Cryptol 16(4):219–237

Buhrman H, Cleve R, Watrous J, de Wolf R (2001) Quantum fingerprinting. Phys Rev Lett 87:167902

Buhrman H, Christandl M, Hayden P, Lo H-K, Wehner S (2008) Possibility, impossibility and cheat-sensitivity of quantum bit string commitment. Phys Rev A 78(32):022316

Carter JL, Wegman MN (1979) Universal classes of hash functions. J Comput Syst Sci 18(2):143–154

Chor B, Rivest RL (1988) A knapsack-type public key cryptosystems based on arithmetic in finite fields. IEEE Trans Info Theory 34:901–909

Cramer R, Shoup V (2003) Design and analysis of practical public-key encryption schemes secure against adaptive chosen ciphertext attack. SIAM J Comput 33(1):167–226

Crépeau C (1988) Equivalence between two flavours of oblivious transfer. In: CRYPTO'87: Advances in cryptology, University of California, Santa Barbara, CA, August 1987. Lecture notes in computer science, vol 293. Springer, New York, pp 350–354

Crépeau C (1994) Quantum oblivious transfer. J Mod Opt 41(12):2445–2454

Crépeau C, Kilian J (1988) Achieving oblivious transfer using weakened security assumptions. In: Proceedings of the 29th annual IEEE symposium on foundations of computer science, IEEE, White Plains, NY, October 1988, pp 42–52

Crépeau C, Savvides G (2006) Optimal reductions between oblivious transfers using interactive hashing. In: EUROCRYPT 2006: Advances in cryptology, St. Petersburg, Russia, May–June 2006. Lecture notes in computer science vol 4004. Springer, Heidelberg, pp 201–221

Crépeau C, Legare F, Salvail L (2001) How to convert the flavor of a quantum bit commitment. In: EUROCRYPT 2001: Advances in cryptology, Innsbruck, Austria, May 2001. Lecture notes in computer science vol 2045. Springer, Berlin, pp 60–77

Crépeau C, Dumais P, Mayers D, Salvail L (2004) Computational collapse of quantum state with application to oblivious transfer. In: Proceedings of the 1st theory of cryptography conference, Cambridge, MA, February 2004. Lecture notes in computer science, vol 2951. Springer, Berlin, pp 374–393

Damgård I (1988) On the randomness of Legendre and Jacobi sequences. In: CRYPTO'88: Advances in cryptology, Santa Barbara, CA, August 1988. Lecture notes in computer science vol 403. Springer, Berlin, pp 163–172

Damgård I, Fehr S, Salvail L (2004) Zero-knowledge proofs and string commitments withstanding quantum attacks. In: CRYPTO 2004: Advances in cryptology, Santa Barbara, CA, August 2004. Lecture notes in computer science, vol 3152. Springer, Berlin, pp 254–272

Diffie W, Hellman ME (1976) New directions in cryptography. IEEE Trans Info Theory 22(5):644–654

Dolev D, Dwork C, Naor M (2000) Non-malleable cryptography. SIAM J Comput 30(2):391–437

Dumais P, Mayers D, Salvail L (2000) Perfectly concealing quantum bit commitment from any quantum one-way permutation. In: EUROCRYPT 2000: Advances in cryptology, Bruges, Belgium, May 2000. Lecture notes in computer science, vol 1807. Springer, Berlin, pp 300–315

Ekert AK (1991) Quantum cryptography based on Bell's theorem. Phys Rev Lett 67:661–663

Even S, Goldreich O, Lempel A (1985) A randomized protocol for signing contracts. Commun ACM 28(6):637–647

Goldreich O, Levin LA (1989) A hard-core predicate for all one-way functions. In: Proceedings of the 21st

ACM symposium on theory of computing, ACM, Seattle, WA, May 1989, pp 25–32

Goldreich O, Micali S, Wigderson A (1987) How to play any mental game or a completeness theorem for protocols with honest majority. In: Proceedings of the 19th ACM symposium on theory of computing, ACM, New York, May 1987, pp 218–229

Goldreich O, Micali S, Wigderson A (1991) Proofs that yield nothing but their validity for all languages in NP have zero-knowledge proof systems. J Assoc Comput Mach 38(3):691–729

Goldwasser S, Micali S (1984) Probabilistic encryption. J Comput Syst Sci 28(2):270–299

Goldwasser S, Micali S, Rackoff C (1989) The knowledge complexity of interactive proof systems. SIAM J Comput 18(1):186–208

Goldreich O, Sahai A, Vadhan S (1999) Can statistical zero knowledge be made non-interactive? Or on the relationship of SZK and NISZK. In: CRYPTO 1999: Advances in cryptology, Santa Barbara, CA, August 1999. Lecture notes in computer science, vol 1666. Springer, Berlin, pp 467–484

Gottesman D, Chuang I (2001) Quantum digital signatures. Available via ArXiv:quant-ph/0103032v2

Grigni M, Schulman LJ, Vazirani M, Vazirani UV (2004) Quantum mechanical algorithms for the nonabelian hidden subgroup problem. Combinatorica 24(1):137–154

Haitner I, Reingold O (2007) Statistically-hiding commitment from any one-way function. In: Proceedings of the 39th ACM symposiom on theory of computing, San Diego, CA, June 2007, pp 1–10

Haitner I, Horvitz O, Katz J, Koo C-Y, Morselli R, Shaltiel R (2005) Reducing complexity assumptions for statistically-hiding commitment. In: EUROCRYPT 2005: Advances in cryptology, Aarhus, Denmark, May 2005. Lecture notes in computer science, vol 3494. Springer, Berlin, pp 58–77

Hallgren S, Russell A, Ta-Shma A (2003) The hidden subgroup problem and quantum computation using group representations. SIAM J Comput 32(4):916–934

Hallgren S, Moore C, Rötteler M, Russell A, Sen P (2006) Limitations of quantum coset states for graph isomorphism. In: Proceedings of the 38th ACM symposium on theory of computing, ACM, Seattle, WA, May 2006, pp 604–617

Hardy L, Kent A (2004) Cheat sensitive quantum bit commitment. Phys Rev Lett 92(15):157901

Håstad J, Impagliazzo R, Levin LA, Luby M (1999) A pseudorandom generator from any one-way function. SIAM J Comput 28(4):1364–1396

Hayashi M, Kawachi A, Kobayashi H (2008) Quantum measurements for hidden subgroup problems with optimal sample complexity. Quantum Info Comput 8:345–358

Kashefi E, Nishimura H, Vedral V (2002) On quantum one-way permutations. Quantum Info Comput 2(5):379–398

Kawachi A, Yamakami T (2006) Quantum hardcore functions by complexity-theoretical quantum list decoding. In: Proceedings of the 33rd international colloquium on automata, languages and programming, Venice, Italy, July 2006. Lecture notes in computer science, vol 4052. Springer, Berlin, pp 216–227

Kawachi A, Koshiba T, Nishimura H, Yamakami T (2005a) Computational indistinguishability between quantum states and its cryptographic application. In: EUROCRYPT 2005: Advances in cryptology, Aarhus, Denmark, May 2005. Lecture notes in computer science, vol 3494. Springer, Berlin, pp 268–284

Kawachi A, Kobayashi H, Koshiba T, Putra RRH (2005b) Universal test for quantum one-way permutations. Theor Comput Sci 345(2–3):370–385

Kempe J, Pyber L, Shalev A (2007) Permutation groups, minimal degrees and quantum computing. Groups Geometry Dyn 1(4):553–584

Kent A (2003) Quantum bit string commitment. Phys Rev Lett 90(23):237901

Kilian J (1988) Founding cryptography on oblivious transfer. In: Proceedings of the 20th ACM symposium on theory of computing, ACM, Chicago, IL, May 1988, pp 20–31

Koashi M, Preskill J (2003) Secure quantum key distribution with an uncharacterized source. Phys Rev Lett 90:057902

Kobayashi H (2003) Non-interactive quantum perfect and statistical zero-knowledge. In: Proceedings of the 14th international symposium on algorithms and computation, Kyoto, Japan, December 2003. Lecture notes in computer science, vol 2906. Springer, Berlin, pp 178–188

Kobayashi H (2008) General properties of quantum zero-knowledge proofs. In: Proceedings of the 5th theory of cryptography conference, New York, March 2008. Lecture notes in computer science, vol 4948. Springer, New York, pp 107–124

Koshiba T, Odaira T (2009) Statistically-hiding quantum bit commitment from approximable-preimage-size quantum one-way function. In: Proceedings of the 4th workshop on theory of quantum computation, communication and cryptography, Waterloo, ON, Canada, May 2009. Lecture notes in computer science, vol 5906. Springer, Berlin, pp 33–46

Lamport L (1979) Constructing digital signatures from a one-way function. Technical Report CSL-98, SRI International

Lo H-K, Chau HF (1997) Is quantum bit commitment really possible? Phys Rev Lett 78(17):3410–3413

Marriott C, Watrous J (2004) Quantum Arthur-Merlin games. In: Proceedings of the 19th IEEE conference on computational complexity, IEEE, Amherst, MA, June 2004, pp 275–285

Mayers D (1996) Quantum key distribution and string oblivious transfer in noisy channels. In: CRYPTO'96: Advances in cryptology, Santa Barbara, CA, August 1996. Lecture notes in computer science, vol 1109. Springer, Berlin, pp 343–357

Mayers D (1997) Unconditionally secure quantum bit commitment is impossible. Phys Rev Lett 78(17):3414–3417

Mayers D, Salvail L (1994) Quantum oblivious transfer is secure against all individual measurements. In: Proceedings of workshop on physics and computation, IEEE, Dallas, TX, November 1994, pp 69–77

Micciancio D, Regev O (2009) Lattice-based cryptography. In Bernstein DJ, Buchmann J, Dahmen E (eds) Post-quantum cryptography. Springer, Berlin, pp 147–191

Naor M (1991) Bit commitment using pseudorandomness. J Cryptol 4(2):151–158

Naor M, Yung M (1989) Universal one-way hash functions and their cryptographic applications. In: Proceedings of the 21st ACM symposium on theory of computing, ACM, Seattle, WA, May 1989, pp 33–43

Naor M, Ostrovsky R, Venkatesan R, Yung M (1998) Perfect zero-knowledge arguments for NP using any one-way permutation. J Cryptol 11(2):87–108

Nguyen M-H, Ong S-J, Vadhan SP (2006) Statistical zero-knowledge arguments for NP from any one-way function. In: Proceedings of the 47th IEEE symposium on foundations of computer science, IEEE, Berkeley, CA, October 2006, pp 3–14

Nguyen PQ, Stern J (2005) Adapting density attacks to low-weight knapsacks. In: ASIACRYPT 2005: Advances in cryptology, Chennai, India, December 2005. Lecture notes in computer science, vol 3788. Springer, Berlin, pp 41–58

Okamoto T, Tanaka K, Uchiyama S (2000) Quantum public-key cryptosystems. In: CRYPTO 2000: Advances in cryptology, Santa Barbara, CA, August 2000. Lecture notes in computer science, vol 1880. Springer, Berlin, pp 147–165

Rabin M (1981) How to exchange secrets by oblivious transfer. Technical Report TR-81, Aiken Computation Laboratory, Harvard University

Rivest RL, Shamir A, Adleman L (1978) A method for obtaining digital signature and public key cryptosystems. Commun ACM 21(2):120–126

Sahai A, Vadhan S (2003) A complete problem for statistical zero knowledge. J ACM 50(2):196–249

Shor PW (1997) Polynomial-time algorithms for prime factorization and discrete logarithms on a quantum computer. SIAM J Comput 26(5):1484–1509

Shor PW, Preskill J (2000) Simple proof of security of the BB84 quantum key distribution protocol. Phys Rev Lett 85:441–444

Sudan M (2000) List decoding: algorithms and applications. SIGACT News 31(1):16–27

van de Graaf J (1997) Towards a formal definition of security for quantum protocols. PhD thesis, Université de Montréal

Watrous J (2002) Limits on the power of quantum statistical zero-knowledge. In: Proceedings of the 43rd IEEE symposium on foundations of computer science, IEEE, Vancouver, BC, Canada, November 2002, pp 459–470

Watrous J (2006) Zero-knowledge against quantum attacks. In: Proceedings of the 38th annual ACM symposium on theory of computing, ACM, Seattle, WA, May 2006, pp 296–305

Wegman MN, Carter JL (1981) New hash functions and their use in authentication and set equality. J Comput Syst Sci 22(3):265–279

Wiesner S (1983) Conjugate coding. SIGACT News 15(1):78–88

Yao AC-C (1995) Security of quantum protocols against coherent measurements. In: Proceedings of the 27th annual ACM symposium on theory of computing, ACM, Las Vegas, NV, May–June 1995, pp 67–75

46 BQP-Complete Problems

Shengyu Zhang
Department of Computer Science and Engineering, The Chinese University of Hong Kong, Hong Kong S.A.R., China
syzhang@cse.cuhk.edu.hk

G. Rozenberg et al. (eds.), *Handbook of Natural Computing*, DOI 10.1007/978-3-540-92910-9_46,
© Springer-Verlag Berlin Heidelberg 2012

Abstract

The concept of completeness is one of the most important notions in theoretical computer science. **PromiseBQP**-complete problems are those in **PromiseBQP** to which all other **PromiseBQP** problems can be reduced in classically probabilistic polynomial time. Studies of **PromiseBQP**-complete problems can deepen our understanding of both the power and limitation of efficient quantum computation. In this chapter we give a review of known **PromiseBQP**-complete problems, including various problems related to the eigenvalues of sparse Hamiltonians and problems about additive approximation of Jones polynomials and Tutte polynomials.

1   Introduction

A celebrated discovery in theoretical computer science is the existence of **NP**-complete problems. A decision problem is **NP**-complete if it is in **NP**, and all other problems in **NP** reduce to it in deterministic polynomial time. After Cook (1971) (and Levin (1973) independently) showed that the Satisfiability problem is **NP**-complete, Karp (1972) found that 21 natural combinatorial problems, mostly problems on graphs, are **NP**-complete as well. Since then, thousands of problems arising from various disciplines such as mathematics, physics, chemistry, biology, information science, etc. have been found to be **NP**-complete (Garey and Johnson 1979).

Why is the notion of **NP**-completeness so important? First, showing a problem to be **NP**-complete means that the problem is very unlikely to be solvable in polynomial time. This makes **NP**-completeness an "important intellectual export of computer science to other disciplines" (Papadimitriou 1997). Second, the number of **NP**-complete problems that arise naturally is huge; thus **NP**-complete problems give a nice classification of **NP** problems. Actually, all but a few **NP** problems are known to be either in **P** or **NP**-complete. Nowadays, whenever people find a new problem in **NP** and cannot quickly find a polynomial-time algorithm, they start to consider whether it is **NP**-complete. Third, the notion of completeness provides a useful method to study a whole class by concentrating on one specific problem. For example, if one wants to separate **NP** from **P**, it is enough to show that any **NP**-complete problem does not have a deterministic polynomial-time algorithm. Of course, if one tries to explore the possibility of **NP** = **P** by designing efficient algorithms for **NP**, it is also sufficient to find an efficient algorithm on any one of the **NP**-complete problems.

Complete problems do not only exist in **NP**, they also exist in other computational classes, although the requirement for the reduction algorithm may vary from case to case. To study a property of a class, it is enough to show the property for one complete problem in the class, as long as the reduction used in the completeness definition does not destroy the property. For example, the fact that Graph 3-Coloring has a (computational) Zero-Knowledge protocol immediately implies that all **NP** problems have a Zero-Knowledge protocol, because for an arbitrary **NP** problem, the prover and verifier can first map the input to a Graph 3-Coloring instance and then run the Zero-Knowledge protocol for the latter (Goldreich et al. 1991).

Quantum computing is a new paradigm rapidly developed since the mid-1990s. Since Shor's fast quantum algorithm for Factoring and Discrete Log caused great excitement in both the physics and computer science communities, many efforts have been put into designing quantum algorithms with exponential speedup over their classical counterparts. However, the progress of this line of research has been disappointingly slower than what people had expected. There are a few quantum algorithms with exponential speedup, such as Hallgren's polynomial-time quantum algorithm for solving Pell's equation (Hallgren 2007) and Kuperberg's $2^{O(\sqrt{n})}$ time quantum algorithm for solving the Hidden Subgroup Problem (Lomont 2004) for a dihedral group (Kuperberg 2005); see a recent survey (Childs and van Dam 2010) for a more comprehensive review. But all the problems are number-theoretical or algebraic. (One exception is the oracle separation for the Glued Tree problem by Childs et al. based on a continuous quantum walk (Childs et al. 2003), but the problem is somewhat artificial.) Why does it seem that fast quantum algorithms are much harder to design and what should one do next? In a survey (Shor 2004), Shor gave a couple of possible reasons, including that one does not have enough intuition and experience in dealing with quantum information, and that there may not be many natural problems in quantum computers that can have speedup. He also suggested to try to first study problems that are solvable on a classical computer, aiming at developing algorithmic designing tools with extensive usefulness. It is hoped that studies of **BQP**-complete problems can shed light on these questions and deepen our understanding of the power and limitation of quantum computation.

BQP is the computational class containing decision problems that are solvable probabilistically on a polynomial-time quantum computer. Since a central problem in quantum computing is the comparison between the computational power of quantum and classical computation, it is natural to see what extra power a quantum computer gives one. In this regard, let **BQP**-completeness be defined with the reduction being **BPP** algorithms, that is, those (classical) probabilistic polynomial-time algorithms. In other words, a problem is **BQP**-complete if it is in **BQP**, and all other **BQP** problems can reduce it by probabilistic polynomial-time classical algorithms. Analogous to the fact that **NP**-complete problems are the "hardest" problems in **NP** and thus capture the computational power of efficient nondeterministic computation, **BQP**-complete problems are the hardest problems in **BQP** and thus capture the computational power of efficient quantum computation.

Like many other "semantic" complexity classes, **BQP** is not known to contain complete problems. What people usually study for completeness, in such a scenario, is the class containing the *promise* problems, that is, those decision problems for which the union of Yes and No input instances is not necessarily the whole set of {0,1} strings. In our quantum case, it is the class **PromiseBQP**, that is, the collection of promise problems solvable in polynomial time on a quantum computer. There are mainly two tracks of **PromiseBQP**-complete problems. The first track contains problems on the eigenvalues of a local or sparse Hamiltonian or unitary matrix (Wocjan and Zhang 2006; Janzing and Wocjan 2007; Janzing et al. 2008). In the first work (Wocjan and Zhang 2006) along this line, Wocjan and Zhang considered the local Hamiltonian eigenvalue sampling (LHES) and local unitary phase sampling (LUPS) problems: Given a classical string $x \in \{0, 1\}^n$ and a $2^n \times 2^n$ dimensional local Hamiltonian $H = \sum_j H_j$ or $U = \prod_j U_j$, where each H_j or U_j are operating on a constant number of qubits, one is approximately sampling the eigenvalues of H or U under the distribution $\langle x|\eta_j\rangle$, where $|\eta_j\rangle$ are the corresponding eigenvectors. These two problems have close connections to other well-known problems in quantum algorithm and complexity theory: LHES is the natural sampling variant of estimating the minimum eigenvalue of a local Hamiltonian, a **QMA**-complete problem (Kitaev et al. 2002; Kempe et al. 2006), and LUPS is the natural sampling variant of estimating the eigenvalue of a unitary on a given eigenvector, a powerful algorithmic tool called phase estimation (Kitaev 1995). Though not defined as a promise problem, it will be shown in this survey that one can easily transform them into promise problems and thus be **PromiseBQP**-complete.

Later, the work was extended in two ways. In Janzing et al. (2008), Janzing, Wocjan, and Zhang showed that even if one restricts the local Hamiltonians in LHES to translationally invariant ones operating on a one-dimensional qudit chain, the problem is still **PromiseBQP**-complete. The other extension, done by Janzing and Wocjan (2007), views the Hamiltonian as a ± 1 weighted graph and lifts the graph to power k, where k is part of the input. Then estimating a diagonal entry is **PromiseBQP**-complete.

The second line of research on **PromiseBQP**-complete problems is about approximating the Jones polynomial and Tutte polynomial (Freedman et al. 2002b, c; Aharonov and Arad 2006; Aharonov et al. 2006, 2007a; Wocjan and Yard 2006). The Jones polynomial is an important knot invariant with rich connections to topological quantum field theory, statistical physics (Wu 1992) and DNA recombination (Podtelezhnikov et al. 1999). The main result for the Jones polynomial in the studies of **PromiseBQP**-completeness is that approximating the Jones polynomial of the plat closure of the braid group at $e^{2\pi i/k}$ to within some precision is **PromiseBQP**-complete. Both the efficient quantum algorithm and the universality for constant k were implicitly given by Freedman et al. (2002b, c), but recent results by Aharonov et al. (2006) and Aharonov and Arad (2006) gave an explicit and simpler algorithm and a hardness proof, which also extend the results to any k bounded by a polynomial of the size of the input braid.

The multivariate Tutte polynomial is more general than the Jones polynomial; the Tutte polynomial has many connections to algebraic graph theory and the Potts model in statistical physics. In Aharonov et al. (2007a), Aharonov, Arad, Eban, and Landau gave an efficient algorithm for additively approximating this general polynomial and also showed that the approximation for some ranges is **PromiseBQP**-complete.

The rest of the chapter is organized as follows. ❷ Section 2 gives precise definitions of **BQP** and **PromiseBQP**. In ❷ Sects. 3 and ❷ 4, the two lines of research on **PromiseBQP**-complete problems are studied. ❷ Section 5 concludes with some questions raised.

2   Preliminaries

A promise problem is a pair (L_{Yes}, L_{No}) of nonintersecting subsets of $\{0, 1\}^*$. A language is the set of Yes instances of a promise (L_{Yes}, L_{No}) satisfying $L_{Yes} \cup L_{No} = \{0, 1\}^*$. For more discussions on promise problems versus languages (and why the former is important and sometimes necessary for complexity theory), the readers are referred to Goldreich's survey (Goldreich 2005).

Definition 1   (PromiseBQP) **PromiseBQP** is the class of promise problems (L_{Yes}, L_{No}) such that there is a uniform family of quantum circuits $\{U_n\}$ operating on $p(n) = \text{poly}(n)$ qubits, with the promise that for any n, applying U_n on $|x, 0^{p(n)-n}\rangle$ and measuring the first qubit gives the outcome 1 with probability at least 2/3 for all $x \in L_{Yes}$, and at most 1/3 for all $x \in L_{No}$.

Definition 2   (BQP) BQP is the class of languages L_{Yes} with $(L_{Yes}, \{0, 1\}^* - L_{Yes})$ in **PromiseBQP**.

Like many other "semantic" classes, **BQP** is not known to have complete problems. However, if one extends languages to promise problems, then **PromiseBQP** has canonical complete problems for any model, such as the quantum Turing machine (Bernstein and Vazirani 1997), quantum circuit (Yao 1993), adiabatic quantum computer (Aharonov et al. 2004), and quantum one-way computer (Browne and Briegel 2006). For the quantum circuit model, for example, the canonical complete problem is the following.

Definition 3 The *canonical complete problem* for **PromiseBQP** in the circuit model has input $(\langle U \rangle, x)$, where $\langle U \rangle$ is the description of a uniform family of quantum circuits $\{U_n\}$ working on $p(n) = \text{poly}(n)$ qubits, with the promise that applying U_n on $|x, 0^{p(n)-n}\rangle$ and measuring the first qubit gives outcome 1 with probability at least 2/3 for all $x \in L_{\text{Yes}}$ and at most 1/3 for all $x \in L_{\text{No}}$. The problem is to distinguish the two cases.

The fact that this is a complete problem for **PromiseBQP** is almost by definition, which also makes the problem not so interesting. We hope to have more "natural" problems that can help us understand the class **PromiseBQP**.

3 Approximate Eigenvalue Sampling

This section provides an overview of the first track of studies on **PromiseBQP**-complete problems, that is on problems about eigenvalues of the local or sparse Hamiltonians. ❷ Section 3.1 starts with the local Hamiltonian eigenvalue sampling (LHES) problem, defined in Wocjan and Zhang (2006), and we show that it is **PromiseBQP**-complete. The problem together with the proof contains the core ideas and techniques for this line of research.

Two extensions are then shown. First, the result in Janzing et al. (2008) is mentioned: even if the Hamiltonians are restricted to be translationally invariant on a one-dimensional qudit chain (with $d = O(1)$), the problem is still **PromiseBQP**-complete. Then ❷ Sect. 3.2 shows the result in Janzing and Wocjan (2007) that estimating an entry in a sparse real symmetric matrix to some power k, where k is part of the input, is also **PromiseBQP**-complete.

The proofs in the three sections have similar ingredients, and an attempt is made to unify the notation and give a consistent treatment.

3.1 Phase Sampling and Local Hamiltonian Eigenvalue Sampling

The problems that one is going to see in this section have close relation to two well-known ones, local Hamiltonian minimum eigenvalue and phase estimation, which will be defined below. The local Hamiltonian minimum eigenvalue (LHME) problem is usually abbreviated as the local Hamiltonian (LH) problem, but here LHME is used to distinguish it from the other related problems, which are defined later.

Definition 4 (Local Hamiltonian minimum eigenvalue (LHME)) A tuple (H, a, b) is given where

1. $H = \sum_j H_j$ is a Hamiltonian operating on n qubits, with j ranging over a set of size polynomial in n, and each H_j operating on a constant number of qubits; it is promised that either $\lambda(H) < a$ or $\lambda(H) > b$, where $\lambda(H)$ is the minimum eigenvalue of H.
2. a and b are two real numbers such that $a < b$ and the gap $b - a = \Omega(1/\text{poly}(n))$.

The task is to distinguish between the case $\lambda(H) < a$ and the case $\lambda(H) > b$.

Definition 5 (Phase estimation (PE)) A unitary matrix U is given by black-boxes of controlled-U, controlled-U^{2^2}, ..., controlled-$U^{2^{t-1}}$ operations, and an eigenvector $|u\rangle$ of U with eigenvalue $e^{2\pi i\varphi}$ with the value of $\varphi \in [0, 1)$ unknown. The task is to output an n-bit estimation of φ.

These two problems are both well studied in quantum computing. The Local Hamiltonian problem was shown by Kitaev et al. (2002) to be **PromiseQMA**-complete when each H_j operates on five qubits; actually it remains **PromiseQMA**-complete even when each H_j operates on only two qubits (Kempe et al. 2006). Kitaev's efficient quantum algorithm for phase estimation (Kitaev 1995) was a powerful tool for quantum algorithm design, such as for factoring (Shor 1997; Nielsen and Chuang 2000) and some recent quantum-walk based algorithms such as the general quantum walk search (Magniez 2007), formula evaluation (Ambainis et al. 2007), and its extension to span-program evaluation (Reichardt and Spalek 2008).

Let us now consider the sampling variant of the above two problems.

Definition 6 A probability distribution q on $\mathbb{R}$ is said to *approximate* another probability distribution p on a discrete set $S \subseteq \mathbb{R}$ with *error* δ and *precision* ε if

$$\mathbf{Pr}_{x\leftarrow q}[s - \varepsilon \leq x \leq s + \varepsilon] \geq (1 - \delta)p(s) \tag{1}$$

for any $s \in S$.

Intuitively, to approximate the probability distribution p, one draws a sample from another distribution q, and the outcome x is ε-close to s at least $(1 - \delta)$ times the correct probability $p(s)$, for each $s \in S$. Now one can define the sampling version of the two problems.

Definition 7 (Local Hamiltonian eigenvalue sampling (LHES)) We are given $(H, \varepsilon, \delta, b)$ where

1. $H = \sum_j H_j$ is a Hamiltonian operating on n qubits, with j ranging over a set of size polynomial in n, and each H_j operating on a constant number of qubits.
2. $\varepsilon = \Omega(1/\text{poly}(n))$ is the required estimation precision.
3. $\delta = \Omega(1/\text{poly}(n))$ is the required sampling error probability.
4. $b \in \{0, 1\}^n$ is a classical n-bit string.

Suppose the eigenvalues and the corresponding eigenvectors of H are $\{(\lambda_k, |\eta_k\rangle): k \in [2^n]\}$ satisfying $|\lambda_k| < \text{poly}(n)$ for each k. Define the probability distribution $D(H, b)$ over the spectrum of H by

$$D(H, b) = \{(\lambda, \mathbf{Pr}(\lambda)) : \mathbf{Pr}(\lambda) = \sum_{\lambda_k = \lambda} |\langle b|\eta_k\rangle|^2\}. \tag{2}$$

The task is to draw a sample from some probability distribution p approximating $D(H, b)$ with error δ and precision ε.

Definition 8 (Local unitary phase sampling (LUPS)) We are given $(H, \varepsilon, \delta, b)$ where

1. U is (the description of) an n-qubit quantum circuit.
2. $\varepsilon = \Omega(1/\text{poly}(n))$ is the required estimation precision.

3. $\delta = \Omega(1/\text{poly}(n))$ is the required sampling error probability.
4. $b \in \{0, 1\}^n$ is a classical n-bit string.

Suppose the eigenvalues of U are $\{\lambda_j = e^{2\pi i \varphi_j}\}_{j=1,\ldots,2^n}$ (where $\varphi_j \in [0, 1)$ for each j), with the corresponding eigenvectors $\{|\eta_j\rangle\}_{j=1,\ldots,2^n}$. The task is to estimate φ_j with error δ and precision ε.

One can immediately see that these two problems are not even promise problems, since the output is a (sampling from a) distribution rather than a Yes/No answer. Nevertheless, they capture the power of efficient quantum computers in the following sense: First, a polynomial-time uniform family of quantum circuits can achieve the sampling requirement. Second, given an oracle for either problem, any **BQP** problem can be solved by a classical polynomial-time algorithm. These facts will be proved in the following two sections. After that, a mention will be made about how to change LHES to a promise problem so that it really becomes **PromiseBQP**-complete.

3.1.1 BQP Algorithm for LHES and LUES

In this section we prove that LHES and LUES are solvable by an efficient quantum circuit. The standard algorithm for phase estimation is first reviewed, then it is observed that the same algorithm actually gives the desired LUES solution. This is then used to show an algorithm for LHES.

Phase estimation can be solved by a quantum algorithm as follows. The working space has two registers. The first register consists of $t = n + \lceil \log(2 + 1/2\delta) \rceil$ qubits and is prepared in $|0\ldots0\rangle$. The second register contains the eigenvector $|u\rangle$. Measuring $\tilde{\varphi}$ in the first register after carrying out the transformations described below gives the desired n-bit estimation of φ with probability of at least $1 - \delta$.

$$|0\rangle^{\otimes t}|u\rangle \tag{3}$$

$$\rightarrow \frac{1}{\sqrt{2^t}} \sum_{j=0}^{2^t-1} |j\rangle|u\rangle \qquad \text{(apply the Fourier transform)} \tag{4}$$

$$\rightarrow \frac{1}{\sqrt{2^t}} \sum_{j=0}^{2^t-1} |j\rangle U^j|u\rangle \qquad \text{(apply the controlled powers of } U\text{)} \tag{5}$$

$$= \frac{1}{\sqrt{2^t}} \sum_{j=0}^{2^t-1} e^{2\pi i j \varphi} |j\rangle|u\rangle \tag{6}$$

$$\rightarrow |\tilde{\varphi}\rangle|u\rangle \qquad \text{(apply the inverse Fourier transform)} \tag{7}$$

The following observation says that the same algorithm actually works for LUES.

Fact If one feeds $|0\rangle|b\rangle$ instead of $|0\rangle|u\rangle$ as input to the above algorithm for the phase estimation problem and $t = \lceil \log \frac{1}{\varepsilon} \rceil + \lceil \log(2 + \frac{1}{2\delta}) \rceil$, then the measurement of the first register gives the desired sampling output. This implies that LUES can be solved by a polynomial-size quantum circuit.

This actually holds not only for $|b\rangle$ but also for a general state $|\eta\rangle$. To see why this is true, write $|\eta\rangle$ as $\sum_{j=1}^{n} \alpha_j |\eta_j\rangle$, then by the linearity of the operations, the final state is $\alpha_j |\tilde{\varphi}_j\rangle |\eta_j\rangle$. For more details, the readers are referred elsewhere (Nielsen and Chuang 2000; Chap. 5). Note that to implement the controlled-U^{2^j} operations for $j = 0, \ldots, 2^t - 1$ in the above algorithm, one needs to run U for 2^t times, which can be done efficiently since $t = \lceil \log \frac{1}{\varepsilon} \rceil + \lceil \log(2 + \frac{1}{2\delta}) \rceil$ and $\varepsilon = \Omega(1/\text{poly}(n))$, $\delta = \Omega(1/\text{poly}(n))$.

Now we can give the efficient algorithm for LHES.

Theorem 1 *LHES can be implemented by a uniform family of quantum circuits of polynomial size.*

Proof By a simple scaling ($H' = H/\Lambda = \sum_j H_j/\Lambda$ where $\Lambda = \max_k |\lambda_k| = \text{poly}(n)$), one can assume that all the eigenvalues λ_k of H satisfy $|\lambda_k| < 1/4$. The basic idea to design the quantum algorithm is to use LUPS on $e^{2\pi i H}$. Note that $e^{2\pi i H}$ is unitary, and if the eigenvalues and eigenvectors of H are $\{\lambda_k |\eta_k\rangle\}$, then those of $e^{2\pi i H}$ are just $\{e^{2\pi i \lambda_k}, |\eta_k\rangle\}$. Therefore, it seems that it is enough to run the LUES algorithm on ($e^{2\pi i H}$, ε, δ, b), and if one gets some $\lambda > 1/2$, then output $\lambda - 1$. However, note that H is of exponential dimension, so $e^{2\pi i H}$ is not ready to compute in the straightforward way. Fortunately, this issue is well studied in the quantum simulation algorithms, and the standard approach is the following asymptotic approximation using the Trotter formula (Trotter 1959; Chernoff 1968; Nielsen and Chuang 2000) or its variants. Here using the simulation technique, one obtains

$$\left(e^{2\pi i \sum_j H_j/m} \right)^m = \left(\prod_j e^{2\pi i H_j/m} \right)^m + O(1/m) \tag{8}$$

Now one runs LUES on (b, ε, $\delta/2$, $e^{2\pi i H}$). Whenever one needs to call $e^{2\pi i H}$, one uses $\prod_j e^{2\pi i H_j/m}$ for m times instead. Note that such substitution yields $O(1/m)$ deviation, so $t = \log \frac{2}{\varepsilon \delta} + O(1)$ calls yield $O(\frac{1}{m\varepsilon\delta}) \leq \frac{c}{m\varepsilon\delta}$ deviation for some constant c. Let $m = \frac{2c}{\varepsilon\delta^2}$, thus the final error probability is less than $\frac{\delta}{2} + \frac{c}{m\varepsilon\delta} = \delta$, achieving the desired estimation and sampling precisions.

From the proof one can see that as long as $e^{2\pi i H}$ can be simulated efficiently from the description of H, one can sample the eigenvalues of H as desired. Since sparse Hamiltonians, which contain local Hamiltonians as special cases, can be simulated efficiently (Aharonov and Ta-Shma 2003; Berry et al. 2007), we know that if we modify the definition of LHES by allowing H to be sparse then it is also **BQP**-complete.

3.1.2 BQP Hardness of LHES

Theorem 2 **PromiseBQP** $\subseteq$ **P**^{LHES}

Proof For any $L \in$ **PromiseBQP**, there is a uniform family of polynomial size quantum circuits with ε-bounded error (for a small constant ε) that decides if $x \in L$ or $x \notin L$. Denote by

U the corresponding quantum circuit (of size n) and suppose the size of U is M, which is bounded by a polynomial in n. Further suppose that the computation is described by $U|x, \mathbf{0}\rangle = \alpha_{x,0}|0\rangle|\psi_{x,0}\rangle + \alpha_{x,1}|1\rangle|\psi_{x,1}\rangle$, where $\mathbf{0}$ is the initial state of the ancillary qubits, and $|\psi_{x,0}\rangle$ and $|\psi_{x,1}\rangle$ are pure states. After the U transform, the first qubit is measured and the algorithm outputs the result based on the outcome of the measurement. The correctness of the algorithm requires that $|\alpha_{x,0}|^2 < \varepsilon$ if $x \in L$, and $|\alpha_{x,1}|^2 < \varepsilon$ if $x \notin L$.

Now, a local Hamiltonian H is constructed encoding the circuit U and a binary string b encoding the inputs x, such that eigenvalue sampling applied to H and b yields significantly different probability distributions for the two cases of $x \in L$ and $x \notin L$.

To this end, the circuit V is constructed as follows: one first applies the original circuit U, then flips the sign for the component with the first qubit being 1, and finally reverses the computation of U. See ❿ Fig. 1.

Suppose the original circuit U is decomposed as the product of elementary gates, that is, $U = U_{m-1} \ldots U_0$, where each U_j is an elementary gate. Let V_j be the j-th gate in V, where $j = 0, 1, \ldots, 2m$. Let $M = 2m + 1$, the number of gates in V. Attach a clock register to the system and define the operator

$$F = \sum_{j=0}^{M-1} V_j \otimes |j+1\rangle\langle j| \tag{9}$$

where the summation in the index is module M. Note that F is an $O(\log N)$-local operator; we will remark how to slightly modify it to be a 4-local operator at the end of the proof. Define

$$|\varphi_{x,j}\rangle = F^j(|x, \mathbf{0}\rangle|0\rangle) \tag{10}$$

for $j \geq 0$, where F^j means that F is applied j times. It is easy to see that $|\varphi_{x,j}\rangle = |\varphi_{x, j \{\mathrm{mod}\}2M}\rangle$ for $j \geq 2M$.

Note that due to the clock register for different j and j' in $\{0, \ldots, 2M-1\}$, $|\varphi_{x,j}\rangle$ and $|\varphi_{x,j'}\rangle$ are orthogonal if $|j-j'| \neq M$. Also note that $U_{m-1}\ldots U_0|x, \mathbf{0}\rangle = U|x, \mathbf{0}\rangle = \alpha_{x,0}|0\rangle|\psi_{x,0}\rangle + \alpha_{x,1}|1\rangle|\psi_{x,1}\rangle$ by definition. Therefore, we have

$$\langle\varphi_{x,j}|\varphi_{x,M+j}\rangle = \langle x, \mathbf{0}|U_0^\dagger \ldots U_{m-1}^\dagger (P_0 - P_1) U_{m-1}\ldots U_0|x, \mathbf{0}\rangle = |\alpha_{x,0}|^2 - |\alpha_{x,1}|^2 \tag{11}$$

for $j = 0, \ldots, M - 1$, where P_0 and P_1 are the projections onto the subspaces corresponding to the first qubit being 0 and 1, respectively. To summarize, we have

$$\langle\varphi_{x,j}|\varphi_{x,j'}\rangle = \begin{cases} 0 & |j-j'| \neq M \\ |\alpha_{x,0}|^2 - |\alpha_{x,1}|^2 & |j-j'| = M \end{cases} \tag{12}$$

◻ **Fig. 1**
Circuit V.

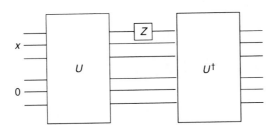

Define the subspace $S_x = \mathrm{span}\{|\varphi_{x,j}\rangle : j = 0, \dots, 2M - 1\}$. The key property here is that though F has an exponentially large dimension, $F|_{S_x}$ is of only polynomial dimension. Depending on the probability, we consider three cases.

Case $|\alpha_{x,0}| = 1$: The dimension of S_x is M, and $F|_{S_x}$ is a shift operator on the basis $\{|\varphi_{x,j}\rangle\}_{j=0,\dots,M-1}$, that is $F|\varphi_{x,j}\rangle = |\varphi_{x,j+1 \bmod M}\rangle$. It is not hard to see that this operator has eigenvalues and the corresponding eigenvectors

$$\lambda_k = \omega_M^k, \quad |\xi_k\rangle = \frac{1}{\sqrt{M}} \sum_{j=0}^{M-1} \omega_M^{-kj} |\varphi_{x,j}\rangle, \quad k = 0, 1, \dots, M - 1 \tag{13}$$

where $\omega_M = e^{2\pi i/M}$.

Case $\alpha_{x,0} = 0$: The dimension of S_x is M, and $F|_{S_x}$ is almost a shift operator on the basis $\{|\varphi_{x,j}\rangle\}_{j=0,\dots,M-1}$: $F|\varphi_{x,j}\rangle = |\varphi_{x,j+1}\rangle$ for all $j = 0, \dots, M - 2$, and $F|\varphi_{x,M-1}\rangle = -|\varphi_{x,0}\rangle$. It is not hard to see that this operator has eigenvalues and the corresponding eigenvectors

$$\mu_k = \omega_M^{k+1/2}, \quad |\eta_k\rangle = \frac{1}{\sqrt{M}} \sum_{j=0}^{M-1} \omega_M^{-(k+1/2)j} |\varphi_{x,j}\rangle, \quad k = 0, 1, \dots, M - 1. \tag{14}$$

Case $0 < |\alpha_{x,0}| < 1$: In this general case, S_x has dimension $2M$. To find the eigenvalues and eigenvectors of $F|_{S_x}$, define

$$|\phi_{x,j}\rangle = \frac{|\varphi_{x,j}\rangle + |\varphi_{x,M+j}\rangle}{\||\varphi_{x,j}\rangle + |\varphi_{x,M+j}\rangle\|}, \quad |\gamma_{x,j}\rangle = \frac{|\varphi_{x,j}\rangle - |\varphi_{x,M+j}\rangle}{\||\varphi_{x,j}\rangle - |\varphi_{x,M+j}\rangle\|} \tag{15}$$

for $j = 0, \dots, M - 1$. Then first, because $\langle \varphi_{x,j} | \varphi_{x,N+j}\rangle = |\alpha_{x,0}|^2 - |\alpha_{x,1}|^2$ is a real number, we have $\langle \phi_{x,j} | \gamma_{x,j}\rangle = 0$. Together with ❷ Eq. 12, we know that

$$\{|\phi_{x,0}\rangle, \dots, |\phi_{x,M-1}\rangle, |\gamma_{x,0}\rangle, \dots, |\gamma_{x,M-1}\rangle\} \tag{16}$$

forms an orthonormal basis of S_x. One can further observe that $F|_{S_x} = F_0 \oplus F_1$, where F_0 and F_1 act on $S_{x,+} = \mathrm{span}\{|\phi_{x,0}\rangle, \dots, |\phi_{x,M-1}\rangle\}$ and $S_{x,-} = \mathrm{span}\{|\gamma_{x,0}\rangle, \dots, |\gamma_{x,M-1}\rangle\}$, respectively, with the matrix representations (in the basis $\{|\phi_{x,j}\rangle\}$ and $\{|\gamma_{x,j}\rangle\}$, respectively) as follows:

$$F_0 = \begin{pmatrix} 0 & 1 & 0 & \dots & 0 & 0 \\ 0 & 0 & 1 & \dots & 0 & 0 \\ \vdots & \vdots & \vdots & & \vdots & \vdots \\ 0 & 0 & 0 & \dots & 0 & 1 \\ 1 & 0 & 0 & \dots & 0 & 0 \end{pmatrix}, \quad F_1 = \begin{pmatrix} 0 & 1 & 0 & \dots & 0 & 0 \\ 0 & 0 & 1 & \dots & 0 & 0 \\ \vdots & \vdots & \vdots & & \vdots & \vdots \\ 0 & 0 & 0 & \dots & 0 & 1 \\ -1 & 0 & 0 & \dots & 0 & 0 \end{pmatrix} \tag{17}$$

So F_0 and F_1 have the same matrix representations as the two operators in the previous two cases, thus having the same eigenvalues and eigenvectors (with respect to different basis vectors though). Precisely, F_0 has eigenvalues λ_k with eigenvectors $|\xi_k\rangle = \frac{1}{\sqrt{M}} \sum_{j=0}^{M-1} \omega_M^{-kj} |\phi_{x,j}\rangle$; F_1 has eigenvalues μ_k with eigenvectors $|\eta_k\rangle = \frac{1}{\sqrt{M}} \sum_{j=0}^{M-1} \omega_M^{-(k+1/2)j} |\gamma_{x,j}\rangle$. (Here one uses the same notation $|\xi_k\rangle$ and $|\eta_k\rangle$ because they are consistent with the previous cases when $|\alpha_{x,0}|^2 = 0$ or 1 respectively.)

By this, it is not hard to find the eigenvalue sampling probabilities:

$$|\langle \varphi_{x,0} | \xi_k\rangle|^2 = |\langle \varphi_{x,0} | \frac{1}{M} \sum_j w_M^{-kj} \phi_{x,j}\rangle|^2 \tag{18}$$

$$= \frac{1}{M} |\langle \varphi_{x,0} | \phi_{x,0}\rangle|^2 \quad \text{(by Eq. 12)} \tag{19}$$

$$= \frac{1}{M} \left| \left\langle \varphi_{x,0} \middle| \frac{\varphi_{x,0} + \varphi_{x,M}}{\| \|\varphi_{x,0}\rangle + |\varphi_{x,M}\rangle \|} \right\rangle \right|^2 \quad \text{(by def of } |\phi_{x,0}\rangle) \quad (20)$$

$$= \frac{|\alpha_{x,0}|^2}{M} \quad \text{(by Eq. 12)} \quad (21)$$

and similarly we have

$$|\langle \varphi_{x,0} | \eta_k \rangle|^2 = \frac{|\alpha_{x,1}|^2}{M} \quad (22)$$

Now the local Hamiltonian is constructed as

$$H = (F + F^\dagger)/2. \quad (23)$$

It is easy to verify that H is a local Hamiltonian. And further, suppose the eigenvalues of F are $\{\kappa_j\}$ with the eigenvectors $\{|\psi_j\rangle\}$, then the eigenvalues of H are just $\{(\kappa_j + \kappa_j^*)/2\}$ with the same corresponding eigenvectors. Thus H has eigenvalues $\cos(2k\pi/M)$ and $\cos((2k+1)\pi/M)$ for $k = 0, 1, \ldots, (M-1)/2$, and the distribution $D(H, |\varphi_{x,0}\rangle)$ is $\{\cos(2k\pi/M)$ with probability $|\alpha_{x,0}|^2/M$, and $\cos((2k+1)\pi/M)$ with probability $|\alpha_{x,1}|^2/M\}$.

Now if $x \in L_{\text{Yes}}$, then $|\alpha_{x,0}|^2$ is very small, thus with high probability the sample gives a random value uniformly chosen from $\{\cos(2k\pi/M): k = 0, \ldots, (M-1)/2\}$. If $x \in L_{\text{No}}$, then $|\alpha_{x,1}|^2$ is very small, thus with high probability the sample gives a random value uniformly chosen from $\{\cos((2k+1)\pi/M): k = 0, \ldots, (M-1)/2\}$. Since any two values chosen from these two sets are at least $\Omega(M^{-2})$ away from each other, an approximator with precision $O(M^{-2})$ suffices to distinguish between these two cases.

Finally, to obtain a 4-local LHES, $|i\rangle$ is replaced by $|e_i\rangle = |0\ldots010\ldots0\rangle$ for $i = 0, \ldots, N-1$, where the only 1 appears at coordinate i. Modify the operator F to be

$$F = \sum_{j=0}^{M-1} V_j \otimes |e_{j+1}\rangle \langle e_j| \quad (24)$$

Note that $|e_j\rangle\langle e_{j-1}|$ and $|e_0\rangle\langle e_{N-1}|$ are 2-local. The remaining proof passes through. This completes the proof for ❯ Theorem 2.

Changing It to a Promise Problem

Now that we have completed both the algorithm and the completeness, it is not hard to see that the sampling problem can be modified to a promise problem as follows. We can derive its **PromiseBQP**-completeness from the above proofs. Since this was not mentioned in the previous work (Wocjan and Zhang 2006), a bit more detail is given below. Let

$$p_0(M) = \text{uniform distribution over } \{\cos(2k\pi/M): k = 0, 1, \ldots, (M-1)/2\};$$
$$p_1(M) = \text{uniform distribution over } \{\cos((2k+1)\pi/M): k = 0, 1, \ldots, (M-1)/2\}.$$

Definition 9 (Local Hamiltonian eigenvalue distribution (LHED)). We are given $(H, \varepsilon, \delta, b, M)$ where

1. $H = \sum_j H_j$ is a Hamiltonian operating on n qubits, with j ranging over a set of size polynomial in n, and each H_j operating on a constant number of qubits.
2. $\varepsilon = \Omega(1/\text{poly}(n))$, $\varepsilon = o(M^{-2})$ is the required estimation precision.

3. $\delta = \Omega(1/\text{poly}(n))$, $\delta < 1/10$ is the required sampling error probability.
4. $b \in \{0, 1\}^n$ is a classical n-bit string.
5. $M = \text{poly}(n)$.

Suppose the eigenvalues and the corresponding eigenvectors of H are $\{(\lambda_k, |\eta_k\rangle) : k \in [2^n]\}$ satisfying $|\lambda_k| < \text{poly}(n)$ for each k. The input has the promise that the probability distribution $D(H, b)$, given by ❷ Eq. 2, approximates exactly one of $p_0(M)$ and $p_1(M)$ with error δ and precision ε. The task is to determine which one, $p_0(M)$ or $p_1(M)$, $D(H, b)$ approximates.

Restricted to One-Dimension Chain

It turns out that even if one restricts the LHES problem to the Hamiltonian on a one-dimensional qudit chain for $d = O(1)$, the problem is still **PromiseBQP**-complete (Janzing et al. 2003). The Hamiltonian is constructed along the same lines as above, with some additional treatment, following the ideas of (Aharonov et al. 2007b), to make it translationally invariant on a one-dimensional qudit chain.

3.2   Lifting the Matrix to a Power

One may find the sampling nature of the previous problems not natural, and would like to consider the average eigenvalue of a local Hamiltonian. (The authors of Wocjan and Zhang (2006) gave the credit for the question about the average eigenvalue to Yaoyun Shi.) If the distribution is, as before, induced by a vector $|b\rangle$, such that the average is $\sum_k |\langle b|\xi_k\rangle|^2 \lambda_k$, then it is not an interesting problem to study for **PromiseBQP**: Note that the value is equal to $\langle b|H|b\rangle$, and further

$$\langle b|H|b\rangle = \langle b| \sum_j H_j |b\rangle = \sum_j \langle b|H_j|b\rangle \tag{25}$$

Since each $\langle b|H_j|b\rangle$ can be easily computed even deterministically, one can obtain the exact average eigenvalue of a local Hamiltonian deterministically in polynomial time.

However, if one raises the matrix to some power, then the problem becomes **PromiseBQP**-complete again. This is the subject of this section, based on the result in Janzing and Wocjan (2007).

Definition 10   The sparse matrix powered entry (SMPE) problem is defined as follows. The input is a tuple $(A, b, m, j, \varepsilon, g)$ where

1. $A \in \mathbb{R}^{N \times N}$ is a symmetric matrix with the operator norm $\|A\| \leq b$, A has no more than $s = \text{polylog}(N)$ nonzero entries in each row, and there is an efficiently computable function f specifying for each given row the nonzero entries and their positions.
2. $m = \text{polylog}(N)$ is a positive integer, $j \in [N]$, $\varepsilon = 1/\text{polylog}(N)$ and $g \in [-b^m, b^m]$.

The input has the promise that either $(A^m)_{jj} \geq g + \varepsilon b^m$ or $(A^m)_{jj} \leq g - \varepsilon b^m$. The task is to distinguish between these two cases.

The main theorem is now given.

Theorem 3 *The problem SMPE is* **PromiseBQP**-*complete*.

The algorithm is very similar to the one in the last section, except that now the efficient quantum algorithm is used to simulate the process e^{-iAt} for general sparse Hamiltonians A (Berry et al. 2007). Next, an account of the proof of the **PromiseBQP**-hardness is given.

From the definition of H in ❷ Eq. 23, one can see that

$$H^t = \sum_k \cos^t(2k\pi/M)|\xi_k\rangle\langle\xi_k| + \sum_k \cos^t((2k+1)\pi/M)|\eta_k\rangle\langle\eta_k| \qquad (26)$$

Thus the distribution $D(H^t, |\varphi_{x,0}\rangle)$ is $\{\cos^t(2k\pi/M)$ with probability $|\alpha_{x,0}|^2/M$, and $\cos^t((2k+1)\pi/M)$ with probability $|\alpha_{x,1}|^2/M\}$, whose average is

$$\begin{aligned}
\langle\varphi_{x,0}|H^t|\varphi_{x,0}\rangle = {} & \frac{1}{M}|\alpha_{x,0}|^2\left(1 + \sum_{k=1}^{(M-1)/2}\cos^t(2k\pi/M)\right) \\
& + \frac{1}{M}|\alpha_{x,1}|^2\left(-1 + \sum_{k=0}^{(M-3)/2}\cos^t((2k+1)\pi/M)\right)
\end{aligned} \qquad (27)$$

When t is large enough, say $t = M^3$, all the cosine terms combined become negligible compared to the 1 or -1 in the summation. Thus, depending on whether $|\alpha_{x,1}|^2$ is close to 1 or 0, the average eigenvalue $\langle\varphi_{x,0}|H^t|\varphi_{x,0}\rangle$ will be close to either $-1/M$ or $1/M$, respectively. Therefore, estimating the average eigenvalue can determine whether $|\alpha_{x,1}|^2$ is close to 1 or 0, solving the starting **PromiseBQP** problem.

4 Additive Approximation of the Jones Polynomials and Tutte Polynomials

A completely different vein of research on **PromiseBQP**-completeness is the study of approximation of Jones polynomials and Tutte polynomials (Freedman 2002b, c; Aharonov et al. 2006, 2007a; Aharonov and Arad 2006; Wocjan and Yard 2008). There are at least two interesting aspects of this line compared to the problems discussed in the previous section. First, the problems look less quantum, at least by the definition. Second, the algorithms do not use quantum Fourier transform as many other quantum algorithms with exponential speedup over their classical counterparts do. What the algorithms use is the homomorphism property of a representation.

Next, this line of research is introduced in ❷ Sect. 4.1 to give some background about the Jones polynomial and Tutte polynomial, mainly about their connections to physics and combinatorics. ❷ Section 4.2 presents the efficient quantum algorithms for approximating Jones polynomials of trace closure of the braid group at some roots of unity. Finally a brief account is given of two subsequent (unpublished) results: the **PromiseBQP**-hardness of approximating Jones polynomials (Aharonov and Arad 2006) in ❷ Sect. 4.3 and the algorithms and complexity of approximating the Tutte polynomials (Aharonov et al. 2007a) in ❷ Sect. 4.4.

4.1 About Jones Polynomials and Tutte Polynomials

A short discussion about the connections of the Jones polynomial and the Tutte polynomial to other fields is given. The definitions are a bit involved and thus deferred to later sections.

In knot theory, the Jones polynomial is a knot invariant discovered by Jones (1985). Specifically, it is, for each oriented knot or link, a Laurent polynomial in the variable $\sqrt{t}$ with integer coefficients. The Jones polynomial is an important knot invariant in low dimensional topology; it is also related to statistical physics (Wu 1992) and DNA recombination (Podtelezhnikov et al. 1999).

The connection between the Jones polynomial and quantum computing is already known. Freedman et al. (2002b, c) showed that a model of quantum computing based on topological quantum field theory and Chern–Simons theory (Freedman 1998; Freedman et al. 2002a) is equivalent to the standard quantum computation model up to a polynomial relation. Note that these results actually already imply an efficient quantum algorithm for approximating the Jones polynomial at $e^{2\pi i/5}$, though the algorithm was not explicitly given. In Aharonov et al. (2006b), Aharonov, Jones and Landau gave a simple and explicit quantum algorithm to approximate the Jones polynomial at all points of the form $e^{2\pi i/k}$, even if k grows polynomially with n. At the universality side, the result in Freedman et al. (2002) implies that approximating the Jones polynomial of the plat closure of a braid at $e^{2\pi i/k}$ is **PromiseBQP**-hard for any constant k. In Aharonov and Arad (2006), Aharonov and Arad generalized this by showing that the problem is **PromiseBQP**-hard also for asymptotically growing k, bounded by a polynomial of size of the braid. These together give a new class of **PromiseBQP**-complete problems.

The usual Tutte polynomial is a two-variable polynomial defined for graphs (or more generally for matroids). It is, essentially, the ordinary generating function for the number of edge sets of a given size and connected components. Containing information about how the graph is connected, it plays an important role in algebraic graph theory. It contains as special cases several other famous polynomials, such as the chromatic polynomial, the flow polynomial and the reliability polynomial, and it is equivalent to Whitney's rank polynomial, Tutte's own dichromatic polynomial, and the Fortuin–Kasteleyn's random-cluster model under simple transformations. It is the most general graph invariant defined by a deletion–contraction recurrence. See textbooks (Bollobás 1998; Biggs 1993; Godsil and Royle 2001) for more detailed treatment.

The result in Aharonov et al. (2007a) focuses on the multivariate Tutte polynomial, which generalizes the two-variable case by assigning to each edge a different variable; see survey (Sokal 2005). This generalized version has rich connections to the Potts model in statistical physics (Wu 1982).

On the complexity side, the exact evaluation of the two-variable Tutte polynomial for planar graphs is #**P**-hard. Both positive and negative results are known for the multiplicative approximation FPRAS (fully polynomial randomized approximation scheme); see Aharonov et al. (2007a) for more details.

4.2 Additive Approximation of the Jones Polynomials

4.2.1 Definitions and the Main Theorems

This section presents an efficient quantum algorithm to additively approximate the Jones polynomial of trace and plat closures of braids at roots of unity. We begin by defining the Jones polynomial. A set of circles embedded in $\mathbb{R}^3$ is called a *link L*. If each circle has a direction, then we say that the link is oriented. A link invariant is a function on links that is invariant under isotopy of links. That is, the function remains unchanged when the links are distorted (without being broken). An important link invariant is the Jones polynomial $V_L(t)$, which is a Laurent polynomial in $\sqrt{t}$ over $\mathbb{Z}$.

Definition 11 (Writhe) The *writhe* $w(L)$ of an oriented link L is the number of positive crossings (l_+ in ❷ *Fig. 2*) minus the number of negative crossings (l_- in ❷ *Fig. 2*) of L.

Definition 12 (Bracket Kauffman polynomial) The *bracket Kauffman polynomial* $\langle L \rangle$ of an oriented link L is

$$\langle L \rangle = \sum_\sigma \sigma(L) \tag{28}$$

where the summation is over all the crossing-breaking states. A crossing-breaking state σ is an n-bit string indicating how to break all crossings by changing each crossing in one of the two ways as shown in ❷ *Fig. 3*. Define $\sigma(L) = A^{\sigma_+ - \sigma_-} d^{|\sigma|-1}$, where σ_+ and σ_- are the numbers of choices of the first and the second case respectively, when breaking all crossings in L, $|\sigma|$ is the number of closed loops in the resulting link, and $d = -A^2 - A^{-2}$.

Definition 13 (Jones polynomial) The *Jones polynomial* of an oriented link L is

$$V_L(t) = V_L(A^{-4}) = (-A)^{3w(L)} \langle L \rangle \tag{29}$$

where $w(L)$ is the writhe of L, and $\langle L \rangle$ is the bracket Kauffman polynomial of L.

There are different ways of forming links. In particular, one can obtain a link from a *braid*. First, an intuitive geometrical definition of a braid is given here. Consider two horizontal bars each with n pegs, one on top of the other; see ❷ *Fig. 4*. The pegs on the upper bar are called the *upper pegs*, and we index them from left to right using $i = 1, \ldots, n$; similarly for the *lower pegs*. Each strand goes from an upper peg, always downwards, to a lower peg, so that finally each peg has exactly one strand attached to it. The *n-strand braid group* B_n is the set of n-strand braids

◻ **Fig. 2**
Positive and negative crossings.

$l+$ $l-$

◻ **Fig. 3**
Two ways to break a crossing.

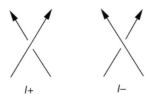

◘ Fig. 4
A braid with 4 strands.

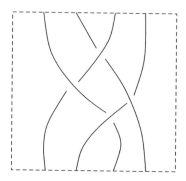

with the multiplication defined by putting the first braid on top of the second. To be more precise, for braids b_1 and b_2, the new braid $b_1 \cdot b_2$ is obtained by putting b_1 on top of b_2, and removing the pegs of the lower bar of b_1 and the upper bar of b_2 with strands on the same peg connected.

The braid group B_n has an algebraic presentation by generators and relations.

Definition 14 Let B_n be the group with generators $\{1, \sigma_1, \ldots, \sigma_{n-1}\}$ with relations

1. $\sigma_i \sigma_j = \sigma_j \sigma_i$ for all $|i - j| \geq 2$
2. $\sigma_i \sigma_{i+1} \sigma_i = \sigma_{i+1} \sigma_i \sigma_{i+1}$

The correspondence between the geometrical pictures and the algebraic presentation is as follows: 1 corresponds to the braid where all strands j go from upper peg j to lower peg j without crossing any other strand; σ_i corresponds to the same braid except that strand i and $i + 1$ have one crossing, with strand i in front. The above two relations are easily verified by this correspondence.

Braids are not links since the strands are not closed circles, but there are different ways of forming links from braids. Two simple ones are the *trace closure* and the *plat closure*. The trace closure B^{tr} of a braid B connects upper peg i and lower peg i without crossing any strand in b. The plat closure B^{pl} of a $2n$-strand braid B connects its upper peg $2i - 1$ and upper peg $2i$ for all $i \in [n]$; similarly for the lower pegs. See **❷** *Fig. 5*.

The main theorem of this section says that there are quantum algorithms additively approximating the Jones polynomials of B^{tr} and B^{pl} at $A^{-4} = e^{2\pi i/k}$ within some precision.

Theorem 4 *There is a quantum algorithm, which on input n-strand m-crossing braid B outputs a complex number c satisfying* $|c - V_{B^{\text{tr}}}(e^{2\pi i/k})| < \varepsilon(-A^2 - A^{-2})^{n-1}$ *with probability* $1 - 1/exp(n, m, k)$ *for some* $\varepsilon = 1/poly(n, m, k)$.

Theorem 5 *There is a quantum algorithm, which on input n-strand m-crossing braid B outputs a complex number c satisfying* $|c - V_{B^{\text{pl}}}(e^{2\pi i/k})| < \varepsilon(-A^2 - A^{-2})^{3n/2}/N$ *with probability* $1 - 1/exp(n, m, k)$ *for some* $\varepsilon = 1/poly(n, m, k)$. *Here*

$$N = \sum_{l=1}^{k-1} \sin(\pi l/k)|P_{n,k,l}| \tag{30}$$

■ Fig. 5
The trace and plat closures of a 4-braid.

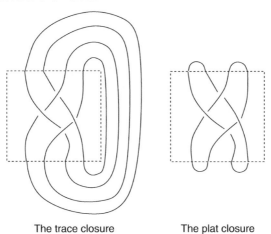

The trace closure   The plat closure

where

$$P_{n,k,l} = \left\{ x \in \{0,1\}^n : 1 \le 2\sum_{i=1}^{j} x_i - j \le k - 1, \forall j \in [n]; 2\sum_{i=1}^{n} x_i - n = l \right\} \quad (31)$$

4.2.2 The Approximation Algorithms

Main Idea and Overview

The main idea of designing the efficient quantum is to design a representation of the braid group and use the homomorphism property of the representation to decompose the computation into ones for each crossing in the braid, which can be done efficiently. This idea was also used to design classical algorithms. Let us recall an example to illustrate the idea. Consider the word problem of the free group generated by $\{a, b\}$. That is, given a sequence of elements from $\{a, b\}$, decide whether the multiplication (in that order) is the identity element of the group. There is a simple linear time algorithm using the stack (or pushdown machine, in other words), but the algorithm also uses linear space. Interestingly, Lipton and Zalcstein showed in 1977 that it can be solved in log space in a very cute manner. Basically, they used the following theorem proved by Ivan Sanov in 1947.

Theorem 6 *There are two integer matrices A, B so that the mapping $a \rightarrow A$ and $b \rightarrow B$ is a faithful representation of the free group on $\{a, b\}$.*

That means that we can replace the word problem by this question: Does a sequence of matrices over $\{A, B, A^{-1}, B^{-1}\}$ equal the identity matrix I? Since multiplication of 2×2 matrices does not need the stack anymore, this basically gives a log space algorithm. (There is actually one more difficulty: during the computation there may be entries that are too large and cannot be stored in log-space. The solution is to do the multiplication mod p for all

p using $O(\log(n))$ bits. Then the correctness is guaranteed by the Chinese Remainder Theorem and the following well-known fact: $\prod_{p \le t} p \ge c^t$, where the product is over primes and $c > 1$.)

The quantum algorithm to approximate the Jones polynomial also uses representation of the braid group. A high level picture is:

$$B_n \xrightarrow{\rho_A} TL_n(d) \xrightarrow{\Phi} \mathbb{C}^{r \times r} \tag{32}$$

Here, $TL_n(d)$ is the *Temperley–Lieb algebra*, which will be defined later. The mapping ρ_A is a homomorphism from B_n to $TL_n(d)$. The mapping Φ is the representation of $TL_n(d)$ by $r \times r$ unitary matrices for some r.

The analysis can be divided into three steps:

1. For the link B^{tr}, the Jones polynomial is

$$V_{B^{tr}}(A^{-4}) = (-A)^{3w(B^{tr})} d^{n-1} tr(\rho_A(B)) \tag{33}$$

The function $tr(\cdot)$ is some mapping from $TL_n(d)$ to $\mathbb{C}$ to be defined later. By this equality, it is enough to approximate $tr(\rho_A(B))$.

2. One can calculate $tr(\rho_A(B))$ by

$$tr(\rho_A(B)) = Tr_n(\Phi \circ \rho_A(B)) \tag{34}$$

where $\circ$ is the standard function composition, and Tr_n is defined as follows. For a matrix $W \in \mathbb{C}^{r \times r}$,

$$Tr_n(\Phi \circ \rho_A(B)) = \frac{1}{N} \sum_{l=1}^{k-1} \sin(\pi l/k) Tr(\Phi \circ \rho_A(B)|_l) \tag{35}$$

where Tr is the standard matrix trace, N is as defined in ❷ Theorem 5, $W|_l$ is W restricted on some subspace. If an orthonormal basis of $\Phi \circ \rho_A(B)|_l$ is $\{|p\rangle\}$, then one can further the calculation by

$$Tr(\Phi \circ \rho_A(B)|_l) = \sum_p \langle p | \Phi \circ \rho_A(B) | p \rangle \tag{36}$$

It turns out that for each l, one can efficiently sample p almost uniformly at random (even on a classical computer). One also needs the following well-known fact in quantum computing.

Fact Given a quantum state $|\psi\rangle$ and a quantum circuit Q, one can generate a random variable $b \in \mathbb{C}$ with $|b| \le 1$ and $E[b] = \langle \psi | Q | \psi \rangle$.

By the above analysis, one can achieve the approximation as long as one can implement $\Phi \circ \rho_A(B)$ efficiently on a quantum computer.

3. This is where the homomorphism comes into the picture. Each braid B can be decomposed into the product of a sequence of basic braids σ_i. The A and Φ are carefully designed such that $\Phi \circ \rho_A$ is a unitary representation. Thus $\Phi \circ \rho_A(B) = \Phi \circ \rho_A(\prod_i \sigma_i) = \prod_i \Phi \circ \rho_A(\sigma_i)$. Therefore, it is enough to implement $\Phi \circ \rho_A(\sigma_i)$ for each basis σ_i, and it turns out that this is not hard to do. This finishes the overview of the whole idea of designing the quantum algorithm.

More Details

Next, more details to carry out the above strategy are given. To start with, the Temperley–Lieb algebra is defined.

Definition 15 (Temperley–Lieb algebra) The *Temperley–Lieb algebra* $\text{TL}_n(d)$ is the algebra generated by $\{1, E_1, \ldots, E_{n-1}\}$ with relations

1. $E_i E_j = E_j E_i, |i - j| \geq 2$
2. $E_i E_{i+1} E_i = E_i E_{i-1} E_i = E_i$
3. $E_i^2 = dE_i$

As the braid group B_n, the Temperley–Lieb algebra also has a nice pictorial description called Kauffman diagrams. A Kauffman n-diagram has an upper bar with n top pegs and a lower bar with n lower pegs as in a braid, but it does not have crossings or loops. A typical Kauffman 4-diagram is as shown in ❯ *Fig. 6*.

The multiplication of two Kauffman n-diagrams $K_1 \cdot K_2$ is obtained similarly for braids: that is, putting K_1 on top of K_2. Note that $K_1 \cdot K_2$ now may contain loops. If there are m loops, then the loops are removed and we add a factor of d^m in the resulting diagram. See ❯ *Fig. 7*. The Kauffman diagrams over $\mathbb{C}$ form an algebra. The correspondence between the pictures and the original definition of Temperley–Lieb algebra is shown in ❯ *Fig. 8*.

The first mapping ρ_A is now defined in ❯ Eq. 32.

◻ **Fig. 6**
A typical Kauffman 4-diagram.

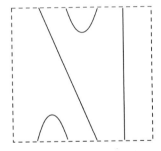

◻ **Fig. 7**
Product of two Kauffman 4-diagrams.

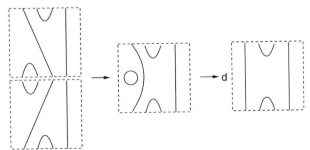

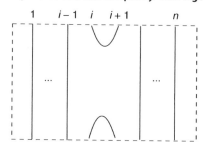

$$1 \quad i-1 \quad i \quad i+1 \quad n$$

$$1 \quad 2 \quad k-2 \quad k-1$$

Definition 16 $\rho_A \colon B_n \mapsto TL_n(d)$ is defined by $\rho_A(\sigma_i) = AE_i + A^{-1}1$.

The following facts are easily verified.

Fact 1 The mapping ρ_A respects the two relations in the definition of the braid group.

Fact 2 If $|A|= 1$ and $\Phi(E_i)$ is Hermitian for each i, then the map $\Phi \circ \rho_A$ is a unitary representation of B_n.

Now the trace function in Step 1 and 2 is defined. It is easier to define the Kauffman diagrams.

Definition 17 The *Markov trace* $tr\colon TL_n(d) \to \mathbb{C}$ on a Kauffman n-diagram K connects the upper n pegs to the lower n pegs of K with nonintersecting curves, as in the trace closure case. Then $tr(K) = d^{a-n}$, where a is the number of loops of the resulting diagram. Extend tr to all of $TL_n(d)$ by linearity.

For this definition, it is not hard to check that the equality in Step 1 holds.

Lemma 1 *For any braid B, one has*

$$V_{B^{tr}}(A^{-4}) = (-A)^{3w(B^{tr})} d^{n-1} tr(\rho_A(B)) \tag{37}$$

Now the second mapping Φ is defined, for which the path model representation of $TL_n(d)$ will be needed. Consider a graph L_{k-1} of a line with $k-1$ points (connected by $k-2$ edges); see ❯ *Fig. 9*. Let $Q_{n,k}$ be the set of all paths of length n on L_{k-1}. That is, $Q_{n,k}$ can be identified with $\{q \in [k-1]^{n+1} \colon |q_i - q_{i+1}| = 1, \forall i \in [n-1]\}$. Let $r = |Q_{n,k}|$. That is, the size of $Q_{n,k}$ is the dimension of matrix $\Phi(K)$; thus one can index the row/column of the matrix by a path in $Q_{n,k}$.

To specify Φ easily, let one use the bit string representation of a path. Each path is specified by a string $p \in \{0, 1\}^n$ such that $p_i = 1$ if the i-th step goes right and $p_i = 0$ if the i-th step goes left. To guarantee that the path is always within the graph, one requires $1 \leq z_j \leq k - 1$ for any $j \in [n]$, where $z_j = 2 \sum_{i=1}^{j} p_i - j$ is the location of the path after first j steps. Let $H_{n,k}$ be the span of all these paths, each treated as a basis state. Denote $p_{<i} = p_1 \cdots p_{i-1}$ and $p_{\geq i} = p_i \cdots p_n$.

To define Φ, it is enough to define its action on each E_i, done as follows, where $\lambda_j = \sin(j\pi/k)$.

$$\Phi(E_i)|p_{<i}00p_{\geq i+2}\rangle = 0 \tag{38}$$

$$\Phi(E_i)|p_{<i}01p_{\geq i+2}\rangle = \frac{\lambda_{z_i-1}}{\lambda_{z_i}}|p_{<i}01p_{\geq i+2}\rangle + \frac{\sqrt{\lambda_{z_i+1}\lambda_{z_i-1}}}{\lambda_{z_i}}|p_{<i}10p_{\geq i+2}\rangle \tag{39}$$

$$\Phi(E_i)|p_{<i}10p_{\geq i+2}\rangle = \frac{\lambda_{z_i+1}}{\lambda_{z_i}}|p_{<i}10p_{\geq i+2}\rangle + \frac{\sqrt{\lambda_{z_i+1}\lambda_{z_i-1}}}{\lambda_{z_i}}|p_{<i}01p_{\geq i+2}\rangle \tag{40}$$

$$\Phi(E_i)|p_{<i}11p_{\geq i+2}\rangle = 0 \tag{41}$$

Theorem 7 *If $d = 2\cos(\pi/k)$, then $\Phi \circ \rho_A$ is a unitary representation of B_n in r-dimensional vector space.*

Now define the subspace $H_{n,k,l}$ of $H_{n,k}$ by $H_{n,k,l} = \text{span}\{|p\rangle : z_n(p) = l\}$. Define Tr_n as in Eq. 35. We are ready to show the algorithm after some final comments: The Fact in Step 2 is covered by the standard Hadamard Test on both the real and the imaginary parts; see Aharonov et al. (2006b) for details. For each basis σ_i, it is not hard to check that $\Phi \circ \rho_A(\sigma_i)$ can be implemented in polynomial time on a quantum computer. The algorithm for approximating the Jones polynomial on B^{tr} is as follows. The averaging over polynomial number of samples at the last step can be shown to give enough approximation by the standard Chernoff bound.

Algorithm 1   Approximate Jones trace closure

1. Repeat for $j = 1$ to $t = \text{poly}(n, m, k)$:
    a. Classically, pick a random path $p \in P_{n,k}$ with probability $\Pr(p) \propto \sin(\pi l/k)$, where l is the index of the site at which p ends.
    b. Use Hadamard Test to output a random variable x_j with $E[x_j] = \text{Re}\langle p|Q(B)|p\rangle$.
2. Use Hadamard Test to output a random variable y_j with $E[y_j] = \text{Im}\langle p|Q(B)|p\rangle$.
3. Let $r = \frac{1}{t}\sum_j (x_j + iy_j)$. Output $(-A)^{3w(B^{tr})}d^{n-1}r$.

The case of plat closure can be reduced to the trace closure case by the following observation. The plat closure of a braid B is isotopic to the trace closure of C, obtained by putting B on top of $n/2$ capcups, where a capcup is a cup on top of a cap. The algorithm is largely the same as the one for trace closure, except that one needs an observation that the state $|\alpha\rangle = |1, 0, 1, 0, \ldots, 1, 0\rangle$ can be used to connect the Jones polynomial of the plat closure and the trace function. To be more precise, one has

Fact 3 $\langle \alpha | \Phi \circ \rho_A | \alpha \rangle = \frac{N}{\sin(\pi/k)d^{n/2}} Tr_n(\Phi \circ \rho_A(C)).$

By this, it is enough to estimate $\langle\alpha|\Phi\circ\rho_A|\alpha\rangle$, which can be done by Hadamard test again like in the trace closure case. The algorithm is as follows.

Algorithm 2   Approximate Jones plat closure

1. Repeat for $j = 1$ to $t = \text{poly}(n, m, k)$:
   a. Generate the state $|\alpha\rangle = |1, 0, 1, 0, \ldots, 1, 0\rangle$.
   b. Use Hadamard Test to output a random variable x_j with $E[x_j] = \text{Re}\langle\alpha|Q(B)|\alpha\rangle$.
2. Use Hadamard Test to output a random variable y_j with $E[y_j] = \text{Im}\langle\alpha|Q(B)|\alpha\rangle$.
3. Let $r = \frac{1}{t}\sum_j(x_j + iy_j)$. Output $(-A)^{3w(B^{tr})}d^{3n/2-1}r\sin(\pi/k)/N$.

4.3   The PromiseBQP-Hardness of Approximating Jones Polynomials

A brief discussion of the idea in Aharonov and Arad (2006) of simulating a quantum circuit by an oracle U approximating the Jones polynomials will be presented here. First, based on a simple procedure similar to the one in ❷ Sect. 3.1.2, it is enough to approximate $\langle 0|U|0\rangle$ for any polynomial size quantum circuit U. So we want to efficiently construct a braid B such that B has polynomially many crossings and $\langle\alpha|\Phi\circ\rho_A(B)|\alpha\rangle \approx \langle 0|U|0\rangle$. (Here the $\approx$ sign means approximation to any desired accuracy.)

Note that in this approach, the working space has to be encoded by paths on graph L_{k-1}. Thus we need to encode the space for the original circuit U by the path space. A simple encoding uses the following four-step paths:

$$|0\rangle \to |1010\rangle, \quad |1\rangle \to |1100\rangle \tag{42}$$

Suppose U is decomposed as $U = U_m \ldots U_1$, where each U_i is an elementary gate acting on at most two qubits. It can be assumed without loss of generality, that each U_i operates on adjacent qubits. Using the path encoding, U_i operates on eight qubits. Note that the path is not arbitrary in $Q_{4n,k}$ since it always returns to the original point every four steps. Denote by S the subspace spanned by all these paths; then it is sufficient if one can efficiently find $B_i \in B_{4n}$ such that $\Phi\circ\rho_A(B_i) \approx U_i$ on the subspace S to the polynomially small accuracy. It turns out that it is doable, as the following density theorem shows for B_8.

Theorem 8   *Suppose $\tilde{U}$ is an encoded two-qubit quantum gate, then for any $\delta > 0$ and $k \geq 11$, one can find a braid $B \in B_8$ with $\text{poly}(k, 1/\delta)$ generators of B_8 such that $\|(\Phi\circ\rho_A(B)|p\rangle - (\tilde{U})|p\rangle\| \leq \delta$ for all $|p\rangle\in H_{n,k,1}$.*

The proof of this theorem is the core technical part needed to show the hardness, but it is a bit far away from the **PromiseBQP**-completeness notion that is being discussed. The readers are referred to Aharonov and Arad (2006) for the details.

4.4   Additive Approximation of the Tutte Polynomials

The results for the Jones polynomials in Aharonov et al. (2006b) and Aharonov and Arad (2006) are generalized to the Tutte polynomials in Aharonov et al. (2007a). The multivariate Tutte polynomial is defined as follows.

Definition 18 (Tutte polynomial) Given an undirected graph $G = (V, E)$ with a weight function on edges $\mathbf{v} = \{v_e : e \in E\}$, the Tutte polynomial is

$$Z_G(q, \mathbf{v}) = \sum_{A \subseteq E} q^{k(A)} \prod_{e \in A} v_e \tag{43}$$

where $k(A)$ is the number of connected components in the subgraph (V, A).

When all v_e's are equal to the same v, then the polynomial becomes

$$Z_G(q, v) = \sum_{A \subseteq E} q^{k(A)} v^{|A|},$$

which is essentially the same as the standard Tutte polynomial

$$T_G(x, y) = \sum_{A \subseteq E} (x - 1)^{k(A) - k(E)} (y - 1)^{|A| + k(A) - |V|} \tag{44}$$

under the change of variables

$$x = 1 + q/v, \quad y = 1 + v \tag{45}$$

$$q = (x - 1)(y - 1), \quad v = y - 1 \tag{46}$$

The main results in Aharonov et al. (2007a) are as follows. First, there is an efficient algorithm to additively approximate the multivariate Tutte polynomial of a given weighted planar graph. Second, there exists a range of complex weights and complex values of q such that the additive approximation of the multivariate Tutte polynomial at those points to within some scale is **PromiseBQP**-hard. The approximation windows in the above two results do not match that nicely in the non-unitary case, in which they modified the definition to get the **PromiseBQP**-complete problems.

The rest of this section will mainly illustrate some ideas of the algorithm. The algorithm actually takes an approach similar to that the Jones polynomial, with generalizations of various objects. First, we start at a planar graph rather than a braid. Like using closures of braids, we also change the planar graph to a knot-like object. Here the *medial graph* is used. For a planar graph G, the medial graph L_G is obtained in the following way. First encircle the facets of G with lines, and then cross the lines that surround each edge by putting a vertex in the middle of the edge. See ❯ *Fig. 10* for an illustration. Note that the resulting graph L_G is 4-regular.

◼ **Fig. 10**
Medial graph of a planar graph.

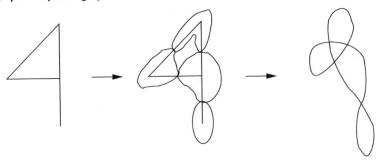

The regions of the medial graph can be {black,white}-colored so that no two adjacent regions have the same color. This coloring is unique up to an overall flip; let the outer region be fixed to be white.

For each crossing, there are two ways to break it, depending on whether one connects the two black or the two white regions. Let σ be an m-bit string indicating the crossing-breaking choices, where $m = |E| = |\sigma|$ is the number of the crossings. Denote by $\text{Black}(\sigma)$ the set of edges, the crossings corresponding to which are broken by σ with two black regions connected. The Kauffman bracket of the medial graph L_G can now be defined by

$$\langle L_G \rangle = \sum_\sigma d^{|\sigma|} \prod_{e \in \text{Black}(\sigma)} u_e \tag{47}$$

where the new variables u_e and the old ones v_e are related by $v_e = d u_e$. The following fact connects the Kauffman bracket of L_G and the Tutte polynomial of G.

Fact 4 $\langle L_G \rangle (d, \mathbf{u}) = d^{-n} Z_G(d^2, d\mathbf{u})$.

Thus it is enough to calculate the Kauffman bracket of the medial graph. To this end, one needs a generalized Temperley–Lieb algebra GTL(d), where one allows an arbitrary number of strands and allows cases with different numbers of upper and lower pegs in one diagram. A diagram does not "change" by adding some trivial strands; that is, those going from an upper peg directly to a lower peg without crossing any other strand. With these relaxations, the product can be defined similarly to the Jones polynomial case (by putting one on top of the other) in a consistent manner.

Using this bridge, generalized versions of the mapping Φ and the path model representation of GTL(d) can be defined. But now the representative does not need to be unitary. Again the homomorphism property of representation reduces the task to approximating each basic tangle diagram.

There is one catch, however: The final scale of the approximation window depends on the decomposition of L_G into some simple structures called *basis tangles*. The optimal decomposition is not known to be easy to obtain. This is also a drawback of the algorithm: the approximation window contains a quantity which is complicated and not directly about the graph itself (but about the layout of it on the two-dimensional plane).

5   Concluding Remarks

5.1   Some Other PromiseBQP-Completeness Related Problems

In Wocjan and Zhang (2006), it is shown that the problem of local unitary phase sampling is also **PromiseBQP**-complete, as is a problem called local unitary average eigenvalue. Even earlier, Knill and Laflamme found that the quadratically signed weight enumerators problem is also **PromiseBQP**-complete (Knill and Laflamme 2001).

In the track of the Jones polynomial and Tutte polynomial, Wocjan and Yard gave new quantum algorithms for approximating HOMFLYPT two-variable polynomials of trace closures of braids (Wocjan and Yard 2008). They also gave algorithms for approximating the Jones polynomial of a general class of closures of braids.

There are some equivalent models of efficient quantum computation, such as adiabatic quantum computation (Aharonov et al. 2004) and one-way quantum computation (Browne and Briegel 2006). One can also view the canonical problems in those models as complete problems for the standard quantum Turing machine or quantum circuit model.

One can interpret the problem of SMPE in ❯ Sect. 3.2 as approximating the total weight of cycles of length t passing a vertex j on a weighted graph. The work (Janzing and Wocjan 2007) can actually show the **PromiseBQP**-completeness even for Hamiltonians with $\{1, -1, 0\}$ entries. However, allowing the weight to be -1 makes weights on cycles cancel, giving the problem a quantum flavor. Childs recently studied the problem with the weight to be chosen only from $\{0, 1\}$ (Childs 2009).

5.2 Future Directions

There are two issues in the state of the art of **PromiseBQP**-completeness. One is that all the known **PromiseBQP**-complete problems are not "natural" enough. In some sense, they are all describing the same class using different languages, though the difficulty of translation may be at different levels. One can say that all completeness results have this feature, but the key reason why **NP**-completeness is so important is that there are so many natural and seemingly unrelated combinatorial problems in theoretical computer science, discrete mathematics, and various other branches of mathematics and science. But not many natural **PromiseBQP**-complete problems are known so far.

Another direction is to try to use the **PromiseBQP**-complete problem to study the classes **PromiseBQP** and **BQP**. For example, one of the main open questions in quantum complexity theory is whether **BQP** is in **PH**, the polynomial hierarchy. The current best known upper bound of **BQP** is **AWPP**, a not-so-natural subclass of **PP**. (See the textbooks (Arora and Barak 2009; Goldreich 2008; and Papadimitriou 1994) and the "complexity zoo" (currently at http://qwiki.stanford.edu/wiki/Complexity_Zoo) for definitions of these complexity classes.) Could the known **PromiseBQP**-complete problems shed any light on the open problem?

References

Aharonov D, Arad I (2006) The **BQP**-hardness of approximating the Jones polynomial. quant-ph/0605181

Aharonov D, Ta-Shma A (2003) Adiabatic quantum state generation and statistical zero knowledge. In: Proceedings of 35th annual ACM symposium on theory of computing (STOC). ACM, New York, pp 20–29

Aharonov D, van Dam W, Kempe J, Landau Z, Lloyd S, Regev O (2004) Adiabatic quantum computation is equivalent to standard quantum computation. In: Proceedings of the 45th annual IEEE symposium on foundations of computer science (FOCS). IEEE Computer Society, Washington, DC, pp 42–51

Aharonov D, Jones V, Landau Z (2006) A polynomial quantum algorithm for approximating the Jones polynomial. In: Proceedings of the 38th annual

ACM symposium on theory of computing (STOC). ACM, New York, pp 427–436

Aharonov D, Arad I, Eban E, Landau Z (2007a) Polynomial quantum algorithms for additive approximations of the Potts model and other points of the Tutte plane. arXiv:quant-ph/0702008

Aharonov D, Gottesman D, Irani S, Kempe J (2007b) The power of quantum systems on a line. In: Proceedings of the 48th annual IEEE symposium on foundations of computer science (FOCS). IEEE Computer Society, Washington, DC, pp 373–383

Ambainis A, Childs AM, Reichardt BW, Spalek R, Zhang S (2007) Any and-or formula of size n can be evaluated in time $n^{1/2+o(1)}$ on a quantum computer. In: Proceedings of the 48th annual IEEE symposium on foundations of computer science

(FOCS). IEEE Computer Society, Washington, DC, pp 363–372

Arora S, Barak B (2009) Computational complexity: a modern approach. Cambridge University Press, Cambridge, UK

Bernstein E, Vazirani U (1997) Quantum complexity theory. SIAM J Comput 26(5):1411–1473

Berry D, Ahokas G, Cleve R, Sanders B (2007) Efficient quantum algorithms for simulating sparse Hamiltonians. Commun Math Phys 270(2):359–371

Biggs N (1993) Algebraic graph theory, 2nd edn. Cambridge University Press, New York

Bollobás B (1998) Modern graph theory. Springer, New York

Browne D, Briegel H (2006) One-way quantum computation - a tutorial introduction. arXiv:quant-ph/0603226

Chernoff P (1968) Note on product formulas for operator semigroups. J Funct Anal 2:238–242

Childs A (2009) Universal computation by quantum walk. Phys Rev Lett 102:180501

Childs A, van Dam W (2010) Quantum algorithms for algebraic problems. Rev Mod Phys arXiv:0812.0380, 82:1–52

Childs A, Cleve R, Deotto E, Farhi E, Gutmann S, Spielman D (2003) Exponential algorithmic speedup by a quantum walk. Proceedings of the 35th annual ACM symposium on theory of computing (STOC). ACM Press, New York, pp 59–68

Cook S (1971) The complexity of theorem-proving procedures. In: Proceedings of 3rd annual ACM symposium on theory of computing (STOC). ACM Press, New York, pp 151–158

Freedman M (1998) **P/NP**, and the quantum field computer. Proc Natl Acad Sci 95(1):98–101

Freedman M, Kitaev A, Larsen M, Wang Z (2002a) Topological quantum computation. Bull Amer Math Soc 40(1):31–38

Freedman M, Kitaev A, Wang Z (2002b) Simulation of topological field theories by quantum computers. Commun Math Phys 227(3):587–603

Freedman M, Larsen M, Wang Z (2002c) A modular functor which is universal for quantum computation. Commun Math Phys 227:605–622

Garey M, Johnson D (1979) Computers and intractability: a guide to the theory of NP-completeness. W.H. Freeman, New York

Godsil C, Royle G (2001) Algebraic graph theory. Springer, New York

Goldreich O (2005) On promise problems (a survey in memory of Shimon Even [1935–2004]). Electronic colloquium on computational complexity (ECCC). TR05–018

Goldreich O (2008) Computational complexity: a conceptual perspective. Cambridge University Press, Cambridge, UK

Goldreich O, Micali S, Wigderson A (1991) Proofs that yield nothing but their validity. J ACM 38(3):690–728

Hallgren S (2007) Polynomial-time quantum algorithms for Pell's equation and the principal ideal problem. J ACM 54(1):1–19

Janzing D, Wocjan P (2007) A simple PromiseBQP-complete matrix problem. Theor Comput 3(1):61–79

Janzing D, Wocjan P, Zhang S (2008) Measuring energy of basis states in translationally invariant nearest-neighbor interactions in qudit chains is universal for quantum computing. New J Phys 10:093004

Jones V (1985) A polynomial invariant for knots via von Neumann algebras. Bull Amer Math Soc 12(1):103–111

Karp R (1972) Reducibility among combinatorial problems. In: Thatcher JW, Miller RE (eds) Complexity of computer computations. Plenum Press, New York

Kempe J, Kitaev A, Regev O (2006) The complexity of the local Hamiltonian problem. SIAM J Comput 35(5):1070–1097

Kitaev A (1995) Quantum measurements and the Abelian stabilizer problem. arXiv:quant-ph/9511026

Kitaev A, Shen A, Vyalyi M (2002) Classical and quantum computation. American Mathematical Society, Providence, RI

Knill E, Laflamme R (2001) Quantum computing and quadratically signed weight enumerators. Infor Process Lett 79(4):173–179

Kuperberg G (2005) A subexponential-time quantum algorithm for the dihedral hidden subgroup problem. SIAM J Comput 35(1):170–191

Levin L (1973) Universal search problems (in Russian). Problemy Peredachi Informatsii 9(3):265–266

Lipton R, Zalcstein Y (1977) Word problems solvable in logspace. J ACM 24(3):522–526

Lomont C (2004) The hidden subgroup problem – review and open problems. arXiv:quant-ph/0411037

Magniez F, Nayak A, Roland J, Santha M (2007) Search via quantum walk. In: Proceedings of the 39th annual ACM symposium on theory of computing (STOC). ACM Press, New York, pp 575–584

Nielsen M, Chuang I (2000) Quantum computation and quantum information. Cambridge University Press, Cambridge, UK

Papadimitriou C (1994) Computational complexity. Addison-Wesley, Reading

Papadimitriou C (1997) NP-completeness: a retrospective. In: Proceedings of the 24th international colloquium on automata, languages and programming (ICALP), Lecture notes in computer science, vol. 1256. Springer, Berlin, pp 2–6

Podtelezhnikov A, Cozzarelli N, Vologodskii A (1999) Equilibrium distributions of topological states in circular DNA: interplay of supercoiling and knotting. Proc Natl Acad Sci USA 96(23):12974–12979

Reichardt B, Spalek R (2008) Span-program-based quantum algorithm for evaluating formulas. In: Proceedings of 40th annual ACM symposium on theory of computing (STOC). ACM Press, New York, pp 103–112

Shor P (1997) Polynomial-time algorithms for prime factorization and discrete logarithms on a quantum computer. SIAM J Comput 26:1484–1509

Shor P (2004) Progress in quantum algorithms. Quant Inform Process 3:5–13

Sokal A (2005) The multivariate Tutte polynomial (alias Potts model) for graphs and matroids. In: Webb BS (ed) Surveys in combinatorics. Cambridge University Press, Cambridge, UK, pp 173–226

Trotter H (1959) On the product of semigroups of operators. Proc Amer Math Soc 10:545–551

Wocjan P, Yard J (2008) The Jones polynomial: quantum algorithms and applications in quantum complexity theory. Quant Inform Comput 8(1&2): 147–180

Wocjan P, Zhang S (2006) Several natural BQP-complete problems. arXiv:quant-ph/0606179

Wu FY (1982) The Potts model. Rev Mod Phys 54: 235–268

Wu FY (1992) Knot theory and statistical mechanics. Rev Mod Phys 64:1099–1131

Yao A (1993) Quantum circuit complexity. In: Proceedings of the 34th annual symposium on foundations of computer science (FOCS). IEEE Computer Society, Washington, DC, pp 352–361